Protein–Protein Interactions

A *Molecular Cloning* Manual

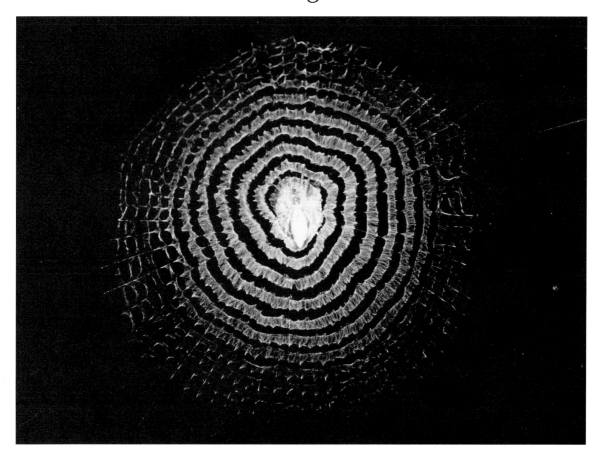

Edited by

Erica Golemis

Fox Chase Cancer Center
Philadelphia, Pennsylvania

COLD SPRING HARBOR LABORATORY PRESS
Cold Spring Harbor, New York

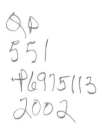

Protein–Protein Interactions
A *Molecular Cloning* Manual

Printed in the United States of America

Developmental Editor	Judy Cuddihy
Project Coordinator	Mary Cozza
Production Editor	Patricia Barker
Desktop Editor	Susan Schaefer
Book Designer	Denise Weiss
Cover Designer	Ed Atkeson

Front cover artwork (paperback edition): The orb web with female spider shown on the cover symbolizes the network of protein interactions that take place in the cell. Simon D. Pollard/Photo Researchers, Inc.

Library of Congress Cataloging-in-Publication Data

Protein-protein interactions : a molecular cloning laboratory manual / edited by Erica Golemis.
  p. cm.
 Includes bibliographical references and index.
 ISBN 0-87969-604-4 (alk. paper) — ISBN 0-87969-628-1 (pbk. : alk. paper)
 1. Protein binding—Laboratory manuals. I. Golemis, Erica.
 QP551 .P6975113 2001
 572′.6—dc21
                                                                      2001047388

10 9 8 7 6 5 4 3 2

Protein–Protein Interactions

A *Molecular Cloning* Manual

Contents

v

Section 4   Recent Developments and Other Tools: Reviews and Short Protocols

Section 5 | **Protein–Protein Interactions: Computation and the Future**

Preface

Under the guidance of general editor Joe Sambrook, Cold Spring Harbor Laboratory Press is launching a new series of advanced laboratory manuals for biological research that build on the 3rd edition of *Molecular Cloning: A Laboratory Manual.*

The major aims of *Molecular Cloning* are to provide researchers with protocols that work reproducibly and the background knowledge required to carry out these protocols intelligently. *Protein–Protein Interactions: A Molecular Cloning Laboratory Manual,* one of the first in the new series of advanced manuals, has similar goals: to provide a complete and current account of technical and theoretical issues in the study of protein associations, for an audience ranging from early graduate students to experienced investigators.

This book owes its existence to the valuable contributions of many talented people. First and foremost, I thank Cold Spring Harbor Laboratory Press for their interest in developing an advanced manual devoted to the topic of protein interactions, and then for allowing me to take an editorial role in the project. In this context, I gratefully acknowledge Patricia Barker, Mary Cozza, Judy Cuddihy, and most of all, Kaaren Janssen. Their collaboration in all aspects of this work cannot be overestimated. Separately, I thank Jan Argentine, John Inglis, and Joe Sambrook for setting such an inspiring precedent in *Molecular Cloning.*

It was a privilege to work with the many authors of this volume, and I greatly appreciate their efforts in constructing thoughtful contributions marked by creativity and clarity. I appreciatively acknowledge my past mentors, Nancy Hopkins and Roger Brent, for years of helpful discussion of issues related to organismal complexity, protein interactions, and the relationships between the two. I thank the members of my laboratory, past and present, for stimulating scientific discussions. I particularly thank Ilya Serebriiskii, whose oversight of ongoing protein interaction benchwork allowed me the time to concentrate on protein interaction editing.

I also acknowledge the rich intellectual and collegial environment at the Fox Chase Cancer Center, which has contributed immeasurably to the thinking behind this project. More pragmatically, Kathy Buchheit and Sarah Costello-Berman were of great assistance in many aspects of organization throughout the editorial process, which would have been much more difficult without their efforts. Finally, I particularly thank Michael and Ian Ochs, and Marion and Emanuel Golemis, for their constant and spirited support.

E.G.

1 | Toward an Understanding of Protein Interactions

Erica Golemis

Division of Basic Science, Fox Chase Cancer Center, Philadelphia, Pennsylvania 19111

The goal of this book is to provide an overview and practical explication of critical technologies in the field of protein–protein interactions, describing both concepts of current importance in this area of research and technologies that enable the characterization and manipulation of proteins. Therefore, the goal of this introduction is to emphasize why an understanding of protein–protein interactions is an essential requirement of the modern molecular biologist, and the manner in which this volume aims to address this need.

SPECIFICS

Proteins control and mediate many of the biological activities of cells. Although some proteins act primarily as single monomeric units (for instance, enzymes that catalyze changes in small-molecule substrates), a significant percentage, if not the majority, of all proteins function in association with partner molecules, or as components of large molecular assemblies. Hence, to gain an understanding of cellular function, the function of proteins must be understood both in isolation and in the context of other interactive proteins. Complicating this process is the fact that a cell is not static: Both internally and externally generated signals induce changes in shape, division, viability, metabolism, and other intrinsic properties. Interpretation of, and response to, these signals requires changes in the protein composition of cells and the pattern of association between cellular proteins. Further complicating the analysis of cellular function is the fact that all cells are not equivalent. Within a single species, such as *Homo sapiens*, liver versus lymphoid versus neural cells contain very different populations of proteins and respond differently to identical external stimuli. Crossing species (for instance, *H. sapiens* to *Mus musculus* to *Gallus gallus*), even a given cell type, such as a lymphoid B cell, possesses significant differences in protein composition and functional properties based on both differences in the genetic code specifying the proteins and changes in the expression levels of discrete proteins. Thus, the situation is highly complex.

To address this complexity, it is useful to generate multiple different classes of information about proteins. For any given protein, these classes of knowledge would include (1) the structural and sequence properties of the protein, including its possession of defined motifs that would allow assignment to a functional class, or be predictive of a certain pattern of protein–protein interaction; (2) the evolutionary history and pattern of conservation of the protein, to allow identification of critical residues; (3) the expression profile of the protein, including issues such as cell-type specificity and abundance, and changes in this profile in response to dynamic processes such as cell cycle progression or growth factor response; (4) the intracellular localization of the protein, and association with specific organelles or structures; (5) the forms of posttranslational regulation to which a protein is subject (phosphorylation, ubiquitination, acylation, and others); and (6) the other cellular proteins with which the protein associates. In fact, all of the first five points together contribute to the determination of the sixth; and given that proteins of linked function tend to associate in transient or stable complexes, determination of a profile of protein–protein interactions is an extremely important step toward the ultimate goal of identifying the functional significance of the activity of any given protein in a cell. Chapters 2 and 3 in Section 1—The Biological Context for Protein Interaction Studies—are essays focused on current topics in signal transduction and human genetics/pharmacogenomics, exemplifying analysis of protein interactions as an organizing principle for these disciplines.

Techniques to provide the classes of information listed above fall into three groups. In one class, the goal is to identify every possible interacting set of proteins for protein X of interest; in this case, physiological significance is temporarily downplayed in an effort to cast a broad net. In the second class, where interacting proteins of interest have been defined, the goal is to detail the biological function and impact of their interactions, i.e., to establish physiological significance. In this case, it is essential to be able to study the interaction under conditions that correspond as closely as possible to the endogenous situation. In the third class, an interaction has been identified, and validated as physiological, and is reasonably well understood; here, the goal is to devise high-throughput methodologies to identify agents that modulate the interaction in desirable ways. No single technique is optimal to address all the points of interest; however, by combining techniques, it is possible to make considerable progress. The techniques described herein bear on all three described classes. Although a complete review of the origins and nature of protein-interactive techniques is beyond the scope of this book, one excellent source of references to this end is provided in Phizicky and Fields (1995), along with an extensive discussion of parameters that can be critical in determining the ability of a specific technique to detect particular protein interactions. Finally, although we have tried to demonstrate the breadth of creativity and options in the protein interaction field, in the interest of space we have not addressed every possible interaction technology. For example, some traditional biochemical methodologies, such as column copurification and differential centrifugation, have been omitted, as have isothermal titration calorimetry and a number of recently developed technologies.

Following the introductory section noted above, this book has been organized into four methodological sections.

Section 2—Standard Technologies Used to Identify and Characterize Protein–Protein Interactions—presents six well-established approaches to the analysis of protein–protein interactions that exemplify standard molecular biological, biochemical, and microbiological approaches used in many laboratories. The use of in vitro GST-fusions in pull-down and far western overlay assays is described in Chapter 4, followed by the complementary methodologies required for coimmunoprecipitation analysis, in Chapter 5. Chapter 6 provides a thorough discussion of the issues involved in the empirical establishment of conditions suitable for protein cross-linking. Chapter 7 is devoted to yeast and bacterial two-hybrid approaches that use transcription-based readout. Chapter 8 details the uses of phage display to identify high-affinity interactions at high-throughput, and Chapter 9 describes classic genetic approaches to discerning protein association based on functional interaction of gene products.

Section 3—Biophysical Approaches to Identify and Characterize Protein–Protein Interactions—in contrast, is devoted to technologies that have biophysical and cell biological roots. The technologies of FRET and PRIM, discussed in Chapters 10 and 11, allow analysis of the association of proteins in situ in either fixed cells or over real time in living cells, providing a unique insight into the changes in protein–protein association dependent on reaction to biological stimulus. Mass spectrometry has been an essential tool for proteomic analyses: Chapter 12 describes issues involved in using this technique to discern interactions between specific proteins in multicomponent complexes. Finally, Chapters 13 (atomic force microscopy), 14 (Biacore), and 15 (QCM biosensor) detail methodologies to obtain specific measurements of the forces involved in protein–protein or in some cases, protein–small molecule ligand interactions.

Section 4—Recent Developments and Other Tools: Reviews and Short Protocols—is devoted to innovative technologies, for the most part developed quite recently, that are either imaginative developments of "themes" related to the core technologies described in Sections 2 and 3, or are unique approaches to the analysis of protein interactions. Although the topics included in this section are quite diverse, they fall into several subgroups. The technique of protease fingerprinting described in Chapter 16 is complementary to protein cross-linking in its use of in vitro biochemical detection methods to identify essential contacts between associating proteins. Chapters 17–20 are related to GST pull-down and immunoprecipitation methodologies and describe means to facilitate purification based on in vitro physical association of proteins with target proteins of interest. Chapters 21–27 are devoted to different forms of two-component systems—in bacteria, yeast, and mammalian cells—for proteins expressed in the nucleus, cytoplasm, or at membranes. Although the ancestral concepts sparking the development of these systems are arguably the two-hybrid system and bacterial α-complementation paradigms, the actual manifestations of protein interaction detection machinery are highly evolved, with each offering unique advantages for different classes of proteins. Chapter 27, describing the screening of peptide aptamers, is a crossover chapter from an orientation of studying protein–protein interactions per se to using protein interaction technologies to develop tools that can be used to regulate protein interactions in vivo. Chapter 28, describing incremental truncation; Chapter 29, describing catalytic antibodies; and Chapter 30, describing protein bundling, suggest alternative technologies that might be valuable toward these ends. Chapters 31 and 32 describe high-throughput methodologies that facilitate the rapid screening of protein interaction pairs.

Finally, the chapters in the last section of the book are forward-looking, based on the assumption that as we begin to acquire a comprehensive understanding of the complement of existing protein and their functions, we can constructively model this information, or employ it to other means. Hence, Chapters 33 and 34 are devoted to descriptions of computational tools and integrative approaches that enhance and organize many of the wet bench techniques described earlier. Chapter 35 describes some experimental efforts to apply protein–protein interactions as tools to manipulate biological systems via creation of synthetic transcriptional control systems in gene therapy applications. Such approaches are likely to become widely adapted in coming years and to allow elegant control of gene function (for other examples, see Gardner et al. 1998, 2000; Elowitz and Liebler 2000).

CONTEXT

Continuing this last theme, following the successful completion of the human genome sequence, the genomic decade of the 1990s spawned a series of derivative "-omic" disciplines, including analysis of the transcriptome (Velculescu et al. 1997), metabolome (Tweedale et al. 1998), and, most relevantly for this book, proteome (Wasinger et al. 1995). As of mid-2001, a search of

Medline using variants of the term "proteomics" identifies in excess of 1000 references, with more than 90% of these references derived from papers published within the last 2 years. The goal of these endeavors is to provide a complete and systematic description of the complete DNA, RNA, and protein content of living organisms to enable understanding of the dynamic regulation of these components. These are very large-scale endeavors and are frequently portrayed as setting the agenda for a new way of doing biological science in the 21st century. Balanced against these global approaches, a search of Medline for the term "protein" per se reveals in excess of 400,000 entries over the same 2-year period. Rather than dealing with global analysis of protein expression and function, these studies usually address the intrinsic mechanisms involved in the creation, modification, interaction, and function of either individual proteins of interest or small protein sets. On the basis of this ratio (1000:400,000), it is clear that whatever the future potential of global approaches, either there are still very significant issues in understanding protein activity on an individual basis, or, alternatively, the general mass of the scientific establishment is still working on a paradigm that will shortly be superseded. Hence, a question of direct relevance to this book is: What is the value of presenting techniques for studying the interactions of individual proteins in the year 2002?

On a fundamental level, biologists want to know first, what we are and how we got here; second, how we relate to a bewilderingly complex biosphere; and third, what actions can be taken to improve the duration and quality of life. Whereas the goal of scientific discovery has generally remained focused on answering these three basic questions, there have historically been very different intellectual approaches employed toward this end, which can be summarized as systemization (or synthesis) versus explication of mechanism (or reductionism). Scientists interested in systemization have focused efforts on the organization of very large quantities of observational results. For example, botanists and zoologists have directed efforts to achieving exact, ordered family relationships of existing flora and fauna. Building on this work, evolutionary biologists additionally delved into the fossil record to determine how life advances from simple to complex forms. In contrast to the historically more ancient systematic approach, the post-Enlightenment biological scientific establishment has attempted to intelligently probe these compendia of observations about nature by formulating and testing specific hypotheses. As one example, by creating targeted crosses of peas and then using the observed results of these crosses to discriminate between opposing models for heredity of traits, Mendel established a paradigm for subsequent work in genetics. As the modern molecular biological disciplines emerged, an earlier descriptive approach to viewing the world lost appeal in the face of new insights into the organism as machine, with key functions controlling inheritance, daily function, and disease recognized to be dependent on the action of biological molecules—nucleic acids, lipids, carbohydrates, and proteins.

Although the systematic (non-hypothesis-driven) and mechanistic (hypothesis-driven) approaches are sometimes portrayed as oppositional, they are more correctly perceived as complementary. Without an ample data set, there is nothing on which to base or interpret experiments, and without rigorous formulation and test of hypothesis, gathered data exist only as information, and fail to lead to knowledge or, ideally, wisdom. In a historical context, the "-omic" studies represent the work of a direct line of descendents of the Systematists, finally reestablishing their importance alongside the Mechanists. Through a combination of systemic and mechanistic analysis, in the next decades we may begin to gain a reasonable understanding of the composition and function of the human organism and other organisms, making progress on the first two fundamental questions noted above.

There remains the third question: How do we improve the quality of life? Inevitably, addressing this question will require the addition of a third element to the knowledge-gathering approaches described above: the science of productively manipulating biological systems. Although biological engineering has sometimes been dismissed as less challenging than "pure science" aimed solely at generating knowledge, in fact, the bioengineer exists in unique dialog with the pure scientist. Because the goal of this discipline is to be able to dissect and manipulate biological systems, of necessity, bioengineering starts with the study of naturally occurring control mechanisms through approaches such as those described in this work. This understanding is then creatively translated in two ways. One way is to adapt naturally occurring biological agents to make tools that can be applied for desirable goals. A considerable number of the protein–protein interaction systems described in this book have such origins. The second way is to identify critical control points utilized in biological regulatory systems and to generate de novo means to perturb them specifically to achieve desirable effects. This latter objective describes the general ambition of applied scientists, including gene therapists, pharmacogenomicists, and clinicians, and is facilitated by possession of a well-stocked toolbox to allow probing of protein interactions in as many ways as possible. In the year 2001, it is hoped that the apparent success of the protein-targeted anticancer drug Gleevac (Druker et al. 1996) may represent one of the first of many such intelligently designed therapies. Given the advancements in our understanding of the basic organismal building materials, the next decades hold promise for a great burst of ingenious exploitation.

In this contextual light, this book has three primary objectives. First, the intent is to introduce current critical issues in analysis of protein–protein interactions. Second, the book presents a number of widely differing approaches to the study of protein–protein interactions, providing insight into the range of tools available to address differing issues of protein interaction and demonstrating the dynamic interplay between the identification and manipulation of protein interactions. Third, and most practically, the included chapters provide clear and detailed technical description of the various protein interaction techniques. However, following in the tradition of the *Molecular Cloning* series, it seems reasonable to propose that a technique can best be practiced when presented in a context of historical development and most appropriate applications. Such a presentation has been attempted by the various authors who have taken part in this project.

A prime challenge for the future is to conduct targeted studies of proteins of interest while considering the larger context of whole organismal function; and conversely, to carefully validate macro-models of organismal function through individual test cases. The sensibility involved in this process is particularly compatible with studies of protein–protein interactions. It is hoped that the sum of these efforts will be to advance the pragmatic needs of the bench practitioner, while conveying a broader sense of the interest and importance of the field of protein interaction-based studies.

REFERENCES

Druker B.J., Tamura S., Buchdunger E., Ohno S., Segal G.M., Fanning S., Zimmermann J., and Lydon N.B. 1996. Effects of a selective inhibitor of the Abl tyrosine kinase on the growth of Bcr-Abl positive cells. *Nat. Med.* **2:** 561–566.

Elowitz M.B. and Leibler S. 2000. A synthetic oscillatory network of transcriptional regulators. *Nature* **403:** 335–338.

Gardner T.S., Cantor C.R., and Collins J.J. 2000. Construction of a genetic toggle switch in *Escherichia coli. Nature* **403:** 339–342.

Gardner T.S., Dolnik M., and Collins J.J. 1998. A theory for controlling cell cycle dynamics using a reversibly binding inhibitor. *Proc. Natl. Acad. Sci.* **95:** 14190–14195.

Phizicky E.M. and Fields S. 1995. Protein-protein interactions: Methods for detection and analysis. *Microbiol. Rev.* **59:** 94–123.

Tweeddale H., Notley-McRobb L., and Ferenci T. 1998. Effect of slow growth on metabolism of *Escherichia coli,* as revealed by global metabolite pool ("metabolome") analysis. *J. Bacteriol.* **180:** 5109–5116.

Velculescu V.E., Zhang L., Zhou W., Vogelstein J., Basrai M.A., Bassett Jr., D.E., Hieter P., Vogelstein B., and Kinzler K.W. 1997. Characterization of the yeast transcriptome. *Cell* **88:** 243–251.

Wasinger V.C., Cordwell S.J., Cerpa-Poljak A., Yan J.X., Gooley A.A., Wilkins M.R., Duncan M.W., Harris R., Williams K.L., and Humphery-Smith I. 1995. Progress with gene-product mapping of the Mollicutes: Mycoplasma genitalium. *Electrophoresis* **16:** 1090–1094.

2 Signal Transduction and Mammalian Cell Growth: Problems and Paradigms

Jonathan Chernoff

Fox Chase Cancer Center, Philadelphia, Pennsylvania 19111

INTRODUCTION

To divide or to differentiate, to attach or to move, to survive or to die—these are among the key decisions cells must make during the development and adult life of a metazoan organism. Such decisions must be accurate and well coordinated and are dictated both by factors external to the cell and by internal cues. The process by which cells carry out these decisions is termed signal transduction. This chapter reviews emerging principles that govern signaling pathways germane to cell growth and division, with particular emphasis on the role of protein–protein interactions. In so doing, the crucial role of such protein interactions in mitogenic signal transduction, and the importance of emerging technology for their detection, will become apparent.

Mitogenic signaling pathways are complicated. With the completion of the human genome sequence, the number of recognizable signaling proteins will certainly increase, and the models of these pathways are apt to become more complicated still. Although many signaling pathways seem formidably complex when viewed as a whole, at closer inspection these pathways often can be described in terms of a series of simple interactions of one protein with another. Indeed, so fundamental are these interactions that it is not an exaggeration to say that they form the basis of all signal transduction machinery.

Why are protein–protein interactions so important in mitogenic signaling? The binding of one signaling protein to another can have a number of consequences. For one, such binding can serve to recruit a signaling protein to a location where it is activated and/or where it is needed to

7

carry out its function. A relevant example that illustrates this phenomenon is the behavior of the protein kinase Raf, which, upon cell stimulation, is recruited from the cytoplasm to the plasma membrane by binding to the GTPase Ras (Avruch et al. 1994). A second consequence of protein interactions is that binding of one protein to another can induce conformational changes that affect activity or accessibility of additional binding domains, permitting additional protein interactions. Such is the case for signaling proteins such as p21-activated kinase, which, upon binding the GTPases Cdc42 or Rac, undergoes a profound conformational change that dislodges an autoinhibitory domain and thereby activates the kinase (Lei et al. 2000). Of course, stimulant-induced changes in protein location and conformation are not mutually exclusive. In many instances, recruitment to a signaling complex results in both relocation and enzymatic activation, as is the case, for example, with the protein tyrosine phosphatase SHP2, which is recruited from the cytoplasm to activated receptor tyrosine kinases (RPTKs) at the plasma membrane and is at the same time activated by the engagement of its SH2 domains to phosphotyrosine residues in the RPTK (Barford and Neel 1998; Hof et al. 1998).

THE SCAFFOLDING OF SIGNAL TRANSDUCTION

Binary protein–protein interactions are the cornerstone of signal transduction; however, an emerging theme in this research area is that higher-order assemblages are also critical for efficient transmission of signals. Scaffolds, adapters, insulators, and inhibitors are superimposed on the basic framework of the mitogenic signaling machinery. These additional layers of complexity have changed the way we look at signal transduction, and point to new ways to consider the organization of such pathways. For example, it has been known for some time that multicomponent complexes are assembled at activated RPTKs at the plasma membrane (Schlessinger 2000), whereas other complexes assemble at gene promoters on chromatin in the nucleus (Lee and Young 2000). It has now become clear that such multicomponent complexes also play a significant role in signaling in the cytoplasm and are critical for the regulation of mitogenesis. As one biologically important example, our understanding of the central Ras-Raf-Mek-mitogen-activated protein kinase (MAPK) pathway has evolved with the discovery that many of the elements of this complex are not only physically associated with one another, but also segregated from other cytoplasmic signaling proteins by several distinct scaffolding proteins (Garrington and Johnson 1999; Kolch 2000). These scaffolding proteins usually, but not always, lack catalytic function; however, they play key roles in signal transduction by virtue of their ability to complex with two or more elements of the Ras/MAPK pathway.

The use of scaffolding proteins in signal transduction is an evolutionarily conserved strategy. Examples of scaffolds for MAPK signaling modules have been found in all commonly studied eukaryotic organisms. At present, the function of these proteins is best understood in yeast (Fig. 1). In *Saccharomyces cerevisiae*, scaffold proteins function to segregate various common elements of MAPK modules. For example, the MAPK Ste11p participates in three distinct MAPK signaling modules: (1) the Ste5p scaffold coordinates components of the pheromone-response MAPK signaling module; (2) the Pbs2p scaffold coordinates components of an osmoregulatory MAPK signaling module; and (3) Ste11p also participates in a MAPK signaling module that regulates filamentation. In this last case, a scaffold protein has not been identified. In the absence of Pbs2p function, osmotic stress induces inappropriate activation of both the filamentous growth pathway and the mating pathway (O'Rourke and Herskowitz 1998; Davenport et al. 1999). Thus, in yeast it seems clear that one function of these scaffold proteins is to specify and insulate the signaling functions of MAPK modules.

In mammalian cells, scaffolds for MAPK modules include kinase suppressor of Ras (KSR) (Downward 1995), growth factor receptor-binding protein 10 (Grb10) (Nantel et al. 1998), and

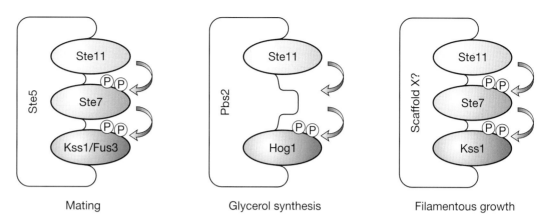

FIGURE 1. Scaffolds in budding yeast. *S. cerevisiae* uses similar signaling proteins toward different ends. The kinase Ste11p is involved in at least three distinct signaling cascades: mating, glycerol synthesis in response to hyperosmotic shock, and filamentous growth. It is thought that scaffolding proteins play a key role in signaling specificity, insulating Ste11p and downstream components from inadvertent activation by inappropriate stimuli. For the mating pathway, Ste5p binds three kinases, Ste11p, Ste7p, and Fus3p (and possibly Kss1p). In the case of hyperosmotic shock, the Ste11p target Pbs2p itself provides a scaffold function, binding both its upstream activator Ste11p and its downstream target Hog1p. It is not known whether the signaling machinery activated during filamentous growth requires an adapter protein.

Mek partner 1 (MP1) (Fig. 2) (Schaeffer et al. 1998). KSR binds to all three members of the canonical MAPK cascade: Raf, Mek, and Erk. Grb10 binds RPTKs, Raf, and Mek, but probably not at the same time, because a single SH2 domain in the carboxyl terminus of Grb10 mediates all these interactions. MP1 tethers the MAPK Erk1 to its activator MEK, and similar scaffolds (e.g,. JIP-1) are known for the stress-activated protein kinase Jnk and its upstream activators MKK7 and MLK (Yassuda et al. 1999). As in budding yeast, such a design may ensure efficient signal transmission and may also serve to prevent excessive interference, or cross-talk, from other signaling pathways in mammalian cells.

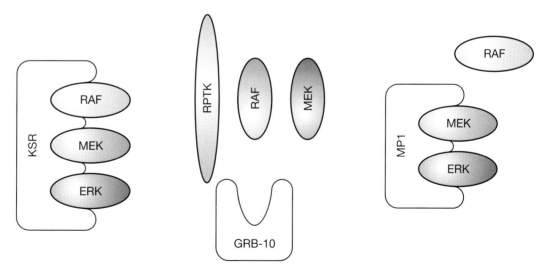

FIGURE 2. Examples of scaffolds in mammalian cells. As in yeast, mammalian signaling pathways also employ scaffolding proteins. KSR is functionally similar to budding yeast Ste5p, in that this protein can bind three members of a MAPK cascade. Grb10 represents a different type of adapter. This protein binds MAPK cascade members as well as RPTKs, but does so via a single binding domain and thus is unlikely to bind all three partners simultaneously. MP1 represents a third type of scaffold, linking together two elements of a MAPK cascade.

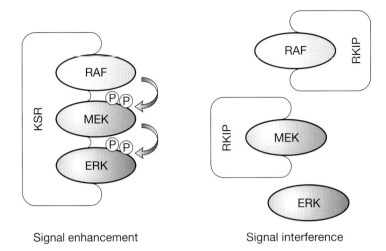

Signal enhancement   Signal interference

FIGURE 3. Antiscaffolds prevent efficient transmission of signals. Scaffold proteins such as KSR are thought to enhance signal transduction by physically coupling enzymes and their substrates. In contrast, antiscaffolds, such as RKIP, may interfere with signaling by sequestering these elements from one another.

Scaffold-like proteins are involved not only in signal transmission, but also in signal interference (Fig. 3). RKIP, a protein initially identified as a binding partner for Raf, can also bind Mek and Erk, suggesting that it might function in a manner analogous to KSR (Yeung et al. 1999, 2000). However, unlike KSR, RKIP inhibits, not enhances, signaling from Raf to Mek. The structural basis for this phenomenon probably lies in the fact that the binding elements on RKIP for Raf and Mek overlap substantially. Because it is unlikely that RKIP can simultaneously bind Raf and Mek, RKIP may act as a sequestering agent for both these proteins, preventing their effective interaction with one another. It would not be surprising if other antiscaffold proteins should be found that disrupt signaling modules in a similar manner.

In mammalian cells, other classes of cytoplasmic scaffolding proteins have been found that link proteins operating in common signaling pathways. Such proteins are sometimes termed anchoring scaffolds. Examples of such anchoring scaffolds that are relevant for mitogenic pathways include the A-kinase anchoring proteins (AKAPs), the cytoplasmic domains of many RPTKs, insulin receptor substrate 1 (IRS1), the T-cell protein linker for activation of T cells (LAT), and Src homology 2 domain-containing protein of 76 kD (SLP76) (Burack and Shaw 2000). These types of scaffolds differ from those involved in the Raf-Mek-MAPK pathway in that the complexed proteins do not act enzymatically upon one other (e.g., Raf phosphorylates Mek, which phosphorylates MAPK), but rather are all involved in a particular signaling function. The anchoring function of this class of scaffold protein may be important in the assembly of signaling elements, imposing a spatial and temporal restriction on their distribution.

PROTEIN–PROTEIN INTERACTIONS AND THE REGULATION OF SIGNALING CROSS-TALK

Although the existence of multicomponent signaling modules in the cytoplasm of mammalian cells is no longer in dispute, their purpose is not completely clear. Is it, as is apparently the case in yeast, to provide insulation against cross-talk from competing signaling pathways? Is it to increase the efficiency of signal transmission by bringing the relevant components in proximity with one another? The prevalence and relevance of cross-talk in mammalian signal transduction have been hotly debated items in the signaling literature. At issue is whether signaling pathways in mammalian cells are best described in terms of linear flow diagrams, or as interconnected webs (Tucker et al. 2001). In simple eukaryotes such as budding yeast, *Drosophila*, and *Caenorhabditis elegans*, the majority of signaling pathways appear to operate as discrete, linear entities, perhaps due to the presence of insulating adapter proteins. This facet of signal transduction can be clearly seen in the mating pathway in *S. cerevisiae*. Activation of the pheromone receptor sets off a chain of events that can be likened to a bucket brigade, with the signal passing from the pheromone receptor to a set of protein kinases arranged in series (and with some components held together by the Ste5p scaffold protein) to a transcription factor (Posas et al. 1998). Experimentally, the essential linearity of the system can be appreciated by the fact that the transcriptional response to mating pheromone is completely dependent on a single transcription factor, Ste12p (Roberts et al. 2000). If significant cross-talk occurred, pheromone signals would be expected to diverge and induce expression of additional, non-Ste12p-dependent genes.

Whether mitogenic signaling pathways operate as discrete linear units or as interconnected webs in higher eukaryotes, it is known that stimulation of a given receptor can activate multiple pathways, and stimulation of similar receptors can yield different outcomes. A classic example of the latter phenomenon is seen in PC12 cells, where switching on the epidermal growth factor (EGF) receptor leads to proliferation, whereas stimulation of the nerve growth factor (NGF) receptor causes growth arrest and differentiation (Marshall 1995). Many of the same signaling proteins are activated by these two receptors, so any description of the differing outcomes of EGF versus NGF signaling will have to invoke either spatial or temporal issues or subtle differences in the identity of the effectors that are recruited to the receptors. We know that EGF leads to transient MAPK activation, whereas NGF leads to prolonged MAPK activation (Qui and Green 1992; Traverse et al. 1992), which is likely to account for their opposing cellular effects; however, the molecular basis for these temporal differences remains uncertain. One strong possibility is that differential use of scaffolds is a key factor. Landreth and colleagues have shown that transient activation of the MAPKs by EGF is a consequence of the formation of a short-lived complex assembling on the EGF receptor itself, comprising the adapter Crk, the exchange factor C3G, the GTPase Rap1, and the protein kinase B-Raf (Kao et al. 2001). In contrast, NGF stimulation of PC12 cells results in the phosphorylation of a scaffold protein, FRS2, on which is assembled a stable complex of Crk, C3G, Rap1, and B-Raf, resulting in the prolonged activation of the MAPKs. Thus, although the same set of signaling proteins is recruited by NGF and EGF, alterate scaffolding arrangements are induced by these different stimuli and may result in the observed unique outcomes.

TECHNOLOGY AND SIGNAL TRANSDUCTION—A VIEW OF THE NEAR FUTURE

Given the importance of scaffolds and antiscaffolds in regulating cross-talk in mitogenic signal transduction, and the fact that the diverse structures of these proteins make their identification by sequence inspection or complementation difficult, emerging technologies for finding proteins of this type could be useful. For example, the multiple-bait interaction trap systems (see Serebriiskii et al. 1999), in which pairs of baits are examined for binding to a common partner, could be used to identify proteins that bind both Raf and Mek, or any other pair of signaling proteins in which scaffolds are suspected to play a role. Because overexpression of scaffolds often inhibits signal transduction (Dickens et al. 1997), probably by altering the stoichiometry of the endogenous signaling components, it is also conceivable that proteins with scaffold or antiscaffold function could be identified by expression screens for inhibitors of mitogenesis. Finally, the rapid evolution of mass spectrometry techniques may make it likely that the components of cytoplasmic mitogenic signaling complexes will be increasingly amenable to direct identification in immunoprecipitates (Pandey and Mann 2000).

If mitogenic signaling in higher eukaryotes in fact operates in such a complex manner, then to understand signaling pathways it will be necessary to develop a means to analyze the operation and interaction of large networks of proteins rather than to consider the behavior of a few individual components in isolation. However, at present we lack good tools to perform such multi-component analyses. Until now, our understanding of signaling pathways has been built mainly on the backbone of biochemistry, the careful and often painstaking analysis of single proteins in vitro. Is a new analytic paradigm for signal transduction required, and, if so, what will it be?

Clearly, data from array analyses of mitogenic signals are telling us something, but the messages are not easy to decipher. Serum stimulation of fibroblasts in culture induces expression of a bewildering array of genes (Iyer et al. 2000). Activated Ras induces or impedes expression of scores, if not hundreds, of genes (Zuber et al. 2000). How does one begin to make sense of these data? Some have suggested that proteins whose expression is coordinately regulated might work together, affecting similar processes. In other words, temporal concordance of expression might imply functional similarities. Perhaps the best case for this theory can be made in the budding yeast mating pathway, where many of the genes involved are coordinately up-regulated. This may also be true in mammalian systems, but there is not much evidence to support it in a general way.

Beyond temporal concordance, it is, in principle, possible to impose a second level of organization of the data, that of spatial location. If we knew where all these proteins were located in the cell, we could begin to sort out what proteins interact with what others. The colocalization of candidate binding proteins can be established by fluorescence resonance transfer methods and related techniques (Bastiaens and Squire 1999; Pollok and Heim 1999). General schemes for localizing proteins have also been published, usually involving screening libraries of green fluorescent protein (GFP) fusions (Ding et al. 2000; Misawa et al. 2000). On a genome-wide scale, it could be useful to combine such localization data with those obtained by independent means.

A similar approach, whose utility has recently been validated, is to determine the composition of signaling networks by combining biochemical data with two-hybrid analyses to analyze protein–protein interactions. In budding yeast, a meta-analysis of existing two-hybrid and biochemical data has confirmed the presence of several clusters of interacting proteins and, in some cases, has suggested functions for proteins whose roles were previously obscure (Schwikowski et al. 2000). Finally, evolutionary relationships may also provide a key to understanding signaling complexes, because conserved proteins that operate together in model eukaryotes such as yeast, worms, or flies are often good indicators of a similar design in humans.

We are entering a new era, and our old ideas about mitogenic signaling are undergoing a rapid revision. It has become abundantly clear not only that protein–protein interactions form the basis for signal transduction, but also that these interactions occur on a scale and with a degree of complexity not previously suspected. It will be a great challenge to integrate the onslaught of data from the genome sequencing projects to determine the organization of signaling proteins in mitogenesis. However, meeting this challenge might allow us to learn how to manipulate protein–protein interactions in the treatment of human diseases such as cancer, where the proper regulation of signal transduction has by definition gone awry.

REFERENCES

Avruch J., Zhang X.F., and Kyriakis J.M. 1994. Raf meets Ras: Completing the framework of a signal transduction pathway. *Trends Biochem. Sci.* **19:** 279–283.

Barford D. and Neel B.G. 1998. Revealing mechanisms for SH2 domain mediated regulation of the protein tyrosine phosphatase SHP-2. *Structure* **6:** 249–254.

Bastiaens P.I.H. and Squire A. 1999. Fluorescence lifetime imaging microscopy: Spatial resolution of biochemical processes in the cell. *Trends Cell Biol.* **9:** 48–52.

Burack W.R. and Shaw A.S. 2000. Signal transduction: Hanging on a scaffold. *Curr. Opin. Cell Biol.* **12:** 211–216.

Davenport K.D., Williams K.E., Ullmann B.D., and Gustin M.C. 1999. Activation of the *Saccharomyces cerevisiae* filamentation/invasion pathway by osmotic stress in high-osmolarity glycogen pathway mutants. *Genetics* **153:** 1091–1103.

Dickens M., Rogers J.S., Cavanagh J., Raitano A., Xia Z., Halpern J.R. Greenberg M.E., Sawyers C.L., and

Davis R.J. 1997. A cytoplasmic inhibitor of the JNK signal transduction pathway. *Science* **277:** 693–696.

Ding D.Q., Tomita Y., Yamamoto A., Chikashige Y., Haraguchi T., and Kiraoka Y. 2000. Large-scale screening of intracellular protein localization in living fission yeast cells by the use of a GFP-fusion genomic DNA library. *Genes Cells* **5:** 169–190.

Downward J. 1995. KSR: A novel player in the RAS pathway. *Cell* **83:** 831–834.

Garrington T.P. and Johnson G.L. 1999. Organization and regulation of mitogen-activated protein kinase signaling pathways. *Curr. Opin. Cell Biol.* **11:** 211–218.

Hof P., Pluskey S., Dhe-Paganon S., Eck M.J., and Shoelson S.E. 1998. Crystal structure of the tyrosine phosphatase SHP-2. *Cell* **92:** 441–450.

Iyer V.R., Eisen M.B., Ross D.T., Schuler G., Moore T., Lee J.C., Trent J.M., Staudt L.M., Hudson J., Boguski M.S., Lashkari D., Shalon D., Botstein D., and Brown P.O. 1999. The transcriptional program in the response of human fibroblasts to serum. *Science* **283:** 83–87.

Kao S., Jaiswal R.K., Kolch W., and Landreth G.E. 2001. Identification of the mechanisms regulating the differential activation of the MAP kinase cascade by epidermal growth factor and nerve growth factor in PC12 cells. *J. Biol. Chem.* **276:** 18169–18177.

Kolch W. 2000. Meaningful relationships: The regulation of the Ras/Raf/MEK/ERK pathway by protein interactions. *Biochem. J.* **351:** 289–305.

Lee T.I. and Young R.A. 2000. Transcription of eukaryotic protein-coding genes. *Annu. Rev. Genet.* **34:** 77–137.

Lei M., Lu W., Merg W., Parrini M.-C., Eck M.J., Mayer B.J., and Harrison S.C. 2000. Structure of PAK1 in an autoinhibited conformation reveals a multistage activation switch. *Cell* **102:** 387–397.

Marshall C.J. 1995. Specificity of receptor tyrosine kinase signaling: Transient versus sustained extracellular signal-regulated kinase activation. *Cell* **80:** 179–185.

Misawa K., Nosaka T., Morita S., Kaneko A., Nakahata T., Asano S., and Kitamura T. 2000. A method to identify cDNAs based on localization of green fluorescent protein fusion products. *Proc. Natl. Acad. Sci.* **97:** 3062–3066.

Nantel A., Mohammad-Ali K., Sherk J., Posner B.I., and Thomas D.Y. 1998. Interaction of the Grb10 adapter protein with the Raf1 and MEK1 kinases. *J. Biol. Chem.* **273:** 10475–10484.

O'Rourke S.M. and Herskowitz I. 1998. The Hog1 MAPK prevents cross talk between the HOG and pheromone response MAPK pathways in *Saccharomyces cerevisiae*. *Genes Dev.* **12:** 2874–2886.

Pandey A. and Mann M. 2000. Proteomics to study genes and genomes. *Nature* **405:** 837–846.

Pollok B.A. and Heim R. 1999. Using GFP in FRET-based applications. *Trends Cell Biol.* **9:** 57–60.

Posas F., Takekawa M., and Saito H. 1998. Signal transduction by MAP kinase cascades in budding yeast. *Curr. Opin. Microbiol.* **1:** 175–182.

Qui M.S. and Green S.H. 1992. PC12 cell neuronal differentiation is associated with prolonged p21ras activity and consequent prolonged ERK activity. *Neuron* **9:** 705–717.

Roberts C.J., Nelson B., Marton M.J., Stoughton R., Meyer M.R., Bennett H.A., He Y.D., Dai H., Walker W.L., Hughes T.R., Tyers M., Boone C., and Friend S.H. 2000. Signaling and circuitry of multiple MAPK pathways revealed by a matrix of global gene expression profiles. *Science* **287:** 873–880.

Schaeffer H.J., Catling A.D., Eblen S.T., Collier L.S., Krauss A., and Weber M.J. 1998. MP1: A MEK binding partner that enhances enzymatic activation of the MAP kinase cascade. *Science* **281:** 1668–1671.

Schlessinger J. 2000. Cell signaling by receptor tyrosine kinases. *Cell* **103:** 211–225.

Schwikowski B., Uetz P., and Fields S. 2000. A network of protein-protein interactions in yeast. *Nature Biotech.* **18:** 1257–1261.

Serebriiskii I., Khazak V., and Golemis E.A. 1999. A two-hybrid dual bait system to discriminate specificity of protein interactions. *J. Biol. Chem.* **274:** 17080–17087.

Traverse S., Gomez N., Paterson H., Marshall C., and Cohen P. 1992. Sustained activation of the mitogen-activated protein (MAP) kinase cascade may be required for differentiation of PC12 cells. Comparison of the effects of nerve growth factor and epidermal growth factor. *Biochem. J.* **288:** 351–355.

Tucker C.L., Gera J.F., and Uetz P. 2001. Towards an understanding of complex protein networks. *Trends Cell Biol.* **11:** 102–106.

Yasuda J., Whitmarsh A.J., Cavanagh J., Sharma M., and Davis R.J. 1999. The JIP group of mitogen-activated protein kinase scaffold proteins. *Mol. Cell Biol.* **19:** 7245–7254.

Yeung K., Janosch P., McFerran B., Rose D.W., Mischak H., Sedivy J.M., and Kolch W. 2000. Mechanism of suppression of the Raf/MEK/extracellular signal-regulated kinase pathway by the raf kinase inhibitor protein. *Mol. Cell Biol.* **20:** 3079–3085.

Yeung K., Seitz T., Li S., Janosch P., McFerran B., Kaiser C., Fee F., Katsanakis K.D., Rose D.W., Mischak H., Sedivy J.M., and Kolch W. 1999. Suppression of Raf-1 kinase activity and MAP kinase signalling by RKIP. *Nature* **401:** 173–177.

Zuber J., Tchernitsa O.I., Hinzmann B., Schmitz A.C., Grips M., Hellriegel M., Sers C., Rosenthal A., and Schafer R. 2000. A genome-wide survey of RAS transformation targets. *Nat. Genet.* **24:** 144–152.

3 Impact of Protein Interaction Technologies on Cancer Biology and Pharmacogenetics

Rebecca Blanchard Raftogianis[1] and Andrew K. Godwin[2]

Departments of [1]Pharmacology and [2]Medical Oncology, Fox Chase Cancer Center, Philadelphia, Pennsylvania 19111

SEQUENCE VARIATIONS: SUSCEPTIBILITY TO CANCER AND TREATMENT BASED ON GENOTYPE-SPECIFIC PROTEIN INTERACTIONS

New genomic and proteomic technologies and the completion of the human genome sequence promise a revolution in our understanding of human disease. These developments are expected to transform dramatically the study of diseases such as cancer and to allow the molecular classification of such diseases. Furthermore, the identification and functional characterization of genetic variation in the human genome will be a major scientific effort in the post-genomics era. Identification of genetic variation is the current focus of several genome projects and is proceeding at a rapid pace (Service 2000). Many diverse sources of data have shown that any two individuals are more than 99.9% identical in sequence, which means that the differences among individuals in our own species that can be attributed to genes falls in a mere 0.1% (or ~2,900,000 nucleotides) of the genome (Venter et al. 2001). Characterization of the functional significance of those variants will be laborious and will likely continue to be a scientific focus for many years. Two areas of particular interest are genetic variants that predispose individuals to disease and variants that affect individual response to drugs (pharmacogenetics/pharmacogenomics). Genetic variation can impart a functional phenotype via a myriad of mechanisms, including the introduction or interruption of protein–protein interactions or protein–ligand interactions. The focus of this chapter is to discuss the application of the techniques described in this book to the study of functional significance of human genetic variation in the context of human disease susceptibility and pharmacogenetics.

CANCER AND THE GENOME

Cancer is an important public health concern in the United States and around the world. After heart disease, it is the second leading cause of death, accounting for 23.3% of deaths in the United States (Greenlee et al. 2000). In the year 2000, about 1,220,100 new cases of invasive cancer were expected to be diagnosed in the United States. In addition, ~1.3 million cases of basal and squamous cell skin cancer are diagnosed annually (Greenlee et al. 2000). An estimated 552,000 Americans were predicted to have died of cancer in 2000, which translates into more than 1500 people a day. Among men, the most common cancers are cancers of the prostate (29%), lung and bronchus (14%), and colon and rectum (10%). Among women, the three most commonly diagnosed cancers are cancers of the breast (30%), lung and bronchus (12%), and colon and rectum (11%) (Greenlee et al. 2000). In 1987, lung cancer surpassed breast cancer as the leading cause of cancer death in women and is expected to account for at least 25% of all female cancer deaths in the coming years. The good news is that following more than 70 years of increases, the recorded number of total cancer deaths among men in the United States has declined for the first time, from a peak of 281,898 in 1996 to 281,110 in 1997. In contrast, among women, the recorded number of total cancer deaths continues to increase, although the rate of increase has diminished in recent years. These somewhat encouraging trends are primarily associated with improved screening techniques and the subsequent increase in diagnosis at an early stage when most, but not all, cancers are more successfully treated. Unfortunately, most current cancer therapies have limited efficacy in curing late-stage disease. Therefore, there continues to be a great and immediate need to develop new approaches to (1) diagnose cancer early in its clinical course, (2) more effectively treat advanced stage disease, (3) better predict a tumor's response to therapy prior to the actual treatment, and (4) ultimately prevent disease from arising through the use of chemopreventive strategies. These goals can only be accomplished through a better understanding of how certain genes and their encoded proteins contribute to disease onset and tumor progression and how they influence the response of patients to drug therapies. Innovations in genetic, biological, and biochemical approaches are necessary to realize these goals.

Genetic Risk Factors

The risk of developing high-incidence cancers (e.g., lung, breast/ovarian, colon, prostate) is not uniformly distributed throughout the population. In part, this lack of uniformity may be explained by different environmental, occupational, and recreational exposure histories (e.g., ultraviolet light from the sun, inhaled cigarette smoke, incompletely defined dietary factors). These carcinogens can affect one or multiple stages of carcinogenesis through both genetic and epigenetic mechanisms (Shields and Harris 2000; Rothman et al. 2001). However, only a small fraction of exposed individuals ultimately develop cancer. Moreover, the risk among individuals with similar exposures is unevenly distributed. Individuals in certain families have been observed to have greater risk and earlier onset of cancer. Families ascertained through a cancer proband (i.e., from the Latin propositus [male] or proposita [female], the individual through whom the family is ascertained) also show significant excess of other cancers in their family members. This excess in cancer incidence persists even when rare Mendelian forms segregating in families are removed. These observations suggest that common cancer susceptibility may, in part, be determined by host genetic factors in addition to Mendelian factors identified to date.

At least two classes of genes have been identified that may determine the risk of developing human cancer. The first class is composed of a restricted number of genes that are directly involved in tumorigenesis, such as proto-oncogenes, tumor suppressor genes, and DNA mismatch repair genes. These can be thought of as the cancer-causing genes (Fearon 1997; Godwin et al. 1997). The risk associated with germ-line mutations in this class of genes, e.g., tumor suppressor genes (including *RB1, TP53, CDKN2, WT1, NF1, NF2, TSC1, TSC2, VHL, PTCH, PTEN, LKB1, SMAD4, APC, MEN1, CDH1, TGFBR2, EXT1, EXT2*), oncogenes (*RET, MET, KIT, CDK4*), DNA mismatch repair genes (*MSH2, MLH1, PMS1, PMS2, MSH6*), and two genes related to DNA repair and tumor suppression (*BRCA1, BRCA2*) is high, but the risk alleles are rare in the population (Fig. 1). For example, the risk of developing a cancer in a RB mutation carrier is extremely high (>1000-fold), whereas the "at risk" allele is infrequent (~1 in 20,000 live births) in the general population. These genes are also frequent targets for mutations in sporadic forms of the disease. These cancer susceptibility genes encode proteins that perform diverse cellular functions, including transcription, cell cycle control, DNA repair, and apoptosis. The proteins encoded by these genes often function through protein–protein interactions, and thus their characterization is particularly amenable to the use of techniques described in this book. The second class is broader and not completely defined (Rothman et al. 2001). The most frequently studied group in this case is composed of genes involved in metabolic detoxification pathways and steroid and amino acid metabolism (Evans and Relling 1999). The risk associated with the second class is significantly lower (maybe two- to tenfold) but is likely to be of greater public health significance as a consequence of the higher frequency (as great as 50% in certain populations) of the risk-modifying factors (Fig. 1). For example, the cytochromes P450 (CYP) evolved to catalyze the metabolism of numerous structurally diverse exogenous and endogenous molecules. Approximately 55 different CYP genes are present in the human genome and are classified into different families and subfamilies on the basis of sequence homology. Members of the CYP3A subfamily, for example, catalyze the oxidative, peroxidative, and reductive metabolism of structurally diverse endobiotics, drugs, and protoxic or procarcinogenic molecules (Rendic and Di Carlo 1997). Moreover, because CYP3A metabolizes estrogens to 2-hydroxyestrone, 4-hydroxyestrone, and 6α-hydroxylated estrogens, all of which have been implicated in estrogen-mediated carcinogenicity, variation in CYP3A may influence the circulating levels of these estrogens and the risk of breast cancer. CYP3A4, a member of the CYP3A family, has been shown to be associated with oxidative deactivation of testosterone. Rebbeck and colleagues demonstrated that a single base change in the *CYP3A4* gene was significantly associated with higher clinical stage and grade in men with prostate tumors (Rebbeck et al. 1998), indicating that mutations in *CYP3A4* may influence prostate carcinogenesis. Kuehl and colleagues have recently shown that a single-nucleotide polymorphism (SNP) in *CYP3A5* leads to alternative splicing and protein truncation and results in the absence of CYP3A5

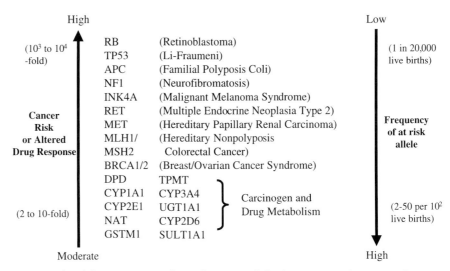

Germline Mutations or Variants

High		Low
(10^3 to 10^4 -fold)	RB (Retinoblastoma) TP53 (Li-Fraumeni) APC (Familial Polyposis Coli) NF1 (Neurofibromatosis) INK4A (Malignant Melanoma Syndrome) RET (Multiple Endocrine Neoplasia Type 2) MET (Hereditary Papillary Renal Carcinoma) MLH1/ (Hereditary Nonpolyposis MSH2 Colorectal Cancer) BRCA1/2 (Breast/Ovarian Cancer Syndrome)	(1 in 20,000 live births)
Cancer Risk or Altered Drug Response		**Frequency of at risk allele**
(2 to 10-fold)	DPD TPMT CYP1A1 CYP3A4 } Carcinogen and CYP2E1 UGT1A1 } Drug Metabolism NAT CYP2D6 GSTM1 SULT1A1	(2-50 per 10^2 live births)
Moderate		High

FIGURE 1. Example of the inverse correlation between allelic frequency and cancer risk associated with inherited cancer susceptibility genes versus the cancer risk or change in drug response associated with variations in drug-metabolizing genes. The cancer syndrome associated with germ-line mutations in each of the cancer susceptibility genes listed is shown in parentheses. (DPD) Dihydropyrimidine dehydrogenase; (CYP1A1) cytochrome P450 1A1; (CYP2E1) cytochrome P450 2E1; (NAT) *N*-acetyltransferase; (GSTM1) glutathione-*S*-transferase, Mu 1; (TPMT) thiopurine methyltransferase; (CYP3A4) cytochrome P450 3A4; (UGT1A1) UDP-glucuronosyltransferase 1A1; (CYP2D6) cytochrome P450 2D6; (SULT1A1) sulfotransferase 1A1.

from tissues in some homozygous individuals (Kuehl et al. 2001). Because CYP3A5 represents at least 50% of the total hepatic CYP3A content in people polymorphically expressing CYP3A5, *CYP3A5* may be the most important genetic contributor to interindividual differences in CYP3A-dependent drug clearance and in response to many medicines.

Molecular genetic methods are being combined with genomic science approaches to facilitate the identification and large-scale characterization of constitutional DNA variation in these kinds of candidate loci. The actual number of genes present in the human genome that contribute to cancer susceptibility as well as response to drugs is unknown and will require many years of additional research to uncover the normal function of the encoded proteins and how these carefully regulated activities are disrupted in cancer. New protein interaction technologies will be key to uncovering how sequence variations as small as a single-nucleotide substitute in any given gene can potentially alter the function of the protein through its interaction with other proteins and/or substrates. Regarding cancer, genes involved in cancer susceptibility, cancer progression, and response to drug therapy will be key. Each of these classes is discussed in some detail below.

CANCER GENES

Despite the prevalence of cancer, little is known about the molecular events that occur during its development. Cancer, both hereditary and nonhereditary, is a multistep process that involves alterations in many specific genes. The normal cell has multiple independent mechanisms that regulate its growth and differentiation, and several separate events are required to override these control mechanisms. The fundamental mechanisms underlying the genetic basis of cancer are being defined and involve, as indicated above, alterations in genes that have been classified into three general categories: (1) proto-oncogenes, which are involved in growth promotion and whose defects leading to cancer are a gain of function; (2) tumor suppressor genes, which are negative regulators of growth and whose loss of function contributes to cancer; and (3) DNA repair genes, which maintain the integrity of the genome and whose loss of function causes increased accumulation of mutations in other critical cancer-causing genes. Progress is being made in isolating these genes and the proteins they encode, determining the normal cellular functions of the proteins, and investigating the mechanisms of tumorigenesis. Recently, it was predicted that ~1.3% (or roughly 351) of the genes present in the human genome are likely to be proto-oncogenes, whereas as many as 130 human DNA repair genes have already been identified (Lander et al. 2001; Wood et al. 2001). The number of classic tumor suppressor genes is relatively small, 20–30; however, loss of heterozygosity (LOH) studies suggest dozens of additional tumor suppressors have yet to be classified (discussed below). A combination of somatic mutations and chromosomal alterations affecting these genes and other genes is thought to be the driving force behind tumorigenesis. Much work remains to uncover how these proteins function normally to understand how they contribute to cancer. However, a number of these have been drawn into nets of interactions with other proteins and are beginning to provide hints to their activity.

Oncogenes

Proto-oncogenes/oncogenes are genes whose action promotes cell proliferation in a positive manner. The normal, nonmutant versions are properly called proto-oncogenes, and the mutant versions, the oncogenes, are excessively or inappropriately active. Mutations converting normal proto-oncogenes into oncogenes are gain-of-function mutations, and these mutations include point mutations; structural alterations such as insertions, deletions, inversions, and translocations; gene amplification; and hypomethylation. Oncogenes were originally identified on the basis of their similarity to retroviral sequences, which were known to be able to transform cells. Many cellular oncogenes have now been identified; however, their involvement in cancer has yet to be fully elucidated. The most commonly studied oncogenes are those of the SRC, RAS, MYC, and ERBB family of proteins (Godwin et al. 1997). Although the functions of these genes initially appeared to be unrelated, elucidation of protein interactions has in some cases shown commonalities of effect. Furthermore, studies of these proteins share some areas with studies of signal transduction (see Chapter 2).

Tumor Suppressor Genes

The concept that genes could suppress cell growth came from early studies of somatic cell fusion (Harris et al. 1969; Stanbridge 1976). In those experiments, fusion of tumorigenic cells with normal cells resulted in hybrids that could continue to grow in culture but were no longer tumorigenic in animals. Moreover, when some of the hybrid cells re-expressed the tumorigenic phenotype, that re-expression correlated with specific loss of chromosomes derived from the normal, nontumorigenic parental cell. In 1969, de Mars proposed that in certain familial cancers, gene carriers might be heterozygous for a recessive mutation and that the cancer appears because of subsequent somatic mutation causing an individual cell to become homozygous for the cancer-causing gene (DeMars 1969). Epidemiological analysis of retinoblastoma and Wilms' tumor provided important evidence for this theory (Knudson 1971; Knudson and Strong 1972) and led to what is now referred to as Knudson's "two-hit" hypothesis. Two forms of the disease were observed: patients with a family history of disease who often presented with multiple bilateral lesions and patients without family history who typically presented with single unilateral lesions and at a later age than those with familial disease. Knudson and Strong speculated that the familial form of disease represented the inheritance of a mutation predisposing to neoplasia and that only one additional rate-limiting genetic event was necessary for tumorigenesis. Sporadic tumors, on the other hand, required two independent genetic lesions and therefore would be slower to develop. Because two independent lesions occurring in the same cell would be rare, only unilateral tumors would be expected. This hypothesis was confirmed at the molecular level with the isolation of the *RB1* gene (Cavenee et al. 1983; Dunn et al. 1989). The development of retinoblastoma has been shown to involve loss of both alleles at a single locus, with individuals with the hereditary form of retinoblastoma having one mutated allele in the germ line (Marshall 1991). Retinoblastoma is one of several dozen prototypic tumor suppressor genes (e.g., *TP53, APC, WT1, NF1, PTEN*) that conform to the classic definition of Knudson's two-hit hypothesis.

Increasingly, different mechanisms of mutation are being implicated in the inactivation of tumor suppressor genes leading to neoplasia (both hereditary and nonhereditary forms). These include point mutations, deletions, insertions, hypermethylation, alterations in genomic imprinting, loss of genetic material leading to haploinsufficiency, dominant negative mutations, and homozygous deletions (Fearon 2000; Hanahan and Weinberg 2000; Prowse et al. 2001). Based on the idea of tumor suppressors as components of complexes, these genetic lesions predict multiple modes of action, including maintenance of genomic stability, programmed cell death (apoptosis), DNA repair, and cell cycle control (Fearon 1997; Godwin et al. 1997; Lundberg and Weinberg 1999; Hanahan and Weinberg 2000). Furthermore, as more and more individuals elect to undergo diagnostic testing at the DNA level for disease susceptibility, the range of known mutations in these genes will continue to grow. This scenario is already giving rise to a new set of problems, namely, delineating deleterious mutations from benign polymorphisms. Protein technology that will help to assess the functional consequences of a single residue change in a protein will be key. For example, at present, over 860 and 880 different sequence variants (deleterious mutations, naturally occurring polymorphisms, and unclassified variants) have been identified in *BRCA1* and *BRCA2*, respectively, and more than 50% have been reported only once (see Breast Cancer Information Core, www.nhgri.nih.gov/ Intramural_research/Lab_transfer/Bic/index.html). Although a few missense mutations within these genes have been identified as conferring a predisposition to breast cancer, most such mutations have not been associated with a known phenotype. Like nonsense mutations, missense mutations result from the substitution of a single nucleotide within a coding codon. Unlike nonsense mutations, the substitution results in a functional codon but encodes a different amino acid at that position. The problem with missense changes is that it is not always simple to determine whether the amino acid substitution will adversely affect the protein's function and thereby contribute to the disease phenotype. If the missense change is commonly found in control populations (ethnically matched disease-free individuals with no family history of breast and/or ovarian cancer), it is deemed to be a naturally occurring polymorphism. However, many of these variants are found in only a limited number of families and are referred to as variants of unknown significance. The frequent discovery of variants of unknown significance is a major problem from a clinical standpoint because many patients who undergo genetic testing are left to interpret these ambiguous results while trying to make important health-care decisions. Therefore, new protein technologies will be necessary to complement genetic studies and to help define the consequence of these single-base changes in regard to protein function and, thus, cancer risk.

Finally, LOH studies are the most frequently used method to define a region that may harbor a tumor suppressor gene. Whole-genome allelotyping studies of nearly every type of common cancers revealed that LOH has been observed on every chromosome arm. These studies suggest that a large number of tumor suppressor genes may exist, most of which have yet to be identified and their functions characterized. As the technology continues to improve, functional genomics and high-throughput screening methods will provide powerful means to identify the genetic components that influence human health and disease. Proteomic approaches will further expand the molecular profile of disease and define the functional pathways.

DNA Repair Genes

The human genome, like other genomes, encodes information to protect its own integrity (Lindahl and Wood 1999). DNA repair enzymes continuously monitor chromosomes to correct damaged nucleotide residues generated by exposure to carcinogens and cytotoxic compounds. Genomic instability caused by the great variety of DNA-damaging agents would be an overwhelming problem for cells and organisms if it were not for DNA repair. Proteins for base excision repair, nucleotide excision repair, and mismatch excision repair have all been identified, and defects in some of these proteins have been shown to be closely associated with cancer development. For example, proteins encoded by mismatch repair genes correct occasional errors of DNA replication as well as heterologies formed during recombination. The bacterial *mutS* and *mutL* genes encode proteins responsible for identifying mismatches, and there are numerous homologs of these genes in the human genome, of greater variety than those found in yeast and nematodes. Some of these proteins are specialized for locating distinct types of mismatches in DNA, some are specialized for meiotic and/or mitotic recombination, and some have functions yet to be determined. Of the 130-plus DNA repair genes, at least 11 are mismatch excision repair genes (Wood et al. 2001), and some of these genes (e.g., *hMSH2, hMLH1, hPMS1,* and *hPMS2*), when mutated, have been shown to predispose to hereditary nonpolyposis colorectal cancer (HNPCC) or Lynch syndrome (Lynch and Smyrk 1996), as well as sporadic forms of the disease. *hMLH1* and *hMSH2* are believed to account for most HNPCC families with an identifiable deleterious mutation (Peltomaki and Vasen 1997). Studies of these proteins from bacteria to mammals emphasize their assembly into recombination "machines." Furthermore, study of the locations of the mutations in the context of the assembled structure of components should both provide insight into basic mechanisms and suggest targets for therapeutic intervention as discussed below.

The study of protein–protein interactions has provided important insights into the functions of many of the known oncogenes, tumor suppressors, and DNA repair proteins. These studies have also demonstrated that members of each class of protein participate in overlapping pathways and, when mutated, each theoretically could contribute to cancer. A strong example of this is the human *TP53* gene. *TP53* codes for a protein product (referred to as p53) that has an important biological function as a cell cycle checkpoint. p53, originally detected by virtue of its ability to form a stable complex with the SV40 large-T antigen (Fig. 2), has been a constant source of fasci-

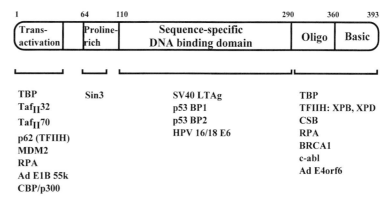

FIGURE 2. Schematic of p53 protein and its interacting proteins. The various protein domains and the proteins that interact with each of these regions are shown (Mansur et al. 1995; Dobner et al. 1996; Gottlieb and Oren 1996; Ko and Prives 1996; Levine 1997; Lill et al. 1997; Zhang et al. 1998; Nie et al. 2000; Zilfou et al. 2001). (This figure was provided courtesy of Dr. Maureen Murphy, Fox Chase Cancer Center, Philadelphia.)

nation since its discovery over a decade ago (for reviews, see Gannon and Lane 1990; Soussi et al. 1990). The gene encoding this 53-kD nuclear phosphoprotein was initially considered to be a cellular oncogene because introduction of expression vectors containing mutant *TP53* cDNA clones by transfection could transform recipient cells in concert with an activated *ras* gene. Subsequently, however, several convergent lines of research indicated that normal (wild-type) p53 actually functioned as a tumor suppressor.

The *TP53* gene is one of the most commonly altered genes identified in human tumors (e.g., sporadic osteosarcomas, soft-tissue sarcomas, brain tumors, leukemias, and carcinomas of the breast, colon, lung, and ovary) occurring in a large fraction (perhaps even half) of the total cancers in the United States (Hollstein et al. 1996; Levine 1997; Hainaut et al. 1998). Unlike many tumor suppressor genes, missense mutations represent a high proportion (>70%) of *TP53* mutations (http://www.iarc.fr/p53/homepage.html). Furthermore, in contrast to the retinoblastoma gene *RB*, where the hereditary syndrome served as the basis for identification of the causal gene, *TP53* was discovered and subsequently found to have a role in hereditary cancer. In 1990, Li and colleagues identified germ-line *TP53* mutations in a series of families with Li-Fraumeni syndrome (LFS), which features diverse childhood cancers as well as early-onset breast cancers (Malkin et al. 1990).

The p53 exists at low levels in virtually all normal cells. Wild-type p53 acts as a negative regulator of cell growth that is induced following DNA damage and mediates cell cycle arrest in late G_1. In some contexts, wild-type p53 can induce apoptosis (programmed cell death) and, in the absence of the wild-type protein, leads to resistance to ionizing radiation and chemotherapeutic agents. For example, in normal cells with DNA damaged by ultraviolet or γ irradiation, progression through the cell cycle is blocked at G_1, coincident with a sharp rise in the levels of p53. During the subsequent arrest of growth, repair of DNA is completed before the cells proceed into S phase. If, however, genomic damage is excessive, the cell undergoes apoptosis, which requires normally functioning p53. Cells can escape apoptosis in the absence of a functional p53 protein, thus allowing the cell to survive and replicate its damaged DNA, which in turn leads to the propagation of the mutation. Therefore, p53 has been described as the "guardian of the genome" because it prevents entry into S phase unless, or until, the genome has been cleared of potentially damaging mutations. In addition, because many chemotherapeutic drugs are believed to kill tumor cells by inducing apoptosis, loss of p53 function may also directly decrease the cells' sensitivity to such cytotoxic agents, enhancing the emergence of drug-resistant populations of cancer cells.

The biochemical mechanisms by which p53 acts in regulating cell proliferation are not fully understood; however, through multiple approaches, p53 has been shown to mediate growth suppression in part through its specific DNA-binding and transcriptional regulatory abilities (El-Deiry et al. 1994; Ko and Prives 1996). Many of these functional studies have relied on defining functional domains of p53 through their interaction with specific binding partners (Mansur et al. 1995; Dobner et al. 1996; Gottlieb and Oren 1996; Ko and Prives 1996; Levine 1997; Lill et al. 1997; Zhang et al. 1998; Nie et al. 2000). As shown in Figure 2, p53 interacts with a host of proteins, including the tumor suppressor, BRCA1, and the proto-oncogene, c-ABL. Recently, Zilfou and colleagues demonstrated through protein interaction studies that the corepressor SIN3 directly interacts with the proline-rich domain of p53 and protects p53 from proteosome-mediated degradation (Zilfou et al. 2001). This interaction is particularly noteworthy because deletion of the proline-rich domain of p53 renders this protein capable of functioning as an activator of transcription, but incapable of inducing programmed cell death, or apoptosis. Elucidation of the functional consequences of the p53–SIN3 interaction is likely to reveal new mechanisms of tumor suppression and apoptosis induction by p53 (Zilfou et al. 2001).

Many oncogenes, tumor suppressors, and DNA repair proteins appear to be part of much larger complexes. For example, Wang and colleagues recently reported that a set of proteins associates with BRCA1 to form a large megadalton protein complex, referred to as BASC (*BRCA1-a*ssociated genome *s*urveillance *c*omplex). This complex includes tumor suppressors; the DNA damage repair proteins MSH2, MSH6, MLH1, ATM, and BLM; and the RAD50–MRE11–NBS1 protein complex (Wang et al. 2000). In addition, DNA replication factor C (RFC), a protein complex that facilitates the loading of PCNA onto DNA, is also part of the BASC. The association of BRCA1 with MSH2 and MSH6, both of which are required for transcription-coupled repair, provides a possible explanation for the role of BRCA1 (and BRCA2) in this pathway. These interactions are consistent with a role of both proteins in some aspects of transcription-coupled DNA repair and DNA recombination.

Studies of the BRCA1 protein following DNA damage (discussed below) have also uncovered a role for other tumor suppressors and oncogenes (i.e., ATM, ATR, CHK2, and CDK2) in regulating BRCA1 activity. BRCA1 exists in nuclear foci but is hyperphosphorylated and disperses after DNA damage (Scully et al. 1997a; Thomas et al. 1997). This dose-dependent change in the state of BRCA1 phosphorylation is accompanied by a specific loss of the BRCA1-containing nuclear foci during S phase. After BRCA1 dispersal, BRCA1, BRCA2, BARD1, and RAD51 accumulate focally on PCNA$^+$ replication structures, implying an interaction of BRCA1/BRCA2/BARD1/RAD51-containing complexes with damaged, replicating DNA (Fig. 3). Lee and colleagues have reported that the human CHK2 kinase (CDS1/CHK2) regulates BRCA1 function after DNA damage by phosphorylating Ser-988 of BRCA1 (Lee et al. 2000). CHK2 and BRCA1 were shown to interact and colocalize within discrete nuclear foci but to separate after γ irradiation. Phosphorylation of BRCA1 at Ser-988 was shown to be required for the release of BRCA1 from CDK2. ATR, a mammalian homolog of yeast S-phase checkpoint gene products, in part controls BRCA1 phosphorylation following hydroxyurea treatment (Tibbetts et al. 2000). Recently, ATR was found to colocalize with BRCA1 in somatic cells, both before and after replication arrest (Tibbetts et al. 2000). BRCA1 has also been shown to be phosphorylated by CDK2 at Ser-1497, concordant with the G_1/S-specific increase in BRCA1 phosphorylation, independent of DNA damage (Ruffner et al. 1999). Phosphorylation of BRCA1 in response to DNA damage has been shown to be dependent on ATM (Cortez et al. 1999). Falck and colleagues recently demonstrated, through various protein studies, a functional link between ATM, CHK2, the phosphatase CDC25A, and CDK2. These proteins have been implicated in radioresistant DNA synthesis, and defects in some have been shown to predispose to, or promote, tumorigenesis (Galaktionov et al. 1995; Bell et al. 1999; Rotman and Shiloh 1999; Falck et al. 2001). Together, these studies emphasize how various approaches to study protein function are uncovering the ways in which key pathways converge and how defects in these components can contribute to cancer.

PHARMACOGENETICS: CURRENT STATUS AND FUTURE GOALS

Whereas the description of cancer genetics above details genetic changes that lead to onset of disease, pharmacogenetics deals with individual response to the environment, whether in susceptibility to mutagenic insult or in ability to respond to therapeutic drugs. Pharmacogenetics has been studied historically in the context of variable drug metabolism (Weber 1997). Several functionally significant genetic polymorphisms within genes that encode drug-metabolizing enzymes have been identified. Classically, those discoveries have been made after observation of variable human response to a drug, often with the variant phenotype manifesting as a toxic drug reaction. For example, ~1 in 300 Caucasian individuals are deficient in the ability to metabolically inactivate the chemotherapeutic agent 6-mercaptopurine (Weinshilboum and Sladek 1980). Individuals with such a deficiency are at risk for profound myelosuppression and ultimately death if they are prescribed standard doses of 6-mercaptopurine (Lennard 1997; Krynetski and Evans 1998). The deficient phenotype is encoded by one of several genetic polymorphisms in the human thiopurine methyltransferase (TPMT) gene (Krynetski et al. 1996; Otterness et al. 1997). Clinicians are now able to phenotype or genotype individuals for the TPMT polymorphism(s) prior to initiation of 6-mercaptopurine therapy (Fig. 4). Functionally significant polymorphisms

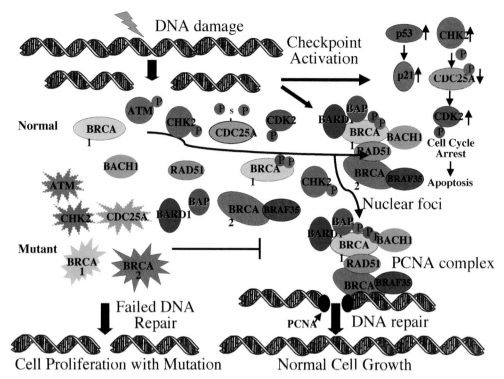

FIGURE 3. Role of BRCA1/2 in DNA replication and repair. Following DNA damage, cell cycle checkpoints are activated, including *p53*. p53 induces *p21* and *CHK2* expression, and signals cell cycle arrest until the damaged DNA can be repaired. The CDC25A phosphatase activates the cyclin-dependent kinase 2 (CDK2), which is also needed for DNA synthesis, but becomes degraded in response to DNA damage. Ionizing radiation-induced destruction of CDC25A requires both ATM and the CHK2-mediated phosphorylation of CDC25A on Ser-123. Failure to repair leads to apoptosis, or programmed cell death (as depicted in the upper right-hand corner). BRCA1 and BRCA2 participate in DNA repair by an unknown mechanism, which is thought to involve transcription-coupled repair. BRCA1 is known to be phosphorylated by ATM and CDK2 in a damage-free cell. BRCA1 is phosphorylated by CHK2 and ATR after damage; phosphorylated BRCA1 associates with each of the proteins BARD1, RAD51, BRCA2, and PCNA during the processes of repair and replication (*Normal*). DNA damage is successfully repaired and normal cell growth proceeds. Mutations in BRCA1, BRCA2, ATM, CHK2, or CDC25A proteins (*Mutant*) may cause dysfunction in this repair/replication complex/system due to their failure to interact with other protein. In this case, DNA repair fails and proliferation continues with damaged DNA. The role of other proteins such as BAP1 (BRCA1-associated protein-1), BRAF35 (BRCA2-associated factor 35), and BACH1 (BRCA1-associated C-terminal helicase) in repair is not known and will require additional studies.

have also been identified in several other drug-metabolizing genes, some of which are associated with variable clinical response to drugs (Ingelman-Sundberg et al. 1999; Iyer et al. 1999; Rettie et al. 1999; Wormhoudt et al. 1999).

In addition to detoxifying and eliminating drugs and metabolites, drug-metabolizing enzymes are often required for activation of prodrugs. For example, many opioid analgesics are activated by CYP2D6 (Poulsen et al. 1996), rendering the 2–10% of the population who are homozygous for nonfunctional CYP2D6 mutant alleles relatively resistant to opioid analgesic effects. Thus, it is not surprising that there is remarkable interindividual variability in the adequacy of pain relief when uniform doses of codeine are widely prescribed.

More recently, pharmacogenetic research has expanded to include the study of drug transporters, drug receptors, and drug targets. As the Human Genome Project progresses, scientists will have a map of hundreds of thousands of common genetic polymorphisms. As this scope expands, genetic profiles, defined by multiple gene loci, will be identified that are associated with particu-

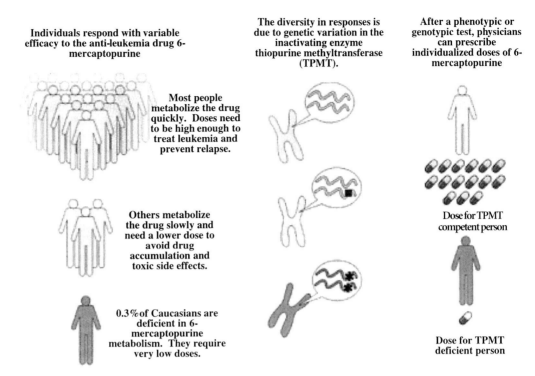

FIGURE 4. Thiopurine methyltransferase (TPMT) deficiency and pharmacogenetics of thiopurine therapy. The TPMT paradigm is used to illustrate a classic pharmacogenetic case in which variation in drug response is caused by a single gene defect. The first panel illustrates phenotypic differences in thiopurine metabolism and response. The second panel indicates the associated genetic variation (a black square marks a heterozygous deficient allele and the asterisks mark deficient alleles expressed homozygously). The third panel illustrates modifications in thiopurine dosing based on TPMT genetics.

larly favorable or adverse responses to drugs. It is hoped that this information will allow clinicians to use genotype information to individualize the prescribing of drugs. As more genetic variants are associated with clinical syndromes and responses, increased efforts will be necessary to understand the molecular mechanisms underlying these phenotypic changes, some of which undoubtedly involve protein–protein interactions.

Many functionally significant polymorphisms alter encoded amino acids that are not involved in the catalytic regions of enzymes. The functional significance of these changes suggests effects with regard to the interaction of these proteins with other species, including enzymes, regulators, and drugs. In fact, a common mechanism by which polymorphisms cause phenotypic change is via altered protein stability. Typically, the amino acid change results in increased degradation of the variant protein, resulting in a lower steady-state level of protein within the cell. For example, two of the TPMT allozymes associated with TPMT deficiency, TPMT*2 and TPMT*3A, are degraded an order of magnitude more rapidly than the wild-type protein (Tai et al. 1997). The specific protein interactions contributing to this regulatory change are not known, however, and application of the techniques such as those described in this book would elucidate this mechanism further.

As described above, study of protein interactions has contributed greatly to the general understanding of mechanisms of cancer through a convergence of fundamental knowledge about signal transduction and DNA repair. The study of protein interactions within the context of classic pharmacogenetics has thus far been relatively limited because most pharmacogenetic investigation has focused on the interaction of proteins with small molecules. However, as the focus of pharmacogenetic research expands to include classes of proteins that function as protein complexes, the tools described in this book will become integrated into pharmacogenetic research. For example, tumor resistance to antiestrogenic therapy of breast cancer has long been recognized but not well understood at the mechanistic level. The past 2 years have given rise to significant advances in our understanding of estrogen receptor (ER)-mediated cell proliferation and mechanisms of antiestrogenic response. Those studies have revealed great complexity in ER signaling and transcriptional activation. It is clear that many coregulators are capable of forming protein complexes with ligand-bound ER (Clarke et al. 2001). The specific coregulators that comprise a given complex determine whether the complex activates or represses transcription of target genes. It has also been recognized that genetic variation in ERs or in proteins that interact with ERs might contribute to antiestrogen resistance (MacGregor and Jordan 1998; Clarke et al. 2001). This level of understanding sets the stage for elucidation of the mechanistic underpinnings of resistance to antiestrogen therapy, and those studies will undoubtedly involve the characterization of protein interactions.

Response to Anticancer Therapies

As indicated above, new clinical applications relating to DNA repair and drug-metabolizing genes are certain to emerge. Tumor cells often acquire resistance to radiation or therapeutic drugs. Genomics approaches, such as array technology, will be used to define any DNA repair genes that may be overexpressed in this context. Furthermore, it will be important to find ways to manipulate DNA repair and drug metabolism specifically or transport selectively to avoid precancerous lesions while promoting the activity of chemotherapeutic drugs. Genetic polymorphisms in relevant genes will be identified and efforts will be made to correlate them with effects on activity of the respective proteins, with response to particular therapies and with clinical outcomes. Although a number of polymorphisms in DNA repair and drug-metabolizing genes are being reported, there is little functional information on the consequence of the attendant amino acid changes. It will be important to find out which polymorphisms actually affect protein functions and then concentrate on these epidemiological and clinical studies. For example, homozygosity for a specific polymorphism in the DNA ligase subunit XRCC1 is associated with higher sister chromatid exchange frequencies in smokers, suggesting an association of this allele with a higher risk for tobacco- and age-related damage (Duell et al. 2000). Furthermore, with the use of gene and protein array techniques, it should be possible to compare expression profiles of DNA repair and drug-metabolizing genes in normal and tumor cells—information that could eventually lead to individually tailored therapies with chemicals and radiation. For example, tumors with low levels of nucleotide excision repair should be more susceptible to treatment with cisplatin (Claij and te Riele 1999). In comparison, mismatch-repair-deficient cells are highly tolerant to alkylating chemotherapeutic drugs (Koberle et al. 1999). The rapidly expanding knowledge of the human genome, coupled with automated methods for detecting gene polymorphisms, provides the tools to elucidate these polygenic determinants of drug effects, thus fueling the burgeoning field of pharmacogenomics. However, methods to evaluate the functional relevance of many of these polymorphisms on protein function will be paramount.

USE OF PROTEIN–PROTEIN INTERACTION STUDIES TO ANALYZE FUNCTIONAL CONSEQUENCES OF GENETIC VARIATION

As described above, study of protein interactors has contributed greatly to the general understanding of mechanisms of cancer genetics by helping to elucidate the function of many cancer-associated genes. Substantial investments are now being made within the pharmaceutical and biotechnology industries to use genomic and proteomic strategies for the discovery of therapeutic targets. It is anticipated that, over the next decade, DNA and protein array technologies, high-throughput screening systems, enhanced mass spectrometry methods, and advanced bioinformatics will be merged to permit rapid elucidation of complex genetic components of human disease. The techniques described in this book are particularly important for correlating genotypic changes with specific biochemical or biophysical phenotypic changes associated with the variant proteins (genotype/phenotype correlation). Some techniques are amenable to the study of DNA interactions and have been applied as genotyping methodologies. The next pages provide examples of how these techniques have been applied or may be applied to the study of functional aspects of genetic variation.

Tagged Fusion Proteins

GST-fusion proteins, or similarly "tagged" proteins, are used extensively in pull-down assays, affinity purification, and far-western analyses (Chapter 4). These classic techniques have been used extensively in the study of protein–protein interactions, including proteins that represent drug targets and are genetically polymorphic. For example, the p53 tumor suppressor gene is one of the most frequently mutated genes in human tumors (Levine 1997). A common Pro72Arg polymorphism has been identified in the polyproline domain of the human p53 protein (Harris et al. 1986). Using GST pull-down assays, Thomas et al. (1999) have shown that the Pro-72 allozyme binds the transcriptional activators TAFII32 and TAFII70 with significantly greater affinity than does the Arg-72 allozyme. These authors further demonstrated that the Pro-72 allozyme is a better *trans*-activator of p53 *trans*-activated genes than the Arg-72 allozyme. These studies begin to determine a molecular mechanism by which the Pro72Arg polymorphism might confer a phenotypic change to the protein. These observations have allowed investigators to further hypothesize that this common polymorphism might influence the biology of the p53 protein.

Coimmunoprecipitation

Like the GST-fusion assays, coimmunoprecipitation (Chapter 5) has become a classic technique for the study of protein–protein interactions. This technique is a rigorous method of establishing physiologically relevant protein interactions. Coimmunoprecipitation and yeast two-hybrid approaches (as discussed below) have been the driving force behind the functional characterization of many of the so-called "orphan" proteins. As indicated above, the *BRCA1* and *BRCA2* breast/ovarian cancer susceptibility genes were identified in 1994 and 1995, respectively (Miki et al. 1994; Wooster et al. 1995; Tavitigian et al. 1996). The primary amino acid sequence of BRCA2 showed a weak similarity to BRCA1 over a restricted region (Wooster et al. 1995; Tavitigian et al. 1996), and only a low level of homology was seen among other known proteins. Not until Scully and colleagues demonstrated a direct interaction using coimmunoprecipitation between BRCA1 and RAD51 and BRCA2 and RAD51 were their potential roles in the cellular response to DNA damage realized (Scully et al. 1997b; Wong et al. 1997). Eukaryotic RAD51 proteins are homologs of bacterial RecA and are required for recombination during mitosis and meiosis and for recombinational repair of double-strand DNA breaks (Shinohara et al. 1992). Through yeast two-hybrid and biochemical assays, they demonstrated that the RAD51 protein interacts specifically with the eight evolutionarily conserved BRC motifs encoded in exon 11 of BRCA2 (Wong et al. 1997). Further coimmunoprecipitation studies using smaller portions of BRCA2 defined at least two additional RAD51-binding domains, residues 982–1066 and 1139–1266 (Katagira et al. 1998). These studies indicated that BRCA2 may interact with RAD51 through multiple sites of BRCA2 and that BRCA2 and RAD51 interact and colocalize in a BRCA1–BRCA2–RAD51 complex (Chen et al. 1997). BRCA2 has also been shown to form in vivo complexes with p53 (Marmorstein et al. 1998). Exogenous BRCA2 expression in cancer cells inhibits p53's transcriptional activity, and RAD51 coexpression enhances BRCA2's inhibitory effects. These findings, made possible using protein interaction technologies, demonstrate that BRCA2 physically and functionally interacts with two key components of cell cycle control, one involving apoptosis and one involving DNA repair. These approaches have led to the identification of more than a dozen proteins that associate with specific domains of BRCA1 and/or BRCA2 (Fig. 5).

These studies are also helping to predict which missense changes may be deleterious mutations on the basis of their location in the protein and possible effects on protein–protein interactions. Understanding protein–protein interactions may also lead to a better understanding of cancer risks. For example, BRCA2 proteins have several significant repeated motifs that are not found in BRCA1 but are conserved in all mammalian BRCA2 proteins that have been sequenced. Eight internal repeats of 30–80 amino acids, known as BRC motifs, are encoded in exon 11 of the human BRCA2 gene (Fig. 5). Each repeat is variably conserved, suggesting that the core sequence was duplicated eight times during evolution, but that many of the repeats are now redundant (Chen et al. 1995). The BRC motifs were later demonstrated to be involved in BRCA2-mediated DNA repair through their interactions with RAD51 (Wong et al. 1997). The BRC repeats lie within a large region spanning exon 11 that has been deemed the ovarian cancer cluster region (OCCR), because mutations in this region have been associated with an increased frequency of ovarian cancer (Gayther et al. 1997; Neuhausen et al. 1998).

Although coimmunoprecipitation can be cumbersome for the detection of novel protein interactions, it is ideally suited for studying genetic variation causing loss or gain of function within known protein interactions. For example, Marin et al. (2000) have used this technique to demonstrate that a mutant form of p53 with Arg at codon 72 has a higher affinity for p73 (a p53 homolog and binding protein) than does the mutant Pro-72 allozyme. These observations again suggest biological significance of the common p53 polymorphism, interestingly, within the context of independent p53 mutations. Furthermore, the ability of mutant p53 to block the chemotherapeutic activity of etoposide correlates with the ability of the protein to bind p73 (Blandino et al. 1999), and p73 plays a role in cisplatin-induced apoptosis (Gong et al. 1999). Collectively, these observations suggest that the common Pro72Arg p53 polymorphism might be associated with chemosensitivity of selected tumors (Marin et al. 2000).

Yeast Two-hybrid

Although most often used to identify novel partners for a known protein, the yeast two-hybrid system (Chapter 7) can be adapted to accommodate the study of variant protein interactions. For example, Maier et al. (1998) have used the yeast two-hybrid approach to analyze the genotype–phenotype correlation among mouse aromatic hydrocarbon receptor (AHR) alleles. The AHR is a transcriptional activator of several drug-metabolizing genes. Translocation of the AHR to the nucleus depends on its interaction with a small-molecule ligand and the ARNT protein (Gu et al. 2000). Affinity of the AHR allozymes for the ARNT protein, in the presence of an AHR ligand such as dioxin, was quantified with a β-galactosidase reporter system (Maier et al. 1998). These authors reported a 15-fold difference in the ligand affinity of two mouse AHR alleles, suggesting biological significance of the polymorphisms. Subsequently, the functional significance of human AHR genetic variation has been demonstrated in studies associating AHR genotype with level of expression of AHR target genes (Smart and Daly 2000).

More recent innovations to the yeast two-hybrid system offer additional promise for pharmacogenetic research. For example, the three-hybrid system includes "hook," "bait," and "fish" components for detecting small ligand–protein receptor interactions (Licitra and Liu 1996). This represents a system in which the genotype–phenotype correlation might be analyzed for small molecule binding proteins such as the drug-metabolizing enzymes. In addition, the reverse two-hybrid system, which selects for interference of a specific protein–protein interaction, may facilitate the detection of variant proteins in which there is a loss of function (Leanna and Hannink 1996).

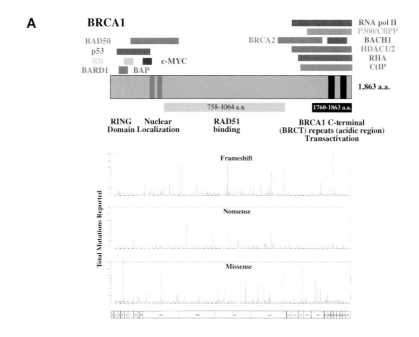

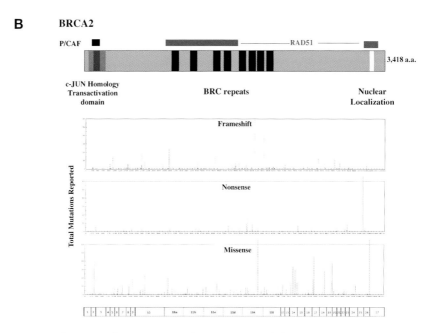

FIGURE 5. Schematic of BRCA1 (*A*) and BRCA2 (*B*) and their interacting proteins. Transcription-coupled repair appears to involve BRCA1 through its association with many proteins, including RNA pol II, acetyl transferases, and deacetylases (P300/CPB, P/CAF, ACRT/SCR-1). During DNA replication, BRCA1 is associated with the replication protein PCNA (see Fig. 3) and with Rad50 and Rad51 during homologous recombinational repair. (BARD) BRCA1-associated RING domain 1 protein; (BAP1) BRCA1-associated protein-1 (ubiquitin conjugating enzymes); (BRCT) BRCA1 C-terminal repeats; (HDAC) histone deacetylase 1 and 2; (RHA) RNA helicase A; (CtIP) CtBp-interacting protein; (BACH1) BRCA1-associated C-terminal helicase, a member of the DEAH helicase; (P/CAF) p300/CBP-associated factor (with histone acetylase activity). The lower panels represent the position and the number of times each of the disease-associated mutations (frameshift, nonsense, and missense) has been reported by BIC (see www.nhgri.nih.gov/Intramural_research/Lab_transfer/Bic/index.html).

FRET

Fluorescence resonance energy transfer (FRET) measures molecular proximity by detecting the excited state of acceptor and donor fluorophore-tagged molecules. This technique, as described in Chapter 10, has been applied to measure protein–protein interactions by alternatively labeling separate proteins of interest with donor or acceptor fluorophores and allows the physiological validation of proposed protein–protein interactions in real time. Intriguingly, the application of this technique to pharmacogenetic research is best exemplified by the dye-terminator incorporation genotyping assay, which detects oligonucleotide proximity rather than protein proximity (Chen and Kwok 1999). In this approach, an allele-specific, donor dye-labeled oligonucleotide is extended by DNA polymerase in PCR using acceptor dye-labeled dNTPs. Once the donor and acceptor dyes become part of a new molecule (via allele-specific PCR amplification), intramolecular FRET is detected and quantified. Another FRET-associated genotyping assay, the Invader assay, was recently developed (Lyamichev et al. 1999; Mein et al. 2000). The Invader assay relies on the specificity of recognition and cleavage, by a flap endonuclease (FEN), of overlapping "signal" and "invader" oligonucleotides that hybridize to target DNA containing a polymorphic site. In the presence of a match between the signal oligonucleotide and the DNA template, the signal oligonucleotide is cleaved to drive a secondary cleavage reaction with a FRET label. The signal is detected with a fluorescence plate reader. An advantage of this technology is the high-throughput flexibility of the assay as well as its sensitivity.

Biacore

Surface plasmon resonance (Chapter 14) enables the detection and quantification of bonding events between molecules by monitoring changes in the refractive index of molecular surfaces upon intermolecular interaction. This technology is useful in obtaining kinetic data regarding the interaction between proteins, lipids, nucleotides, and small molecules. Because of this versatility, there is great potential for the application of this technique to the study of phenotypic changes associated with genetic variation. For example, Baynes et al. (2000) have used Biacore technology to study the binding characteristics of wild-type and mutant human p85α proteins to lipid substrates. p85α is a PI3 kinase involved in insulin-stimulated glucose disposal. These investigators identified a common p85α polymorphism as well as one mutation in subjects with severe insulin resistance. Cells expressing the polymorphic variant of p85α maintained similar PI3 kinase activity toward lipid substrates as cells expressing the wild-type protein. However, cells expressing the mutant p85α exhibited significantly lower levels of PI3 kinase activity. These authors then used the Biacore technique to determine whether the mutant protein was defective in binding lipid substrates. Their data demonstrated that wild-type, polymorphic, and mutant p85α each have similar binding capacity, thus suggesting that the mechanism by which the mutant protein exerted a phenotypic change was not via altered substrate binding.

Atomic Force Microscopy

AFM (Chapter 13) maps the topography of a surface by detecting minute changes in bonding forces at the surface of a molecule. This technique is often used to study ligand–receptor interactions. Yip et al. (1998) have used AFM to demonstrate that a mutant form of the human insulin protein forms much weaker self-associated oligomeric structures than does wild-type insulin. Consequently, in solution, a higher percentage of mutant insulin adopted monomeric conformations as opposed to hexamers. These observations subsequently contributed to the development of insulin therapies with better bioavailability because the efficacy of commercial insulin preparations is dependent on the dissolution of insulin hexamers to monomers.

Similar to Biacore technologies, AFM has been applied as a technique for high-throughput genotyping. One of the most difficult aspects of genotyping is the assignment of haplotypes

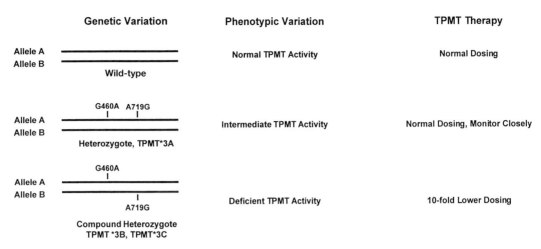

FIGURE 6. Haplotype–phenotype correlation with thiopurine methyltransferase. The figure illustrates the importance of being able to distinguish technically between individuals who carry one copy of the TPMT*3A allele (and hence would not need dose modification) and those who are compound heterozygotes expressing one copy of the *3B and one of the *3C allele (and hence would need very low doses of thiopurines). The *3A allele is defined by two SNPs, a G→A transition at nucleotide 460 and an A→G transition at nucleotide 719 in the cDNA. These two positions are separated by 13 kb of sequence on human chromosome 6. The *3B allele is defined only by the SNP at position 460, and the *3C allele by the SNP at position 719.

(Service 2000). The ability to assign a haplotype is sometimes paramount to clinical decisions. For example, the TPMT polymorphic phenotype described previously in this chapter is encoded by several *TPMT* alleles. One of those alleles, *3A, is defined by two separate SNPs at loci within the gene that are separated by more than 13 kb of DNA sequence. However, each of these SNPs also defines separate alleles (*3B and *3C, respectively) that each encode deficient TPMT enzyme. In the clinical setting, it would be critical to distinguish between an individual who is heterozygous for the *3A allele and one who is a compound heterozygote (*3B/*3C). That is true because in the former case, the individual would be predicted to have an "intermediate" TPMT phenotype and could be prescribed standard doses of mercaptopurine drugs. However, the individual who is a *3B/*3C compound heterozygote would be TPMT deficient and would likely experience profound myelosuppression with that same dose (Fig. 6). Woolley et al. (2000) have described modification of the AFM technique, using carbon nanotube probes, to determine genomic haplotypes. This approach involves the hybridization of fluorescently labeled, allele-specific oligonucleotide probes to target DNA fragments, followed by detection of the presence and spatial location of the labels by AFM. A major advantage of this methodology is the ability to codetect SNPs as far as several kilobases (perhaps 100 kb) apart and its amenability to high-throughput genotyping (Service 2000).

CONCLUSIONS

Susceptibility to cancer and the development of effective approaches to treat these diseases will depend greatly on our ability to decipher the function of hundreds, maybe thousands, of proteins. With the sequencing of the human genome and the identification of thousands of previously uncharacterized genes now a reality, the work to elucidate the function of the proteins they encode is just beginning. The number of proteins directly or indirectly involved in cancer susceptibility, cancer progression, and response to drug therapy is difficult, if not impossible, to predict.

If one were to consider just the oncogenes, tumor suppressor genes, and DNA repair genes as the sole genes associated with cancer,the number might approach more than 500. However, it is clear that not all of these genes have a role in human disease and that they are clearly not the only ones that contribute to the risk of disease. The class of genes involved in metabolic detoxification pathways and steroid and amino acid metabolism, for example, is quite large and is likely to be intimately involved in many facets of cancer. Furthermore, it is likely that a small but significant fraction of the several million SNPs present in any genome will result in multiple forms of the same protein with slightly modified biochemical and biological properties. Therefore, new genomic and proteomic technologies will be necessary to identify important genomic risk factors and translate these findings into a functional reality. In this book, the authors have presented new and exciting advances in the area of protein interaction technologies that will help to move forward this area of research. Specific protein–protein and protein–ligand interactions are central to most biological processes and are the focus of many avenues of research to develop small molecule-based therapies that will disrupt these essential interactions. We are only beginning the post-genomics era. Therefore, we can look forward to an equally exciting decade of great discoveries and challenges that will be necessary to define the function of thousands of new proteins as well as their potentially important variant forms and allozymes.

REFERENCES

Baynes K.C., Beeton C.A., Panayotou G., Stein R., Soos M., Hansen T., Simpson H., O'Rahilly S., Shepherd P.R., and Whitehead J.P. 2000. Natural variants of human p85 alpha phosphoinositide 3-kinase in severe insulin resistance: A novel variant with impaired insulin-stimulated lipid kinase activity. *Diabetologia* **43:** 321–331.

Bell D.W., Varley J.M., Szydlo T.E., Kang D.H., Wahrer D.C., Shannon K.E., Lubratovich M., Verselis S.J., Isselbacher K.J., Fraumeni J.F., Birch J.M., Li F.P., Garber J.E., and Haber D.A. 1999. Heterozygous germ line hCHK2 mutations in Li-Fraumeni syndrome. *Science* **286:** 2528–2531.

Blandino G., Levine A.J., and Oren M. 1999. Mutant p53 gain of function: Differential effects of different p53 mutants on resistance of cultured cells to chemotherapy. *Oncogene* **18:** 477–485.

Cavenee W.K., Dryja T.P., Phillips R.A., Benedict W.F., Godbout R., Gallie B.L., Murphree A.L., Strong L.C., and White R.L. 1983. Expression of recessive alleles by chromosomal mechanisms in retinoblastoma. *Nature* **305:** 779–784.

Chen H., Lin R., Schiltz R., Chakravarti D., Nash A., Nagy L., Privalsky M., Nakatani Y., and Evans R. 1997. Nuclear receptor coactivaor ACTR is a novel histone acetyltransferase and forms a multimeric activation complex with P/CAF and CPB/p300. *Cell* **90:** 569–580.

Chen X. and Kwok P.Y. 1999. Homogeneous genotyping assays for single nucleotide polymorphisms with fluorescence resonance energy transfer detection. *Genet. Anal.* **14:** 157–163.

Chen Y., Chen C.-F., Riley D.J., Allred D.C., Chen P.-L., von Hoff D., Osborne C.K., and Lee W.-H. 1995. Aberrant subcellular localization of BRCA1 in breast cancer. *Science* **270:** 789–791.

Claij N. and te Riele H. 1999. Microsatellite instability in human cancer: A prognostic marker for chemotherapy? *Exp. Cell Res.* **246:** 1–10.

Clarke R., Leonessa F., Welch J.N., and Skaar T.C. 2001. Cellular and molecular pharmacology of antiestrogen action and resistance. *Pharmacol. Rev.* **53:** 25–71.

Cortez D., Wang Y., Qin J., and Elledge S. 1999. Requirement of ATM-dependent phosphorylation of brca1 in the DNA damage response to double-strand breaks. *Science* **286:** 1162–1166.

DeMars R. 1969. Genetics concepts and neoplasia; a collection of papers. In *23rd Annual Symp. Fundamental Cancer Research* (ed. W.A.Wilkins), pp. 105–106, Baltimore.

Dobner T., Horikoshi N., Rubenwolf S., and Shenk T. 1996. Blockage by adenovirus E4orf6 of transcriptional activation by the p53 tumor suppressor. *Science* **272:** 1470–1473.

Duell E.J., Wiencke J.K., Cheng T.J., Varkonyi A., Zuo Z.F., Ashok T.D., Mark E.J., Wain J.C., Christiani D.C., and Kelsey K.T. 2000. Polymorphisms in the DNA repair genes XRCC1 and ERCC2 and biomarkers of DNA damage in human blood mononuclear cells. *Carcinogenesis* **21:** 965–971.

Dunn J.M., Phillips R.A., Zhu X., Becker A., and Gallie B.L. 1989. Mutations in the RB1 gene and their effects on transcription. *Mol. Cell. Biol.* **9:** 4596–4604.

El-Deiry W.S., Harper J.W., O'Conner P.M., Velculescu V.E., Canman C.E., Jackman J., Peintenpol J.A.,

Burrell M., Hill D.E., Wang Y., Wilman K.G., Mercer W.E., Kastan M.B., Konh K.W., Elledge S.J., Kinzler K.W., and Vogelstein B. 1994. WAF1/CIP1 is induced in p53-mediated G1 arrest and apoptosis. *Cancer Res.* **54:** 1169–1174.

Evans W.E. and Relling M.V. 1999. Pharmacogenomics: Translating functional genomics into rational therapeutics. *Science* **286:** 487–491.

Falck J., Mailand N., Syljuasen R.G., Bartek J., and Lukas J. 2001. The ATM-Chk2-Cdc25A checkpoint pathway guards against radioresistant DNA synthesis. *Nature* **410:** 842–847.

Fearon E.R. 1997. Human cancer syndromes: Clues to the origin and nature of cancer. *Science* **278:** 1043–1050.

———. 2000. BRCA1 and E-cadherin promoter hypermethylation and gene inactivation in cancer-association or mechanism? *J. Natl. Cancer Inst.* **92:** 515–517.

Galaktionov K., Lee A.K., Eckstein J., Draetta G., Meckler J., Loda M., and Beach D. 1995. CDC25 phosphatases as potential human oncogenes. *Science* **269:** 1575–1577.

Gannon J. and Lane D. 1990. Interactions between SV40 T antigen and DNA polymerase alpha. *New Biol.* **2:** 84–92.

Gayther S., Mangion J., Russell P., Seal S., Barfoot R., Ponder B., Stratton M., and Easton D. 1997. Variation of risks of breast and ovarian cancer associated with different germline mutations of the BRCA2 gene. *Nat. Genet.* **15:** 103–105.

Godwin A.K., Schultz D.C., Hamilton T.C., and Knudson A.G. 1997. Oncogenes and tumor suppressor genes. In *Gynecologic oncology: Principles and practice* (ed. W.J. Hoskins et al.), pp. 107–148. Lippencott, Philadelphia.

Gong J.G., Costanzo A., Yang H.Q., Melino G., Kaelin W.G., Jr., Levrero M., and Wang J.Y. 1999. The tyrosine kinase c-Abl regulates p73 in apoptotic response to cisplatin-induced DNA damage (see comments). *Nature* **399:** 806–809.

Gottlieb T.M. and Oren M. 1996. p53 in growth control and neoplasia. *Biochim. Biophys. Acta* **1287:** 77–102.

Greenlee R.T., Murray T., Bolden S., and Wingo P.A. 2000. Cancer statistics, 2000. *CA Cancer J. Clin.* **50:** 7–33.

Gu Y.Z., Hogenesch J.B., and Bradfield C.A. 2000. The PAS superfamily: Sensors of environmental and developmental signals. *Annu. Rev. Pharmacol. Toxicol.* **40:** 519–561.

Hainaut P., Hernandez T., Robinson A., Rodriguez-Tome P., Flores T., Hollstein M., Harris C.C., and Montesano R. 1998. IARC Database of p53 gene mutations in human tumors and cell lines: Updated compilation, revised formats and new visualisation tools. *Nucleic Acids Res.* **26:** 205–213.

Hanahan D. and Weinberg R.A. 2000. The hallmarks of cancer. *Cell* **100:** 57–70.

Harris H., Miller O.J., Klein G., Worst P., and Tachibana T. 1969. Suppression of malignancy by cell fusion. *Nature* **223:** 363–368.

Harris N., Brill E., Shohat O., Prokocimer M., Wolf D., Arai N., and Rotter V. 1986. Molecular basis for heterogeneity of the human p53 protein. *Mol. Cell. Biol.* **6:** 4650–4656.

Hollstein M., Shomer B., Greenblatt M., Soussi T., Hovig E., Montesano R., and Harris C.C. 1996. Somatic point mutations in the p53 gene of human tumors and cell lines: Updated compilation. *Nucleic Acids Res.* **24:** 141–146.

Ingelman-Sundberg M., Oscarson M., and McLellan R.A. 1999. Polymorphic human cytochrome P450 enzymes: An opportunity for individualized drug treatment. *Trends Pharmacol. Sci.* **20:** 342–349.

Iyer L., Hall D., Das S., Mortell M.A., Ramirez J., Kim S., Di Rienzo A., and Ratain M.J. 1999. Phenotype-genotype correlation of in vitro SN-38 (active metabolite of irinotecan) and bilirubin glucuronidation in human liver tissue with UGT1A1 promoter polymorphism. *Clin. Pharmacol. Ther.* **65:** 576–582.

Katagira T., Saito H., Shinohara A., Ogawa H., Kamada N., Nakamura Y., and Miki Y. 1998. Multiple possible sites of BRCA2 interacting with DNA repair protein RAD51. *Genes Chromosomes Cancer* **21:** 217–222.

Knudson A.G. 1971. Mutation and cancer: Statistical study of retinoblastoma. *Proc. Natl. Acad. Sci.* **68:** 820–823.

Knudson A.G. and Strong L.C. 1972. Mutation and cancer: A model for Wilms' tumor of the kidney. *J. Natl. Cancer Inst.* **48:** 313–324.

Ko L.J. and Prives C. 1996. p53: Puzzle and paradigm. *Genes Dev.* **10:** 1054–1072.

Koberle B., Masters J.R., Hartley J.A., and Wood R.D. 1999. Defective repair of cisplatin-induced DNA damage caused by reduced XPA protein in testicular germ cell tumours. *Curr. Biol.* **9:** 273–276.

Krynetski E.Y. and Evans W.E. 1998. Pharmacogenetics of cancer therapy: Getting personal. *Am. J. Hum. Genet.* **63:** 11–16.

Krynetski E.Y., Tai H.L., Yates C.R., Fessing M.Y., Loennechen T., Schuetz J.D., Relling M.V., and Evans W.E. 1996. Genetic polymorphism of thiopurine S-methyltransferase: Clinical importance and molecular mechanisms. *Pharmacogenetics* **6:** 279–290.

Kuehl P., Zhang J., Lin Y., Lamba J., Assem M., Schuetz J., Watkins P.B., Daly A., Wrighton S.A., Hall S.D.,

Maurel P., Relling M., Brimer C., Yasuda K., Venkataramanan R., Strom S., Thummel K., Boguski M.S., and Schuetz E. 2001. Sequence diversity in CYP3A promoters and characterization of the genetic basis of polymorphic CYP3A5 expression. *Nat. Genet.* **27:** 383–391.

Lander E.S., Linton L.M., Birren B., Nusbaum C., Zody M.C., Baldwin J., Devon K., Dewar K., Doyle M., FitzHugh W., et al. 2001. Initial sequencing and analysis of the human genome. *Nature* **409:** 860–921.

Leanna C.A. and Hannink M. 1996. The reverse two-hybrid system: A genetic scheme for selection against specific protein/protein interactions. *Nucleic Acids Res.* **24:** 3341–3347.

Lee J.-S., Collins K.M., Brown A.L., Lee C.-H., and Chung J.H. 2000. hCds1-mediated phosphorylation of BRCA1 regulates the DNA damage response. *Nature* **404:** 201–204.

Lennard L. 1997. Methyl transferases. In *Comprehensive toxicology* (ed. F.P. Guengerich), pp. 437–454, Pergamon, London.

Levine A.J. 1997. p53, The cellular gatekeeper for growth and division. *Cell* **88:** 323–331.

Licitra E.J. and Liu J.O. 1996. A three-hybrid system for detecting small ligand-protein receptor interactions. *Proc. Natl. Acad. Sci.* **93:** 12817–12821.

Lill N.L., Grossman S.R., Ginsberg D., DeCaprio J., and Livingston D.M. 1997. Binding and modulation of p53 by p300/CBP coactivators. *Nature* **387:** 823–827.

Lindahl T. and Wood R.D. 1999. Quality control by DNA repair. *Science* **286:** 1897–1905.

Lundberg A.S. and Weinberg R.A. 1999. Control of the cell cycle and apoptosis. *Eur. J. Cancer* **35:** 1886–1894.

Lyamichev V., Mast A.L., Hall J.G., Prudent J.R., Kaiser M.W., Takova T., Kwiatkowski R.W., Sander T.J., de Arruda M., Arco D.A., Neri B.P., and Brow M.A. 1999. Polymorphism identification and quantitative detection of genomic DNA by invasive cleavage of oligonucleotide probes. *Nat. Biotechnol.* **17:** 292–296.

Lynch H. and Smyrk T. 1996. Hereditary nonpolyposis colorectal cancer (Lynch syndrome): An updated review. *Cancer* **78:** 1149–1167.

MacGregor J.I. and Jordan V.C. 1998. Basic guide to the mechanisms of antiestrogen action. *Pharmacol. Rev.* **50:** 151–196.

Maier A., Micka J., Miller K., Denko T., Chang C.Y., Nebert D.W., and Alvaro P. 1998. Aromatic hydrocarbon receptor polymorphism: Development of new methods to correlate genotype with phenotype. *Environ. Health Perspect.* **106:** 421–426.

Malkin D., Li F., Strong L., Fraumeni J., Nelson C., Kim D., Kassel J., Gryka M., Bisehoff F., Tainsky M., and Friend S. 1990. Germ line p53 mutations in a familial syndrome of breast cancer, sarcomas, and other neoplasms. *Science* **250:** 1233–1238.

Mansur C.P., Marcus B., Dalal S., and Androphy E.J. 1995. The domain of p53 required for binding HPV 16 E6 is separable from the degradation domain. *Oncogene* **10:** 457–465.

Marin M.C., Jost C.A., Brooks L.A., Irwin M.S., O'Nions J., Tidy J.A., James N., McGregor J.M., Harwood C.A., Yulug I.G., Vousden K.H., Allday M.J., Gusterson B., Ikawa S., Hinds P.W., Crook T., and Kaelin W.G., Jr. 2000. A common polymorphism acts as an intragenic modifier of mutant p53 behaviour. *Nat. Genet.* **25:** 47–54.

Marmorstein L.Y., Ouchi T., and Aaronson S.A. 1998. The *BRCA2* gene product functionally interacts with p53 and RAD51. *Proc. Natl. Acad. Sci.* **95:** 13869–13874.

Marshall C.J. 1991. Tumor suppressor genes. *Cell* **64:** 313–326.

Mein C.A., Barratt B.J., Dunn M.G., Siegmund T., Smith A.N., Esposito L., Nutland S., Stevens H.E., Wilson A.J., Phillips M.S., Jarvis N., Law S., de Arruda M., and Todd J.A. 2000. Evaluation of single nucleotide polymorphism typing with invader on PCR amplicons and its automation. *Genome Res.* **10:** 330–343.

Miki Y., Swensen J., Shattuck-Eidens D., Futreal P., Harshman K., Tavtigian S., Liu Q., Cochran C., Bennett L., Ding W., et al. 1994. A strong candidate for the breast and ovarian cancer susceptibility gene *BRCA1*. *Science* **266:** 66–71.

Neuhausen S.L., Godwin A.K., Gershoni-Baruch R., Schubert E., Garber J., Stoppa-Lyonnet D., Olah E., Csokay B., Serova O., Lalloo F., et al. 1998. Haplotype and phenotype analysis of nine recurrent BRCA2 mutations in 111 families: Results of an international study. *Am. J. Hum. Genet.* **62:** 1381–1388.

Nie Y., Li H.H., Bula C.M., and Liu X. 2000. Stimulation of p53 DNA binding by c-Abl requires the p53 C terminus and tetramerization. *Mol. Cell. Biol.* **20:** 741–748.

Otterness D., Szumlanski C., Lennard L., Klemetsdal B., Aarbakke J., Park-Hah J.O., Iven H., Schmiegelow K., Branum E., O'Brien J., and Weinshilboum R. 1997. Human thiopurine methyltransferase pharmacogenetics: Gene sequence polymorphisms. *Clin. Pharmacol. Ther.* **62:** 60–73.

Peltomaki P. and Vasen H. 1997. Mutations predisposing to hereditary nonpylyposis colorectal cancer: Database and results of a collaborative study. The International Collaborative Group on Hereditary Nonpolyposis Colorectal Cancer. *Gastroenterology* **113:** 1146–1158.

Poulsen L., Brosen K., Arendt-Nielsen L., Gram L.F., Elbaek K., and Sindrup S.H. 1996. Codeine and morphine in extensive and poor metabolizers of sparteine: pharmacokinetics, analgesic effect and side effects. *Eur. J. Clin. Pharmacol.* **51:** 289–295.

Prowse A., Frolov A., and Godwin A.K. 2001. The genetics of ovarian cancer. In *American Cancer Society*

Atlas of clinical oncology. B.C. Decker, Hamilton, Ontario. (In press.)

Rebbeck T.R., Jaffe J.M., Walker A.H., Wein A.J., and Malkowicz S.B. 1998. Modification of clinical presentation of prostate tumors by a novel genetic variant in CYP3A4. *J. Natl. Cancer Inst.* **90:** 1225–1229.

Rendic S. and Di Carlo F.J. 1997. Human cytochrome P450 enzymes: A status report summarizing their reactions, substrates, inducers, and inhibitors. *Drug Metab. Rev.* **29:** 413–580.

Rettie A.E., Haining R.L., Bajpai M., and Levy R.H. 1999. A common genetic basis for idiosyncratic toxicity of warfarin and phenytoin. *Epilepsy Res.* **35:** 253–255.

Rothman N., Wacholder S., Caporaso N.E., Garcia-Closas M., Buetow K., and Fraumeni J.F. 2001. The use of common genetic polymorphisms to enhance the epidemiologic study of environmental carcinogens. *Biochim. Biophys. Acta* **1471:** C1–10.

Rotman G. and Shiloh Y. 1999. ATM: A mediator of multiple responses to genotoxic stress. *Oncogene* **18:** 6135–6144.

Ruffner H., Jiang H., Craig A., Hunter T., and Verman I. 1999. BRCA1 is phosphorylated at serine 1497 in vivo at a cyclin-dependent kinase 2 phosphorylation site. *Mol. Cell. Biol.* **19:** 4843–4854.

Scully R., Chen J., Ochs R.L., Keegan K., Hoekstra M., Feunteun J., and Livingston D.M. 1997a. Dynamic changes of BRCA1 subnuclear location and phosphorylation state are initiated by DNA damage. *Cell* **90:** 425–435.

Scully R., Chen J., Plug A., Xiao Y., Weaver D., Feunteun J., Ashley T., and Livingston D.M. 1997b. Association of BRCA1 with Rad51 in mitotic and meiotic cells. *Cell* **88:** 265–275.

Service R.F. 2000. DNA imaging. Getting a feel for genetic variations (news). *Science* **289:** 27–28.

Shields P.G. and Harris C.C. 2000. Cancer risk and low-penetrance susceptibility genes in gene-environment interactions. *J. Clin. Oncol.* **18:** 2309–2315.

Shinohara A., Ogawa H., and Ogawa T. 1992. Rad51 protein involved in repair and recombination in *Saccharomyces cerevisiae* is a recA-like protein. *Cell* **69:** 457–470.

Smart J., and Daly A.K. 2000. Variation in induced CYP1A1 levels: Relationship to CYP1A1, Ah receptor and GSTM1 polymorphisms. *Pharmacogenetics* **10:** 11–24.

Soussi T., Caron de Fromentel C., and May P. 1990. Structural aspects of the p53 protein in relation to gene evolution. *Oncogene* **5:** 945–952.

Stanbridge E.J. 1976. Suppression of malignancy in human cells. *Nature* **260:** 17–20.

Tai H.L., Krynetski E.Y., Schuetz E.G., Yanishevski Y., and Evans W.E. 1997. Enhanced proteolysis of thiopurine S-methyltransferase (TPMT) encoded by mutant alleles in humans (TPMT*3A, TPMT*2): Mechanisms for the genetic polymorphism of TPMT activity. *Proc. Natl. Acad. Sci.* **94:** 6444–6449.

Tavitigian S., Rommens J., Couch F., Shattuck-Eidens D., Neuhausen S., Merajver S., Thorlacius S., Offit K., Stoppa-Lyonnet D., Belanger C., et al. 1996. The complete *BRCA2* gene and mutations in chromosomes 13q-linked kindreds. *Nat. Genet.* **12:** 333–337.

Thomas J.E., Smith M., Tonkinson J., Rubinfeld B., and Polakis P. 1997. Induction of phosphorylation on *BRCA1* during the cell cycle and after DNA damage. *Cell Growth Differ.* **8:** 801–809.

Thomas M., Kalita A., Labrecque S., Pim D., Banks L., and Matlashewski G. 1999. Two polymorphic variants of wild-type p53 differ biochemically and biologically. *Mol. Cell. Biol.* **19:** 1092–1100.

Tibbetts R., Cortez D., Brumbaugh K.M., Scully R., Livingston D., Elledge S.J., Abraham R.T. 2000. Functional interactions between BRCA1 and the checkpoint kinase ATR during genotoxic stress. *Genes Dev.* **14:** 2989–3002.

Venter J.C., Adams M.D., Myers E.W., Li P.W., Mural R.J., Sutton G.G., Smith H.O., Yandell M., Evans C.A., Holt R.A., et al. 2001. The sequence of the human genome. *Science* **291:** 1304–1351.

Wang Y., Cortez D., Yazdi P., Neff N., Elledge S.J., and Qui J. 2000. BASC, a super complex of BRCA1-associated proteins involved in the recognition and repair of aberrant DNA structures. *Genes Dev.* **14:** 927–939.

Weber W.W. 1997. *Pharmacogenetics.* Oxford Press, New York.

Weinshilboum R.M. and Sladek S.L. 1980. Mercaptopurine pharmacogenetics: Monogenic inheritence of erythrocyte thiopurine methyltransferase activity. *Am. J. Hum. Genet.* **32:** 651–662.

Wong A., Pero R., Ormonde P., Tavigian S., and Bartel P. 1997. Rad51 interacts with the evolutionarily conserved BRC motifs in the human breast cancer susceptibility gene *BRCA2. J. Biol. Chem.* **272:** 31941–31944.

Wood R.D., Mitchell M., Sgouros J., and Lindahl T. 2001. Human DNA repair genes. *Science* **291:** 1284–1289.

Woolley A.T., Guillemette C., Li Cheung C., Housman D.E., and Lieber C.M. 2000. Direct haplotyping of kilobase-size DNA using carbon nanotube probes (see comments). *Nat. Biotechnol.* **18:** 760–763.

Wooster R., Bignell G., and Lancaster J. 1995. Identification of the breast cancer susceptibility gene BRCA2. *Nature* **378:** 789–792.

Wormhoudt L.W., Commandeur J.N., and Vermeulen N.P. 1999. Genetic polymorphisms of human N-acetyltransferase, cytochrome P450, glutathione-S-transferase, and epoxide hydrolase enzymes: Relevance to xenobiotic metabolism and toxicity. *Crit. Rev. Toxicol.* **29:** 59–124.

Yip C.M., Brader M.L., DeFelippis M.R., and Ward M.D. 1998. Atomic force microscopy of crystalline insulins: The influence of sequence variation on crystallization and interfacial structure. *Biophys. J.* **74:** 2199–2209.

Zhang H., Somasundaram K., Peng Y., Tian H., Zhang H., Bi D., Weber B.L., and El-Deiry W.S. 1998. BRCA1 physically associates with p53 and stimulates its transcriptional activity. *Oncogene* **16:** 1713–1721.

Zilfou J.T., Hoffman W.H., Sank M., George D.L., and Murphy M. 2001. The co-repressor Sin3 directly interacts with the proline-rich domain of p53 and protects p53 from proteosome-mediated degradation. *Mol. Cell. Biol.* **21:** 3974–3985.

4

Identification of Protein–Protein Interactions with Glutathione-S-Transferase Fusion Proteins

Margret B. Einarson[1] and Jason R. Orlinick[2]

[1]Fox Chase Cancer Center, Philadelphia, Pennsylvania 19111; [2]Brigham & Women's Hospital, Boston, Massachusetts 02115

INTRODUCTION

Glutathione-S-transferase (GST) fusion proteins have had a wide range of applications since their introduction as tools for synthesis of recombinant proteins in bacteria (Smith and Johnson 1988), and they are routinely used for antibody generation, protein–protein interaction studies, and biochemical analysis. This chapter describes the use of GST-fusion proteins as probes for the identification of protein–protein interactions. These techniques are fundamentally similar to analogous techniques developed for antibody detection of proteins. For example, in western blots, an antibody is used to detect a query protein on a membrane. In contrast, in a far western (also known as an overlay assay), the antibody is replaced by a recombinant GST-fusion protein produced and purified from bacteria, and the interaction of this protein with a target protein on the membrane is assayed.

This chapter is adapted and expanded from Chapter 18, Protocol 2, in Sambrook and Russell 2001, *Molecular Cloning*.

Far-western analysis was originally developed to screen protein expression libraries with a ^{32}P-labeled GST-fusion protein (Blackwood and Eisenman 1991; Kaelin et al. 1992); however, this protocol can also be used for membranes generated by transfer following SDS-polyacrylamide gel electrophoresis (PAGE), allowing simultaneous interaction detection and information about the size of interacting proteins. Similarly, immunoprecipitation is based on the ability of the antibody to bind to its antigen in solution and the subsequent purification of the immunocomplex by collection on protein A- or G-coupled beads. The GST pull-down (Kaelin et al. 1991) is an affinity purification of an unknown protein from a pool of proteins in solution by its interaction with the GST-fusion probe protein and isolation of the complex by collection of the interacting proteins through the binding of GST to glutathione-coupled beads.

Advantages of Far Westerns and Pull-down Techniques

These techniques can allow a large-scale screen of a library with a novel protein to identify new interactions, as well as identify specific regions of proteins that mediate these interactions. Moreover, these approaches can be initiated prior to the availability of antibodies to the protein of interest, or when antibodies to the protein have been found to interfere with its protein–protein interactions. Although the protocols described in this unit produce results that are qualitative, GST-fusion proteins can be used in highly quantitative and sophisticated assays (see, e.g., Posern et al. 1999; McDonald et al. 1999). It should be noted that these methods characterize in vitro interactions, which should subsequently be substantiated in vivo by independent means such as coimmunoprecipitation (see Chapter 5). The flexibility of these techniques permits their application to a wide range of biological questions.

Alternative Approaches, Similar Paradigm

The protocols described below use GST-fusion proteins. There are many systems available for production of recombinant proteins that can be substituted. The most commonly used protein fusion systems include: a polyhistidine tag using this system (His-tag; Gentz et al. 1989); maltose-binding protein fusions (MBP; Bedouelle and Duplay 1988; Guan et al. 1988; Maina et al. 1988), fusions that become biotinylated in vivo (Promega), as well as tags for defined antigenic regions such as the Myc epitope. His-tag fusion proteins incorporate a polyhistidine stretch that can be purified by its high-affinity binding to metal ions (Hochuli et al. 1987). This system has the advantage that the fusion protein can be denatured during the purification while the His-tag is still competent to bind to metal ions, allowing a denatured fusion protein to be affinity-purified. This is a distinct advantage for insoluble proteins. Many of these fusion protein vectors are available for bacterial, *Drosophila*, and mammalian expression systems. The two most commonly used systems are GST and His-tag fusion proteins.

The development of vectors and antibodies to support GST-fusion protein applications has made them a popular choice. However, each of these fusion moieties permits different methods of affinity purification that may suit the particular protein under study. Therefore, it is in the best interest of the investigator to weigh the advantages of each system for a particular protein of interest.

OUTLINE OF PROCEDURE

Far-western Analysis

A far-western experiment is outlined in Figure 1. The probe protein is synthesized and purified from bacteria (see Preparation of GST-fusion proteins, Protocol 1). In this example, the probe protein is a GST-fusion containing the GST moiety followed by a protease cleavage site and a tar-

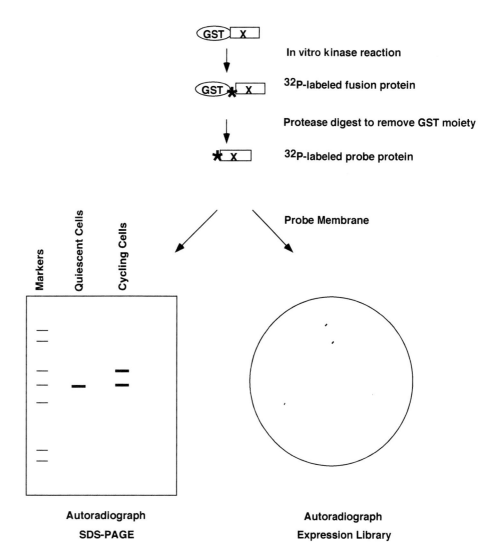

FIGURE 1. Outline of a far-western assay. The recombinant purified GST-fusion protein is radioactively labeled at a kinase consensus site contained in the fusion portion of the protein. The ^{32}P-labeled probe protein is separated from the GST moiety by protease digestion. The protein can be incubated with membranes generated subsequent to SDS-PAGE (*left*) or membranes generated by plating an expression library (*right*). Variations on this procedure are described in the text. (Redrawn, with permission, from Sambrook and Russell 2001, *Molecular Cloning*.)

get site for a known kinase translated in-frame with the protein of interest. The purified fusion protein is bound to glutathione beads and is labeled radioactively with ^{32}P using a commercially available kinase. Unincorporated nucleotide is removed from the fusion protein by washing, and the radioactively labeled protein is cleaved with protease (Factor X or thrombin) or eluted with glutathione to release the probe protein. The sample is prepared for probing by resolving the target proteins on an SDS-PAGE gel and transferring the proteins to a membrane, or plating a protein expression library and transferring the proteins to a membrane. The membranes are washed and blocked prior to addition of the radioactive probe protein. After incubation with the probe protein, the membrane(s) is washed and subjected to autoradiography. The method described above can be used for screening an expression library (Macgregor et al. 1990; Blackwood and Eisenman 1991; Kaelin et al. 1991; Blanar and Rutter 1992; Ayer et al. 1993; Einarson and Chao 1995) or proteins resolved by SDS-PAGE and transferred to a membrane (see, e.g., Grgurevich et al. 1999; Hunter et al. 1999; Posern et al. 1999).

Detection Methods Used in a Far Western

Three methods are generally used to detect an interaction: radioactive labeling of the fusion protein, detection by anti GST antibodies, and biotinylation of the fusion protein. Each of these methods has advantages. Radioactively labeling the fusion protein is rapid and easy and generally has little impact on the subsequent activity of the protein because the phosphorylation site is in the fusion portion of the protein. A variety of protein kinase sites have been integrated into fusion protein vectors (Blanar and Rutter 1992; Kaelin et al. 1992; Ron and Dressler 1992) to allow ^{32}P-labeling of recombinant proteins expressed in bacteria. An early variation on this technique is using endogenous phosphorylation sites in the protein of interest as the site of ^{32}P phosphorylation (Skolnick et al. 1991; Ayer et al. 1993). This strategy can only be employed with protein probes that contain such sites and when the phosphorylation of those residues is beneficial (at least, nondeleterious) to the experimental design (in terms of impact on protein interaction). Biotinylation of proteins is now relatively easy due to the availability of kits (from Amersham Pharmacia Biotech, Inc.); however, biotinylation can also have an impact on the interaction of the protein. Finally, using anti-GST antibodies that are commercially available (from Amersham Pharmacia, Santa Cruz Biotech, Upstate Biotechnology, Convance/Babco) is a good option for detecting interactions on membranes containing cell lysates. The latter two methods have the advantage of being nonradioactive. However, if a library screen is planned, the cost of the anti-GST antibody and detection reagents may be excessive.

Design of the Bait Fusion Protein

The design of the fusion protein depends on the chosen method of detection. If the option to ^{32}P-label the protein is desired, it is important to use a vector that contains a kinase target site. Another issue to consider is using cleaved (GST-removed) versus full-length GST-fusion proteins in actual screening. The GST moiety can generate background; therefore, most fusion vectors contain cleavage sites to separate the GST portion from the protein of interest. However, if the choice is to cleave the fusion protein, the interaction cannot be detected with anti-GST antibodies. Furthermore, in choosing which cleavage site to use, it is important to scan the bait protein and ensure that it does not contain the recognition site of the protease. The region of the protein of interest that is expressed will be determined by what type of screen is being planned. For a library screen, inclusion of as much of the protein as possible detects the greatest variety of interactors. In contrast, if known interaction domains of a protein are being tested to confirm predicted interactions, a fusion protein containing only these motifs is preferred, because it is less prone to false positive interactions.

Other Considerations

Being able to synthesize the fusion protein without excess degradation and insolubility is one of the most important considerations in planning a far western. If the fusion protein is excessively degraded during purification, the probe protein will be a mixture of different products of degradation. This may result in variable results from different fusion protein preparations. Therefore, it is important to monitor the status of the fusion protein after purification and storage prior to use.

Probing proteins immobilized on a membrane with a GST-fusion protein is analogous to western analysis. However, one variable specific to protein–protein probing is whether to probe the membrane with or without a cycle of denaturation/renaturation, which is designed to allow misfolded proteins to refold into their native conformations. In general, proteins transferred to a membrane from an SDS-polyacrylamide gel do not require denaturation/renaturation. However, many investigators who probe filters displaying an expression library do find it necessary. If probing proteins transferred to a membrane from SDS-PAGE is unsuccessful, adding the denaturation/renaturation cycle in Support Protocol 1 (p. 52) may yield positive results.

Controls

The following issues should be considered in designing controls for far-western experiments:

- If the GST moiety is retained on the fusion protein, the experiment must be replicated with GST alone. In the case of a library screen where this would be impractical, it is important to test positive plaques after quaternary purification, prior to clone characterization.

- Generating a probe protein with the predicted interaction domain mutated, and demonstrating loss of interaction in an experiment performed in parallel to the actual probe, is the best control for specificity of the interaction.

- A negative control of GST alone or a nonspecific protein can be included on an SDS-PAGE gel with the target proteins. If the protein of interest is a member of a conserved family of proteins, the interaction of the protein of interest can be compared to other proteins from the same gene family.

GST Pull-down Analysis

The GST pull-down protocol was developed to probe interactions between a fusion protein and potential interactors in solution (Kaelin et al. 1991). A GST pull-down experiment is outlined in Figure 2. First, the probe GST-fusion protein is expressed and affinity-purified from bacteria. At the same time, a cell lysate (which can be ^{35}S-labeled or unlabeled) is prepared. Then, the GST-fusion probe and cell lysates are mixed in the presence of glutathione-Sepharose beads and incubated to allow protein associations to occur. Centrifugation is used to collect the GST-fusion probe protein and any associated molecules. The complexes are washed to remove nonspecifically adhering proteins. Excess free glutathione is used to elute the complexes from the beads, or they are boiled directly into an SDS-PAGE sample buffer. The proteins are resolved on an SDS-PAGE gel and processed for further analysis (western, autoradiography, or staining). This technique is particularly useful to detect protein–protein interactions in solution that might not be detected in a membrane-based assay.

Applications of the Pull-down Technique

There are two general uses of the GST pull-down experiment. One is to identify novel interactions between a fusion protein and unknown proteins (Kaelin et al. 1991; Orlinick and Chao 1996); another is to identify interactions between the fusion protein and a known protein (see, e.g., Grgurevich et al. 1999; Hunter et al. 1999; Posern et al. 1999; Sun et al. 1999) that is a suspected interactor. These two experiments are designed and executed differently.

Identifying an Unknown Interactor

Many of the variables that determine the success or failure of this type of experiment must be empirically determined at the outset of the experiment. Primarily, what is the appropriate source of interactor proteins to probe with the bait? It is important to choose the source of the interacting protein carefully. This experiment requires that the potential interactor be present in the protein source being probed at a high enough level that it is within the limits of the detection method chosen. For this reason, radioactively labeled lysates are most frequently used as the source of protein. Some important questions to ask before choosing the interactor source are: Is the protein of interest expressed in that particular cell or tissue type? (If not, the physiologically relevant interactors may not be present either.) Are different types of cell populations (i.e., quiescent versus cycling or growth-factor-treated versus untreated) being compared?

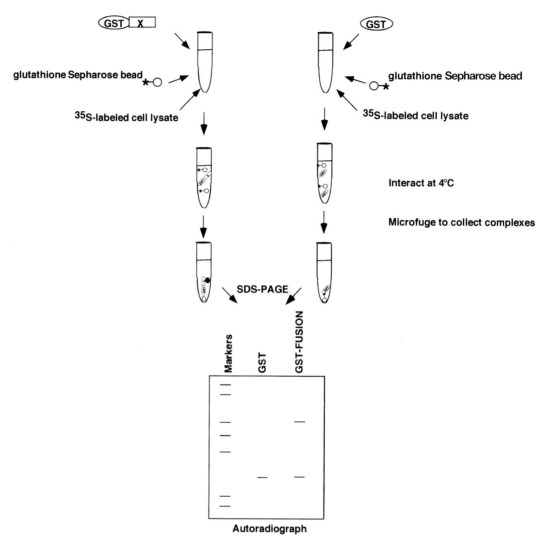

FIGURE 2. Outline of a GST pull-down assay. The recombinant GST-fusion protein, or control GST, is incubated with a cell lysate in the presence of glutathione-Sepharose beads. The proteins are allowed to incubate with end-over-end mixing at 4°C. The reaction is centrifuged to collect the GST or GST-fusion proteins and associated proteins. The proteins are resolved on an SDS-PAGE gel and subjected to autoradiography. The bands on the gel that associate uniquely with the GST-fusion protein and not GST indicate novel interacting proteins. Variations on this protocol are described in the text. (Redrawn, with permission, from Sambrook and Russell 2001, *Molecular Cloning.*)

Characterizing a Predicted or Known Interaction

The experimental design is more straightforward if a predicted interaction is being tested, as a wide variety of protein sources can be generated to identify and map the interaction. Detection of the interaction is determined by the availability of antibodies to the predicted interactor. If none is available, ^{35}S-labeled in-vitro-translated protein can be used, or the predicted interactor can be epitope-tagged. In addition, cells can be transfected to increase the abundance of the predicted interactor. However, it is important to control for the effects of mass action (aggregation) and to ensure the specificity of the association. The best control to test specificity is to include a GST-fusion protein with a mutated interaction domain and demonstrate loss of the interaction, and to perform a comparison to the GST moiety alone.

Other Considerations

Some optimization for each protein complex being assayed is required for this technique. The buffer in which the interaction will take place is one of the more important variables to be considered. This is often the buffer in which the potential interactors are prepared and can vary from a cell lysis buffer (such as RIPA) to an in vitro translation reaction mixture. Although a wide variety of buffers work, each interaction may be more or less detectable depending on the buffers employed. A second variable is the amount of target protein mixed with the fusion protein. The abundance of the potential interactor and the affinity of the interaction (both variables are generally unknown when the experiment is initiated) determine the amount of material used. A third variable that can be optimized is wash conditions. In this protocol, the cell lysis buffer was used; however, buffers of varying salt and detergent concentrations can be used to eliminate nonspecific interactions.

Controls

The following controls should be considered:

- Perform a pull-down experiment with GST alone and compare to the GST-fusion protein.
- Create a mutation in the putative interaction domain and demonstrate loss of the interaction.

Add increasing amounts of excess fusion protein minus the GST moiety (or the interaction domain alone) to demonstrate competition. Perform this competition in parallel with a mutant of the putative interaction domain.

Preparation of GST-fusion Proteins

This protocol is designed for IPTG-inducible bacterial expression vectors.

MATERIALS

CAUTION: See Appendix for appropriate handling of materials marked with <!> .

Buffers and Solutions

Coomassie Blue
Elution buffer
20 mM reduced glutathione in 50 mM Tris-Cl (pH 8.0)
Isopropylthio-β-D-galactoside (IPTG) <!>
LB broth
Lysis buffer
PBS + 1% Triton X-100 supplemented with protease inhibitors (e.g, 2 µg/ml aprotinin,
1 µg/ml leupeptin, 25 µg/ml phenylmethylsulfonyl fluoride <!>)
Phosphate-buffered saline
150 mM NaCl
20 mM sodium phosphate (pH 7.4) <!>

Antibiotics

Appropriate antibiotic (see step 1)

Bacterial Strains

Bacterial strain transformed with GST and GST-fusion expression plasmids

Special Equipment

Centrifuge
Eppendorf tubes
Disposable chromatography column (Econo-column; Bio-Rad)
Glutathione-Sepharose beads
SDS-polyacrylamide gel electrophoresis equipment and reagents <!>
Sonicator

METHOD

1. Inoculate an overnight bacterial culture of each construct (GST alone and GST-fusion proteins) in 5 ml of LB + antibiotic.

   Although variations of GST-fusion protein expression vectors are available, the most commonly used versions (available from Amersham Pharmacia) include the GST moiety separated from the protein of interest by a multiple cloning site; an IPTG-inducible promoter; the ampicillin resistance gene; the *lacI* gene for expression control; and a bacterial origin of replication. Many bacterial strains can be used, including those commonly used for cloning. Alternatively, protease-deficient strains such as CAG456 (grows at 30°C) can be used for proteins that are produced poorly at 37°C.

2. Inoculate 1 liter of LB + antibiotic selection with 5 ml of the overnight culture from step 1.

3. Grow to an OD_{600} of 0.5–1.0.

4. Induce expression of the protein by adding IPTG to a final concentration of 0.1 mM.

   Different proteins are optimally produced using different concentrations of IPTG. If protein expression is problematic, first titrate the amount of IPTG added to determine the optimal conditions for protein induction.

5. Grow for an additional 3 hours at 37°C with shaking.

6. Centrifuge the bacterial culture at 3500*g* for 20 minutes at 4°C.

7. Discard the supernatant.

   At this point, pellets can be stored frozen at –20°C if necessary.

8. Resuspend pellet in 20 ml of lysis buffer plus protease inhibitors.

9. On ice, sonicate the pellet in short 10-second bursts alternated with 10 seconds resting on ice. Three cycles of sonication are usually sufficient.

   Oversonication can result in degradation and denaturation of the fusion protein. It can also result in contamination with bacterial host proteins. If these problems occur in the preparation of the protein of interest, titration of the time of sonication is required to release the protein.

10. Centrifuge the lysate at 3500*g* for 15 minutes at 4°C.

11. Transfer the supernatant to a fresh tube.

12. Add 5 ml of a 50:50 slurry of glutathione-Sepharose beads.

    Commercially available glutathione-Sepharose beads are often provided in a solution containing alcohols or other ingredients. Prior to use, these resins should be washed with lysis buffer and stored as a 50:50 (v/v) slurry at 4°C.

13. Incubate for 30 minutes at 4°C, rotating the tube end over end to ensure mixing.

14. Centrifuge at 750*g* at 4°C to pellet the beads.

15. Wash the beads with 5 ml of ice-cold PBS plus protease inhibitors.

16. Centrifuge at 750*g* at 4°C to pellet the beads.

17. Add 5 ml of ice-cold PBS plus protease inhibitors. Resuspend the beads by gentle mixing.

18. Centrifuge at 750*g* at 4°C to pellet the beads.

    The fusion protein can be stored on the beads at 4°C. This is appropriate if the protein is to be labeled or used in a GST pull-down experiment.

19. Add 5 ml of ice-cold PBS plus protease inhibitors. Resuspend the beads with gentle mixing.

20. Pour the slurry into a commercially available disposable chromatography column (such as Bio-Rad Econo-columns).

21. Allow the PBS to run out of the column. Wash with 5 ml of ice-cold PBS plus protease inhibitors.

22. While the column is flowing, prepare a rack of 10 Eppendorf tubes labeled 1–10.

23. Add 5 ml of cold 20 mM reduced glutathione in 50 mM Tris (pH 8.0) to elute the fusion protein.

24. Collect ~0.5-ml fractions of the eluate in each Eppendorf tube.

25. Store the eluate at 4°C prior to use.

    The column can be stored in PBS at 4°C properly sealed to prevent desiccation and contamination.

    The eluted proteins are in a solution containing 20 mM glutathione. In most instances, it is optimal to remove the glutathione. This can be accomplished by dialysis against the buffer that is most compatible with the assay in which the protein will be utilized.

26. Perform a protein assay on the eluted fractions. Run the protein on an SDS-PAGE gel and stain with Coomassie Blue.

TROUBLESHOOTING

An excellent reference for affinity purification of GST-fusion proteins can be found in the Amersham Pharmacia Web Site: http://www.apbiotech.com/product/publication/brochure/brochures_recombinant.html.

LOW YIELD

If the yield is low, repeat protein purification on a smaller scale and remove aliquots at the following steps of the procedure: step 3 prior to addition of IPTG; step 9; step 11 supernatant; step 14 supernatant; eluate fractions.

1. To compare the presence of the fusion protein present in the fractions collected, load the following fractions on a mini SDS-PAGE gel. Volumes are based on a starting volume of a liter of culture; adjust accordingly.

    15 µl of uninduced culture from step 3
    15 µl of induced culture from step 5
    0.2% of supernatant at step 9 (total cell lysate): 40 µl
    0.2% of supernatant at step 11 (soluble lysate): 40 µl
    0.2% of supernatant at step 14 (lysate after incubation with beads): 10 µl
    0.2% of aliquot of an eluate fraction: 1 µl
    2% of aliquot of an eluate fraction: 10 µl

2. Combine with the SDS sample buffer and run on an SDS-polyacrylamide gel. The GST moiety adds ~27 kD molecular mass.

3. Stain the gel with Coomassie Blue. Analysis of these fractions will tell:

    a. whether the protein was induced (uninduced versus induced culture).

    b. whether the protein is soluble (compare aliquot step 9 versus aliquot step 11).

    c. whether a substantial portion of the fusion protein did not bind to the beads (compare aliquot step 14 to aliquot step 11.

    d. the yield, and the integrity of the protein.

FAILURE TO INDUCE PROTEIN EXPRESSION

To determine whether the induction conditions are working, prepare a culture of GST in parallel with the GST-fusion protein. If GST is produced, optimization of induction of the fusion protein is necessary. Titrate the amount of IPTG added, alter the OD_{600} when the IPTG is added, lower the temperature, and/or induce for longer or shorter times.

DEGRADATION OF THE PROTEIN

This can be addressed by using protease-deficient (*lon⁻*) bacterial strains. There are many commercially available bacterial strains designed for protein expression with appropriate genetic backgrounds to minimize protein degradation (BL21, CAG456). Degradation can also result from oversonication. Determine the appropriate sonication time for the protein of interest by performing small-batch preparations and test different times of sonication.

INSOLUBILITY

This can be ameliorated by inducing at lower temperatures and for longer times. In addition, using different detergents during lysis can help (an excellent reference for this problem is Frangoni and Neel 1993). However, this can be an insurmountable difficulty. To continue analysis of the protein by the methods described in this chapter, transferring the protein of interest to a His-tagged vector (which allows affinity purification of denatured proteins) may be an appropriate alternative.

FAILURE TO BIND THE GLUTATHIONE-SEPHAROSE

This can occur when the fusion protein is denatured excessively. This is often the result of oversonication. To address this possibility, decrease the time and intensity of sonication. Addition of DTT to 5 mM final concentration prior to lysis can also increase the binding of some fusion proteins (see Amersham Pharmacia Web Site: http://www.apbiotech.com/product/publication/brochure/brochures_recombinant.html.

Far Western: Labeling GST-fusion Proteins

This protocol is for proteins containing a protein kinase A phosphorylation site in the fusion protein (Ron and Dressler 1992).

MATERIALS

CAUTION: See Appendix for appropriate handling of materials marked with <!> .

Buffers and Solutions

2x PK buffer
    100 mM potassium phosphate (KPO_4) (pH 7.15) <!>
    20 mM magnesium chloride ($MgCl_2$) <!>
    10 mM sodium fluoride (NaF) <!>
    9 mM dithiothreitol (DTT) <!>

Reagents

$[\gamma\text{-}^{32}P]$ATP (6000 Ci/mmole) <!>
20 mM Tris (pH 8)
20 mM Reduced glutathionine
GST-fusion protein (see below)
Protein kinase A (Sigma)
    Prepare fresh at each use as per manufacturer's instructions.

Special Equipment

G50 Sephadex Spin Column (optional; Boehringer-Mannheim)
Glutathionine-Sepharose beads
Microcentrifuge
Microcentrifuge tubes

METHOD

1. Prepare the following reaction mix in an Eppendorf tube:

   5 µl of $[\gamma\text{-}^{32}P]$ATP (6000 Ci/mmole)
   1 unit/µl protein kinase A (Sigma) (made up fresh each use)
   1–3 µg of GST-fusion protein on glutathione-Sepharose beads
   12.5 µl of 2x PK buffer
   dH_2O to 25 µl

   The amount of protein used depends on the application (i.e., western blot versus library screen). A good starting point is a final concentration of 1–5 nM during the incubation with the membrane(s).

Optional: The protein can be protease-cleaved prior to the labeling reaction. This is generally described in detail by manufacturers' protocols related to supplied vectors and should be done as specified therein. Substitute cleaved protein for protein bound to glutathione-Sepharose beads in the labeling reaction and proceed to step 2.

2. Incubate for 30 minutes at 37°C.

3. After the labeling reaction is complete, wash the beads in 200 μl of 1x PK buffer. Centrifuge for 1 minute at 13,000 rpm. Discard the supernatant containing the free nucleotide appropriately. Repeat the wash once. At this step, the labeled protein can be protease-cleaved (if desired) or directly eluted in 20 mM reduced glutathione in 50 mM Tris (pH 8.0). After the protein is released from the beads, it is ready to be used for probing. The probe should be used the same day it is made and stored at 4°C prior to use.

   Alternatively: If the labeled protein was cleaved from the GST moiety prior to the labeling reaction (see "optional" note in step 1), it is loaded onto a prepared Sephadex G50 column (Boehringer-Mannheim) equilibrated with 1x PK buffer to remove the free nucleotide. Use columns according to the manufacturer's instructions. Once the probe protein is separated from the free nucleotide, it is ready to be used for probing. The probe should be used the same day it is made and stored at 4°C prior to use.

   Because the probe is diluted in the next protocol, the final volume of the probe at this step is not fixed. The most important variable to note at this stage is the total microgram or molar input quantity of the probe. The final concentration of probe when probing the membranes is 1–5 nM. Prior to labeling, the protein concentration should be determined by protein assay or by detection with Coomassie Blue or silver staining of an SDS-PAGE gel as compared to a standard of known concentration. If a comparison is being made between two proteins (GST versus GST-fusion or a wild-type versus mutant fusion protein), it is important to ensure that each probe is at the same final concentration during probing.

Far Western: Probing Membranes

Support protocols to deal with lack of protein interaction after transfer and a nonradioactive approach for detection of an interaction follow this protocol.

MATERIALS

CAUTION: See Appendix for appropriate handling of materials marked with <!> .

Buffers and Solutions

Basic buffer
   20 mM HEPES (pH 7.5)
   50 mM potassium chloride (KCl) <!>
   10 mM magnesium chloride (MgCl$_2$) <!>
   1 mM dithiothreitol (DTT) <!>
   0.1% Nonidet P-40
Blocking buffer
   5% nonfat dry milk in basic buffer
Denaturation buffer
   6 M guanidine hydrochloride <!> prepared in basic buffer (optional)
Interaction buffer
   1% nonfat dry milk in basic buffer + 5% glycerol
Wash buffer 1
   PBS + 0.2% Triton X-100
Wash buffer 2
   PBS + 0.2% Triton X-100 + 100 mM KCl

Reagents

Labeled GST-fusion protein
Potential targets immobilized on nylon or nitrocellulose membrane

Special Equipment

Plastic containers that can accommodate membranes
X-ray film

METHOD

1. Prepare a membrane for probing by transfer according to standard techniques. Proteins transferred from an SDS-PAGE gel can generally be probed directly. If proteins are being

transferred from an expression library, perform Support Protocol 1 (p. 52) before proceeding to step 2.

2. Wash the membrane in enough basic buffer to completely cover the membrane(s) for 10 minutes at 4°C with gentle agitation.

3. Incubate the membrane in enough blocking buffer to cover the membrane completely at 4°C with gentle agitation for 4 hours to overnight.

4. Mix 1–3 μg (1–5 nM final concentration) of the labeled fusion protein in interaction buffer. The probe solution should contact the entire surface of the membrane(s) evenly. Incubate for 4–5 hours at 4°C with gentle agitation.

5. Discard the radioactive probe solution appropriately. Wash the membrane with enough wash buffer 1 to cover the membrane(s) completely for 10 minutes at 4°C with gentle agitation. Repeat three times. Discard the wash solutions appropriately.

6. Wash the membrane in enough wash buffer 2 to cover membrane(s) completely for 10 minutes at 4°C with gentle agitation. Repeat once.

7. Wrap the membrane carefully in plastic wrap and expose it to X-ray film.

TROUBLESHOOTING

NO SIGNAL

1. Ensure that the protein is being radioactively labeled by analyzing the labeled protein on an SDS-polyacrylamide gel; dry and expose to X-ray film.

2. Determine whether the membrane contains protein (i.e., was the library induced correctly or did the proteins transfer properly).

3. Perform a denaturation–renaturation cycle (Support Protocol 1, p. 52) prior to probing.

4. The probe protein is limiting. Perform a titration adding increasing amounts of probe protein to obtain the optimal probe concentration.

5. The source of interacting proteins may not contain a partner. Try the assay with a different source of interacting proteins.

6 If one is available, include a positive control. However, it is always possible that no signal could be the correct result for the source of interacting proteins being probed.

EXCESSIVE SIGNAL

1. The probe protein is degraded or impure. Check the probe protein integrity by analysis on an SDS-polyacrylamide gel.

2. Background is due to the presence of the GST moiety. Remove the GST moiety by proteolytic cleavage prior to probing. If this is impractical, add an excess of unlabeled GST prior to blocking or during probe incubation (this approach cannot be used if anti-GST antibodies are being used to detect the interaction).

3. The probe is in excess. This can be controlled by performing a titration of the probe to determine optimal probe concentration.

4. The incubation time can be shortened to reduce nonspecific interactions.

Denaturation–Renaturation in Guanidine Hydrochloride

In some instances, the proteins on the membrane are not able to interact after transfer. This is possibly due to improper folding. Thus, denaturation and slow renaturation of the proteins may be necessary. This is accomplished by washing the membrane in a denaturation buffer, which is serially diluted to permit slow renaturation on subsequent washing. This is most efficiently accomplished by starting with a stock solution of 6 M guanidinium hydrochloride made in basic buffer and serially diluting this stock solution as the membrane is washed.

After transfer, the membrane is washed in denaturation buffer for 10 minutes at 4°C with gentle agitation. Repeat. The denaturation buffer just used is diluted 1:1 with basic buffer. The membrane is incubated in diluted denaturation buffer as before. The denaturation buffer is serially diluted 1:1 each time for a total of five dilutions. For example:

1. Wash the membrane in 50 ml of denaturation buffer for 10 minutes at 4°C with gentle agitation.

2. Repeat with a fresh aliquot of 50 ml of denaturation buffer.

3. Remove the denaturation buffer from the container and retain 25 ml. Add 25 ml of basic buffer for a total volume of 50 ml. This is the first 1:1 dilution of the denaturation buffer. Return the diluted denaturation buffer to the container and wash the blot for 10 minutes at 4°C with gentle agitation.

4. Repeat this dilution and wash cycle four more times. The final wash contains 175 mM guanidinium hydrochloride.

   At this point, proceed with the protocol (Probing Membranes) at step 2.

Detection with Anti-GST Antibodies

A nonradioactive approach is to interact the membrane with an unlabeled GST-fusion protein and to detect the interaction using commercially available anti-GST antisera.

1. Subsequent to generating the membrane with potential interacting proteins (step 1, Protocol 3), proceed through step 6 of Protocol 3.

2. Subsequent to step 6 of Protocol 3, incubate the membrane with anti-GST antibodies according to manufacturer's specifications.

Using this approach when screening a library is substantially more expensive (purchase of antibody and detection reagents) than using radioactively labeled proteins. In addition, care must be taken to control for nonspecific interactions with GST alone, and a signal resulting from antibody cross-reactivity.

Protocol 4

GST Pull-down

This protocol is designed to use a ^{35}S-labeled cell lysate as the source for interacting proteins.

MATERIALS

CAUTION: See Appendix for appropriate handling of materials marked with <!>.

Buffers and Solutions

Lysis buffer
   20 mM Tris (pH 8.0)
   200 mM NaCl
   1 mM EDTA (pH 8.0)
   0.5% Nonidet P-40
   2 µg/ml aprotinin
   1 µg/ml leupeptin
   0.7 µg/ml pepstatin
   25 µg/ml phenylmethylsulfonyl fluoride (PMSF) <!>

Reagents

Coomassie Blue
GST protein
GST-fusion protein
Reduced glutathione (Sigma)
^{35}S-labeled cell lysate
Silver nitrate
Tris (50 mM, pH 8.0)

Special Equipment

End-over-end sample rotator
Equipment for SDS-polyacrylamide gel electrophoresis
Glutathione-Sepharose beads
   Beads are often supplied by commercial vendors in solutions containing alcohols. It is important to wash the beads thoroughly in lysis buffer and to generate a 50/50 slurry of beads in lysis buffer prior to use.
Microcentrifuge
Microcentrifuge tubes
X-ray film

METHOD

1. Incubate the cell lysate with 50 μl of glutathione-Sepharose beads (50/50 slurry in lysis buffer) and 25 μg of GST for 2 hours at 4°C with end-over-end mixing. The amount of lysate needed to detect an interaction is highly variable. Start with lysate equivalent to 1×10^6 to 1×10^7 tissue culture cells (Orlinick and Chao 1996; Spector et al. 1998).

    It is important to keep in mind that the following experiment compares GST versus GST-fusion protein, so it is necessary to prepare enough lysate to allow for equal amounts of lysate in each reaction. Allow enough volume to permit liberal mixing; 500 μl to 1 ml is a good starting point.

    This step is designed to preclear proteins from the lysate that interact nonspecifically with the GST moiety or the beads alone. It is not strictly necessary to preclear the lysates with GST or glutathione-Sepharose beads if the interaction will be detected primarily with antibodies directed to a candidate interacting protein. However, when ^{35}S-labeled cell lysates are used to identify novel protein–protein interactions, these steps can aid in reducing background. When detecting the interactor with antibodies to the interacting protein, it is important to include "GST + beads" and "beads alone" controls. Additionally, if the interacting protein of interest is known to be confined to a specific cellular compartment (e.g., the nucleus), a fraction of the cell lysate corresponding to that compartment (e.g., a nuclear extract) can be used as the source of potential interactors.

2. Centrifuge at 13,000 rpm for 2 minutes at 4°C in a microcentrifuge.

3. Remove the supernatant to a fresh tube.

4. Incubate the cell lysate with GST and GST-probe proteins, separately. Set up two tubes containing equal amounts of precleared cell lysate. Add 50 μl of glutathione-Sepharose beads (50/50 slurry in lysis buffer) and add GST protein to one tube and the GST-fusion probe to the other (~10 μg each).

    The amount of protein added should be equimolar in the two reactions (i.e., the final molar concentration of GST should be the same as that of the GST-probe protein).

5. Incubate the tubes for 2 hours at 4°C with end-over-end mixing.

6. Centrifuge the samples at 13,000 rpm for 2 minutes in a microcentrifuge.

7. Save the supernatant in a fresh microfuge tube (for SDS-PAGE analysis only).

8. Wash the beads four times with 1 ml of ice-cold lysis buffer. Discard the washes.

9. At this point, the fusion protein and any proteins bound to it can be eluted with 50 μl of 20 mM reduced glutathione in 50 mM Tris (pH 8.0).

    Alternatively, the beads can be mixed with an equal volume of 2x SDS-PAGE sample buffer and boiled prior to gel loading.

    It is important to include a "glutathione-Sepharose beads only" control if the sample is to be boiled off the beads. Proteins bound nonspecifically to the beads can appear as bound to the fusion protein even in comparison to GST alone.

10. Run the proteins on an SDS-PAGE gel. For detection of interaction:

    a. If the goal is to detect the ^{35}S-labeled proteins associated with the fusion protein after SDS-PAGE, the gel is dried on a gel dryer and exposed to X-ray film.

    b. If the goal is to detect specific partners, after SDS-PAGE the proteins should be transferred to a membrane and western analysis performed.

    c. If the goal is to determine the size and abundance of proteins associated with the fusion protein from a nonradioactive lysate subsequent to SDS-PAGE, the gel should be stained with Coomassie or silver nitrate.

TROUBLESHOOTING

DETERMINING OPTIMAL CONDITIONS

If it is difficult to determine whether the conditions for the GST pull-down experiment are optimal, an initial troubleshooting step is to load equal volume percentage aliquots of each fraction generated during the protocol.

For example, analysis of a 1% volume aliquot of the following samples on the final SDS-PAGE gel can aid in troubleshooting:

Total cell lysate (step 1)
Eluate (step 9)
Beads post-elution (step 9)
Supernatant saved at step 7
"Beads + GST" eluate
"Beads alone" eluate

Harvest these samples at the indicated steps, add the appropriate amount of SDS-PAGE sample buffer, and flash-freeze in a dry-ice ethanol bath for future analysis.

Aliquot from step 1: These samples will inform you of the prevalence of the novel interactor in the total cell lysate.

Aliquot from step 9 prior to elution: How much is bound to the GST-fusion protein.

Aliquot of eluate from step 9 post-elution: How much GST-fusion protein + associated proteins were eluted from the beads.

Beads remaining from step 9: How much of the novel interactors remained bound. GST-plus-beads and beads-alone controls will identify nonspecific interactions.

LOW SIGNAL

If an interaction is yielding a low signal, even though the interactor is abundant in the total cell lysate, this may indicate that the binding conditions are not optimal. A change in salt and detergent concentrations, in addition to increasing the time allowed for association, may improve the result.

In contrast, the complex may be retained on the glutathione-Sepharose beads due to inefficient elution. This can be determined by SDS-PAGE comparison of the eluate versus the "fusion-protein-beads post-elution" fraction. If this is the problem, it may be remedied by pooling multiple elutions or increasing the time for elution.

NONSPECIFIC BACKGROUND

Preclearing a lysate with GST or beads alone can help to minimize nonspecific interactions. Decreasing the amount of lysate added and increasing the stringency of the wash conditions can also reduce background.

IMPORTANT CONTROL

If you are analyzing the interaction by western blot, it is important to reprobe the membrane with anti-GST antibodies subsequent to probing for the candidate interactor. This will determine whether all samples were incubated with the same amount of GST-fusion protein, and it will help determine whether the fusion protein is undergoing degradation while incubated with the cell lysate.

REFERENCES

Ayer D.E., Kretzner L., and Eisenman R.N. 1993. Mad: A heterodimeric partner for Max that antagonizes Myc transcriptional activity. *Cell* **72:** 211–222.

Bedouelle H. and DuPlay P. 1988. Production in *Escherichia coli* and one-step purification of bifunctional hybrid proteins which bind maltose. *Eur. J. Biochem.* **171:** 541–549.

Blackwood E.M. and Eisenman R.N. 1991. Max: A helix-loop-helix zipper protein that forms a sequence-specific DNA-binding complex with Myc. *Science* **251:** 1211–1217.

Blanar M.A. and Rutter W.J. 1992. Interaction cloning: Identification of a helix-loop-helix zipper protein that interacts with c-Fos. *Science* **256:** 1014–1018.

Einarson M.B. and Chao M.V. 1995. Regulation of Id1 and its association with basic helix-loop-helix proteins during nerve growth factor-induced differentiation of PC12 cells. *Mol. Cell. Biol.* **15:** 4175–4183.

Frangoni J.V. and Neel B.G. 1993. Solubilization and purification of enzymatically active glutathione-S-transferase (GEX) fusion proteins. *Anal. Biochem.* **210:** 179–187.

Gentz R., Chen C.H., and Rosen C.A. 1989. Bioassay for trans-activation using purified human immunodeficiency virus tat-encoded protein: Trans-activation requires mRNA synthesis. *Proc. Natl. Acad. Sci.* **86:** 821–824.

Grgurevich S., Mikhael A., and McVicar D.W. 1999. The Csk homologous kinase, Chk, binds tyrosine phosphorylated paxillin in human blastic T cells. *Biochem. Biophys. Res. Commun.* **256:** 668–675.

Guan C., Li P., Riggs P.D., and Inouye H. 1988. Vectors that facilitate the expression and purification of foreign peptides in *Escherichia coli* by fusion to maltose binding protein. *Gene* **67:** 21–30.

Hochuli E., Dobeli A., and Schacher A. 1987. New metal chelate adsorbent selective for proteins and peptides containing neighbouring histidine residues. *J. Chromatogr.* **411:** 177–184.

Hunter S., Burton E.A., Wu S.C., and Anderson S.M. 1999. Fyn associates with Cbl and phosphorylates tyrosine 731 in Cbl, a binding site for phosphatidylinositol 3-kinase. *J. Biol. Chem.* **274:** 2097–2106.

Kaelin W.G., Pallas D.C., DeCaprio J.A., Kaye F.J., and Livingston D.M. 1991. Identification of cellular proteins that can interact specifically with the T/E1A binding region of the retinoblastoma gene product. *Cell* **64:** 521–532.

Kaelin W.G., Krek W., Sellers W.R., DeCaprio J.A., Ajchenbaum F., Fuchs C.S., Chittenden T., Li Y., Farnham P.J., Blanar M.A., Livingston D.M., and Flemington E.K. 1992. Expression cloning of a cDNA encoding a retinoblastoma-binding protein with E2F-like properties. *Cell* **70:** 351–364.

Macgregor P.F., Abate C., and Curran T. 1990. Direct cloning of leucine zipper proteins: Jun binds cooperatively to the CRE with CRE-BP1. *Oncogene* **5:** 451–458.

Maina C.V., Riggs P.D., Grandea A.G. III, Slatko B.E., Moran L.S., Tagliamonte J.A., McReynolds L.A., and Guan C. 1988. A vector to express and purify foreign proteins in *Escherichia coli* by fusion to, and separation from, maltose binding protein. *Gene* **74:** 365–373.

McDonald O.B., Chen W.J., Ellis B., Hoffman C., Overton L., Rink M., Smith A., Marshall C.J., and Wood E.R. 1999. A scintillation proximity assay for the Raf/MEK/ERK kinase cascade: high-throughput screening and identification of selective enzyme inhibitors. *Anal. Biochem.* **268:** 318–329.

Orlinick J.R. and Chao M.V. 1996. Interactions of the cellular polypeptides with the cytoplasmic domain of the mouse Fas antigen. *J. Biol. Chem.* **271:** 8627–8632.

Phizicky E.M. amd Fields S. 1995. Protein-protein interactions: Methods of detection and analysis. *Microb. Rev.* **59:** 94–123.

Posern G., Zheng J., Knudsen B.S., Kardinal C., Muller K.B., Voss J., Sishido T., Cowburn D., Cheng G., Wang B., Kruh G.D., Burrell S.K., Jacobson C.A., Lenz D.M., Zamborelli T.J., Adermann K., Hanafusa H., and Feller S. 1998. Development of highly selective SH3 binding peptides for Crk and CRKL which disrupt Crk-complexes with DOCK180, SoS, and C3G. *Oncogene* **16:** 1903–1912.

Ron D. and Dressler H. 1992. pGSTag—A versatile bacterial expression plasmid for enzymatic labeling of recombinant proteins. *BioTechniques* **13:** 866–869.

Sambrook J. and Russell D. 2001. *Molecular cloning: A laboratory manual,* 3rd edition. Cold Spring Harbor Laboratory Press, Cold Spring Harbor, New York.

Skolnick E.Y., Margolis B., Mohammadi M., Lowenstein E., Fischer R., Drepps A., Ullrich A., and Schlessinger J. 1991. Cloning of PI3 kinase associated p85 utilizing a novel method for expression/cloning of target proteins for receptor tyrosine kinases. *Cell* **65:** 83–90.

Smith D.B. and Johnson K.S. 1988. Single-step purification of polypeptides expressed in *Escherichia coli* as fusions with glutathione-S-transferase. *Gene* **67:** 31–40.

Spector D.L., Goldman R.D., and Leinwand L.A. 1998. *Cells: A laboratory manual.* Cold Spring Harbor Laboratory Press, Cold Spring Harbor, New York.

Sun Y., Liu X., Ng-Easton E., Lodish H.F., and Weinberg R.A. 1999. SnoN and Ski protooncoproteins are rapidly degraded in response to transforming growth factor β signaling. *Proc. Natl. Acad. Sci.* **96:** 12442–12447.

5 | Identification of Associated Proteins by Coimmunoprecipitation

Peter D. Adams,[1] Steven Seeholzer,[1] and Michael Ohh[2]

[1]Fox Chase Cancer Center, Philadelphia, Pennsylvania 19111; [2]Dana-Farber Cancer Institute and Harvard Medical School, Cambridge, Massachusetts 02115

INTRODUCTION

Many of the protein–protein associations that exist within the intact cell are conserved when a cell is lysed under nondenaturing conditions. This approach takes advantage of this fact to detect and identify physiologically relevant protein–protein interactions. As illustrated in Figure 1, if protein X is immunoprecipitated with an antibody to X, then protein Y, which is stably associated with X in vivo, may also precipitate (Fig. 1, complex 1). This precipitation of protein Y, based on a physical interaction with X, is referred to as coimmunoprecipitation. Some of the earliest examples of this approach used antibodies to SV40 T antigen and adenovirus E1a to determine the host cellular proteins that interact with these viral transforming oncoproteins. Two interacting proteins of particular note are the tumor suppressor proteins, p53 and pRB (Lane and Crawford 1979; Yee and Branton 1985; Harlow et al. 1986).

This chapter is adapted and expanded from Chapter 18, Protocol 4, in Sambrook and Russell 2001, *Molecular Cloning*.

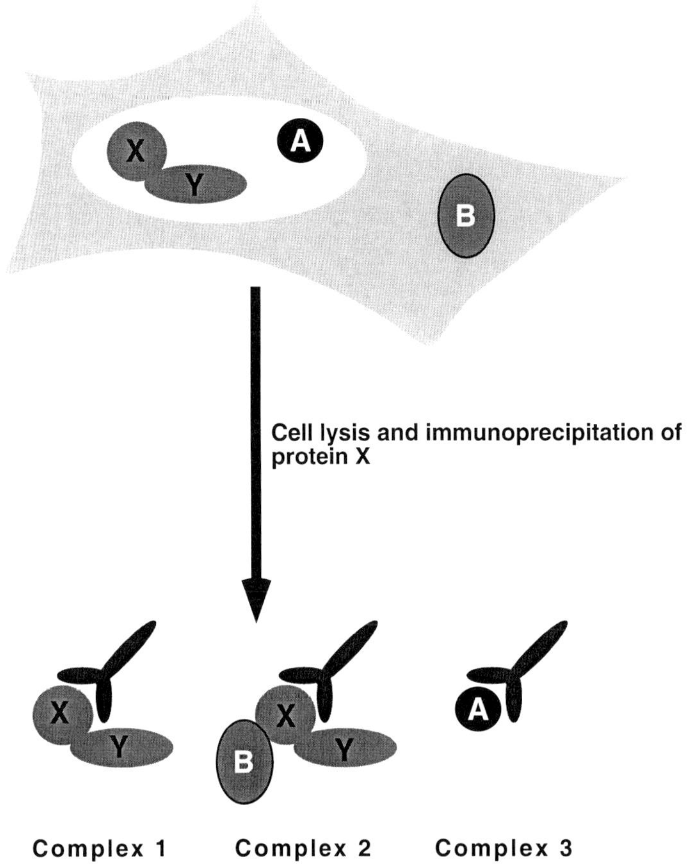

FIGURE 1. Principle and pitfalls of detection of proteins by coimmunoprecipitation. In the intact cell, protein X is present in a complex with protein Y. This complex is preserved after cell lysis and allows protein Y to be coimmunoprecipitated with protein X (complex 1). However, the disruption of subcellular compartmentalization allows artifactual interactions to occur between some proteins, e.g., protein X and protein B (complex 2). Furthermore, the antibody that is used for the immunoprecipitation will nonspecifically cross-react with other proteins, e.g., protein A (complex 3). The key to identification of protein–protein interactions by coimmunoprecipitation is to perform the proper controls so as to identify protein Y but not proteins A and B. (Adapted from Sambrook and Russell 2001, *Molecular Cloning*.)

A variety of methods are available to detect and then identify Y (see below). This approach is most commonly used to test whether two proteins of interest are associated in vivo. However, it can also be used to objectively identify novel interacting partners of a particular protein, including, for example, p21cip1 as a protein that interacts with cyclin D/cdk4 kinase (Xiong et al. 1993); cul2, elongins B and C, and fibronectin as proteins that interact with the von Hippel-Lindau tumor suppressor protein (pVHL) (Duan et al. 1995; Kibel et al. 1995; Pause et al. 1997; Lonergan et al. 1998; Ohh et al. 1998); and transformation/transcription domain-associated protein (TRRAP) as a protein that interacts with the E2F1 and c-*myc* transcription factors (McMahon et al. 1998).

Detection of an interaction by this method requires that the protein–protein complex remain intact through a series of wash steps. Therefore, low-affinity and transient interactions that exist in the cell in a state of dynamic equilibrium may not be observed with this method. Moreover, this approach is only applicable to proteins that persist in physiological complexes after they have been solubilized from the cell. Thus, it may not be appropriate for detection of protein–protein interactions that make up large, insoluble macromolecular structures of the cell, e.g., the nuclear and extracellular matrices (see Cell Lysis and Immunoprecipitation for a discussion of lysis conditions).

Testing an Interaction between Two Known Proteins

Perhaps the most rigorous demonstration of a physiological in vivo interaction between two proteins is their coimmunoprecipitation from cell extracts. For example, a putative interaction might first be identified by use of a powerful, high-throughput approach such as a yeast two-hybrid screen and then subsequently shown by coimmunoprecipitation of the two proteins to be an in vivo physiological interaction. Detection of an interaction in this way can often be facilitated by ectopic expression of the proteins concerned; e.g., by transient transfection of the cells with plasmids encoding the relevant proteins. However, if possible, it is obviously desirable to show such an interaction without recourse to such methods because overexpression may drive unphysiological interactions.

Identification of Novel Protein–Protein Interactions

Coimmunoprecipitation can also be used to search for novel proteins that interact with a known protein of interest. The major advantage of this approach, as compared to many other methods to identify associated proteins, is that it can be a particularly powerful way to identify physiological protein–protein interactions that exist within the intact cell. The major disadvantage of this approach is that it can be relatively laborious and time-consuming, and it often requires a very large number of cultured cells (although frozen cell pellets derived from many liters of cultured suspension cells can now be obtained commercially, e.g., from the National Cell Culture Center in the United States [http://www.nccc.com]).

OUTLINE OF PROCEDURE

Whether the aim is to test for a specific interaction between two known proteins or to identify novel proteins that interact with a known protein, the principle is the same. The cells are harvested and lysed under conditions that preserve protein–protein interactions, a protein of interest is specifically immunoprecipitated from the cell extracts, and the immunoprecipitates are then fractionated by polyacrylamide gel electrophoresis (PAGE). In the past, this has generally been by one-dimensional SDS-PAGE. However, two-dimensional isoelectric focusing (IEF)/SDS-PAGE gels offer improvements in both resolution and sensitivity, and the development of "proteomics" technologies is making such gels more commonplace.

Coimmunoprecipitation of a protein of known identity is most commonly detected by western blotting with an antibody directed against that protein. Alternatively, if the cells are metabolically labeled with [^{35}S]methionine prior to lysis, coprecipitating proteins can be detected by autoradiography. The identity of known or suspected radiolabeled proteins can be confirmed using immunological techniques (e.g., re-immunoprecipitation of the protein in question with a relevant antibody; Beijersbergen et al. 1994; Ginsberg et al. 1994; Vairo et al. 1995) or comparative peptide mapping. In the latter approach, the ^{35}S-labeled protein is digested with a protease, e.g., V8 protease or chymotrypsin, and the peptides are fractionated by SDS-PAGE. The ^{35}S-peptide map is compared with the map that is obtained from a ^{35}S-labeled protein derived from in vitro transcription and translation of a known cDNA. If the two proteins are identical, they will result in identical peptide maps (Xiong et al. 1992, 1993; Kibel et al. 1995).

If the aim is to identify novel interacting proteins, the putative associated protein is first detected in the gel, usually as a ^{35}S-labeled or silver-stained band. Subsequently, the coimmunoprecipitation is carried out at a large enough scale so as to obtain sufficient material for identification of the associated protein by Edman sequencing- or mass spectrometry (MS)-based methods.

The purpose of this chapter is not to repeat protocols for cell lysis, immunoprecipitation, and related techniques that are very well described elsewhere (Harlow and Lane 1988). Rather, the aim is to describe the pitfalls of this approach and the controls that should be performed to ensure that a coimmunoprecipitating protein is truly an in-vivo-associated protein which is likely to be physiologically relevant.

Cell Lysis and Immunoprecipitation

When lysing the cells, the major consideration is that the lysis conditions solubilize and extract the protein to be immunoprecipitated, but do not disrupt all of the protein–protein interactions that exist within the cell. Thus, it is best to determine in advance the mildest lysis conditions that efficiently solubilize the majority of the protein of interest. These pilot experiments can generally be performed on a small scale with extract derived from 1×10^6 to 1×10^7 cells per immunoprecipitation. In general, a higher salt concentration (200–1000 mM NaCl) and the presence of an ionic detergent (e.g., 0.1–1% SDS or sodium deoxycholate [DOC]) are more disruptive than lower salt (120 mM NaCl) and nonionic detergent (e.g., 0.1–1% NP-40 or Triton X-100). Mechanical processes, e.g., sonication, also tend to denature and disrupt protein–protein interactions. However, it is important to note that the lysis conditions that solubilize and yet do not dissociate a particular protein complex should be empirically determined. EBC (50 mM Tris [pH 8], 120 mM NaCl, 0.5% NP-40) and RIPA (50 mM Tris [pH 8], 150 mM NaCl, 1% NP-40, 0.5% DOC, 0.1% SDS) are two commonly used buffers. Each of these buffers should also be supplemented with a cocktail of protease and phosphatase inhibitors (e.g., 500 µg/ml phenylmethylsulfonyl fluoride, 50 µg/ml leupeptin, 100 µg/ml aprotinin, 100 mM NaF, 0.2 mM sodium orthovanadate).

Once the lysis conditions have been determined, the cleared cell lysate is subjected to a standard immunoprecipitation with an antibody directed against the protein of interest. It is important to perform a control immunoprecipitation in parallel with the relevant control antibody (see Optimization and Controls). When determining the conditions for cell lysis and protein solubilization, one should bear in mind that high concentrations of ionic detergent (e.g., greater than 0.2% SDS) and reducing agents (e.g., dithiothreitol [DTT] and mercaptoethanol) will tend to denature the antibody and interfere with the immunoprecipitation. After the immunoprecipitation has been performed, the proteins are generally fractionated by one-dimensional SDS-PAGE and detected by the methods described below.

Detection of Associated Proteins

If the purpose of the experiment is to test whether two particular proteins interact in vivo, the presence of the second protein is usually detected by western blotting or, if the cells were metabolically labeled with [^{35}S]methionine before lysis, by autoradiography. The former is straightforward and, assuming that the antibody used for the western blot is well characterized (see below), relatively unambiguous. The latter has the advantage that the ^{35}S-labeled protein of interest can be compared by tryptic mapping with ^{35}S-labeled protein derived by in vitro transcription and translation of a defined cDNA, resulting in unambiguous confirmation of the protein's identity (Xiong et al. 1992, 1993; Kibel et al. 1995).

If the purpose is to identify novel associated proteins, these can be detected by direct staining of the proteins in the gel. At this stage, prior to scaling up of the procedure, only the more sensitive stains such as silver staining and imidazole-zinc negative staining (as opposed to Coomassie Blue, for example) generally have the required level of sensitivity (Matsui et al. 1999). Recently, a fluorescent stain, SYPRO Ruby, has been developed that is as sensitive as silver and imidazole-zinc negative staining and is readily compatible with mass spectrometry-based methods of protein identification (see below) (Berggren et al. 2000). Alternatively, if cells are labeled prior to lysis with

[^{35}S]methionine, radiolabeled associated proteins can be detected by autoradiography or with a phosphorimager.

Optimization and Controls

The major pitfall associated with identification of associated proteins by this approach is identification of false positives. These false positives arise from the presence of proteins in the washed immunoprecipitate that are not normally associated with the protein of interest in the intact cell (Fig. 1). Such contaminants result from formation of nonphysiological protein–protein interactions after cell lysis (Fig. 1, complex 2), or cross-reactivity and nonspecific binding of the antibody to other cellular proteins (Fig. 1, complex 3). Fortunately, much can be done to eliminate or control for such confounding interactions.

Characterization of Antibodies

Any antibody that is used for the immunoprecipitation should be well defined (Hu et al. 1991; Marin et al. 1998). That is, it should be definitively demonstrated that it immunoprecipitates, from crude cell extracts, the protein against which it was raised. There are a number of ways to demonstrate this: (1) show that multiple antibodies, independently raised to the same protein, recognize the same polypeptide; (2) show that the antibodies fail to detect their target protein in a cell line that lacks that protein. Cell lines lacking a particular protein are sometimes available as human cell lines derived from individuals with cancer predisposition syndromes (e.g., 786-O renal carcinoma cells lack the pVHL tumor suppressor protein) or from mice with a targeted knockout of a particular gene; (3) compare the peptide map that results from proteolytic digestion of ^{35}S-labeled protein immunoprecipitated from metabolically labeled cells with that derived from ^{35}S-labeled in vitro translation of a defined cDNA. If the protein immunoprecipitated from the cell extracts is identical to the protein encoded by the cDNA, the peptide maps should be identical (Xiong et al. 1992, 1993; Kibel et al. 1995).

Control Antibodies

Even mouse monoclonal antibodies interact nonspecifically with proteins that are distinct from the immunogen (Fig. 1, complex 3). Any putative associated protein that immunoprecipitates with a control as well as the specific antibody is, by definition, nonspecifically binding to the antibody. Obviously, the more closely matched the control antibody and the specific antibody, the less the chance of erroneously identifying a nonspecifically interacting protein. For a mouse monoclonal antibody, the proper control is another monoclonal antibody of the same subclass, for a rabbit serum it is the preimmune serum from the same rabbit, and for a purified rabbit polyclonal it is another purified rabbit polyclonal antibody.

Multiple Antibodies

A protein that is associated with the protein that is the primary target of the antibody, rather than cross-reacting with the antibody (Fig. 1, complex 3), is likely to be coimmunoprecipitated by more than one antibody directed to the immunoprecipitated protein. Thus, the level of confidence in a particular putative associated protein is raised if it is observed to coimmunoprecipitate with more than one antibody to a single protein. Likewise, it should be possible to detect a physiological interaction between two proteins by immunoprecipitation of either protein in the complex. A caveat here is that different antibodies recognize different epitopes, some of which might be obscured in a particular protein complex. Conversely, some antibodies might disrupt particular protein–protein interactions. Thus, not all antibodies should be expected to coimmunoprecipitate the same panel of associated proteins.

Use Cell Lines Lacking the Target Protein

A protein that is associated with the immunoprecipitated protein, rather than nonspecifically binding to the antibody (Fig. 1, complex 3), will not coimmunoprecipitate in a cell line which lacks the protein that is the primary target of the antibody (Kibel et al. 1995). Thus, if possible, control immunoprecipitates should be performed from cell lines lacking that protein (see Characterization of Antibodies, above).

Test Biologically Relevant Mutants

A protein that is associated with the immunoprecipitated protein in a functionally significant way may fail to interact with biologically inactive mutants of the protein. Thus, if possible, the ability of the putative coimmunoprecipitating protein to interact with biologically inactive mutants of the immunoprecipitated protein should be tested. Such mutants might be naturally expressed in certain human cell lines (e.g., tumor-derived cell lines containing mutant versions of a protein) or, alternatively, can be ectopically expressed as epitope-tagged proteins by transient transfection of cells. The ectopically expressed mutant proteins can then be specifically immunoprecipitated with antibodies directed against the epitope tag. For example, the cullin family member, Cul2, and a number of other physiologically relevant pVHL-binding proteins, e.g., elongins B and C and fibronectin, failed to interact with a tumor-derived, naturally occurring point mutant of pVHL (Fig. 2).

Test whether Association Occurs before or after Cell Lysis

Cell lysis involves a massive disruption of cellular compartmentalization, bringing into proximity proteins that might never normally be in the same subcellular location. This provides an opportunity for nonphysiological complexes to form after cell lysis (Fig. 1, complex 2). In principle, it is possible to compete out such post-lysis interactions by inclusion of the putative interacting protein in the cell lysis buffer. For example, Ohh et al. (1998) included purified, nonradiolabeled fibronectin in the cell lysis buffer and showed that it was unable to prevent coimmunoprecipitation with pVHL of the endogenous ^{35}S-labeled fibronectin that was present in the cell prior to lysis. From this, and other cell lysate mixing experiments, they concluded that the ^{35}S-labeled fibronectin was associated with pVHL before cell lysis.

Reduce the Backgound of Nonspecific Proteins

The number of proteins that coimmunoprecipitate by virtue of nonspecific interaction with the antibody (Fig. 1, complex 3) or unphysiological association with the protein to which the antibody is raised (Fig. 1, complex 2) can be reduced by careful optimization of the conditions for the immunoprecipitation. Three procedures are commonly adopted to this end.

1. Increase the ionic strength of the immunoprecipitate wash buffer. Just as a higher concentration of salt in the lysis buffer is generally more disruptive of protein–protein associations, a higher concentration of salt in the wash buffer will, in general, reduce nonspecific protein–protein interactions. Titrate the salt concentration in the wash buffer from 120 to 1000 mM NaCl. However, high concentrations of salt can perturb the SDS-PAGE. Therefore, the immunoprecipitates should be washed with a standard salt buffer (120 mM NaCl) immediately prior to loading on the gel.

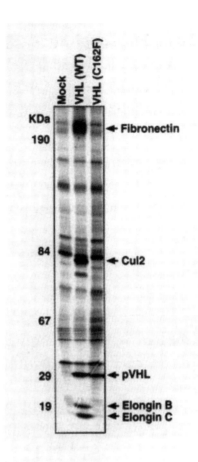

FIGURE 2. Coimmunoprecipitation of pVHL-associated proteins. 786-O renal carcinoma cells stably transfected with a backbone expression plasmid (*left lane*), a plasmid encoding HA-epitope-tagged wild-type pVHL (*middle lane*), or a plasmid encoding HA-epitope-tagged tumor-derived pVHL (C162F) were labeled with [^{35}S]methionine, lysed, and immunoprecipitated with anti-HA antibody. Coimmunoprecipitates were washed with NETN + 900 mM NaCl. Bound proteins were resolved by electrophoresis in a 7.5–15% discontinuous SDS-PAGE and detected by fluorography. (Reprinted, with permission, from Sambrook and Russell 2001, *Molecular Cloning*.)

2. Decrease the amount of primary antibody in the immunoprecipitation. Decreasing the concentration of the primary antibody in the immunoprecipitate will decrease the total number of proteins that coprecipitate nonspecifically. The concentration of the antibody should be decreased to a point where the signal obtained from any specific protein is maximized relative to any nonspecific protein.

3. Preclear the cell lysate with the control antibody. Before performing the immunoprecipitation with the specific antibody directed against the protein of interest, carry out an immunoprecipitation of the cell lysate with an excess of the control antibody and protein A–Sepharose (Harlow and Lane 1988).

Protein Identification

If the purpose is to identify novel associated proteins, the next step is to determine the identity of the coimmunoprecipitating protein(s). Proteins can be identified by direct Edman sequencing (Aebersold et al. 1987; Kamo and Tsugita 1999) or, alternatively, by MS-based approaches (Courchesne and Patterson 1999; Jensen et al. 1999; Wilkins et al. 1999; Yates et al. 1999). In the past, the partial sequence information obtained by Edman sequencing was used to design nucleic acid probes to screen a library for the corresponding cDNA. More recently, and increasingly in the future, MS-based approaches have been and will be used to obtain sequence-dependent information that allows positive identification in a database search of known proteins, ESTs, or proteins predicted from genomic sequencing projects. MS-based approaches do not always provide primary sequence data (see below).

Generally, a significant scale-up of the purification is required to obtain enough material for protein identification. The minimum amount of protein required to obtain useful sequence information by either Edman sequencing or MS depends on a number of factors, including the specific protein concerned and the method of sample preparation. In general, Edman sequencing-based methods require approximately 10 pmoles of protein for internal sequencing (500 ng of a 50-kD protein). "State-of-the-art" MS-based methods are more sensitive and can be used for sequencing at low picomole or subpicomole levels. A 1-pmole amount of a 50-kD protein is approximately 50 ng of protein. Thus, MS-based methods can, in principle, be used to identify proteins stained with silver, imidazole-zinc, or SYPRO Ruby that are not detectable with Coomassie Blue. However, it should be noted that the glutaraldehyde commonly used in silver-staining protocols is not compatible with protein identification by MS. Modified protocols have been developed that omit the glutaraldehyde (Shevchenko et al. 1996a).

Mass Spectrometry-based Approaches

For several reasons, MS-based approaches have become the methods of choice for protein identification. First, as stated above, MS-based approaches are more sensitive than chemical sequencing methods. Second, MS-based methods are greatly facilitated by information obtained from recent genome sequencing projects. Third, MS-based approaches can identify individual proteins present within mixtures of proteins. Fourth, MS can also yield information on posttranslational modification of proteins. Fifth, MS-based approaches are faster than chemical sequencing methods. For these reasons, although still an emerging technology, MS-based approaches promise to revolutionize the practice of protein identification. A brief review of MS-based approaches for protein identification follows. Because the technology is complicated and still being developed, detailed protocols for sample preparation, handling, and analysis should be devised after discussion with an individual experienced in these methods.

Overview

MS can be divided into three steps (Kuster and Mann 1998; Gygi et al. 1999; Yates 2000). First, the sample is ionized. Second, the ions are analyzed according to their mass/charge (m/z) ratio. Third, the ions are detected.

Two methods are commonly used to ionize biological samples. In matrix-assisted laser desorption ionization (MALDI), ionization is achieved by irradiating a crystalline light-absorbing matrix, doped with the biological analytes, with a short laser pulse. In electrospray ionization (ESI), a continuous flow of ions is generated by ejecting the biological analyte in solution from a capillary needle that is held at high electrical potential.

After ionization, the resulting ions can be analyzed in one of several ways. In a time of flight (TOF) mass analyzer, the ions are accelerated in an electric field over a fixed distance toward a detector. The time taken to reach the detector depends on the m/z ratio of the ion. Accordingly, TOF is commonly used to determine the masses of proteolytic peptides in a mixture (or the mass of the intact protein). However, it yields very little information on peptide sequence. In contrast, triple quadrupole and ion trap MS mass analyzers can be used to perform tandem mass spectrometry (MS/MS), which generates sequence-dependent information. In such machines, the first MS stage measures the m/z ratio and hence the mass of individual peptides in a mixture.

Subsequently, a peptide of a defined *m/z* ratio can be selected and subjected to collision-induced dissociation (CID), which fragments the starting peptide at amide bonds to produce two complementary series of product ions (referred to as y-series and b-series) whose masses provide detailed sequence-dependent information.

MALDI TOF

This particular combination of ionization and analysis procedures is commonly used to determine a so-called "peptide mass map" (Fig. 3). The protein of interest is excised from the gel and digested with a sequence-specific protease. The peptides are extracted and analyzed by MALDI TOF to determine the mass of each peptide in the digest and so generate a "fingerprint" unique to that protein. By comparing this fingerprint with the predicted fingerprints of proteins in databases, it is

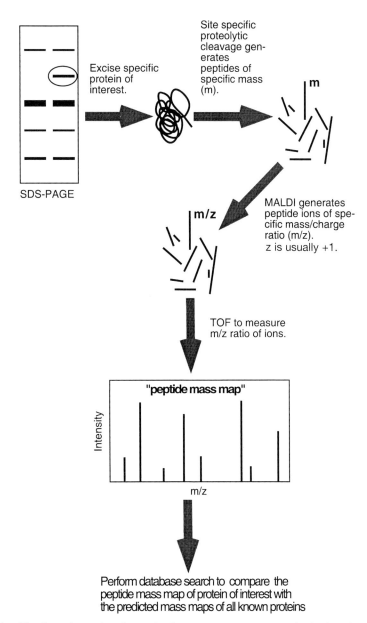

FIGURE 3. Identification of associated proteins by MALDI TOF. See text for further details.

possible to identify the protein. Once an organism's genome has been entirely sequenced, this MALDI TOF fingerprint can be compared with the fingerprint of every protein predicted to be encoded by the genome. As validation of this approach, it was used to identify 90% of the proteins in *Saccharomyces cerevisiae* and 65% of the proteins in *Haemophilus influenzae* that were excised from two-dimensional polyacrylamide gels (Shevchenko et al. 1996b; Fountoulakis et al. 1997).

ESI-MS/MS

This combination of ionization and analysis systems is used to obtain both peptide mass information and, most importantly, peptide sequence information (Fig. 4). This information is generally sufficient to identify a single protein in a database, particularly when mass and sequence infor-

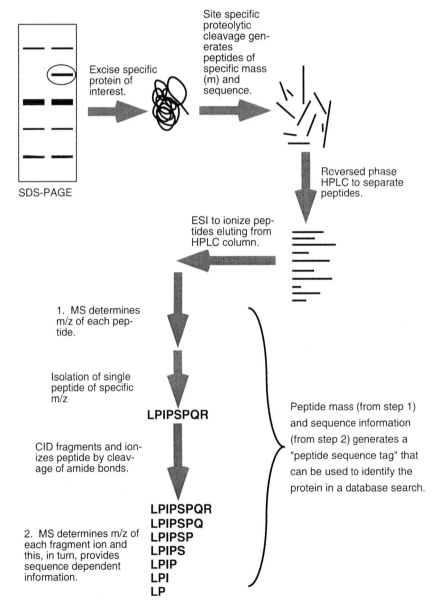

FIGURE 4. Identification of associated proteins by ESI MS/MS. See text for further details.

mation of multiple peptides from a single protein are obtained. The protein of interest is commonly excised from a polyacrylamide gel, digested with a sequence-specific protease (typically trypsin) and, while in solution, subjected to ESI-MS/MS. Since the sample is analyzed in solution, it is possible to couple this approach to high-performance liquid chromatography (HPLC) or capillary electrophoresis (CE) fractionation of the peptides in the digest (LC-MS/MS or CE-MS/MS), thus simplifying the mixture of peptides being analyzed at any one time and making data analysis more straightforward.

This approach has two advantages over MALDI TOF. First, because the MS/MS data contain short stretches of direct sequence information from within individual peptides, rather than being merely a mass map or fingerprint of the whole protein, it can readily be used to search translated EST databases that do not provide sequence information of the entire protein. Second, because sequence information is obtained on a single peptide of a defined *m/z* ratio, it is possible to sequence peptides within a mixture of peptides derived from a mixture of proteins. Potentially, this removes the requirement for prior purification of proteins by one-dimensional or two-dimensional polyacrylamide gels (McCormack et al. 1997; Link et al. 1999; Tong et al. 1999). For example, Yates and coworkers used CE-MS/MS and LC-MS/MS to analyze purified yeast ribosomes after tryptic digestion and successfully identified virtually all of the 78 proteins known to be present in the complex (Link et al. 1999; Tong et al. 1999) Moreover, they were able to identify 71 of the 78 ribosomal proteins in a total yeast cell lysate (Link et al. 1999).

Characterization of Multiprotein Complexes by MS

The particular advantages of MS-based approaches outlined above have made it possible to identify protein components of purified multicomponent complexes. Aebersold and coworkers used LC-MS/MS to identify a large number of proteins that coimmunoprecipitated with the apoptotic inhibitor protein, bclX$_L$ (Gygi et al. 1999). Mann and coworkers used ESI-MS/MS to identify proteins in yeast and human RNA spliceosomes (Neubauer et al. 1997, 1998; Ajuh et al. 2000). The splicing complexes were purified through a combination of gel filtration and affinity chromatography. ESI-MS/MS was also used to identify subunits of the immunopurified yeast anaphase-promoting complex (APC), a protein–ubiquitin ligase complex required for progression through and exit from mitosis (Zachariae and Nasmyth 1999).

One subunit, Apc2p, was found to be similar to the cullin, Cdc53p, that is found in another ubiquitin–protein ligase complex, SCF (*Skp1*, *Cdc53*, *F*-box), that is required for DNA replication. This suggests an evolutionary link between the APC and the SCF (Zachariae et al. 1998; Koepp et al. 1999). Recently, two groups used LC-MS/MS to identify proteins associated with BRCA1 in vivo. In one approach, BRCA1 was directly immunoprecipitated from cell extracts and LC-MS/MS was used to identify coimmunoprecipitating proteins. BRCA1 was found to be present in a high-molecular-weight complex (the BRCA1-associated genome surveillance complex, BASC) containing tumor suppressor and DNA damage repair proteins such as MSH2, MSH6, MLH1, ATM, and BLM (Wang et al. 2000). In another approach, a BRCA1-containing complex was purified by ion exchange and immunoaffinity chromatography, and LC-MS/MS identified members of the SWI/SNF chromatin remodeling family as BRCA1-associated proteins (Bochar et al. 2000).

Identification of pVHL-associated Protein, Cul2

The approach discussed above has been used to identify novel proteins associated with von Hippel-Lindau tumor suppressor protein. von Hippel-Lindau disease is a hereditary cancer syndrome that is characterized by the development of renal carcinomas, pheochromocytomas, and vascular tumors of the central nervous system and retina (Ohh and Kaelin 1999). Tumors result from functional inactivation of both copies of the *VHL* gene. The protein product of the *VHL* gene, pVHL, has been shown to interact with a number of cellular proteins, e.g., elongins B and C, Rbx1, and a member of the Cullin family of proteins, Cul2 (Duan et al. 1995; Kibel et al. 1995; Pause et al. 1997; Lonergan et al. 1998; Kamura et al. 1999b). Recent work suggests that the pVHL–elongin B and C–Rbx1–Cul2 complex serves as an E3 ubiquitin ligase and targets certain proteins for degradation (Iwai et al. 1999; Kamura et al. 1999a,b; Lisztwan et al. 1999; Tyers and Willems 1999). This activity of pVHL is thought, at least in part, to underlie its tumor suppressor function. The pVHL protein has also been shown to stably interact with the extracellular matrix protein, fibronectin (Ohh et al. 1998). Below is an outline of the protocol that was used to identify Cul2 and fibronectin as pVHL-associated proteins. The protocol was performed on 786-O renal carcinoma cells that were stably transfected with a plasmid expressing hemagglutinin (HA)-tagged pVHL (786-O–HA-pVHL). Control cells, which were stably transfected with the empty vector alone, were processed in parallel (Fig. 4). This protocol was optimized for purification of the pVHL-associated proteins in question. Optimum conditions for identification of other associated proteins will probably vary.

MATERIALS

CAUTION: See Appendix for appropriate handling of materials marked with <!>.

Buffers and Solutions

Acetonitrile (50%) <!>
EBC lysis buffer
    50 mM Tris (pH 8)
    120 mM sodium chloride
    0.5% Nonidet P-40 (NP-40)
    5 µg/ml leupeptin <!>
    10 µg/ml aprotinin <!>
    50 µg/ml phenylmethylsulfonyl fluoride <!>
    0.2 mM sodium orthovanadate <!>
    100 mM sodium fluorine <!>
    20 mM Tris (pH 8)
EDTA (1 mM)
1x Laemmli sample buffer
NP-40 (0.5%)

NETN
  20 mM Tris (pH 8)
  1 mM EDTA
  0.5% NP-40
Phosphate-buffered saline
Protein A–Sepharose
Sodium chloride (90 mM)
Tris (20 mM, pH 8)
Trypsin

Cells and Antibodies

12CA5, anti-HA mouse monoclonal antibody
786-0/HA-pVHL cells, 6×10^7

Gels and Dyes

Acetonitrile (50%) <!>
Coomassie Blue R250
SDS-PAGE gel

Special Equipment

ABI 477A or 494A sequence machine
HPLC equipment

METHOD

1. Wash 30 x 10-cm plates of asynchronously growing 786-O/HA-pVHL cells (a total of approximately 6×10^7 cells) in phosphate-buffered saline and scrape each plate of cells into 1 ml of ice-cold EBC lysis buffer.

2. Centrifuge cell lysate at 14,000*g* for 15 minutes at 4°C.

3. Pool the soluble fraction (30 ml) and add 30 μg of the anti-HA mouse monoclonal antibody, 12CA5. Rock the immunoprecipitate for 1 hour at 4°C.

4. Add 0.9 ml of a 1:1 slurry of protein A-Sepharose (in 20 mM Tris, pH 8, 1 mM EDTA, 0.5% NP-40). Rock the immunoprecipitate for another 30 minutes at 4°C.

5. Wash the protein A-Sepharose four times in NETN + 900 mM NaCl and then once in NETN.

6. Aspirate the protein A-Sepharose beads to dryness, add 800 μl of 1x Laemmli sample buffer, and boil for 4 minutes.

7. Prepare the SDS-PAGE gel. The separation gel (a discontinuous 7.5–15% gel at pH 8.8) should be 20–30 cm long, the stacking gel (5% at pH 6.8) should be 10 cm long, and the well itself should be 10 cm deep. Pour the stacking gel without a multiwell comb and construct the wells by inserting spacers vertically between the glass plates so that they form a well that is large enough for the sample. Load the sample into the well and run at 10 mA constant current overnight.

8. To detect the protein bands, stain with Coomassie Blue R-250. Excise the band of interest from the gel, place it in an Eppendorf tube, and wash two times for 3 minutes in 1 ml of 50% acetonitrile. Digest the protein with trypsin while it is still in the gel, electroelute the peptides, and fractionate by narrow-bore HPLC. Subject the collected peptides to automated Edman degradation sequencing on an ABI 477A or 494A machine (Lane et al. 1991).

> Procedures for detection and digestion of the protein vary greatly. In particular, digestion can be performed while the protein is still in the polyacrylamide gel or after it has been transferred to a nitrocellulose or PVDF membrane. Processing of the samples for Edman sequencing or MS analysis also varies and this is best discussed with the scientist operating the machinery. A review of methods is given in Matsudaira (1993) and Link (1999).

Identification of coimmunoprecipitating proteins is a powerful, albeit relatively laborious, method of identifying physiologically relevant associated proteins. Future increases in the sensitivity of MS-based methods of protein identification will complement the wealth of sequence information available from the Human Genome Project to promote the application of this method.

ACKNOWLEDGMENT

We thank Bill Kaelin for his review of the manuscript.

REFERENCES

Aebersold R.H., Leavitt J., Saavedra R.A., Hood L.E., and Kent S.B. 1987. Internal amino acid sequence analysis of protein separated by one- or two-dimensional gel electrophoresis after in situ protease digestion on nitrocellulose. *Proc. Natl. Acad. Sci.* **84:** 6970–6974.

Ajuh P., Kuster B., Panov K., Zomerdijk J.C., Mann M., and Lamond A.I. 2000. Functional analysis of the human CDC5L complex and identification of its components by mass spectrometry. *EMBO J.* **19:** 6569–6581.

Beijersbergen R.L., Kerkhoven R.M., Zhu L., Carlee L., Voorhoeve P.M., and Bernards R. 1994. E2F-4, a new member of the E2F gene family, has oncogenic activity and associates with p107 *in vivo. Genes Dev.* **8:** 2680–2690.

Berggren K., Chernokalskaya E., Steinberg T.H., Kemper C., Lopez M.F., Diwu Z., Haugland R.P., and Patton W.F. 2000. Background-free, high sensitivity staining of proteins in one- and two-dimensional sodium dodecyl sulfate-polyacrylamide gels using a luminescent ruthenium complex. *Electrophoresis* **21:** 2509–2521.

Bochar D.A., Wang L., Beniya H., Kinev A., Xue Y., Lane W.S., Wang W., Kashanchi F., and Shiekhattar R. 2000. BRCA1 is associated with a human SWI/SNF-related complex: Linking chromatin remodeling to breast cancer. *Cell* **102:** 257–265.

Courchesne P.L. and Patterson S.D. 1999. Identification of proteins by matrix-assisted laser desorption/ionization mass spectrometry using peptide and fragment ion masses. *Methods Mol. Biol.* **112:** 487–511.

Duan D.R., Pause A., Burgess W.H., Aso T., Chen D.Y., Garrett K.P., Conaway R.C., Conaway J.W., Linehan W.M., and Klausner R.D. 1995. Inhibition of transcription elongation by the VHL tumor suppressor protein. *Science* **269:** 1402–1406.

Fountoulakis M., Langen H., Evers S., Gray C., and Takacs B. 1997. Two-dimensional map of *Haemophilus influenzae* following protein enrichment by heparin chromatography. *Electrophoresis* **18:** 1193–1202.

Ginsberg D., Vairo G., Chittenden T., Xiao Z.X., Xu G., Wydner K.L., DeCaprio J.A., Lawrence J.B., and Livingston D.M. 1994. E2F-4, a new member of the E2F transcription factor family interacts with p107. *Genes Dev.* **8:** 2665–2679.

Gygi S.P., Han D.K.M., Gingras A.-C., Sonenberg N., and Aebersold R. 1999. Protein analysis by mass spectrometry and sequence database searching: Tools for cancer research in the post-genomic era. *Electrophoresis* **20:** 310–319.

Harlow E. and Lane D. 1988. *Antibodies: A laboratory manual.* Cold Spring Harbor Laboratory Press, Cold Spring Harbor, New York.

Harlow E., Whyte P., Franza B.R., and Schley C. 1986. Association of adenovirus early-region 1A proteins with cellular polypeptides. *Mol. Cell. Biol.* **6:** 1579–1589.

Hu Q.J., Bautista C., Edwards G.M., Defeo-Jones D., Jones R.E., and Harlow E. 1991. Antibodies specific for the human retinoblastoma protein identify a family of related polypeptides. *Mol. Cell. Biol.* **11:** 5792–5799.

Iwai K., Yamanaka K., Kamura T., Minato N., Conaway R.C., Conaway J.W., Klausner R.D., and Pause A. 1999. Identification of the von hippel-lindau tumor-suppressor protein as part of an active E3 ubiquitin ligase complex. *Proc. Natl. Acad. Sci.* **96:** 12436–12441.

Jensen O.N., Wilm M., Shevchenko A., and Mann M. 1999. Peptide sequencing of 2D gel isolated nanoelectrospray tandem mass spectrometry. *Methods Mol. Biol.* **112:** 571–588.

Kamo M., and Tsugita A. 1999. N-terminal amino acid sequencing of 2-DE spots. *Methods Mol. Biol.* **112:** 461–466.

Kamura T., Conrad M.N., Yan Q., Conaway R.C., and Conaway J.W. 1999a. The Rbx1 subunit of SCF and VHL E3 ubiquitin ligase activates Rub1 modification of cullins Cdc53 and Cul2. *Genes Dev.* **13:** 2928–2933.

Kamura T., Koepp D.M., Conrad M.N., Skowyra D., Moreland R.J., Iliopoulos O., Lane W.S., Kaelin Jr., W.G., Elledge S.J., Conaway R.C., Harper J.W., and Conaway J.W. 1999b. Rbx1, a component of the VHL tumor suppressor complex and SCF ubiquitin ligase. *Science* **284:** 657–661.

Kibel A., Iliopoulos O., DeCaprio J.A., and Kaelin Jr., W.G. 1995. Binding of the von Hippel-Lindau tumor suppressor protein to Elongin B and C. *Science* **269:** 1444–1446.

Koepp D.M., Harper J.W., and Elledge S.J. 1999. How the cyclin became a cyclin: Regulated proteolysis in the cell cycle. *Cell* **97:** 431–434.

Kuster B. and Mann M. 1998. Identifying proteins and post-translational modifications by mass spectrometry. *Curr. Opin. Struct. Biol.* **8:** 393–400.

Lane D.P. and Crawford L.V. 1979. T antigen is bound to a host protein in SV40-transformed cells. *Nature* **278:** 261–263.

Lane W.S., Galat A., Harding M.W., and Schreiber S.L. 1991. Complete amino acid sequence of the FK506 and rapamycin binding protein, FKBP, isolated from calf thymus. *J. Protein Chem.* **10:** 151.

Link A.J., ed. 1999. *Methods in molecular biology*, volume 112. *2D Proteome analysis protocols.* Humana Press, Totowa, New Jersey.

Link A.J., Eng J., Schieltz D.M., Carmack E., Mize G.J., Morris D.R., Garvik B.M., and Yates III, J.R. 1999. Direct analysis of protein complexes using mass spectrometry. *Nat. Biotechnol.* **17:** 676–682.

Lisztwan J., Imbert G., Wirbelauer C., Gstaiger M., and Krek W. 1999. The von Hippel-Lindau tumor suppressor protein is a component of an E3 ubiquitin-protein ligase activity. *Genes Dev.* **13:** 1822–1833.

Lonergan K.M., Iliopoulos O., Ohh M., Kamura T., Conaway R.C., Conaway J.W., and Kaelin Jr., W.G. 1998. Regulation of hypoxia-inducible mRNAs by the von Hippel-Lindau tumor suppressor protein requires binding to complexes containing elongins B/C and Cul2. *Mol. Cell. Biol.* **18:** 732–741.

Marin M.C., Jost C.A., Irwin M.S., DeCaprio J.A., Caput D., and Kaelin Jr., W.G. 1998. Viral oncoproteins discriminate between p53 and the p53 homolog p73. *Mol. Cell. Biol.* **18:** 6316–6324.

Matsui N.M., Smith-Beckerman D.M., and Epstein L.B. 1999. Staining of preparative 2D gels. *Methods Mol. Biol.* **112:** 571–588.

Matsudaira P., ed. 1993. *A practical guide to protein and peptide purification for microsequencing.* Academic Press, Harcourt Brace, New York.

McCormack A.L., Schieltz D.M., Goode B., Yang S., Barnes G., Drubin D., and Yates J.R. 1997. Direct analysis and identification of proteins in mixtures by LC/MS/MS and database searching at the low-femtomole level. *Anal. Chem.* **69:** 767–776.

McMahon S.B., Van Buskirk H.A., Dugan K.A., Copeland T.D., and Cole M.D. 1998. The novel ATM-related protein TRRAP is an essential cofactor for the c-Myc and E2F oncoproteins. *Cell* **94:** 363–374.

Neubauer G., Gottschalk A., Fabrizio P., Seraphin B., Luhrmann R., and Mann M. 1997. Identification of the proteins of the yeast U1 small nuclear ribonucleoprotein complex by mass spectrometry. *Proc. Natl. Acad. Sci.* **94:** 385–390.

Neubauer G., King A., Rappsilber J., Calvio C., Watson M., Ajuh P., Sleeman J., Lamond A., and Mann M. 1998. Mass spectrometry and EST-database searching allows characterization of the multi-protein spliceosome complex. *Nat. Genet.* **20:** 46–50.

Ohh M. and Kaelin Jr., W.G. 1999. The von Hippel-Lindau tumour suppressor protein: New perspectives. *Mol. Med. Today* **5:** 257–263.

Ohh M., Yauch R.L., Lonergan K.M., Whaley J.M., Stemmer-Rachamimov A.O., Louis D.N., Gavin B.J., Kley N., Kaelin Jr., W.G., and Iliopoulos O. 1998. The von Hippel-Lindau tumor suppressor protein is required for proper assembly of an extracellular fibronectin matrix. *Mol. Cell* **1:** 959–968.

Pause A., Lee S., Worrell R.A., Chen D.Y., Burgess W.H., Linehan W.M., and Klausner R.D. 1997. The von Hippel-Lindau tumor-suppressor gene product forms a stable complex with human CUL-2, a member of the Cdc53 family of proteins. *Proc. Natl. Acad. Sci.* **94:** 2156–2161.

Shevchenko A., Wilm M., Vorm O., and Mann M. 1996a. Mass spectrometric sequencing of proteins silver-stained polyacrylamide gels. *Anal. Chem.* **68:** 850–858.

Shevchenko A., Jensen O.N., Podtelejnikov A.V., Sagliocco F., Wilm M., Vorm O., Mortensen P., Shevchenko A., Boucherie H., and Mann M. 1996b. Linking genome and proteome by mass spectrometry: Large-scale identification of yeast proteins from two dimensional gels. *Proc. Natl. Acad. Sci.* **93:** 14440–14445.

Tong W., Link A., Eng J.K., and Yates J.R. 1999. Identification of proteins in complexes by solid-phase microextraction/multistep elution/capillary electrophoresis/tandem mass spectometry. *Anal. Chem.* **71:** 2270–2278.

Tyers M. and Willems A.R. 1999. One ring to rule a superfamily of E3 ubiquitin ligases. *Science* **284:** 601, 603–604.

Vairo G., Livingston D.M., and Ginsberg D. 1995. Functional interaction between E2F-4 and p130: Evidence for distinct mechanisms underlying growth suppression by different retinoblastoma protein family members. *Genes Dev.* **9:** 869–881.

Wang Y., Cortex D., Yazdi P., Neff N., Elledge S.J., and Qin J. 2000. BASC, a super complex of BRCA1-associated protein involved in the recognition and repair of aberrant DNA structures. *Genes Dev.* **14:** 927–939.

Wilkins M.R., Gasteiger E., Bairoch A., Sanchez J.-C., Williams K.L., Appel R.D., and Hochstrasser D.F. 1999. Protein identification and analysis tools in the ExPASy server. *Methods Mol. Biol.* **112:** 531–552.

Xiong Y., Zhang H., and Beach D. 1992. D type cyclins associate with multiple protein kinases and the DNA replication and repair factor PCNA. *Cell* **71:** 505–514.

Xiong Y., Hannon G.J., Zhang H., Casso D., Kobayashi R., and Beach D. 1993. p21 is a universal inhibitor of cyclin kinases. *Nature* **366:** 701–704.

Yates J.R. 2000. Mass spectrometry from genomics to proteomics. *Trends Genet.* **16:** 5–8.

Yates J.R., Carmack E., Hays L., Link A.J., and Eng J.K. 1999. Automated protein identification using micro-column liquid chromatography-tandem mass spectrometry. *Methods Mol. Biol.* **112:** 553–569.

Yee S.P. and Branton P.E. 1985. Detection of cellular proteins associated with human adenovirus type 5 early region 1A polypeptides. *Virology* **147:** 142–153.

Zachariae W. and Nasmyth K. 1999. Whose end is destruction: Cell division and the anaphase-promoting complex. *Genes Dev.* **13:** 2039–2058.

Zachariae W., Shevchenko A., Andrews P.D., Ciosk R., Galova M., Stark M.J., Mann M., and Nasmyth K. 1998. Mass spectrometric analysis of the anaphase-promoting complex from yeast: Identification of a subunit related to cullins. *Science* **279:** 1216–1219.

6 Chemical Cross-linking in Studying Protein–Protein Interactions

Owen W. Nadeau and Gerald M. Carlson

Division of Molecular Biology and Biochemistry, School of Biological Sciences, University of Missouri, Kansas City, Missouri 64110-2499

INTRODUCTION

Cross-linking is a technique for covalently linking distinct chemical functionalities. In the case of proteins, either nucleophilic side chains or the termini of the polypeptide chains are the functional groups that are linked. This technique has been used for studying protein–protein interactions for nearly 50 years, making it one of the oldest such techniques. Cross-linking is also one of the most powerful and versatile techniques available for studying the interactions of proteins. It can be used to study the interactions of proteins within a stable complex or between reversibly interacting proteins. Within a hetero-oligomeric complex, cross-linking is used to identify neighboring subunits or proteins (Dey et al. 1998), to give the maximal distance separating the proteins in question (Hajdu et al. 1977), and even to identify specific adjacent amino acid residues of the interacting proteins (Hartman and Wold 1967). Cross-linking can be used to determine the minimal subunit stoichiometry in such a complex, even in the case of insoluble, membranous complexes (Heymann and Mentlein 1980). It can be used to detect conformational changes within a complex, such as those induced by an effector, through alterations in subunit cross-linking patterns (Nadeau et al. 1997a,b), and it can be used to lock an oligomer into a particular conformational state (Benesch and Kwong 1991). In the case of reversibly interacting proteins, cross-linking can identify the partners, regardless of whether both are soluble, or one is soluble and the other is membranous (e.g., a ligand and its membrane receptor; Ji 1977), or both are insoluble within a membrane. Thus, cross-linking has the versatility of ranging from screening to supplying relative and absolute structural information for interacting proteins (Ji 1977). Of course, one of its common current uses is to provide corroboration for the results of two-hybrid screens, pull-downs, and immunoprecipitations.

In this chapter, we consider only chemical cross-linking; i.e., that covalent cross-linking brought about by addition of a chemical reagent. This usually results in the incorporation of a portion of the molecule of cross-linking reagent between the nucleophilic groups of the proteins that become cross-linked; however, in the case of zero-length cross-linking, none of the reagent that induces cross-linking is incorporated into the conjugated product. Cross-linking can also be achieved enzymatically through use of enzymes such as transglutaminase; although enzymatic cross-linking is not covered in this chapter, most of the variables and concerns are the same as in chemical cross-linking. For expanded treatises on cross-linking that are thorough, scholarly, and experimentally useful, the reader should consult the books by Wong (1993) or Hermanson (1996). Another useful source of information is the product catalog of the Pierce Chemical Company, a large commercial source of cross-linking reagents.

There are two paramount points, which are interrelated, that cannot be overemphasized regarding the use of chemical cross-linking to study protein–protein interactions. First, cross-linking is a completely empirical process. One cannot predict which proteins will be cross-linked by which reagents under which conditions. A diverse range of experimental conditions and reagents should be screened to achieve optimal cross-linking. Second, extreme caution should be used if one attempts to interpret the absence of cross-linking. This is an excellent example of the dictum that "The absence of evidence is not necessarily evidence of absence." As is described in this chapter, for two interacting proteins to be cross-linked, the appropriate functional groups, e.g., the side chains, of each must be properly positioned and sufficiently reactive to be modified by a cross-linking reagent having the appropriate physico-chemical characteristics.

IN VIVO VERSUS IN VITRO CROSS-LINKING

Besides the traditional and widely used in vitro approach, cross-linking can also be performed in vivo, although there are far fewer examples in the literature of in vivo cross-linking. Obviously, the great advantage of in vivo cross-linking is that the protein–protein interactions of interest are occurring under conditions that are close to natural; i.e., they are potentially perturbed only by introduction of the cross-linker. Because the complexity and concentration of the protein components are so much greater in an in vivo system, specificity in detecting targets that genuinely interact is presumably enhanced by using short cross-linkers so that only proximal proteins will be conjugated. For example, formaldehyde, a lipid-soluble reagent with a very short cross-linking span (2–3 Å), has been used for in vivo cross-linking (Jackson 1987). One of the primary difficulties of this approach is, in fact, controlling the specificity of cross-linking. In the general application of in vivo cross-linking, lipid-soluble, hydrophobic cross-linkers (e.g., formaldehyde, CS_2, or C_2N_2) are used to penetrate the cell membrane (Valentine et al. 1993; Alaedini and Day 1999; Jackson 1999). Determining or controlling the conditions required to optimally cross-link specific targets within a cell is a difficult process at best. Besides the proteins of interest, the cross-linker can react with numerous other targets, including cell-surface proteins, integral membrane proteins, cytosolic proteins, and nuclear proteins. Moreover, metabolites and other non-protein components are also potential competitors for the cross-linker, particularly if they contain reactive groups that are homologous to protein side-chain groups, such as amines, carboxylates, or thiols. Generally, it is difficult both to control the conditions of cross-linking and to alter protein–protein interactions within the cell. The large number of potential targets further complicates interpretation of the data.

In vitro cross-linking offers a far greater potential for controlling both the specificity and conditions of cross-linking. For example, water- and lipid-soluble cross-linkers can be used to modify, respectively, hydrophilic and hydrophobic regions of interacting targets. The pH, concentration of reactants, and temperature can all be easily manipulated as well. In vitro cross-linking also allows the level of purity to be optimized for proteins and their complexes, so that the cross-linking patterns of protein conjugates are more easily interpreted. Moreover, the actual regions of cross-linking are more easily determined for well-resolved conjugates than for those in complex mixtures of proteins. Another advantage of the in vitro approach is the large variety of cross-linking reagents and procedures that are available. The obvious disadvantage of in vitro cross-linking is that interacting proteins are conjugated under nonphysiological conditions. Even when cross-linking is carried out in cell lysates, dissolution of the membrane with detergents abolishes compartmentalization of proteins and their complexes and potentially disrupts protein–membrane and protein–protein interactions.

The type of protein target is an important determinant in choosing the appropriate cross-linking approach. Elucidating the interactions of integral membrane proteins might be best served, for example, through using in vivo cross-linking. The interactions of cytokine receptors are commonly studied by cross-linking the receptors on cell surfaces in the presence of activating ligands, such as interferon (Nadeau et al. 1999b). Cytosolic protein–protein interactions are amenable to analysis by either the in vivo or in vitro approach, but proteins that can be easily purified in sufficient quantities are natural candidates for an in vitro approach.

CROSS-LINKING PROCEDURES

Contributions of Proteins as Reactants in Cross-linking

Cross-linking is simply an extension of the general chemical modification of proteins. Both processes are empirical and are influenced by several variables, including the concentrations of reactants, time, temperature, and pH. Controlling the extent of cross-linking and selectivity for amino acid side chains is important in minimizing the formation of unwanted products, because the process of cross-linking is ongoing; i.e., product formation can progress from monosubstituted proteins, to cross-linked proteins, and, ultimately, to insoluble, extensively cross-linked aggregates. Besides causing technical difficulties, the presence of the latter products in the reaction mixture also diminishes the probability of the cross-linking being specific. Even when conditions are optimized, there are at least three different forms of covalent modification possible in a cross-linking reaction: monosubstitution, intramolecular cross-linking, and intermolecular cross-linking. In addition, there are combinations of these three (Fig. 1A).

1. Monosubstituted linkages occur when proteins are modified by only one reactive group of the cross-linker. The remaining free active group does not react with a proximal protein side

A.

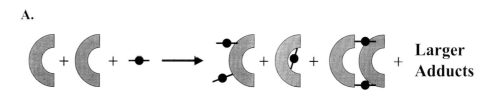

B.

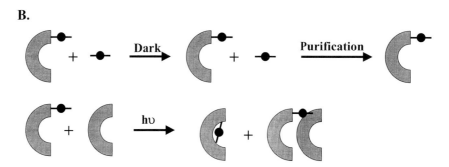

FIGURE 1. (*A*) One-step method of cross-linking. Proteins react with bifunctional reagents to form mono-substituted, intramolecularly, and intermolecularly cross-linked forms of the interacting proteins. Subclasses within each form can simultaneously occur as a result of different combinations of monosubstitution and intramolecular cross-linking. In the continuous process of cross-linking, adducts of increasing size are also formed, progressing from trimers, tetramers, etc., to large, extensively cross-linked polymers. (*B*) Two-step method of cross-linking. A monosubstituted bait protein is first formed by its selective chemical modification with the reactive arm of a heterobifunctional cross-linker (photoactivatable in this example), followed by subsequent purification steps to remove excess non-bound cross-linker. In the second step, the modified bait protein is combined with the target protein. Cross-linking is then induced by activating the latent functional group on the "free" arm of the cross-linker with light of an appropriate wavelength. In contrast to the one-step method in *A*, the formation of large adducts is essentially averted by reducing the number of potential reactants and time of reaction in the cross-linking step.

chain, because it cannot adapt the required orientation for such a reaction, or because a second reactive nucleophilic polypeptide side chain does not happen to be nearby, or because it undergoes competing reactions with the solvent or with non-protein solutes.

2. Intramolecular cross-linking within a single polypeptide chain is generally favored when the concentrations of either the protein or cross-linker are low or the length of the cross-linker is sufficient only for bridging the distance between neighboring side chains of the same protein, rather than between adjacent proteins.

3. Intermolecular cross-linking takes place when the geometry, flexibility, span, and chemistry of the cross-linker are suitable for covalently linking reactive side chains between two separate proteins. Although several screens comparing different cross-linkers and reaction conditions are usually necessary to achieve cross-linking of the desired targets, a general understanding of the nature of the components in a typical protein cross-linking experiment lessens the number of experiments required to optimize the reaction.

In cross-linking, it is useful to think of a protein as a reactant that contains a number of potentially reactive amino acid side chains that may be targeted selectively by distinct classes of chemical cross-linkers. The side chains of the amino acids vary in size, shape, polarity, charge, and nucleophilicity. Based on hydrophobicity, there are two general classes of amino acids: (1) nonpolar residues that contain hydrogen (glycine), aliphatic (alanine, valine, leucine, isoleucine, methionine, and proline), or aromatic (phenylalanine and tryptophan) side chains and (2) polar residues that contain side chains which are generally either nonionizable (asparagine, glutamine, serine, and threonine) or ionizable (arginine, tyrosine, lysine, cysteine, histidine, aspartic acid, and glutamic acid). With only a few exceptions, it is the nucleophilic side chains of this last group of amino acids that undergo cross-linking. The reactivity of any given type of side-chain functional group is influenced by its location in the protein, i.e., its microenvironment (accessibility, polarity, noncovalent interactions with other residues or solvent, etc.). Thus, not all the primary amines of lysyl residues, for example, will have the same reactivity, even on the surface of a protein. These differences in reactivity are part of the reason that cross-linking must be considered as an empirical procedure.

The Effects of pH

Most cross-linking reactions are nucleophilic substitution reactions in which the attacking nucleophilic amino acid side chain donates a pair of electrons to the cross-linker, resulting in the direct displacement of a leaving group from the cross-linker and formation of a covalent bond between the side chain and the cross-linker. The rate of all such nucleophilic displacement reactions is dependent on the ease with which the leaving group can be displaced (i.e., the chemical nature of the cross-linker) and the nucleophilicity of the attacking side chain. The most potent nucleophilic side chain normally found in proteins is R-S$^-$, the anionic deprotonated thiolate form of cysteine. Protonation decreases the nucleophilicity of this and the other ionizable, reactive side chains by decreasing their effectiveness as electron-pair donors. Thus, pH is one of the most important variables affecting cross-linking reactions.

The Henderson-Hasselbalch equation predicts that as one raises the pH from a value equal to its pK_a to 2 pH units above its pK_a, the percent deprotonation of a side chain increases from 50% to 99%, and its reactivity increases accordingly. Although the typical side-chain pK_a values listed in textbooks provide a useful reference point, the actual pK_a of any given side-chain functional group is, like its reactivity, dependent on its microenvironment. The ionic strength and temperature of the reaction mixture can also affect the pKa value. Thus, pH must also be treated as an empirical factor that can control the rate, extent, and, most importantly, selectivity of cross-linking reactions. Technical problems occur, especially in identification of the protein components, when cross-linking is too extensive; however, this can often be avoided by judiciously lowering the pH of the cross-linking reaction. When optimizing reaction conditions, it is prudent to screen a number of different pH values.

Types of Cross-linkers

A remarkable variety of cross-linkers is commercially available from companies such as Pierce Chemical Company, Sigma-Aldrich, and Molecular Probes (Table 1). The most common chemical cross-linkers available are bifunctional reagents, i.e., compounds containing two chemically reactive groups that react with two side chains of protein targets to form a cross-linked complex. The part of the cross-linker that bridges the two reactive groups is referred to as the spacer or linker. In addition to determining the distance between the two reactive groups, the spacer can impart other important properties. For example, functional groups such as disulfides can be included within the spacer, thus making it susceptible to chemical cleavage by reductants; such compounds are termed cleavable cross-linkers. Other cleavable cross-linkers contain covalent bonds that can be cleaved by oxidants or bases. The spacer also contributes to the geometry of the cross-linker, as well as the reagent's solubility. Homobifunctional and heterobifunctional are terms used for bifunctional reagents that contain either identical or different reactive functional groups, respectively. Heterobifunctional cross-linkers have the advantages of being able to select for different classes of nucleophilic side chains with each functional group and of potentially allowing the two steps of the cross-linking reaction to be performed separately and under different conditions. One popular class of heterobifunctional reagents, photoactivatable cross-linkers, combines chemically

TABLE 1. Cross-linkers

Structure and Name	Abbreviation	Water-soluble	Reactive groups	Selectivity
Homobifunctional				
1. *N,N'-p*-phenylenebismaleimide	*p*-PDM	no	maleimide	sulfhydryl
2. Dimethyl pimelimidate•2HCl	DMP	yes	imido ester	amine
3. Disuccinimidyl glutarate	DSG	no	*N*-hydroxy succinimide/ maleimide	amine
4. 4,6-Dibromomethyl-3,7 dimethyl-1,5-diazabicyclo [3.3.0]octa-3,6-diene-2,8-dione	bBBr (bromo-bimane)	moderately	alkyl halide	sulfhydryl
Heterobifunctional				
5. *m*-Maleimidobenzoyl-*N*-hydroxy-succinimide ester	MBS	no	maleimide/N-hydroxy-succinimide ester	sulfhydryl/amine
6. Succinimidyl *N*-14-(2-hydroxybenzoyl)-*N*-11-(4-azido-benzoyl)-9-oxo-8,11,14-triaza-4,5-dithiatetradecanoate	SHAD	no	aryl azide/N-hydroxy-succinimide ester	broad/amine
Trifunctional				
7. β-[Tris(hydroxymethyl)phosphino] propionic acid	THPP	yes	hydroxymethyl phosphino	amine
Zero-length				
8. 1-Ethyl-3-[3-dimethylaminopropyl] carbodiimide hydrochloride	EDC	yes	carbodiimide	carboxyl/amine

reactive and photoreactive groups. The latter groups are activated by exposure to light of an appropriate wavelength.

Trifunctional cross-linkers, a more recently developed class of cross-linking reagents, are composed of molecules containing three reactive groups in a trigonal arrangement. An example of this class is tris(hydroxymethyl)phosphine (Table 1), whose three chemically reactive arms can potentially cross-link three different targets. A different class of reagents that is also sometimes referred to as trifunctional cross-linkers contains two reactive groups with cross-linking potential plus a specific ligand to aid in affinity purification of the cross-linked targets (Alley et al. 2000).

Zero-length cross-linkers constitute another class of cross-linkers. These reagents activate side-chain functional groups on targeted proteins, allowing covalent bonds to form between the side chains without insertion of an exogenous spacer, thus the name zero-length. The carbodiimides, probably the most widely used zero-length cross-linkers, activate carboxylate groups to form active acylisourea intermediates, which are, in turn, modified by incoming amines, resulting in an amide linkage (isopeptide bond) between the carboxyl and amine groups. Through a mechanism similar to carbodiimide-induced amidation of carboxyl and amine side chains, cyanogen (ethanedinitrile) covalently links salt-bridged lysyl and acidic residues in interacting proteins (Simon and Day 2000). Because cyanogen is a gaseous reagent that readily permeates cell membranes, it can also be used to perform in vivo cross-linking. The transglutaminases are enzymatic zero-length cross-linkers that catalyze the formation of isopeptide bonds between adjacent glutamyl and lysyl residues of interacting proteins.

Site-specific photocrosslinkers are photolabile amino acid analogs that can be used to replace natural amino acids through site-directed mutagenesis at a desired position within a protein (Fischer et al. 2000). The altered proteins can then be incubated with target proteins and irradiated to facilitate site-specific cross-linking.

Specificity and Selectivity of Cross-linking

Throughout this chapter, we use the term "specificity" to refer to the observed potential interactions between proteins that are detected by cross-linking; "selectivity," on the other hand, is used to denote the preference of a given class of cross-linking reagent for a particular type of amino acid side chain. Specificity and selectivity are directly interrelated, because less selective cross-linking results in a greater extent of modification, which in turn can lead to less specific cross-linking. Although cross-linkers are often chosen on the basis of the selectivity of their reactive functional groups for a given type of nucleophile (Table 1), it must be remembered that few, if any, reagents used to chemically modify proteins are absolutely specific for a given type of amino acid side chain in all proteins. Microenvironmental effects exerted on a particular nucleophile in a protein can alter its properties to mimic the reactivity of another nucleophilic species. Nevertheless, high selectivity can often be achieved empirically by varying the reaction time, concentration of reactants, type of cross-linker, and the pH of cross-linking. Selecting an appropriate pH and cross-linker goes hand-in-hand in the overall design of an experiment. For example, if lysyl ε-amines are the preferred target on the protein, an appropriate pH to deprotonate at least a subpopulation of these nucleophiles and an amine-selective cross-linker should be chosen. Of course, cross-linkers with functional groups that are selective for side chains other than amines are also available, including the *bis*-maleimides and *bis*-haloacetylamides, which are selective for sulfhydryl groups. A number of cross-linkers are selective for sugars, prosthetic groups, and other components of proteins, and are the subject of several excellent reviews (see, e.g., Wong 1993; Hermanson 1996).

Controlling the specificity of cross-linking is perhaps one of the most critical variables in the use of this method in determining interacting proteins. In one-step cross-linking procedures, where all the reactants are included in a single step, excessive cross-linking can readily occur under certain conditions, potentially leading to the conjugation of noninteracting species, i.e., false positives. For instance, cross-linking at pH values above 8 can promote deprotonation and activation of many classes of nucleophiles on the protein's surface, so that the protein effectively becomes a sponge for cross-linking reagents and, ultimately, solutes that collide with it during the reaction. The same scenario occurs when the concentration of the cross-linker is in great excess of the concentrations of the target proteins. Reaction conditions that can alter a protein's native structure may expose buried regions, leading to nonspecific reactions with either the cross-linker or other components of the reaction. Because cross-linkers are chemical reagents, they also have the potential to alter a protein's higher-order structure, which can result in loss of activity or formation of aggregates. Carrier solvents, which are used to introduce insoluble hydrophobic cross-linkers into a reaction, also have the potential to alter the structure of proteins and to create false positives. Besides potentially influencing the rate of the organic cross-linking reaction, temperature and ionic strength can further influence the specificity of cross-linking by directly affecting the interactions between proteins. The choice of buffer can also affect the extent and specificity of cross-linking. Buffers such as Tris that contain free amines, and other buffers containing other nucleophiles, can effectively compete with the nucleophilic side chains of proteins for the cross-linker, especially given that their concentrations are far in excess of the protein targets.

Another problem related to specificity is the presence of nonspecific cross-linked products that do not result from false positives, but instead from the too-extensive cross-linking of genuinely interacting proteins. This can be a particular problem with large, stable hetero-oligomeric complexes, in which virtually every polypeptide chain can eventually be linked directly or indirectly to all others. Such nonspecific cross-linking, even to a modest extent, makes analysis and interpretation of products progressively more difficult as the size of the adduct increases. For many hetero-oligomeric complexes, one of the greatest experimental challenges is to control the extent of cross-linking. By far, the easiest cross-linked adducts to analyze and interpret are bimolecular.

A two-step cross-linking process, which exploits the different properties of the reactive groups of heterobifunctional cross-linkers, can be used to increase specificity of cross-linking. In these procedures, the purified protein of interest is commonly modified with the chemically reactive group (e.g., a maleimido or succinimidyl group) of a photoreactive heterobifunctional cross-linker. Excess noncovalently linked cross-linker is then removed from the modified protein. In the second step, the mono-derivatized protein is combined with its potential protein partner(s) and exposed to an activating wavelength of light to induce cross-linking (Nadeau et al. 1999a). A scheme for such a two-step cross-linking procedure using a photoactivatable cross-linker is shown in Figure 1B. An alternative to the chemical modification procedure in the first step is to generate mutants of the protein that actually incorporate photolabile amino acid analogs into the polypeptide (Fischer et al. 2000). The advantage of this latter technique is that the conjugating moiety is inserted into the protein at a precise position, rather than randomly. The ability to choose sites on a molecule that are implicated in protein–protein interactions by other methods significantly increases the probability of successfully and specifically cross-linking it to its interacting partner.

Cross-linking reagents are derived from monofunctional chemical modifiers, sometimes called blocking reagents, which contain the same types of reactive groups shown in Table 1. The actions of cross-linkers can therefore be attenuated by first modifying a protein with a monofunctional blocking reagent that reacts selectively with a particular type of amino acid side chain. For example, formaldehyde selectively and stably methylates the ε-amine of lysine in the presence of NaCNBH$_3$ to form N,N-dimethyl-lysine, which is a poor nucleophile compared to its parent form. Because the methyl groups are small and the charge on the amine remains unchanged (Jentoft and Dearborn 1983), the overall structure and activity of the protein generally remains unaltered. Nucleophiles from less abundant residues (e.g., cysteine) on the derivatized protein can then be targeted to promote more specific, controlled cross-linking. An alternative approach is to convert the reactive group of an amino acid side chain to a different chemical functionality having a different reactivity. For instance, conversion of thiols to amines can be carried out with either ethylenimine or 2-bromoethyl amine. Reagents are available for the conversion of many of the major functional groups of side chains (Wong 1993).

The solubility of a cross-linker is an important determinant in the types of proteins that it targets. Hydrophobic cross-linkers are routinely used to cross-link membrane-spanning regions of integral membrane proteins, or hydrophobic regions in oligomeric complexes (Sanders et al. 1992). Nonpolar cross-linkers are also more likely to react with buried reactive side-chain groups than are polar water-soluble cross-linkers; accordingly, they are also more likely to promote intramolecular cross-linking of a protein. Water-soluble cross-linking reagents are favored to react with solvent-accessible side chains, and thus are often used to achieve intermolecular cross-linking of soluble cytosolic proteins.

Specificity and Distance in Cross-linking (Molecular Rulers and Nearest Neighbors)

Because side-chain nucleophiles cannot be cross-linked if they are separated by a distance greater than the length of the cross-linker, the specificity of cross-linking is dependent on the distance separating the reactive groups of the cross-linker being used. A basic method for determining the maximal distance between reactive amino acid residues, especially in an oligomeric complex, is to use a homologous series of cross-linkers with spacers of differing lengths (molecular rulers) or geometry (reviewed extensively in Wong 1993). The identity of the reactive groups of the cross-linker is maintained in the series of homologous rulers to simplify the interpretation of results by eliminating differences in chemistry among the cross-linkers; even so, the caveat against interpretation of negative results (the absence of cross-linking) still holds. The distance between reacting residues determined using this molecular ruler approach is defined as maximal because residues separated by a distance less than the extended length of the cross-linker may still be cross-linked, owing to the potential flexibility of the cross-linker (Green et al. 2001). Protein breathing could potentially further complicate determination of the distance separating cross-linked residues.

The nearest-neighbor approach is a method that is used to determine the topography, with respect to relative subunit locations, of hetero-oligomeric complexes. The basic premise of this approach is that proteins in a complex are cross-linked only if they are in sufficiently close proximity to accommodate the selected span of the cross-linker. A topographical model can be deduced through identification of the components of multiple cross-linked adducts. An excellent example of the successful use of this approach is the mapping of the 30S ribosomal subunit (Lambert et al. 1983). Interpretation of results is easiest when there is but a single copy of each subunit type, although the approach is still feasible with complexes that have more than one copy of a given subunit. A variation of this approach is to analyze changes in subunit interactions within an oligomer that are induced by substrates or effectors and that can be detected by alterations in the subunit cross-linking patterns. Different activation states of the hexadecameric phosphorylase kinase complex have been distinguished using this variation (Nadeau et al. 1997a,b).

General Protocols

Selecting an appropriate cross-linker is the first step in determining the optimal conditions for cross-linking interacting proteins. Considering the potential differences in the reactivities of protein side chains, a screen of several different types of cross-linkers is generally carried out to obtain detectable quantities of the desired conjugate(s). A screen comparing both hydrophilic and hydrophobic classes of cross-linkers is a good starting point. Secondary screens using the most promising class are then performed by altering both the chemistry (varying functional groups selective for different amino acid side chains) and span of the cross-linkers. In the secondary screens, the concentration of reactants, pH, temperature, and time of cross-linking may also be varied. A flowchart showing the various steps and strategies of screening is shown in Figure 2.

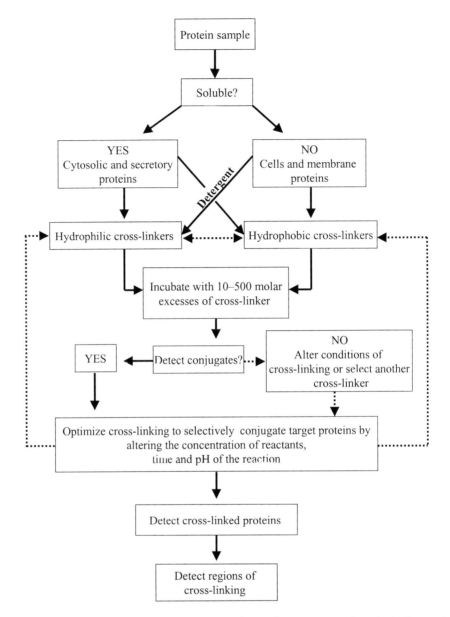

FIGURE 2. Flowchart for cross-linking screening. This scheme depicts a general method of screening cross-linkers for potential target proteins and illustrates the variables that must be considered to achieve positive outcomes (*solid arrows*) in the selection of a suitable cross-linker. The paths indicated by the dotted arrows illustrate the empirical nature of the cross-linking approach. For example, if a desired conjugate is not detected after incubating the target proteins with a large excess of the cross-linker being tested, then either different conditions of cross-linking (pH, time, protein concentration, etc.) must be tested, or a new cross-linker with different chemistries must be evaluated.

The protocol outlined above is applicable to a broad range of sample formulations, from those containing only the purified targets to those containing complex mixtures of proteins from cell lysates. Note that the protocol is for one-step cross-linking; i.e., both cross-linking reactions occur under the same conditions. For ease of discussion, we will assume that the targets of interest are in sufficient quantity for detection either by Coomassie staining of SDS-PAGE gels or by western blotting. Gradient polyacrylamide gels are often used for observing the patterns of cross-linking because they provide greater flexibility in resolving reaction products of widely different masses, from the parent monomers to extensively cross-linked aggregates. One precaution in resolving cross-linked proteins on SDS-PAGE gels is related to intramolecular cross-linking. A protein that is intramolecularly cross-linked cannot unfold completely in the presence of denaturants, leading to a decrease in its Stokes radius, which in turn causes it to migrate faster than its non-cross-linked counterpart. It is not uncommon to obtain two well-resolved adducts with different apparent molecular weights, but composed of the identical polypeptides in the same stoichiometry. This possibility obviously complicates data interpretation.

Sample Preparation

For the in vivo approach, the first step is to determine the amount of cells (50–100 million/ml is generally sufficient) that is necessary to detect the target proteins by either gel staining or immunochemical techniques. Alternatively, subconfluent cultures can be labeled with mCi amounts of [^{35}S]methionine prior to cross-linking to detect small amounts of proteins that are cross-linked to an overexpressed target (De Gunzburg et al. 1989). Before adding the cross-linker, subject the cells to several exchanges in phosphate-buffered saline (PBS) to remove any cell medium, which contains peptides and other components that may potentially react with the cross-linker.

For in vitro approaches, dissolve or dialyze the proteins to a final concentration of 0.05–0.5 mg/ml in a buffer that is unlikely to react with the reactive groups of the cross-linker. A pH value of 7.5 is convenient for the initial screening.

Screening Reaction Conditions

Add the cross-linker to the protein sample. If the cross-linker is hydrophobic, it must first be dissolved in a suitable carrier solvent (e.g., acetonitrile, acetone, or methanol) and added in small quantities so that the volume of the solvent is not excessive in the reaction (generally ~1% of the total reaction volume, but certainly less than the amount necessary to affect the target protein's activity). In separate experiments, increase the molar ratio of cross-linker to protein in desired increments from 10 to 500. Initially, cross-linking for 30 minutes is a convenient starting point. The effects of both temperature and pH can be tested using the same conditions. In general, cross-linking below neutral pH will favor selection of protein sulfhydryl groups (particularly with thiol-selective cross-linkers such as phenylenebismaleimide), whereas cross-linking at alkaline pH will include greater amounts of amine-targeted cross-linking.

Time Course of Cross-linking

Using the ratio of concentrations of cross-linker to protein that resulted in the maximum extent of formation of the desired conjugate, further optimize the reaction by varying the time of cross-linking. It is prudent to avoid long reaction times because of potential instability of reactants.

If there is concern over the stability of the cross-linker because of pH, temperature, hydrolysis, etc., the cross-linker can be added in multiple aliquots at different times to increase the cross-linking yield. The ultimate goal is usually to maximize the ratio of desired conjugate(s) formed to reactants consumed, while maintaining specificity, i.e., avoiding formation of too many conjugates or conjugates that are too large or complex. When using cross-linking as a probe of effector-induced conformational changes in an oligomeric complex, it must be established that the product adduct of interest is in the linear range of its rate of formation, thus allowing either increases or decreases in its formation to be observed (Nadeau et al. 1997a).

Quenching of Cross-linking

The final step in the cross-linking reaction is its quenching. Excess amounts of nucleophiles (e.g., glycine, Tris) or other reagents that compete for the reactive groups of the cross-linker are commonly added to quench additional protein cross-linking. Purify protein conjugates from the other reaction components by such techniques as chromatography, ultrafiltration, or dialysis. An alternative approach is simultaneously to quench and denature with detergents. This latter technique is useful for the rapid screening of cross-linkers by SDS-PAGE or western blotting. The typical SDS-sample buffer used in PAGE contains both the denaturant SDS, which perturbs protein interactions, and two nucleophiles (Tris and β-mercaptoethanol), which compete for the cross-linker's reactive groups. Cross-linking reactions can be rapidly quenched by adding the SDS-sample buffer, vortexing, and heating for 5 minutes at 90°C. Reactions can also be quenched by precipitating the protein with trichloroacetic acid or other precipitants, including polyethylene glycol or ammonium sulfate. Take care when using protein precipitants, because the aggregate formation they promote can allow residual nonspecific cross-linking. A better strategy is first to add nucleophilic competitors of cross-linking, before adding the precipitant.

DETECTION AND PURIFICATION OF CROSS-LINKED PROTEINS

The biochemistry of a protein helps in choosing an appropriate method for detecting both the identity of conjugate partners and the specific regions of cross-linking between protein side chains. For example, if the function of an enzyme or receptor is known, then the substrates or ligands of these proteins can be used to purify conjugates of the target protein and its interacting partners by affinity chromatography. If there are no available probes to identify a desired target, cleavable cross-linkers may be used to identify protein conjugates of the target by diagonal methods. Alternatively, radiolabeled or immunoreactive cross-linkers can be selected to detect cross-linked species, as can cross-linkers containing a specific ligand for affinity binding. If the protein of interest can be expressed, one can potentially generate mutant proteins that contain either immunoreactive tags or metal-binding ligands, which can be used to purify conjugates of these proteins by affinity-based chromatographic methods.

Cleavable Cross-linkers

Bifunctional reagents containing labile bonds that can be chemically cleaved following cross-linking are termed cleavable cross-linkers. Two cleavable moieties that are routinely incorporated into the spacer regions of cross-linkers are disulfides and vicinal diols, with the former cleaved by reduction with thiols and the latter by oxidation with periodate. Conjugate pairs of initially cross-linked adducts can be identified in complex mixtures by diagonal methods (Traut et al. 1989), which combine cleavable cross-linkers and two-dimensional gels. Following cross-linking with the cleavable reagent, proteins are resolved by SDS-PAGE in the absence of the cleavage reagent. A vertical lane is excised from the gel and exposed to the cleavage reagent to disrupt the cross-links. The gel strip is then placed horizontally across the top of an SDS-PAGE slab gel. The proteins are resolved again by mass in the second dimension. Proteins that were not cross-linked will migrate along a diagonal, progressing from high to low mass. Each of the cleaved monomers that was originally in a cross-linked adduct that was resolved in the first dimension will now migrate to different distances below the diagonal, based on its individual mass. This approach is especially useful in studying a complex system having multiple protein components.

Labeled Cross-linkers

There are two general classes of labels that are incorporated into cross-linkers to aid in identifying cross-linked adducts. The first class comprises groups that can be detected by spectroscopic methods, including fluorophores (Kosower and Kosower 1995), UV-VIS chromaphores (Sigrist et al. 1982), and spin labels (Gaffney et al. 1983). The second class of reporter groups contains radiolabels. Radiolabeled iodine (^{125}I) is routinely incorporated into phenolic cross-linkers with IODO-GEN-(1,3,4,6-tetrachloro-3α,6α-diphenyl glycouril) (Fraker and Speck 1978), and reagents necessary for synthesizing this type of ^{125}I-labeled cross-linker are commercially available from Pierce Chemical Company. A common problem encountered when using labeled cross-linkers to detect cross-linked proteins is monosubstitution, where the protein is labeled by the cross-linker without being cross-linked.

The problem of false positives generated by monosubstitution can be partially alleviated through the use of label transfer techniques (for review, see Traut et al. 1989; Fancy 2000). The first step is to couple a bait protein with a cleavable, labeled, heterobifunctional, photoactivatable cross-linker (e.g., SHAD in Table 1). Then purify the labeled protein from excess reagent. Target proteins are then added and cross-linking is induced by photolysis, which results in direct covalent attachment of the labeled group to the interacting target partner. Subsequent cleavage of the cross-linker releases the now labeled partner, which can be purified through tracking of the label.

Several difficulties may be encountered when using ^{125}I-labeled aryl azides, which are among the most common photoreactive groups used in this approach. Upon activation, aryl azides form highly reactive nitrenes, which reportedly undergo transformations to form azepines and other reactive intermediates (Li et al. 1988), ultimately leading to loss of the radioactive label by unknown mechanisms (Watt et al. 1989). Separating the radiolabel from the reactive azido group, as in the case of SHAD, minimizes release of the label (Koch et al. 1994). The transient nature of the reactive photo-induced intermediate(s) in theory also diminishes the yield of label transfer; i.e., the lifetime of the reactive intermediate must exceed the time of association between the labeled bait and the target protein to obtain significant yields of specific intermolecular cross-linking. Transient associations between interacting proteins may, therefore, favor either intramolecular cross-linking of the tethered probe with endogenous residues of the bait protein or preferential reaction with solvent to form monosubstituted product.

Bromobimanes (Table 1), which are available from Molecular Probes, form fluorescently labeled cross-linked complexes with proteins. Although not intrinsically fluorescent themselves, bromobimanes contain alkyl bromide groups that alkylate sulfhydryls to form fluorescent conjugates. Several advantages are inherent in the use of these reagents. First, the thiol-selectivity of these reagents makes them less susceptible to nucleophilic attack by the more abundant lysyl ε-amines, increasing the specificity of cross-linking. Second, the stability of these reagents significantly reduces the probability of their reacting with solvent molecules to form monosubstituted proteins. Third, only cross-linking forms fluorescent products, which greatly simplifies interpretation and product analysis.

Affinity Detection and Purification of Cross-linked Proteins

The recently developed so-called trifunctional cross-linkers contain bifunctional reactive groups for cross-linking, plus a covalently attached ligand for detection or purification of cross-linked products by affinity methods (Hermanson 1996). The attached affinity ligand is often biotin, which targets conjugates for binding by streptavidin or avidin (Wedekind et al. 1989). Cross-linked species are purified by passing the reaction mixture over a streptavidin column, followed by washing and elution with biotin. Analytical scale variants of the above procedure can be used to screen different conditions of cross-linking. Cross-linking reactions are carried out in microcentrifuge tubes (final volumes 50–300 µl). Following cross-linking, suspensions of streptavidin immobilized on agarose are added directly to the cross-linking mixture and incubated at ambient temperature with gentle mixing. Nonspecifically bound components are removed by successive washing and centrifugation. The pellets are then boiled in SDS-buffer and resolved by SDS-PAGE, with protein staining or immunochemical detection.

A related method, first described by Ballmer-Hofer et al. (1982), involves the use of antibodies directed toward the cross-linker. After cross-linking, conjugates are isolated by immunoprecipitation with anti-cross-linker polyclonal antibodies. As is the case with labeled cross-linkers, monosubstitution and intramolecular cross-linking can complicate both detection and purification by affinity methods.

IDENTIFICATION OF CROSS-LINKED RESIDUES

By far the most difficult challenge in using cross-linking to study protein–protein interactions is determining the specific amino acid residues that are cross-linked to each other within a complex. Besides the expected problems associated with peptide isolation and identification, there is invariably the confounding factor of monosubstitution. Regardless of whether a protein molecule is also cross-linked, monosubstitution greatly complicates detection, purification, and characterization using most available methods. A peptide that is modified by the cross-linking reagent, but not cross-linked, nearly always contains that very part of the cross-linker that serves as the basis for detection and purification. Thus, false positives are present and must be eliminated one by one. Similarly, intramolecular cross-linking must be characterized before it can be distinguished from intermolecular cross-linking.

Peptide Mapping by HPLC

The traditional approach to determining regions of cross-linking between proteins is peptide mapping. This technique requires the use of HPLC, and its chance for success is enhanced if the proteins composing the conjugate are small and readily purified. The cross-linked peptides are detected either directly by the specific properties of the cross-linker or indirectly by difference peptide maps. In peptide maps, modified peptides are identified on the basis of their disappearance from digests of the native parent protein(s) and their appearance as new species in the digests of the cross-linked target. Methods such as two-dimensional PAGE have also been used to map CNBr digests of proteins (Dukan et al. 1998) and have potential use in the peptide mapping of cross-linked proteins.

Peptide Mapping by Mass Spectrometry

With the advent of proteomics, new approaches for rapid peptide mapping by mass spectrometry (MS) are becoming the methods of choice. MS methods of peptide mapping eliminate many of the steps previously required to analyze complex digests of proteins and, more importantly, require far less sample (often only pmole amounts). This high sensitivity allows digests of single spots of resolved, cross-linked adducts from either gel sections (Sechi and Chait 1998) or blots (Binz et al. 1999) to be examined using matrix-assisted laser desorption/ionization time-of-flight mass spectrometry (MALDI-TOF-MS) (Bennett et al. 2000). The combination of microbore HPLC and electrospray ionization mass spectrometry (ESI-MS) is perhaps the most powerful technique for analyzing the digests of native and modified proteins (Kosaka et al. 2000). Peptides are first resolved by microbore HPLC, and a fraction of the eluate is shunted to the MS, where peptide masses are determined directly. Peptides are identified through comparing their experimentally determined masses against the theoretical masses that would be obtained from digestion of the protein in question with the same cleavage reagent. There are now many Web Sites (ExPASy, SWISS-PROT, NCBI, and other databases) that contain search engines and predictive programs that generate theoretical peptide maps of proteins using numerous cleavage reagents, including trypsin, chymotrypsin, and CNBr.

Although the above techniques are generally used simply to identify proteins, new refined techniques are emerging that directly address the identification of cross-linked peptides. For example, Chen et al. (1999) have combined isotopic derivatization and ESI-MS into a universal technique to isolate cross-linked peptides. Proteins are first cross-linked using any desired cross-linker. To prevent their further reaction with subsequent reagents, the lysyl ε-amine groups of the proteins are blocked by reductive methylation with formaldehyde in the presence of $NaCNBH_3$. The proteins are then digested with proteases specific for groups other than lysine. The α-amines of the liberated peptides, the only free amines now present, are modified with an equimolar mixture of isotopically labeled dinitrofluorobenzene (DNFB) and unlabeled DNFB. Cross-linked peptides contain two α-amine targets versus only one for peptide monomers, which allows the modified peptides to be fractionated into mono-DNP and bis-DNP pools by chromatography over phenyl columns. Each pool is further resolved by HPLC and analyzed using ESI-MS. The cross-linked peptides are identified by their signature mass spectrum, resulting from the distribution of label afforded by their bis-substitution with labeled DNFB. Other combinations of MS and labeling techniques can be envisioned, given the large number and variety of labeled cross-linkers.

Affinity Purification of Cross-linked Peptides

The use of trifunctional cross-linkers can be extended from identifying and purifying cross-linked proteins, as discussed in the section on Affinity Detection and Purification of Cross-linked Proteins, p. 87), to determining specific residues involved in cross-linking. Following their affinity purification, the cross-linked proteins are digested with appropriate reagents. The digest is passed over a column that is specific for the affinity ligand attached to the cross-linker. Following elution, the identity of the cross-linked peptide(s) is determined using MS or gas phase sequencing. Antibodies directed against a specific cross-linker can be used in a similar manner. Instead of streptavidin or other ligand-binding proteins, immobilized antibodies are used to purify both the conjugate and its cross-linked peptides. Monosubstitution again contributes complexity to both of these affinity-based approaches.

REFERENCES

Alaedini A. and Day R.A. 1999. Identification of two penicillin-binding multienzyme complexes in *Haemophilus influenzae*. *Biochem. Biophys. Res. Commun.* **264:** 191–195.

Alley S.C., Ishmael F.T., Jones A.D., and Benkovic S.J. 2000. Mapping protein-protein interactions in the bacteriophage T4 DNA polymerase holoenzyme using a novel trifunctional photo-cross-linking and affinity reagent. *J. Am. Chem. Soc.* **122:** 6126–6127.

Ballmer-Hofer K., Schlup V., Burn P., and Burger M.M. 1982. Isolation of *in situ* crosslinked ligand-receptor complexes using an anticrosslinker specific antibody. *Anal. Biochem.* **126:** 246–250.

Benesch R.E. and Kwong S. 1991. Hemoglobin tetramers stabilized by a single intramolecular cross-link. *J. Prot. Chem.* **10:** 503–510.

Bennett K.L., Matthiesen T., and Roepstorff P. 2000. Probing protein surface topology by chemical surface labeling, crosslinking, and mass spectrometry. In *Mass spectrometry of proteins and peptides* (ed. J.R. Chapman), pp. 113–131. Humana Press, Totowa, New Jersey.

Binz P.-A., Muller M., Walther D., Bienvenut W.V., Gras R., Hoogland C., Bouchet G., Gasteiger E., Fabbretti R., Gay S., Palagi P., Wilkins M.R., Rouge V., Tonella L., Paesano S., Rossellat G., Karmime A., Bairoch A., Sanchez J.-C., Appel R.D., and Hochstrasser D.F. 1999. A molecular scanner to automate proteomic research and to display proteome images. *Anal. Chem.* **71:** 4981–4988.

Chen X., Chen Y.H., and Anderson V.E. 1999. Protein cross-links: Universal isolation and characterization by isotopic derivatization and electrospray ionization mass spectrometry. *Anal. Biochem.* **273:** 192–203.

De Gunzburg J., Riehl R., and Weinberg R.A. 1989. Identification of a protein associated with p21[ras] by chemical crosslinking. *Proc. Natl. Acad. Sci.* **86:** 4007–4011.

Dey D., Bochkariov D.E., Jokhadze G.G., and Traut R.R. 1998. Cross-linking of selected residues in the N- and C-terminal domains of *Escherichia coli* protein L7/L12 to other ribosomal proteins and the effect of elongation factor Tu. *J. Biol. Chem.* **273:** 1670–1676.

Dukan S., Turlin E., Biville F., Bolbach G., Touati D., Tabet J.C., and Blais J.C. 1998. Coupling 2D SDS-PAGE with CNBr cleavage and MALDI-TOFMS: A strategy applied to the identification of proteins induced by a hypochlorous acid stress in *Escherichia coli*. *Anal. Chem.* **70:** 4433–4440.

Fancy D.A. 2000. Elucidation of protein-protein interactions using chemical cross-linking or label transfer techniques. *Curr. Opin. Chem. Biol.* **4:** 28–33.

Fischer K.D., Helms J.B., Zhao L., and Wieland F.T. 2000. Site-specific photocrosslinking to probe interactions of Arf1 with proteins involved in budding of COPI vesicles. *Methods* **20:** 455–464.

Fraker P.J. and Speck Jr., J.C. 1978. Protein and cell membrane iodinations with a sparingly soluble chloroamide, 1,3,4,6-tetrachloro-3a, 6a-diphenylglycoluril. *Biochem. Biophys. Res. Commun.* **80:** 849–857.

Gaffney B.J., Willingham G.L., and Schepp R.S. 1983. Synthesis and membrane interactions of spin-label bifunctional reagents. *Biochemistry* **22:** 881–892.

Green N.S., Reisler E., and Houk K.N. 2001. Quantitative evaluation of the lengths of homobifunctional protein cross-linking reagents used as molecular rulers. *Protein Sci.* **10:** 1293–1304.

Hajdu J., Wyss S.R., and Aebi H. 1977. Properties of human erythrocyte catalases after crosslinking with bifunctional reagents. Symmetry of the quaternary structure. *Eur. J. Biochem.* **80:** 199–207.

Hartman F.C. and Wold F. 1967. Cross-linking of bovine pancreatic ribonuclease A with dimethyl adipimidate. *Biochemistry* **6:** 2439–2448.

Hermanson G.T. 1996. *Bioconjugate techniques.* Academic Press, San Diego.

Heymann E. and Mentlein R. 1980. Cross-linking experiments for the elucidation of the quaternary structure of carboxylesterase in the microsomal membrane. *Biochem. Biophys. Res. Commun.* **95:** 577–582.

Jackson V. 1987. Deposition of newly synthesized histones: New histones H2A and H2B do not deposit in the same nucleosome with new histones H3 and H4. *Biochemistry* **26:** 2315–2325.

———. 1999. Formaldehyde cross-linking for studying nucleosomal dynamics. *Methods* **17:** 125–139.

Jentoft N. and Dearborn D.G. 1983. Protein labeling by reductive alkylation. *Methods Enzymol.* **91:** 570–579.

Ji T.H. 1977. A novel approach to the identification of surface receptors. The use of photosensitive heterobifunctional cross-linking reagent. *J. Biol. Chem.* **252:** 1566–1570.

Koch T., Suenson E., Korsholm B., Henriksen U., and Buchardt O. 1994. Pitfalls in characterization of protein interactions using radioiodinated crosslinking reagents. Preparation and testing of a novel photochemical ^{125}I-label transfer reagent. *Bioconjug. Chem.* **5:** 205–212.

Kosaka T., Takazawa T., and Nakamura T. 2000. Identification and C-terminal characterization of proteins from two-dimensional polyacrylamide gels by a combination of isotopic labeling and nanoelectrospray fourier transform ion cyclotron resonance mass spectrometry. *Anal. Chem.* **72:** 1179–1185.

Kosower E.M. and Kosower N.S. 1995. Bromobimane probes for thiols. *Methods Enzymol.* **251:** 133–148.

Lambert J.M., Boileau G., Cover J.A., and Traut R.R. 1983. Cross-links between ribosomal proteins of 30S subunits in 70S tight couples and in 30S subunits. *Biochemistry* **22:** 3913–3920.

Li Y.-Z., Kirby J.P., George M.W., Poliakoff M., and Schuster G.B. 1988. 1,2-Didehydroazepines from the photolysis of substituted aryl azides: Analysis of their chemical and physical properties by time-resolved spectroscopic methods. *J. Am. Chem. Soc.* **110:** 8092–8098.

Nadeau O.W., Sacks D.B., and Carlson G.M. 1997a. Differential affinity cross-linking of phosphorylase kinase conformers by the geometric isomers of phenylenedimaleimide. *J. Biol. Chem.* **272:** 26196–26201.

———. 1997b. The structural effects of endogenous and exogenous Ca^{2+}/calmodulin on phosphorylase kinase. *J. Biol. Chem.* **272:** 26202–26209.

Nadeau O.W., Traxler K.W., Fee L.R., Baldwin B.A., and Carlson G.M. 1999a. Activators of phosphorylase kinase alter the cross-linking of its catalytic subunit to the C-terminal one-sixth of its regulatory alpha subunit. *Biochemistry* **38:** 2551–2559.

Nadeau O.W., Domanski P., Usacheva A., Uddin S., Platanias L.C., Pitha P., Raz R., Levy D., Majchrzak B., Fish E., and Colamonici O.R. 1999b. The proximal tyrosines of the cytoplasmic domain of the beta chain of the Type I interferon receptor are essential for signal transducer and activator of transcription (Stat) 2 activation. *J. Biol. Chem.* **274:** 4045–4052.

Sanders S.L., Whitfield K.M., Vogel J.P., Rose M.D., and Schekman R.W. 1992. Sec61p and BiP directly facilitate polypeptide translocation into the ER. *Cell* **69:** 353–365.

Sechi S. and Chait B.T. 1998. Modification of cysteine residues by alkylation. A tool in peptide mapping and protein identification. *Anal. Chem.* **70:** 5150–5158.

Sigrist H., Allegrini P.R., Kempf C., Schnippering C., and Zahler P. 1982. 5-Isothiocyanato-1-naphthalene azide and *p*-azidophenylisothiocyanate: Synthesis and application in hydrophobic heterobifunctional photoactive cross-linking of membrane proteins. *Eur. J. Biochem.* **125:** 197–201.

Simon M.J. and Day R.A. 2000. Improved resolution of hydrophobic penicillin-binding proteins and their covalently linked complexes on a modified C18 reversed phase column. *Anal. Lett.* **33:** 861–867.

Traut R.R., Casiano C., and Zecherle N. 1989. Cross-linking of protein subunits and ligands by the introduction of disulphide bonds. In *Protein function: A functional approach* (ed. T.E. Creighton), pp. 101–133. IRL Press, Oxford.

Valentine W.M., Graham D.G., and Anthony D.C. 1993. Covalent cross-linking of erythrocyte spectrin by carbon disulfide *in vivo. Toxicol. Appl. Pharmacol.* **121:** 71–77.

Watt D.S., Kawada K., Leyva E., and Platz M.S. 1989. Exploratory photochemistry of iodinated aromatic azides. *Tetrahedron Lett.* **30:** 899–902.

Wedekind F., Baer-Pontzen K., Bala-Mohan S., Choli D., Zahn H., and Brandenburg D. 1989. Hormone binding site of the insulin receptor: Analysis using photoaffinity-mediated avidin complexing. *Biol. Chem. Hoppe-Seyler* **370:** 251–258.

Wong S.S. 1993. *Chemistry of protein conjugation and cross-linking.* CRC Press, Boca Raton.

7 Yeast and Bacterial Two-hybrid Selection Systems for Studying Protein–Protein Interactions

Ilya Serebriiskii[1] and J. Keith Joung[2]

[1]*Fox Chase Cancer Center, Philadelphia, Pennsylvania 10111;* [2]*Howard Hughes Medical Institute/Massachusetts Institute of Technology, Cambridge, Massachusetts 02139 and Massachusetts General Hospital, Boston, Massachusetts 02114*

INTRODUCTION

The yeast two-hybrid system has provided a powerful tool for studying protein–protein interactions. This genetic method, based on the reconstitution of a functional transcriptional activator in yeast (Fields and Song 1989), has now been used extensively both to identify novel protein–protein interactions and to analyze known interactions. Many extensions to the original two-hybrid system have also greatly expanded its utility (reviewed most recently in Serebriiskii et

This chapter is adapted and expanded from Chapter 18, Protocol 1, in Sambrook and Russell 2001, *Molecular Cloning.*

al. 2001). Recently, a bacterial genetic selection system analogous to the yeast two-hybrid has also been described (Dove et al. 1997; Dove and Hochschild 1998; Joung et al. 2000). This bacterial-based system offers two significant advantages over its yeast counterpart: It permits the analysis of very large libraries ($>10^8$ in size) and it provides an alternative approach for eukaryotic proteins that are not amenable to analysis in the yeast-based system (e.g., self-activating baits, proteins toxic to yeast, or proteins that have undesired interactions with endogenous yeast proteins). Together, the yeast and bacterial two-hybrid systems provide powerful methods for analyzing protein–protein interactions.

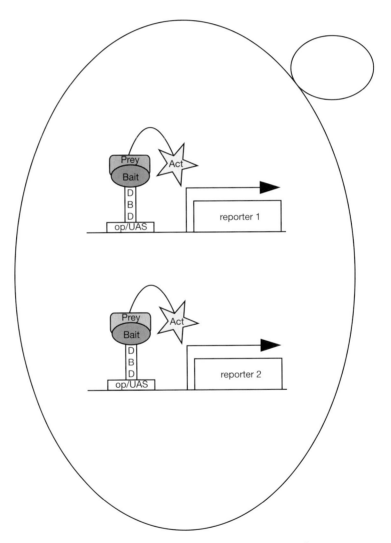

FIGURE 1. Schematic of the basic two-hybrid system to detect interactions between two proteins in a yeast cell. As shown, a DNA-binding domain (DBD)-fused bait protein of interest interacts with an activation domain (Act) used to a partner protein (Prey), either known or selected from a cDNA library. The interacting pair binds a specific sequence motif (op [operator] or UAS [upstream activating sequence], depending on whether LexA or GAL4 is used as a DBD), activating transcription of two separate reporter genes. (Reprinted, with permission, from Sambrook and Russell 2001, *Molecular Cloning.*)

Protocol: Part 1

A Yeast Two-hybrid System Based on Transcriptional Activation

An important element in the characterization of the function of a protein is the identification of other proteins with which it interacts. Detection of interacting proteins can be accomplished by multiple approaches, and classically has involved the biochemical copurification and microsequencing of components of protein complexes, followed by subsequent cloning. During the last several years, a powerful genetic screening strategy utilizing yeast, termed the two-hybrid system (Fig. 1) (Fields and Song 1989), has gained wide popularity because of the relative ease and speed with which novel interacting proteins can be isolated. A number of variants of the two-hybrid system have been developed for library screening (Chien et al. 1991; Durfee et al. 1993; Gyuris et al. 1993; Vojtek et al. 1993), all using a similar theoretical paradigm, but with distinct and generally noninterchangeable sets of reagents (see http://www.fccc.edu/research/labs/golemis/InteractionTrapInWork.html for details). This section provides a detailed discussion of screening for novel proteins by use of the interaction trap variant of the two-hybrid system (Gyuris et al. 1993).

The term "two-hybrid" derives from the two classes of chimeric proteins used in a screen, the first a fusion of a protein of interest "x" to a *DNA binding domain* (DBD-x), the second a fusion of a cDNA library "y" to a transcriptional *activation domain* (AD-y). Multiple classes of proteins can be used in yeast two-hybrid screens, with the main requirements being that the protein can be effectively transported to the nucleus, and not activate transcription. In a general two-hybrid strategy (Fig. 1), the DBD-x fusion protein (generally referred to as a Bait) is transformed into a suitable strain of *Saccharomyces cerevisiae* with a dual reporter system that contains binding sites for the DBD. In the interaction trap, the DBD used is the bacterial protein LexA. A first reporter contains *lexA* operator sites (DNA-binding sites) upstream of the *lacZ* gene, encoding β-galactosidase, and a second contains *lexA* operators upstream of *LEU2*, a required gene in the yeast leucine biosynthetic pathway or, alternatively, another biosynthetic gene such as *HIS3*. Yeast containing the nontranscriptional activator bait do not turn on either of these reporters, and so in the absence of LacZ or Leu2, protein products are white on medium containing X-Gal, and fail to grow on medium lacking leucine. An AD-cDNA library is introduced into such yeast (either by mating with pretransformed partner strain or by direct transformation) and plated on selective media. In cases where a library-encoded protein interacts with the bait, the AD is brought to the *lexA* sites flanking the *lacZ* and *LEU2* reporters, resulting in colony growth on medium lacking leucine, and blue color on medium containing X-Gal.

The following protocols divide the execution of an interaction trap/two-hybrid screen into three stages. In the first, characterization of a novel bait is described, with attention to controls to increase the likelihood that it will function effectively in a two-hybrid screen. In the second, introduction of a cDNA library and selection of positive interactors are detailed. In the third, a series of first-order and subsequent control experiments designed to establish whether an interacting protein is biologically significant are outlined. A flowchart illustrating the order and approximate time necessary to perform various steps is shown in Figure 2.

Stage 1: Constructing and Characterizing a Bait Protein

The first step in an interactor hunt is to construct a plasmid that expresses LexA fused to the protein of interest. This construct is cotransformed with a *lexAop-lacZ* reporter plasmid into a yeast

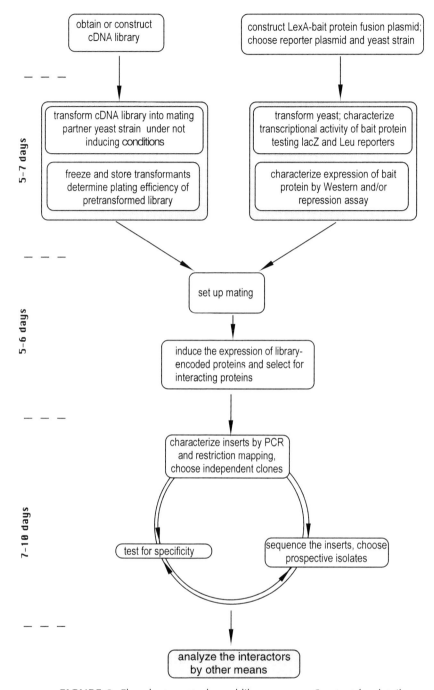

FIGURE 2. Flowchart, controls, and library screen. See text for details.

strain containing a chromosomally integrated *lexAop-LEU2* reporter gene. A series of control experiments is performed to establish that the Bait protein is made as a stable protein in yeast, that it is capable of entering the nucleus and binding *lexA* operator sites (Fig. 3, repression assay), and that it does not appreciably activate transcription of the *lexA* operator-based reporter genes. Depending on the result with these controls, the bait may be used directly to screen a library using the initial test conditions; alternatively, different combinations of reporter strains/plasmids can be used, or the bait can be modified according to provided guidelines.

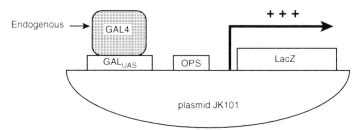

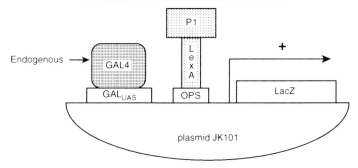

FIGURE 3. Repression assay for DNA binding (Brent and Ptashne 1984). The plasmid JK101 contains the upstream activating sequence (UAS) from the GAL1 gene followed by LexA operators upstream of the *lacZ* coding sequence. Thus, yeast containing pJK101 will have significant β-galactosidase activity when grown on medium in which Gal is the sole carbon source because of binding of endogenous yeast GAL4 to the UAS_{GAL} (*top*). LexA-fused proteins that are made, enter the nucleus, and bind the lexA operator sequences will block activation from the UAS_{GAL}, repressing β-galactosidase activity three- to fivefold (*bottom*). Note that on glucose X-Gal medium, yeast containing JK101 should be white because UAS_{GAL} transcription is repressed. (Reprinted, with permission, from Golemis and Serebriiskii 1998.)

The following protocols use basic yeast media and transformation procedures. As a general rule, it is good practice to move expeditiously through the characterization steps described below. While plasmids will be retained for extended periods of time in yeast maintained on Parafilm-wrapped selective plates placed in a 4°C refrigerator, expressed protein levels will gradually drop, and results may become somewhat variable after more than ~2 weeks of such maintenance. If delays are foreseen, the best options are either to repeat transformations with bait protein and *lexAop-lacZ* reporter before moving on to library screening, or to freeze at –70°C a stock of yeast transformed with bait and reporter that can be thawed prior to library screening.

MATERIALS

CAUTION: See Appendix for appropriate handling of materials marked with <!>.

Antibodies

Monoclonal antibody to LexA (CLONTECH) or polyclonal antibody to LexA (Invitrogen) or specific antibody to the fusion domain of the target protein (if available)

Buffers and Solutions

For components of stock solutions, buffers, and reagents, see Sambrook and Russell 2001, *Molecular Cloning*. Dilute stock solutions to the appropriate concentrations.

Chloroform <!>

Dimethylsulfoxide (DMSO) <!>

Ethanol <!>

β-Glucuronidase lysis solution

    50 mM Tris-HCl (pH 7.5)

    10 mM EDTA

    0.3% (v/v) β-mercaptoethanol (prepare fresh)

    1:50 β-glucuronidase type HP-2, crude solution from *H. pomatia* <!>(Sigma)

Lysis solution

    Zymolase 100T <!> dissolved 2–5 mg/ml in rescue buffer

    OR

    β-glucuronidase <!> 100,000 units/ml (Sigma) 1:50 in rescue buffer

       Prepare fresh solution for each use.

Sterile glycerol solution for freezing transformants

    65% sterile glycerol

    0.1 M magnesium sulfate ($MgSO_4$) <!>

    25 mM Tris-HCl (pH 8.0)

    20 mg/ml X-Gal

Rescue buffer

    50 mM Tris-HCl (pH 7.5)

    10 mM EDTA

    0.3% β-mercaptoethanol <!>

       Prepare fresh solution for each use.

SDS (10% w/v) <!>

2x SDS gel-loading buffer

    100 mM Tris-HCl (pH 6.8)

    200 mM dithiothreitol <!>

    4% SDS (electrophoresis grade) <!>

    0.2% bromophenol blue <!>

    20% glycerol <!>

       SDS gel-loading buffer lacking dithiothreitol can be stored at room temperature. Add dithiothreitol from a 1 M stock just before the buffer is used.

TE (pH 7.5) (sterile)

TE (pH 7.5) containing 0.1 M lithium acetate <!>

TE (pH 7.5) containing 40% PEG 4000 <!> and 0.1 M lithium acetate <!> (sterile)

Centrifuges and Rotors

Sorvall GSA rotor or equivalent

Sorvall RT6000 centrifuge and H10000B rotor or equivalent

Enzymes and Buffers

Restriction endonucleases *Eco*RI, *Xho*I, *Hae*III

Gels

Agarose gel

Agarose gel (2%)

Low-melting-temperature agarose gel (0.7%)

Low-melting-temperature agarose (1%) in 100 mM potassium hydrogen phosphate ($KHPO_4$) <!> (pH 7.0), melted and cooled to 60°C

SDS <!>-polyacrylamide gel

> For preparation of SDS-polyacrylamide gels used in the separation of proteins, see Appendix 8 in Sambrook and Russell 2001, *Molecular Cloning*.

Media

For components of yeast media, see Appendix 2 of Sambrook and Russell (2001)

CM selective media

> To estimate the amount of media required, use Table 3 (Sambrook and Russell 2001) and to prepare the necessary selective media, use Tables 18-8 and 18-9 (Sambrook and Russell 2001).

> Yeast nitrogen base without amino acids (YNB) is sold either with or without ammonium sulfate <!>. Table 18-8 (Sambrook and Russell 2001, *Molecular Cloning*) assumes that the YNB contains ammonium sulfate. If the bottle of yeast nitrogen base instructs that 1.7 g/liter be added to make the medium, then it does not contain ammonium sulfate, and 5 g of ammonium sulfate per liter of medium should be added.

YPD medium

> 20 g of peptone
>
> 10 g of yeast extract
>
> 20 g of glucose
>
> 20 g of agar (if for plates)
>
> > Add 1 liter of distilled H_2O and autoclave the medium for 20 minutes. Before pouring plates, cool the autoclaved medium to 55°C.

Yeast selective agar plates (for recipes, see Table 18-4 in Sambrook and Russell 2001)

> CM(Glu) –Ura –His
>
> CM(Glu) –Ura
>
> CM(Gal/Raf) –Ura (use the analogous recipe containing glucose, but substitute 20 g of galactose and 10 g of raffinose)
>
> CM(Gal/Raf) –Ura –His –Leu

Nucleic Acids and Oligonucleotides

Carrier DNA

> Sheared salmon sperm DNA is typically used as carrier. It is extremely important that this DNA be of very high quality; use of a poor-quality preparation can drop transformation frequencies one to two orders of magnitude. For a simple procedure for generating high-quality salmon sperm DNA, see Schiestl and Gietz (1989) and Sambrook and Russell 2001, *Molecular Cloning*; alternatively, a number of companies sell such preparations commercially.

Library to be screened for interaction

> Interaction trp libraries are available commercially from, e.g., CLONTECH (MatchMaker System), Invitrogen, OriGene, Display, MoBiTech (for a comprehensive listing of libraries from various species and tissue sources, see www.fccc.edu/research/labs/golemis/lib_sources.html).

Oligonucleotides primers (20 mM in TE [pH 8.0])

| Forward primer (FP1) | 5′-CTG AGT GGA GAT GCC TCC |
| Reverse primer (FP2) | 5′-CTG GCA AGG TAG ACA AGC CG |

Target DNA encoding the protein of interest (bait)

TE (pH 8.0)

Special Equipment

Culture plates (24 x 24 cm) for selective media (see Table 18-3, Sambrook and Russell 2001)
Falcon tubes (50 ml, sterile)
Glass beads (0.45-mm diameter, sterile; Sigma)
Glass beads (150–212 μM in TE [pH 8.0])
Heat block preset to 42°C
Inoculating manifold/frogger (e.g., Dankar Scientific, or cut in half Bel-Blotter, Bel-Art Products)

> A frogger for the transfer of multiple colonies can be purchased or easily homemade; it is important that all of the spokes have a flat surface and that the spoke ends are level. The frogger can be sterilized by autoclaving or by flaming in alcohol.

Microtiter plate (96-well)
Multichannel pipettor or inoculating manifold/frogger (e.g., Dankar Scientific)
Reagents for analyzing yeast colonies by PCR (see Sambrook and Russell 2001, Chapter 4, Protocol 13)
Reagents for purifying DNA fragments from low-melting-temperature agarose (see Sambrook and Russell 2001, Chapter 5, Protocol 6 or 7)
Repeating pipettor
Thermocycler
Thin-walled microfuge tubes (0.5 ml, sterile)

Vectors and Yeast Strains

S. cerevisiae strains for selection and propagation of vectors (see Table 18-6, Sambrook and Russell 2001)
Vectors carrying LexA (see Table 18-1, Sambrook and Russell 2001) and activation domain fusion sequences (see Table 18-2, Sambrook and Russell 2001) and LacZ reporter plasmids (see Table 18-5, Sambrook and Russell 2001)

METHOD

1. Using standard subcloning techniques, insert the DNA encoding the protein of interest into the polylinker of pMW103 (Fig. 4) or other LexA-fusion plasmid (Table 1) to make an in-frame protein fusion, incorporating a translational stop sequence at the carboxy-terminal end of the desired bait sequence.

> When deciding how to construct a bait, the following points are important. The assay depends on the ability of the bait to enter the nucleus, and it requires the bait to be a transcriptional nonactivator. Obvious sequences that confer attachment to membranes or sequences that are transcriptional activation domains should be removed from the chosen protein. It is not entirely clear whether or not the use of two-hybrid systems to find associating partners for proteins that are normally extracellular is a generally successful strategy; thus, the procedure should be regarded as extremely high-risk.

2. Transform (by scaling down ~1:30 transformation protocol in Stage B below) the SKY48 *lexAop-LEU2* selection strain of yeast using the following combinations of LexA-fusion and *lexAop-lacZ* plasmids (see Table 1):

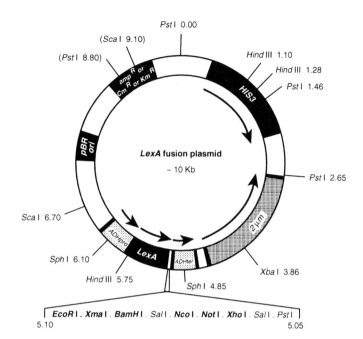

FIGURE 4. LexA fusion vector. The strong ADH promoter is used to express bait proteins as fusions to the DNA-binding protein LexA. A number of restriction sites are available for insertion of coding sequences: Those shown in bold type are unique. The reading frame for insertion is GAA TTC CCG GGG ATC CGT CGA CCA TGG CGG CCG CTC GAG TCG ACC TGC AGC. The sequence CGT CAG CAG AGC TTC ACC ATT G can be used to design a primer to confirm the correct reading frame for LexA fusions. The plasmid contains the *HIS3* selectable marker and the 2μ origin of replication to allow propagation in yeast, and an antibiotic-resistance gene and the pBR origin of replication to allow propagation in *E. coli*. Sequence data are available for pEG202; in pMW101 and pMW103, ApR was replaced with CmR and KmR, respectively (Watson et al. 1996). Only unique (*in bold*) and some selected sites are shown; sites in parentheses are specific for the ApR gene. Maps, sequences, and polylinkers for interaction trap-compatible plasmids are given on the Web site: http://www.fccc.edu/research/labs/golemis/InteractionTrapInWork.html. (Reprinted, with permission, from Golemis and Serebriiskii 1998.)

a. pBait + pMW109 (test for activation)

b. pEG202-hsRPB7+ pMW109 (weak positive control for activation)

c. pSH17-4 + pMW109 (strong positive control for activation)

d. pEG202– Ras + pMW109 (negative control for activation)

e. pBait + pJK101 (test for repression)

f. pEG202– Ras + pJK101 (positive control for repression)

g. pGKS3 + pJK101 (negative control for repression)

> PGKS3 can be substituted for any other yeast HIS3 plasmid not expressing LexA.
>
> A description of the function of each of the plasmids is provided in Table 1.

Use of the two LexA fusions (pEG202– Ras, pSH17-4) as strong positive and negative controls allows a rough assessment of the transcriptional activation profile of LexA bait proteins. pMW103 itself (or related plasmid pEG202) is not a good negative control because the peptide encoded by the uninterrupted polylinker sequence is itself capable of very weakly activating transcription.

pMW109 is a very sensitive *lacZ* reporter and will detect any potential ability to activate *lacZ* transcription. However, the *LEU2* reporter in SKY48 is even more sensitive than the pMW109 reporter for some baits, so it is possible that a bait protein that gives little or no signal in a β-galactosidase assay may nevertheless permit some level of growth on –Leu medium.

TABLE 1. Interaction Trap-compatible Two-hybrid System Plasmids and Strains

Plasmid name/source	Selection in yeast	in E.coli	Number of operators	Comment/Description
LexA fusion plasmids				
pEG202 (pMW101, 103)	HIS3	Ap^R		ADH1 promoter expresses LexA followed by polylinker; basic plasmids to clone bait as LexA-fusion. n.b.: E.coli marker for pMW101 is Cm^R, for pMW103, Km^R
pJK202	HIS3	Ap^R		pEG202 derivative, incorporating nuclear localization sequences between LexA and polylinker (enhanced ability to translocate bait to nucleus)
pNLexA	HIS3	Ap^R		polylinker is upstream of LexA; allows fusion of LexA to carboxyl terminus of bait, leaving amino-terminal residues of bait unblocked
pGilda	HIS3	Ap^R		GAL1 promoter expresses LexA followed by polylinker, for use with baits whose continuous presence is toxic to yeast
pEG202I	HIS3	Ap^R		pEG202 derivative, which can be integrated into yeast HIS3 gene after digestion with KpnI; ensures lower levels of bait expression
Reporter plasmids				
pMW111	URA3	Km^R	1 lexA	lexA operators direct transcription of the lacZ gene: sensitivity to transcriptional activation is a function of operator number
pMW109	URA3	Km^R	2 lexA	
pMW112	URA3	Km^R	8 lexA	
Activation domain fusion plasmids				
pJG4-5, pYesTrp2	TRP1	Ap^R		library or prey expression plasmids; GAL1 promoter provides efficient expression of a gene fused to a cassette consisting of nuclear localization sequence, transcriptional activation domain, and HA or V5 epitope tags

	Genotype			
LEU2/LYS2 selection strains				
SKY48 (MATα)			6 lexA	stringent selection for interaction partners of cI-fused baits; most
	trp1 ,his3, ura3,		3 cI	sensitive lexA-responsive LEU2 reporter
SKY191 (MATα)	lexAop-LEU2,		2 lexA	most stringent lexA-responsive LEU2 reporter; and more sensitive cI-
	cIop-LYS2		3 cI	responsive LYS2 reporter versus SKY48
SKY473 (MATa)			4 lexA	to be used as mating partner for SKY48 and SKY191 strains
			3 cI	

	Selection in yeast	in E.coli		
Control set of plasmids: Testing specificity				
pEG202-Ras	HIS3	Ap^R		expresses LexA-Ras fusion protein; use as negative control for activation assay, positive control in repression assay, and as test bait in interaction assay
pEG202-hsRPB7	HIS3	Ap^R		expresses LexA-hsRPB7 fusion protein; use as weak positive control for activation assay
pGKS3	HIS3	Ap^R		does not express any LexA-fusion protein; use as negative control for repression assay
pSH17-4	HIS3	Ap^R		expresses LexA-GAL4 fusion protein; use as strong positive control for activation assay
pJG4-5-Raf1	TRP1	Ap^R		Raf interacts with Ras; control in interaction assay

See text for details.

We note that SKY48 is described rather the EGY48 as host strain; this strain possesses additional features allowing the potential of adapting a two-hybrid screen to a two-bait two-hybrid screen, as described in Serebriiskii et al. (1999). Additional information on interaction trap-compatible reagents is provided on the Web site: http://www.fccc.edu/research/labs/golemis/InteractionTrapInWork.html.

3. Plate each transformation mixture on a Glu/CM –Ura, –His plate. Incubate 2–3 days at 30°C to select for yeast that contain plasmids.

> If colonies are not apparent within 3–4 days, or if only a very small number (less than 20) of colonies are obtained, results obtained in characterization experiments may not be typical, and it is suggested that transformation be repeated.

4. Use the replica plating technique (Fig. 5) to assess the phenotype of the yeast colonies. Typically, multiple independent colonies are assayed for each combination of plasmids. This is important, because for some baits, protein expression level is heterogeneous between independent colonies, with accompanying heterogeneity of apparent ability to activate transcription of the two reporters.

Earlier versions of these protocols have provided alternative means to assess *lacZ* activation (see, e.g., Golemis and Serebriiskii 1998; Golemis et al. 1997), which have included the use of toothpicks to streak colonies to plates with X-Gal incorporated in the medium. Although these approaches are certainly acceptable, we currently prefer the replicator-based overlay approach described here for a number of reasons. First, this approach is significantly faster than the other approaches, in terms both of transfer of colonies between plates and of time of color development in the assay (hours with an overlay versus 2–3 days on plates). Second, the assay is more sensitive, allowing the reliable detection of lower levels of β-galactosidase activity (Serebriiskii and Golemis 2000). Third, use of the overlay assay rather than growth on plates is less likely to result in the detection of false positives (Serebriiskii et al. 2000b). Finally, the use of an array format allows easy transition to automated scoring of β-galactosidase assays, for example, through the use of plate readers (Serebriiskii et al. 2000a).

5. Dispense 50–75 μl of sterile H$_2$O to each well of one-half (6 x 8) of a 96-well microtiter plate (e.g., using a syringe-based repeater). Place an insert grid from a rack of pipette tips (Rainin RT series, 200-μl capacity) over the top of the microtiter plate and attach it with tape: The holes in the insert grid should be placed exactly over the wells of the microtiter plate. Using sterile plastic pipette tips, pick 6 yeast colonies (1–2 mm diameter) from each of the transformation plates **a**–**g**, and insert tips into the water-filled wells through the holes in the insert grid. Tips will be supported in near-vertical position by the insert grid, so leave them there until all colonies have been picked. Swirl the plate gently to mix the yeast into suspension, then remove the sealing tape and lift the insert grid, thereby removing all of the tips at once.

6. Put a suitable replicator in the plate. If yeast has already sedimented, shake the replicator in a circular movement or vortex the whole plate at medium speed. Lift the replicator (which will now carry drops of liquid on its spokes), and put it on the surface of the solidified medium. Tilt it slightly in a circular movement, then lift the replicator and put it back in the plate. (Be sure to keep the correct orientation!) Check that all of the drops are left on the surface and are of approximately the same size. If only 1 or 2 drops are missing, it is easy to correct by dropping about 3 μl of yeast suspension on the missing spots from the corresponding wells. If many drops are missing, make sure that all of the spokes of the replicator are in good contact with liquid in the microtiter plate (it may be necessary to cut off the side protrusions on the edge spokes of the plastic replicator) and redo the whole plate. Continue replicating by shuttling back and forth between microtiter and medium plates, vortexing briefly as before between replications to prevent sedimentation.

7. Transfer yeast suspension on the following plates:

Glu/CM –Ura, –His	2
Gal/Raf/CM –Ura, –His	1
Gal/Raf/CM –Ura, –His, –Leu	1
Gal/Raf/CM –Ura, –His, –Trp	1

Let the liquid adsorb to the agar before putting the plates upside down in the incubator.

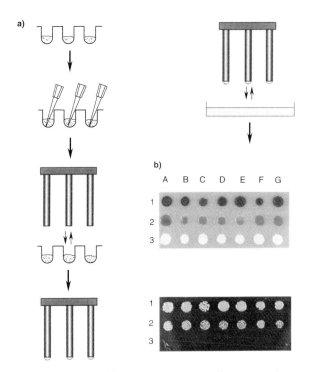

FIGURE 5. (a) Replica technique/ gridding yeast. From transformation plates, pick each yeast colony (1–2 mm diameter) to be tested, and resuspend it in 50–75 µl of sterile distilled H₂O in a well of a 96-well microtiter plate. If sterile toothpicks are used for picking colonies, they need to be removed immediately after resuspension of a colony, to prevent absorption of the liquid. Plastic pipette tips can be also used; placing an insert from a rack of pipette tips (e.g., Rainin RT series) over the top of the microtiter plate (not shown) helps to shake and remove all tips at once. A frogger for the transfer of multiple colonies can be purchased or easily homemade; it is important that all of the spokes have a flat surface, and that spoke ends are level. The frogger can be sterilized by autoclaving or by alcohol/flame sterilization. When making prints on a plate, put the frogger on the surface of the solidified medium and tilt slightly in a circular movement, then lift the frogger. Make sure the drops left on the surface are of approximately the same size. Let the liquid adsorb to the agar before putting the plates upside down in the incubator. In a subsequent transfer between the master plates, invert the frogger on the lab bench, and then place the master plate upside down on the spokes, making sure that a proper alignment of the spokes and the colonies is made. Lift the plate and insert the frogger into a microtiter plate with ~50 µl of sterile H₂O per well. Let the plate sit for 5–10 minutes, shaking from time to time to resuspend the cells left on the spokes. When all yeast are resuspended, print as described in the main text on the appropriate plates. Alternatively, rather than using extended shaking, it is possible to wash the frogger and flame-sterilize it before printing to ensure that none of the spokes retains excess yeast. (b) Typical results. Patches obtained after printing of the yeast suspension on Gal/Raf/CM –Ura –His –Trp –Leu (bottom) and CM/Gal/Raf/XGal –Ura (top) plates. (1A-1G) SKY48 plus strongly activating LexA fusion; (2A-2G) SKY48 plus moderately activating LexA fusion; (3A-3G) SKY48 plus nonactivating LexA fusion. (Reprinted, with permission, from Golemis and Serebriiskii 1998.)

8. Incubate at 30°C overnight (–Ura –His plates) before assaying for β-galactosidase activity, or for 3–4 days (–Ura, –His, –Leu and –Ura, –His, –Trp plates) monitoring for growth. Be sure to save one Glu/CM –Ura, –His plate (this will be the master plate). At 48 hours after plating to Gal/Raf/CM –Ura –His –Leu, **c** should be as well grown as it is on a Glu/CM –Ura –His master plate, whereas **a**, **b**, and **d–g** should show no growth. Ideally, **a** will still display no apparent growth at 96 hours after plating. An optional –Ura, –His, –Trp plate is a negative control for contamination: Nothing should grow there.

9. Assay β-galactosidase activity of the transformants by a chloroform overlay assay (adapted from Duttweiler 1996). To perform the assay, take one Gal/Raf/CM –Ura –His plate and one Glu/CM –Ura, –His plate with spotted yeast transformants.

a. Gently overlay each plate with chloroform, pipetting slowly in from the side so as not to smear colonies. Leave colonies completely covered for 5 minutes, and then remove the chloroform by decanting.

b. Gently rinse the plates with another ~5 ml of chloroform and let dry for another 5 minutes upside down in a chemical fume hood.

c. Prepare 1% low-melting agarose in 100 mM $KHPO_4$ (pH 7.0). For each plate, take ~7 ml of agarose cooled to ~60°C and add X-Gal to 0.25 mg/ml. Overlay the plate, making sure that all yeast spots are completely covered.

> Plates will be chilled after chloroform evaporation, so it will be difficult to spread less than 7 ml of top agarose.

d. Incubate at 30°C and monitor for color changes. It is generally useful to check the plates after 20 minutes, and again after 1–3 hours. For the activation assay, strong activators such as the LexA–GAL4 control (pSH17-4) will produce a blue color in 5–10 minutes, and a bait protein (LexA-fusion protein) that does so is likely to be unsuitable for use in an interactor hunt. Weak activators will produce a blue color in 1–6 hours (cf. negative control pEG202-Ras and weak positive control pEG202-hsRPB7). Any bait activating stronger than pEG202-hsRPB7 most likely will not be suitable.

The repression assay should be monitored within 1–2 hours if using the overlay assay, because the high basal LacZ activity will make differential activation of JK101 impossible to see with longer incubations. A good result (i.e., real repression) will generally reflect a two- to threefold reduction in the degree of blue color detected for JK101+ bait versus JK101 alone on plates containing galactose.

10. A table of expected results for a well-behaved bait is provided below (Table 2). In an optimal result, all six colonies assayed for each transformation would possess approximately equivalent phenotypes in activation and repression assays. For a small number of baits, this is not the case. The most typical deviation is that of eight colonies assayed for a new bait; some fractions are white on X-Gal and do not grow on –Leu medium, whereas the remaining fraction displays some degree of blueness and growth. Do not select the white, nongrowing colonies as a starting point in a library screen; generally, these colonies possess the phenotypes they do because they are synthesizing little or no bait protein (as can be assayed by western blot, see below). The reasons for this are not clear; however, it appears to be a bait-specific phenomenon and may be linked to some degree of toxicity of continued expression of particular proteins in yeast. It will be necessary to adjust sensitivity levels to allow work with blue, growing colonies (see below).

Note also that, for a small percentage of baits, the repression assay does not work, although the bait protein is clearly made to high levels, and there is no reason to believe it is not nuclear. In these cases, it is generally reasonable to go ahead with the library screen.

Detection of Bait Protein Expression

An important step in characterization of a bait protein is the direct assay of whether the bait is detectably expressed and whether the bait is of the correct size. In most cases, both of the above will be true; however, some proteins (especially where the fusion domain is ~60–80 kD or larger) will either be synthesized at low levels, or be posttranslationally clipped by yeast proteases. Either of these two outcomes can lead to problems in library screens. Proteins expressed at low levels, and apparently inactive in transcriptional activation assays, can be up-regulated to much higher levels under the leucine-deficient selection and suddenly demonstrate a high background of transcriptional activation. Where proteins are proteolytically clipped, screens might inadvertently be performed with LexA fused only to the amino-terminal end of the larger intended bait. To antic-

TABLE 2. Bait Characterization: Expected Results

| | Plasmids to transform | | Anticipated results | | |
| | | | X-Gal plates | | Growth Leu– |
Assay	reporter	LexA-fusion	Glu	Gal/Raf	Gal/Raf
Activation					
test	pMW112	pBait	white/light blue	white/light blue	(no)
negative control	pMW112	pEG202- Ras	white	white	no
weak positive control	pMW112	pEG202-hsRPB7	light blue	bluish	no
strong positive control	pMW112	pSH17-4	dark blue	blue	yes
Repression					
test	pJK101	pBait	(white)	(lighter blue)	
negative control	pJK101	— (pGKS3*)	white	blue	no
positive control	pJK101	pEG202- Ras	white	lighter blue	

			Test by western blot	
Expression				
test	use clones from	pBait	(single band)	
positive control 1	activation	pEG202- Ras	single band	
positive control 2	assay	pSH17-4	single band	

Adapted from Table 18-7, Sambrook and Russell 2001, *Molecular Cloning*.

ipate and forestall these potential problems, western analysis of lysates of yeast containing LexA-fused baits is helpful (see Table 2).

1. Inoculate at least two primary transformants for each novel bait construct assayed, and for a positive control for protein expression such as pEG202-Ras and/or pSH17-4. Use a sterile toothpick to pick colonies from the Glu/CM –Ura –His master plate into Glu/CM –Ura –His liquid medium.

   If wearing gloves while holding the toothpick, it can generally be dropped into the culture tube and left there without contamination.

2. Grow overnight cultures on a roller drum or other shaker at 30°C. In the morning, dilute saturated cultures into fresh tubes containing ~2.5 ml of Glu/CM –Ura –His, with a starting density of OD_{600} ~0.2 and grow again as before.

   Mark which colonies are being assayed for protein expression on the master plate, and use the same colony that has been assayed and shown to express bait appropriately as the founder to grow up for library introduction.

3. When the culture has reached OD_{600} ~0.45–0.7 (~4–6 hours), remove 1.5 ml to an Eppendorf tube and spin cells in a microfuge at full speed for 3–5 minutes. When pellet is visible, decant or aspirate the supernatant.

   For some LexA-fusion proteins, the levels of the protein drop off rapidly in cultures approaching stationary phase. This is caused by a combination of the diminishing activity of the *ADH1* promoter in late growth phases and the relative instability of particular fusion domains. Thus, it is not necessarily a good idea to let cultures become saturated in the hopes of getting a higher yield of protein to assay (although for some proteins, this does work). It may be helpful to freeze duplicate samples at this stage, if more than one round of assay is anticipated (and in case of accidents).

4. Working rapidly, add 50 µl of 2x Laemmli sample buffer to the tube, vortex, and place the tube on dry ice. Samples may be frozen at this stage at –70°C, and are stable for extended periods (at least 4–6 months).

5. When ready to run a polyacrylamide gel (SDS-PAGE) prior to western analysis, remove samples from freezer directly to a boiling water bath or to a heating block or PCR machine set to 100°C. Boil for 5 minutes.

6. Chill on ice and centrifuge for 5–30 seconds in a microfuge to pellet large cell debris. Use ~20–50 µl to load gel lanes.

7. Perform PAGE, blotting, and western analysis by standard protocols (Harlow and Lane 1988; Sambrook et al. 1989; Sambrook and Russell 2001, *Molecular Cloning*). LexA fusions can be detected using antibody to the fusion domain, if available; alternatively, use antibody to LexA (commercially available from CLONTECH and Invitrogen). It is preferable to use antibodies to LexA because this allows comparison of expression levels of the bait protein under test with other standard bait proteins, e.g., LexA–Ras.

8. Note which colonies on the master plate express bait appropriately, and use one of these colonies as the founder to grow up for library introduction.

    At this time, all of the alternative bait expression vectors carry the *bla* (Amp^R) gene and thus should be selected on ampicillin when transformed in bacteria. Hence, with these vectors, the investigator may need to use a KC8 bacterial passage to facilitate isolation of the library plasmid following a library screen.

TROUBLESHOOTING AND MODIFICATIONS OF BAITS

There are three basic problems that can be identified and potentially corrected before screening.

1. *The bait activates transcription.* This problem can be addressed in several ways. One is to make a series of truncations of the protein, in an attempt to eliminate the activation domain. If the bait activates transcription very strongly, that is, as well as the positive control, this step will be necessary.

    If the bait activates moderately, the simplest approach is to re-do the control experiments using less stringent reporter plasmids and strain (see Tables 2 and 3) and to see whether activation becomes minor. Alternatively, the protein can be truncated, as for a strong activator. Finally, it is possible to use an integrating form of bait vector (see Table 2), which will result in a stable reduction of protein levels.

    If the bait is a weak activator, one option is to use less stringent reporter plasmids; alternatively, the investigator may choose to proceed with the tested reagents, assuming that a small background of false positives may be identified. In general, it is a good idea to use the most sensitive screening conditions possible; in some cases, use of very stringent interaction strains eliminates detection of biologically relevant interactions (Estojak et al. 1995).

2. *The bait plasmid produces an inappropriate level or size of protein.* If the protein being poorly expressed is very large, it may be simplest to subdivide into two or three overlapping constructs, each of which can be tested independently. Alternatively, the vector pJK202 incorporates a nuclear localization sequence, and will increase the general concentration of the bait in the nucleus, where it is required for the assay. Such an approach may make the best of a situation where a protein cannot readily be expressed at higher levels.

3. *Very few transformants containing the bait plasmid express the bait protein, or yeast expressing the bait protein grow noticeably more poorly than control yeast in the absence of any selection.* These results would suggest that the bait protein is somewhat toxic to the yeast. Because toxicity can cause difficulties in performing library screening, it may be desirable to reclone the protein of interest into pGilda, which allows inducible expression of the protein from the GAL1 promoter, thus limiting time of expression of the protein to actual selection. Note, however, that tests with a pGilda construct should be performed on medium containing galactose as a carbon source.

Suggested modifications for enhancing bait performance are summarized below (Table 3).

TABLE 3. Possible Modifications to Enhance Bait Performance in Specific Applications

Response \ Bait Problem	Strongly activating	Weakly activating	Not transported to the nucleus, or low expression level	Continuous expression of LexA-fusion is toxic to yeast	Bait protein requires unblocked amino-terminal end for function	Bait protein expressed at high levels, unstable, or interacts promiscuously	Potential new problem[a]
Truncate / modify bait	+	–	–	–	–	–	It may be necessary to subdivide bait into two or three overlapping constructs, each of which must be tested independently.
Use more stringent strain/reporter combination	+	+	–	–	+?	+?	Use of very stringent interaction strains may eliminate detection of biologically relevant interactions.
Fuse to nuclear localization sequence pJK202	–	–	+	–	–	–	
Put LexA-fused protein under GAL1-inducible promoter pGilda	+?	–	–	+	+?	+?	Can no longer use GAL-dependence of reporter phenotype to indicate cDNA-dependent interaction.
Fuse LexA to the carboxyl terminus of the bait pNLexA	–	–	–	–	+	–	Generally, LexA poorly tolerates attachment of the amino-terminal fusion domain; only ~60% of constructs are expressed correctly.
Integrate bait, reduce concentration pEG202I	+?	–	–	+?	+?	+	Reduced Bait protein concentration may lead to reduced assay sensitivity.

(+) Would usually help; (+?) may help; (–) will not help.

[a]All of the alternative bait expression vectors remain on an Amp^R selection for bacteria. If using them as is, the investigator may need to use a KC8 bacterial passage to isolate the library plasmid after a library screen.

Transforming and Characterizing the Library

A list of some of the libraries currently available for use with this system is found at http://www.fccc.edu/research/labs/golemis/InteractionTrapInWork.html. Currently, the most convenient source of libraries suitable for the interaction trap is commercial, and can be viewed at the Web sites of the companies supplying these reagents. The protocol outlined below describes the steps used to perform a screen that should saturate a cDNA library derived from a genome of mammalian complexity. Fewer plates will be required for screens with libraries derived from lower eukaryotes with less complex genomes, and researchers should scale back accordingly.

Because a useful control in subsequent steps will involve the calculation of false positive frequencies (see below), in parallel to the large-scale library transformation, it is helpful to perform a small-scale transformation of the mating strain SKY473 with the library vector only.

> Currently, we recommend mating in the library against the bait of interest (see Finley and Brent 1994) as the most convenient strategy. The main advantage of this approach is that if the investigator wishes to use the same library to screen multiple baits, only a single large-scale transformation is required, followed by relatively easy mating steps. Alternatively, with minor modifications, this protocol can be used to directly transform the library into yeast containing the bait, when only a single bait is to be screened. In the latter case, the library transformation should be done as part of Stage II below, replacing the mating.

1. Select a colony of an appropriate mating partner yeast strain (such as SKY473) and grow an ~20-ml culture in liquid YPD medium overnight at 30°C on a shaker.

   It is important to use a fresh (thawed from –70°C and streaked to a single colony less than ~7 days previously) colony and to maintain sterile conditions throughout all subsequent procedures.

2. Dilute the ~20-ml overnight culture into approximately 300 ml of Glu/CM –Ura, –His liquid dropout medium such that the diluted culture has an OD_{600} ~0.15. Incubate at 30°C on an orbital shaker until the culture has reached OD_{600} ~0.50–0.7.

3. Transfer the culture to sterile 6 x 50-ml Falcon tubes, and spin at 1000–1500g for 5 minutes at room temperature. Gently resuspend the pellets in ~5 ml of sterile H_2O each, and combine all slurries in one of the Falcon tubes. Add sterile H_2O to the top of the tube and mix.

4. Respin cells again at 1000–1500g for 5 minutes. Pour off H_2O and resuspend yeast in 1.5 ml of TE buffer/0.1 M lithium acetate.

   In the following protocol, it is generally a good idea to mate new bait strains additionally with a control strain. The control strain is the same strain used for the library (e.g., SKY473) but containing the library vector with no cDNA insert. Use an aliquot of competent yeast from step 4 to transform the empty library plasmid (pJG4-5 or pYesTrp2). Plate on a 100-mm plate, and collect the transformed cells as for the library (protocol outlined below). This control strain can be safely reamplified in liquid medium.

5. Mix 30 µg of library DNA and 1.5 mg of freshly denatured carrier DNA in an Eppendorf tube, and then add the DNA mix to the yeast. Mix gently and aliquot DNA/yeast suspension into 30 microfuge tubes (~60 µl each).

6. To each tube, add 300 µl of sterile 40% PEG 4000, 0.1 M lithium acetate, and TE buffer (pH 7.5). Mix by gently inverting the tubes a number of times (do not vortex). Place the tubes for 30–60 minutes at 30°C.

7. To each tube, add ~40 µl of DMSO, and again mix by inversion. Place the tubes in a heat block set to 42°C for 10 minutes.

8. Pipette contents of each tube onto a 24 x 24-cm Glu –Trp dropout plate. Spread the cells evenly using one to two dozen sterile glass beads, invert the plates, and incubate at 30°C until colonies appear (usually 2–3 days). Although it is possible to throw away the beads after

spreading, it is acceptable and efficient to keep the glass beads on the lids while incubating the plates; glass beads will be needed to harvest the library transformants (see below).

9. Select a few representative transformation plates, draw a 23 x 23-mm square (1% of plate bottom surface) over an average density spot, count colonies in each grid section, and recalculate for the whole transformation. A good transformation done by this protocol should yield ~20,000–40,000 colonies per large plate and represent ~10^5 transformants per microgram of library DNA.

Doing transformations in small aliquots reduces the likelihood of contamination and, for reasons that are not clear, provides significantly better transformation efficiency than scaled-up versions. Do not use excess transforming library DNA per aliquot of competent yeast cells, or each competent cell may take up multiple library plasmids, complicating subsequent analysis. Under the conditions described here, less than 10% of yeast will contain two or more library plasmids.

If the library is directly transformed against the bait, the only modifications of the protocol above are:

a. Use the characterized bait strain instead of native yeast.

b. In steps 1 and 2, grow it in Glu/–Ura –His liquid medium instead of YPD.

c. In step 8, use Glu/–Ura –His –Trp plates rather than Glu/–Trp.

Collect Primary Transformant Cells

The procedure described below creates a slurry in which cells derived from primary transformants are homogeneously dispersed. A precalculated number of these cells are used to perform mating with each bait strain.

1. Inspect all transformation plates. If visible molds or other contaminants (colonies of unusual color, shape, etc.) are observed on the plates, carefully excise them and a region around them using a sterile razor blade prior to harvesting library transformants.

2. Pour 10 ml of sterile H_2O on each of five 24 x 24-cm plates containing transformants. If needed, pour sterile glass beads. Stack the five plates on top of each other. Holding tightly, shake the stack until all colonies are resuspended (1–2 minutes). Using a sterile pipette, collect 5 ml of yeast slurry from each plate (tilt the plate) and pool into sterile 50-ml conical tubes. Proceed to the next five plates. The final result of this step should be a total volume of up to 150 ml of liquid spread across three 50-ml tubes.

> The same glass beads can be transferred to the fresh plates or new beads can be used. More than five plates can be washed off at once (as many as can be held in the hands and shaken). Music with a good beat is helpful!

3. Fill each conical tube containing the yeast to the top with sterile TE buffer or H_2O, and vortex or invert to suspend the cells. Spin the tube in a low-speed centrifuge at 1000–1500g for ~5 minutes at room temperature, and discard the supernatant. Wash the cells by adding sterile TE buffer or H_2O. Resuspend the cells and centrifuge at 1000–1500g for ~5 minutes at room temperature, and then discard the supernatant. Repeat the wash.

> After the second wash, the cumulative pellet volume should be ~25 ml of cells derived from 1.5 x 10^6 transformants.

4. Resuspend the pellet in 1 volume of glycerol solution. Combine the contents of the different Falcon tubes and mix well. Freeze in 1-ml aliquots at –70°C (These aliquots are stable for at least 1 year.) If proceeding directly with mating, leave one aliquot unfrozen and assume that viability of the culture is 100%.

> If the library was directly transformed against the bait, leave one aliquot unfrozen and proceed directly to Stage 2, Selecting an Interactor.

Determine Replating Efficiency

In general, the viability for yeast frozen for less than 1 year will be >90%. However, refreezing a thawed aliquot results in loss of viability; therefore, many frozen aliquots (0.2–1.0 ml) should be made (especially if interaction mating is intended). A series of limiting dilutions on Glu/CM –Trp (or Glu/CM –Trp –His –Ura for a direct library transformation) should be performed to determine survival rate, expressed as number of colony-forming units per milliliter (cfu/ml).

Stage 2: Selecting an Interactor

Mating the Bait Strain and the Pretransformed Library

Once the bait strain has been made and characterized and the library strain has been transformed and frozen in aliquots, the next step is to mate the two strains. To mate the two strains, the bait strain is grown in liquid culture and then mixed with a thawed aliquot of the pretransformed library strain. The mixture is then plated on YPD and grown overnight. During this time, individual cells of the bait strain fuse with individual cells of the library strain to form diploid cells. The mixture of diploids and unmated haploids is then collected and plated on media to select for interactors. In practice, the diploid/haploid mixture is generally frozen in a few aliquots to allow titering and repeated platings at various dilutions.

It is generally a good idea to mate new bait strains with a control strain. The control strain is the same strain used for the library, but containing the library vector with no cDNA insert. Mating with the control strain can be performed at the same time as the library mating, and both matings can be treated identically in the next step, selecting interactors. This control will provide a clear estimate of the frequency of cDNA-independent false positives, a frequency that is important to know when deciding how many positives to pick and characterize.

1. Start a 30-ml Glu/CM –Ura –His liquid culture of the bait strain from the Glu/CM –Ura –His master plate. Grow with shaking at 30°C to late-log phase (OD_{600} = 1.0–2.0).

2. Collect the cells by centrifugation at 1000g for 5 minutes. All manipulations can be performed at room temperature. Resuspend the cell pellet in 1 ml of sterile H_2O and transfer to a sterile 1.5-ml microfuge tube. Measure the optical density of 1:100 diluted aliquot to ensure that the OD_{600} of the undiluted suspension is 30–50. This will correspond to about 1×10^9 cfu/ml.

3. Thaw an aliquot of the pretransformed library strain at room temperature. Mix 200 µl of the bait strain with ~10^8 cells of the pretransformed library strain. In parallel, mix the bait strain with a similar number of cells containing library vector only. Process separately: Obviously, fewer plates need to be screened, as the goal is not to obtain library representation.

4. Centrifuge the mixture of cells at 1000g for 5 minutes and discard the supernatant. Resuspend the cell pellet in 200 µl of YPD medium. Plate on a single 100-mm-diameter YPD plate and incubate for 12–15 hours at 30°C.

5. Add 1–2 ml of sterile H_2O to the surface of the YPD plate and resuspend the cells while shaking with sterile glass beads. Transfer to sterile collection tubes and vortex gently for 2 minutes. Collect the cells by centrifugation at 1000g for 5 minutes and resuspend in 1 volume of sterile glycerol solution. Distribute into 200-µl aliquots and freeze at –80°C. As with the frozen pretransformed library strain, the mated yeast should not be thawed and refrozen. Only one or a few of the aliquots will be needed to represent the library; therefore, thawed aliquots can be discarded after use.

6. Titer the mated cells by thawing an aliquot and plating serial dilutions to Glu/CM –Trp –His –Ura plates. Unmated haploids will not grow on this medium. Count the colonies that grow after 2–3 days and determine the plating efficiency of the frozen mated cells.

This count will basically show how successful your mating was and how many plates you would have to use to get full representation of the library. For example, imagine you have obtained 1×10^6 clones in your library transformation. It is desirable that for actual selection, each primary colony obtained from the transformation be represented on the selection plate by 3–10 individual yeast cells (see explanation below). Therefore, you have to plate a total of about 5×10^6 diploid cells on selection plates. If you have determined that the titer of diploids is 5×10^7 cfu/ml, you will have to plate about 100 μl of your slurry. However, the total number of cells you plate will be much higher, because the efficiency of mating typically is 1–10%. An approximate titer can be recalculated from the OD reading. Realistically, the same 100 μl of your slurry, which contains 5×10^6 cfu/ml of diploids, would contain about 1×10^8 cells. To avoid cross-feeding, put only around 10^6 cells on one 90-mm plate. Therefore, to screen the library exhaustively, you would have to plate about 50 plates. (See the note to the next protocol.)

We recommend plating 3–10 times more diploids than the number of colonies obtained from the transformation, because not all cells that contain interacting proteins plate at 100% efficiency on –Leu medium (Estojak et al. 1995) and because the created slurry may not be completely homogeneous. This will in some cases lead to multiple isolations of the same cDNA; however, it will increase the likelihood that all primary transformants are represented by at least one cell on the selective plate, and, in fact, resolution of an identical cDNA among a relatively small set of "positives" can be taken as one sign of a specific interaction.

Screen for Interacting Proteins

In the next step, interactors will be selected by plating the mated cells onto selection plates lacking leucine. It will be important to know how many viable diploids were plated to these selection plates to gain a sense of how much of the library has been screened and to determine the false-positive frequency (see discussion in the following protocol). Step 6 above will determine the number of viable diploids per unit volume of frozen mated cells, expressed as cfu/ml. To save time, the titering can be done at the same time as the selection for interactors in the protocol below. Because the titer of diploids will likely be within an order of magnitude of 10^8 cfu/ml, you can plate two or three different dilutions onto the selection plates to select for interactors, while at the same time plating dilutions to determine the exact titer as in step 6 above.

1. Thaw one aliquot of the mated yeast (or cells containing pBait that were transformed directly with the library DNA) and dilute 1:100 with Gal-Raff/CM –Ura, –His, –Trp dropout medium. Incubate with shaking 4–6 hours at 30°C to induce the GAL1 promoter on the library. If the frozen culture was not previously titered, plate serial dilutions now onto Glu/CM –Ura –His –Trp plates. Process mated yeast containing bait and library vector only in parallel.

2. Plate ~1×10^6 cells (or 50 μl of a culture at $OD_{600} = 1.0$) on each of ten 100-mm Gal/Raff/CM –Ura, –His, –Trp, –Leu dropout plates. For the bait/library vector mating, use only 1 or 2 plates.

The value of ~10^6 cells per plate has not been randomly chosen. It is easiest to visually scan for Leu$^+$ colonies using cells plated at ~10^6 cfu per 100-mm plate. Plating at much higher density may contribute to cross-feeding between yeast, resulting in spurious background growth. However, plating at higher densities also greatly reduces the number of plates to pour and inoculate. Depending on the bait background, up to 10^7 cfu per 90-mm plate may be used. Therefore, we suggest initially plating mated yeast at two different cell densities and comparing the results.

Plate ~5×10^6 cells on each of 10 additional 100-mm plates.

3. Incubate the plates 5 days at 30°C. Good candidates for positive interactors will generally produce colonies over this time period, depending on the individual bait used. The most common appearance of colonies should be at 2–4 days after plating.

4. Compare the selection plates seeded with lower and higher densities. The number of colonies should be roughly proportional to the seeding density.

> Success is no background growth on the more densely seeded plates. Disproportionally more colonies seen on the more densely seeded plates (especially sitting on the thin "lawn") is probably background due to cross-feeding. In this case, take another aliquot of frozen mating, repeat induction, and plate at 1×10^6 cells per plate on as many plates as are necessary for full representation of the calculated number of diploids.

> It is best to observe the plates daily, but not necessarily to pick colonies immediately. Instead, the first-day colonies are visible by eye; mark their location on the plate with a dot of a given color from a lab marker (for example, Day 3 = red). Each day, mark further colonies arising with distinctive colors (Day 4 = blue, etc).

5. At day 5, create a master plate (Glu/CM –Ura –His –Trp) in which colonies are grouped by day of appearance.

> If a large number of apparent positives are present, it might be necessary to have separate plates with colonies obtained on day 2, on day 3, and on day 4. It is appropriate to generate a negative control for subsequent steps, picking at random three to five colonies from the viability determination plates (above) and streaking them in parallel on the master plates to be tested. If a manifold or frogger is used for subsequent assay steps (see below), make sure the streaks are gridded in such a way that colonies will correspond to the spokes on a frogger (Fig. 5).

> If no colonies appear within a week, those arising later are likely to be artifactual. In tests of master plates (below), true interactors tend to appear at a time specific for a given bait, with false positives clustering at a different time point. Hence, pregrouping by date of growth facilitates the decision of which clones to analyze first. If using a replica technique, it is worthwhile to make an extra Glu master plate (see below). The number of Leu$^+$ colonies to pick and characterize should be based on the number of cDNA-independent false positives that arise on the Leu$^+$ plates for the control mating with empty library vector. The higher the frequency of false positives, the more Leu$^+$ colonies should be picked to find rare true positives. Because the frequency of true positives will be unknown at this step, the goal is to pick through all of the false positives that are expected in the number of library transformants being screened. For example, if the number of library transformants was 10^6, and more than this was mated to ensure that all 10^6 are represented in diploids, the goal will be to pick through the number of false positives expected in 10^6 diploids. If the cDNA-independent false-positive frequency is 1 Leu$^+$ colony in 10^4 cfu plated, it will be necessary to pick at least 100 Leu$^+$ colonies to find a true positive that exists at a frequency of 1 in 10^6 cfu plated.

> If contamination has occurred at an earlier step, this is generally reflected by the growth of a very large number of colonies (more than 500/plate) within 24–48 hours after plating on selective medium. Compare the smell of the plates and morphology of the colonies to that on the titer plates (dissecting microscope can be helpful). In the latter case, if contamination can be confirmed to be bacterial (rather than fungal), one option is to recreate the selective plates adding tetracycline (15 μg/ml), and repeat library induction and plating.

6. Incubate at 30°C until patches or colonies form.

First Confirmation of Positive Interactions

The following steps test for galactose-dependent transcriptional activation of both the *lexAop-LEU2* and *lexAop-lacZ* reporters. Simultaneous activation of both reporters in a galactose-specific manner generally indicates the transcriptional phenotype is attributable to expression of library-encoded proteins, rather than derived from mutation of the yeast. The test for β-galactosidase activity and for Leu requirement is described above in Stage 1, Constructing and Characterizing a Bait Protein (p. 95).

TABLE 4. Interpretation of the Phenotype of Primary Candidates

Phenotype				Explanation		Suggestion
−Leu growth		X-Gal color				
Glu	Gal	Glu	Gal	traditional	optimistic	
−	+	−	+	very good sign		work with those clones first
(+)	+	(+)	+	bait is up-regulated/mutated to a high background of transcriptional activation	• GAL1 promoter is slightly leaky • both proteins are very stable • interaction occurs with high affinity	take one clone for confirmation of interaction; store the rest.
−	+	−	−	yeast mutation occurred that favors growth or transcriptional activation on galactose medium	some bait–interactor combinations are known to activate *lacZ* versus *LEU2* preferentially, or vice versa	if all other candidates winnow out, check these clones
All other phenotypes				contamination/plasmid rearrangements/mutations	something really new	trash

See text for details.

1. Invert the replicator or frogger on the lab bench. Then place the master plate upside down on the spokes, making sure that a proper alignment of the spokes and the colonies is made. Lift the plate and insert the frogger into a microtiter plate that already contains 50–75 µl of sterile H_2O per well. Let the apparatus sit for 5–10 minutes, shaking from time to time to resuspend the cells left on the spokes. When all yeast are resuspended, print (as described in the Bait Characterization protocol) on the plates listed below.

   Transfer yeast from the microtiter plate to each of the following plates:

Glu/CM −Ura −His −Trp	2
Gal/Raff//CM −Ura −His −Trp	1
Glu/CM −Ura −His −Trp −Leu	1
Gal/Raff/CM −Ura −His −Trp−Leu	1

2. Incubate for 3–4 days at 30°C. After 1 day of incubation, take out all three −Ura −His −Trp plates. One of the Glu/CM −Ura −His −Trp can be kept as a fresh master plate, while the second one should be used along with Gal/Raf/CM −Ura −His −Trp plate for β-galactosidase assay. Continue to monitor growth. Differential growth on leucine will generally be apparent between 48 and 72 hours on the −Ura −His −Trp −Leu plates.

3. Interpretation of the phenotype of primary candidates is given in Table 4.

At this point, colonies and the library plasmids they contain are tentatively considered positive if X-Gal analysis indicates a blue color following culture on Gal/Raf/CM plates, but white, or only faintly blue, following culture on Glu/CM plates, and if they grow on Gal/Raff/CM −Ura −His −Trp −Leu but not on Glu/CM −Ura −His −Trp −Leu plates.

In very rare cases, if an interaction occurs with high affinity, and both proteins are stable in yeast, a weak enhancement of growth and X-Gal activity will occur on glucose as well as galactose medium. This is because the very low levels of transcription from GAL1 promoter on glucose medium allow accumulation of sufficient levels of strongly interacting, stable AD fusions for an interaction to be detected.

In some cases, strong activation of one of the two *lexAop* reporters will be observed, but only poor activation of the other. Some bait–interactor combinations clearly and preferentially activate *lacZ* versus *LEU2*, or vice versa (Estojak et al. 1995). In assessing whether these are likely to be real, we note that, in general, the confidence level in such interactors is substantially lower than in interactors that strongly activate both reporters (particularly in cases where strong growth is observed on medium lacking leucine, but virtually no blue color is observed with X-Gal). Whether these are pursued is a decision for the investigator.

Isolation of Library Plasmid Insert, and Second Confirmation of Positive Interactions

How many positives are obtained will vary drastically from bait to bait. How they are processed subsequently will depend on the number initially obtained and on the preference of the individual investigator. If a reasonably small number (1–48) are obtained, the steps detailed below can be followed essentially as written. However, sometimes searches will yield large numbers of colonies (50–300 or more). In this case, there are various options.

The first option is to "warehouse" the majority of the positives and work up the first 24–48 arising, following the observation that those growing fastest are frequently the strongest interactors. These can be checked for specificity of interaction; that is, for their ability to bind unrelated baits in addition to the original one. At the same time, restriction digests can be used to establish whether they are all independent cDNAs or represent multiple isolates of a limited number of cDNAs. If the former is true, it may be advisable to repeat the screen in a less sensitive strain/plasmid background, because obtaining many different interactors can be a sign of a low-affinity, nonspecific background. The second major option is to work up large numbers of positives to get a complete profile of isolated interactors. The approach taken depends on the goals and resources of the investigator.

The protocol described below is useful when both PCR and automated sequencing are readily available and affordable. Under these conditions, the protocol saves time by minimizing the number of library isolates that must be subjected to specificity testing (thus reducing numbers of both *Escherichia coli* and yeast transformations). This protocol has been successfully used in many labs and is gradually replacing an older approach, which emphasizes thorough characterization of specificity of bait–library protein interaction prior to sequencing.

Finally, in some cases, no positives are obtained for a given bait. There may be several reasons for this. One is that the library source was inappropriate or that an insufficient number of colonies were screened. A second reason is that the bait does not, in fact, interact with any single partner protein with sufficient affinity to be detected (Estojak et al. 1995). A third reason is that the bait is not in native conformation, either because of a disfavored truncation or because of steric interactions with the fusion domain, or that the bait is not interacting well with the *lexA* operator because some moiety on the fusion protein is sequestering the protein (Golemis and Brent 1992). Sometimes it is possible to modify bait and screening conditions to obtain interactors; however, some baits simply do not appear to generate interactors. The investigator must decide how much effort to apply.

1. Starting from a Glu/CM –Ura –His –Trp master plate, use a multicolony replicator/frogger to resuspend yeast in 25 µl of β-glucuronidase solution in a 96-well microtiter plate (same technique as in First Confirmation of Positive Interactions, step 1). Seal the wells using tape, and incubate on a horizontal shaker for 1.5–3.5 hours at 37°C.

   Alternatively, pipette tips or toothpicks can be used to transfer yeast in the wells (approximately the volume of one middle-sized yeast colony, or 2–3 µl of packed pellet). Do not take more, or the quality of isolated DNA will suffer. The master plate does not have to be absolutely fresh: Plates that have been stored for 5 days at 4°C have been used successfully. Zymolase has also worked very well at this step as a substitute for β-glucuronidase.

2. Remove the tape, and add about 25 µl of glass beads (150–212 µm; e.g., Sigma G-1145) to each well, and seal again. Attach the microtiter plate to a vortex with a flat-topped surface (e.g., using rubber bands) and mix vigorously for 5 minutes.

3. Add 100 µl of sterile distilled H_2O to each well. Take 1 –2 µl as a template for each PCR. Reseal the plate with tape, and keep the remainder frozen at –70°C.

   In many cases, PCR product can be obtained directly from the yeast colonies even without β-glucuronidase treatment (e.g., by introducing a 10-minute, 94°C step at the beginning of the PCR program). However, the crude yeast lysate obtained in this protocol (step 3) also can be used as a source of plasmid DNA (use 1–5 µl for electroporation into *E. coli*), instead of the more time-consuming protocols (making it worth doing the more thorough preparation).

4. Perform a PCR amplification in ~30 μl volume using the following program:

2 minutes at 94°C
45 seconds at 94°C ⎤
45 seconds at 56°C ⎬ 35 cycles
45 seconds at 72°C ⎦

Use primers specific for the library plasmid used.

For the JG4-5 plasmid:

Forward primer (FP1) 5′-CTG AGT GGA GAT GCC TCC

Reverse primer (FP2) 5′ CTG GCA AGG TAG ACA AGC CG

Modified versions of this protocol with extended elongation times were also found to work; however, the variant given above has amplified fragments of as much as 1.8 kb in pretty fair quantity. Primers for pYesTrp2 plasmid are available from Invitrogen.

To interpret the results of the PCR, it is helpful to have the following control templates:

a. Empty library plasmid (about 0.1 ng).

b. Yeast from the positive control colonies, treated along with experimental clones as above.

c. Same amounts of library plasmid and positive control yeast, mixed together.

Interpretation of possible PCR outcomes is summarized in Table 5.

5. Take 10 μl of the PCR product for the HaeIII digestion (below), and run out the remainder of the PCR (about 20 μl) on a 0.7% agarose gel. Identify fragments that appear to be of the same size; HaeIII digests of these fragments should be run side by side. Place the gel in a refrigerator until you are ready to isolate fragments.

6. Perform a restriction digest of 10 μl of the PCR product with HaeIII in a total volume of 20 μl. To group apparently similar clones, rearrange the loading order according to the results

TABLE 5. Interpretation of PCR Outcome

Template			Possible outcomes			
Plasmid prep[a]	Control yeast[b]	Test clones				
√			−	+	+	+
	√		−	−	−	+/−
√	√		−	+	−	+/−
		√	−	−	−	+/−
Interpretation			bad mastermix/ wrong settings/ faulty amplifier	not enough template	lysed yeast inhibited PCR	uneven template load/digestion
Recommendation			double-check/ repeat	add more template/ improve lysis	add less template	adjust template load/re-PCR from obtained bands

Double bands may NOT be observed on the PCR done on the mixture of the plasmid prep with yeast, because smaller product from the empty vector may out-compete the bigger one from the plasmid.

[a]JG4-5–Raf1 is a suitable template; be sure to use as control library plasmid (with the insert) of the same type as you have in our library.

[b]PCR from the empty vector yields a product of ~130 bp for JG4-5 (FP1 and FP2 primers); ~185 bp for YesTrp (YesTrp forward and reverse primers).

obtained with nondigested PCR, and load the digestion products on a 1.5% agarose gel. Run out the DNA a sufficient distance to get good resolution of DNA products in the 200- to 1000-bp size range.

> This will generally yield distinctive and unambiguous groups of inserts, confirming whether multiple isolates of a small number of cDNAs have been obtained. As mentioned above, rarely a single yeast will contain two or more different library plasmids. If this happens, it will be immediately revealed by PCR; therefore, after bacterial transformation, an increased number of clones should be checked to avoid the loss of the "real" interactor.

7. Purify undigested fragments from the agarose gel using standard molecular biology techniques. In cases where a very large number of isolates of a small number of cDNAs have been obtained, the investigator may choose to sequence the PCR product directly.

> The PCR product obtained in this step will be used not only for sequencing, but also for second confirmation of interaction (see protocol below). To get enough PCR products, reamplify if necessary.

> Only the forward primer, FP1, works reliably in sequencing of PCR fragments; the reverse primer will only work in sequencing from purified plasmid. In general, the TA-rich nature of the ADH terminator sequences downstream of the polylinker in the pJG4-5 vector make it difficult to design high-quality primers in this region.

Stage 3: Specificity Testing

The next step is to determine whether isolated cDNAs are repeatable and specific to the pBait of interest, and to exclude library-encoded cDNAs that interact with LexA (instead of the bait), "sticky" proteins that interact with the pBait in a nonspecific manner, and clones isolated because of mutations in the initial SKY48 strain that render growth and transcriptional activation nonspecifically possible. This can be done using a PCR-recombination approach (derived from Petermann 1998) in a single step, after which confirmed specific positive clones can be worked up through conventional plasmid purification.

MATERIALS

Plasmids and PCR Products

Enough PCR products (~1 µg) from the yeast colonies to be assessed for second confirmation (from the previous step); reamplify if necessary
High-quality digested library plasmid (see steps 1–4 below)
Yeast with the following sets of plasmids (obtain by a recent transformation)
   pMW109 + your original bait (pBait)
   pMW109 + pEG202– Ras
   pMW109 + another unrelated bait (optional, but highly recommended)

METHODS

1. Digest an empty library plasmid with two enzymes producing incompatible ends in the polylinker region (e.g., *Eco*RI and *Xho*I for the JG4-5 plasmid). Make sure that the restriction enzyme sites are in the region flanked by the priming sites.

2. Perform a PCR from a pJG4-5-Raf1 control plasmid using the FP1 and FP2 primers as before, and purify the PCR product.

3. Transform SKY48 containing pEG202– Ras and pMW109 with:

   a. digested library plasmid (50–100 ng)

   b. digested library plasmid (50–100 ng) and Raf1 control PCR product (0.5–1 µg)

   c. uncut library plasmid (50–100 ng)

   Save the digested library plasmid and the Raf1 PCR product for step 13.

4. Plate the transformations on Glu/CM –Ura, –His, –Trp dropout plates and incubate at 30°C until colonies grow (2–3 days). Store at 4°C.

   This control experiment is an indicator of the degree of digestion of the library plasmid. The background level of colonies transformed with digested empty library plasmid (*a*) should be minimal. In case the background is high, make sure that the digestion of the empty library plasmid is full and not partial by increasing the digestion incubation time or the restriction enzyme concentration.

   When transformed together, the PCR-amplified cDNA fragment from the Raf1 PCR product and the digested library plasmid will undergo homologous recombination in vivo in up to 97% of the transformants that acquired both vector and insert. This is due to the identity between the cDNA PCR fragment and the plasmid at the priming sites. If transformation efficiency in *b* is better than in *a* by 5- to 20-fold, it is acceptable to proceed to the next steps. *c* is a positive control for the transformation.

5. Use digested library vector in combination with selected PCR products (use same ratios as in step 3*a*) to transform the following:

   SKY48 containing pMW109 and:

   a. pBait

   b. pEG202–Ras

   c. another bait

   In parallel, transform the digested library plasmids and Raf1 control PCR product into SKY48 containing pMW109 and *a–c*.

   In the event that the pBait used in the screen shows weak transcriptional activity on its own, it is advisable to choose a nonspecific control bait that can weakly activate transcription on its own to have comparable background levels of transcriptional activation. This is highly recommended because baits that have transcriptional activation capacity have greater difficulties with false-positive background in general.

6. Plate each transformation mix on Glu/CM –Ura, –His, –Trp dropout plates and incubate at 30°C until colonies grow (2–3 days).

7. Prepare a master plate for each library plasmid being tested. Each plate should contain at least 10 colonies of the transformed PCR-insert/digested plasmid into each of *a–c*.

8. Test for β-galactosidase activity and for leucine requirement exactly as described for First Confirmation of Positive Interactions, above. True positives should show a Leu+ LacZ+ phenotype with *a*, but not with *b* and *c*. Clones transformed with Raf control PCR will provide both positive and negative controls: *a* and *c* should be negative whereas *b* should be positive when assayed for β-galactosidase and growth on –Leu plates.

9. Proceed with sequencing and biological characterization. Because with this step you are about to leave working with the two-hybrid system, you may be required to recover the yeast plasmid for archival purposes. (For all subsequent cloning steps, PCR product would more than suffice.) To do so, you can transform selected positives into *E. coli*, using 2–4 µl of the β-glucuronidase-treated frozen yeast (from the Rapid Screen for Interaction Trap Positives protocol). Use of electroporation is highly recommended. For yeast plasmid isolation, a number of commercially available kits exist (e.g., CLONTECH, Zymo Research), or an old phenol-chloroform technique can be used.

A major strength of the protocol given above is that it will identify redundant clones prior to plasmid isolation and bacterial transformation, which in some cases greatly reduces the amount of work required. However, accurate records should be maintained as to how many of each class of cDNA are obtained; if any ambiguity is present as to whether a particular cDNA is part of a set or unique, investigators should err on the side of caution.

A database of common false-positives has been compiled and made available on the World Wide Web, at http://www.fccc.edu/research/labs/golemis/Interaction TrapInWork.html. For cDNAs only isolated a single time, or that do not appear to make biological sense in the context of the starting bait, it may be helpful to consult the database to ensure the clone has not been reported by multiple additional groups.

A Bacterial Two-hybrid System Based on Transcriptional Activation

The bacterial two-hybrid system described here is based on the observation that two interacting proteins X and Y can trigger transcriptional activation of a weak promoter in *E. coli*. As shown in Figure 6, transcriptional activation occurs if two fusion proteins are expressed in the cell. One of the fusions consists of protein X covalently linked to a DNA-binding protein that in turn binds to a specific DNA site positioned near the weak promoter. The other fusion consists of protein Y linked to a subunit of the *E. coli* RNA polymerase (RNAP). In this configuration, X is tethered near the weak promoter (via the DNA-binding-protein part of the fusion) and, because of the interaction between X and Y, recruits RNAP to the weak promoter and thereby activates transcription. In theory, any interacting protein–protein (X–Y) or protein–DNA pair should be able to mediate this transcriptional activation, and a number of experiments have demonstrated that this is generally true. Interacting protein–protein and protein–DNA pairs from prokaryotes (Dove et al. 1997, 2000), yeast (Dove and Hochschild 1998; Joung et al. 2000), and mammals (Joung et al. 2000; Shaywitz et al. 2000; J.K. Joung and A. Hochschild; J.K. Joung et al., both unpubl.) have all been shown to activate transcription in this system. In addition, the affinity of the interacting partners correlates with the magnitude of transcriptional activation observed (Dove et al. 1997). The encouraging results of these experiments have given support to the initial proposal (Dove et al. 1997) that this system could form the basis of a bacterial two-hybrid system.

Identifying Protein–Protein Interactions

To use the bacterial two-hybrid system to identify protein–protein interactions, one could tether a bait protein near the weak promoter by fusing it to a DNA-binding protein (e.g., λcI protein) that recognizes a nearby DNA site. Much as in the yeast two-hybrid system, proteins that interact with the bait could then in theory be identified from a library of candidates fused to an RNAP subunit (e.g., the α subunit). Note that the placement of the bait and prey is also potentially reversible; that is, one could also fuse the bait to an RNAP subunit and create a library of preys fused to the DNA-binding domain. Fusing the bait to the DNA-binding domain is preferable because one can readily assess the stability and activity of such a fusion (see Methods); in contrast, assays to check the stability and activity of RNAP fusions are not currently available.

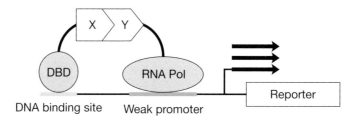

FIGURE 6. Schematic showing the components of the bacterial two-hybrid system. Protein X is fused to a DNA-binding domain (DBD) that in turn binds to a specific DNA-binding site positioned near a weak promoter. Protein Y is fused to a subunit of the *E. coli* RNA polymerase (RNA Pol). Interaction between proteins X and Y mediates the recruitment of RNA polymerase to the weak promoter, thereby activating expression of a downstream reporter gene.

Identifying Protein–DNA Interactions

Protein–DNA interactions can also be identified using the bacterial two-hybrid system. As shown in Figure 7, in this use of the system, the "bait" is a DNA-binding site of interest positioned near the weak promoter. One can then identify "prey" DNA-binding domains that bind to the bait DNA-binding site by one of two approaches. In the first method, a library of preys is fused to a constant protein X, which in turn interacts with a protein Y linked to a subunit of the RNAP (two-hybrid method). Alternatively, in another approach, a library of prey DNA-binding domains can be fused directly to a subunit of the RNAP (one-hybrid method). In either case, productive interaction of a "prey" DNA-binding domain with the "bait" DNA-binding site recruits RNAP to the weak promoter and activates transcription. Prokaryotic (Dove et al. 1997) and eukaryotic (Joung et al. 2000; J.K. Joung and C.O. Pabo, unpubl.) DNA-binding domains have been shown to activate transcription in both the one- and two-hybrid configurations.

Selectable Marker Genes Used with the Bacterial Two-hybrid System

Two different selectable marker genes have been used with the activation-based bacterial two-hybrid system to identify desired interaction partners from libraries:

1. *The yeast HIS3 gene* (Pabo and colleagues). *E. coli* cells bearing a *hisB* gene deletion do not grow on medium lacking histidine (His-deficient medium). One can use the yeast *HIS3* gene as a selectable marker in such cells because expression of *HIS3* complements the *hisB* defect, permitting growth on His-deficient medium (Struhl et al. 1976; Struhl and Davis 1977). The stringency of this selection can be easily raised or lowered by altering the concentration of 3-aminotriazole (3-AT)<!>, a competitive inhibitor of the *HIS3* enzyme, in the medium (Brennan and Struhl 1980). Two characteristics of the selectable *HIS3* system make it useful for finding rare candidates from very large ($>10^8$ in size) libraries: It exhibits a low spontaneous background frequency ($\sim 3 \times 10^{-8}$ breakthrough colonies with 20 mM 3-AT) (Joung et

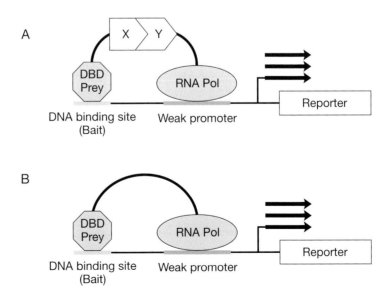

FIGURE 7. Identifying protein–DNA interactions with the bacterial two-hybrid system. Two strategies for isolating DNA-binding proteins from a library of preys. (*A*) Two-hybrid strategy. (*B*) One-hybrid strategy. See text for details.

al. 2000), and ~10^9 candidates can be plated on a single 245 x 245-mm agar plate. This system has been used successfully to identify zinc finger variants with altered DNA-binding specificities from a large randomized library (Joung et al. 2000).

2. *The β-lactamase* (bla) *gene* (Hochschild, Greenberg, and colleagues). This system uses the ampicillin-resistance gene β-*lactamase* (*bla*) as a selectable marker. The higher breakthrough frequency (~ 2 x 10^{-6}) in this system (Shaywitz et al. 2000) and the ability to plate no more than 10^6 transformants per agar plate limit its use to analysis of relatively smaller libraries. However, the medium used in the *bla* system is somewhat simpler to prepare than the histidine-deficient medium required for the *HIS3* system. Hochschild, Greenberg, and colleagues have used this selection system to identify a CREB-binding protein (CBP) mutant with enhanced affinity for CRE-binding protein (CREB) (Shaywitz et al. 2000).

Although the *HIS3* selection system has been used to identify protein–DNA interactions and the *bla* system has been used to identify an altered protein–protein interaction, the only significant difference between the two systems is the selectable reporter used. Theoretically, it should be possible to use either selection system to study both protein–DNA and protein–protein interactions. The primary consideration in determining which selection system to use should be the size of the library to be assessed. Experiments requiring the analysis of very large libraries (e.g., randomized peptide libraries, targeted mutagenized or randomized protein libraries, or rare candidates from cDNA libraries) should use the *HIS3* system, whereas experiments with smaller libraries could use either selection system.

An additional distinction between the two selection systems is the presence of different secondary reporter genes. As shown in Figure 8, the reporter strains in both selection systems also harbor a secondary reporter gene that is expressed co-cistronically with the primary (*HIS3* or *bla*) selectable marker. These secondary reporters can provide a rapid means to verify potential positives that come through the initial selection (see Methods below). The selectable bacterial *aadA* gene (conferring resistance to the antibiotics streptomycin and spectinomycin) is present in the *HIS3* system, and the β-galactosidase gene (*lacZ*) is present in the *bla* system. An advantage of the *lacZ* reporter is that its expression can be measured quantitatively, and thus one can readily assess the magnitude of activation seen with potential positives. The different characteristics of the secondary reporter may influence the choice of which system to use.

The following section provides detailed protocols for using the bacterial two-hybrid to analyze protein–protein interactions with the *HIS3* selection system or the *bla* selection system. Both of these selection systems have been used successfully to isolate candidates of interest from randomized and/or mutagenized libraries. These protocols should also be applicable to the isolation of proteins from cDNA libraries, although as yet, these systems have not been used for this purpose.

Stage 1 details methods to construct and characterize a selection strain harboring a "bait" fusion protein. Stage 2 describes methods for introducing a library of prey fusion proteins into the selection strain and protocols for performing the selection. Finally, Stage 3 details additional experiments for validating potential positives from the selection. A flowchart giving an overview of the various stages, as well as estimated time frames for each step, is shown in Figure 9.

A. HIS3 Selection System

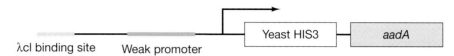

B. *bla* Selection System

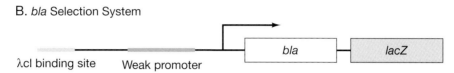

FIGURE 8. Reporters in the different bacterial two-hybrid selection systems. (*A*) Expression of the yeast *HIS3* gene can be selected for in *E. coli* strains lacking a functional *hisB* gene. The bacterial *aadA* gene encodes for streptomycin and spectinomycin antibiotic resistance. (*B*) The *bla* gene encodes resistance to β-lactam antibiotics (e.g., carbenicillin). The *lacZ* gene encodes the readily quantifiable enzyme β-galactosidase.

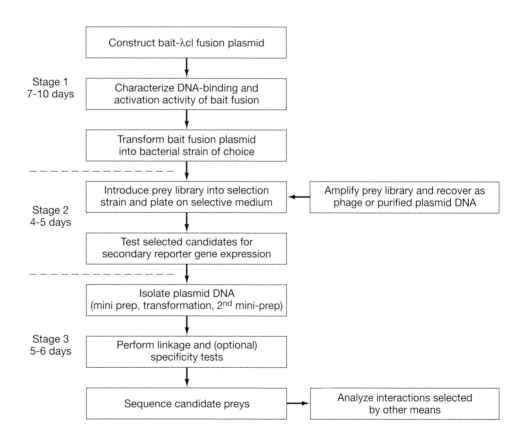

FIGURE 9. Flowchart of the bacterial two-hybrid system. See text for details.

Stage 1: Construction of a Selection Strain Harboring a Bait

The protocol for this stage details construction of a bait–DNA-binding protein fusion for use in identifying protein–protein interactions. The end goal of this protocol is the creation of a selection strain (harboring a protein bait) into which a library of prey fusion proteins can then be introduced.

MATERIALS

CAUTION: See Appendix for appropriate handling of materials marked with <!>.

Buffers and Solutions

Amino acid mixture: 17 different amino acids (no His, Met, or Cys)

Make the following six mixtures first:

  i. Phe 0.99%, Lys 1.1%, Arg 2.5% in H_2O
  ii. Gly 0.2%, Val 0.7%, Ala 0.84%, Trp 0.41% in H_2O
  iii. Thr 0.71%, Ser 8.4%, Pro 4.6%, Asn 0.96% in H_2O
  iv. Asp 1.04%, Gln 14.6% in 3% HCl
  v. Glu 18.7%, Tyr 0.36% in H_2O with 4 g of NaOH
  vi. Ile 0.79%, Leu 0.79% in H_2O

Mix together equal volumes of solutions i–vi and filter-sterilize through a 0.2-μm nylon filter. Store at 4°C, wrap in foil to protect from light.

3-Aminotriazole (3-AT) (1 M) <!>

Dissolve in H_2O and filter through 0.2-μm nylon filter.

IPTG (1 M) (filter-sterilized) <!>
10x M9 salts (Miller recipe)

Media

LB agar plates (see Sambrook and Russell 2001, *Molecular Cloning*, p. A2.5 for recipe)
LB/C/K plates – LB plates with chloramphenicol <!> and kanamycin
Liquid LB medium (see Sambrook and Russell 2001, *Molecular Cloning*, p. A2.2)
Liquid NM medium

For 500 ml, mix:

ddH$_2$O	418 ml
10x M9 salts (Miller recipe)	50 ml
20% glucose	10 ml
20 mM adenine HCl	5 ml
amino acid mixture (see above)	15 ml
1 M MgSO$_4$ <!>	0.5 ml
10 mg/ml thiamine	0.5 ml
10 mM ZnSO$_4$	0.5 ml
100 mM CaCl$_2$ (always add this last)	0.5 ml

Filter-sterilize through a 0.2-μm filter.

LB/T/K plates – LB plates with tetracycline and kanamycin
LB/T/C/K plates – LB plates with tetracycline, chloramphenicol, and kanamycin
LB/T/C/K/I plates – LB plates with tetracycline, chloramphenicol, kanamycin, and IPTG

TABLE 6. Concentrations of Antibiotics and Other Supplements Used in Agar Plates

Antibiotic	Stock solution	Final concentration in plates
Carbenicillin	50 mg/ml in H_2O	for *bla* selection plates: 250 to 1000 µg/ml
		for *HIS3* selection plates: 100 µg/ml
Chloramphenicol	30 mg/ml in ethanol	30 µg/ml
Kanamycin	10 mg/ml in H_2O	50 µg/ml
Tetracycline	10 mg/ml in 80% ethanol	15 µg/ml
IPTG	1 M in H_2O	variable but typically 50 µM
X-Gal	20 mg/ml in dimethylformamide	80 µg/ml
Phenyethyl-β-D-thio-galactoside	10 mM in dimethylformamide	0.1 mM

LB/T/C/K/I/XG plates – LB plates with tetracycline, chloramphenicol, kanamycin, IPTG, X-Gal, and phenyethyl-β-D-thiogalactoside

NM agar plates

> For 500 ml of plates, autoclave 418 ml of ddH₂O with 7.5 g of Bacto-Agar and a stir bar. Allow agar to cool to 65–70°C and then add the same basic components as for the liquid NM medium (premixed): and antibiotics, IPTG, and 3-AT as needed.

NM/K plates – NM plates with kanamycin

NM/K/5 mM 3-AT plates–NM plates with kanamycin and 5 mM 3-AT

> See Table 6 for concentrations of supplements for specialized plates.

Special Equipment

Sterile toothpicks

Vectors and Bacterial Strains

See Tables 7 through 9.

METHOD

In this stage, the bait protein of interest is fused to the DNA-binding protein bacteriophage λ repressor (λcI protein). Two functional tests of this bait fusion protein are then performed: (1) a phage immunity assay to check for stability and DNA-binding capability inside a bacterial cell and (2) an activation assay to verify that the bait fusion does not activate the reporter promoter on its own.

Construction of a Bait-λcI Fusion Protein

1. Insert the bait-encoding DNA into the polylinker of the λcI-encoding expression vector, pBT, to create the bait-λcI fusion protein. A translation stop codon should be created in-frame at the end of the bait sequence. This plasmid will be referred to as pBait in this protocol.

Characterization of DNA-binding and Activation Activity: Assay for Phage Immunity and *lacZ* Expression

Steps 2–6 test the ability of the bait-λcI fusion protein to bind specifically the DNA-binding site of λcI in the cell (DNA-binding assay) and to activate transcription on its own from the weak reporter promoter (activation assay). Checking DNA-binding activity is critical, because if the

TABLE 7. Fusion Protein Expression Vectors for the Bacterial Two-hybrid System

Plasmid name	Antibiotic marker	Plasmid origin	Comment/Description	Contact/Sources
pBT	CAM	p15A	basic plasmid to clone bait as a fusion to λcI protein[a]; expression driven by *lac*UV5 promoter	Stratagene
pTRG	TET	pBR322	basic plasmid to clone prey as fusion to *E. coli* RNAP α subunit residues 1–248[b]; expression driven by tandem *lpp/lac*UV5 promoters	Stratagene
pBR-UV5–(αLP)	AMP	pBR322	basic plasmid to clone prey as a fusion to *E. coli* RNAP α subunit residues 1–248; expression driven by tandem *lpp/lac*UV5 promoters; has f1 ori of replication	Joung, Hurt, and Pabo (unpubl.)
pBR-UV5–λcI(LP)	AMP	pBR322	basic plasmid to clone prey as a fusion to λcI protein; expression driven by UV5 promoter; has f1 ori of replication	Joung, Hurt, and Pabo (unpubl.)
pKJ1323	CAM	p15A	basic plasmid to clone bait as fusion to *E. coli* RNAP α subunit residues 1–248; expression driven by tandem *lpp/lac*UV5 promoters	Joung and Pabo (unpubl.)
pKJ1514	AMP	pBR322	basic plasmid for fusing a DNA-binding prey to a fragment of Gal11P; expression driven by a *lac*UV5 promoter, has f1 ori of replication	Joung, Hurt, and Pabo (unpubl.)
pACL-αGal4	CAM	p15A	expresses αGal4 fusion protein (interacts with Gal11P fusions); expression driven by tandem *lpp/lac*UV5 promoters	Joung et al. (2000)
pTRG-Gal11P (control)	TET	pBR322	Gal11P fragment cloned into pTRG backbone; expresses an α-Gal11P fusion protein driven by tandem *lpp/lac*UV5 promoters	Stratagene
pBT-LGF2 (control)	CAM	p15A	Gal4 fragment cloned into pBT backbone; expresses a λcI-Gal4 fusion protein driven by a *lac*UV5 promoter	Stratagene
pACλcI-Gal11p (control)	CAM	p15A	control plasmid that expresses a λcI-Gal11P fusion protein driven by a *lac*UV5 promoter	available from Joung and Pabo[c]
pBR-αGal4 (control)	AMP	pBR322	control plasmid that expresses an α-Gal4 fusion protein driven by tandem *lpp/lac*UV5 promoters	available from Joung and Pabo[c]
pAC-UV5 (control)	CAM	p15A	control plasmid that does not express any fusion protein; has a *lac*UV5 promoter	available from Joung and Pabo[d]

[a]A derivative of pACλ cI32 (S.L. Dove et al., unpubl.) (see Hu et al. 2000) but with a longer linker and additional cloning sites between λcI and the bait.
[b]A derivative of pBRstar (Shaywitz et al. 2000) except with a longer linker and additional cloning sites between α-coding sequence and the prey.
[c]Joung, Dove, and Hochschild, unpublished.
[d]Joung and Hochschild, unpublished.

TABLE 8. Bacterial Strains and Phage

Strain	Genotype	Comments/Description	Contact/Sources
Bacteriomatch reporter	Δ(*mcrA*)183 Δ(*mcrCB-hsdSMR-mrr*)173, *endA1 supE44, thi-1, recA1, gyrA96, relA1, lac*, [F′ *lacI*q *bla lacZ Kan*R]	reporter strain in which the expression of the *bla* and *lacZ* reporter genes is directed by a weak promoter bearing an upstream λcI DNA-binding site[a]	Stratagene
KJ1567	Δ*hisB463*, Δ(*gpt-proAB-arg-lac*)XIII, *zaj::Tn10* [F′ *lacI*q HIS3 *aadA Kan*R]	reporter strain in which the expression of the HIS3 and aadA reporter genes is directed by a weak promoter bearing an upstream λcI DNA-binding site	Joung, Hurt, and Pabo (unpubl.)
KJ1C (control)	F-, Δ*hisB463*, Δ(*gpt-proAB-arg-lac*)XIII, *zaj::Tn10*	parental strain to KJ1567	Joung et al. (2000)

[a]Derived from strain US3F′3.1 (Shaywitz et al. 2000) and XL1-Blue MR (Stratagene).

TABLE 9. Bacteriophage

Phage	Description	Contact/Sources
λcI⁻ phage	bacteriophage λ with a mutation in the cI gene; grows only lytically; used for immunity assay	available from Joung and Pabo
M13KO7 or VCS-M13 helper phage	M13 helper phage with p15A origin of replication and Kan^R cassette inserted into phage origin of replication	New England Biolabs, Stratagene

fusion cannot bind to its specific binding site, it is not appropriate for use in selections. In addition, bait fusions that are self-activating cannot be used for selections.

DNA-binding capability is checked using a phage immunity assay. Cells expressing the bait λcI fusion are tested for their immunity to superinfection by λcI⁻ phage. Cells expressing a fusion protein capable of binding to a λcI-binding site should be immune to infection and lysis by λcI⁻ phage.

Transcriptional activation activity of the bait fusion on its own is checked by assaying expression of a *lacZ* gene reporter directed by a weak promoter bearing an upstream λcI-binding site. Desirable fusions should NOT activate expression of *lacZ* in the absence of the RNAP–prey fusion.

2. Double-transform the Bacteriomatch reporter strain with the following combinations of plasmids:

   a. pBait + pTRG (test for activation and repression)

   b. pBT-LGF2 + pTRG-Gal11P (positive control for activation and repression)

   c. pAC-UV5 + pTRG (negative control for activation and repression)

3. Plate each transformation on LB/T/C/K plates and incubate for 12 –18 hours at 37°C.

4. For phage immunity (DNA binding) tests, stripe ~100 μl of high-titer λcI⁻ phage down the center of an LB T/C/K/I plate and let dry. Pick at least two individual colonies from each transformation plated in step 2 for immunity challenge. Touch colonies with a sterile toothpick and, starting from one side of the plate, drag the toothpick perpendicularly across the phage stripe. Incubate plates for 16–18 hours at 37°C.

   Positive control cells should grow through the phage stripe, whereas negative control cells should grow robustly only up until the streak reaches the phage stripe. A desirable bait should resemble the positive control cells in phenotype.

5. For transcriptional activation tests:

   a. Pick two individual colonies from each transformation (from step 2) and resuspend in 100 μl of LB medium.

   b. Serially dilute each resuspension 1:10 twice (i.e., dilute to 10⁻²) in LB medium.

   c. Spot 5 μl of each resuspension and dilution on LB/T/C/K/I/XG plates. Incubate for 16–18 hours at 37°C.

   d. Examine the color of well-isolated colonies for each transformant. Positive controls should yield blue colonies, negative controls should yield white to pale blue colonies, and cells with desirable bait fusions should resemble the negative controls.

TROUBLESHOOTING FOR BAITS WITH UNDESIRABLE CHARACTERISTICS

- If a bait fails to confer immunity, consider increasing (or decreasing) the concentration of IPTG used in the test plates. In addition, consider lengthening the linker between λcI and the bait. Expression of the bait protein can be also monitored by western blot using cI antibodies available from Invitrogen.

- If a bait confers immunity but activates the promoter by itself, consider fusing the bait to the RNAP α subunit instead (see alternative protocol below).

Construction of the Bait-containing Selection Strain

6. If a suitable bait is obtained, transform the pBait plasmid into the Bacteriomatch strain (if using the *bla* selection) or into strain KJ1567 (if using the HIS3 selection) and plate on LB/C/K plates. Incubate at 37°C for 16–18 hours (for Bacteriomatch reporter) or 12–14 hours (for strain KJ1567).

7. If using the *bla* selection system, make the Bacteriomatch reporter cells harboring pBait competent and freeze these cells at –80°C. If using the *HIS3* selection system, proceed using the cells directly from the transformation plate; if, however, there will be a delay of longer than a week before proceeding to the next stage, the KJ1567 strain cells harboring pBait should be stored as a frozen glycerol stock at –80°C.

Fusion of Bait to RNA Polymerase α Subunit

In certain situations, it may be desirable to fuse the bait protein to the RNAP α subunit (e.g., if the bait is self-activating when fused to the λcI DNA-binding protein). Unlike the case of fusion to λcI, however, bait fusions made to the RNAP α subunit cannot be easily checked for stability or function, nor for incorporation into the RNAP complex. The only test that can be performed is to ensure that the bait fusion does not spuriously cause increased transcription from the reporter promoter.

METHOD

Construction of α-Bait Fusion Protein

1. Insert the bait-encoding DNA into the polylinker of an α-encoding expression vector. If the *bla* system will be used, use plasmid pTRG. If the *HIS3* system will be used, use plasmid pKJ1323 to create the α-bait fusion protein. A translation stop codon should be created in-frame at the end of the bait sequence. This plasmid will be referred to as pαBait in this protocol.

Test for Activation Activity of the α-Bait Fusion Protein

2. Transform the Bacteriomatch reporter strain with each of the following plasmids:

   a. for *HIS3* system: pαBait + pTRG

      for *bla* system: pαBait + pBT(test for spurious bait activation)

   b. pAC-UV5 + pTRG (negative control for activation)

   c. pBT-LGF2 + pTRG-Gal11P (positive control for activation)

3. Plate each transformation on LB/T/C/K plates and incubate for 12 –18 hours at 37°C.

4. For transcriptional activation tests:

   a. Pick two individual colonies from each transformation (from step 2) and resuspend in 100 µl of LB medium.

   b. Prepare 1:100 dilution of each resuspension (e.g., by two subsequent 1:10 dilutions) in LB medium.

   c. Spot 5 µl of each resuspension and dilutions on LB/T/C/K/I/XG plates. Incubate for 16–18 hours at 37°C.

   d. Examine the color of well-isolated colonies for each transformant. Positive controls should yield blue colonies, negative controls should yield white to pale blue colonies, and cells with desirable bait fusions should resemble the negative controls.

5. A desirable α-bait fusion should not cause spurious transcriptional activation.

Construction of the α-Bait-containing Selection Strain

6. If using the *HIS3* selection system, transform strain KJ1567 with the pαBait plasmid and plate on LB/C/K plates. Incubate for 12–14 hours at 37°C. Proceed using the cells directly from the transformation plate; if, however, there will be a delay of longer than a week before proceeding to the next stage, the cells should be stored as a frozen glycerol stock at –80°C.

7. If using the *bla* selection system, transform Bacteriomatch reporter cells with the pαBait plasmid and plate on LB/T/K plates. Incubate at 37°C for 16–18 hours. Before proceeding to the next stage, make the transformed Bacteriomatch reporter cells competent and freeze them at –80°C.

Stage 2: Introducing the Library into Selection Strain Cells and Selecting Potential Interactors

The methods for introducing prey libraries into the selection strain cells (prepared in Stage 1) differ depending on the selection system used. For the *HIS3* system, the prey plasmid library is electroporated into high transformation efficiency cells and converted into a library of infectious transducing phage; the library is then introduced into selection strain cells by simple infection with the phage; these additional steps permit the reproducible plating of $>10^9$ library members on a single His-deficient medium plate. For the *bla* system, direct transformation of a prey library into selection strain cells followed by plating on LB plates containing carbenicillin works well. Alternatively, one may amplify the library by passage through a standard cloning strain. Protocols for all of these manipulations are given in this section.

Following plating and growth on selective plates, potential candidates are subjected to an initial confirmatory test that rechecks for increased expression not only of the primary selectable marker, but also of the secondary reporter. Prey candidates that pass this initial test are carried through to additional testing in Stage 3.

MATERIALS

CAUTION: See Appendix for appropriate handling of materials marked with <!>.

Buffers and Solutions

5x PEG/NaCl solution
   17.5% PEG 8000 <!> and 12.5% NaCl in H_2O. Filter-sterilize through a 0.2-μm filter.
Sterile 50% glycerol (sterilize by filtration through a 0.2-μm filter)

Nucleic Acids and Oligonucleotides

Library of prey fusion proteins
Libraries suitable for use with the *bla* selection system
   These are commercially available from Stratagene.

Media

LB/C plates – LB plates with chloramphenicol <!>
LB/C/C/K plates – LB plates with carbenicillin (100 μg/ml), chloramphenicol, and kanamycin

LB/C/C/K/I plates – LB plates with carbenicillin (100 µg/ml), chloramphenicol, kanamycin, and IPTG <!>

LB/C/K plates – LB plates with chloramphenicol and kanamycin

LB/T plates – LB plates with tetracycline

LB/T/C plates – LB plates with tetracycline and chloramphenicol

LB/T/C/K plates – LB plates with tetracycline, chloramphenicol, and kanamycin

LB/T/C/K/C plates – LB plates with tetracycline, chloramphenicol, kanamycin, and carbenicillin (250–1000 µg/ml)

LB/T/C/K/C/XG plates – LB plates with tetracycline, chloramphenicol, kanamycin, carbenicillin (250–1000 µg/ml), X-Gal, and phenyethyl-β-D-thiogalactoside

NM/C/C/K/I/20 mM 3-AT – NM plates with carbenicillin (100 µg/ml), chloramphenicol, kanamycin, IPTG, and 20 mM 3-AT <!>

2× YT liquid medium (see Sambrook and Russell 2001, *Molecular Cloning*, p. A2.4 for recipe)

Special Equipment

Culture plates (245 × 245-mm)

Glass beads

Microtiter plate (96-well)

Multichannel micropipettor

Vectors and Bacterial Strains

See Tables 6 through 9.

METHOD

Construction and Amplification of the Prey Fusion Protein Library

1. To construct a prey fusion protein library, clone the prey-encoding inserts into:

   a. pTRG plasmid (for protein–protein experiments with the *bla* system and bait fused to λcI)

   or

   b. pBR-UV5-αLP phagemid (for protein–protein experiments with the HIS3 system and bait fused to λcI)

   or

   c. pBR-UV5-λcILP phagemid (for protein–protein experiments with the HIS3 system and bait fused to RNAP α subunit)

   or

   d. pBT plasmid (for protein–protein experiments with the *bla* system and bait fused to RNAP α subunit)

2. Amplify the library.

   For libraries to be used with the *HIS3* selection:

   a. Electroporate the library ligation into a high transformation efficiency, F⁺ cloning strain (e.g., XL-1 Blue, Stratagene). This strain should be sensitive to kanamycin (permits one to select cells infected by helper phage; see below) and should have a selectable antibiotic resistance marker on the F′ strain.

b. Pool the electroporations and allow the cells to recover in 2x YT (use 10x the volume of the pooled transformations) for 1 hour with shaking at 37°C.

c. Serially dilute by 1:10 a small aliquot of the transformed cells at least 8 times (to 10^{-8}) in 2x YT medium. Perform this dilution in triplicate. This can be easily performed using small 100-µl volumes (dilute 10 µl of sample into 90 µl of LB) in a 96-well titer plate with a multichannel pipettor. Spot 5 µl of each dilution three times on

> LB/T/C plates (to assess number of transformants)
> LB/C/K plates (control; should not yield any colonies)

These titers represent the preamplification library size.

d. Inoculate the remainder of the transformed cells into 2x YT (use 10x the volume of the recovered cell culture) supplemented with tetracycline (selects for presence of the F′) and carbenicillin (selects for presence of the library plasmids). Shake for 2 hours at 37°C.

e. Serially dilute by 1:10 a small aliquot of the transformed cells at least 8 times (to 10^{-8}) in 2x YT medium. Perform this dilution in triplicate. Spot 5 µl of each dilution three times on

> LB/T/C plates (to assess number of amplified transformants)
> LB/C/K plates (control, should not yield any colonies)

Incubate plates for 16–18 hours at 37°C.

f. Pellet cells in the remainder of the amplified culture by centrifugation. Resuspend the cell pellet in 2x YT with 15% (v/v) glycerol (use 1/100th the volume of the amplified culture). Divide into at least four aliquots and freeze in a dry ice/ethanol bath. Store at –80°C.

g. Assess the library size by determining the transformation efficiency from the number of colonies on the LB/T/C plates of step *b*. Assess the fold-amplification by comparing the number of colonies on the LB/T/C plates of step *d* with those of step *b*. A typical amplification is four- to eightfold. Confirm that no colonies grow on the LB/C/K plates of steps *b* and *d*.

h. Thaw one or more aliquots of the amplified cells frozen in step *f*. Use enough cells based on the calculated titers to oversample the original library size by at least threefold. Add the thawed cells to 2x YT (use 10x volume of the thawed cells) supplemented with carbenicillin and tetracyline. Shake for 1 hour at 37°C.

i. Infect the cells with helper phage (either M13KO7 or VCS-M13) at a multiplicity of infection of >100:1. Add the phage, mix the culture well, and allow the phage to adsorb to cells by leaving the culture at room temperature without shaking for 30 minutes.

j. Shake the culture for 1.5 hours at 37°C. This permits expression of the kanamycin-resistance gene on the helper phage genome.

k. Add kanamycin to the culture to a final concentration of 70 µg/ml. The presence of kanamycin ensures that only phage-infected cells will grow. Allow the phage culture to grow for 18 hours at 37°C or, alternatively, for 24 hours at 30°C.

During growth of the helper phage-infected culture, single-stranded versions of the library plasmids are produced (due to the presence of an f1 phage origin of replication on the plasmid) and packaged as infectious M13 particles. The plasmids inside these phage can be reintroduced into any F$^+$ cell by simply allowing the phage particle to infect the recipient cell. The single-stranded plasmid is converted back to a double-stranded plasmid by the recipient cell and then begins to replicate as a plasmid again. Thus, simple infection by these "pseudo-phage" provides a means for "transforming" a recipient cell with the packaged plasmid.

l. To harvest the phage library, pellet the cells twice by centrifugation and save the supernatant (containing the phage).

m. Determine the titer of the phage library by quantifying the ampicillin-transducing units (ATUs) in the stock as follows:

   i. Serially dilute the phage stock 1:10 at least eight times (i.e., to a 10^{-8} dilution) in 2x YT (can be done conveniently in a microtiter plate).

   ii. Add 10 µl of each dilution to 50 µl of an overnight culture of XL-1 Blue (or other F^+, ampicillin-sensitive strain) in microtiter plate wells. Allow phage to adsorb by leaving at room temperature without shaking for 10–15 minutes.

   iii. Add 190 µl of 2x YT to each well and leave for 2 hours at 37°C without shaking.

   iv. Spot 5 µl of each infection in triplicate (i.e.,15 µl total) onto LB/T/C plates. Allow spots to dry and incubate for 16–18 hours at 37°C.

   v. Calculate ATUs in the original phage stock by counting ampicillin-resistant colonies.

n. The concentration of ATU in the phage stock should be sufficiently high so that the number of candidates one wishes to assess can be found in 500 µl or less of the stock. If this is not the case, then a PEG precipitation method should be used to concentrate the phage.

   i. Add 4 volumes of phage stock to 1 volume of 5x PEG/NaCl solution.

   ii. Place on ice and incubate for >2 hours.

   iii. Centrifuge at 6000*g* for 40 minutes.

   iv. Remove supernatant. Phage will typically be present as a large pellet along the side of the centrifuge tube or bottle. Resuspend this pellet in a small volume of 2x YT with 15% glycerol.

o. Titer the concentrated phage library stock as in step *m*.

p. Aliquot the phage library and freeze at –80°C in cryotubes.

Alternate step 2: For libraries to be used with the *bla* selection (optional):

a. Transform (by electroporation or chemical methods) a standard cloning strain (e.g., XL-1 Blue MR [Stratagene] or DH5 [GIBCO BRL]) with the library ligation.

b. Serially dilute by 1:10 a small aliquot of the transformation at least 6 times (to 10^{-6}) in LB medium. Perform this dilution in triplicate. This can be easily performed using small 100-µl volumes (dilute 10 µl of sample into 90 µl of LB) in a 96-well titer plate with a multichannel pipettor. Spot 5 µl of each dilution three times on

   LB/T plates (for libraries based on the pTRG plasmid)
   LB/C plates (for libraries based on the pBT plasmid)

c. Plate the remainder of the transformation on large

   LB/T plates (for libraries based on the pTRG plasmid)
   LB/C plates (for libraries based on the pBT plasmid)

d. Incubate all plates for 18–24 hours at 3°C.

e. Calculate library size by counting colonies from the serial dilution plates.

f. Harvest bacteria from the larger plates by scraping the cells into liquid LB medium.

g. Isolate plasmid DNA using commercially available alkaline-lysis mini-prep kits on half of the cell resuspension. This DNA is the amplified library and will be used for subsequent transformation of the selection strain.

h. Add glycerol to a final concentration of 15% to the remaining half of the cell resuspension and store frozen at –80°C.

Introducing the Library into Selection Cells and Performing the Selection

3. Introduce the library into the selection strain prepared in Stage 1 as follows:

For *HIS3* Selection:

a. Inoculate a 10-ml culture of the selection strain from Stage 1 using NM medium with chloramphenicol, kanamycin, and 50 μM IPTG. Use a 125-ml flask and shake gently (125–150 rpm) for 20–24 hours at 37°C.

b. Check the optical density (600 nm) of the selection strain culture. A saturated culture should yield an $OD_{600} > 2.5$.

c. Transfer 5 ml of the selection strain culture to a sterile 125-ml flask and then add the phage library prepared in step 2. Up to 10^9 ATUs can be added to the cells.

   i. Briefly swirl the culture to mix the phage with the cells, and allow the culture to stand at room temperature without shaking for 25–30 minutes. The phage adsorb to the cells during this time.

   ii. Add 20 ml of NM medium with chloramphenicol, kanamycin, and 50 μM IPTG (prewarmed to at least room temperature). Shake gently (125–150 rpm) for 1.5 hours at 37°C to allow the expression of the ampicillin-resistance gene and prey fusion protein encoded on the incoming plasmid.

   iii. Transfer the culture to a sterile 50-ml conical tube and spin in a tabletop centrifuge at 2000*g* for 30 minutes at room temperature.

   iv. Drain the medium and resuspend the cell pellet in 2.5 ml of NM medium with chloramphenicol, kanamycin, and 50 μM IPTG.

d. Serially dilute by 1:10 a small aliquot of the transduced cells at least eight times (to 10^{-8}) in NM medium. Perform this dilution in triplicate. Spot 5 μl of each dilution three times (i.e., 15 μl total) on:

   LB/C/K plates (to assess the total number of cells)
   LB/C/C/K plates (to assess the number of transduced cells)
   NM/C/C/K/I plates (to assess the number of transduced cells capable of growing on His-deficient medium)

   Incubate plates for 16–18 hours at 37°C.

e. Place the remainder of the transduced cells onto a NM/C/C/K/I/20 mM 3-AT plate. (Up to 10^9 transductants can be plated on a single 245-mm × 245-mm plate.) Add about a dozen sterile glass beads and gently agitate (do not shake) plates to distribute the cells evenly on the plate. Allow to air-dry, invert, and incubate for 24 hours at 37°C, and then let sit 12 hours at room temperature.

f. Count the colonies that grow on the titer plates from step *h* to verify the total number of cells transduced with the prey plasmids that were plated on the selection plate. Note that the total number of cells plated (transduced and not transduced) should exceed the number of transduced cells plated by a factor of at least 5. This ratio ensures that most cells were transduced/infected with only a single prey plasmid.

g. Inspect 3-AT-selection plates for the appearance of colonies.

i. If a sufficient number of candidates have been plated but no colonies appear on the selection plates, the selection can be redone using lower concentrations of 3-AT (as low as 5 mM). Note, however, that the appearance of background colonies increases with lower concentrations of 3-AT.

ii. If too many candidates appear on the selection plates, redo the selection using higher concentrations of 3-AT (as high as 40 mM). Caution should be used, however, because concentrations of 3-AT above 30 mM may decrease the plating efficiency of positive candidates.

h. Colonies that grow on the selective plates should be checked immediately for first confirmation (step 4).

For *bla* Selections:

It is preferable first to perform a small-scale transformation to determine the transformation efficiency of the selection cells with the library.

a. Transform chemically competent selection strain cells from Stage 1 with 0.1 µg of either the library ligation from step 1 or the amplified DNA library from step 2.

b. Serially dilute a small aliquot of the transformation 1:10 at least eight times (i.e., to a 10^{-8} dilution) in LB. Spot 5 µl of each dilution three times on LB/T/C/K. Let the spots dry and incubate overnight at 30°C.

c. From the titer plates of step *b*, calculate the number of cells in the transformation that contain prey plasmids.

Next, scale up the transformation to perform the selection.

d. Use enough library to obtain the desired number of transformants and transform the selection strain cells. Remove a small aliquot of the transformation and serially dilute and plate as in step *b*. Plate the remainder of the transformation on LB/T/C/K plates with carbenicillin at 250 µg/ml. No more than ~10^5 transformants should be placed on a 100-mm agar plate.

e. Incubate plates for 18–24 hours at 30°C. Do not incubate plates longer than 24 hours.

f. Inspect titer plates to ensure that a sufficient number of prey plasmid-transformed cells have been plated. Also inspect carbenicillin-selection plates for the appearance of colonies.

i. If a sufficient number of candidates have been plated but no colonies appear on the selection plates, the selection can be redone using lower concentrations of carbenicillin (as low as 150 µg/ml).

ii. If too many candidates appear on the selection plates, the selection can be redone using higher concentrations of carbenicillin (as high as 1000 µg/ml). Caution should be used, however, because concentrations of carbenicillin above 250 µg/ml may decrease the plating efficiency of positive candidates. However, more candidates can be plated when using higher concentrations of carbenicillin.

g. Colonies that grow on the selective plates should be checked immediately for first confirmation (step 4).

First Confirmation of Positive Interactions: Test for Secondary Reporter Activity

One rapid method to verify potential positive colonies from step 3 is to examine the expression of the secondary reporter present in the selection strains. True positives, due to increased transcription from the weak promoter, should also show increased secondary reporter expression. Step 4 details how to assess increased *aadA* expression (for the HIS3 selection strain) and increased *lacZ* expression (for the *bla* selection strain).

4. Assays for increased expression of the secondary reporter gene.

For the *aadA* Gene (*HIS3* Selections):

   a. Pick potential positive colonies directly from the selection plate and resuspend in 100 μl of NM medium in a microtiter plate well.

   b. Serially dilute each colony resuspension 1:10 in NM medium two times (i.e., to 10^{-2}).

   c. Spot 5 μl of each resuspension and dilutions on the following plates:

   NM/C/C/K/I/20 mM 3-AT (to reconfirm growth)
   NM/C/C/K/I/20 mM 3-AT/50 μg/ml spectinomycin (to check *aadA* expression)

   d. Incubate the plates 24 hours at 37°C. Inspect the plates for growth. If necessary, allow the plates to incubate an additional 12–18 hours at room temperature. If none of the candidates grow on the spectinomycin plate, consider replating the serial dilutions of step *b* on plates with lower concentrations of spectinomycin.

   e. Colonies that grow on the spectinomycin plates should be designated as first-round positives and carried forward to Stage 3 for further analysis. This plate will also serve as the master plate for these candidates.

For the *lacZ* Gene (*bla* Selections):

   a. Pick potential positive colonies directly from the selection plate and resuspend in 100 μl of LB medium in a microtiter plate well.

   b. Serially dilute each colony resuspension 1:10 in LB medium two times (i.e., to 10^{-2}).

   c. Spot 5 μl of each resuspension and dilutions on the following plates:

   LB/T/C/K/C (to reconfirm growth)

   LB/T/C/K/C/XG (to check for increased *lacZ* expression)

   d. As a negative control, also serially dilute and spot the selection strain obtained in Stage 1 harboring the pTRG plasmid.

   e. Incubate the plates for 12–18 hours at 37°C.

   f. Inspect the plates. Examine the color of well-isolated colonies for each candidate and compare the color of well-isolated colonies of the negative control. Desirable candidates should be more blue than the negative control. These candidates should be carried forward to Stage 3 for further analysis.

Stage 3: Second Confirmation of Potential Positive Candidates

In this stage, candidates initially confirmed as positives in Stage 2 are tested to determine whether the increased reporter gene expression is linked to the expression of the specific prey isolated from the library. To perform this linkage analysis, the plasmid encoding the prey fusion is isolated and reintroduced into naive selection strain cells. If the ability to grow on selection plates is linked to the prey plasmid, then the insert is sequenced.

MATERIALS

CAUTION: See Appendix for appropriate handling of materials marked with <!>.

Buffers and Solutions

Amino acid mixture (see Stage 1)
3-Aminotriazole (1 M) (3-AT) <!>
> Dissolve in H_2O and filter through a 0.2-μm nylon filter.

IPTG (1 M) (filter sterilized) <!>
10x M9 salts (Miller recipe)

Media

LB agar plates (see Sambrook and Russell 2001, *Molecular Cloning*, p. A2.5 for recipe)
LB/C plates – LB plates with chloramphenicol <!>
LB/Carb plates – LB plates with 100 μg/ml carbenicillin
LB/C/C/K/I plates – LB plates with carbenicillin (100 μg/ml), chloramphenicol, kanamycin, and IPTG <!>
LB/T plates – LB plates with tetracycline
LB/T/C/K plates – LB plates with tetracycline, chloramphenicol, and kanamycin
Liquid LB medium (see Sambrook and Russell 2001, *Molecular Cloning*, p. A2.2 for recipe)
NM liquid medium (see Stage 1)
NM agar plates (see Stage 1)
NM/C/C/K/I/20 mM 3-AT – NM plates with carbenicillin (100 μg/ml), chloramphenicol, kanamycin, IPTG, and 20 mM 3-AT
> See Table 6 for concentrations of supplements for specialized plates.

Vectors and Bacterial Strains

See accompanying tables.

METHOD

Isolation of Positive Plasmids

1. Isolation of purified prey plasmid from potential positive candidates.

   a. For each potential candidate, inoculate a well-isolated colony from the master plates created in Stage 2 (p. 130) into 2 ml of LB supplemented with 100 μg/ml of carbenicillin (if using the *HIS3* selection) or into LB supplemented with 15 μg/ml of tetracycline (if using the *bla* selection). Grow with agitation for 12 hours at 37°C.

   b. Isolate the plasmid DNA from 1.5 ml of the overnight culture using a standard miniprep isolation procedure. Resuspend or elute DNA in a final volume of 100 μl.

   The plasmid DNA isolated by this method will include not only the prey plasmid but also the bait plasmid. Thus, another round of transformation is necessary to separate the prey from the bait.

   c. Use 1 μl of the DNA from step *b* to transform a standard cloning strain not resistant to either carbenicillin or tetracycline (e.g., XL1-Blue MR [Stratagene]). These transformations can be performed easily in a 96-well format. Spot 1/20th of the transformation on a LB/Carb (for *HIS3* selections) or LB/T (for *bla* selections) plate. Incubate 16–18 hours at 37°C.

d. Pick two colonies from each transformation and patch to:

LB/C and then a LB/Carb plate (if using the *HIS3* selection)
LB/C and then a LB/T plate (if using the *bla* selection)

Let patches grow 6–8 hours at 37°C.

e. Transformants harboring only the prey plasmid should fail to grow on the chloramphenicol (LB/C) plate. For each transformant, pick one of the candidates that meet this criterion from the LB/Carb or LB/T plate patch made in step *d* and inoculate a 10-ml culture in LB supplemented with 100 µg/ml carbenicillin (if using the *HIS3* selection) or LB supplemented with 15 µg/ml tetracycline (if using the *bla* selection). Grow for 16–18 hours at 37°C with agitation.

f. Isolate plasmid DNA from the 10-ml cultures using standard commercially available alkaline lysis/column purification methods. Utilize procedures for low-copy-number plasmids and perform all extra wash steps.

g. This DNA is the purified prey plasmid.

Second Confirmation of Positive Interactions: Linkage Testing

2. Use 1 µl of purified prey plasmid DNA to transform selection strain cells constructed in Stage 1 (p. 124).

3. Serially dilute each transformation 1:10 two times (i.e., to 10^{-2}) in NM medium (if isolated from a HIS3 selection) or in LB (if isolated from a *bla* selection).

4. Spot 5 µl of each transformation and its dilutions on the indicated plates.

a. For candidates selected with the HIS3 system:

NM/C/C/K/I plates (positive control for transformation)
NM/C/C/K/I/20 mM 3-AT plates (tests linkage)

Incubate for 18–24 hours at 37°C.

b. For candidates selected with the *bla* system:

LB/T/C/K plates (positive control for transformation)
LB/T/C/K plates supplemented with 250 µg/ml carbenicillin (tests linkage)

Incubate for 12–18 hours at 37°C.

5. Analyze growth on the plates. Candidates that grow again on selective medium should have their prey inserts sequenced (use the purified prey plasmid DNA isolated in step *1g*).

Optional Step: Specificity Test for Positive Candidates

One additional test that may be performed before sequencing the prey-encoding plasmids is to check that the candidate preys interact specifically with the bait in the two-hybrid assay. This is accomplished by testing whether the prey can activate the weak promoter reporter in the absence of the bait fusion protein.

6. Perform the specificity test as follows:

For Candidates Selected Using the *HIS3* System:

  a. Double-transform strain KJ1567 with the following combinations of plasmids:

   i. Prey candidate plasmid + pBT (specificity test)

   ii. Prey candidate plasmid + pBait (specificity test – positive control)

   iii. pBR-αGal4 + pACλcI-Gal11P (positive control)

  b. Serially dilute each transformation 1:10 two times (i.e., to 10^{-2}) in NM medium.

  c. Spot 5 μl of each transformation and its dilutions on the indicated plates.

> NM/C/C/K/I plates (verifies transformation efficiency)
> NM/C/C/K/I/20 mM 3-AT plates (checks activation of the reporter)

Incubate for 18–24 hours at 37°C.

  d. Analyze growth on the plates. Candidates that grow on selective medium only in the presence of the bait fusion protein likely encode preys that interact specifically with the bait.

For Candidates Selected Using the *bla* System:

  a. Double-transform the Bacteriomatch reporter strain with the following combinations of plasmids:

   i. Prey candidate plasmid + pBT (specificity test)

   ii. Prey candidate plasmid + pBait (specificity test – positive control)

   iii. pTRG-Gal11P + pBT-LGF2 (positive control)

  b. Serially dilute each transformation 1:10 two times (i.e., to 10^{-2}) in LB medium.

  c. Spot 5 μl of each transformation and its dilutions on the indicated plates:

> LB/T/C/K plates (verifies transformation efficiency)
>
> LB/T/C/K plates with 250 μg/ml carbenicillin (checks activation of the reporter)

Incubate for 18–24 hours at 37°C.

  d. Analyze growth on the plates. Candidates that grow on selective medium only in the presence of the bait fusion protein likely encode preys that interact specifically with the bait.

 7. For additional characterization of positive isolates, please see the discussion on Subsequent Characterization from the Yeast Two-hybrid section above.

Additional Method: For Protein–DNA Interactions

This section provides an overview of a protocol using the bacterial two-hybrid system to study protein–DNA interactions. The intended goal of this type of experiment is to identify proteins (preys) from a library that bind specifically to a DNA site of interest (the bait). The protocol described here uses the *HIS3* selection system and two fusion proteins containing fragments of the yeast Gal11P and Gal4 proteins, known to interact with each other (Farrell et al. 1996). The Gal4 fragment is fused to the RNAP α subunit, and the Gal11P fragment is fused to a library of prey proteins. In this configuration, if a prey fusion protein binds to the bait DNA-binding site, RNAP is recruited to the weak reporter promoter via the Gal11P–Gal4 interaction. This approach has worked previously in selections to identify DNA-binding proteins with new specificities (Joung et al. 2000). Theoretically, although not described here, one could also use the *bla* selection system or a single fusion protein ("one hybrid") approach (with preys fused directly to RNAPα) as well.

Construction of a Bait Reporter Strain

1. A DNA site of interest (the bait) is positioned upstream of the weak promoter that controls expression of the selectable reporter. This modified reporter is referred to hereafter as the "bait reporter." The bait reporter is then introduced into the Δ*hisB* strain KJ1C by bacterial mating.

   The technical details of the method to accomplish this are beyond the scope of this chapter; however, the reader is referred to references that provide additional information (Whipple 1998; Joung et al. 2000).

Activation Test of the Bait Reporter

After introducing the bait DNA sequences upstream of the weak promoter, check that these new sequences do not affect basal transcription of the promoter. To do this, compare the level of *HIS3* expression from the new bait reporter with that from a reporter that lacks the upstream bait DNA sequences. *HIS3* expression can be tested by examining growth of cells on His-deficient medium. If the level of *HIS3* is not decreased or elevated, the new bait reporter is suitable for use in selection experiments.

2. Inoculate cultures of KJ1567 (control) and the KJ1C strain harboring the bait reporter in 3 ml of NM medium supplemented with kanamycin (50 μg/ml). Grow for 16–18 hours at 37°C.

3. Serially dilute each culture 1:10 eight times (i.e., dilution to 10^{-8}) in NM medium.

4. Spot 5 μl of the 10^{-3} through 10^{-8} dilutions for each culture on NM/K and NM/K/5 mM 3-AT plates. Allow the spots to dry and incubate 12–18 hours at 37°C.

5. Identify the dilution spots with single well-isolated colonies and score growth. KJ1567 should express sufficient *HIS3* to permit growth on NM plates lacking 3-AT but should be unable to grow on NM plates with 5 mM 3-AT. The growth pattern of the KJ1C strain with the bait reporter should be similar to that of KJ1567 on the two types of plates.

Construction of a Selection Strain Harboring the Bait Reporter

6. Transform strain KJ1C harboring the bait reporter with the pACL-αGal4 plasmid and plate on LB/C/K plates. Proceed using the cells directly from the transformation plate; however, if there will be a delay of longer than a week before proceeding, the cells should be stored as a frozen glycerol stock at –80°C.

   From this point forward, use the same *HIS3* system protocols detailed in Stages 2 and 3 above (for protein–protein interactions). The one difference is that when making the prey fusion protein library, the prey-encoding inserts should be cloned into phagemid pKJ1514 (see Tables).

ACKNOWLEDGMENTS

We thank Scot Wolfe, Jeff Miller, and Matt Rhoades for helpful comments on the manuscript; Jessica Hurt and Elizabeth Ramm for expert help in constructing bacterial 2Hyb plasmids and strains; Carl Pabo for his resources, support, and encouragement; and Ann Hochschild and Simon Dove (Harvard Medical School) and Bonnie Wu (Stratagene) for providing advice and protocols for the *bla* selection method.

REFERENCES

Brennan M.B. and Struhl K. 1980. Mechanisms of increasing expression of a yeast gene in *Escherichia coli. J. Mol. Biol.* **136:** 333–338.

Brent R. and Ptashne M. 1980. The lexA gene product represses its own promoter. *Proc. Natl. Acad. Sci.* **77:** 1932–1936.

———. 1984. A bacterial repressor protein or a yeast transcriptional terminator can block upstream activation of a yeast gene. *Nature* **312:** 612–615.

Chien C.T., Bartel P.L., Sternglanz R., and Fields S. 1991. The two-hybrid system: A method to identify and clone genes for proteins that interact with a protein of interest. *Proc. Natl. Acad. Sci.* **88:** 9578–9582.

Dove S.L. and Hochschild A. 1998. Conversion of the omega subunit of *Escherichia coli* RNA polymerase into a transcriptional activator or an activation target. *Genes Dev.* **12:** 745–754.

Dove S.L., Huang F.W., and Hochschild A. 2000. Mechanism for a transcriptional activator that works at the isomerization step. *Proc. Natl. Acad. Sci.* **97:** 13215–13220.

Dove S.L., Joung J.K., and Hochschild A. 1997. Activation of prokaryotic transcription through arbitrary protein-protein contacts. *Nature* **386:** 627–630.

Durfee T., Becherer K., Chen P.L., Yeh S.H., Yang Y., Kilburn A.E., Lee W.H., and Elledge S.J. 1993. The retinoblastoma protein associates with the protein phosphatase type 1 catalytic subunit. *Genes Dev.* **7:** 555–569.

Duttweiler H.M. 1996. A highly sensitive and non-lethal beta-galactosidase plate assay for yeast. *Trends Genet.* **12:** 340–341.

Estojak J., Brent R., and Golemis E.A. 1995. Correlation of two-hybrid affinity data with in vitro measurements. *Mol. Cell. Biol.* **15:** 5820–5829.

Farrell S., Simkovich N., Wu Y., Barberis A., and Ptashne M. 1996. Gene activation by recruitment of the RNA polymerase II holoenzyme. *Genes Dev.* **10:** 2359–2367.

Fields S. and Song O. 1989. A novel genetic system to detect protein-protein interaction. *Nature* **340:** 245–246.

Finley R.L., Jr. and Brent R. 1994. Interaction mating reveals binary and tertiary interactions between *Drosophila* cell cycle regulators. *Proc. Natl. Acad. Sci.* **91:** 12980–12984.

Golemis E.A. and Brent R. 1992. Fused protein domains inhibit DNA binding by LexA. *Mol. Cell. Biol.* **12:** 3006–3014.

Golemis E. and Serebriiskii I. 1998. Two-hybrid systems/interaction trap. In *Cells: A laboratory manual* (ed. D.L. Spector et al.), vol. 3, pp. 69.1–69.40. Cold Spring Harbor Laboratory Press, Cold Spring Harbor, New York.

Golemis E.A., Serebriiskii I., Gyuris J., and Brent R. 1997. Interaction trap/two-hybrid system to identify interacting proteins. In *Current protocols in molecular biology* (ed. F.M. Ausubel et al.), vol. 3, pp. 20.1.1–20.1.35. John Wiley, New York.

Gyuris J., Golemis E.A., Chertkov H., and Brent R. 1993. Cdi1, a human G1 and S phase protein phosphatase that associates with Cdk2. *Cell* **75:** 791–803.

Harlow E. and Lane D. 1988. *Antibodies: A laboratory manual.* Cold Spring Harbor Laboratory, Cold Spring Harbor, New York.

Hu J.C., Kornacker M.G., and Hochschild A. 2000. *Escherichia coli* one- and two-hybrid systems for the analysis and identification of protein-protein interactions. *Methods* **20:** 80–94.

Joung J.K., Ramm E.I., and Pabo C.O. 2000. A bacterial two-hybrid selection system for studying protein-DNA and protein-protein interactions. *Proc. Natl. Acad. Sci.* **97:** 7382–7387.

Petermann R., Mossier B.M., Aryee D.N., and Kovar H. 1998. A recombination based method to rapidly assess specificity of two-hybrid clones in yeast. *Nucleic Acids Res.* **26:** 2252–2253.

Sambrook J. and Russell D. 2001. *Molecular cloning: A laboratory manual,* 3rd ed. Cold Spring Harbor Laboratory Press, Cold Spring Harbor, New York.

Sambrook J., Fritsch E.F., and Maniatis T. 1989. *Molecular cloning: A laboratory manual,* 2nd ed. Cold Spring Harbor Laboratory. Cold Spring Harbor, New York.

Schiestl, R.H. and Gietz, R.D. 1989. High efficiency transformation of intact yeast cells using single stranded nucleic acids as a carrier. *Curr. Genet.* **16:** 339–346.

Serebriiskii I.G. and Golemis E.A. 2000. Uses of lacZ to study gene function: Evaluation of beta-galactosidase assays employed in the yeast two-hybrid system. *Anal. Biochem.* **285:** 1–15.

Serebriiskii I., Khazak V., and Golemis E.A. 1999. A two-hybrid dual bait system to discriminate specificity of protein interactions. *J. Biol. Chem.* **274:** 17080–17087.

———. 2001. Redefinition of the yeast two-hybrid system in dialog with changing priorities in biological research. *BioTechniques* **30:** 634–636, 638, 640.

Serebriiskii I.G., Toby G.G., and Golemis E.A. 2000a. Streamlined yeast colorimetric reporter assays, using scanners and plate readers. *BioTechniques* **29:** 278–279, 282–284, 286–288.

Serebriiskii I., Estojak J., Berman M., and Golemis E.A. 2000b. Approaches to detecting two-hybrid false positives. *BioTechniques* **28:** 328–336.

Shaywitz A.J., Dove S.L., Kornhauser J.M., Hochschild A., and Greenberg M.E. 2000. Magnitude of the CREB-dependent transcriptional response is determined by the strength of the interaction between the kinase-inducible domain of CREB and the KIX domain of CREB-binding protein. *Mol. Cell. Biol.* **20:** 9409–9422.

Struhl K., Cameron J.R., and Davis R.W. 1976. Functional genetic expression of eukaryotic DNA in *Escherichia coli. Proc. Natl. Acad. Sci.* **73:** 1471–1475.

Stuhl K. and Davis R.W. 1977. Production of a functional eukaryotic enzyme in *Escherichia coli:* Cloning and expression of the yeast structural gene for imidazole glycerolphosphate dehydratase (his3). *Proc. Natl. Acad. Sci.* **74:** 5255–5259.

Vasavada H.A., Ganguly S., Germino F.J., Wang Z.X., and Weissman S.M. 1991. A contingent replication assay for the detection of protein-protein interactions in animal cells. *Proc. Nat. Acad. Sci.* **88:** 10686–10690.

Vojtek A.B., Hollenberg S.M., and Cooper J.A. 1993. Mammalian Ras interacts directly with the serine/threonine kinase Raf. *Cell* **74:** 205–214.

Watson M.A., Buckholz R., and Weiner M.P. 1996. Vectors encoding alternative antibiotic resistance for use in the yeast two-hybrid system. *BioTechniques* **21:** 255–259.

Whipple F.W. 1998. Genetic analysis of prokaryotic and eukaryotic DNA-binding proteins in *Escherichia coli. Nucleic Acids Res.* **26:** 3700–3706.

8 Phage-display Approaches for the Study of Protein–Protein Interactions

Carl S. Goodyear and Gregg J. Silverman

Department of Medicine, University of California, San Diego, La Jolla, California 92093-0663

INTRODUCTION

In recent years, phage display has evolved into a powerful tool providing opportunities to define natural protein–protein interactions and to mold novel ligand receptors. The essential advantages of phage-display approaches originate in the incorporation of the protein and genetic components into a single phage particle (Smith 1985). By providing a direct physical link between the expressed protein and the encoding genetic information, one can efficiently perform iterative rounds of selection of clones with desirable functional capacities followed by amplification of the selected sublibrary. Hence, during library selection (or panning), specific phage clones are progressively enriched on the basis of their specificity and affinity for ligand. Thus, relatively rare ligand-binding clones can be rescued rapidly and efficiently from large libraries. As these expression cloning systems have matured, versatile selection methods have been reported that are based on the functional properties of displayed proteins in diverse immunochemical and biological settings.

143

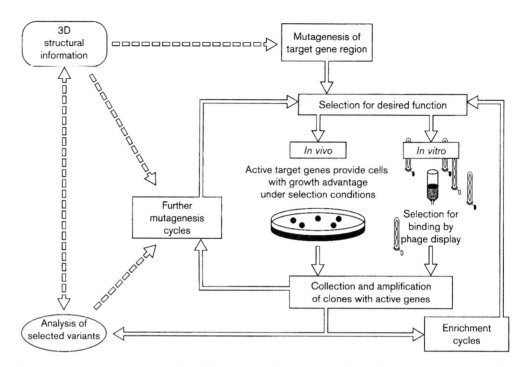

FIGURE 1. Engineering of proteins based on structural information. Three-dimensional structural information is an essential guide for targeting randomizing mutagenesis to regions of proteins presumed to be critical for function. This scheme depicts two examples of successful strategies for in vivo and in vitro selection of desired variants from large combinatorial libraries, with the dashed arrows indicating the flow of information. The outcome of such experiments will also yield new information about structure and protein interactions. (Reprinted, with permission, from Kast and Hilvert 1997, © Excerpta Medica Inc.)

To appreciate the special advantages of phage display, one must understand the associated features that distinguish it from other expression cloning methods. In many of these other approaches (Helfman et al. 1983; Young and Davis 1983; Seed and Aruffo 1987; Sikela and Hahn 1987; Singh et al. 1988; Fields and Song 1989; Germino et al. 1993), the gene encoding a protein of interest is also identified from complex cloned libraries based on expression in prokaryotic hosts. The central difference lies in how specific binders are isolated. In these earlier expression systems, screening requires the immobilization of the expressed library of protein translation products onto a solid support, such as a transfer membrane, which potentially results in altered protein conformations that compromise the structural and ligand-binding properties of these recombinant molecules. Moreover, in these alternative approaches, only a limited number of individual clones can be examined easily, and it is difficult to scale up these methods to increase the size of the library that can be thoroughly screened. Hence, in these libraries, the identification of genes represented at low frequencies is technically demanding and labor-intensive. In contrast, in phage display, the direct linkage of the phenotypic properties of a clone with encoding genetic information enables efficient selection and enrichment on the basis of a variety of functional interactions, and thereafter, this library subset can be readily expanded.

A wide variety of molecules of divergent size and character have been successfully displayed on the surface of filamentous phage. These include enzymes (McCafferty et al. 1990), antibodies (Burton and Barbas 1993; Winter et al. 1994), cytokines (Gram et al. 1993), protein fragments (Petersen et al. 1995), and peptides (Smith 1985; Cwirla et al. 1990; Scott and Smith 1990). These studies document the fact that recombinant proteins directed to the periplasmic space of *Escherichia coli* often refold into conformations that convey functional properties similar to those occurring in their native environment. Therefore, this evolving technology has offered a means to

TABLE 1. Genes and Gene Products of the f1 Bacteriophage

Gene	Function	No. of amino acids	Protein MW
II	DNA replication	410	46,137
X	DNA replication	111	12,672
V	binding ssDNA	87	9,682
VIII	major capsid protein	50	5,235
III	minor capsid protein	406	42,522
VI	minor capsid protein	112	12,342
VII	minor capsid protein	33	3,599
IX	minor capsid protein	32	3,650
I	assembly	348	39,502
IV	assembly	405	43,476
XI	assembly	108	12,424

The number of amino acids and the molecular weight are for the mature proteins. The initiating methionine is included in proteins that do not contain an amino-terminal signal sequence. (Reprinted, with permission, from Webster 2001.)

harness the insights from advanced structural analyses (Fig. 1) and the explosion of DNA sequences from host genomes. This technology has also been used for the discovery and characterization of novel ligand receptor-binding interactions, even in cases where a detailed understanding of structure–function relationships is not available (for review, see O'Neil and Hoess 1995; Katz 1997). In fact, by combination of rational design and a combinatorial approach, phage-display technology can be used to create novel small proteins with properties and functional capacities that exceed those of naturally occurring proteins.

OUTLINE OF PROCEDURE

Overview of the Biology of Phage Display

Phage-display technology was made possible through the in-depth understanding of the genetics and physiology of the highly related f1 and M13 filamentous bacteriophage (Fig. 2) (for review, see Sambrook and Russell 2001, *Molecular Cloning*, pp. 3.1–3.49; Webster 2001). Although there are several different potential variations of approach, many investigators have used cloning systems in which the DNA sequences encoding the coat protein–target peptide fusions are carried within a plasmid, or phagemid (Sambrook and Russell 2001, *Molecular Cloning*, pp. 3.42–3.49), that also contains a filamentous phage origin of DNA replication. Consequently, when permissive *E. coli* strains containing these plasmids are later infected (or rescued) with helper phage carrying the other native phage proteins, the resulting phagemid constructs are efficiently packaged into phage particles. The helper phage contains an attenuated origin of replication; therefore, its genome is relatively inefficiently packaged in comparison to the phagemid sequence, which enhances the representation of useful clones in a library. Compared to phage vectors, the phagemid vector systems offer the advantage of simple manipulation of libraries in plasmid form. There are also implications for the efficiency and potential valency of display (as discussed below).

The properties of the selection system are greatly affected by the choice of the phage coat protein used for fusion with the recombinant protein(s) of interest. There are only 11 gene products encoded by the genome of the filamentous phage (Table 1). To date, the most commonly adopted approach employs cloning into mature or truncated phage coat proteins III or VIII. For pVIII, there are an estimated 2700 copies per individual virion. Fusion proteins involving pVIII can afford high multivalent display, which enables selection methods that yield clones on the basis of avidity effects. For example, murine anti-ganglioside-specific antibody clones could be selected

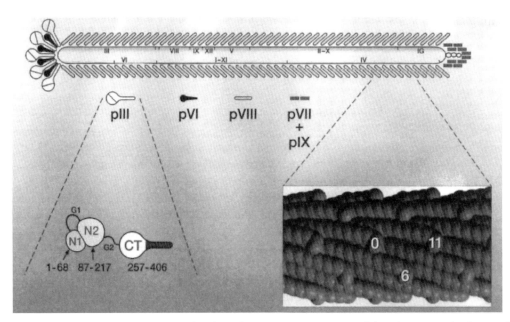

FIGURE 2. The Ff bacteriophage particle. This schematic representation of the phage particle depicts the location of the capsid protein and the orientation of the phage DNA. At the lower left is a schematic of the structure of the pIII, in which N1, N2, and CT refer to domains, and G1 and G2 refer to glycine-rich regions. At the lower right is a representation of the orientation of the VIII molecules along the cylindrical portion of the phage. Three neighboring pVIII molecules are labeled as 0, 6, and 11. (Adapted, with permission, from Webster 2001; Marvin 1998, © Excerpta Medica Inc.)

as pVIII-fusion proteins from a phage-display library, but not from the equivalent pIII library, presumably because the monomeric pIII display was below some threshold of binding affinity necessary for selection (Qiu et al. 1999). There is further evidence to suggest that pVIII systems facilitate selection on the basis of lower-affinity interactions that might not be recovered in gene III-based systems (Folgori et al. 1994). However, there is only relatively inefficient interclonal competition, and thus the highest-affinity clones may not be as efficiently recovered. pVIII systems also may be limited in the size of the cloned insert due to impaired phage assembly, because fusions of polypeptides greater than 10 amino acids to pVIII are generally not well tolerated (Iannolo et al. 1995; Malik et al. 1996). This is an especially important limitation in phage systems that do not allow mixing of wild-type coat protein from helper phage. In contrast, pIII has more often been used for displaying polypeptides encoded by cDNA and genomic sequences, because it can readily accommodate fusion with larger polypeptides (Crameri and Suter 1993; Shanmugavelu et al. 2000). For pIII, there are only five copies per virion, and these are localized to one tip of the phage particle. Hence, vectors that create fusions to pIII may endow more oligovalent (or even monovalent) interactions, which can be an advantage for the selection of uncommon species of phage from large libraries on the basis of affinity.

Other filamentous phage coat proteins have also been used to display foreign proteins, including pVI (Jespers et al. 1995; Fransen et al. 1999) or pVII and pIX (Gao et al. 1999). These systems may also be modified to construct cDNA or genomic expression libraries (Sambrook and Russell 2001, *Molecular Cloning*, pp. 11.1–11.124, 14.1–14.51). Several other display systems have also been developed using nonfilamentous phage. These include fusions with the bacteriophage λ capsid protein D, which accepts large fusions at the amino or carboxyl termini (Sternberg and Hoess 1995; Santini et al. 1998), and the tail protein V of the same virus (Maruyama et al. 1994; Dunn 1995; Kuwabara et al. 1997). In addition, the bacteriophage T4 has also been exploited for displaying peptides and larger protein domains (Efimov et al. 1995; Ren et al. 1996; Houshmand et al. 1999).

Potential Limitations of Phage Display

A major goal of protein engineering studies is to improve upon the functional capacity of a domain by altering the composition and, at times, the size of the polypeptide. These studies require systems capable of efficient and productive in vitro expression. Several decades of investigation have provided impressive successes for in vitro protein expression. However, all too often, the expression of a functional eukaryotic protein has proven to be incompatible with bacterial systems. These limitations often derive from the inherent inability of bacteria to perform many common posttranslational modifications. In other cases, unpredictable pitfalls have been encountered that are attributed to problems in protein folding, because mammalian protein folding into a functional conformation is often dependent on chaperone proteins. These chaperones perform essential functions that are as yet poorly understood. Local structural context can also affect protein folding; the fusion of a gene sequence to a foreign gene can interfere with the native functional activity. From this perspective, it was almost unexpected that mammalian proteins could be displayed fused to the coat protein of a filamentous bacteriophage yet still show their native functional capacity. The list of reports of successfully displayed proteins is now quite extensive (for reviews, see Lowman et al. 1991; Chiswell and McCafferty 1992; Clackson and Wells 1994; O'Neil and Hoess 1995).

From one perspective, a functional domain from a biologic system represents the end product of a highly directed evolutionary process. Unlikely as it may seem, it has now been firmly established in a number of studies that phage-display approaches can yield variant domains with properties that have been highly improved for binding affinity, or for fine specificity, or even for structural refinements that greatly enhance stability (Roberts et al. 1992).

Considerations for the Design of Phage-display Library Approaches

In engineering studies of a domain, the first standard step in developing phage-display libraries is a proof-of-principle demonstration that the native domain retains functionality when displayed on a phage surface. Only thereafter can phage-display methods be applied to investigate the minimal structural requirements for the defined functional activity.

When preparing a library for phage display, it is an essential requirement for expression that the encoding sequences are directionally cloned to maintain the proper 5′ and 3′ orientation in the display vector. Equally important is the correct reading frame of the encoding gene that must be maintained in its position between the carboxyl terminus of a bacterial export signal sequence and the downstream pIII or pVIII filamentous coat protein gene. Although effective strategies can readily be designed for the cloning of libraries of relatively homogeneous genes, like antibody genes or oligonucleotides encoding defined peptides, in more diverse libraries like those of genomic DNA fragments and cDNA, the 5′ and 3′ ends may be heterogeneous and/or undefined. For these types of genes, potentially only 1 of every 18 clones created (2 × 3 × 3) will contain a fusion that is optimal. It necessitates both the correct orientation (i.e., twofold variations) and the correct reading frame with both the signal sequence that is required for expression of a soluble protein directed to the periplasmic space (i.e., threefold variation) and the phage coat protein that is required for display on the phage surface (i.e., threefold variations). These problems greatly reduce the representation of potentially functional clones that can be selected from the library.

During the generation of cDNA libraries, there is also an additional complication, stemming from the presence of translational stop codons at the 3′ end of the mRNA of cloned eukaryotic genes. In these cases, a stop codon occurring before the amino terminus of the phage coat proteins prematurely truncates the fusion product, effectively preventing the display of their expressed polypeptides. In a recent report, the fusion of the library of variant domains to the carboxyl terminus of coat proteins, pIII and pVIII, was used to avoid this problem (Fuh and Sidhu 2000; Fuh et al. 2000).

Success in the application of phage-display methods requires an understanding of the conformational and functional integrity of the targeted proteins in the cDNA or genomic expression libraries. Efforts to alter a functional activity are greatly advanced by an understanding of the specific amino acids that are the contact residues responsible for the functional properties, or which stabilize the overall structure of the protein (also termed the framework). Structural crystallographic or nuclear magnetic resonance (NMR) studies can identify these specific contact sites. Methods like alanine scanning (Sambrook and Russell 2001, *Molecular Cloning*, pp. 13.81) can confirm the relative contribution of these positions to functional activity and structure. From these insights, a more limited number of critical residues can be identified, and the codons responsible for these residues can be randomized (also termed variegated) to generate a library that provides the means to isolate a variant with functional properties altered by design.

It must also be appreciated that by using currently available methods, there are limitations to the size and diversity that can be created within any combinatorial library, especially when sequence diversity is increased experimentally by codon randomization. To randomize (or variegate) the amino acids expressed at critical positions, oligonucleotides must be synthesized that have degenerate codons which have at most 64 different nucleic acid sequences ($4 \times 4 \times 4$). Hence, for the variegation of a gene at N codon positions, a library of up to 64^N members is required. To decrease the required codon diversity, several nucleotide doping strategies are often used. These include the degenerate codon, NNK (in which N is any nucleotide, and K is either G or T), which allows for all 20 amino acid combinations within 32 codons. Alternatively, the degenerate codon VNS (in which V represents A, C, or G, and S represents G or C), allows for 24 codons, omitting Tyr, Phe, Cys, and all stop codons. These types of strategies are highly desirable due to the strict numerical relationship size of library (or number of independent transformants) and the maximum number of positions that can be completely randomized. Hence, with the most commonly used cloning and transfection methods, libraries up to about 10^8–10^9 independent members represent a practical limit. Consequently, practical considerations dictate that at most 6–8 positions can be randomized (representing 20^6–20^8) (Clackson and Wells 1994). However, there is ample evidence that it may not be necessary to generate "complete libraries" with representation of all possible sequence variations to obtain successful isolation of valuable clones. As discussed below, Nord et al. reported the successful isolation of functional clones from a library generated with only 4×10^7 independent clones, even though 13 codons were randomized (which would require approximately 4×10^{19} variations to be complete; Nord et al. 1997), demonstrating that an underrepresented library can be sufficient for the isolation of specific binders.

Greater diversity can also be created using methods that enable the generation of very large libraries, including the use of λ packaging systems (Sambrook and Russell 2001, *Molecular Cloning*, pp. 2.1–2.117) to boost transformation efficiencies (Alting-Mees and Short 1993), or by Cre-lox-based in vivo recombination (Sambrook and Russell 2001, pp. 4.82–4.85) of duplexed libraries (Fisch et al. 1996). These approaches have not yet been widely adopted. In a recent report (Sidhu et al. 2000), Sidhu and coworkers describe a bacterial strain that greatly improves the capacity to scale up the size of libraries obtained from electroporation (Sambrook and Russell 2001, pp. 16.33–16.36) approaches.

Although initial reports focused on selection against ligands immobilized onto a solid phase, there is now ample evidence that ligand-specific clones can be efficiently selected in a variety of assay formats without the need to isolate or purify the target ligand of interest. Of the alternatives available, enzyme-linked immunosorbent assay (ELISA) plates, magnetic beads, and even cell surfaces have been used successfully. Popular opinion states that the conformational and functional properties of a protein may be affected by the format or immobilization setting with which it is associated. Of special note are the reports in which direct immobilization of purified proteins onto plastic surfaces can at times partially denature the protein, altering the conformation and potentially accessible binding epitopes. This topic has been reviewed at length (Silverman 2001), and it is important to consider that it may be advisable to design selection conditions that mostly closely emulate the setting in which the properties of the isolated clone will be used in future work. Otherwise, one may end up selecting clones that do not have desirable (or useful) binding properties.

Phage Display and a Model Domain

There are reports of many different types of methodological approaches that use the power of phage display to recover variant or gene-fragmented clones with desirable properties. In identifying a useful example, we opted to provide detailed methods for the generation of libraries of variant domains that contain a limited number of variegated codons. The adapted methods presented below have been validated in our laboratory by efforts now in progress that seek to isolate variant domains based on a single domain of the virulence factor, protein A, produced by clinical isolates of *Staphylococcus aureus* (SpA).

In its native form, SpA is a 45-kD secreted membrane protein that contains five highly homologous 56- to 61-amino-acid domains that each demonstrate two separate immunoglobulin (Ig) binding activities—IgG Fc (*F*ragment that *c*rystallizes) specificity and Fab (*F*ragment *a*ntigen *b*inding) specificity. Bacterial Ig-binding proteins have proven to be an attractive model for understanding the functional capacities of defined domains. These proteins have become of interest, in part, due to their highly ordered secondary structure, which imparts great structural stability without the involvement of internal disulfide bonds. In reports from two independent groups, single domains of SpA displayed as fusions at the amino terminus of the gIII coat protein were shown to retain Fc binding activity, and IgG binding could be used for efficient selection from dilute libraries (Djojonegoro et al. 1994; Kushwaha et al. 1994).

Wells and colleagues used the detailed understanding of structural basis for the IgG Fc-SpA-binding interaction to design a strategy to create much smaller protein derivatives with the same functional capacity (Fig. 3A). In the native molecule, only two of the three α helices in the 59-amino-acid domain B of SpA are involved in Fc binding, whereas the third appears to stabilize a hydrophobic core and may be involved in the separate Fab-binding activity. To optimize an Fc-binding domain, a library based on the first 38 amino acids was created, and, following several rounds of randomization of several codons, a variant was selected in which the hydrophobic interaction of helix 1 and helix 2 was stabilized, while Fc-binding activity was retained (Fig. 3B–D). Hence, starting from a domain of 59 amino acids that folds into a 3-α-helical bundle that binds the Fc fragment of IgG with a K_D of 10 nM, a 38-residue form (termed Z38) that binds with a K_D of 43 nM was characterized (Braisted and Wells 1996). In a later study (Starovasnik et al. 1997), NMR analysis demonstrated that a synthetic peptide of the Z38 sequence is composed of two antiparallel α helices in a tertiary structure remarkably similar to that observed for the first two helices of the B-domain in the B-domain/Fc complex (Deisenhofer 1981). A 34-residue analog (Z34C) was designed in which an interhelical disulfide bond provided dramatically enhanced stability, and ninefold higher affinity IgG-binding activity. By NMR analysis, Z34C, like Z38, was shown to have a structure virtually identical to the equivalent region from SpA domains. This stabilized, minimized, two-helix peptide, which is about half the size and has one-third of the remaining residues altered, accurately mimics both the structure and function of the native domain (Braisted and Wells 1996; Starovasnik et al. 1997). This small engineered peptide analog, Z34C, currently represents one of the smallest functional peptides with features typical of a folded protein.

The well-ordered triple α-helical bundle structure of SpA domains has also been successfully exploited to derive efficient and stable domains with novel binding activities. Using a solid-state synthesis strategy, the codons for 13 noncontiguous surface-exposed residues of helix 1 and helix 2 of a synthetic 58-amino-acid SpA domain were variegated, and nonexhaustive libraries of less than 10^8 members were created (Nord et al. 1995, 1997). Seven of these 13 positions are directly involved in the native Fc-binding activity. Residues involved in stabilizing the hydrophobic core

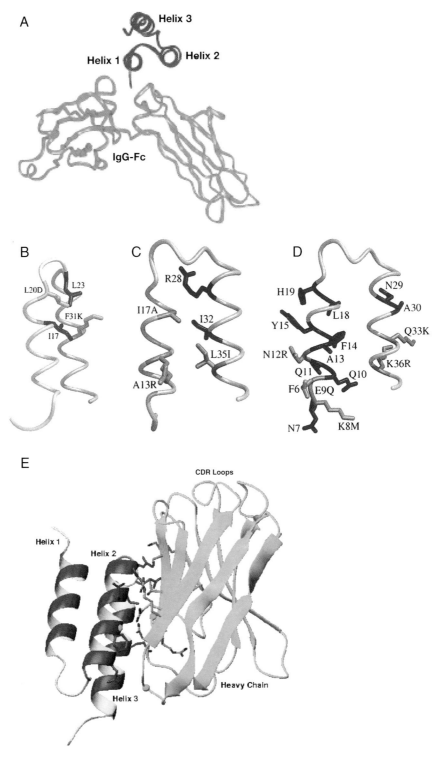

FIGURE 3. (*See facing page for legend.*)

were conserved. From phage-display libraries, "affibodies" of diverse specificity were subsequently recovered after 3–5 rounds of selection, including binders to *Taq* DNA polymerase, human insulin, and human apolipoprotein. Clones expressed in *E. coli* were evaluated by SDS-PAGE, biosensor studies, and CD spectroscopy to demonstrate the conservation of the parental α-helical structure and specific micromolar binding activities. These binding affinity constants (i.e., dissociation constants, K_D) ranged from ~2 × 10⁻⁶ M for *Taq* DNA polymerase, ~3 × 10⁻⁵ M for human insulin, and ~3 × 10⁻⁶ M for apolipoprotein. These studies document potential engineering opportunities that use the stable secondary structure of the SpA domain to build novel binding activities upon a well-ordered scaffolding. In these studies, the exceptional stability and high solubility of SpA were significant assets, and the possible utility of expressing these affibodies as intracellular proteins with therapeutic activities was also suggested.

Our group has been primarily interested in understanding the Fab-binding specificity of SpA. In earlier studies, we evaluated the binding interaction of SpA with human monoclonal Ig proteins of known variable region usage (Sasso et al. 1989, 1991). To characterize the interaction further, we used immobilized SpA to select binders from IgM and IgG Fab phage-display libraries generated from the peripheral lymphoid cells of a healthy donor (Sasano et al. 1993; Silverman et al. 1996). These demonstrated that the SpA-binding activity was restricted to antibodies expressing variable-region genes from the human VH3 family, and binding was unaffected by antibody light-chain usage. In more recent studies, crystallographic analysis of a cocrystal of a human VH3 IgM Fab and a domain of SpA elucidated the molecular basis of this special type of binding interaction. These structural analyses demonstrated that contacts were mediated by conserved VH3 family-specific residues in framework 1 and framework 3 of the Fab, without involvement of the variable-region β loops, termed complementarity-determining regions (CDR), that mediate conventional antigen binding (Fig. 3E) (Graille et al. 2000). Although the overwhelming involvement of Fab β-strand contacts has not been reported in any type of antigen receptor-binding interaction, other features of this interaction are reminiscent of the properties of microbial superantigens for T lymphocytes. It is also notable that the SpA-specific VH-binding motif is highly conserved in the repertoires of almost all mammalian B-cell repertoires, and it appears to have first arisen at least as early as the amphibian immune repertoire (Cary et al. 2000).

Within investigations of the murine response to in vivo challenge, we have demonstrated that SpA can induce large-scale lymphocyte clonal deletion that is specific for B cells expressing the conserved VH-binding motif for SpA (Silverman et al. 1998, 2000). We have developed the following protocol in our efforts to determine whether phage-display technology can be used to develop SpA variant domains with optimized B-cell superantigen activities.

FIGURE 3. Immunoglobulin binding interactions of a SpA domain. Each domain of SpA uses residues in helix 1 and helix 2 for Fc binding and residues in helix 2 and helix 3 for Fab binding. The triple α-helical structure is stabilized by a hydrophobic core. (*A*) Ribbon diagram of helix 1 and helix 2 of domain B of SpA (dark) in a complex with an IgG-Fc fragment (light). (*B, C*, and *D*) Ribbon diagrams of the truncated Z-domain, a double helical variant, selected for Fc binding from exoface (*B*), intraface (*C*), and interface (*D*) libraries produced by helix 2. Colored residues represent randomized sites in each library; residues that were conserved as wild type are blue and residues that were selected as residue other than wild type are in yellow. (*E*) Schematic representation of the complex between SpA domain D and 2A2 Fab V_H region from a human IgM. This side view shows the interaction of the helix 2 and helix 3 of SpA domain D (*red*) with residues in the β strands of the framework region 1 and framework region 3 V_H subdomains of the heavy chain (*blue*). (Provided, with permission, by E. Stura, M. Graille, and J.-P. Charbonnier, DIEP, CEA, Saclay, France.) (*A, B, C*, and *D* were reprinted, with permission, from Braisted and Wells 1996, © National Academy of Sciences U.S.A.)

Generation of Variant Domain Libraries

The following protocol, depicted as a flowchart in Figure 4, can be used for the generation and/or engineering of protein domains. It was adapted in our laboratory for the development of variant SpA domains with novel Fab-binding specificities. Using the structural information outlined above, we used this protocol to generate libraries of domains in which we randomized 6–8 codons that define the natural VH3 specificity of SpA. In our effort, these codons were variegated using the NNK or VNS doping strategy, as described above (p. 148). Toward this goal, three large (60- to 88-residue) oligonucleotides with targeted degenerate codons were designed. To facilitate construction of the composite domain library, these oligonucleotides had overlapping complementary stretches of at least 12 nucleotides. Of course, for application to other domains, the number and composition of the required oligonucleotides are dependent on the specific variant domains to be generated.

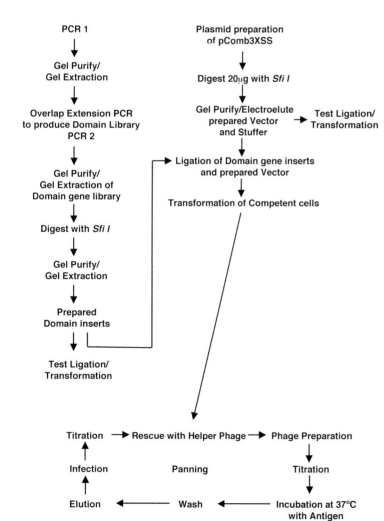

FIGURE 4. Flowchart depicting domain gene library construction and selection. The initial construction of a domain library begins with the annealing and amplification of oligonucleotides to generate the degenerate library, followed by the preparation of these inserts (*left*). During the preparation process, pComb3X vector is also prepared and tested for ligation efficiency. Prepared inserts are ligated into the phagemid vector and transformed into competent bacterial cells. Phage are rescued by the addition of helper phage, and the panning/selection process is started using the rescued phage. Panning is a cyclic procedure that is carried out in consecutive rounds over the course of several days.

MATERIALS

CAUTION: See Appendix for appropriate handling of materials marked with <!>.

Buffers and Solutions

Agarose (GIBCO-BRL 15510-027)

Blocking buffer

    3% (w/v) bovine serum albumin diluted in PBS

    1% (w/v) bovine serum albumin diluted in PBS

Carbenicillin stock solution

    50 mg/ml in dH_2O, filter-sterilized (Sigma C1389)

10x DNA gel loading dye

    3.9 ml of glycerol

    500 µl of 10% SDS <!>

    200 µl of 0.5 M EDTA

    0.025 g of bromophenol blue <!>

    0.025 g of xylene cyanol <!>

        Bring to 10 ml total volume with H_2O.

dNTP mix (2.5 mM) (dATP/dCTP/dGTP/dTTP set, 100 mM, Amersham Pharmacia Biotech 27-2035-02)

DNA molecular weight marker (100 bp; Amersham Pharmacia Biotech 27-4001-01)

DNA molecular weight marker (1-kb; GIBCO-BRL 15615-024)

Dynabeads M-280 streptavidin (Dynal 112.05)

Ethanol <!>

Ethanol (70%), diluted in dH_2O

Kanamycin stock solution, 50 mg/ml (Sigma K 0254)

LB Agar plates with 50 µg/ml carbenicillin

    Combine 32 g of LB Agar (GIBCO-BRL 22700-041) with 1 liter of H_2O. Stir and autoclave for 15 minutes at 121°C. When cooled to 45–50°C, add 1 ml of carbenicillin stock solution. Pour into petri dishes and allow to solidify. Store at 4°C.

Low-molecular-weight mass ladder (GIBCO-BRL 10068-013)

Nuclease-free dH_2O (Promega P 1193)

Phosphoric acid (1 M) (H_3PO_4) <!>

5x Polyethylene glycol <!>/NaCl

    200 g of PEG-8000 (Sigma P 2139)

    150 g of sodium chloride

        Bring to 1 liter total volume with dH_2O and stir until dissolved. Sterilize by autoclaving at 121°C for 20 minutes at 15 psi on liquid cycle.

Phosphate-buffered saline (PBS)

SB medium

    30 g of tryptone (BD Biosciences, Difco 0123-17-3)

    20 g of yeast extract (BD Biosciences, Difco 0127-17-9)

    10 g of MOPS (3[*N*-morpholino] propanesulfonic acid; Sigma M 8899) <!>

        Bring to 1 liter total volume with dH_2O, stir until dissolved, and titrate to pH 7. Sterilize by autoclaving at 15 psi on liquid cycle for 20 minutes at 121°C.

SOC medium

    20 g of tryptone

    5 g of yeast extract

    0.5 g of NaCl

186 mg of KCl <!>

Bring to 1 liter total volume with dH$_2$O, stir until dissolved, and titrate to pH 7. Sterilize by autoclaving on liquid cycle for 20 minutes at 121°C at 15 psi. When cooled, add 10 ml of sterile 1 M MgCl$_2$ and 20 ml of filter-sterilized 1 M glucose. Using aseptic conditions, aliquot into 10-ml portions and store at room temperature.

Sodium acetate (3 M, pH 5.2), sterilized by autoclaving <!>
1x TAE electrophoresis running buffer
48.4 g of Tris base
11.4 ml of acetic acid <!>
20 ml of 0.5 M EDTA

Bring to 1 liter with dH$_2$O.

Tetracycline stock solution
5 mg/ml in ethanol
TMB microwell peroxidase substrate system (2-C) (Kirkegaard & Perry Laboratories 50-76-00)
VCSM13 helper phage (Stratagene 200251)
Washing buffer
PBS with 0.5% Tween-20

Antibodies

Anti-M13 antibody (Pharmacia Biotech 27-9410-01)
Horseradish peroxidase (HRP)-anti-M13 conjugate (Pharmacia Biotech 27-9411-01)

Enzymes and Buffers

AmpliTaq DNA polymerase (Perkin Elmer N801-0533)
10x PCR buffer (Perkin Elmer, supplied with polymerase)
10x Restriction digest buffer (supplied with enzyme)
*Sfi*I restriction enzyme (Roche Molecular Biochemicals, 40 units/μl, 1 288 059)
T4 DNA ligase (GIBCO-BRL, 1 unit/μl, 15224 090)
5x T4 DNA ligase buffer (supplied with enzyme)

Plasmid and Bacterial Cells

pComb3XSS (available from Prof. Carlos Barbas III, The Scripps Research Institute, La Jolla, California)
ER2738 glycerol culture (New England Biolabs E4104S)
XL1-Blue MRF′ electrocompetent-competent cells (Stratagene 200158)

Specific Reagents and Special Equipment

Dynal magnetic stand MPC-S (Dynal 120.20)
Electroporation apparatus (e.g., Gene Pulser II, Biorad)
Electroporation cuvettes (2-mm, BTX 10-000295-01)
ELISA plate (96-well, Corning Coster 3690)
Linbro plate sealer with adhesive back (ICN Biochemicals 76-401-05)
Oligonucleotides
QIAEX II gel extraction kit (Qiagen 20021)
Quik Pik electroelution capsules (Stratagene 400855)
Specific oligonucleotide primers

Stage 1: Generating Libraries

METHOD

First Round of Overlap PCR

The initial PCR amplification requires the mixing of equimolar amounts of two oligonucleotides to form a template for the subsequent PCR. Figure 5 shows a schematic representation of the entire overlap PCR steps. On the basis of experience, each 50-μl reaction produces approximately 2–4 μg of impure product. To ensure that adequate amounts of purified PCR product of the correct size are obtained, 10 single reactions are usually assembled and later pooled.

1. Assemble 10 reactions for the amplification.

   Each reaction consists of the following:

oligonucleotide 1 (antisense)	100 ng
oligonucleotide 2 (sense)	100 ng
primer 1 (sense)	60 pmoles
primer 2 (anti-sense)	60 pmoles
10× PCR buffer	5 μl
2.5 mM dNTPs	8 μl
Taq DNA polymerase	0.5 μl

2. Add nuclease-free dH_2O to a final volume of 50 μl.

3. Perform the PCR under the following conditions: 94°C for 5 minutes followed by 30 cycles of 94°C for 15 seconds, 56°C for 15 seconds, 72°C for 90 seconds, followed by 72°C for 10 minutes.

4. Evaluate 5–10 μl of the PCRs on a 2% TAE/agarose gel using DNA gel loading dye and appropriate molecular-weight markers.

   From these amplifications the size of the PCR product that has been produced should be equivalent to the calculated size of the oligonucleotides used as a template.

Isolation of the PCR Product

5. Pool the 10 PCRs and precipitate with ethanol.

6. Add 10% (v/v) 3 M sodium acetate (pH 5.2) and mix.

7. Add 2.2 volumes of 100% ethanol and mix.

8. Incubate for 1 hour to overnight at –70°C.

9. Centrifuge at 14,000 rpm for 30 minutes at 4°C.

10. Decant the ethanol and wash the pellet in 70% ethanol.

11. Centrifuge at 14,000 rpm for 15 minutes at 4°C.

12. Decant the 70% ethanol and dry the pellet on the benchtop for 5–10 minutes.

13. Resuspend in an appropriate volume of nuclease-free dH_2O (50–100 μl).

14. Run the products on a 2% TAE/agarose gel and excise the correct-size band with a scalpel.

15. Purify either with electroelution (e.g., Quik Pik columns, Stratagene) or with resin binding (e.g., QIAEX II gel Extraction Kit, QIAGEN).

16. Quantitate yields on a 2% TAE/agarose gel with an appropriate mass ladder or by reading the optical density at 260 nm (1 O.D. unit = 50 μg/ml).

    If the first-round reaction is below a specific size, certain purification techniques cannot be used. Check the manufacturer's protocols.

Second Round of Overlap PCR

17. To generate the total domain, perform a second PCR extension. Mix the first-round PCR product at an equimolar amount with a third oligonucleotide. Add specific external primers, with encoded *Sfi*I restriction sites, to generate the full-length product.

18. Assemble ten 50-μl reactions. Each reaction should include:

first-round product	100 ng
oligonucleotide 3 (sense)	100 ng
primer 3 (sense)	60 pmoles
primer 4 (anti-sense)	60 pmoles
10x PCR buffer	5 μl
2.5 mM dNTPs	8 μl
Taq DNA polymerase	0.5 μl

    Add water to a final volume of 50 μl.

19. Perform PCR under the following conditions: 94°C for 5 minutes followed by 30 cycles of 94°C for 15 seconds, 56°C for 15 seconds, 72°C for 90 seconds, followed by 72°C for 10 minutes.

20. Evaluate 5–10 μl on a 2% TAE/agarose gel using DNA gel-loading dye and appropriate molecular-weight markers.

Isolate the PCR Product

21. Pool the PCR products, ethanol-precipitate, purify, and quantify the product on a 1–2% TAE/agarose gel as described under Isolation of the PCR Product above.

Restriction Digestion of the Protein Domain PCR Product and the pComb3XSS Vector with *Sfi*I

22. Prepare the PCR product and pComb3XSS (see Fig. 5) for cloning using an *Sfi*I restriction digest. We prefer the higher concentration (40 units/μl) and the 10x buffer M from Roche Molecular Biochemicals.

    a. The digestion of the PCR product should contain:

purified PCR product	10 μg
*Sfi*I (36 units per μg of DNA)	360 units
10x buffer M	20 μl

    Add H$_2$O to a final volume of 200 μl.
    The units of enzyme used are dependent on the size of the domain that is being digested.

    b. The digestion of vector should contain:

pComb3XSS (contains a stuffer fragment between the two *Sfi*I cloning sites)	20 μg
*Sfi*I (6 units per μg of DNA)	120 units
10x buffer M	20 μl

    Add H$_2$O to a final volume of 200 μl.

    c. Incubate the restriction digestions for 5 hours at 50°C.

    d. Ethanol-precipitate as discussed under Isolation of the PCR Product above, and purify the domain on a 1–2% TAE/agarose gel and the vector and stuffer on a 1% TAE/agarose gel.

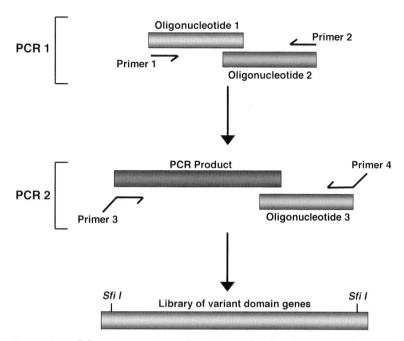

FIGURE 5. Generation of domain gene inserts by PCR overlap for cloning into the pComb3X system. In PCR 1, the first set of oligonucleotides is fused using complementary sequences, thereby allowing the amplification of the first PCR product by a specific set of primers. PCR 2 fuses the product from PCR 1 with a third oligonucleotide containing a complementary sequence, generating the final domain gene library. The inserts all contain asymmetric *Sfi*I restriction sites on the 5´ and 3´ ends that are used for directional cloning into the pComB3X vector. *Note:* The variegated codons should never be placed in the overlapping sequence portion of the oligonucleotides.

We recommend electroelution for purification, using Quik Pik columns (Stratagene). It is imperative that the digestion and purification are optimal to yield vector and insert with high transformation efficiencies and low background after ligation. The stuffer fragment should also be purified for use in the test ligations to determine vector quality.

TEST LIGATIONS

To verify that the vector and inserts produced are suitable for the creation of a library of sufficient size, first determine the ligation efficiency of the vector and the inserts and determine the background transformation frequency of the vector alone. If a relatively high frequency (>10%) of transformants is attained when the vector sample is ligated without the insert, this indicates that the vector preparation is of poor quality. Hence, only a very limited library will be generated with this preparation, because too much of the "library" will not have cloned inserts. Another aliquot of the vector should then be prepared.

At various points in the following steps, bacterial cultures are grown in the presence of antibiotics. The reasons for antibiotic addition are as follows:

- The XL1-Blue and ER2738 bacterial strains contain a tetracycline-resistance gene. The reasons for using tetracycline in the media are: (a) this will avoid the contamination of cultures with other strains of bacteria; (b) tetracycline promotes the expression of the F-pilus on the bacterial surface, a requirement for phage infection.

- The plasmid used in this protocol (pComb3X) has an ampicillin-resistance gene, and therefore, any bacteria containing this gene will be resistant to the effects of carbenicillin in the medium.

- For the expression of phage particles in cultures, helper phage is added and a kanamycin-resistance gene is present in the helper phage genome. Therefore, the addition of kanamycin selects for bacteria that are co-infected with helper phage.

23. Assemble the following small-scale ligation reactions.

    a. PCR product ligation

pComb3X (*Sfi*I-digested and purified)	140 ng
protein domain PCR product	
(*Sfi*I-digested and purified)	70 ng
5x ligase buffer	4 µl
ligase	1 µl

    Add H$_2$O to a volume of 20 µl.

    b. Control product ligation

pComb3X (*Sfi*I-digested and purified)	140 ng
stuffer fragment (*Sfi*I-digested and purified)	70 ng
5x ligase buffer	4 µl
ligase	1 µl

    Add H$_2$O to a volume of 20 µl.

    c. Vector self-ligation control

pComb3X (*Sfi*I-digested and purified)	140 ng
5x ligase buffer	4 µl
ligase	1 µl

    Add nuclease-free H$_2$O to a volume of 20 µl.

    d. Incubate all reactions overnight at room temperature.

    e. Transform 1 µl of each ligation reaction into 50 µl of ER2738 or XL1-Blue by electroporation (see Test Transformation, below).

    f. Plate 100 µl, 10 µl, and 1 µl onto LB + carbenicillin plates. Incubate the plates overnight at 37°C.

       These results should give an indication of the library size that can be generated with prepared reagents. The potential library size per microgram of vector DNA should be at least 1 x 10^7 transformants before continuing on to the large-scale library. The self-ligation background should be less than 10% (ideally >5%). If either of these requirements is not fulfilled, repeat the vector digestion, purification, and ligations until good-quality vector has been generated.

Test Transformation

To achieve the highest possible transformation efficiencies, we recommend electroporation. Protocols are available for generating electrocompetent cells (Rader et al. 2001), but XLI-Blue cells can be purchased in an electrocompetent state (Stratagene 200158).

24. Perform electroporation as follows:

    a. To an appropriate number of labeled 14-ml polypropylene tubes, add 2 ml of room-temperature SOC medium.

    b. Place 2-mm electroporation cuvettes (stored at –20°C) on ice prior to the transformation (one for each sample).

    c. Set the Gene Pulser II apparatus to 25 µF capacitance, 2.5 kV, and 200 Ω on the pulse controller unit.

    d. For each of the 20-µl test ligation samples, aliquot 1 µl into a fresh Eppendorf tube and store on ice. Gently thaw the electrocompetent cells in your hand. Add 50 µl of cells to the 1-µl test ligation sample, and mix by stirring with the pipette tip.

e. Transfer the DNA–cell mixture to a chilled electroporation cuvette and tap the sample to the bottom of the cuvette, avoiding air bubbles.

f. Place the cuvette in the Gene Pulser II apparatus and apply one pulse, which should result in a time constant of 4–5 milliseconds.

g. Immediately add 1 ml of SOC medium to the cuvette and gently resuspend the cells. Any delay in adding the SOC medium will result in decreased transformation efficiency (1 minute of delay is equivalent to a threefold decrease).

h. Transfer the cell suspension to a labeled polypropylene tube containing an additional 2 ml of SOC medium.

i. Incubate the cultures for 1 hour at 37°C while shaking at 250 rpm.

j. Plate 100 μl, 10 μl, and 1 μl onto LB + carbenicillin plates and incubate overnight at 37°C. (Serial dilution of the culture should be performed so that a volume of 100 μl is plated.)

25. To determine the total transformants, count the colonies on each plate:

Total transformants = [number of colonies × culture volume (μl)/ plating volume (μl)] × [total ligation volume (μl)/ligation volume transformed (μl)]

Multiply the total number of transformants in the test ligation by a factor of 7.14 to determine the size of a library that would result from transforming 1 μg of a library ligation or by a factor of 10 to determine the library ligation. Ascertain the number of library ligations that would be required to produce a library size of 5×10^7.

Library Ligation and Transformation

26. Assemble enough reactions to produce at least 5×10^7 transformants:

pComb3X (*Sfi*I-digested and purified)	1.4 μg
protein domain PCR product	
(*Sfi*I-digested and purified)	700 ng
5× ligase buffer	40 μl
ligase	10 μl

Add H_2O to a volume of 200 μl.

27. Incubate the ligation reactions overnight at room temperature.

28. Perform an ethanol precipitation as described under Isolation of the PCR Product (p. 155).

29. Resuspend the ethanol-precipitated ligation reaction in 15 μl of sterile H_2O.

Library Transformation

> Contamination of the library with phage from various sources can be disastrous. Therefore, it is important to use a 10% bleach solution to clean the bench areas, centrifuges, bottles, and pipettors to avoid contamination from alternative phage particles (helper, environmental, or prior libraries) throughout all of the following steps.

30. Remove a 2-mm electroporation cuvette from –20°C and place it on ice.

31. Thaw the electrocompetent cells (ER2738) by hand and add 300 μl of cells per library transformation. Once the cells are thawed, add them directly to the 15 μl of DNA and mix by gentle pipetting.

32. Transfer the DNA–cell mixture into the chilled cuvette and electroporate in the Gene Pulser II apparatus, applying one pulse (time constant of ~4 msec).

33. Immediately add 1 ml of room-temperature SOC medium, and gently mix the cells. Transfer the 1 ml to a labeled 14-ml polypropylene tube. Flush the cuvette twice with 1 ml of SOC and combine in the polypropylene tube. Incubate for 1 hour at 37°C, shaking at 200–250 rpm.

34. Add 6 ml of prewarmed (37°C) SB medium containing 20 μg/ml carbenicillin (3.6 μl of the 50 mg/ml carbenicillin stock solution) and 10 μg/ml tetracycline (18 μl of the tetracycline stock solution) to the 3-ml culture. For titering of the library, remove 20 μl of the culture and add it to 180 μl (1:10 dilution), mix by gentle vortexing, and serially dilute as above to have 1:10, 1:100, and 1:1000 dilutions. Plate 100 μl of each dilution on LB + carbenicillin plates and incubate overnight at 37°C.

    The total number of transformants = number of colonies x (culture volume/plating volume)

35. Incubate the 9-ml culture for 1 hour at 37°C with shaking at 250 rpm.

36. Increase the carbenicillin concentration to 50 μg by the addition of 5.4 μl of the carbenicillin stock solution and incubate for a further 1 hour at 37°C, shaking at 250 rpm.

37. Add 1 ml of helper phage to the 9-ml culture and increase the volume to 100 ml with 90 ml of prewarmed SB + carbenicillin/tetracycline (50 μg/ml/10 μg/ml concentration, respectively) in a sterile 250-ml conical flask. Incubate for a further 2 hours at 37°C, shaking at 250 rpm.

38. Add 140 μl of the kanamycin stock solution to the 100-ml culture and incubate overnight at 37°C, shaking at 250 rpm.

Precipitation Using Polyethylene Glycol

39. Centrifuge the culture at 3000g (e.g., 4000 rpm in a Beckman JA-10 rotor) for 15 minutes at 4°C.

40. Decant the supernatant into a sterile 250-ml centrifuge bottle, add 25 ml of 5x PEG/NaCl, and incubate for 30 minutes on ice.

41. Centrifuge the supernatant at 15,000g (e.g., 9000 rpm in a Beckman JA-10 rotor) for 30 minutes and discard the supernatant. Turn the centrifuge bottle upside down on a paper towel to drain excess liquid off the phage pellet.

42. Resuspend the phage pellet in 2 ml of 1% BSA/PBS and filter-sterilize with a 0.2-μm syringe filter.

    The phage suspension can be stored at 4°C indefinitely, but if panning is intended, the phage should be used straight away.

Stage 2: Panning

There are many methodological approaches that can be used for library selection, but in this section we discuss the use of streptavidin-coated magnetic beads (for alternative methods, see Rader et al. 2001). The strategy used to recover desirable clones should be chosen on the basis of the type of panning strategy employed and the properties of the ligand that has been targeted. Of the several alternatives, we have compared the common recovery strategies of low pH, trypsinization of bound libraries, and competition with soluble antigen. When considering which option to employ, it is worth knowing how this may affect the type of binder that will be recovered. Low pH will dissociate most or all binders, whether they are specific or nonspecific for the selection antigen, whereas trypsin targets the fusion area between the protein and gene III and specifically cleaves it. As a result, for the latter, only phage that bind via the protein will be recovered. In contrast, soluble antigen will provide specific binders of relatively high affinity because the elution will work in a competitive manner.

We recommend an empiric approach to determine which is best for the system. For the low pH strategy, we usually use 50 μl of 100 mM glycine-HCl (pH 2.2) for 10 minutes followed by neutralization with 3 μl of 2 M Tris base. The trypsin strategy uses 200 μl of a freshly prepared 10 mg/ml trypsin solution (diluted in 1× PBS) followed by an incubation at 37°C. Finally, the antigen elution strategy usually will be a 50–100-μg solution of the unbiotinylated target protein (diluted in 1× PBS) to act as a competitive binder for 30 minutes at 37°C with mixing.

1. In preparation of the amplification of the panned phage, first set up a bacterial culture for infection. The electrocompetent bacteria used for the library transformation are normally used. Prior to the panning protocol, make 5-μl aliquots and store them at –70°C.

2. Inoculate 5 ml of SB medium + tetracycline (10 μg/ml) in a 15-ml polypropylene tube with 5 μl of the electrocompetent cells and grow for 1.5–2.5 hours at 37°C, 250 rpm, until an optical density (OD) at 600 nm is ~1.

3. Mix 100 μl of the PEG-precipitated phage suspension with 5–10 μg of biotinylated target protein (see Dynal protocols to estimate the amount of protein that will associate with the amount of beads added to the protocol). Add 10 μl of blocking buffer and increase the volume to 200 μl with 1× PBS. Incubate for 2 hours at 37°C with mixing (end over end).

4. Wash 1 mg of the Dynabeads (100 μl) with blocking buffer three times to remove preservatives. In brief, aliquot 100 μl of beads into an Eppendorf tube and place the Eppendorf tube in a Dynal magnetic stand. Leave for 2 minutes to allow Dynabeads to settle next to the magnet. Aspirate the liquid and wash with 200 μl of blocking buffer. Repeat three times. Leave the final 200 μl of blocking buffer on the beads with mixing for 1 hour at room temperature.

5. At this point, prepare the phage suspension for input titering. Add 20 μl of the phage suspension to 180 μl of SB to give a 10^{-1} dilution, and continue the serial dilution to give 10^{-6}, 10^{-7}, and 10^{-8} phage suspensions.

6. Remove the blocking buffer from the beads and resuspend the Dynabeads in the biotinylated protein:phage solution. Incubate for 30 minutes at 37°C with mixing.

7. Wash the Dynabeads five times with 200 μl of 1× PBS/0.5% Tween-20. Place the Eppendorf tube in the Dynal magnetic stand and, once the Dynabeads have settled, aspirate the liquid. Add 200 μl of wash buffer and vortex lightly, replace on the magnetic stand, and remove the wash buffer. Repeat five times.

   Over subsequent rounds of panning, the number and stringency of washings can be increased to remove more nonspecific or weak-affinity binders.

8. After the final wash, resuspend the Dynabeads:biotinylated protein:phage in an appropriate elution buffer (use either low pH, trypsin, or soluble antigen) and incubate for 10–30 minutes at 37°C with mixing.

9. Remove the elution buffer from the beads using the magnetic stand, and use this buffer to infect bacteria for the amplification of the selected phage.

10. Add the phage solution to 2 ml of the prepared bacterial culture and incubate for 15 minutes at room temperature.

11. Add 6 ml of prewarmed (37°C) SB, 3.2 μl of 50 mg/ml carbenicillin, and 16 μl of 5 mg/ml tetracycline and transfer the culture to a 50-ml polypropylene tube.

12. For the output titering, remove 20 μl of the sample into 180 μl of SB and plate 100 μl of the 1:10, 1:100, and 1:1000 dilutions (to make the 1:100 and 1:1000 dilutions, serially dilute 20 μl of the first 1:10 dilution into 180 μl of SB medium). Place the 8-ml culture for 1 hour at 37°C with shaking at 250 rpm.

13. At the same time, the input titers can be used for infection. Add 2 μl of the 10^{-6}, 10^{-7}, and 10^{-8} phage suspensions to separate Eppendorf tubes containing 98 μl of prepared bacterial culture. Incubate for 15 minutes at room temperature.

14. Plate the input (50 μl) and output (100 μl) onto LB agar + carbenicillin plates and incubate overnight at 37°C.

    To calculate the input and output:

    Input (cfu/ml) = colonies × (1/dilution) × 1000

    Output (cfu) = colonies × (culture volume/plating volume)

> Determination of the input and output titers allows the ongoing evaluation of the efficiency of selection of the library. In most cases, the titer of the input phage should be consistent from round to round. However, the output phage will be dependent on the number of specific phage present in the library and the efficiency of their recovery. In theory, the initial library will contain a small number of specific binders, and a low output titer will be expected. With each sequential round of selection, the number of specific phage in the resulting library should increase (enrichment), and the output number should also increase. Hence, the calculation of the input and output titers provides a simple means for monitoring library panning efficiency, which is especially helpful if a progressive pattern of increased output is demonstrated.

15. After the 8-ml culture has been incubated for 1 hour, add 4.8 μl of 50 μg/ml carbenicillin and incubate for 1 hour at 37°C with shaking at 250 rpm.

16. Add 1 ml of VCSM13 helper phage (10^{12} to 10^{13} pfu) to the 8-ml culture and increase the volume to 100 ml in a 250-ml sterile conical flask with 91 ml of prewarmed SB medium containing 92 μl of 50 μg/ml carbenicillin and 184 μl of 5 mg/ml tetracycline. Incubate for 2 hours at 37°C with shaking at 250 rpm.

17. Add 140 μl of 50 mg/ml kanamycin and incubate overnight at 37°C with continued shaking (250 rpm).

18. Precipitate with PEG as described above and continue panning for 3–6 rounds. The number of rounds is dependent on strategy, protein interactions, washing stringency, and library size.

Stage 3: Binding Analysis

After panning, the rounds of selection should be analyzed to evaluate the enrichment for binders. To do this, the PEG-precipitated phage from each round can be tested in an assay, be it ELISA-based or flow cytometric. The assay should include all rounds of phage samples from the initial unselected phage to the last round. The method chosen to evaluate whole phage libraries can also be used as a system of evaluating single clones from the appropriate round. The appropriate round is the one that yields a library exhibiting the best binding characteristics (or best signal in the detection assay) associated with the protein–protein interaction. This section outlines a basic ELISA-based analysis strategy.

1. Precoat ELISA plate wells with 25 μl of the appropriate protein(s) diluted in coating buffer (depending on the protein, e.g., 1× PBS or 0.1 M $NaHCO_3$, pH 8.0). The protein(s) should be at a concentration of 0.5–5 μg/ml. Anti-M13 antibody should also be precoated in parallel, at 5 μg/ml. This will enable the determination of the phage levels in the samples. Cover the ELISA plates with a Linbro plate sealer and incubate overnight at 4°C.

2. Shake out the precoat and add 150 μl of blocking buffer to the wells. Cover with a plate sealer and incubate for 1 hour at 37°C.

3. Discard the blocking buffer and add 50 μl of the phage samples at a 1:5 dilution, serially diluting the phage down the entire plate at 1:10 dilutions. Incubate for 2 hours at 37°C (all phage dilutions are in a 1% BSA/PBS buffer). Wells should include samples tested in duplicate to assess binding to all relevant precoats, including anti-M13.

4. Discard the phage solutions and wash the wells five times with 1× PBS / 0.05% Tween-20. Fill the wells completely with the wash buffer.

5. Prepare a 1:3000 dilution of HRP-conjugated anti-M13 antibody in 1% BSA/PBS. Add 50 μl of the diluted secondary antibody to each well and incubate for 1 hour at room temperature.

6. Wash the wells as described above and develop with 50 μl of TMB microwell substrate system (Kirkegaard and Perry Laboratories, Gaithersburg, Maryland).

7. Read the plates at an optical density of 405 nm on an ELISA reader at appropriate time points, dependent on the signal intensity. Stop the reaction with 50 μl of 1 M H_3PO_4 when a strong signal intensity is achieved with a negligible background signal. If the reactions are already stopped, read the plates at an optical density of 450 nm.

8. Determine the best round of panning on the basis of comparative binding using ELISA (i.e., highest relative OD signal) or an alternative binding assay. Analyze single clones as described above. Use the titration plates made to evaluate the input and output number as a source of single clones to inoculate 5–100 ml of SB medium with 50 μg/ml of carbenicillin.

9. Incubate the single clones at 37°C for 4–8 hours, depending on the size of the culture, and add helper phage in a comparable volume to the culture size (refer to the volume added to the amplification step in the panning protocol, p. 160).

10. After 2 hours, add kanamycin to give a concentration of 70 μg/ml and incubate overnight at 37°C with shaking at 250 rpm.

11. Precipitate the cultures using PEG as described above (p. 160), taking into account the volumes used, and analyze.

After positive clones have been identified, large-scale preparations can be done. To produce soluble proteins without the gene III fragment in *E. coli*, single clones must be transferred into an alternative *E. coli* nonsuppressor strain. This allows the amber stop codon in pComb3X (see Fig. 6) to be used and soluble protein to be produced. Detailed protocols for the production of soluble proteins are provided in Barbas et al. (2001).

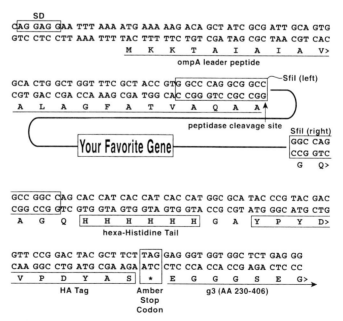

FIGURE 6. Details of the pComb3X cloning site. The *Sfi*I sites have different sequences, thereby allowing the directional cloning that is needed for the display of cloned proteins on the surface of phage. The site also contains an *ompA* signal peptide for the transport of the protein to the periplasm, two tags (hexahistidine and hemagglutinin-specific decapeptide [HA]), and an amber stop codon. (Adapted, with permission, from Scott and Barbas 2001.)

ACKNOWLEDGMENTS

This presentation was modified from its original form, which appeared as part of a much more complete compilation of contributions on the biology of filamentous phage and their exploitation in diverse phage-display applications (Silverman 2001). For more detailed presentations on the background and methods used in phage display, the reader is referred to this text.

REFERENCES

Alting-Mees M.A. and Short J.M. 1993. Polycos vectors: A system for packaging filamentous phage and phagemid vectors using lambda phage packaging extract. *Gene* **137:** 93–100.

Barbas III, C.F.I., Burton D.R., Scott J., and Silverman G.J. 2001. *Phage display: A laboratory manual.* Cold Spring Laboratory Harbor Press, Cold Spring Harbor, New York.

Braisted A.C. and Wells J.A. 1996. Minimizing a binding domain from protein A. *Proc. Natl. Acad. Sci.* **93:** 5688–5692.

Burton D.R. and Barbas C.F. 1993. Human antibodies to HIV-1 by recombinant DNA methods. *Chem. Immunol.* **56:** 112–126.

Cary S.P., Lee J., Wagenknecht R., and Silverman G.J. 2000. Characterization of superantigen-induced clonal deletion with a novel clan III-restricted avian monoclonal antibody: Exploiting evolutionary distance to create antibodies specific for a conserved V_H region surface. *J. Immunol.* **164:** 4730–4741.

Chiswell D.J. and McCafferty J. 1992. Phage antibodies: Will new "coliclonal" antibodies replace monoclonal antibodies? *Trends Biotechnol.* **10:** 80–84.

Clackson T. and Wells J.A. 1994. In vitro selection from protein and peptide libraries. *Trends Biotechnol.* **12:** 173–184.

Crameri R. and Suter M. 1993. Display of biologically active proteins on the surface of filamentous phages: A cDNA cloning system for selection of functional gene products linked to the genetic information responsible for their production. *Gene* **137:** 69–75.

Cwirla S.E., Peters E.A., Barrett R.W., and Dower W.J. 1990. Peptides on phage: A vast library of peptides for identifying ligands. *Proc. Natl. Acad. Sci.* **87:** 6378–6382.

Deisenhofer J. 1981. Crystallographic refinement and atomic models of a human Fc fragment and its complex with fragment B of protein A from S*taphylococcus aureus* at 2.9- and 2.8-Å resolution. *Biochemistry* **20:** 2361–2370.

Djojonegoro B.M., Benedik M.J., and Willson R.C. 1994. Bacteriophage surface display of an immunoglobulin-binding domain of *Staphylococcus aureus* protein A. *Bio/Technology* **12:** 169–172.

Dunn I.S. 1995. Assembly of functional bacteriophage lambda virions incorporating C-terminal peptide or protein fusions with the major tail protein. *J. Mol. Biol.* **248:** 497–506.

Efimov V.P., Nepluev I.V., and Mesyanzhinov V.V. 1995. Bacteriophage T4 as a surface display vector. *Virus Genes* **10:** 173–177.

Fields S. and Song O.K. 1989, A novel genetic system to detect protein-protein interactions. *Nature* **340:** 245–246.

Fisch I., Kontermann R.E., Finnern R., Hartley O., Soler-Gonzalez A.S., Griffiths A.D., and Winter G. 1996. A strategy of exon shuffling for making large peptide repertoires displayed on filamentous bacteriophage. *Proc. Natl. Acad. Sci.* **93:** 7761–7766.

Folgori A., Tafi R., Meola A., Felici F., Galfré G., Cortese R., Monaci P. and Nicosia A. 1994. A general strategy to identify mimotopes of pathological antigens using only random peptide libraries and human sera. *EMBO J.* **13:** 2236–2243.

Fransen M., Van Veldhoven P.P., and Subramani S. 1999. Identification of peroxisomal proteins by using M13 phage protein VI phage display: Molecular evidence that mammalian peroxisomes contain a 2,4-dienoul-CoA reductase. *Biochem. J.* **340:** 561–568.

Fuh G. and Sidhu S.S. 2000. Efficient phage display of polypeptides fused to the carboxy-terminus of the M13 gene-3 minor coat protein. *FEBS Lett.* **480:** 231–234.

Fuh G., Pisabarro M.T., Li Y., Quan C., Lasky L.A., and Sidhu S.S. 2000. Analysis of PDZ domain-ligand interactions using carboxyl-terminal phage display. *J. Biol. Chem.* **275:** 21486–21491.

Gao C., Mao S., Lo C.H., Wursching P., Lerner R.A., and Janda K.D. 1999. Making artificial antibodies: A format for phage display of combinatorial heterodimeric arrays. *Proc. Natl. Acad. Sci.* **96:** 6025–6030.

Germino F.J., Wang Z.X., and Weissman S.M. 1993. Screening for in vivo protein-protein interaction. *Proc. Natl. Acad. Sci.* **90:** 933–937.

Graille M., Stura E.A., Corper A.L., Sutton B., Taussig M., Charbonnier J.-B., and Silverman G.J. 2000. Crystal structure of a *Staphylococcus aureus* protein A domain complexed with the Fab fragment of a human IgM antibody: Structural basis for recognition of B-cell receptors and superantigen activity. *Proc. Natl. Acad. Sci.* **97**: 5399–5404.

Gram H., Strittmatter U., Lorenz M., Gluck D., and Zenke G. 1993. Phage display as a rapid gene expression system: Production of bioactive cytokine-phage and generation of neutralizing monoclonal antibodies. *J. Immunol. Methods* **161**: 169–176.

Helfman D.M., Fiddes J.R., Thomas G.P., and Hughes S. 1983. Identification of clones that encode chicken tropomyosin. *Proc. Natl. Acad. Sci.* **80**: 31–35.

Houshmand H., Fröman G., and Magnusson G. 1999. Use of bacteriophage T7 displayed peptides for determination of monoclonal antibody specificity and biosensor analysis of the binding reaction. *Anal. Biochem.* **268**: 363–370.

Iannolo G., Minenkova O., Petruzzelli R., and Cesarini G. 1995. Modifying filamentous phage capsid: Limits in the size of the major capsid protein. *J. Mol. Biol.* **248**: 835–844.

Jespers L.S., Messens J.H., De Keyser A., Eeckhout D., Van Den Brande I., Gansemans Y.G., Lauwereys M.J., Vlasuk G.P., and Stanssens P.E. 1995. Surface expression and ligand-based selection of cDNAs fused to filamentous phage gene VI. *Bio/Technology* **13**: 378–382.

Kast P. and Hilvert D. 1997. 3D structural information as a guide to protein engineering using genetic selection. *Curr. Opin. Struct. Biol.* **7**: 470–479.

Katz B.A. 1997. Structural and mechanistic determinants of affinity and specificity of ligands discovered or engineered by phage display. *Annu. Rev. Biophys. Biomol. Struct.* **26**: 27–45.

Kushwaha A., Chowdhury P.S., Arora K., Abrol S., and Chaudhary V.K. 1994. Construction and characterization of M13 bacteriophages displaying functional IgG-binding domains of staphylococcal protein A. *Gene* **151**: 45–51.

Kuwabara I., Maruyama H., Mikawa Y.G., Zuberi R.I., Liu F.T., and Maruyama I.N. 1997. Efficient epitope mapping by bacteriophage λ surface display. *Nat. Biotechnol.* **15**: 74–78.

Lowman H.B., Bass S.H., Simpson N., and Wells J.A. 1991. Selecting high-affinity binding protein by monovalent phage display. *Biochemistry* **30**: 10832–10838.

Malik P., Terry T.D., Gowda L.R., Petukhov A.L.S.A., Symmons M.F., Welsh L.C., Marvin D.A., and Perham R.N. 1996. Role of capsid structure and membrane protein processing in determining the size and copy number of peptides displayed on the major coat protein of filamentous bacteriophage. *J. Mol. Biol.* **260**: 9–21.

Maruyama I.N., Maruyama H.I., and Brenner S. 1994. λfoo: A lambda phage vector for the expression of foreign proteins. *Proc. Natl. Acad. Sci.* **91**: 8273–8277.

Marvin D.A. 1998. Filamentous phage structure, infection and assembly. *Curr. Opin. Struct. Biol.* **8**: 150–158.

McCafferty J., Griffiths A.D., Winter G., and Chiswell D.J. 1990. Phage antibodies: Filamentous phage displaying antibody variable domains. *Nature* **348**: 552–554.

Nord K., Nilsson B., Uhlen M., and Nygren P.A. 1995. A combinatorial library of an alpha-helical bacterial receptor domain. *Protein Eng.* **8**: 601–608.

Nord K., Gunneriusson E., Ringdahl J., Stahl S., Uhlen M., and Nygren P.A. 1997. Binding proteins selected from combinatorial libraries of an alpha-helical bacterial receptor domain. *Nat. Biotechnol.* **15**: 772–777.

O'Neil K.T. and Hoess R.H. 1995. Phage display: Protein engineering by directed evolution. *Curr. Opin. Struct. Biol.* **5**: 443–449.

Petersen G., Song D., Hugle-Dorr B., Oldenburg I., and Bautz E.K. 1995. Mapping of linear epitopes recognized by monoclonal antibodies with gene-fragment phage display libraries. *Mol. Gen. Genet.* **249**: 425–431.

Qiu J.X., Kai M., Padlan E.A., and Marcus D.M. 1999. Structure-function studies of an anti-asialo GM1 antibody obtained from a phage display library. *J. Neuroimmunol.* **97**: 172–181.

Rader C., Steinberger P., and Barbas C.F. III. 2001. Selection from antibody libraries. In *Phage display: A laboratory manual* (ed. C.F. Barbas III et al.), pp. 10.1–10.20. Cold Spring Harbor Laboratory Press, Cold Spring Harbor, New York.

Ren Z.J., Lewis G.K., Wingfield P.T., Locke E.G., Steven A.C., and Black L.W. 1996. Phage display of intact domains at high copy number: A system based on SOC, the small outer capsid protein of bacteriophage T4. *Protein Sci.* **5**: 1833–1843.

Roberts B.L., Markland W., Ley A.C., Kent R.B., White D.W., Guterman S.K., and Ladner R.C. 1992. Directed evolution of a protein: Selection of a potent neutrophil elastase inhibitor. *Gene* **121**: 9–15.

Sambrook J. and Russell D. 2001. *Molecular cloning: A laboratory manual*, 3rd edition. Cold Spring Harbor Laboratory Press, Cold Spring Harbor, New York.

Santini C., Brennan D., Mennuni C., Hoess R.H., Nicosia A., Cortese R., and Luzzago A. 1998. Efficient display of an HCV cDNA expression library as C-terminal fusion to the capsid protein D of bacteriophage lambda. *J. Mol. Biol.* **282:** 125–135.

Sasano M., Burton D.R., and Silverman G.J. 1993. Molecular selection of human antibodies with an unconventional bacterial B cell antigen. *J. Immunol.* **151:** 5822–5839.

Sasso E.H., Silverman G.J., and Mannik M. 1989. Human IgM molecules that bind staphylococcal protein A contain VHIII H chains. *J. Immunol.* **142:** 2778–2783.

———. 1991. Human IgA and IgG F(ab´)2 that bind to staphylococcal protein A belong to the VHIII subgroup. *J. Immunol.* **147:** 1877–1883.

Scott J.K. and Barbas C.F.I. III. 2001. Phage-display vectors. In *Phage display: A laboratory manual* (ed. C.F. Barbas III et al.), pp. 2.1–2.19. Cold Spring Harbor Laboratory Press, Cold Spring Harbor, New York.

Scott J.K. and Smith G.P. 1990. Searching for peptide ligands with an epitope library. *Science* **249:** 386–390.

Seed B. and Aruffo A. 1987. Molecular cloning of the CD2 antigen, the T-cell erythrocyte receptor, by a rapid immunoselection procedure. *Proc. Natl. Acad. Sci.* **84:** 3365–3369.

Shanmugavelu M., Baytan A.R., Chesnut J.D., and Bonning B.C. 2000. A novel protein that binds juvenile hormone esterase in fat body tissue and pericardial cells of the tobacco hornworm *Manduca sexta L. J. Biol. Chem.* **275:** 1802–1806.

Sidhu S.S., Lowman H.B., Cunningham B.C., and Wells J.A. 2000. Phage display for selection of novel binding proteins. *Methods Enzymol.* **328:** 333–363.

Sikela J.M. and Hahn W. 1987. Screening an expression library with a ligand probe: Isolation and sequence of a cDNA corresponding to a brain calmodulin binding protein. *Proc. Natl. Acad. Sci.* **84:** 3038–3042.

Silverman G.J. 2001. Functional domains and scaffolds. In *Phage display: A laboratory manual* (ed. C.F. Barbas III et al.), pp. 5.1–5.24. Cold Spring Harbor Laboratory Press, Cold Spring Harbor, New York.

Silverman G.J., Pirès R., and Bouvet J.P. 1996. An endogenous sialoprotein and a bacterial B-cell superantigen compete in their VH family-specific binding interactions with human immunoglobulins. *J. Immunol.* **157:** 4496–4502.

Silverman G.J., Cary S.P., Dwyer D.C., Linda L., Wagenknecht R., and Curtiss V.E. 2000. A B-cell superantigen induced persistent "hole" in the B-1 repertoire. *J. Exp. Med.* **192:** 87–98.

Silverman G.J., Nayak J.V., Warnatz K., Cary S., Tighe H., and Curtiss V.E. 1998. The dual phases of the response to neonatal exposure to a VH family-restricted staphylococcal B-cell superantigen. *J. Immunol.* **161:** 5720–5732.

Singh S.J.H., LeBowitz A.S., Baldwin J., and Sharp, P.A. 1988. Molecular cloning of an enhancer binding protein: Isolation by screening of an expression library with a recogntion site DNA. *Cell* **52:** 415–423.

Smith G.P. 1985. Filamentous fusion phage: Novel expression vectors that display cloned antigens on the virion surface. *Science* **228:** 1315–1317.

Starovasnik M.A., Braisted A.C., and Wells J.A. 1997. Structural mimicry of a native protein by a minimized binding domain. *Proc. Natl. Acad. Sci.* **94:** 10080–10085.

Sternberg N. and Hoess R.H. 1995. Display of peptides and proteins on the surface of bacteriophage λ. *Proc. Natl. Acad. Sci.* **92:** 1609–1613.

Webster R.E. 2001. Filamentous phage biology. In *Phage display: A laboratory manual* (ed. C.F. Barbas III et al.), pp. 1.1–1.37. Cold Spring Harbor Laboratory Press, Cold Spring Harbor, New York.

Winter G., Griffiths A.D., Hawkins R.E., and Hoogenboom H.R. 1994. Making antibodies by phage display technology. *Annu. Rev. Immunol.* **12:** 433–455.

Young R.A. and Davis R.W. 1983. Efficient isolation of genes using antibody probes. *Proc. Natl. Acad. Sci.* **80:** 1194–1198.

9 Genetic Approaches to Identify Protein–Protein Interactions in Budding Yeast

Randy Strich

Fox Chase Cancer Center, Philadelphia, Pennsylvania 19111

INTRODUCTION

Model organisms such as yeast, flies, and nematodes have provided many of the current paradigms regarding basic cellular processes found in mammals. For example, cell cycle control and checkpoint mechanisms found in yeast can be directly translated to vertebrates. Similarly, developmental pathways uncovered in *Drosophila melanogaster* and *Caenorhabditis elegans* have provided blueprints for embryonic patterning and organogenesis in humans. One outcome of these studies has been the important role that protein–protein interactions play in controlling the transition from one cellular state to the next. The relationships between important players in these regulatory cascades, as well as between potential direct interactors, have been predicted using classic genetic approaches. For example, genetic studies have been exploited to probe protein interactions within large biochemical machines such as the DNA replication initiation complex (Homesley et al. 2000), the 26S proteasome (Smyth and Belote 1999), and the RNA polymerase holoenzyme (Mortin 1990). In addition, simpler contacts such as those required for muscle function in *C. elegans* have been dissected using genetic approaches (Levin and Horvitz 1993). Therefore, when considering identifying the full spectrum of protein–protein interactions that define a particular pathway, the use of genetic approaches provides a powerful in vivo supplement to modern methods that rely on artificial protein fusions or in vitro assays.

Suppressor analysis is the most common genetic strategy used to identify potential protein–protein interactors. Suppressors are defined as a mutation in a second gene that restores the wild-type phenotype to a mutant individual. An example is illustrated in Figure 1. This hypothetical process requires the interaction between two proteins, A and B. A mutation in A that disrupts the interaction with B abrogates the pathway, resulting in a mutant phenotype. However, a second mutation in B, which permits association with A, restores the pathway, and returns the wild-type phenotype. More recent technologies have allowed the application of another type of suppression via overexpression. This strategy relies on the premise that the reduced affinity caused by the mutation in A can be overcome by increasing the effective concentration of B (Fig. 1B). This chapter describes classic suppressor analysis approaches as they have been applied to the budding yeast *Saccharomyces cerevisiae*. In addition, the ability to use yeast as a surrogate genetic system to transplant a biological system from another organism is discussed.

Why Use the Genetic Approach?

As with any experimental approach, there are advantages and disadvantages to using suppressor analyses. One advantage is the inherent physiological nature of the process. There are no gene fusions or forced localization of a protein to a specific compartment of the cell, as is required in two-hybrid systems. Moreover, all of the reactions take place inside the cell and are therefore not subject to artifacts derived from extract preparation or in vitro binding experiments. However, when the goal is to obtain a protein that directly interacts with a specific target, potential problems exist if a genetic approach is chosen. For example, suppressors may function several steps upstream or downstream of the target protein or in a separate pathway altogether. Although this property is of great utility if the entire pathway is being studied, it can be a source of frustration if only direct binding partners of the target protein are being sought. In addition, a successful suppressor analysis may rely heavily on selecting the correct mutant allele in the target protein. Certain alleles will work whereas others will not, regardless of the effort. These concerns can be minimized by the careful selection of mutant alleles and the judicial use of secondary screens described below. A flowchart outlining the basic considerations for a suppressor hunt is shown in Figure 2. Details for each step are described below.

OUTLINE OF PROCEDURE

Preparation for a Suppressor Hunt

Choice of a Phenotype

Phenotypes come in two "flavors," selections and screens. A selectable phenotype uses life and death to distinguish between a wild-type and a mutant individual, respectively. For example, a mutation in the target gene may result in cell death under a particular growth condition (Fig. 3, top panel). A suppressor of this mutation would restore growth under the test condition. Therefore, only individuals able to grow are interesting. This approach allows the rapid examination of a large number of candidates with relatively little effort. The second general phenotypic class is a screen. A screen differs from a selection in that each individual must be inspected for the wild-type or mutant phenotype. For example, the original mutation may cause the cell to assume an abnormal shape when grown under a particular set of conditions (open colonies, Fig. 3, lower panel). The suppressor returns the shape to wild type (filled colony). Although time-consuming, screens have proven to be very helpful in *C. elegans* and *D. melanogaster* as well as in yeast. As with

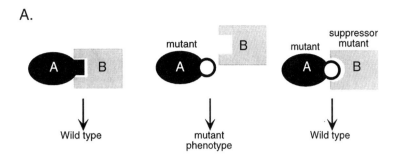

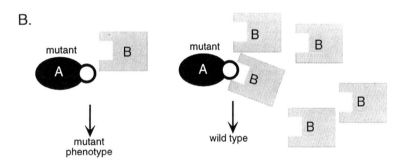

FIGURE 1. Suppressor strategies. (*A*) Mutational approach. The interaction of *A* and *B* generates a wild-type phenotype (*left* panel). The mutation in *A* disrupts this interaction, causing a mutant phenotype (*middle* panel). A new mutation in *B* restores this interaction, thereby reconstituting the pathway and generating a wild-type phenotype (*right* panel). (*B*) High-copy suppression. As in panel *A*, the mutation in protein A reduces the binding affinity to B, leading to a mutant phenotype. Increasing the effective concentration of B compensates for the lower binding ability producing wild-type activity.

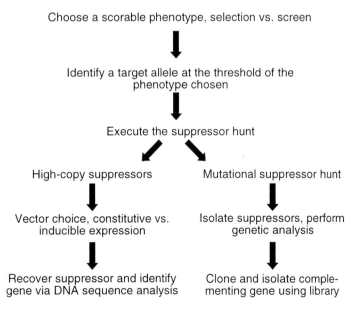

FIGURE 2. Suppressor hunt flowchart.

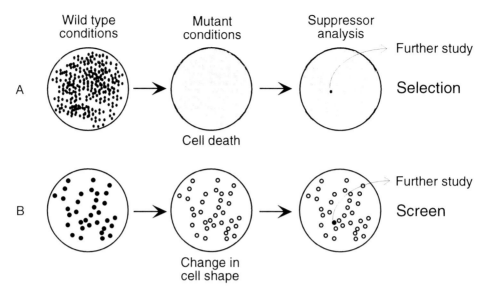

FIGURE 3. Mutant phenotypes for suppressor analysis. (*A*) Selection. Transfer of growing yeast colonies (*filled circles*) to the test medium results in cell death (*shaded circles*). Suppression of the mutation restores cell growth. (*B*) Screen. Under test conditions, the cell shape changes from wild type (*filled circles*) to mutant (*open circles*). Suppression of the mutant phenotype restores normal cell shape (*filled circle, right panel*).

most genetic approaches, it usually comes down to numbers. Because more individuals can be tested using a selective protocol, the likelihood that even rare suppressors are recovered is increased. Therefore, whenever possible, the use of a selective protocol is recommended to both save time and saturate the system.

Choice of the Target Allele

The selection of the target allele to initiate the suppressor hunt is critical for enhancing the chances of success. Because the goal is to identify a protein that interacts directly with the target, mutant alleles that are deletions or that cause a significant truncation of the protein should be avoided. Rather, mutations that alter protein–protein interacting domains (e.g., coiled coil, helix-loop-helix) should be given high priority. In particular, conditional mutations (cold-sensitive or heat-sensitive) have been successfully employed. Regardless of the allele selected, the presence of the mutant target protein should be verified by western blot analysis. Mutations that severely destabilize the target protein will reduce the chances of isolating an interacting suppressor.

In the new genomic era, genes are frequently identified in a gene of interest by homology with a gene family rather than by a classic mutant hunt. How then does one generate a mutation that could serve as the starting point for a suppressor hunt? David Botstein and coworkers replaced clusters of charged residues in actin with alanines and recovered conditional alleles at relatively high efficiency (Wertman et al. 1992). The rationale for this protocol is that a highly charged region of the protein will probably be exposed on the outside of the protein rather than buried within. These exposed residues may be involved in establishing protein interactions through their charged side groups. This strategy has also been successfully employed with proteins that do not make highly ordered structures (Diamond and Kirkegaard 1994). Therefore, the target protein should be visually inspected to identify clusters of charged residues (three out of five). Substitution of all the charged residues with alanine can be accomplished using standard oligonucleotide-directed mutagenesis protocols. The mutant alleles should then be tested for activity under normal conditions and for temperature-sensitive (ts) or cold-sensitive (cs) conditional behavior. Alleles demonstrating tight conditional behavior should be chosen for the next round of analysis.

Setting the Phenotypic Threshold

The suppression of a mutant phenotype is often partial in nature. Therefore, it is best to use initial conditions that provide the most sensitivity for suppression. For example, if a *ts* allele is being used, reduce the temperature in 0.5° increments and reexamine the phenotype. Choose the lowest temperature that still provides a workable phenotype. This approach will increase the likelihood that second mutations causing even mild suppression can be identified. However, if the threshold is set too low, the hunt may produce a large number of suppressors that are due to nonspecific events. Therefore, if a large number of suppressors is obtained (e.g., >50), the strongest suppressor should be identified for further study. With a *ts* target allele, this is accomplished by simply elevating the temperature and identifying the suppressors that still function as the target protein becomes increasingly more inactivated. This method is effective except in the case of bypass suppressors (see below). If the target mutation is not conditional, choose a stronger allele and either assay the suppressors already identified or repeat the hunt.

Executing the Suppressor Hunt

High-copy Suppressors

High-copy suppression uses increased gene dosage, or expression from a strong heterologous promoter, to increase the concentration of a protein that binds to the mutated target. The rationale for high-copy suppression is that a mutation in the target protein reduces its ability to bind another protein causing a mutant phenotype. The increased concentration of the second protein will compensate for this reduced binding affinity (Fig. 1). High-copy suppression has three advantages over the mutagenesis approach described below. First, although the same care should be given to selecting the target allele, only a yeast transformation is required to start the analysis. Second, because the library plasmid can be rescued back into *Escherichia coli*, the gene or genes providing suppressor activity can be quickly identified. Finally, the source of the library can be derived from any organism, as long as it contains yeast expression and plasmid maintenance determinants. This option may be especially important if a heterologous system is being established in yeast (see below). For these reasons, high-copy suppressors are normally sought first.

Choice of Vector

The high-copy suppression approach demands the elevated expression of the interacting protein. This can be accomplished in two ways. First, random yeast genomic fragments are inserted into a vector containing a 2μ origin of replication but lacking a centromere (a centromere would reduce the copy number to approximately one). The copy number of such a plasmid is usually 20–30 per cell (Nelson and Fangman 1979), and several have been engineered with convenient cloning sites and selectable markers (Christianson et al. 1992). These plasmids are well maintained and segregate to daughters, unlike other high-copy centromere-lacking plasmid systems (e.g., Yrps; Strich et al. 1986). A yeast genomic library with an insert size of 8–12 kbp and a high ligation efficiency (>85%) requires approximately 4000 independent clones to provide coverage of all the genes. If heterologous cDNA expression is desired, the GALSET series of plasmids (Enomoto et al. 1998) is a good choice for a high-copy plasmid backbone. In addition to providing a high-copy-number plasmid, expression is also enhanced through use of the strong *GAL1* promoter (Johnston and Davis 1984). This promoter is induced by growth on medium containing galactose but repressed by the presence of dextrose. Constructing a *GAL1*–cDNA library from higher organisms requires the analysis of a greater number of independent clones (~100,000 to 1,000,000), because the complexity of the genome increases while the efficiency of producing a productive fusion gene decreases.

Mutation-induced Suppressors

In standard suppressor hunt protocols, a haploid strain harboring the target mutation is used to allow the recovery of recessive, loss-of-function alleles. However, if the identification of interacting proteins is the ultimate goal of the study, a diploid strain homozygous for the target mutation should be employed. The use of a diploid strain will restrict the suppressors recovered to only dominant, gain-of-function alleles that will function even in the presence of the wild-type allele. Many types of mutagens have been successfully employed to increase the mutation rate in yeast (e.g., chemical mutagens, UV irradiation). An ethylmethyl sulfonamide (EMS) mutagenesis protocol is described in the Protocols section. This mutagenesis protocol increases the mutation rate to approximately 10^{-3}. The spontaneous mutation rate in yeast is approximately 10^{-6}. Regardless of the mutagen chosen, the execution of the suppressor hunt itself is essentially the same. The number of survivors tested depends on whether a screen or selection is employed (see below).

Execution of Suppressor Hunt

The use of a selection permits either the transformants or the mutagenized survivors to be plated at high density (~5000–10,000 per plate). If a screen is required, then ~400 colony-forming units should be plated. It should be determined empirically what percentage of the transformation reaction needs to be used to obtain either number. Moreover, because mutagenesis reduces cell viability, the plating efficiency of the mutagenized culture should be determined prior to starting the suppressor hunt. Moreover, these cells must be spread carefully. Good spacing of the colonies may allow the plates to be directly replicated to the phenotype-testing medium, bypassing the time-consuming requirement to make master plates by picking colonies with toothpicks. If master plates are required, two possible templates are shown in Figure 4. The number of individual colonies to screen depends on many factors, but 5,000–10,000 is a normal starting target. Finally, if a reporter plasmid is used to express the phenotype, it must be determined whether the suppressor obtained is due to the high-copy library/host mutation or an alteration on the reporter gene itself, the classic *cis, trans* test. A mutation on a reporter gene will behave identically to real suppressors; i.e., it will possess dominant characteristics and suppress the mutant phenotype. Therefore, reporter plasmids must be cured from the yeast strain by growth under conditions that

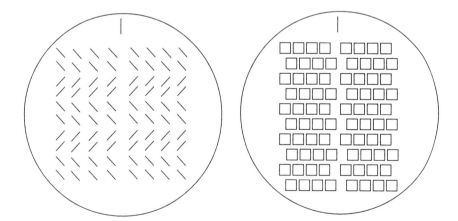

FIGURE 4. Yeast master templates. Two templates are illustrated for generating master plates for suppressor screen protocols. The left template is designed for plus/minus tests (e.g., no growth, reporter gene activation/inactivation). The right template allows mastering at lower cell density on the agar. This method is helpful when examining partial phenotypes such as increased drug resistance or loss of viability due to exposure to xenotoxic compounds.

do not select for plasmid maintenance (e.g., YPDA). Then an unmutagenized source of the reporter must be reintroduced.

Analysis of Suppressors

If the goal of this approach is to identify proteins that directly interact with the target protein, the suppressors will normally possess the following characteristics: (1) The new mutation or high-copy plasmid will *not* be able to suppress a null allele of the target gene. (2) The suppression will be allele-specific.

The first characteristic is absolutely essential. This criterion demands that the target protein be present in the cell before the wild-type phenotype can be generated by either type of suppressor. To test this criterion, disrupt the target allele using gene replacement (Rothstein 1991). For high-copy suppressors, simply retransform the rescued plasmid into a strain harboring a null allele of the target gene. Suppressors that still function in the presence of a null allele most likely represent genes that function either downstream or independent of the target gene. If the null mutant is inviable, the mutation must be covered by the introduction of a wild-type allele of the target gene on a plasmid carrying either the *URA3* or *LYS2* gene for maintenance selection. The chromosomal allele of the target gene can now be disrupted in the presence of the plasmid. To test whether suppression occurs with the null allele, simply transfer the strain to medium containing 5-fluoroorotic acid (5-FOA) or aminoadipate, which selects against the maintenance of the *URA3* or *LYS2* plasmid, respectively (Boeke et al. 1984; Zaret and Sherman 1985). The ability of the strain to grow in the absence of the wild-type target gene indicates that a bypass suppressor has been isolated. Although bypass suppressors may be of interest to the overall goal of the research program, they most likely do not represent interacting proteins.

The second condition of allele specificity is the gold standard for direct interacting suppressors. Allele specificity suggests that the change in the suppressor protein is able to compensate specifically for the target mutation. This information is invaluable for delimiting interacting domains for both proteins simply by sequencing the mutant alleles and finding where the mutations lie. For suppressor mutations, outcrossing to a different target allele is necessary. For high-copy suppressors, the connection between allele specificity and direct interactors is less well documented. However, the high-copy suppressor can be tested for allele specificity by simply transforming the suppressor-harboring plasmid into a different mutant strain.

Heterologous Systems

Suppressor analysis has long been part of the experimental arsenal of researchers working in genetically tractable organisms. However, similar approaches have proven difficult to recapitulate in mammalian cell systems. One approach to circumvent this problem has been to establish a genetic system in yeast using mammalian target proteins of interest. This method takes advantage of the considerable homology observed between basic biological processes in yeast and mammals. For example, the glucocorticoid receptor/transcription factor has been shown to activate reporter gene expression in yeast (Schena and Yamamoto 1988; Garabedian and Yamamoto 1992). Moreover, G-coupled protein receptors have been modified to activate the mating pheromone mitogen-activated protein (MAP) kinase cascade (Klein et al. 1998; Reilander and Weiss 1998). In addition, the BRCA1 tumor suppressor (Hu et al. 2000), when fused to LexA, functions as a transcription factor in both yeast and human cells. Finally, systems have been established analyzing apoptosis, a process that does not formally exist in yeast (Xu and Reed 1998; Kang et al. 1999; Wright et al. 2000). The functionality of these proteins in yeast obviously requires them to interact with yeast proteins to perform their function. Therefore, the identification of yeast interactors may provide valuable clues as to the nature of associating proteins in the organism of interest.

Synthetic Phenotypes

There are two approaches, loss or gain of function, to establish a genetic system in yeast based on a foreign target gene. The loss-of-function method requires that a functional homolog of the heterologous protein of interest be identified in yeast. The complete sequence of the yeast genome has provided all the available open reading frames. If a potential homolog is identified, the published information on the protein(s) in question can obtained in two excellent World Wide Web sites, the Saccharomyces genome database (SGD, http://genome-www.stanford.edu/Saccharomyces/) and the yeast proteome database (YPD, http://www.proteome.com/databases/index.html). The yeast homolog can be examined as outlined above for a phenotype to use in a suppression analysis. If no obvious homolog exists, another path can be taken to identify a yeast protein that has a similar function to the target. This method involves a modification of the synthetic lethal screen developed a decade ago (Kranz and Holm 1990; Bender and Pringle 1991). Briefly, this strategy involves identifying a yeast mutation that makes the target gene of interest essential for growth. This method may allow the identification of a functional homolog that does not contain sufficient sequence similarity to be culled from the database by current algorithms. This method is time-consuming and probably has the least chance for success versus other methods. Therefore, this avenue of investigation should be used only after exhausting other methods. Once a putative homolog has been identified by sequence comparison, the physiological significance of the similarity can be probed by making a mutant in the yeast gene and assaying for expected phenotypes. For example, if the target is involved in ion homeostasis, the yeast mutant may be sensitive to medium containing either too much or too little calcium. Complementation of the mutation by its mammalian counterpart would add significant credence to the hypothesis that the two proteins represent functional homologs. However, cross-species complementation is not always successful, even with closely related proteins.

The second avenue that can be pursued is to generate a phenotype in yeast through overexpression of the target protein from a heterologous promoter. Many vectors and promoters are available for expressing foreign genes in yeast (for review, see Buckholz 1993). The use of conditional promoters, such as *GAL1* or *GAL10* of the galactose utilization pathway, may be best suited for this study because the overexpression of the target gene may produce a lethal phenotype. Death itself may not provide much information, but it does provide a strong phenotype for suppressor analysis.

Mutagenesis

Once the appropriate choices of phenotype and target allele have been made, this protocol is used to perform the mutagenesis.

MATERIALS

CAUTION: See Appendix for appropriate handling of materials marked with <!>.

Buffers and Solutions

Ethylmethyl sulfonamide (EMS)
NaPO$_4$ buffer, (0.1 M, pH 7.0) <!>
   57.7 ml of Na$_2$HPO$_4$
   42.3 ml of NaH$_2$PO$_4$,
     Dilute to 1000 ml with dH$_2$O.
Synthetic dextrose (SD)
   2% dextrose
   0.17% yeast nitrogen base
   0.5% ammonium sulfate <!>
YPDA
   1% yeast extract
   2% dextrose
   2% peptone, supplemented with 10 mg/liter adenine

Yeast Strains

Yeast host strain

METHOD

1. Inoculate a 25-ml culture of the yeast host strain in the appropriate medium. If no plasmid maintenance is required, rich medium (YPDA) should be used, because the cells will grow to a higher density. If plasmid selection is required to elaborate the target phenotype, a synthetic dextrose medium (SD) supplemented with the required amino acids or nucleotides should be used.

   General guides to the maintenance and propagation of yeast are available (Sherman 1974 #306; Burke et al. 2000) and should be consulted prior to starting this protocol.

   For galactose-responsive promoter induction, substitute raffinose (2%) to grow cultures under nonrepressing, but noninducing, conditions. These promoters can be activated by addition of galactose (1%) to the medium.

2. Grow the culture for 2 days, with shaking, at the permissive temperature. It is important that the culture be grown to saturation so that very few (<5%) budded cells remain in the population, because mutagenizing a mother but not the bud will mask any phenotype in the colony. Harvest cells by centrifugation (6000g) and wash once in one culture volume of 0.1 M NaPO$_4$ buffer (pH 7.0).

3. Resuspend the cell pellet in 1/10 culture volume of NaPO$_4$ buffer. If a screen is to be employed, lightly sonicate the culture (lowest level, 50% duty cycle, 30 seconds) to ensure that the colonies will arise from a single cell. Transfer the prepared culture to a sterile 25-ml flask and add 0.05 ml of EMS. Gently shake the flask in a chemical fume hood on a rotating table for 180 minutes.

4. Remove 0.1 ml of the mutagenized culture and add to 9.9 ml of 5% sodium thiosulfate and 0.1 M NaPO$_4$ (pH 7.0) to quench the EMS. This sample can be plated directly and stored at 4°C for 1 week without significant loss in viability. Quench the remaining culture with 100 ml of sodium thiosulfate and store at 4°C for reserve if necessary. Proceed to phenotypic screening or selection.

Protocol 2

Total Yeast DNA Preparation

Total yeast DNA preparations are required to rescue high-copy suppressor plasmids from yeast back into *E. coli* and for the construction of genomic libraries to clone dominant suppressors.

MATERIALS

CAUTION: See Appendix for appropriate handling of materials marked with <!>.

Buffers and Solutions

Ethanol (100% [0.8 ml] and 70% [0.5 ml]) <!>
PCI (0.2 ml) <!>
    25:24:1 phenol (distilled):chloroform:isoamyl alcohol <!>
YRLB (0.2 ml) <!>
  0.5 M NaCl
  0.2 M Tris-HCl (pH 7.5)
  10 mM EDTA
  1% SDS

Yeast Strains

Yeast host strain culture (see previous protocol)

Special Equipment

Glass beads (0.5 mm)

METHOD

1. Grow a 15-ml overnight culture in defined medium selecting for plasmid maintenance (if appropriate).

2. Harvest cells by centrifugation ($6000g$), resuspend the pellet with 1 ml of H_2O, and transfer to a microfuge tube.

3. Add 1 cell pellet volume of 0.5-mm glass beads, 0.2 ml of YRLB, and 0.2 ml of PCI.

4. Vortex vigorously for 2 minutes. Add 0.3 ml of YRLB and 0.2 ml of PCI. Repeat vortexing.

5. Centrifuge at $12,000g$ for 5 minutes.

6. Remove the aqueous phase, extract with 0.4 ml of PCI, and collect the aqueous phase.

7. Add 0.8 ml of 100% ethanol and incubate for 15 minutes at room temperature.

8. Centrifuge for 10 minutes. Wash the pellet with 0.5 ml of 70% ethanol.

9. Resuspend the dried pellet in 100 µl of H_2O.

To rescue putative high-copy suppressors, transform 1–2 µl of total DNA preparation into competent *E. coli*.

Protocol 3

Yeast Transformation

Several excellent protocols have been published detailing high-efficiency transformation of budding yeast. The most detailed protocol is provided by Gietz and coworkers (http://www.umanitoba.ca/faculties/medicine/biochem/gietz/Trafo.html).

REFERENCES

Bender A. and Pringle J.R. 1991. Use of a screen for synthetic lethal and multicopy suppressor mutants to identify two new genes involved in morphogenesis in *Saccharomyces cerevisiae*. *Mol. Cell. Biol.* **11:** 1295–1305.

Boeke J.D., LaCroute F., and Fink G.R. 1984. A positive selection for mutants lacking orotidine-5′-phosphate decarboxylase activity in yeast: 5-Fluoro-orotic acid resistance. *Mol. Gen. Genet.* **197:** 345–346.

Buckholz R.G. 1993. Yeast systems for the expression of heterologous gene products. *Curr. Opin. Biotechnol.* **4:** 538–542.

Burke D., Dawson D., and Stearns T. 2000. *Methods in yeast genetics: A Cold Spring Harbor Laboratory course manual.* Cold Spring Harbor Laboratory Press, Cold Spring Harbor, New York.

Christianson T.W., Sikorski R.S., Dante M., Shero J.H., and Hieter P. 1992. Multifunctional yeast high-copy-number shuttle vectors. *Gene* **110:** 119–122.

Diamond S.E. and Kirkegaard K. 1994. Clustered charged-to-alanine mutagenesis of poliovirus RNA-dependent RNA polymerase yields multiple temperature-sensitive mutants defective in RNA synthesis. *J. Virol.* **68:** 863–876.

Enomoto S., Chen G., and Berman J. 1998. Vectors for expressing T7 epitope- and His$_6$ affinity-tagged fusion proteins in *S. cerevisiae*. *Bio/Technology* **24:** 782–786.

Garabedian M.J. and Yamamoto K.R. 1992. Genetic dissection of the signaling domain of a mammalian steroid receptor in yeast. *Mol. Biol. Cell* **3:** 1245–1257.

Homesley L., Lei M., Kawasaki Y., Sawyer S., Christensen T., and Tye B.K. 2000. Mcm10 and the MCM2-7 complex interact to initiate DNA synthesis and to release replication factors from origins. *Genes Dev.* **14:** 913–926.

Hu Y.F., Miyake T., Ye Q., and Li R. 2000. Characterization of a novel trans-activation domain of BRCA1 that functions in concert with the BRCA1 C-terminal (BRCT) domain. *J. Biol. Chem.* **275:** 40910–40915.

Johnston M. and Davis R.W. 1984. Sequences that regulate the divergent *GAL1-GAL10* promoter in *Saccharomyces cerevisiae*. *Mol. Cell. Biol.* **4:** 1440–1448.

Kang J.J., Schaber M.D., Srinivasula S.M., Alnemri E.S., Litwack G., Hall D.J., and Bjornsti M.A. 1999. Cascades of mammalian caspase activation in the yeast *Saccharomyces cerevisiae*. *J. Biol. Chem.* **274:** 3189–3198.

Klein C., Paul J.I., Sauve K., Schmidt M.M., Arcangeli L., Ransom J., Trueheart J., Manfredi J.P., Broach J.R., and Murphy A.J. 1998. Identification of surrogate agonists for the human FPRL-1 receptor by autocrine selection in yeast. *Nat. Biotechnol.* **16:** 1334–1337.

Kranz J.E. and Holm C. 1990. Cloning by function: An alternative approach for identifying yeast homologs of genes from other organisms. *Proc. Natl. Acad. Sci.* **87:** 6629–6633.

Levin J.Z. and Horvitz H.R. 1993. Three new classes of mutations in the *Caenorhabditis elegans* muscle gene sup-9. *Genetics* **135:** 53–70.

Mortin M.A. 1990. Use of second-site suppressor mutations in *Drosophila* to identify components of the transcriptional machinery. *Proc. Natl. Acad. Sci.* **87:** 4864–4868.

Nelson R.G. and Fangman W.L. 1979. Nucleosome organization of the yeast 2 micron DNA plasmid: A eukaryotic minichromosome. *Proc. Natl. Acad. Sci.* **76:** 6515–6519.

Reilander H. and Weiss H.M. 1998. Production of G-protein-coupled receptors in yeast. *Curr. Opin. Biotechnol.* **9:** 510–517.

Rothstein R. 1991. Targeting, disruption, replacement, and allele rescue: Integrative DNA transformation in yeast. *Methods Enzymol.* **194:** 281–301.

Schena M. and Yamamoto K.R. 1988. Mammalian glucocorticoid receptor derivatives enhance transcription in yeast. *Science* **241:** 965–967.

Smyth K.A. and Belote J.M. 1999. The dominant temperature-sensitive lethal DTS7 of *Drosophila melanogaster* encodes an altered 20S proteasome beta-type subunit. *Genetics* **151:** 211–220.

Strich R., Woontner M., and Scott J.F.. 1986. Mutations in *ARS1* increase the rate of simple loss of plasmids in *Saccharomyces cerevisiae*. *Yeast* **2:** 169–178.

Wertman K.F., Drubin D.G., and Botstein D. 1992. Systematic mutational analysis of the yeast ACT1 gene. *Genetics* **132:** 337–350.

Wright M.E., Han D.K., and Hockenbery D.M. 2000. Caspase-3 and inhibitor of apoptosis protein(s) interactions in *Saccharomyces cerevisiae* and mammalian cells. *FEBS Lett.* **481:** 13–18.

Xu Q. and Reed J.C. 1998. Bax inhibitor-1, a mammalian apoptosis suppressor identified by functional screening in yeast. *Mol. Cell.* **1:** 337–346.

Zaret K.S. and Sherman F. 1985. alpha-Aminoadipate as a primary nitrogen source for *Saccharomyces cerevisiae* mutants. *J. Bacteriol.* **162:** 579–583.

10 | Imaging Protein Interactions by FRET Microscopy

Peter J. Verveer, Ailsa G. Harpur, and Philippe I.H. Bastiaens

Cell Biology and Cell Biophysics Programme, European Molecular Biology Laboratory, D-69117 Heidelberg, Germany

This chapter is adapted and expanded from Chapter 18, Protocol 5, in Sambrook and Russell 2001, *Molecular Cloning.*

INTRODUCTION

GFP: Impact on Molecular and Cellular Biology and Application to Protein Interactions

Following the discovery and cloning of the green fluorescent protein (GFP) from the jellyfish, *Aequorea victoria* (Prasher et al. 1992), many opportunities for the analysis of gene function and protein–protein interactions have emerged. Such opportunities range from the visual tracking of proteins in a spatial and temporal manner to the quantitation of the intrinsic photophysical properties of the GFPs themselves, allowing them to be used as indicators or sensors of biological activities/processes and protein interactions (for an extensive review, see Wouters et al. 2001).

GFP is intrinsically fluorescent and does not require exogenous cofactors or substrates, a property that renders it highly useful as a genetically encoded reporter tag. This is particularly valuable for analysis of gene expression in embryonic and intact organisms using either endogenously regulated (gene trapping) or tissue-specific promoters (Chalfie et al. 1994; Moss et al. 1996). The utility of GFP has also been demonstrated in establishing the subcellular localization of fusion proteins (Chalfie et al. 1994; Moss et al. 1996; Arnone et al. 1997). Organelle-specific localization signals have been successfully fused to GFP mutants, thus targeting GFP to locations such as the endoplasmic reticulum, the Golgi apparatus (Dayel et al. 1999), mitochondria (Niwa et al. 1999), and the plasma membrane (Okada et al. 1999). In addition to localization of proteins, GFP has been used as an intracellular sensor of biological activity by exploiting its photophysical properties. For example, the utility of GFP has been extended to that of an indicator of protein proximity by detection of the excited-state reaction, *fluorescence resonance energy transfer* (FRET). To date, this has been employed in calcium-sensitive constructs (Miyawaki et al. 1997, 1999), sensors for cyclic AMP (Zaccolo et al. 1999) or cyclic GMP (Honda et al. 2001), protease substrates (Heim and Tsien 1996; Mitra et al. 1996; Xu et al. 1998; Mahajan et al. 1999), and for detecting the interaction of a number of proteins (Day 1998; Mahajan et al. 1998). Additionally, FRET between GFP and Cy3 has been used to image cellular processes such as the phosphorylation of intracellular and transmembrane proteins (Ng et al. 1999; Wouters and Bastiaens 1999; Verveer et al. 2000). This chapter gives a detailed description of the latter type of experiment using several approaches for imaging FRET in a microscope.

Fluorescence Resonance Energy Transfer

Photophysical Principles of FRET

Until recently, only a limited number of techniques that rely heavily on the use of chemical cross-linking agents and antibodies were available for molecular and cellular biologists to address the question of specific protein–protein interactions. However, the difficulty with such techniques as immunoprecipitation and affinity chromatography is that they do not preserve the physiological conditions under which proteins may normally interact in the cell. Additionally, such approaches do not provide information on the spatial distribution of the interacting proteins. Immunocytochemical colocalization provides indirect evidence of the presence of two particular proteins within the same cellular compartment; however, it fails to confirm whether these particular proteins actually interact directly. FRET measurements can address such questions of specificity in intact cells (Bastiaens and Squire 1999).

FRET, a nonradiative, dipole–dipole coupling process, transfers energy from an excited donor fluorophore to an acceptor fluorophore in very close proximity (typically within 10 nm) (Förster 1948; Herman 1989; Clegg 1996). Thus, excitation of the donor produces a sensitized emission from the acceptor that in the absence of FRET ordinarily would not occur. Proteins can be fused to genetically encoded GFP variants or chemically modified by covalent attachment of synthetic

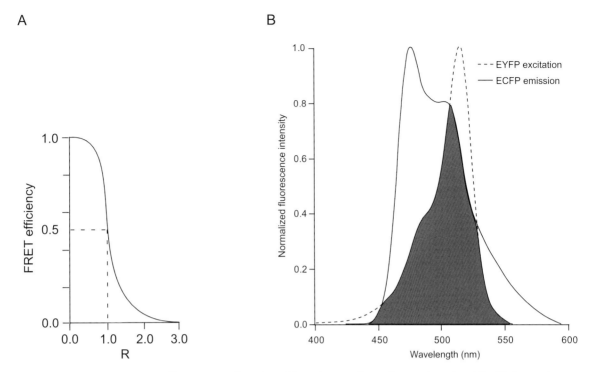

FIGURE 1. (A) FRET efficiency as a function of interchromophore distance in units of R_0. (B) Spectral overlap (*shaded area*) between the fluorescence emission spectrum of ECFP and the absorption spectrum of EYFP. (Reprinted, with permission, from Sambrook and Russell 2001, *Molecular Cloning*.)

fluorophores (Bastiaens and Jovin 1998; Griffin et al. 1998), and the molecular interaction of the proteins in question can then be inferred by FRET between the fluorophores. The rate (k_t; ns^{-1}) at which Förster-type energy transfer occurs is given by the equation:

$$k_t = k_D \left(\frac{R_0}{r}\right)^6 \tag{1}$$

where r (nm) is the actual distance between the fluorophores, k_D (ns^{-1}) is the rate at which the excited donor returns to its ground state in absence of an acceptor (equal to the inverse of τ_D (ns), the fluorescence lifetime of the donor), and R_0 (nm) is the distance at which 50% energy transfer takes place (Fig. 1), defined as:

$$R_0 = \left(\kappa^2 \cdot J(\lambda) \cdot n^{-4} \cdot Q\right)^{\frac{1}{6}} \cdot 9.7 \cdot 10^2 \tag{2}$$

Finally, the efficiency of energy transfer E is defined by the ratio of the Förster-type energy transfer rate to the total sum of rates for all processes by which the excited donor can return to its ground state:

$$E = \frac{k_t}{k_t + k_D} = \frac{R_0^6}{R_0^6 + r^6} \tag{3}$$

This equation shows a steep dependency of the energy transfer efficiency on the distance between the donor and acceptor. The energy transfer efficiency E characterizes the efficiency of nonradiative energy transfer between a single donor/acceptor pair at a given distance r. Factors such as the quantum yield (Q) of the donor, the relative orientation of the transition dipoles of the fluorophores (κ^2), the refractive index of the intervening medium (n), and the overlap integral ($J(\lambda)$; cm^6 mole^{-1}) of the donor emission and acceptor absorption spectra will affect R_0 and therefore E (Clegg 1996). The overlap integral ($J(\lambda)$;cm^6 mole^{-1}) is given by:

$$J(\lambda) = \frac{\int_0^{\infty} F(\lambda)\varepsilon(\lambda)\lambda^4 d\lambda}{\int_0^{\infty} F(\lambda) d\lambda} \tag{4}$$

where $\varepsilon(\lambda)$ (M^{-1}cm^{-1}) is the extinction coefficient of the acceptor, and $F(\lambda)$ is the fluorescence intensity of the donor, at wavelength λ. In selecting a suitable chromophore pair for FRET-based protein interaction detection, two main factors must be considered to optimize R_0: (1) the quantum yield of the donor Q and (2) the spectral overlap between donor emission and acceptor absorption determined by the overlap integral $J(\lambda)$, which is dependent on the acceptor extinction coefficient $\varepsilon(\lambda)$. In addition, the choice also depends on the chromophore pair being spectrally distinct enough to facilitate unambiguous detection of the donor with an appropriate band pass filter. The final considerations are practicalities, such as the feasibility of the construction of functional fusion proteins or fluorophore conjugates and their delivery into cells. Therefore, the optimal R_0 must be considered bearing the examined protein system in mind.

The measurement of FRET between GFP mutants (for review, see Heim and Tsien 1996; Tsien 1998; Matz et al. 1999; Pollock and Heim 1999) (these variants are available commercially from CLONTECH) has been facilitated by the engineering of blue and red-shifted variants of GFP, such as cyan (CFP) and yellow (YFP) fluorescent protein. The enhanced cyan fluorescent protein (ECFP) as a donor to the enhanced yellow fluorescent protein (EYFP) is an example of a genetically encodable, spectrally compatible, chromophore combination. The R_0 value for this pair is 4.9 nm (Table 1), which enables FRET detection over distances of $\sim1.6 \times R_0 = 7.8$ nm corresponding to a detection limit of 5% FRET efficiency (Fig. 1). Genetically encoded fluorescent FRET pairs allow the analysis of protein interactions without the need to conjugate fluorophores to purified proteins and reintroduce them into cells. If, however, the use of sulfoindocyanine dye conjugates is feasible, the large extinction coefficient and spectral overlap makes Cy3 (Amersham Pharmacia Biotech) a suitable acceptor for energy transfer from the donor EGFP (Ng et al. 1999; Wouters and Bastiaens 1999; Verveer et al. 2000). The R_0 of this pair is 6 nm, enabling efficient FRET detection over a distance up to 9.6 nm. Obviously, the higher the R_0, the greater the distance over which FRET can be detected. However, even under optimal donor–acceptor spectral conditions, R_0 will generally not exceed 7 nm, which confers an upper distance limit of ~11 nm for the detection of FRET. In practical terms, these factors may be considered in relation to average protein sizes. Assuming that the specific volume of a protein is ~0.74 cm^3/gram, the radius of a 100-kD molecular-mass globular protein can be calculated by using the following equation, where R is the radius (nm):

$$R = 6.76 \cdot 10^{-2} \cdot \sqrt[3]{MW} \tag{5}$$

resulting in a radius of 3.1 nm. A protein of 200 kD has a calculated radius of 3.95 nm. These calculations of radii demonstrate that FRET is likely to occur between fluorophores on two interacting proteins with molecular masses below 100 kD, whereas the likelihood of FRET between dyes on globular proteins of molecular masses above 200 kD is low.

Imaging Protein Phosphorylation with FRET

FRET can be used as a highly specific indicator of protein modifications (such as phosphorylation) through the interaction of a target protein with a reagent (such as an antibody) that recognizes the modified residues. Signal specificity for protein modification is achieved by FRET between chromophores present on the phosphorylated protein and an antibody. Typically, this

TABLE 1. R_0 Values (nm) for Green Fluorescent Protein and Cy Dyes

Acceptor ⇒ Donor ⇓	ECFP	EGFP	EYFP	Cy3	Cy3.5	Cy5
ECFP	n.d.	4.65	4.87	n.d.	n.d.	n.d.
EGFP		4.61	5.47	6.0	5.7	n.d.
EYFP			5.05	n.d.	n.d.	n.d.
Cy3				4.4	5.1	5.3
Cy3.5					4.6	6.4

The values were calculated as described previously (Bastiaens and Jovin 1996), using published extinction coefficients for GFP variants (Tsien 1998) and those supplied by the manufacturer for the Cy dyes (Amersham Pharmacia Biotech). (n.d.) Not determined. (Reprinted, with permission, from Sambrook and Russell 2001, *Molecular Cloning.*)

type of experiment is carried out using a GFP moiety, fused to the phosphoprotein as the donor, and Cy3 conjugated to an antibody as the acceptor. The antibody–Cy3, or Fab fragment–Cy3, conjugate is introduced by either microinjection into live cells or incubation on permeabilized, fixed cells. In this approach, FRET can be detected by changes in the quantum yield of the donor, typically measured by changes in fluorescence lifetime or, in fixed samples, by acceptor photobleaching. *F*luorescence *l*ifetime *i*maging *m*icroscopy (FLIM) is particularly suited to live-cell FRET measurements because the acceptor-bearing antibody is present in large molar excess over the donor fusion protein. In addition, the antibody does not necessarily need to be uniquely specific for the epitope, as is often the case. Semipurified antibody preparations are accommodated that possess minor quantities of specific IgG molecules. We describe the detection of interactions with (1) one genetically encoded donor component and one microinjected, labeled acceptor component and (2) two labeled donor/acceptor components. These protocols are equally compatible with the detection of interactions between two genetically encoded probes such as ECFP and EYFP or EGFP and DsRed. Although we present here a specific protocol for measuring the activation state of protein kinase Cα (PKCα) by following the phosphorylation of Thr-250 (Ng et al. 1999), the approach is generic for measuring protein covalent modification(s) and can also be applied to the detection of protein interactions per se. Examples are described for both a live- and a fixed-cell FLIM experiment using EGFP-tagged PKCα and a labeled antibody specific for phosphorylated Thr-250, and also for a fixed-cell FLIM experiment using two labeled antibodies (one to PKCα and the other to phospho-Thr-250). In addition to these, protocols are given for sensitized emission measurements and FRET detection by acceptor photobleaching. The latter is illustrated with an example of the detection of the phosphorylation state of GFP-tagged ErbB1 with a generic antibody against phosphotyrosine.

OUTLINE OF PROCEDURE

FRET Detection Methods

Methods of measuring FRET in a microscope can be broadly divided into intensity-based methods and fluorescence-decay-kinetics-based methods (see Wouters et al. 2001). Excitation of a donor fluorophore in a FRET pair leads to quenching of the donor fluorescence and, in an increased and sensitized acceptor, emission. Intensity-based or steady-state fluorescence intensity measurements make use of these effects (Bastiaens and Jovin 1998; Gordon et al. 1998; Nagy et al. 1998). Excited-state decay kinetics of the donor is measured in kinetic-based approaches, either by measuring the fluorescence lifetime of the donor chromophore (Gadella et al. 1993; Bastiaens and Squire 1999) or by measuring the photobleaching kinetics of the donor (Jovin and Arndt-Jovin 1989a,b).

Sensitized Emission Measurements

Preferential excitation of a donor fluorophore that is located in molecular proximity to an appropriate acceptor fluorophore will result in sensitized emission from the acceptor and quenching of the donor fluorescence. The change in the ratio of the donor and acceptor emissions following an event that induces protein association can be measured (Adams et al. 1991; Miyawaki et al. 1997, 1999). This measurement has a number of limitations despite its practicality. For instance, because the ratio is not an absolute determination of the FRET efficiency, such measurements must be performed over time to detect change that can be ascribed to protein association. Because fluorescence intensity is proportional to concentration, the donor–acceptor intensity ratio will depend on local concentrations of labeled proteins. Differential redistribution of the fluorescent proteins as a result of biological activity during the measurement makes this aspect particularly problematic. Ratiometric concentration effects are not an issue with the use of reporter probes that encode the donor and acceptor GFPs on the same polypeptide, as is the case with molecules such as the calcium-sensing "chameleons" (Miyawaki et al. 1997, 1999). These reporter probes consist of a donor and acceptor flanking an "activity"-specific module, in which the stoichiometry of the donor–acceptor is constant for all pixels, and therefore, any change in emission ratio can be ascribed to changes in FRET efficiency. However, one should take care to ensure that observed ratio changes are not due to pH changes that differentially affect the intensity of GFP variants (Llopis et al. 1998). Alternatively, the detection of acceptor-sensitized emission alone can be used as a measure of FRET (Mahajan et al. 1998; Xu et al. 1998), an approach that is useful in situations where donor intensities are low and/or there is contamination with high background (auto) fluorescence in the donor channel. However, absorption spectra characteristically exhibit long tails in the shorter wavelength (blue) region, which may result in the direct excitation of the acceptor molecule in addition to that of the donor, thus resulting in mixing of direct and sensitized emission. Conversely, fluorescence emission tends to tail into the red part of the spectrum, causing donor fluorescence bleedthrough into the acceptor detection channel. Corrections for these effects have been documented (Gordon et al. 1998; Nagy et al. 1998) and involve the acquisition of fluorescence images of samples containing the donor, the acceptor, and both for three different filter settings. The images taken with the donor filter set (I^D), the acceptor filter set (I^A), and the FRET filter set (I^F) can be written in terms of the concentrations of free donor-tagged molecules ([D]), free acceptor-tagged molecules ([A]), and the concentration of interacting molecules ([DA]):

$$I^D = \varepsilon_D^D Q_D^D([D] + (1-E)[DA])$$ (6)

$$I^A = \varepsilon_A^A Q_A^A([A] + [DA])$$ (7)

$$I^F = \varepsilon_D^D Q_D^A([D] + (1-E)[DA]) + \varepsilon_A^A Q_A^A([A] + [DA]) + \varepsilon_D^D Q_A^A E[DA]$$
$$= \frac{Q_D^A}{Q_D^D} I^D + \frac{\varepsilon_A^D}{\varepsilon_A^A} I^A + \varepsilon_D^D Q_A^A E[DA]$$ (8)

where E is the true energy transfer efficiency of the donor/acceptor pair (as defined by Eq. 3). Here ε_D^D is the excitation efficiency of the donor molecules through the donor excitation filters (donor filter set and FRET filter set), and ε_A^A and ε_A^D are the excitation efficiencies of the acceptor through the acceptor and donor excitation filters, respectively. These efficiencies include factors such as the extinction coefficients of the donor/acceptor molecules and illumination intensity. The detection efficiency of the donor fluorescence through the donor and acceptor emission filters is given by Q_D^D and Q_D^A, respectively. The detection efficiency of the acceptor fluorescence through the acceptor filter is given by Q_A^A. These detection efficiencies include such factors as the quantum yields of the donor and acceptor molecules and exposure time.

The sensitized emission follows from Equation 8:

$$I^{sens} = \varepsilon_D^D Q_A^A E[DA] = I^F - \frac{Q_D^A}{Q_D^D} I^D - \frac{\varepsilon_A^D}{\varepsilon_A^A} I^A \qquad (9)$$

The ratio Q_D^A/Q_D^D can be found by two measurements of a sample containing donor molecules only, using the FRET filter set and the donor filter set, and dividing the integrated intensities. Similarly, the ratio $\varepsilon_A^D/\varepsilon_A^A$ can be found using a sample with acceptor molecules only by dividing the integrated intensities detected with the FRET and the acceptor filter sets.

By dividing the sensitized emission (Eq. 9) by the acceptor emission (Eq. 7), an apparent energy transfer efficiency can be calculated in each pixel that is proportional to the relative concentration of bound acceptor molecules:

$$E_i^{A,sens} = \frac{I^{sens}}{I^A} = \frac{\varepsilon_D^D}{\varepsilon_A^A} \cdot E \frac{[DA]}{[A] + [DA]} \qquad (10)$$

where we added a subscript i to indicate that this measure is different in each pixel. A similar equation can be derived that is proportional to the relative concentration of donor molecules in bound state. To find the total concentration of donor molecules one calculates

$$I^{D,na} = I^F - I^D(Q_D^A/Q_D^D - Q_A^A/Q_A^D) - I^A \varepsilon_A^D/\varepsilon_A^A = \varepsilon_D^D Q_A^A([D] + [DA]) \qquad (11)$$

resulting in an apparent energy transfer efficiency equal to

$$E_i^{D,sens} = \frac{I^{sens}}{I^{D,na}} = E \frac{[DA]}{[D] + [DA]} \qquad (12)$$

Unfortunately, Equation 11 contains the ratio Q_A^A/Q_D^D, that, unlike the ratios encountered before, cannot be easily measured using samples containing donor or acceptor molecules only.

ENERGY TRANSFER EFFICIENCY

Historically, measures such as Equations 10 and 12 have been called "energy transfer efficiency," but they should not be confused with the energy transfer efficiency E, as defined by Equation 3. The latter is a single number that characterizes the efficiency of transfer of energy from the donor to the acceptor and is dependent on the fixed geometry of the donor/acceptor pair in a complex. The pixel-by-pixel apparent energy-transfer efficiencies that can be defined for most FRET measurements in the microscope do depend on E, but, more importantly, also on the concentrations of free and bound donor- or acceptor-tagged molecules. Thus, the physical interpretation of these kinds of energy-transfer-efficiency calculations is that they provide a measure for the fraction of molecules that are interacting in each pixel, with respect to the total concentration of donor- or acceptor-tagged molecules. Note that, in all cases, these measures are not absolute values, but only proportional to the relative concentrations of bound molecules. However, pixel-by-pixel comparisons or comparisons between different images are possible, as long as the excitation efficiencies ε, the detection efficiencies Q, and the true energy-transfer efficiency E, are invariant.

Acceptor Photobleaching

FRET detection is also possible by examining quenching of the donor emission only (Bastiaens and Squire 1999; Wouters et al. 2001). From Equation 6, we see that if no FRET occurs ($E = 0$), I^D is proportional to the total donor concentration. Experimentally, this can be measured on the same specimen by specifically photobleaching the acceptor by excitation at its absorption maximum and acquiring another image using the donor filter set (Bastiaens et al. 1996; Bastiaens and Jovin 1998; Wouters et al. 1998):

$$I^{D,pb} = \varepsilon_D^D Q_D^D([D] + [DA]) \qquad (13)$$

In practice, this approach involves (1) the acquisition of a donor image, (2) photobleaching of the acceptor, and then (3) a second donor image acquisition. An apparent energy transfer efficiency can then be calculated in each pixel that is proportional to the relative concentration of bound donor molecules:

$$E_i^{D,pb} = \frac{I^{D,pb} - I^D}{I^{D,pb}} = E \frac{[DA]}{[D] + [DA]} \tag{14}$$

If an additional measurement is made with an acceptor filter set, an apparent energy transfer measurement can be defined that is proportional to the relative concentration of bound acceptor molecules:

$$E_i^{A,pb} = \frac{I^{D,pb} - I^D}{I^A} = \frac{\varepsilon_D^D Q_D^D}{\varepsilon_A^A Q_A^A} E \frac{[DA]}{[A] + [DA]} \tag{15}$$

In contrast to sensitized emission measurements, photobleaching can be performed with high selectivity of the acceptor because absorption spectra tend to tail in the blue part of the spectrum but are steep at their red edge. In using acceptor photobleaching FRET measurements, care should be taken that the photochemical product of the bleached acceptor does not have residual absorption at the donor emission and, more importantly, that it does not fluoresce in the donor spectral region. Because of mass movement of protein during the extended time required for photobleaching (typically 1–20 minutes), it is preferable to perform this type of FRET determination on fixed cell samples. Live-cell FRET measurements based on this approach are more feasible using fluorescence lifetime imaging because lifetimes are independent of probe concentration and light path length.

Fluorescence Lifetime Measurements

FRET reduces the fluorescence lifetime of the donor fluorophore (τ; a measure of the excited-state duration) because it depopulates its excited state (Clegg and Schneider 1996), an effect that is also manifest as a reduction in the donor *quantum yield* (the ratio of the number of fluorescence photons emitted compared to the number of photons absorbed). FRET can be detected purely via donor fluorescence because τ is an intrinsic fluorescence parameter, that is, it is independent of probe concentration and light path length, but dependent on excited-state processes. The specificity of the acceptor probe is not required because its fluorescence is not detected. This enables the use of saturating amounts of acceptor molecules or pools of acceptor-labeled protein with nonfunctional subpopulations. It is possible to calculate apparent FRET efficiencies at each pixel of an image. As in the case of intensity-based images, this apparent energy transfer efficiency is dependent on the true efficiency of the complex E and the concentrations of free and bound donor-tagged molecules. If we assume that the measured lifetime is equal to the sum of the lifetimes of bound and free donor weighted by their relative concentration, then we find

$$E_i^{D,lt} = 1 - \frac{\tau_i}{\tau_D} = 1 - \frac{[D]\tau_D + [DA]\tau_{DA}}{([D] + [DA])\tau_D} = E \frac{[DA]}{[D] + [DA]} \tag{16}$$

where τ_D is the lifetime of the free donor and τ_{DA} the lifetime of the donor in a complex. The lifetime of the free donor can be obtained at the end of an experiment by acceptor photobleaching or from a different sample that has no acceptor molecules. We note that, in practice, the lifetime that is measured with the instruments described below is a not-weighted sum of the true lifetimes τ_D and τ_{DA}. Thus, formally, the energy transfer efficiency calculated by Equation 16 cannot be interpreted to be proportional to the fraction of bound donor-tagged molecules. However, in most cases this is a reasonable approximation.

FLIM enables the measurement of fluorescence lifetimes on a pixel-by-pixel basis (Lakowicz and Berndt 1991; Clegg et al. 1992; Gadella et al. 1993, 1994) and the calculation of a FRET efficiency map from that image (Wouters and Bastiaens 1999). Broadly speaking, determination of fluorescence lifetimes can be performed by two general approaches.

1. *The time domain.* Here a delayed, gated measurement of the fluorescence intensity is made following excitation of the sample with a short pulse of light (Sytsma et al. 1998). Sampling of the fluorescence intensity is performed at sequential time points along the exponential fluorescence decay. In cases where a sample contains a single fluorescence species, the fluorescence lifetime is given as the time over which the fluorescence intensity falls to about 37% of its initial value. In this scenario, repetition of excitation ensures an adequate signal-to-noise ratio. In practice, a lifetime (τ_i) can be calculated in each pixel from the ratio of the integrated intensities (R_i) for two time-gated segments of equal length such that

$$\tau_i = \frac{\Delta t}{\ln R_i} \tag{17}$$

where Δt (ns) is the time delay between the two gates.

2. *The frequency domain* (Lakowicz and Berndt 1991; Gadella et al. 1993). If a sample is excited by a sinusoidally modulated light source, the resultant fluorescence emission is also sinusoidally modulated at the same frequency as the excitation source but is phase-shifted and has a reduced modulation depth. The phase (τ_i^ϕ) and modulation (τ_i^M) fluorescence lifetimes can be calculated from this phase shift and demodulation according to

$$\tau_i^\phi = \tan(\Delta\phi_i/\omega) \tag{18}$$

$$\tau_i^M = \sqrt{1/M_i^2 - 1/\omega} \tag{19}$$

where $\Delta\phi_i$ and M_i represent the phase shift and demodulation, respectively, at each pixel i, and ω is the angular modulation frequency.

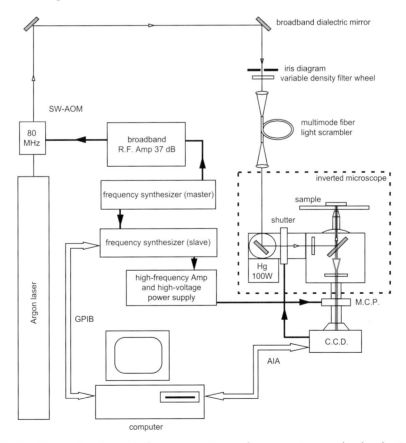

FIGURE 2. Schematic of a single frequency FLIM configuration. See text for details. (Reprinted, with permission, from Sambrook and Russell 2001, *Molecular Cloning.*)

AN EXPERIMENTAL SETUP FOR FREQUENCY DOMAIN FLIM

The following are key elements of a frequency domain fluorescence lifetime imaging system: (1) a standing wave acousto-optic modulator (SW-AOM) that modulates the intensity of the excitation light source at high frequency and (2) a frequency mixing device such as an image intensifier to perform phase-sensitive detection of the fluorescence emission. Figure 2 shows a single frequency FLIM configuration. These components are based around a vibrationally isolated, inverted microscope (e.g., Zeiss Axiovert 135 TV). The basic light source requirement is an argon laser with sufficient output power (>100 mW per line). The 457.9-nm , 488.0-nm, and 514.5-nm lines are ideal for excitation of the GFP variants, ECFP, EGFP, and EYFP, respectively. The output of this light source is modulated sinusoidally by the SW-AOM (Intra-Action Corp. Belwood). A minimal setup requires a single SW-AOM with a central resonant frequency at 80 MHz (optimal for GFP fluorescence lifetimes around 2 ns). An iris diaphragm placed about 1.5 meters from the SW-AOM is employed to isolate the central (zero-order) modulated beam selectively from the higher-order diffracted beams. To minimize thermal phase drift in the SW-AOM, this apparatus is coupled to a thermostatic water bath/circulator and the microscopy room is temperature-stabilized, keeping the SW-AOM to within $\pm 0.1°C$ of a set temperature that is essential for phase coherence between the fluorescent signal and the modulated gain of the detector.

The modulated central light beam is then directed into a 1.5-m step index silica fiber having a 1-mm core and numerical aperture of 0.37 (Technical Video Ltd.) using a 12-cm focal-length lens. The coherence of the laser light is disrupted by vibrating the fiber at frequencies of ~100 Hz, resulting in a randomly moving speckled illumination of the specimen, which is integrated during detection. Koehler illumination at the sample plane is achieved by incorporating an achromatic lens (7.6-cm focal length) just before the epi-illumination port of the microscope to collect and collimate the laser light. This focal length is sufficient for the fiber core to appear as a point source, thus giving a collimated beam resulting in flat illumination in the sample plane. A 100 W mercury arc lamp (Zeiss HBO 100 W/2) attached to the second epi-illumination port provides an alternate illumination source. A rotating mirror facilitates rapid switching between the laser and the lamp. The laser illumination is used for donor fluorescence lifetime imaging, and the lamp is employed for imaging and bleaching the acceptor.

A large part of the microscope is enclosed within a Perspex thermostatic chamber heated to 37°C to facilitate live-cell experiments. By incorporating most of the microscope in this chamber, temperature gradients between the microscope components (such as objectives) and the sample are eliminated. Providing a CO_2 source to the chamber is optional. A TV port is situated directly below the sample and objectives and is used for coupling the detector, thus providing the shortest route for fluorescence emission with minimal losses. Light from the sample is imaged onto the photocathode of a high-frequency-modulated image intensifier head (Hamamatsu C5825 or LaVision Picostat HR), which employs proximity-focusing of photoelectrons ejected from the photocathode onto the face of a microchannel plate (MCP). The electron image at the output of the MCP strikes a phosphor screen to generate an intensified light image. For homodyne detection, the effective gain of the image intensifier is modulated at a frequency equal to that of the SW-AOM (80 MHz) by the application of a biased sinusoidal voltage to the photocathode. The amplified, phase-locked outputs from high-frequency synthesizers (2023 Marconi) provide highly stable sinusoidal voltage sources for modulating both the excitation field via the SW-AOM and the gain of the image intensifier unit.

Using a telescopic lens with a magnification of 0.5, the amplified image at the phosphor screen of the MCP is projected onto the chip of a scientific grade CCD camera (Quantix, Photometrics). The magnification of 0.5 matches the full surface area of the CCD chip with that of the phosphor on the MCP. The 12-bit CCD camera houses a Kodak KAF1400 chip with a 1317 x 1035 array of 6.8-μm square pixels. Typically, two-by-two binning is applied during image acquisition on the CCD to accommodate the lower resolution (12 lp/mm) of the MCP. A phase-dependent signal at each pixel of the image is achieved by sequential phase-stepping the gain source and recording a series of images throughout an entire cycle (0–360°). A Fourier analysis is applied to each pixel in the image sequence to obtain the phase and modulation images, from which the fluorescence lifetimes are calculated. Typically, 16 phase-dependent images of ~300 x 300 pixels are acquired in a FLIM sequence. With a maximal readout rate of 5 Mpixels s^{-1}, each phase-dependent image can be read in ~20 msec.

Photobleaching is minimized by illuminating the sample only during image acquisition. This is achieved with an external, high-speed shutter (Uniblitz VS25 and D122 shutter and driver, Vincent Associates) located between the filter block and the epi-illumination port of the microscope. Synchronous triggering of the shutter is controlled with a BNC output on the CCD, indicating shutter status. The phase setting of the frequency synthesizer modulating the image intensifier gain is precisely stepped via commands sent over a GPIB interface housed in a PC. The incorporation of extensions into IPLab Spectrum (Signal Analytics Corp.) for phase-stepping control and downloading images from the CCD provides the software interface for the collection of FLIM data.

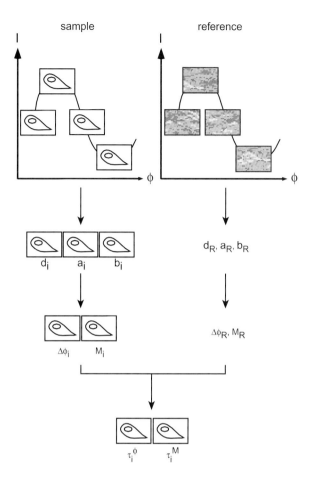

FIGURE 3. Image processing flow diagram. See text for details. (Reprinted, with permission, from Sambrook and Russell 2001, *Molecular Cloning*.)

Frequency Domain FLIM Data Analysis

FLIM data consist of a sequence of phase-dependent images over a full cycle of 2π radians (Fig. 3). From these data, the phase shifts ($\Delta\phi_i$) and modulations (M_i) must be extracted at each pixel i in order to calculate the fluorescence lifetimes. The phase-dependent raw FLIM data $D_i(k)$ can be written as a Fourier series.

$$D_i(k) = d_i + a_i \cos(k\Delta\psi) + b_i \sin(k\Delta\psi) \tag{20}$$

This signal is sampled by acquiring an image at each sequential step ($k = 0,1,2,3...K-1$) of $\Delta\psi$ in the phase of the image intensifier modulation. The parameters d_i, a_i, and b_i can be calculated using discrete sine and cosine transformations:

$$D_i = \frac{1}{K}\sum_{k=0}^{K-1} D_i(k) \tag{21}$$

$$a_i = \frac{2}{K}\sum_{k=0}^{K-1} D_i(k)\cos(k\Delta\psi) \tag{22}$$

$$b_i = \frac{2}{K}\sum_{k=0}^{K-1} D_i(k)\sin(k\Delta\psi) \tag{23}$$

The phase shifts and modulations can be found from these parameters:

$$\Delta\phi_i = \arctan(b_i/a_i) - \Delta\phi_R \tag{24}$$

$$M_i = \sqrt{a_i + b_i}/(d_i M_R) \tag{25}$$

where $\Delta\phi_R$ and M_R are the calculated phase shift and modulation, respectively, of a reflective reference sample (foil) that has a zero lifetime value. The phase and modulation lifetime images can then be calculated by substitution of the phase shifts and demodulations into Equations 18 and 19, respectively.

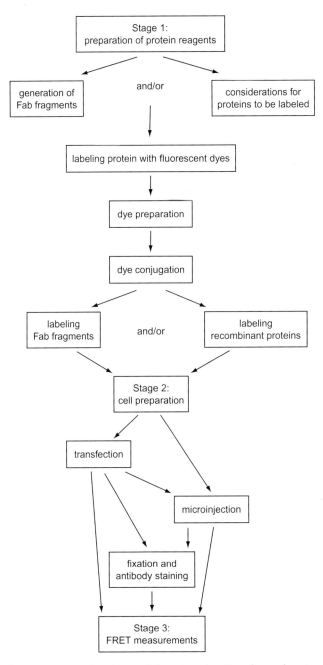

FIGURE 4. Flow diagram representing the possible protocol options for performing a FLIM-FRET experiment (see text for full details). (Modified, with permission, from Sambrook and Russell 2001, *Molecular Cloning*.)

FRET Experiments

To perform any FRET experiment, sample and reagent preparation follow a basic order. There are various alternatives at each point, and these are outlined in a flowchart (Fig. 4). The first stage considers the preparation of labeled proteins and components. A specific protocol for the generation of Fab fragments and then generic protocols for labeling these or other recombinant proteins are provided. Begin with Stage 2: Cell Preparation for FRET Analysis (p. 197) if your experiment does not require exogenous reagents. We describe the necessary steps to generate either live or fixed cells for analysis. For live cells (e.g., analysis of cells transfected with proteins fused to two distinct GFP-derived fluorophores), cDNA transfection or microinjection, with or without protein/antibody microinjection, may be required. For fixed cells, cDNA transfection or microinjection prior to antibody staining and/or fixation may be involved. Each step is outlined independently and can be applied to a specific experimental regime. Finally, a detailed description of the image collection and analysis for the different FRET measurement approaches follows in Stage 3: FRET Measurement. We give the following protocols: (1) sensitized emission measurements using the CFP/YFP donor/acceptor pair; (2) acceptor photobleaching FRET measurements using a EGFP donor and a Cy3 acceptor; (3) FLIM measurements using EGFP or Oregon Green as a donor fluorophore. Note that these protocols can all be adapted for other fluorophores by adjusting excitation wavelengths and filter sets.

Stage 1: Labeling Proteins with Fluorescent Dyes

This stage provides a specific protocol for generating Fab fragments as well as generic protocols for labeling these or other recombinant proteins.

MATERIALS

CAUTION: See Appendix for appropriate handling of materials marked with <!>.

Buffers and Solutions

Bicine (1 M, pH 7.5 and pH 9.0–9.5) adjusted with NaOH
Digestion buffer
  20 mM sodium phosphate <!>
  10 mM EDTA
  20 mM cysteine/HCl buffer (pH 7.0)
N,N-Dimethylformamide (DMF) <!>
Phosphate-buffered saline (PBS)

Antibodies

Purified protein such as antibody (MC5 monoclonal or T250 polyclonal)

Labeled Compounds

Cy3 and Cy5 OSu monofunctional sulfoindocyanine succinimide esters (Amersham)

Special Equipment

Centricon concentrators, YM10, YM30, and YM100 (Amicon)
Econopac 10DG disposable prepacked gel filtration/size exclusion chromatography
  columns (Bio-Rad)
UV/visible absorption spectrophotometer

METHODS

Generation and Purification of Fab Fragments

If Fab fragments are required, they must be produced prior to fluorescence labeling. In the particular case demonstrated here, Fabs of the polyclonal phospho-Thr-250 antibody (T[p] 250) were generated.

> Fab fragments are preferred for FRET experiments in live cells to prevent potential cross-linking of antigen by whole antibodies.

1. Concentrate the antibodies by centrifugation to 15–20 mg/ml in 20 mM sodium phosphate and 10 mM EDTA (pH 7.0) with a disposable concentrator such as a Centricon YM100 (100-kD molecular-weight cutoff; Amicon).

2. Add 500 µl of digestion buffer and 500 µl of a 50% slurry (in digestion buffer) of papain immobilized to agarose beads (Sigma) to 250 µl of concentrated antibody preparation (~4–5 mg). Digest for 6–10 hours at 37°C with vigorous shaking (400 rpm in a shaking incubator).

> **CAUTION:** Digestion in excess of 16 hours will result in the cleavage of the Fab fragments into smaller polypeptides. Therefore, first perform a pilot experiment where small aliquots of the digestion (~5 µl) are retained at 1-hour intervals, for analysis by reducing SDS-PAGE, to determine the progress of the reaction. Under reducing conditions, undigested antibody migrates as 50 kD and 25 kD for the heavy- and light-chain IgG fragments, respectively, whereas the digested antibody will migrate at around 25 kD.

3. Purify the Fab fragments from the Fc portion and nondigested antibodies using a protein A–Sepharose column equilibrated by washing with 20 bed volumes of 20 mM sodium phosphate. Apply the digestion mixture to the equilibrated column and immediately begin collecting 1-ml fractions of the flowthrough.

> Protein A will bind the Fc portion of the antibody, thereby removing both the whole antibody and the Fc fragments from the digestion mixture. The flowthrough should contain Fab fragments alone. Note that a number of subclasses of IgG molecules do not bind to protein A; therefore, make sure that the antibodies are compatible. Also consider using Protein G when mouse monoclonal antibody Fab fragments are generated.

4. Determine the OD$_{280}$ to identify which fraction contains the purified Fab fragments. In general, the Fab fragments will be present in three or four fractions; therefore, pool the highest peak fractions.

5. Concentrate the pooled fractions by centrifugation in a Centricon YM10 (10-kD molecular-weight cutoff; Amicon) to a volume below 0.5 ml.

6. Label the Fab fragments by exchanging them into a buffer that does not contain free amino groups (i.e., it is necessary to remove the free cysteine from the digestion buffer). Apply the sample to a low-molecular-weight (10-kD exclusion) gel filtration chromatography column (Bio-Rad, Econo-Pac 10 DG, 10-ml columns) equilibrated in PBS, and elute with the same buffer (according to the manufacturer's instructions).

7. Concentrate the Fab fragments following elution and determine the protein concentration by Bradford assay. Add 1/10 volume of 1 M Bicine (pH 9.0). Label the concentrated, exchanged Fabs with fluorescent sulfoindocyanine (Cy) dyes.

> The Fab concentration determined by Bradford assay, using BSA as a standard, must be multiplied by a factor of 2.

Dye Preparation

Succinimide esters of sulfoindocyanine (Cy) dyes covalently bind to free amino groups (the α-amino-terminal or ϵ-amino groups on lysine side chains). Cy dyes (Amersham) are provided as lyophilized samples that must be maintained in a desiccated environment at all times, because they react with water. The amount of dye in each vial must be quantitated individually because there can be batch variations. Note that the succinimide ester of Oregon Green (Molecular Probes) was used to label the MC5 PKCα antibody in the example provided.

1. Resuspend a single vial of lyophilized dye in 20 μl of dry DMF, which will give a Cy solution with a concentration of ~10 mM.

> Water is extracted from DMF with the addition of approximately one-third volume of hygroscopic resin to the storage vessel. For example, AG 501-X8 mixed-bed resin from Bio-Rad Laboratories can be placed in the stock DMF bottle and maintained long term. If there is any doubt as to the "dryness" of the DMF, either add further fresh resin or replace with fresh resin. This issue is a consideration should there be problems labeling proteins to a desired ratio, because water reacts with succinimide esters.

2. Dilute 1 μl of Cy solution (from step 1) 1:10,000 in PBS and calculate the dye concentration from the visible peak absorption (A). The extinction coefficients (ϵ) for Cy3 and Cy5 are 150 mM^{-1} and 250 mM^{-1} at 554 nm and 650 nm, respectively. Use the equation $c = A/\epsilon$ (for 7-cm path length cuvettes).

3. The dye solution is now ready for labeling the protein of interest.

Dye Conjugation

This amino residue labeling protocol is equally applicable to Fab fragments, whole antibodies, or purified proteins. In each case, the labeling reaction must be performed in a buffer free of amine groups. Many commercial antibody preparations are provided with stabilizing agents that contain free amino groups (such as BSA or gelatin) that will compete for the Cy dye. A number of suppliers will provide the antibody preparations free of these compounds upon request. In this case, request that they are provided at 1 mg/ml in PBS.

1. To label the proteins, they must be in a buffer that does not contain free amino groups. If they are not in an appropriate buffer, perform steps 6 and 7 in Generation and Purification of Fab Fragments (p. 194).

   If the protein has a special buffer requirement for stability, the gel filtration column can be equilibrated with that particular buffer. Care should be taken that no buffer component has free amino groups or factors that are inhibitory to the labeling reaction (e.g., Tris should be avoided). Also note that reducing agents such as dithiothreitol (DTT) or β-mercaptoethanol will interfere with the labeling reaction. However, if a reducing agent must be included to maintain biological/protein function, β-mercaptoethanol has a lower interference. Both T(p) 250 and MC5 were exchanged into PBS prior to labeling in 0.1 M Bicine, pH 9.0.

   It is advisable to maintain an alkaline pH (while maintaining protein integrity) in the labeling reaction to promote deprotonation of ε-amino groups of lysines and thereby efficient coupling.

2. React the sample (typically 0.1–0.5 ml at 1–10 μM) with a 10- to 40-fold molar excess of the sulfoindocyanine fluorescent dye, Cy3 (Amersham) for 30 minutes at room temperature (see above for preparation of Cy dyes). Add the dye very slowly while simultaneously stirring the solution with the pipette tip.

   The volume of Cy3/DMF added must not exceed 10% of the total volume to avoid protein denaturation by DMF.

   The determination of the optimal molar excess and reaction time is ultimately an empirical one, such that the final labeling ratio is between 1 and 3 dye molecules per protein molecule (see step 7, below, for labeling ratio criteria). This will vary greatly with individual proteins or antibodies. It is possible to relabel the antibody if the final labeling ratio is too low. Both T(p) 250 and MC5 were labeled with a dye molar excess of 40-fold.

3. Terminate the labeling reaction by adding free amine-containing buffer such as Tris to a final concentration of 10 mM.

4. Remove the excess unreacted dye by exchanging the buffer once more with gel filtration column chromatography (10 DG, Bio-Rad), as in step 6, Preparation of Fab fragments, and elute the protein into PBS.

   a. Equilibrate the column with 5 bed volumes of PBS or buffer of choice (i.e., 50 ml).

   b. Load the labeling reaction mixture directly onto the resin to maximize separation. Take care to apply it to as small an area as possible.

   c. Wash with 2.5 ml of buffer (~3.3 ml is void) and discard this.

      The labeled product will run as the leading front and should be visible (with free dye retarded).

   d. Add a further 2 ml of buffer and collect the visible protein fraction, discarding the remainder.

5. Concentrate the labeled protein once again using a Centricon (appropriate molecular-weight cutoff).

6. Verify successful covalent labeling of the protein by SDS–PAGE. Following electrophoresis, examine the gel directly on a UV transilluminator (302 nm) and detect the labeled product.

   A fluorescent band should migrate at the expected molecular weight of the protein. There should be no fluorescence from free dye at the migration front.

   UV excitation at 302 nm is suboptimal for most dyes. Thus, if possible, consider shorter wavelength excitation if no bands are visible upon gel illumination.

7. Following the labeling reaction, the labeling ratio is calculated by the following formula

   $$A_{\lambda} \cdot M / [(A_{280} - f A_{\lambda}) \cdot \varepsilon(\lambda)]$$

where A_λ is the absorption of the dye at its absorption maximum at wavelength λ, A_{280} is the absorption of the protein at 280 nm, M is the molecular weight of the protein in kD, and $\varepsilon(\lambda)$ is the molar extinction coefficient of the dye at wavelength λ in $mM^{-1}cm^{-1}$. The equation also corrects for absorption of the dye at 280 nm. This factor f is the ratio between absorption of the dye at 280 nm and its maximal visible absorption at wavelength λ. For example, the labeling ratio formula for Cy3-labeled antibody becomes

$$A_{554} \times 170/[(A_{280} - 0.05 \times A_{554}) \times 150]$$

The specific activity of the labeled product should be compared with that which is unlabeled to verify that dye coupling has not damaged the biological function of the Fab, whole antibody, or protein. Each labeled product will be affected to a different extent by the labeling reaction. The presence of damaged material among a population of functional protein is tolerable; however, the percentage of functional, labeled protein must be determined. This is critical because it is possible that if only a small proportion of the protein is functional, this is not the labeled component. In this case, steps must be taken to modify the labeling regime until there is functional, labeled protein.

Stage 2: Cell Preparation for FRET Analysis

GFP Cloning Vectors

A wide range of GFP variants are commercially available from Clontech Laboratories, Inc. (details can be found on their Web site, http://www.clontech.co.uk/techinfo/vectors/catlc.html), including various spectrally optimized versions (enhanced green, yellow, and cyan fluorescent protein), a red-shifted variant from the coral species, *Anthozoa*, and forms fused to intracellular compartment targeting sequences (such as the farnesylation sequence of H-ras for plasma membrane localization, a nuclear localization sequence, and Golgi complex or endoplasmic reticulum retention sequences). These cloning vectors contain promoter sequences compatible with high-level mammalian expression, and the green fluorescent protein (GFP) variants have been modified to encode humanized codon preferences for further enhancement of expression in mammalian cells. Standard molecular cloning protocols (see Chapter 1 in Sambrook and Russell 2001, *Molecular Cloning*) may be employed to generate amino- or carboxy-terminal fusions of the protein of interest with the various GFP mutants (see Chapter 17 in Sambrook and Russell 2001, *Molecular Cloning*).

MATERIALS

CAUTION: See Appendix for appropriate handling of materials marked with <!>.

Buffers and Solutions

Gelatin solution (0.1%)

Dissolve 0.5 g of gelatin (Porcine, Sigma) in 500 ml of 1x PBS. Sterilize by autoclaving.

Paraformaldehyde fixative solution (4%) <!>

Dissolve 4 g of paraformaldehyde (Sigma) in 50 ml of distilled, deionized H_2O, and then add 1 ml of 1 M NaOH solution. Stir gently on a heating block (~65°C) until the paraformaldehyde is dissolved. Then add 10 ml of 10x PBS and allow to cool to room temperature. Adjust the pH to 7.4 with 1 M HCl <!> (~1 ml) and then adjust the final volume to 100 ml with ddH$_2$O. Filter the solution through a 0.45-μm membrane filter to remove any particulate matter, and store in aliquots at –20°C for several months. Avoid repeated freeze-thawing of the paraformaldehyde solution.

Phosphate-buffered saline

> To prepare a 10x stock, dissolve 68 g of NaCl, 18.8 g of Na_2HPO_4 <!>, and 2 g of KH_2PO_4 <!> in 1 liter of ddH_2O.

Poly-L-lysine

> This is available commercially from Sigma as a 0.1% w/v solution in H_2O; dilute 1:10 in H_2O prior to use.

Quenching solution

> 50 mM Tris (pH 8.0)
> 100 mM NaCl

Media

Low-background fluorescence CO_2-independent medium

> This is commercially available from GIBCO BRL, or alternatively, adjust the standard formulation of Dulbecco's modified Eagle's medium omitting each of the following: pH indicator phenol red, antibiotics penicillin and streptomycin, folic acid, and riboflavin. Immediately prior to use, supplement with 50 mM HEPES adjusted with NaOH to pH 7.4.

Mowiol mounting medium

> This is the desired mounting medium because it does not quench GFP fluorescence. Mix 6 ml of glycerol, 2.4 g of Mowiol 4-88 (Calbiochem), and 6 ml of distilled, deionized H_2O. Initially shake for 2 hours at room temperature, then add 12 ml of 200 mM Tris-HCl (pH 8.5) and incubate at 50°C with occasional mixing until the Mowiol dissolves (~3 hours). Filter through a 0.45-μm membrane filter and store in aliquots at 4°C for weeks or –20°C for months.

Other Reagents

Transfection reagent (e.g., Superfect, Qiagen; Lipofectamine, GIBCO-BRL; Fugene 5, Boehringer Mannheim)

Microinjection Equipment and Supplies

Cells of interest seeded onto glass-bottomed cell culture vessels
GELloader tips (Eppendorf)
HEPES/NaOH (1 M, pH 7.4)
High-purity (e.g., HPLC grade) H_2O
Inverted microscope with 10x and 40x air objectives
Ion-exchange columns (QIagen midiprep or maxiprep)
Microinjector (Eppendorf model 5246)
Micromanipulator (Eppendorf model 5171)
Micropipette bubble meter (Clark Electromedical Instruments, Reading, United Kingdom)
Micropipettes (Eppendorf, Femtotip)
Millex-GV4 0.22-mm syringe filtration units
Needle-pulling device or commercially produced microinjection needles (Femtotip)
Plasmid DNA or labeled protein

Special Equipment

Glass chamber slides or coverglass chamber slides (Labtek, Nunc)
Tissue culture dishes (glass-bottomed 35-mm; MatTek Corporation, Ashland, Massachusetts)
Tissue culture plates (6- or 12-well; Nunc)

METHODS

Preparation of Live Cells for a FRET Experiment

The only consideration for the choice of cell type, other than their suitability for the system to be investigated, is that there is the facility to immobilize them in some manner (see below for suggestions).

1. Seed the cells onto an appropriate surface for microscopy, such as glass-bottomed dishes (MatTek Corporation, Ashland, Massachusetts) or coverglass chamber slides (Labtek, Nunc) for live-cell experiments. For fixed-cell preparations, cells are seeded onto glass coverslips placed in 6- or 12-well tissue culture dishes. For detecting PKCα activity in MCF10A cells, plate at a density such that they are approximately 40% confluent the following day.

   The final choice of vessel depends on whether the cells will need to be microinjected; if microinjection is planned, the shape and clearance space for needle access will need to be considered.

   For suspension cells, media such as gelatin (0.1% w/v in PBS; autoclaved) or poly-L-lysine (0.01% w/v in H_2O) can be used to facilitate immobilization. In either case, first coat the surface by covering either the wells, coverslips, or coverglass slides for ~30 minutes with the chosen medium and then aspirate the excess. Seed the cells directly onto the coated surface. The cells can also be maintained in suspension throughout the transfection procedure and then immobilized immediately prior to imaging. Once the cells are immobilized, microinjection is also possible.

2. Transfect the cells with the cDNA encoding the GFP-tagged protein of interest, using any one of the many transfection methods available (see Chapter 16 in Sambrook and Russell 2001, *Molecular Cloning*; for example, calcium phosphate precipitation or lipofection with a range of lipid-based products such as Lipofectamine, GIBCO BRL; Fugene 5, Boehringer Mannheim). Alternatively, cDNA can be introduced into the cells using nuclear microinjection (protocol described below, p. 200). For monitoring PKCα activity in live cells, transfect 2.5 μg of DNA (Lipofectamine, GIBCO-BRL) for 6 hours. Allow cells to express protein for 16–24 hours.

   In some cases, the level of protein expression needs to be controlled accurately. If this is the case, you may consider optimizing transfection efficiencies as well as the elapsed time prior to the FLIM measurement, or alternatively, using nuclear microinjection.

3. Following transfection and plasmid expression for the optimized time period, and immediately before performing an experiment at the microscope, replace the culture medium with either CO_2-independent medium (GIBCO-BRL Life Technologies) or Dulbecco's modified Eagle's medium lacking the constituents phenol red, riboflavin, penicillin, streptomycin, and folic acid, and buffered with 50 mM HEPES (pH 7.4).

   The imaging medium must not contain components that autofluoresce, and the above medium fulfills this criterion. Additionally, should the experimental design require serum starvation of the cells prior to the relevant treatment, this should be performed for the prescribed time frame prior to imaging. If, however, the cells undergo apoptosis in response to serum deprivation, maintain the cells in 0.5% fetal calf serum for this period and replace with the serum-free medium described in step 3 above immediately prior to imaging.

Cell Preparation for Fixed Cell FRET Experiments

This is a protocol for the preparation of cells to perform FRET experiments on fixed cells. The reagents used have been optimized to minimize the quenching of GFP mutants and fluorescent dyes. This protocol can easily be applied to the preparation of fixed cell samples that do not require treatment with labeled antibodies; i.e., samples that have both the donor and acceptor fluorophores expressed as fluorescent fusion proteins. If this is the case, omit the Triton X-100 permeabilization step (steps 4 and 5, below) and proceed with the remainder of the protocol.

1. Seed the cells onto either glass coverslips or glass chamber slides and transfect, if necessary, as described above.

2. Following transfection and adequate expression time (optimized as required), treat the cells with the desired stimulus.

3. Wash the cells twice in PBS, aspirate, and then add 4% paraformaldehyde. Incubate at room temperature for 10 minutes to fix the cells. Quickly wash the cells with 50 mM Tris (pH 8.0) + 100 mM NaCl to rinse off the excess fixative, and wash again for 5 minutes to quench the remaining aldehyde groups.

   Paraformaldehyde is the fixative of choice when performing FRET experiments because it is a cross-linking agent that preserves the integrity of protein interactions and does not precipitate cellular proteins (cf. methanol fixation).

4. Permeabilize the cells in 0.1% Triton X-100 (Sigma)/PBS for 15–20 minutes.

5. Wash the cells twice with PBS to remove the permeabilization solution.

6. Incubate the coverslips with Cy3-labeled Fab fragments prepared in Stage 1 or whole antibody diluted in 1% BSA in PBS for 1 hour at room temperature in a humidified chamber. Typical concentration ranges for incubation are between 1 and 100 nM, corresponding to 0.15–15 µg/ml for a whole antibody. It is advisable to perform a titration to determine the optimal dilution that results in epitope saturation. To detect the activation state of PKCα, we incubated MCF10A cells with 10 µg/ml Cy3-labeled T250 antibody (αT250-Cy3) and 10 µg/ml Oregon Green-labeled anti-PKCα (MC5-OG) in PBS containing 1% BSA.

   To minimize the quantity of antibody or Fab used, coverslips can be placed cell-side-down on 25-µl droplets of the antibody solution. This can be performed upon a sheet of Parafilm or other clean, nonstick surface. To retrieve the coverslips at the end of the incubation, place a 100-µl droplet of PBS at the edge of the coverslip; this will raise the coverslip sufficiently to grasp the edge with jeweler's forceps.

7. Wash the coverslips five times in PBS.

8. Blot off excess PBS with a tissue at the edge of the coverslip. Do not allow the cells to dry out.

9. Mount the coverslips on 10 µl of Mowiol on glass microscope slides, and allow this medium to harden overnight at 4°C. The Mowiol must solidify before imaging or the coverslip must be secured by rubber cement or molten agarose (1% in H_2O). Do not use nail varnish because this has been shown to quench GFP fluorescence.

   If glass chamber slides have been used to prepare the cells, proceed as above and place 10-µl drops of Mowiol over each well of cells and then place a long, square coverslip over the entire slide, taking care to minimize air bubbles.

10. The slides are now ready to image.

Microinjection of DNA or Labeled Proteins or Antibodies

The introduction of DNA into cells by microinjection is the method of choice when high-level, controlled protein expression is required. (For a detailed technical description of microinjection, see Pepperkok et al. 1998.) When live-cell FRET experiments are to be performed, labeled proteins such as Fab fragments or recombinant proteins must also be microinjected. To facilitate easier injections, inject labeled proteins into the cytoplasm at its highest point near the nucleus. Conversely, DNA is injected directly into the nucleus. Successful microinjection will depend on the purity of the DNA used or the removal of aggregates from protein solutions.

1. Purify DNA either by double-cesium chloride gradients or Qiagen ion-exchange columns (midiprep or maxiprep). Both the DNA and protein solutions should be filtered of particulate matter.

2. Dilute the purified DNA in HPLC-grade H_2O to a concentration appropriate for the experimental design. For optimal expression overnight, dilute to 1 μg/ml, and for expression after a few hours, dilute to 100 μg/ml.

3. Clear the DNA or protein solution of particulate matter to avoid blockage of the microinjection needle.

    We favor the use of 0.22-μm Millipore syringe filter units (Millex-GV4) for their ease of use and low protein-binding characteristics.

4. Place a filter unit within a 0.5-ml microcentrifuge tube, and place this inside a 1.5-ml microcentrifuge tube. Filter the DNA–protein solution by centrifugation at 13,000 rpm in a standard microcentrifuge for 1 minute.

    As little as 10 μl may be filtered this way. If filter units are unavailable, clarification may also be performed by centrifugation of the DNA solution at 25,000g for 20 minutes.

5. Pull needles that have an opening of ~0.25 μm in diameter using a needle-pulling device. This will be influenced by the filament temperature setting, and can be determined empirically using a syringe-operated bubble meter by measuring the air pressure required to expel air bubbles from the pipette into a liquid. Alternately, Eppendorf supplies commercially produced microinjection needles (Femtotip).

6. Load the glass needle with 2 μl of DNA or protein solution using GELloader pipette tips (Eppendorf). To avoid the introduction of air bubbles, place the end of the pipette as close to the needle opening as possible when loading the solution. Air bubbles will interrupt and prevent microinjection.

7. Fit the needle directly into the holder of the microinjection device.

    The optimal pressure and injection time settings vary with cell type and needle diameter; however, a typical range for these values is 0.3 second and 150–400 hPa injection pressure. To prevent medium from entering the needle by capillary action, a backpressure of ~20 hPa is needed. During optimization of injection conditions, aim to achieve injection without disruption of the integrity of the cellular structure (i.e., without excessive pressure). Excessive pressure during nuclear injection is apparent as separation of the nucleus from the surrounding cellular material visible by a light ring around the nucleus. Ideally, injection is performed on an inverted microscope using either 40x or 10x air objectives.

8. Perform injection at room temperature in standard CO_2-dependent growth medium and only for intervals of ~10 minutes to avoid acidification of the medium. If prolonged periods of injection are necessary, substitute the medium for one that is CO_2-independent.

9. At this stage:

    a. If cDNA has been injected for a period of time necessary to allow expression of the protein of interest (this time frame is determined empirically for the cDNA in question but starting estimates are given in step 2 above), return the cells to a CO_2 incubator.

    or

    b. If protein or antibody has been injected, immediately place the cells into CO_2-independent imaging medium in preparation for FRET imaging.

    The number of cells to be injected will again need to be determined empirically for the experiment concerned. If single cells are being imaged in a time course, only a few cells will need to be injected; however, allowances must be made for unsuccessful injections and cell damage or death as a result of injection.

Stage 3: FRET Measurements

SENSITIZED EMISSION

This protocol describes the steps involved for a sensitized emission measurement of the CFP/YFP donor–acceptor pair.

MATERIALS

Special Equipment

CCD (e.g., Photometrics Quantix with Kodak KAF 1400 chip)
Filters
    Acceptor (YFP) filterset (HQ500/20, Q515LP, HQ535/30; Chroma)
    Donor (CFP) filterset (D436/20, 455DCLP, D480/40; Chroma)
    FRET filter set (D436/20, Q515LP, HQ535/30; Chroma)
Image processing software (e.g., IPLab Spectrum; SignalAnalytics)
Inverted microscope (Zeiss 135TV)
Mercury arc lamp (100 W; Zeiss, HBO 100 AttoArc)
Oil objective (100x/1.4 NA; Zeiss Fluar)
Shutter driver (Vincent Associates, Uniblitz D122)
Shutter (high-speed; Vincent Associates, Uniblitz VS25)

METHOD

1. Make reference measurements to obtain the ratios of detection and excitation efficiencies. These need to be done only once at the beginning or the end of the experiment:

    a. Acquire an image of a sample with donor-tagged molecules *only* using the FRET filter set.

    b. Acquire an image of the same sample using the donor filter set.

    c. Perform background correction to both images acquired in steps *a* and *b*:

        i.   Select a small region in the background.

        ii. Calculate the mean intensity in this region.

        iii. Subtract this mean intensity from the image.

    d. Divide the total intensity of the image acquired in step *a* by the total intensity of the image acquired in step *b*, to obtain the ratio Q_D^A/Q_D^A.

        To only add the intensities within cells, it is advisable to set all pixels below a preset threshold value equal to zero. Because in each image the threshold must be chosen individually, make sure that they are chosen such that in both images the same area of the cells is left, e.g., by only excluding background.

    e. Acquire an image of a sample with acceptor-tagged molecules *only* using the FRET filter set.

    f. Acquire an image of the same sample using the acceptor filter set.

    g. After background correction as described above in *c*, divide the total intensity of the image acquired in step *e* by the total intensity of the image acquired in step *f* to obtain the ratio $\varepsilon_A^D/\varepsilon_A^A$.

The following steps are done for the samples that have both donor and acceptor-tagged molecules:

1. Acquire an image with the donor filter set (I^D).

2. Acquire an image with the FRET filter set (I^F).

3. Acquire an image with the acceptor filter set (I^A).

CAUTION: To calculate the results from the measured images, they must be in register, because the calculations are done on a pixel-by-pixel basis. Thus, the pixels in the different images should correspond to the same spatial position in the image. Shifts can easily occur, because the images are obtained at different wavelengths and/or with different emission filters and simply because drift may occur during acquisition. These conditions should be verified and, in the case of movement during data acquisition, a correction should be applied. This can be done manually or can be automated as described in step 5 below.

---

5. Register the FRET image I^F to the donor image I^D:

    a. Calculate the Fourier transform of I^D.

    b. Calculate the Fourier transform of I^F.

    c. Multiply the conjugate of the Fourier transform of I^D with the Fourier transform of I^F.

    d. Take the inverse Fourier transform of the result to get the correlation image.

    e. Find the position of the maximum in the correlation image. The shifts (Δx, Δy) between the images are equal to the horizontal and vertical distances between the peak and the origin of the image.

    > The origin of the image is in this case defined by the implementation of Fourier transform. It is usually in the upper left corner of the image, or in the middle of the image. A simple way to find the position of the origin is to perform the calculations above using two identical images. The position of the maximum in the result indicates the origin.

    f. Shift I^F accordingly.

6. Register I^A to I^D as described in the previous step.

7. After background correction, as in step 1c, calculate the apparent energy transfer efficiency using Equation 10 with the correction factors obtained as described above in step 1.

CAUTION: Calculation of Equation 10 (p. 187) involves a pixel-by-pixel division. Division by pixels that have a very low value will lead to noise amplifications. In the case of Equation 10, this means that pixels where I^A is low do not yield a good result. Therefore, for all pixels where I^A is lower than a preset threshold, do not calculate the result in that pixel, or else set the corresponding pixel in the result equal to zero.

ACCEPTOR PHOTOBLEACHING

The following protocol describes the steps involved in an acceptor photobleaching experiment using a wide-field microscope. The use of a confocal microscope makes it possible to perform the measurement in a limited region in the cell by selecting a region of interest (ROI), where the acceptor is photobleached. The result can then be compared to regions in the same cell that are not photobleached. Wide-field microscopes do not allow the limiting of the region of irradiation to an arbitrary ROI. However, it is possible to use the field diaphragm to limit the illumination to

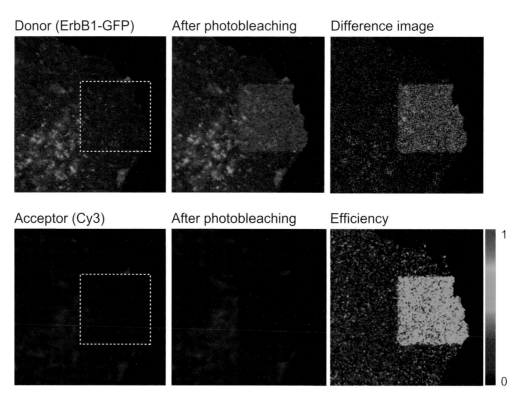

FIGURE 5. Acceptor photobleaching experiment on fixed MCF7 cells detecting the phosphorylation state of ErbB1-GFP receptors using an anti-phosphotyrosine antibody labeled with Cy3. Fixation and antibody incubation were performed as described by Wouters and Bastiaens (1999). (*Upper row*) ErbB1-GFP images before and after photobleaching of the acceptor within the indicated region and their difference image. (*Lower row, left and middle*) Acceptor intensities before and after photobleaching in the indicated region. (*Lower row, right*) Apparent energy transfer efficiency calculated from the donor intensities before and after photobleaching the acceptor.

the center of the image. Alternatively, acceptor destruction can be performed on the whole image. An example of a FRET measurement using acceptor photobleaching is given in Figure 5, where the phosphorylation state of ErbB1-GFP receptors was detected using an anti-phosphotyrosine antibody labeled with Cy3. Here a confocal microscope was used for illustrative purposes, because it facilitates easy selection of a region of interest to photobleach the acceptor Cy3.

MATERIALS

Special Equipment

CCD (e.g., Photometrics Quantix with Kodak KAF 1400 chip)
Filters
    Cy3 (HQ545/30, Q565LP, HQ610; Chroma)
    EGFP filterset (HQ480/20, Q495LP, HQ510/20; Chroma)
Image-processing software (e.g., IPLab Spectrum; SignalAnalytics)
Inverted microscope (Zeiss 135TV)
Mercury arc lamp (100 W; Zeiss, HBO 100 AttoArc)
Oil objective (100x/1.4 NA; Zeiss Fluar)
Shutter driver (Vincent Associates, Uniblitz D122)
Shutter (high-speed; Vincent Associates, Uniblitz VS25)

METHOD

1. Acquire an image with the donor filter set (I^D). Take care to minimize photobleaching of the donor!

2. Acquire an image with the acceptor filter set (I^A).

3. Photodestruct the acceptor in an ROI or in the whole image by continuous illumination with the acceptor filter set until fluorescent signals are below background levels.

   For an accurate determination of the apparent energy transfer, it is essential that the acceptor be completely destroyed by a sufficiently long period of illumination of high intensity.

4. Acquire a donor image ($I^{D,pb}$) as in step 1.

> **CAUTION:** Due to drift during image acquisition and especially during acceptor photo-bleaching, the donor images that are acquired in steps 1 and 4 must be registered.

5. Apply image registrations as described in step 5 of the protocol for sensitized emission measurements.

6. Apply background corrections as described in step 1c of the protocol for sensitized emission measurements.

7. Calculate the apparent energy transfer efficiency in every pixel using Equation 14 or 15 (see p. 188).

   See the caution about divisions by low values in point 7 of the preceding protocol for sensitized emission.

   Step 3 usually requires an extended period of illumination during which mass movement of proteins may occur in living cells. Therefore, this type of analysis is preferably performed on fixed cell samples.

FLIM MEASUREMENTS

This image acquisition protocol is a basic plan for taking a FLIM series using the microscopy setup described in this chapter. The protocol specifically describes the data acquisition for EGFP or Oregon Green as a donor fluorophore. It can be adapted for other chromophore systems by adjusting the excitation wavelength and filter sets (Table 2).

MATERIALS

CAUTION: See Appendix for appropriate handling of materials marked with <!>.

Special Equipment

Amplifier (ENI 403LA or Intra-action PA-4)
Argon laser (Coherent, Innova 70C) <!>

Broad-band dielectric mirrors (2) (Newport Corporation)
Detectors
    MCP (Hamamatsu C5825 or LaVision Picostat HR)
    CCD (Photometrics Quantix with Kodak KAF 1400 chip)
Filters
    GFP, OG (Q495LP, HQ510/20; Chroma)
    Cy3 (HQ545/30, Q565LP, HQ610/75; Chroma)
Half-silvered mirror (Zeiss)
Flexure mirror mounts (2), lens holders (2), post and post holders (5) (Newport Corporation)
Frequency synthesizers (2) (IFR 2023)
Inverted microscope (Zeiss 135TV)
IPLab Spectrum (SignalAnalytics)
Iris diaphragm (Comar)
Lenses
    12- and 7.6-cm focal length (Newport Corporation)
    2.5- and 3.8-cm focal length (Newport Corporation)
Mercury arc lamp (100 W; Zeiss, HBO 100 AttoArc)
Multimode fiber step indexed 1-mm core (Newport)
Oil objective (100x/1.4 NA; Zeiss Fluar)
Optical breadboard (2 x 1 m; TMC)
Power meter and head (Ophir, Nova Display and 2A-SH)
Power PC (Macintosh) equipped with PCI-GPIB card (National Instruments)
Shutter driver (Vincent Associates, Uniblitz D122)
Shutter, high-speed (Vincent Associates, Uniblitz VS25)
Standing wave acousto-optic modulator (Intra-action, 80 MHz)
Variable density filter wheel (Laser Components)
Water circulating cooler (Grant)

METHOD

1. Adjust the wavelength selector prism to select the 488-nm excitation wavelength on the argon laser. Fine-tune the position of the high-reflector mirror using the control knobs at the back of the laser to optimize the output power.

2. Set the frequency synthesizer driving the AOM to around 40 MHz at a resonance frequency (for the experiments described herein, a driving frequency of 40.112 MHz was employed). This produces intensity oscillations in the laser light beam at twice the driving frequency (80.224 MHz).

3. Adjust the angle of incidence and monitor the intensity of the undiffracted zero order beam with a power meter to optimize the diffraction in the AOM and thereby the modulation depth. The optimal angle of diffraction (corresponding to maximal diffraction) results in a minimum in the output power of the zero order beam.

4. Turn on the MCP and the CCD. Set the bias of the photocathode voltage to −2 V and adjust the gain to match the full dynamic range on the CCD. This setting is dependent on the fluorescence intensity of the sample and must be determined empirically. Ideally, the gain should

TABLE 2. Filter Specifications for FRET Microscopy Using Selected Donor and Acceptor Pairs

FRET pair: Donor and acceptor	excitation laser line/ exc. filter	dichroic	emission	excitation	dichroic	emission
	Donor filter set			Acceptor filter set		
ECFP and EYFP	457.9 nm D436/20	455DCLP	D480/40	HQ500/20	Q515LP	HQ535/30
EGFP and DsRed	488 nm HQ480/20	Q495LP	HQ510/20	HQ546/12	HQ560LP	HQ605/75
Cy3 and Cy5	514.5 nm HQ545/30	HQ565LP	HQ610/75	HQ620/60	HQ660L	HQ700/75
EGFP and Cy3	488 nm HQ480/20	Q495LP	HQ510/20	HQ545/30	HQ565LP	HQ610/75

The excitation of the donor is either with an argon laser (FLIM), or using a excitation filter. All filters are available from Chroma. Table adapted from Table 18-11, Sambrook and Russell 2001, *Molecular Cloning*.

be kept as low as possible to reduce noise. Typically the gain is set at 1 for the Hamamatsu C5825, with the gain of the Photometrics Quantix CCD set at 3. Set the readout of the CCD to 2 × 2 binning.

5. Set the master frequency synthesizer so that the MCP is driven to a value exactly double (for the above example: 80.224 MHz) that driving the AOM.

6. Choose the most suitable objective for the experiment. A Zeiss Fluar 100x/1.4 NA oil objective was used for this example.

7. To obtain a zero lifetime reference image, record a cycle of 16 phase-dependent images each separated by 22.5° from a strong scatterer, for example, a small piece of aluminum foil placed on the imaging surface of a coverslip or glass-bottom dish.

   a. Exchange the fluorescence filter set for a half-silvered mirror and reduce the intensity of the incident beam to a minimum with a variable density filter wheel. Adjust the focus on the foil surface.

CAUTION: When setting up the foil image, take extreme care not to look directly into the microscope ocular until the incident laser source has been reduced to a minimum.

   b. Take a single image of the aluminum foil with an exposure time of ~100 msec. This image is taken to select an ROI and to gauge the likely exposure time required. Because the phase of the master frequency synthesizer may not be at a maximum, and to avoid saturation of the detector, select an exposure time that generates ~1000 counts.

   c. A cycle of 16 phase images is recorded, each separated by 22.5°. Determine the phase setting at which maximum intensity is reached in the image series. Reset this phase to 0° on the master frequency synthesizer.

   d. Record another cycle of 16 phase images from the foil to save as a zero lifetime reference. While performing lifetime imaging, phase stability should be monitored over time. In our setup, the phase is stable within 0.3° over a period of 1 hour. For our setup, a reference foil sequence is recorded and saved each hour.

8. Restore the incident excitation source to a maximum (using the variable density filter wheel). The system is now ready for the acquisition of cell imaging data.

9. Acquire a donor image using the GFP filter set (Dichroic, Q495LP; emission, HQ510/20). Select an exposure time such that 75% of the dynamic range of the CCD (3000 counts for a 12-bit CCD) is filled. Select a region of interest in the image on which to measure fluorescence lifetimes.

10. Take two contiguous series of 16 phase-dependent images (45° phase-stepped), one forward and one reverse cycle to correct for photobleaching. The set exposure time for each phase image should be as optimized in step 9. Take an additional image in the absence of sample illumination. This background offset image is subtracted from all phase images in the series. Each of the pairs of equivalent phase images of the forward and reverse cycles is then summed to first-order correct for bleaching of the donor. From these data and the zero lifetime reference (aluminum foil), a donor fluorescence lifetime map can be calculated as described in the section Frequency domain FLIM data analysis (see Eqs. 20–25, pp. 191–192).

11. Change to the 100 W Mercury arc lamp (Zeiss Attoarc) as a source of illumination to image and photobleach the acceptor. Move the Cy3 filter set (excitation, HQ545/30; dichroic, Q565L;, emission, HQ610/75 chroma) into the detection path. Take an image of the acceptor by optimizing the exposure time to occupy the full dynamic range of the CCD. Close the detector port and illuminate the acceptor until there is no discernible Cy3 fluorescence.

FLIM Experiment on Live Cells

This is an example of a FLIM-FRET experiment to monitor the activation status of PKCα in live cells via its phosphorylation on Thr-250 (Fig. 6). The donor in this experiment is EGFP fused to the amino terminus of PKCα and the acceptor is Cy3 conjugated to Fab fragments against phosphorylated Thr-250 (αT(p)250-Cy3). This protocol can be adapted to other protein systems in live cells.

1. Prepare cells as described above for Live Cell FRET experiments (see p. 199).

2. Identify cells that are expressing the transfected GFP-tagged PKCα and microinject selected cells with αT(p)250-Cy3 (0.5 mg/ml).

3. Following microinjection, acquire a phase image series of the aluminum foil as a zero lifetime reference.

4. Proceed to acquire fluorescence phase image sequences of the donor (GFP filter set) as described above, before and after treatment with the desired stimulus (100 nM PMA for PKCα). Include at least one uninjected cell in the imaged region of interest as a donor reference lifetime measurement in the absence of acceptor.

5. Proceed to calculate fluorescence lifetime maps as described in Frequency domain FLIM data analysis.

FLIM Experiment on Fixed Cells

Here we describe an example of a fixed cell FLIM-FRET experiment to monitor the activation status of PKCα via its phosphorylation on Thr-250 (Fig. 7). This experiment involves the donor and acceptor being present on two different antibodies; the donor Oregon Green is conjugated to the PKCα-specific monoclonal antibody (MC5-OG), and the acceptor Cy3 is conjugated to the polyclonal antibody directed against phosphorylated Thr-250 (αT(p)250-Cy3). This protocol can be adapted to other protein systems in fixed cells.

1. Prepare cells as described for Fixed Cell Experiments (see pp. 199–200), completing the desired stimulus and time regime. Modify the antibody incubation to incorporate 1:1 (10 μg/ml) mix of donor (MC5-OG) and acceptor-labeled (αT(p)250-Cy3) antibodies.

2. Acquire a phase image series of the foil as a zero lifetime reference.

3. Acquire an Oregon Green donor phase image sequence for a region of interest using the GFP filter set (Dichroic, Q495LP; emission, HQ510/20), as described above.

4. Image and photobleach the acceptor by changing to the 100 W mercury arc lamp (Zeiss Attoarc) as a source of illumination. Move the Cy3 filter set (excitation, HQ545/30; dichroic, Q565LP; emission, HQ610/75 chroma) into the detection path. Take an image of the acceptor by optimizing the exposure time to occupy the full dynamic range of the CCD. Close the detector port and illuminate the acceptor until there is no discernible Cy3 fluorescence.

5. Return to the donor filter set and acquire a second, post-bleach donor-phase image series. This is the donor reference lifetime image, in the absence of acceptor.

6. Calculate fluorescence lifetime maps as described in Frequency Domain FLIM Data Analysis (see p. 191).

7. Calculate FRET efficiency maps by taking the ratio (R) of the lifetime maps acquired after and before acceptor photobleaching. The FRET efficiency equals $1 - R_i$.

   When performing acceptor photobleaching experiments, it is important to establish that there is no influence from the photoproduct of the photobleached acceptor.

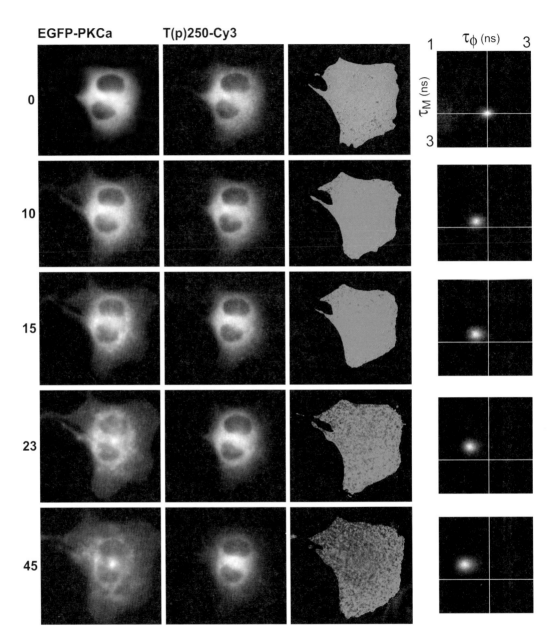

FIGURE 6. FLIM experiment on live cells monitoring the activation status of PKCα following PMA treatment. COS-7 cells transfected with donor EGFP-PKCα were microinjected with acceptor antibodies, T(p)250-Cy3. FLIM data at 101.118 MHz were acquired before (0 minutes), and after (10, 15, 23, 45 minutes) treatment with 100 nM PMA. Average fluorescence lifetime maps were calculated (τ) for each cell, and these are depicted as a pseudo-color map corresponding to average lifetimes ranging from 1.3 nsec (*dark blue*) to 2.5 nsec (*red*). Phase and modulation lifetimes (before and after treatment) are also plotted as 2D histograms. The change from higher lifetimes (*green* and *red*, 0 minutes) to lower lifetimes (*blue*, 45 minutes) in this cell demonstrates the progressive phosphorylation of Thr-250, and hence an induction of FRET resulting in a reduction in the donor fluorescence lifetimes. (Reprinted, with permission, from Sambrook and Russell 2001, *Molecular Cloning*.)

acceptor photobleaching

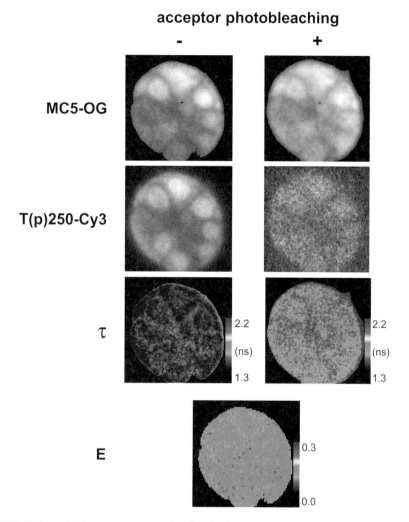

FIGURE 7. FLIM experiment on fixed cells demonstrating the activation status of PKCα. MCF 10A cells were fixed and incubated with the donor, Oregon Green (OG)-labeled anti-PKCα antibodies (MC5-OG, 10 μg/ml) and the acceptor, Cy3-labeled PKCα phospho-Thr-250 antibodies (T(p)250-Cy3, 10 μg/ml). FLIM data at 80.244 MHz were acquired from which average fluorescence lifetime maps (τ) were calculated before (*left* panels) and after (*right* panels) acceptor photobleaching, and a calculated FRET efficiency image map (E) is shown. The data presented herein represent a cluster of cells, and the nuclei of these are visible in the Cy3 and OG images. It is apparent in the lifetime maps (τ) that each cell prior to acceptor photobleaching has lower fluorescence lifetimes than following acceptor photobleaching, thus demonstrating the presence of FRET as a result of phosphorylation of Thr-250. (Reprinted, with permission, from Sambrook and Russell 2001, *Molecular Cloning.*)

ACKNOWLEDGMENTS

We thank Timo Zimmermann (EMBL) for providing the results for Figure 5, and Tony Ng and Sandra Schmidt of the Protein Phosphorylation and Cell Biophysics Laboratories, Imperial Cancer Research Fund, for the provision of the experimental results presented in Figures 6 and 7.

REFERENCES

Adams S.R., Harootunian A.T., Buechler Y.J., Taylor S.S., and Tsien R.Y. 1991. Fluorescence ratio imaging of cyclic AMP in single cells. *Nature* **349:** 694–697.

Arnone M.I., Bogarad L.D., Collazo A., Kirchhamer C.V., Cameron R.A., Rast J.P., Gregorians A., and Davidson E.H. 1997. Green fluorescent protein in the sea urchin: New experimental approaches to transcriptional regulatory analysis in embryos and larvae. *Development* **124:** 4649–4659.

Bastiaens P.I.H. and Jovin T.M. 1996. Microspectroscopic imaging tracks the intracellular processing of a signal transduction protein: Fluorescent-labeled protein kinase C βI. *Proc. Natl. Acad. Sci.* **93:** 8407–8412.

———. 1998. FRET microscopy. In *Cell biology: A laboratory handbook* (ed. J.E. Celis), pp. 136–146. Academic Press, New York.

Bastiaens P.I.H. and Squire A. 1999. Fluorescence lifetime imaging microscopy: Spatial resolution of biochemical processes in the cell. *Trends Cell Biol.* **9:** 48–52.

Bastiaens P.I.H., Majoul I.V., Verveer P.J., Söling H.-D., and Jovin T.M. 1996. Imaging the intracellllar trafficking and state of the AB₅ quaternary structure of cholera toxin. *EMBO J.* **15:** 4246–4253.

Chalfie M., Tu Y., Euskirchen G., Ward W.W., and Prasher D.C. 1994. Green fluorescent protein as a marker for gene expression. *Science* **263:** 802–805.

Clegg R.M. 1996. Fluorescence resonance energy transfer. In *Fluorescence imaging spectroscopy and microscopy* (ed. X.F. Wang and B. Herman), pp. 179–252. John Wiley, New York.

Clegg R.M., and Schneider P.C. 1996. Fluorescence time-resolved imaging microscopy: A general description of lifetime-resolved imaging measurements. In *Fluorescence microscopy and fluorescence probes* (ed. J. Slavik), pp. 15–33. Plenum Press, New York.

Clegg R.M., Feddersen B.A., Gratton E., and Jovin T.M. 1992. Time resolved imaging fluorescence microscopy. *Proc. SPIE* **1640:** 448–460.

Day R.N. 1998. Visualization of Pit-1 transcription factor interactions in the living cell nucleus by fluorescence resonance energy transfer microscopy. *Mol. Endocrinol.* **12:** 1410–1419.

Dayel M.J., Hom E.F., and Verkman A.S. 1999. Diffusion of green fluorescent protein in the aqueous-phase lumen of endoplasmic reticulum. *Biophys. J.* **76:** 2843–2851.

Förster T. 1948. Zwischenmolekulare Energiewanderung und Fluoreszenz. *Ann. Phys.* **2:** 57–75.

Gadella Jr., T.W.J., Clegg R.M., and Jovin T.M. 1994. Fluorescence lifetime imaging microscopy: Pixel-by-pixel analysis of phase-modulation data. *Bioimaging* **2:** 139–159.

Gadella Jr., T.W.J., Jovin T.M., and Clegg R.M. 1993. Fluorescence lifetime imaging microscopy (FLIM)—Spatial resolution of microstructures on the nanosecond time-scale. *Biophys. Chem.* **48:** 221–239.

Gordon G.W., Berry G., Liang X.H., Levine B., and Herman B. 1998. Quantitative fluorescence energy transfer measurements using fluorescence microscopy. *Biophys. J.* **74:** 2702–2713.

Griffin A.B., Adams S.R., and Tsien R.Y. 1998. Specific convalent labeling of recombinant protein molecules inside live cells. *Science* **281:** 269–272.

Heim R. and Tsien R.Y. 1996. Engineering green fluorescent protein for improved brightness, longer wavelengths and fluorescence resonance energy transfer. *Curr. Biol.* **6:** 178–182.

Herman B. 1989. Resonance energy transfer microscopy. *Methods Cell Biol.* **30:** 219–243.

Honda A., Adams S.R., Sawyer C.L., Lev Ram V.V., Tsien R.Y., and Dostmann W.R. 2001. Spatiotemporal dynamics of guanosine 3′,5′-cyclic monophosphate revealed by a genetically encoded, fluorescent indicator. *Proc. Natl. Acad. Sci.* **98:** 2437–2442.

Jovin T.M. and Arndt-Jovin D.J. 1989a. FRET microscopy: Digital imaging of fluorescence resonance energy transfer. In *Cell structure and function by microspectrofluorometry* (ed. E. Kohen and J.H. Hirschberg), pp. 99–115. Academic Press, San Diego.

———. 1989b. Luminescence digital imaging microscopy. *Annu. Rev. Biophys. Biophys. Chem.* **18:** 271–308.

Lakowicz J.R. and Berndt K. 1991. Lifetime-selective fluorescence imaging using an rf phase-sensitive camera. *Rev. Sci. Instrum.* **62:** 1727–1734.

Llopis J., McCaffery J.M., Miyawaki A., Farquhar M.G., and Tsien R.Y. 1998. Measurement for cytosolic, mitochondrial, and Golgi pH in single living cells with green fluorescent proteins. *Proc. Natl. Acad. Sci.* **95:** 6803–6808.

Mahajan N., Harrison-Shostak D.C., Michaux J., and Herman B. 1999. Novel mutant green fluorescent protein protease substrates reveal the activation of specific caspases during apoptosis. *Chem. Biol.* **6:** 401–407.

Mahajan N., Linder K., Berry G., Gordon G.W., Heim R., and Herman B. 1998. Bcl-2 and Bax interactions in mitochondria probed with green fluorescent protein and fluorescence resonance energy transfer. *Nat. Biotechnol.* **16:** 547–552.

Matz M.V., Fradkov A.F., Labas Y.A., Savitsky A.P., Zaraisky A.G., Markelov M.L., and Lukyanov S.A. 1999. Fluorescent proteins from nonbioluminescent Anthozoa species. *Nat. Biotechnol.* **17:** 969–973.

Mitra R.D., Silva C.M., and Youvan D.C. 1996. Fluorescence resonance energy transfer between blue-emitting and red-shifted excitation derivatives of the green fluorescent protein. *Gene* **173:** 13–17.

Miyawaki A., Griesbeck O., Heim R., and Tsien R.Y. 1999. Dynamic and quantitative Ca^{2+} measurements using improved cameleons. *Proc. Natl. Acad. Sci.* **96:** 2135–2140.

Miyawaki A., Llopis J., Heim R., McCaffery J.M., Adams J.A., Ikura M., and Tsien R.Y. 1997. Fluorescent indicators for Ca^{2+} based on green fluorescent proteins and calmodulin. *Nature* **388:** 882–887.

Moss J.B., Price A.L., Raz E., Driever W., and Rosenthal N. 1996. Green fluorescent protein marks skeletal muscle in murine cell lines and zebrafish. *Gene 173: 89–98.*

Nagy P., Vámosi G., Bodnár A., Locket S.J., and Szöllösi J. 1998. Intensity-based energy transfer measurements in digital imaging microscopy. *Eur. Biophys. J.* **27:** 377–389.

Ng T., Squire A., Hansra G., Bornancin F., Prevostel C., Hanby A., Harris W., Barnes D., Schmidt S., Mellor H., et al.. 1999. Imaging protein kinase $C\alpha$ activation in cells. *Science* **283:** 2085–2089.

Niwa Y., Hirano T., Yoshimoto K., Shimizu M., and Kobayashi H. 1999. Non-invasive quantitative detection and applications of non-toxic, S65T-type green fluorescent protein in living plants. *Plant J.* **18:** 455–463.

Okada A., Lansford R., Weimann J.M., Fraser S.E., and McConnell S.K. 1999. Imaging cells in the developing nervous system with retrovirus expressing modified green fluorescent protein. *Exp. Neurol.* **156:** 394–406.

Pepperkok R., Saffrich R., and Ansorge W. 1998. Computer-automated and semiautomated capillary microinjection of macromolecules into living cells. In *Cell biology: A laboratory handbook,* 2nd edition (ed. J.E. Celis), pp. 23–30. Academic Press, New York.

Pollock B.A. and Heim R. 1999. Using GFP in FRET-based applications. *Trends Cell Biol.* **9:** 57–60.

Prasher D.C., Eckenrode V.K., Ward W.W., and Prendergast F.G. 1992. Primariy structure of *Aequorea victoria* green-fluorescent protein. *Gene* **111:** 229–233.

Sambrook J. and Russell D. 2001 *Molecular cloning: A laboratory manual,* 3rd edition. Cold Spring Harbor Laboratory Press, Cold Spring Harbor, New York.

Sytsma J., Vroom J., de Grauw C.J., and Gerritsen H.C. 1998. Time-gated fluorescence lifetime imaging and microvolume spectroscopy using two-photon excitation. *J. Microsc.* **191:** 39–51.

Tsien R.Y. 1998. The green fluorescent protein. *Ann. Rev. Biochem.* **67:** 509–544.

Verveer P.J., Wouters F.S., Reynolds A.R., and Bastiaens P.I.H. 2000. Quantitative imaging of lateral ErbB1 receptor signal propagation in the plasma membrane. *Science* **290:** 1567–1570.

Wouters F.S. and Bastiaens P.I.H. 1999. Fluorescence lifetime imaging of receptor tyrosine kinase activity in cells. *Curr. Biol.* **9:** 1127–1130.

Wouters F.S., Verveer P.J., and Bastiaens P.I.H. 2001. Imaging biochemistry inside cells. *Trends Cell Biol.* **11:** 203–211.

Wouters F.S., Bastiaens P.I.H., Wirtz K.W.A., and Jovin T.M. 1998. FRET microscopy demonstrates molecular association of non-specific lipid transfer protein (nsL-TP) with fatty acid oxidation enzymes in peroxisomes. *EMBO J.* **17:** 7179–7189.

Xu X., Gerard A.L., Huang B.C., Anderson D.C., Payan D.G., and Luo Y. 1998. Detection of programmed cell death using fluorescence energy transfer. *Nucleic Acids Res.* **26:** 2034–2035.

Zaccolo M., De Giorgi F., Cho C.Y., Feng L., Knapp T., Negulescu P.A., Taylor S.S., Tsien R.Y., and Pozzan T. 1999. A genetically encoded, fluorescent indicator for cyclic AMP in living cells. *Nat. Cell Biol.* **2:** 25–29.

11 Detection of Homotypic Protein Interactions with Green Fluorescent Protein Proximity Imaging (GFP-PRIM)

Dino A. De Angelis

Cellular Biochemistry and Biophysics Program, Memorial Sloan-Kettering Cancer Center, New York, New York 10021

INTRODUCTION

Green fluorescent protein (GFP) is a true workhorse of modern cellular biology. GFP and its spectral mutants have been used primarily as passive fluorescent markers to monitor the location of tagged polypeptides within living cells and organisms. A relatively recent but important trend has been the development of GFP-based active indicators of cellular processes and microenvironments (Tsien 1998). It is now possible to measure fluctuations in the concentration of cellular ions (Miyawaki et al. 1997; Romoser et al. 1997; Miesenböck et al. 1998), to monitor the activity of proteases (Pollok and Heim 1999), to assess the translational and rotational motion of polypeptides (Yokoe and Meyer 1996; Swaminathan et al. 1997), to observe misfolding and aggregation of polypeptides (Waldo et al. 1999), to show differences in potential across electrically active membranes (Siegel and Isacoff 1997), and, of interest to readers of this volume, to detect protein–protein interactions.

Fluorescence resonance energy transfer (FRET) between appropriately chosen spectral variants of GFP has been used successfully to detect protein–protein interactions in vivo as well as in vitro (see Chapter 10). FRET is well suited for monitoring heterotypic protein–protein interactions but suffers from some drawbacks when monitoring homotypic protein interactions. This

215

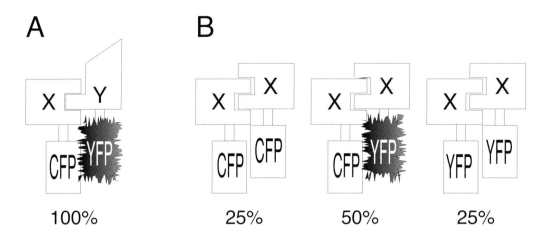

FIGURE 1. Detection of heterotypic protein–protein interactions versus homotypic interactions with FRET. Energy transfer from CFP to YFP is indicated by a YFP with jagged contour. (*A*) Heterotypic protein–protein interactions between protein X and protein Y bring CFP and YFP in close proximity 100% of the time. (*B*) Detection of homotypic protein interactions with FRET between CFP and YFP. Only 50% of homotypic interactions involving protein X will bring CFP and YFP together; the rest will either bring CFP with CFP or YFP with YFP.

chapter introduces GFP-PRIM (*Proximity Im*aging), a technique specializing in the detection of homotypic protein interactions.

Detecting Homotypic Protein Interactions with FRET

FRET requires that two proteins of interest be fused, respectively, to a fluorescence donor (typically cyan fluorescent protein, CFP) and a fluorescence acceptor (typically yellow fluorescent protein, YFP). If the two proteins interact, the close proximity of CFP and YFP facilitates energy transfer from excited CFP to YFP. The efficiency of energy transfer is highly dependent on the distance separating the chromophores and on the relative orientation of their transition dipoles. Energy transfer can be monitored by a decrease in the ratio of cyan to yellow fluorescence emission by the complex following excitation of CFP (see Chapter 10). Heterotypic protein–protein interactions are more readily detected than homotypic interactions when using FRET. If two different proteins respectively labeled with CFP and YFP interact stably and with high affinity, every complex formed will contain the fluorescence donor and acceptor in close proximity (Fig. 1A). Thus, all heterotypic protein interactions can result in a FRET signal if the separation between CFP and YFP is suitable. In contrast, if two versions of a protein capable of stable and high-affinity homodimerization (labeled with CFP and YFP, respectively) are incubated together, proximity of the donor and acceptor fluorophores will occur in 50% at most of complexes formed (Fig. 1B). The remaining 50% will consist of interactions that do not produce FRET (CFP/CFP or YFP/YFP).

Detecting Homotypic Protein Interactions with GFP-PRIM

Recently, we reported that tagging a protein with a GFP variant can provide clues about the protein's oligomeric status (De Angelis et al. 1998). The particular mutant used for these experiments

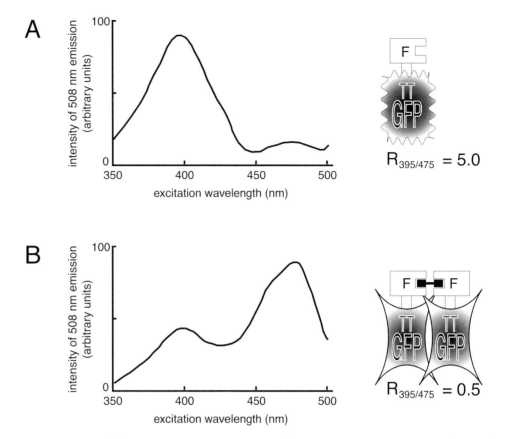

FIGURE 2. Detection of homotypic protein interactions with PRIM. (*A*) Excitation spectrum of ttGFP fused to FK506 binding protein (FKBP, labeled F in the diagram) obtained by reading the intensity of green fluorescence emitted at 508 nm as a function of excitation wavelength from 350 to 500 nm in a Perkin Elmer LB-50 spectrofluorometer. The spectrum displays two excitation peaks, with maxima at 395 nm and 475 nm. The excitation ratio ($R_{395/475}$) of this fusion protein and every ttGFP fusion protein in the monomeric state is 5.0 at 25°C in phosphate-buffered saline. (*B*) Excitation spectrum of ttGFP-FKBP following addition of the dimerizer drug FK1012 (dumbbell-shaped molecule). Dimers isolated from gel permeation chromatography display a $R_{395/475}$ of 0.5, which is a tenfold change from the monomeric ttGFP-FKBP excitation ratio. The change in shape of excitation spectrum is represented by a change in the shape of the ttGFP molecule.

is thermotolerant GFP (ttGFP) (Siemering et al. 1996). ttGFP is essentially wild-type (wt) GFP with two amino acid substitutions (V163A and S175G) that promote efficient folding and maturation of the protein at 37°C, likely by minimizing the aggregation tendency of wt GFP (Yokoe and Meyer 1996; Fukuda et al. 2000). ttGFP has two excitation peaks with respective maxima at 395 nm and 475 nm (see Fig. 2A). Illumination of the protein at wavelengths within either peak results in the emission of green light (508 nm) characteristic of GFP. The ratio of the two excitation maxima (the excitation ratio, also $R_{395/475}$) is the same for any monomeric ttGFP fusion protein when measured in the same buffer and at constant temperature. When ttGFP is fused to a protein that is capable of self-association, the excitation ratio in the homo-oligomeric state can differ from that of the monomeric state in a characteristic way, depending on the attached protein (Fig. 2B). PRIM exploits these spectral changes to detect the extent of self-association of ttGFP-tagged proteins in vitro and in vivo.

Differences between PRIM and FRET

PRIM and FRET are fundamentally very different (outlined in Table 1). The emission ratio changes measured with FRET derive from quantum mechanical interactions between any spectrally matched fluorescence donor and acceptor. Efficient interactions between CFP and YFP always result in a decrease in the CFP/YFP emission ratio; fluorescence changes are therefore unidirectional with respect to proximity. In contrast, PRIM is specific to some GFP mutants displaying both 395-nm and 475-nm excitation peaks. Current evidence suggests that the excitation ratio changes measured by the method result from direct structural interactions between two ttGFP moieties (De Angelis et al. 1998). The changes in PRIM are bidirectional: Homotypic interactions between ttGFP-labeled proteins can result in either increases or decreases from the baseline monomeric $R_{395/475}$. Although further studies are required to elucidate the precise mechanism giving rise to GFP-PRIM, it is easy to see the advantages of this method over FRET for the detection of homotypic protein interactions. FRET requires CFP-labeled and YFP-labeled versions of the protein, and these must be expressed at similar levels. Moreover, only half of the homotypic protein interactions will contribute to a CFP-YFP FRET signal, as mentioned above (Fig. 1). In contrast, PRIM only requires expression of the candidate protein labeled with ttGFP, and every homotypic interaction contributes to the signal.

GFP-PRIM can monitor changes in the degree of self-association of a given protein in vitro and in vivo. Ideal candidates for a GFP-PRIM experiment are proteins that can exist in one of two states (monomeric or self-associated, whether dimeric or trimeric or tetrameric, etc.) and that undergo self-association or dissociation in response to a given stimulus. There are two prerequisites for a successful GFP-PRIM experiment: (1) Experimental conditions must exist under which monomers can be resolved from homo-oligomers and (2) the excitation ratios of ttGFP obtained in each of these states must differ sufficiently. In these cases, PRIM can be a very powerful method to monitor the dynamics of homotypic interactions. The following sections outline the major steps of carrying out GFP-PRIM experiments in vitro by spectrofluorometry and in vivo by excitation ratio imaging of live cells.

TABLE 1. Summary of the Principal Differences between FRET and PRIM

FRET	PRIM
Requires two GFP mutants of different colors. The emission of the FRET donor has to overlap with the excitation of the FRET acceptor.	Requires ttGFP, which has two excitation maxima (395 nm and 475 nm).
Readout is a change in the emission ratio of the donor/acceptor pair upon proximity.	Readout is a change in the excitation ratio of ttGFP upon proximity.
Proximity always results in a decrease of the donor/acceptor emission ratio (unidirectional).	Proximity can result in either an increase or a decrease in the excitation ratio of ttGFP (bidirectional).
Relative orientation of the donor and acceptor within a complex affects the efficiency of the FRET signal.	Relative orientation of the two copies of ttGFP within a complex determines both the efficiency and the direction of the PRIM signal.
Quantum mechanical interaction between matched donor/acceptor fluorophore pair.	Structural interaction between adjacent ttGFP molecules.
Ideal to measure heterotypic protein–protein interactions.	Ideal to measure homotypic protein interactions.

GFP-PRIM In Vitro: Measuring the Excitation Ratio by Spectrofluorometry

Determination of the Dynamic Range of PRIM

Spectral changes seen with GFP-PRIM are not predictable a priori: On dimerization, the $R_{395/475}$ of a ttGFP-labeled polypeptide can either increase, decrease, or stay the same. The outcome ultimately depends on the relative orientations of the ttGFP moieties within a homotypic complex. To maximize the probability of success, it is advisable to construct both amino- and carboxy-terminal ttGFP fusions of the protein of interest and to test them independently. In principle, insertion of ttGFP within the polypeptide is also feasible when one or both termini must remain free to preserve function of the candidate protein. The excitation ratio of the resulting chimeric proteins is measured in a spectrofluorometer by recording the intensity of green light emitted at 508 nm following illumination of the protein solution at 395 nm and 475 nm. The quotient of these intensities is the excitation ratio or $R_{395/475}$. An excitation spectrum, as shown in Figure 2, is obtained by recording the intensity of 508-nm light emitted as a function of increasing excitation wavelength from 350 nm to 500 nm.

If the excitation ratio of the ttGFP fusion protein differs from ttGFP alone, chances are that the protein is capable of a certain degree of self-association. The extent of self-association can be determined biochemically by gel filtration chromatography, a method that separates proteins in the native state according to molecular weight (Fig. 3). Theoretical and practical aspects of this technique, also known as size exclusion chromatography or gel permeation chromatography, have been reviewed previously (Stellwagen 1990). Gel filtration achieves two important goals: (1) to resolve monomers from homo-oligomers, thus allowing their excitation ratios to be independently measured, and (2) to determine the exact proportion of ttGFP–protein X in the monomeric and oligomeric state. When relatively crude extracts are applied on sizing columns, both the size and the excitation ratio of ttGFP–protein X can be determined by spectrofluorometry of the collected fractions. When in the monomeric state, every ttGFP-labeled protein tested so far has the same excitation ratio as ttGFP alone. In most cases, the $R_{395/475}$ of homo-oligomers differs significantly from the monomeric $R_{395/475}$ (De Angelis et al. 1998). The dynamic range of PRIM for a given ttGFP fusion protein—a quantitative measure of the difference in excitation ratio between each state—is obtained by dividing the $R_{395/475}$ of the monomer by that of the homo-oligomer (or the reciprocal if the result is below 1).

Knowing the excitation ratios of 100% homo-oligomeric and 100% monomeric ttGFP fusion proteins allows one to calculate the relative fraction of monomeric protein from a mixed population displaying an intermediate $R_{395/475}$ value, using the equation:

$$R_{395/475 \text{ [mixed population]}} = x \left(R_{395/475 \text{ [monomer]}} \right) + (1 - x) \left(R_{395/475 \text{ [homo-oligomer]}} \right)$$

where x is the relative fraction of monomers and $(1 - x)$ is the relative fraction of homo-oligomers. This can only be calculated if there is a single homo-oligomeric state of the protein (dimeric or trimeric, etc.). A greater dynamic range will enable a more accurate determination of the proportion of the ttGFP–fusion protein in each state. With such information, the kinetics of self-association or dissociation of the protein following stimulation by a test substance can be assayed in a spectrofluorometer.

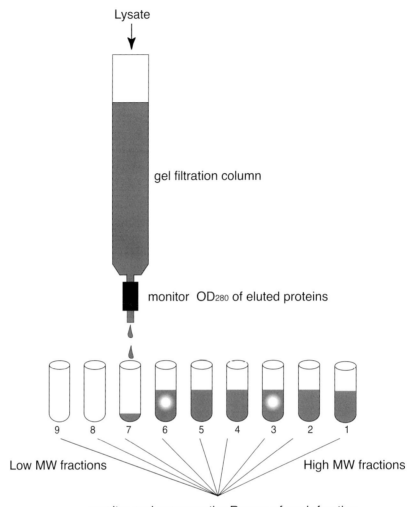

FIGURE 3. Sizing and analysis of ttGFP–fusion proteins in vitro. ttGFP–fusion proteins, fully or partially puri-fied, are separated according to size by gel filtration. Elution of polypeptides from high to low molecular weight is monitored by measuring the absorbance of the column output at 280 nm. The native size of the ttGFP–fusion proteins can be estimated by comparing the position of fluorescent material collected (shown by the white area in fractions 3 and 6) to that of molecular weight standards. The $R_{395/475}$ of the monomer-ic and multimeric ttGFP fusion protein (respectively, fractions 6 and 3) can then be directly compared. If they are sufficiently different, ttGFP-PRIM experiments will be possible. The extent of homo-oligomerization of the fusion protein will be detectable directly in mixtures; moreover, shifts in the equilibrium of the two forms will be detectable in real time.

Sensitivity of the Excitation Ratio to Variables Other Than Proximity

The excitation ratio of wt GFP can vary with changes in temperature, pH, and ionic strength (Ward et al. 1982); this is also true of ttGFP. Although these parameters may not change signifi-cantly over time during the assay of a given sample, care should be exercised in keeping them as uniform as possible from sample to sample.

A more important concern is that under certain illumination conditions, ttGFP undergoes photoisomerization. This is manifested by a decrease in the intensity of the first excitation peak and a reciprocal increase in that of the second peak, resulting in a gradual decrease in excitation ratio over time. Light in the UV range is particularly potent at causing photoisomerization, as is illumination at the 395-nm peak, albeit to a lower extent (Cubitt et al. 1995). This phenomenon can be minimized by (1) reducing the intensity of light applied to the sample and (2) illuminating it for shorter periods of time (by shuttering off excitation between data acquisition). In cases where this is impractical, the excitation ratio change of a test sample must be normalized to that of a control sample that does not undergo changes in self-association under the same illumination conditions. This will distinguish between excitation ratio changes due to photoisomerization and those due to changes in self-association.

Experimental Controls, False Negatives, and False Positives

Whether the assays are conducted with pure protein or crude extracts, several controls should be included in GFP-PRIM experiments. Subjecting ttGFP to the same challenge as the ttGFP–fusion protein should not result in a change in excitation ratio. If possible, a fusion between ttGFP and a non-oligomerizing mutant of the protein should also be included; such a protein should not undergo a change in $R_{395/475}$. Another negative control consists of challenging the ttGFP-labeled protein with an inactive version of the oligomerizing agent. Such negative controls are necessary to correct for the effects of photoisomerization on the excitation ratio (discussed above). Competition between ttGFP-labeled and nonlabeled versions of the polypeptide should also be performed. Inducing self-association in the presence of increasing amounts of the nonlabeled protein should result in a progressive decrease in the PRIM signal.

Similarity between the excitation ratio of ttGFP and the ttGFP-labeled protein does not rule out the possibility of protein homo-oligomerization. False negatives could occur, for instance, if the ttGFP subunits within a homotypic complex are too far apart to interact, or if their relative orientation is such that structural interactions between them fail to generate a PRIM signal. Independently testing two versions of the polypeptide with ttGFP fused at either terminus will help maximize the chances of success.

False positives have not yet been encountered; however, they are conceivable because of the structural (as opposed to spectral) requirement for generating a PRIM signal. One can imagine a ttGFP–fusion protein in which part of the attached protein folds back onto the surface of ttGFP, such that a change in excitation ratio is now caused intramolecularly (between the protein and ttGFP) instead of intermolecularly (between adjacent ttGFP modules).

At high concentrations, purified wt GFP has a tendency to form a specific dimer (Ward et al. 1982; Palm et al. 1997). The dimerization tendency is diminished in so-called "folding" variants of GFP that harbor the V163A point mutation, such as ttGFP (Yokoe and Meyer 1996). In their pioneering work, Ward and colleagues reported increases in the excitation ratio of highly concentrated wt GFP. Interestingly, proximity of ttGFP induced via fusion to certain proteins (such as glutathione-*S*-transferase, GST) also results in an increase in excitation ratio relative to the monomer. The relative geometry of ttGFP moieties in a ttGFP-GST dimer, therefore, might resemble that of the specific wt GFP dimer obtained at high concentrations or seen in some of the crystal structures (see, e.g., Yang et al. 1996). However, fusion of ttGFP to other dimerizing proteins causes decreases in $R_{395/475}$ (e.g., ttGFP-FKBP; Fig. 2). This completely opposite spectral behavior suggests that proximity of two ttGFP molecules does not necessarily lead to the formation of a specific dimer analogous to the one formed at high concentrations of wt GFP. Although the tendency for ttGFP to form specific dimers is lower than that of wt GFP, the protein concentration should be kept constant from assay to assay to avoid possible misinterpretations of changes in $R_{395/475}$.

GFP-PRIM in Live Cells: Imaging Homotypic Protein Interactions

The point of a GFP-PRIM experiment is to show dynamic changes in homotypic interactions brought about by a challenge—usually an externally administered substance. A series of 410-nm and 470-nm image pairs is acquired prior to stimulus application, to determine the baseline excitation ratio (Fig. 4B). During recording, a solution of the test substance can be applied by careful pipetting onto the cells. More complex perfusion setups are also available (see, e.g., Rieder and Cole 1998). Additional image pairs can then be recorded for the estimated duration of the response (the appropriate amount of time will vary for each ttGFP–fusion protein, but can range from a few minutes to several hours). Excitation ratios can be calculated and displayed during image acquisition (on-line), or after all primary images have been collected (off-line). The series of ratio images produced displays the spatiotemporal dynamics of homotypic interactions of ttGFP–protein X in pseudocolor over the recorded interval. (For an example, see De Angelis et al. 1998, Fig. 3.)

MATERIALS

A list of the principal components and a brief description of their function follows (the specific models used in our laboratory are indicated in parentheses).

Instrumentation

Computer equipped with software to control the image acquisition and perform the image analysis (Metamorph/Metafluor 3.0 software; Universal Imaging)

Dichroic mirror and bandpass filter combination to capture the green light emitted from ttGFP while rejecting the 410-nm and 470-nm excitation wavelengths (respectively, 500DCXR and HQ535/550; Chroma Technology)

Experimental chamber with temperature controller (Medical Systems Corp.)

Light source capable of emitting light of the wavelengths to be ratioed, with an output that can be coupled to a microscope (Polychrome II grating monochromator fitted with a 50-W Xenon lamp; Till photonics)

Low-light-level camera (Pentamax 512EFT frame transfer camera with fiber-coupled Gen IV image intensifier; Princeton Instruments cooled 12-bit EEV 37 charge-coupled device chip)

Microscope (Zeiss Axiovert 135TV)

Cuvette assays with intact cells are complicated by the fact that cells have a high level of intrinsic autofluorescence that can interfere with, or even mask, the PRIM signal. This hinders a precise assessment of the extent of self-association in live cells by spectrofluorometry. The spatiotemporal dynamics of homotypic interactions between ttGFP-labeled proteins can be analyzed in single live cells using excitation ratio imaging. In vivo proximity imaging experiments are conceptually similar to in vitro assays using a spectrofluorometer: In both cases, the excitation ratio of a ttGFP-labeled protein is followed over time after application of a stimulus. Using microscopy, additional information can be gathered concerning the subcellular localization of homotypic interactions within live cells.

The light path is more complex in microscopy than in spectrofluorometry: Light needs to travel through several optical components before and after the sample, prior to forming an image that can be digitally acquired, stored, and analyzed. The objectives must be corrected for spherical and chromatic aberrations. The light source must provide flat and even illumination for the whole field of view. A thorough discussion of excitation ratio imaging systems is beyond the scope of this chapter, and the reader is referred to more exhaustive reviews of the subject (Bright et al. 1989; Dunn and Maxfield 1998; Silver 1998).

METHOD

Recording the Excitation Ratio from Live Cells

In a typical experiment:

1. Grow cultured cells expressing a specific protein fused to ttGFP as a monolayer on glass coverslips.

    The fusion protein can be introduced into cells by a variety of means (usually transfection, infection, or microinjection).

2. For imaging, place a coverslip into the experimental chamber in defined medium and identify a field containing one or more positive cells.

3. For every ratio, acquire two digital images by illumination of the sample at 410 nm and 470 nm (Fig. 4A).

    These wavelengths, slightly shifted from the 395 nm and 475 nm used in the spectrofluorometer, were empirically determined to be optimal for imaging ttGFP in our system (Miesenböck et al. 1998). Because each ratio represents one time point in the experiment, the image pair has to be acquired in rapid succession (within a few milliseconds at most).

4. The two images are produced as pixel arrays stored on the computer hard drive; each pixel has a location and an intensity value. Background correction of images is necessary. After subtracting the average intensity of a region devoid of fluorescence from the entire image, the excitation ratio is subsequently obtained by dividing the intensity of the 410-nm image with the 470-nm image on a pixel-by-pixel basis. The resulting ratios are displayed on a third image, typically color-coded from red (high) to blue (low) ratios.

Determination of the Dynamic Range In Vivo

In intact cells, monomeric and multimeric ttGFP–fusion proteins cannot be readily resolved as is the case in vitro (described above), where proteins can be fractionated according to size. For this reason, the respective excitation ratios of the ttGFP–fusion protein in the monomeric and multimeric state should be determined in vitro prior to embarking on an imaging experiment whenever possible. In cases where this is impractical, control conditions that induce maximal and minimal self-association of the candidate protein, respectively, should be known a priori for the cell type under study. The two excitation ratios thus obtained will represent the upper and lower lim-

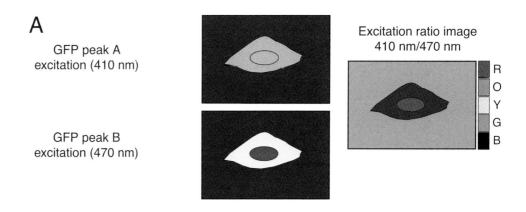

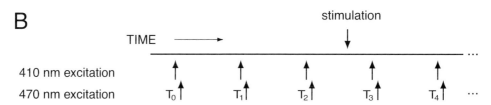

FIGURE 4. Proximity imaging by excitation ratio microscopy. (*A*) Acquisition of a GFP-PRIM ratiometric image. Two images acquired in rapid succession from illumination of the sample at 410 nm and 470 nm. Emitted light is passed through an emission filter that allows green light exclusively through the intensified camera. The two images formed are background subtracted (see text) and then divided on a pixel per pixel basis to generate the $R_{410/470}$ ratio image. The highest ratios (e.g., the nucleus of this cell) are assigned a red color (R) and the lowest ratios (the cytoplasm) a blue color (B), with intermediate values in other colors (Orange, Yellow, Green). (*B*) Time flow of a proximity imaging experiment. Pairs of images are acquired in rapid succession at 410 and 470 nm (pairs of arrows pointing up) prior to the application of a test substance (from T_0 to T_2). The test substance is applied (stimulation) and further excitation ratios can be recorded (T_3, T_4, and beyond).

its of the PRIM signal for this system. Again, a large dynamic range will facilitate the detection of small shifts in the proportion of monomeric versus multimeric ttGFP–fusion protein. The dynamics of self-association initiated by the application of different test substances will be detectable in real time and with spatial resolution.

Experimental Controls and Caveats for In Vivo Imaging

Experimental controls for GFP-PRIM in vivo are the same as in vitro (see above): No change in $R_{410/470}$ should result from treatment of cells with an inactive version of the oligomerizing agent. Cells expressing ttGFP alone, or ttGFP fused to a version of the protein incapable of self-association, should not undergo changes in $R_{410/470}$.

If a ttGFP–fusion protein capable of homodimerization is introduced into cells that endogenously express the protein, interactions between the labeled and nonlabeled versions of the protein will occur. These interactions will not result in a PRIM signal, because two ttGFP moieties are required to produce an excitation ratio shift. To maximize the strength of the PRIM signal, the ttGFP–fusion protein should be expressed at a level higher than that of the endogenous protein. Cells with different expression levels of the ttGFP–fusion protein might display different excitation ratios, even when maximal self-association is induced. In these cases, cells with similarly high expression levels should be selected and analyzed.

The precautions taken with the sample in vitro with spectrofluorometry also apply with in vivo cell imaging. The same medium should be used for imaging, and the temperature should be kept constant. Photoisomerization must be minimized by both (1) shuttering off illumination between image acquisitions and (2) using as little light intensity as possible to illuminate the sample (the use of very sensitive intensified cameras helps reduce the amount of illumination required to produce an image). To correct for the effects of photoisomerization, images from untreated samples must be acquired under the same exact conditions as test samples. The ratio changes due to photoisomerization will then be distinguishable from those due to changes in homo-oligomerization status.

CONCLUSION

PRIM and FRET are two techniques that complement one another in the study of protein–protein interactions. FRET is best suited for the detection of heterotypic protein interactions, whereas PRIM is specialized for the study of homotypic protein interactions.

Conditions for a successful GFP-PRIM experiment are twofold. First, the monomeric state must be experimentally resolvable from the multimeric state. For in vitro experiments by spectrofluorometry, this can be achieved by sizing cellular extracts by gel filtration. For cellular imaging, conditions must be known a priori under which the protein will be maximally monomeric or maximally self-associated in the cells of interest. Second, the excitation ratios of monomers must differ sufficiently from multimers. When these two conditions are met, PRIM is a very powerful method to detect homotypic protein interactions in real time. Structural studies are clearly required to understand fully the basis for excitation ratio changes seen with PRIM. Despite the fact that the magnitude and direction of ratio change cannot be predicted a priori for any protein that can self-associate, the method can nevertheless be useful for monitoring the dynamics of homotypic protein–protein interactions in vivo and in vitro.

REFERENCES

Bright G.R., Fisher G.W., Rogowska J., and Taylor D.L. 1989. Fluorescence ratio imaging microscopy. *Methods Cell Biol.* **30:** 157–192.

Cubitt A.B., Heim R., Adams S.R., Boyd A.E., Gross L.A., and Tsien R.Y. 1995. Understanding, improving and using green fluorescent proteins. *Trends Biochem. Sci.* **20:** 448–455.

De Angelis D.A., Miesenböck G., Zemelman B.V., and Rothman J.E. 1998. PRIM: Proximity imaging of green fluorescent protein-tagged polypeptides. *Proc. Natl. Acad. Sci .* **95:** 12312–12316.

Dunn K. and Maxfield F.R. 1998. Ratio imaging instrumentation. *Methods Cell Biol.* **56:** 217–236.

Fukuda H., Arai M., and Kuwajima K. 2000. Folding of the green fluorescent protein and the cycle3 mutant. *Biochemistry* **39:** 12025–12032

Miesenböck G., De Angelis D.A., and Rothman J.E. 1998. Visualizing secretion and synaptic transmission with pH-sensitive green fluorescent proteins. *Nature* **394:** 192–195.

Miyawaki A., Llopis J., Heim R., McCaffery J.M., Adams J.A., Ikura M., and Tsien R.Y. 1997. Fluorescent indicators for Ca^{2+} based on green fluorescent proteins and calmodulin. *Nature* **388:** 882–887.

Palm G.J., Zdanov A., Gaitanaris G.A., Stauber R., Pavlakis G.N., and Wlodawer A. 1997. The structural basis for spectral variations in green fluorescent protein. *Nat. Struct. Biol.* **4:** 361–365.

Pollok B.A. and Heim R. 1999. Using GFP in FRET-based applications. *Trends Cell Biol.* **9:** 57–60.

Rieder C.L. and Cole R.W. 1998. Perfusion chambers for high-resolution video light microscopic studies of vertebrate cell monolayers: Some considerations and a design. *Methods Cell Biol.* **56:** 253–275.

Romoser V.A., Hinkle P.M., and Persechini A. 1997. Detection in living cells of Ca^{2+}-dependent changes in the fluorescence emission of an indicator composed of two green fluorescent protein variants linked by a calmodulin-binding sequence. A new class of fluorescent indicators. *J. Biol. Chem.* **272:** 13270–13274.

Siegel M.S. and Isacoff E.Y. 1997. A genetically encoded optical probe of membrane voltage. *Neuron* **19:** 735–741.

Siemering K.R., Golbik R., Sever R., and Haseloff J. 1996. Mutations that suppress the thermosensitivity of green fluorescent protein. *Curr. Biol.* **6:** 1653–1663.

Silver R.B. 1998. Ratio imaging: Practical considerations for measuring intracellular calcium and pH in living tissue. *Methods Cell Biol.* **56:** 237–251.

Stellwagen E. 1990. Gel filtration. *Methods Enzymol.* **182:** 317–328.

Swaminathan R., Hoang C.P., and Verkman A.S. 1997. Photobleaching recovery and anisotropy decay of green fluorescent protein GFP-S65T in solution and cells: Cytoplasmic viscosity probed by green fluorescent protein translational and rotational diffusion. *Biophys. J.* **72:** 1900–1907.

Tsien R.Y. 1998. The green fluorescent protein. *Annu. Rev. Biochem.* **67:** 509–544.

Waldo G.S., Standish B.M., Berendzen J., and Terwilliger T.C. 1999. Rapid protein-folding assay using green fluorescent protein. *Nat. Biotechnol.* **17:** 691–695.

Ward W.W., Prenctice H.J., Roth A.F., Cody C.W., and Reeves S.C. 1982. Spectral perturbations of the *Aequorea* green-fluorescent protein. *Photochem. Photobiol.* **35:** 803–808.

Yang F., Moss L.G., and Phillips G.N., Jr. 1996. The molecular structure of green fluorescent protein. *Nat. Biotechnol.* **14:** 1246–1251.

Yokoe H. and Meyer T. 1996. Spatial dynamics of GFP-tagged proteins investigated by local fluorescence enhancement. *Nat. Biotechnol.* **14:** 1252–1256.

12 Characterization of Multiprotein Complexes by Mass Spectrometry

Carol V. Robinson

Oxford Centre for Molecular Sciences, New Chemistry Laboratory, Oxford, OX1 3QT, United Kingdom

INTRODUCTION

Despite the fact that mass spectrometry measures species in the gas phase, it is becoming widely accepted that, under the appropriate conditions, non-covalent features of protein structure can be retained in mass spectra (Loo 1997; Rostom and Robinson 1999). Numerous mass spectrometric examples now exist whereby protein interactions have been maintained from simple dimers (Vis et al. 1998) through to macromolecular complexes such as *Escherichia coli* ribosomes (Rostom et al. 2000). Recent applications have shown that mass spectrometry can provide information not only about the stoichiometry of subunits within a complex (van Berkel et al. 2000; Zhang et al. 1999), but also about subsets of interacting proteins within macromolecular complexes (Rostom et al. 2000), the symmetry of building blocks of insoluble capsular structures (Tito et al. 2000), and the thermal stability and assembly of multiprotein complexes from protein components in real time (Fändrich et al. 2000).

The mass spectrometry approach benefits from the fact that a pure protein sample is not required because spectra can be obtained from mixtures of proteins with different masses

(Fändrich et al. 2000), and heterogeneous dynamic particles of protein such as the ribosome (Rostom et al. 2000) can be examined. In addition, very small quantities are required, typically microliter quantities of solution in micromolar concentrations, and spectra can be acquired in a matter of seconds. The limitations of the technique include the solubility of protein complexes in the appropriate aqueous buffer, a low tolerance of inorganic salts, and difficulty in obtaining a stable electrospray of some protein solutions.

The survival in the gas phase of non-covalent complexes formed in solution has improved dramatically with the introduction of two enabling technologies. The first, nanoflow electrospray, reduces the initial energy of the ions formed during the ionization process (Wilm and Mann 1994). During the conventional electrospray process, solutions are infused at a flow rate of 10 μl/minute, which requires high needle voltages and large volumes of gas to produce a stable spray and to effect the evaporation of droplets to produce gaseous ions. The low flow rates (typically 2–10 nl/minute) and needle voltages associated with nanoflow sample introduction (500 V–1.5 kV compared with 3–4 kV in conventional electrospray) and considerably smaller droplet size remove the need for large volumes of gas in the atmospheric pressure region of the mass spectrometer. The combination of these effects reduces the frequency of high-energy collisions with the gas-phase protein assemblies and enhances their survival (Chung et al. 1999).

Large proteins analyzed from aqueous solutions at neutral pH, however, are not sufficiently charged to lie within the mass-to-charge range of quadrupole mass spectrometers, typically up to 3,000 or 4,000 D/e. The second enabling technology that has overcome this limitation is the introduction of time-of-flight (TOF) mass spectrometers coupled with electrospray (Verentchikov et al. 1994). TOF mass analyzers, although not a new technology, are more readily coupled with matrix-assisted laser desorption ionization (MALDI) because it is possible to synchronize the timing of the laser pulse with the initiation of the TOF (Karas and Hillenkamp 1988). The coupling of a continuous-flow ionization method, such as electrospray, is more difficult but is achieved using an orthogonal TOF mass analyzer with gated ion flow (Dawson and Guilhaus 1989). In this arrangement, accelerator electrodes are pulsed to extract ions orthogonally from the continuous ion beam. In theory, the mass range of TOF analyzers is unlimited; in practice, mass-to-charge ratios up to 30,000 D/e have been reported (Rostom et al. 2000; Tito et al. 2000). Thus, nanoflow electrospray ionization coupled with orthogonal TOF mass analysis enables much larger protein complexes to be studied than conventional electrospray using quadrupole mass spectrometers.

In this chapter, two worked examples have been chosen to demonstrate the type of information that can be obtained from the study of non-covalent complexes by mass spectrometry. The first is that of the transthyretin (TTR) retinol-binding protein (RBP) complex purified from chicken plasma (Rostom et al. 1998), and the second is that of an archaeal GimC/prefoldin homolog termed MtGimC (Leroux et al. 1999). The complexes formed between RBP and TTR were characterized by mass spectrometry, providing information about the overall stoichiometry of the complexes present in the mixture, whereas dissociation experiments gave an insight into the interactions between the subunits (Rostom et al. 1998). In the case of MtGimC, mass spectrometry was used to explore the stoichiometry of interacting subunits as well as the thermal stability and real-time assembly of the MtGimC complex (Fändrich et al. 2000). The assembly of multiprotein complexes from their components is difficult to monitor, primarily because the rate of assembly is often fast and the detection of any intermediates is difficult (Liljas 1999). Mass spectrometry, however, appears to offer an unprecedented opportunity for monitoring the assembly of macromolecular complexes because ions produced from aqueous solution can be recorded within a few milliseconds after their formation (Kebarle and Tang 1993), whereas TOF analysis is capable of resolving complexes differing in single protein subunits. In addition, real-time acquisition of mass spectra enables the time-dependent molecular weight changes to be observed.

Background to Complexes Used as Examples

TTR:RBP Complex

In plasma, RBP is the specific carrier of retinol (vitamin A), transporting it from storage sites in the liver to receptors on the surface of its target cells (Kanai et al. 1968). A second serum protein, TTR, is involved in the plasma transport of thyroid hormones and forms a complex with RBP (Peterson and Berggård 1971). Complex formation by RBP and TTR is thought to prevent filtration in the glomerulus of both the RBP and bound retinol (Raz et al. 1970). TTR:RBP complexes have been identified in a variety of species, including human (Kanai et al. 1968; Peterson and Berggård 1971) and chicken (Mokady and Tal 1974). The RBPs and TTRs of these species have molecular masses of 21 kD and 55 kD, respectively; the TTRs are composed of four identical subunits. The binding affinities of TTR:RBP complexes (typically from 1.1×10^{-7} to 1.5×10^{-7} M) are essentially the same within species and across species in chimeric complexes (Kopelman et al. 1976). Early studies of the protein assemblies, however, presented conflicting results regarding the stoichiometry, showing both one (Kanai et al. 1968; Peterson and Berggård 1971; Raz et al. 1970) and four (van Jaarsveld et al. 1973) molecules of RBP in complex with the TTR tetramer. Studies involving gel filtration, electrophoresis, and circular dichroism reported that one mole of human TTR binds up to a maximum of 1.35 molar equivalents of RBP using protein concentrations up to 20 mg/ml and at pH 7.0 (Heller and Horwitz 1974). Fluorescence polarization studies, under different experimental conditions, indicated that fewer than two molecules of human RBP and up to four molecules of chicken RBP were bound to human and chicken TTR, respectively (Kopelman et al. 1976).

MtGimC

An archaeal counterpart of the molecular chaperone GimC from *Methanobacterium thermoautotrophicum*, termed MtGimC, was found to suppress the aggregation of nonnative proteins in vitro and to stabilize them for subsequent folding by chaperonins (Leroux et al. 1999). MtGimC is assembled from two distinct subunits, MtGimα and MtGimβ, of known molecular mass. At the time of the mass spectrometry experiments, the precise stoichiometry of the complex or the arrangement of the subunits was not known.

OUTLINE OF PROCEDURE

Sample Introduction

The key to success in the analysis is the preparation of protein complex in the appropriate buffer. Small molecules and contaminants are effectively removed prior to analysis by Centricon filtering (Amicon), or by using PD-10 (Pharmacia) or biospin columns (Biorad). The solution for analysis would ideally be presented in aqueous solution at a pH of 6–8 and containing a volatile buffer. Typically, the protein concentration for the analysis would be in the range of 5–20 μM, and 1–2 μl would be consumed. A number of biochemical buffers and solution additives in common use in protein preparation procedures are deleterious to the electrospray process. For example, millimolar concentrations of involatile salts such as sodium chloride and calcium chloride, and viscous additives such as glycerol or glycols, are best avoided. Volatile buffers such as Tris and ammonium acetate and additives such as dithiothreitol (DTT), mercaptoethanol, and EDTA can be tolerated at the low millimolar concentration.

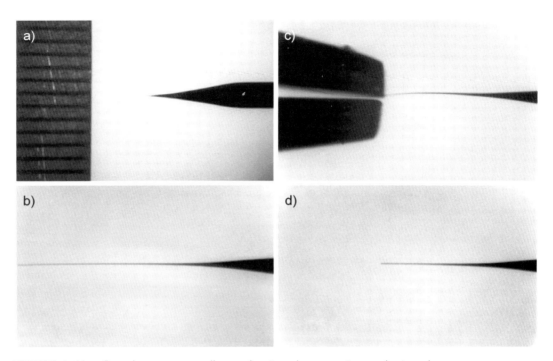

FIGURE 1. Nanoflow electrospray needles used to introduce protein samples into the mass spectrometer. Gold-plated borosilicate glass capillaries are prepared by extrusion under heat and gold sputtered. (*a*) The tip after extrusion. (Magnification, 34x.) Divisions of 5 mm are represented on the left-hand side of the picture. (*b*) Further magnification of the capillary tip. (*c*) Manual breaking of the capillary tip using forceps. (*d*) The capillary tip ready for use in the nanoflow electrospray experiment.

The introduction of the protein-containing solution from an appropriate nanoflow capillary is also a key component of the success of the process. Nanoflow needles are typically prepared from borosilicate glass capillaries using a micropipette puller, with a final gold coating of ~500 Å applied with a sputter coating system. A variety of commercial nanoflow capillaries are available. Finer-tapered capillaries are employed for general applications of peptide sequencing and protein analysis from organic solvents. For the study of protein interactions, a wider internal diameter of the finely drawn capillary is required due to the viscous nature of the aqueous solution. Manually opening the electrospray needle under a stereomicroscope and applying a pressure to the back of the capillary provides a convenient mechanism for the formation of a fine spray (Fig. 1). The preparation of the TTR:RBP complex from chicken serum and the expression of the MtGimα and MtGimβ subunits have been described previously (Rostom et al. 1998; Leroux et al. 1999).

Obtaining Mass Spectra of Individual Proteins and Their Complexes

Conventional electrospray methods rely on the use of organic co-solvents and acids to promote high charge states of unfolded protein molecules. Because these additives are likely to denature proteins and disrupt non-covalent interactions, significantly less charging is observed for tightly folded proteins in aqueous solution. This reduction in charge states requires a much higher *m/z* range instrument than is available with conventional quadrupole mass spectrometers. For this reason, electrospray TOF mass spectrometers are the instruments of choice. A further concept, important for maintaining gas phase macromolecular species, is collisional cooling, a process by which low internal energy is maintained within macromolecular ions. This is achieved by allowing a large number of low-energy collisions with an inert gas such as nitrogen, and in practical

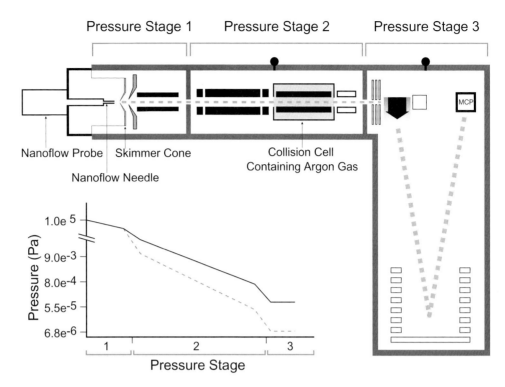

FIGURE 2. Diagram to show the different pressure regimes in the Q-TOF mass spectrometer (Micromass UK Ltd.). Reducing the efficiency of pumping in the source and intermediate pressure regions as well as admitting a collision gas to the collision cell improves the transmission of high-mass species through the instrument. (MCP) Multichannel plate detector.

terms involves increasing the pressure in different regions of the mass spectrometer (Fig. 2). The translational energy that typically exists with highly charged gas-phase ions is imparted to the neutral gas molecule through many low-energy collisions (Krutchinsky et al. 1998). In contrast, fewer high-energy collisions at a lower pressure result in the dissociation of macromolecular complexes to their component subunits. Collisional cooling has been shown to be effective in the transmission of whole particles through the mass spectrometer, allowing them to reach the detector intact (Rostom et al. 2000; Tito et al. 2000). To evaluate the conditions within the mass spectrometer for the preservation of macromolecular complexes, readily available proteins such as hemoglobin or alcohol dehydrogenase can be examined, and conditions can be optimized to obtain maximum intensity for the *m/z* signals at about 5000 *m/z* units.

The complexes formed between RBP and TTR were analyzed under the conditions optimized for the preservation of protein interactions, and the resulting spectrum is shown in Figure 3. The measured masses of the complexes are higher than those calculated from simple addition of the masses of the component subunits. This observation, together with the broad nature of the peaks measured for the higher oligomers, is most likely caused by an association of counterions or water molecules. This is consistent with the observed increase of the measured molecular weights relative to calculated values and has been reported for other systems (Loo 1995; Nettleton et al. 1998). To overcome the difficulty in assigning the charge to these broad peaks, the spectra of the protein complexes were simulated using Gaussian distributions based on the known isotopic content of the individual proteins and a distribution of small molecules (Fig. 3b). The simulations define a distribution of 10 to 90 small molecules in various components consistent with the partial occupancy of the central channel in the TTR tetramer (Rostom et al. 1998).

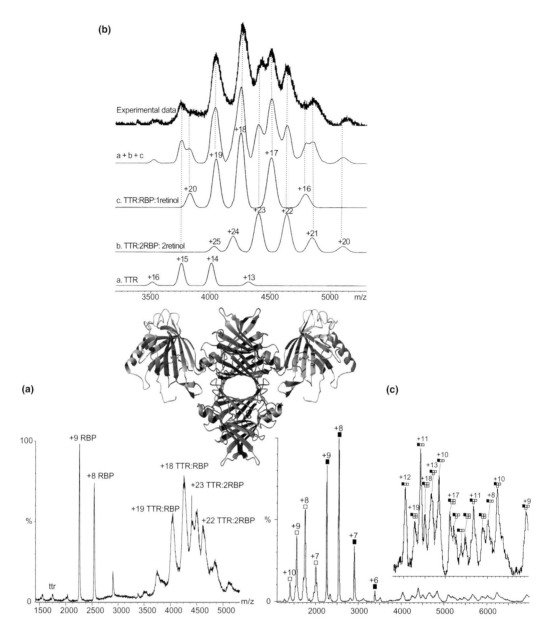

FIGURE 3. (a) Positive ion nanoflow electrospray mass spectrum, recorded at a cone voltage of 100 V, of complexes purified from chicken plasma consisting of transthyretin and retinol-binding protein. The charge states series below m/z 3000 are assigned to *apo* retinol-binding protein and to transthyretin monomer. (b) Expansion of the high-mass region of the spectrum shown in a and comparison with simulations of the charge states for the component protein complexes. Individual charge state simulations for transthyretin tetramer, TTR:2RBP with two retinol molecules and TTR:RBP with single retinol, were summed to give a simulation, similar to that observed in the experimental data. Minor discrepancies arise because the simulation employs a Gaussian distribution centered on the average isotopic composition, whereas the experimental data are more accurately described by a binomial distribution (Chung et al. 1997). Furthermore, no attempt has been made to model the asymmetry of the peak arising from salt adducts. The simulations were carried out in Sigma Plot (Jandell Scientific). The measured masses from the complexes were 76,650 ± 23 and 97,200 ± 30 for the complexes with one and two molecules of RBP, respectively. Good agreement was found between simulation and experiment assuming an average of 27 water molecules or ammonium ions in the TTR tetramer and in complexes with RBP. (c) Analysis of the chicken plasma protein complexes under high-energy conditions, cone voltage 150 V. The monomeric species occur below m/z 3000. Ions occurring above m/z 3000 have been assigned to subunits of the protein complexes and reveal the important interactions that stabilize the protein complexes. The identity of the high m/z ions are RBP:3ttr, RBP:2ttr, RBP:TTR, ■ RBP, and □ttr. (Reprinted, with permission, from Rostom et al. 1998; ©Wiley-Liss.)

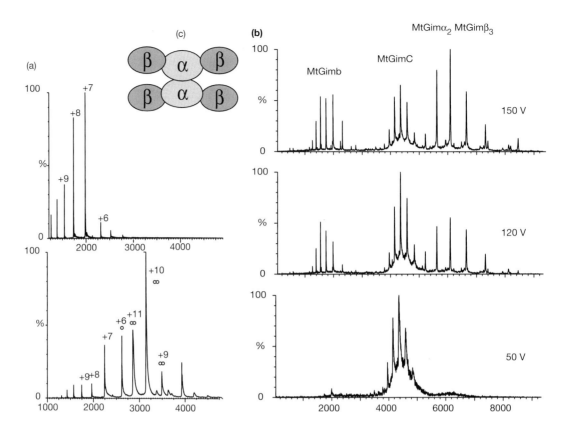

FIGURE 4. Mass spectra of the isolated MtGimα and β subunits (a) and collision-induced dissociation of MtGimC from a solution at 15 μM concentration in 10 mM ammonium acetate at pH 8.0 (b). The spectra recorded for MtGimβ (a) show that the protein is monomeric (measured mass 13,861.9 ± 0.8 D), whereas that of MtGimα (*lower panel*) is consistent with equilibrium between monomeric and dimeric forms (measured masses 15,699.8 ± 1 and 31,399.4 ± 2 D). The spectra in *b* were recorded on the same sample using cone voltages from 50 V to 150 V, as indicated. At 50 V, the charge-state series corresponds to those of the intact complex (a measured mass 86,910 ± 17 D; cf. calculated mass 86,849 D). Increasing the cone voltage to 120 V and 150 V leads to a disruption of the hexamer and the observation of a pentameric species MtGimα$_2$MtGimβ$_3$. Correspondingly, monomeric MtGimβ can be observed at m/z values below 3000. Pressures of 8.7×10^{-6} and 7.1×10^{-8} Pa were used in the source and analyzer, respectively, and a capillary voltage of 1.65 kV was applied. (Reprinted, with permission, from Fändrich et al. 2000; ©National Academy of Sciences, U.S.A.)

The measured masses of the two isolated subunits MtGimα (15,698.9 kD) and MtGimβ (13,862.7 kD) obtained under the normal operating pressure of the TOF mass spectrometer confirm their respective masses. Whereas MtGimα was found to be monomeric by all methods used, under conditions optimized for preserving protein interactions, two series of peaks were observed for MtGimα (Fig. 4). The even-numbered charge states, such as the peak at m/z 3 200, could arise either from monomeric protein with 5 positive charges or from a protein dimer with +10. The presence of odd-numbered charge states in this series (+9 and +11), however, is unique for the protein dimer. These observations showed that the MtGimα subunits are the most favored species under the gentlest conditions, a finding in agreement with the results of cross-linking and gel filtration experiments in showing that MtGimβ exists predominantly as a dimer in solution (Leroux et al. 1999).

The purified complex, MtGimC, analyzed under low-energy conditions (50 V), gives rise to a series of charge states corresponding to a mass of 86,910 ±17 D (Fig. 4b). The measured masses quoted represent the average masses of all the charge states in the spectra. This mass agrees well with measurements of the total mass of MtGimC by sedimentation equilibrium centrifugation (83.7 ± 8 kD) (Leroux et al. 1999) and is consistent with only one possible combination of the two types of protein subunits: a hexamer containing two copies of MtGimα and four copies of MtGimβ (MtGimα$_2$MtGimβ$_4$) giving rise to a calculated mass of 86,848.6 D. Using other biochemical techniques, including analytical ultracentrifugation, it is still a formidable task to rule out the possibility of microscopic constitutional heterogeneities of the protein complexes present in solution. In contrast, on the basis of the precision of the mass spectrometry data, the possibility of MtGimC hexamers containing different subunit stoichiometries could be firmly excluded.

Charge States of Individual Proteins and Their Complexes

The relationship between the observed charge state in the electrospray mass spectrum and the amino acid composition of the protein depends on the interplay of many different factors. These include the pH and ionic strength of the solution from which the protein is analyzed, the protein structure (Chowdhury and Chait 1990), the counterions present (Mirza and Chait 1994), and the cone and needle voltages set in the electrospray interface. From the analysis of the TTR:RBP complex obtained from aqueous buffer solution at pH 7.0, a number of different species are obtained even under the lowest-energy conditions (Fig. 3). The amino acid composition of chicken TTR gives a total of 17 acidic residues, including the carboxyl terminus, and 18 basic groups including the amino terminus, per monomer. Thus, at neutral pH, the overall charge on the protein will be +1, but the observed charges (+7, +8, and +9) on the monomeric form of transthyretin (ttr) demonstrate the charging effects of the electrospray process. Because both transthyretin monomer (ttr) and tetramer (TTR) are observed within the same spectrum, and thus under identical solution, electrospray, and instrument settings, charge states ranging from +28 to +36 might be predicted by simple multiplication of the +7, +8, and +9 charge states observed for the monomer. However, the predominant charge states for the TTR tetramer are +15 and +16, suggesting that 13–20 charges are lost from the component monomers when the tetramer is formed.

The changes in charge states that occur upon binding one molecule of RBP to the TTR were examined in a similar manner. The observed charge states of the 1:1 complex suggest that 5–6 positively charged residues are involved in ion-pairing interactions between the RBP and TTR tetramer. Because binding of retinol to RBP involves hydrophobic burial of the uncharged retinol molecule within RBP, and because the small conformational change associated with binding is unlikely to result in any changes in the burial of charged residues (Cowan et al. 1990), the observed changes must arise from interactions between RBP and TTR. Five basic residues are implicated in the proposed model, and the changes in the charge states observed suggest that an average of five ionic interactions are involved in the TTR:RBP complex. Further support for these ionic interactions comes from the change in charge states upon binding the second RBP to TTR:RBP. The observed charges on the complex suggest that five or six ionic interactions are formed between TTR and the second RBP molecule.

Arrangement of the Subunits within the Complex

Although the relationship between the gas-phase structure of protein complexes and their solution-phase counterparts is the subject of ongoing research, many examples have demonstrated that protein subunits on the periphery of the particle are released selectively under increased energy conditions (Rostom et al. 2000). A slight increase in collision energy coupled with a weakening of solution interactions by reduction of ionic strength and dilution of the protein complex can effect the partial dissociation of complexes (Fändrich et al. 2000). In the case of the TTR:RBP complexes, increasing the energy of collisions with residual gas molecules in the electrospray interface gives rise to a markedly different electrospray mass spectrum from that obtained under conditions chosen to mimic solution phenomena (Fig. 3). The low *m/z* charge state series assigned to dissociated RBP and monomeric ttr now dominate, whereas the intensities of the broad peaks at higher *m/z* are significantly reduced. A series of much weaker high *m/z* species (*m/z* 4000–7000) is also generated, extending beyond the m/z of the protein complexes (*m/z* 3500–5500). These must arise from species carrying fewer charges than the protein complexes or from larger protein assemblies. Analysis of the peaks confirms that they correspond in mass to varying numbers of protein subunits, and reveals that the most intense series of peaks corresponds to 3ttr:1RBP:1retinol. The second highest intensity peaks correspond to 2ttr:1RBP:1retinol, and a lower-intensity series corresponds in mass to TTR:1RBP:1retinol. The precise mechanism for the disruption of multiprotein complexes in the gas phase is the subject of ongoing research, but it seems likely that the role of counterions in stabilizing naked gas-phase assemblies plays a part in their fragmentation (A.A. Rostom et al., in prep.). However, the results of these experiments show that the interactions observed in the X-ray analysis of TTR:2RBP (Monaco et al. 1995) are maintained in the hexameric complex, since each RBP molecule interacts with 3 ttr subunits.

To investigate the quaternary architecture of the intact MtGimC complex and, in particular, to assess whether the two MtGimα subunits interact directly within MtGimC, the complex was partially disrupted within the gas phase (Fig. 4). Starting from conditions optimized to observe intact MtGimC, the complex was dissociated by increasing the collision energy and keeping the pressures within the mass spectrometer constant. The most prominent series of peaks induced by such a process occurs at *m/z* values between 5000 and 8000 and corresponds in mass to the pentamer MtGimα$_2$MtGimβ$_3$. At low *m/z* values, a second series of peaks, corresponding to monomeric MtGimβ, can be seen. A further increase in the cone voltage leads to a greater relative intensity of the peaks assigned to the pentamer, making it the dominant series in the spectrum. The observed loss of a single MtGimβ molecule as an initial dissociation step suggests that these subunits may be located at peripheral positions within MtGimC, thereby facilitating their selective release as the collision energy is increased. In addition to the pentamer MtGimα$_2$MtGimβ$_3$, a number of dissociation products can be observed at very low intensity in these spectra. However, under the high-energy conditions used in this experiment, the possibility that subunit rearrangements may take place in the gas phase, generating such low-intensity species, cannot be fully ruled out.

Probing the Thermal Stability of Protein Interactions

The ability to control the temperature of the nanoflow capillary enables the study of thermal stability of complexes by mass spectrometry. This is achieved by placing the nanoflow needle within an aluminum block and incorporation of a thermoelectric device and a resistive sensor with read-back control (described in detail elsewhere [Tito 2001]). Given that nanoflow droplets at low micromolar concentrations are calculated to contain single molecules per droplet (Wilm and Mann 1994), and because the escape from solution into the gas phase is very fast (Kebarle 2000), it is not possible for protein subunits dissociated in solution to reassociate in the gas phase. An example of this approach is given for MtGimC. Because *M. thermoautotrophicum* is a hyperthermophile growing optimally at 65°C, the effects of elevated temperatures on the structure of MtGimC were examined. The mass spectra were recorded on salt-free aliquots of MtGimC solution heated from 40°C to 70°C (Fig. 5a). At temperatures below 60°C, MtGimC accounts for more than 50% of the total ion intensity, indicating that more than half of the complex remains intact under these conditions. On raising the temperature to 70°C, the relative intensity of the signal assigned to intact MtGimC decreases as that of the dissociation products increases. However, at temperatures between 50°C and 70°C, still below the melting temperature of the MtGimα dimer (Fändrich et al. 2000), MtGimα is evident both as a monomer and a dimer, presumably

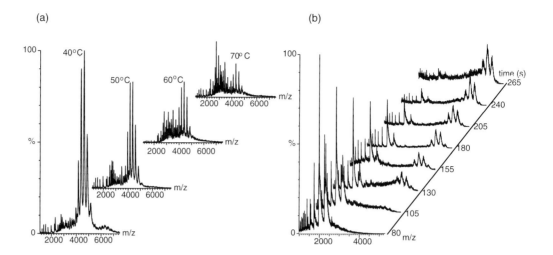

FIGURE 5. Thermal dissociation of MtGimC (*a*) and assembly of the complex from its component α and β subunits (*b*). Spectra of MtGimC were recorded at the various temperatures indicated in the plot *a*, using aliquots from the same solution with identical spectrometer settings. Peaks corresponding to intact MtGimC appear between 4000 and 5000 *m/z* and represent the most intense charge state series in the spectra recorded at low temperature. By raising the temperature, MtGimC dissociated in solution, giving rise to various species including MtGimβ, the dominant series at *m/z* values below 2000, and α₂, β₂, β₃ between *m/z* 2000 and 5000. For each spectrum, MtGimC (12 μM) was placed in a borosilicate glass capillary and incubated in a thermally controlled probe. The temperature within the nanoflow source and the countercurrent drying gas was maintained to be the same as the solution, achieved by a flow of nitrogen gas (140 liter/hour) through a thermally controlled manifold. Capillary and cone voltages were set to 1400 V and 120 V, and a pressure of 2.35×10^{-8} Pa was recorded in the analyzer. Each spectrum in the figure represents the summation of 10 acquisition steps. Full details of the experimental setup are described in detail elsewhere (Tito 2001). (*b*) Mass spectra recorded for a 2:1 mixture of MtGimβ:MtGimα at various time intervals after mixing the two subunits. Each trace represents the average of five acquisitions and has been normalized according to the total ion count in each spectrum. The capillary voltage was set to 1400 V with a cone voltage of 90 V. The analyzer pressure was recorded as 1.3×10^{-7} Pa. (Reprinted, with permission, from Fändrich et al. 2000; ©National Academy of Sciences, U.S.A.)

reflecting the stability of the latter in solution. MtGimβ is observed as a monomer and also as a dimer and a trimer. The latter are likely to represent aggregated denatured species originating from unfolded MtGimβ present at temperatures well above the T_m value of MtGimβ. Therefore, the mass spectrometry data support a model in which the association of MtGimα with MtGimβ to form MtGimC stabilizes the complexed MtGimα subunits against thermal denaturation.

Real-time Assembly of Multiprotein Complexes

To investigate the assembly of complexes from their component subunits, isolated subunits are mixed and the sample is introduced with minimal delay. In the case of MtGimC, a 1:2 molar ratio of MtGimα:MtGimβ was mixed and analyzed immediately. Mass spectra, recorded at various time points, were used to monitor the time-dependent mass changes in the various species (Fig. 5b). The first spectrum was recorded after 80 seconds, the effective dead time required for mixing and sample injection. At this time, charge-state series are evident that correspond in mass to the MtGimβ monomer and to the MtGimα monomer and dimer. In the spectra recorded 130 seconds after mixing, however, a series of broad, high-*m/z* peaks can be seen, with a molecular weight corresponding to intact MtGimC. After 205 seconds, only very low intensity peaks are found for MtGimα, and the peaks assigned to the intact complex now dominate in intensity over those from MtGimβ. After 265 seconds, the spectrum shows that the major species present in solution is MtGimC. No other species can be detected in any of the spectra, suggesting that the assembly of MtGimC is a highly cooperative process.

Outlook

Many aspects of protein interactions are exemplified by mass spectral analysis of these two complexes. Although this application of mass spectrometry is still in its infancy, the physiologically important complexes of TTR and RBP purified from plasma provide an example of a heterogeneous solution of different protein components. In contrast, the subunits and complex from the chaperone MtGimC form a well-defined complex that can be analyzed under a variety of different mass spectrometry conditions. The fact that these multiprotein complexes survive in the gas phase and can be assigned to complexes as well as individual protein subunits enables an evaluation of solution phenomena, such as an estimate of K_d (Rostom et al. 1998). Comparison of the charge states of the monomeric proteins with those of the complexed proteins suggests the number of ionic interactions made by the protein subunits. The opportunity to probe subunit contacts by increasing the energy of collisions to induce dissociation of protein complexes allows the overall architecture of the assembly to be examined. The observed pattern of dissociation products generated by mass spectrometry is consistent with a model for MtGimC in which MtGimα forms a dimeric core with two MtGimβ subunits bound to each MtGimα subunit (Fig. 4c) (Leroux et al. 1999). The high stability of the MtGimα dimer and the presence of four α/β contacts but only one α/α interaction are both consistent with the experimental observation that the loss of one MtGimβ subunit is the primary and most dominant dissociation step. On the basis of its thermal stability, revealed by mass spectrometry, the intact MtGimC hexamer is most likely to represent one structural unit inside its thermophilic host *M. thermoautotrophicum*. Assembly of the complex from isolated subunits to the hexameric complex is consistent with a cooperative process reflecting a highly favorable association of the six protein subunits. More generally, these examples highlight the power of mass spectrometry to probe not just the stoichiometry of multiprotein complexes, but also the pathways of assembly and disassembly of interacting subunits in macromolecular complexes. This detailed information, coupled with the minimal sample requirements (1–2 μl of solution at 5 μM concentration) and tolerance of mixtures, ensures a rich future for mass spectrometry in the elucidation of the structures of protein complexes.

Mass Spectrometry Characterization of Multiprotein Complexes

This protocol outlines the procedure followed for the mass spectra used as examples in this chapter.

MATERIALS

Special Equipment

Mass Spectrometry

Mass spectra used as examples in this chapter were recorded on Q-TOF and LC-TOF mass spectrometers (Micromass UK Ltd.). The Q-TOF and LC-TOF are equipped with Z-spray nanoflow electrospray sources. In the Q-TOF mass spectrometer, ions are focused by a radio frequency (RF) lens before transmission to the quadrupole, which in these experiments was used in the RF-only mode as a wide bandpass filter. Collisional cooling on the Q-TOF instrument was achieved by manually throttling the analyzer roughing pump and applying up to 30 mBa of argon gas in the collision cell. Spectra recorded on the LC-TOF mass spectra were obtained with argon gas introduced into the first hexapole lens. Transmission into the TOF was achieved with an acceleration voltage of 8 kV and 3 kV for the Q-TOF and LCT mass spectrometers, respectively. A 1-GHz TDC was used for both mass spectrometers and the multichannel plate (MCP) set at 2700 V in both cases. Cone voltage was typically set at 150 V unless stated otherwise, and a needle voltage of 1.4 kV was generally applied. Mass spectra were acquired without source heating. Mass Lynx software (version 3.1) was used for all spectra. The spectra represent the raw data with a mean smoothing algorithm applied at peak width at half height.

Nanoflow Capillaries

Borosilicate glass capillaries, external diameter 1.0 mm and internal diameter 0.78 mm (Harvard Apparatus Limited)

Micropipette puller Model P-97 (Flaming/Brown)

SEM coating system (Polaron)

METHOD

1. Tune for optimal mass spectrometry conditions by manipulating the pressure within the mass spectrometer, either by restricting pumping or admitting gases in the different pressure stages.

   Suitable complexes for tuning that are readily available commercially include yeast alcohol dehydrogenase and hemoglobin (Sigma). Both give rise to non-covalent tetramers in solution with ammonium acetate buffer at pH 6.5 at an *m/z* range value of 4000–6000 under the appropriate mass spectrometry conditions. If these non-covalent species cannot be observed, the conditions within the mass spectrometer should be reexamined before proceeding to the next step.

2. Calibrate the mass spectrometer under the conditions appropriate for the experiment. CsI (Sigma) is the most suitable calibrant for high *m/z* ratios because it produces high-molecu-

lar-weight clusters and, because it is monoisotopic, gives rise to sharp peaks at regular intervals. Equine myoglobin (Sigma) is used for lower *m/z* complexes.

3. Prepare the complex for analysis in the appropriate buffer using dialysis, Centricon concentrators (Amicon), or column chromatography for sample cleanup. This procedure typically is carried out at 4°C and is often repeated up to five times to ensure efficient removal of involatile salts.

   If the resulting solution does not produce a stable electrospray, it is likely that further sample cleanup is required.

4. Introduce the sample from the nanoflow capillary and optimize the source and pressure conditions for this complex.

   At this stage, it is possible that the needle will readily block. To overcome this problem, it is advisable to vary the pH or ionic strength of the solution to optimize the overall solubility of the complex.

5. Investigate collision-induced dissociation to probe the topological arrangement of subunits, ideally using cone voltage parameters to perturb protein subunit interactions. It is also possible to reduce the pressure within the mass spectrometer, thus removing collisional cooling, to disrupt the complex partially.

6. Investigate the effects of pH and ionic strength by manipulating solution conditions and reintroduction of the sample.

REFERENCES

Chowdhury S.K. and Chait B.T. 1990. Analysis of mixtures of closely related forms of bovine trypsin by electrospray ionization mass spectrometry—Use of charge state distribution to resolve ions of the different forms. *Biochem. Biophys. Res. Commun.* **173:** 927–931.

Chung E.W., Henriques D., Renzoni D., Ladbury J.E., and Robinson C.V. 1999. Probing the nature of interactions in SH2 binding interfaces—Evidence from electrospray mass spectrometry. *Protein Sci.* **8:** 1962–1970.

Chung E.W., Nettleton E.J., Morgan C.J., Gross M., Miranker A., Radford S.E., Dobson C.M., and Robinson C.V. 1997. Hydrogen exchange properties of proteins in native and denatured states monitored by mass spectrometry and NMR. *Protein Sci.* **6:** 1316–1324.

Cowan S., Newcomer M., and Jones T. 1990. Crystallographic refinement of human serum retinol binding protein at 2A resolution. *Proteins Struc. Func. Genet.* **8:** 44–61.

Dawson J. and Guilhaus M. 1989. Orthogonal-acceleration time-of-flight mass spectrometer. *Rapid Commun. Mass Spectrom.* **3:** 155–159.

Fändrich M., Tito M.A., Leroux M.R., Rostom A.A., Hartl F.U., Dobson C.M., and Robinson C.V. 2000. Observation of the non-covalent assembly and disassembly pathways of the chaperone complex MtGimC by mass spectrometry. *Proc. Natl. Acad. Sci.* **97:** 14151–14155.

Heller J. and Horwitz J. 1974. The binding stoichiometry of human plasma retinol-binding protein to pre-albumin. *J. Biol. Chem.* **249:** 5933–5938.

Kanai M., Raz A., and Goodman D.S. 1968. Retinol-binding protein: The transport protein for vitamin A in human plasma. *J. Clin. Investig.* **47:** 2025–2044.

Karas M. and Hillenkamp F. 1988. Laser desorption ionization of protein with molecular masses exceeding 10,000 daltons. *Anal. Chem.* **60:** 2299–2301.

Kebarle P. 2000. A brief overview of the present status of the mechanisms involved in electrospray mass spectrometry. *J. Mass Spectrom.* **35:** 804–817.

Kebarle P. and Tang L. 1993. From ions in solution to ions in the gas phase. *Anal. Chem.* **65:** 972–986.

Kopelman M., Cogan U., Mokady S., and Shinitzky M. 1976. The interactions between retinol-binding proteins and pre-albumins studied by fluorescence polarization. *Biochim. Biophys. Acta* **439:** 449–460.

Krutchinsky A.N., Laboda A.V., Spicer V.L., Dworschak R., Ens W., and Standing K.G. 1998. Orthogonal injection of matrix-assisted laser desorption/ionization ions into a TOF spectrometer through a collisional damping interface. *Rapid Commun. Mass Spectrom.* **12:** 505–518.

Leroux M.R., Fandrich M., Lunker D., Siegers K., Lupas A.N., Brown J.R., Schiebel E., Dobson C.M., and Hartl F.U. 1999. MtGimC, a novel archaeal chaperone related to the eukaryotic chaperonin cofactor GimC/prefoldin. *EMBO J.* **18:** 6730–6743.

Liljas L. 1999. Virus assembly. *Curr. Opin. Struct. Biol.* **9:** 129–134.

Loo J.A. 1995. Observation of large subunit protein complexes by electrospray ionization mass spectrometry. *J. Mass Spectrom.* **30:** 180–183.

———. 1997. Studying non-covalent complexes by electrospray mass spectrometry. *Mass Spectrom. Rev.* **16:** 1–23.

Mirza U.A. and Chait B.T. 1994. Effects of anion on the positive ion electrospray ionization mass spectra of peptides and proteins. *Anal. Chem.* **66:** 2898–2904.

Mokady S. and Tal M. 1974. Isolation and partial characterization of retinol binding protein from chicken plasma. *Biochim. Biophys. Acta* **336:** 361–366.

Monaco H., Rizzi M., and Coda A. 1995. Structure of the complex of two plasma proteins: Transthyretin and retinol binding protein. *Science* **268:** 1039–1041.

Nettleton E.J., Sunde M., Zhihong L., Kelly J.W., Dobson C.M., and Robinson C.V. 1998. Protein subunit interactions and structural integrity of amyloidogenic transthyretins: Evidence from electrospray mass spectrometry. *J. Mol. Biol.* **281:** 553–564.

Peterson P.A. and Berggård I. 1971. Isolation and properties of a human retinol-transporting protein. *J. Biol. Chem.* **246:** 25–33.

Raz A., Shiratori T., and Goodman D.S. 1970. Studies on the protein-protein and protein-ligand interactions involved in retinol transport in plasma. *J. Biol. Chem.* **245:** 1903–1912.

Rostom A.A. and Robinson C.V. 1999. Disassembly of intact multiprotein complexes in the gas phase. *Curr. Opin. Struct. Biol.* **9:** 135–141.

Rostom A.A., Fucini P., Benjamin D.R., Juenemann R., Nierhaus K.H., Hartl F.U., Dobson C.M., and Robinson C.V. 2000. Selective dissociation of intact ribosomes in a mass spectrometer. *Proc. Natl. Acad. Sci.* **97:** 5185–5190.

Rostom A.A., Sunde M., Richardson S.J., Schreiber G., Jarvis S., Bateman R., Dobson C.M., and Robinson C.V. 1998. Dissection of multi-protein complexes using mass spectrometry: Subunit interactions in transthyretin and retinol-binding protein complexes. *Proteins Struct. Func. Genet.* (suppl.) **2:** 3–11.

Tito M.A. 2001. "Probing macromolecular assemblies using electrospray time of flight mass spectrometry." Ph.D. thesis, University of Oxford.

Tito M.A., Tars K., Valegard K., Hadju J., and Robinson C.V. 2000. Electrospray TOF mass spectrometry of the intact MS2 virus capsid. *J. Am. Chem. Soc.* **122:** 350–351.

van Berkel W. J. H., Van Den Heuvel R. H. H., Versluis C., and Heck A. J. R. 2000. Detection of intact megaDalton protein assemblies of vanillyl-alcohol oxidase by mass spectrometry. *Protein Sci.* **9:** 435-439.

van Jaarsveld P.P., Edelhoch H., Goodman D.S., and Robbins J. 1973. The interaction of human plasma retinol-binding protein and prealbumin. *J. Biol. Chem.* **248:** 4698–4705.

Verentchikov A., Ens W., and Standing K. 1994. A reflecting time of flight mass spectrometer with an electrospray ion source and orthogonal extraction. *Anal. Chem.* **66:** 126–133.

Vis H., Heinemann U., Dobson C.M., and Robinson C.V. 1998. Dectection of a monomeric intermediate associated with dimerization of protein HU by mass spectrometry. *J. Am. Chem. Soc.* **120:** 6427–6428.

Wilm M. and Mann M. 1994. Electrospray and Taylor-cone theory, Dole's beam of macromolecules at last? *Int. J. Mass Spectrom. Ion Proc.* **136:** 167–180.

Zhang Z., Krutchinsky A., Endicott S., Realini C., Rechsteiner M., and Standing K.G. 1999. Proteasome activator 11S REG or PA28: Recombinant REGa/REGb hetero-oligomers are heptamers. *Biochemistry* **38:** 5651–5658.

13 Probing Ligand–Receptor Interactions with Atomic Force Microscopy

Xiaohui Zhang, Aileen Chen, Ewa Wojcikiewicz, and Vincent T. Moy

Department of Physiology and Biophysics, University of Miami School of Medicine, Miami, Florida 33136

INTRODUCTION

In recent years, new technology has been developed to directly measure the forces involved in protein–protein interactions. Previously, studies of ligand–receptor interactions usually involved biochemical methods of binding affinities or rate constants. Although this kind of measurement remains an essential part of protein interaction studies, it lacks information regarding the influence of internal and external forces. For example, blood flow currents constantly perturb migrating leukocytes bound to endothelium. This external force induces the adhesion receptors to undergo cycles of adhesion and de-adhesion as the cell rolls along the blood vessel. Now, however, the field has broadened to include measurements revealing the mechanical properties of biomolecules under applied force. These direct measurements can be used to determine the dynamic strength and characterize the changes in free energy (i.e., energy landscape) during breakage of a protein complex. Techniques that have been employed include the use of microneedles (Kishino and Yanagida 1988), optical tweezers (Ashkin et al. 1990; Svoboda et al. 1993), magnetic beads (Smith et al. 1992), the biomembrane force probe (BFP) (Evans et al. 1995), and atomic force microscopy (AFM) (Burnham and Colton 1989). Initial pioneering work includes force measurements of actin filament (Kishino and Yanagida 1988), DNA (Perkins et al. 1994), and ligand–

241

receptor systems (Florin et al. 1994). Although each of the different methods has its own merits, this chapter focuses on the use of the AFM in measuring ligand–receptor interactions. Also included are detailed methods and experimental protocols that can be applied toward studying most protein–protein interactions.

The AFM was originally designed to be an imaging tool (Binnig et al. 1986). Modified from the design of the scanning tunneling microscope (STM), the AFM acquires topographic images by methodically scanning the specimen with a flexible cantilever that bends according to the contours of the surface. Atomic-level resolution is acquired by translating the deflection of the cantilever into an image map of surface height differences. Because mapping of the surface can be conducted in both air and aqueous environments, imaging studies of biomolecules are possible under near-physiological conditions, thus enabling researchers to examine the subtle details of biological structures such as biomembranes (Zasadzinski et al. 1988), bacteriorhodopsin (Butt et al. 1990), and DNA (Hansma et al. 1992). For excellent reviews on AFM imaging, see Heinz and Hoh (1999) or Engel et al. (1999).

The AFM can also be operated in the force scan mode in which its ultrasensitivity can be used to measure interactions between two apposing surfaces down to the single-molecule level. In studies of ligand–receptor forces, the ligand is immobilized on the surface of a flexible AFM cantilever while the receptor is attached to a suitable substrate. The interaction force is acquired from the deflection of the cantilever during the approach and withdrawal of the cantilever from the substrate. Using this method, Lee et al. (1994b) were able to measure directly the unbinding force of a single ligand–receptor interaction. This novel application led to the use of the AFM as an ultrasensitive force transducer for probing biomolecular interactions. A partial list of the unbinding forces of measured ligand–receptor pairs is given in Table 1. In recent years, this nonimaging AFM technique (often referred to as force spectroscopy) has also been employed to study the unfolding of individual proteins (Rief et al. 1997b; Marszalek et al. 1999; Oesterhelt et al. 2000).

The AFM force measurements of ligand–receptor interactions can be interpreted within the theoretical framework proposed by Bell (1978). In this model, the application of a mechanical force to a ligand–receptor bond is predicted to reduce the activation energy and hence accelerate the dissociation of the bond. Moreover, the unbinding force should increase with the logarithm of the rate at which an external mechanical force is applied toward unbinding of adhesion complexes (i.e., loading rate). Confirming the Bell model, recent studies using the BFP and the AFM

TABLE 1. Summary of Reported Unbinding Force of Ligand–Receptor Bonds

Ligand–receptor pair	Unbinding force (pN)	Loading rate (pN/s)	References
Streptavidin–biotin	120–300	100–5000	Yuan et al. (2000)
	5–170	0.05–60,000	Merkel et al. (1999)
	200	N/A	Wong et al. (1998)
Avidin–biotin	115–170	100–5000	Yuan et al. (2000)
	5–170	0.05–60,000	Merkel et al. (1999)
W120F[a]–biotin	90–170	100–5000	Yuan et al. (2000)
P-selectin–glycoprotein ligand-1	115–165	–[c]	Fritz et al. (1998)
VE-cadherin-FC pair	35–55	–[c]	Baumgartner et al. (2000)
DNP-hapten[b]-antibody (ANO2)	60 ± 30	–[c]	Heymann and Grubmüller (1999)
Cell adhesion proteoglycans	40 ± 15	N/A	Dammer et al. (1995)
Meromyosin–actin	15–25	N/A	Nakajima et al. (1997)
Ferritin–antibody	49	N/A	Allen et al. (1997)
Human serum albumin–antibody	244	N/A	Hinterdorfer et al. (1996)

(pN) Piconewton.
[a]W120F: a streptavidin mutant in which tryptophan 120 was replaced by a phenylalanine.
[b]DNP–hapten: spin-labeled dinitrophenyl hapten.
[c]Authors studied unbinding force vs. different pulling velocities; we were unable to convert the velocity to loading rates.

have shown that increases in loading rate cause an increase in rupture force between individual complexes of streptavidin/biotin (Merkel et al. 1999; Yuan et al. 2000). The same relationship has been shown for fluorescein and its different antibodies (Schwesinger et al. 2000), concanavalin A (ConA)/D-mannose (Chen and Moy 2000), cadherins (Baumgartner et al. 2000), and complementary DNA strands (Lee et al. 1994a; Strunz et al. 1999).

In this review, we discuss ligand–receptor interaction studies and begin by first introducing AFM instrumentation, including a list of potential commercial suppliers and a protocol for calibrating the AFM cantilever. Next, we provide experimental protocols for measuring ligand–receptor interactions by probing receptor protein immobilized on agarose beads or on the surface of living cells. Finally, we conclude by describing exciting advances that have been made in AFM force spectroscopy techniques and discuss future directions for the field.

OUTLINE OF PROCEDURE

AFM Instrumentation

Before discussing the basic mechanisms of the AFM, let us first examine the steps of a simple force scan measurement. At the start of an experiment, a flexible cantilever coated with the ligand of interest is brought in contact with receptor protein attached to an apposing substrate. The cantilever is then subsequently withdrawn until the final bond separates. Throughout the measurement, changes in cantilever deflection are detected by a two-segment photodiode that senses alterations in the reflective path of a laser spot focused on the back of the cantilever. The signals from the photodiode are then amplified and sent to a computer for analysis. Thus, the essential components for an AFM include a piezoelectric translator that displaces the cantilever, a laser and photodiode to detect changes in cantilever deflection, and a computer to control movement of the cantilever as well as to process and record data.

To carry out the above procedures, we use a homemade AFM (Fig. 1) in our laboratory; modification of the standard AFM design used for imaging can improve the quality of the signal acquired in AFM force measurements. For instance, we have found that uncoupling the mechanism for lateral and vertical scans reduced mechanical and electrical noise and improved the sen-

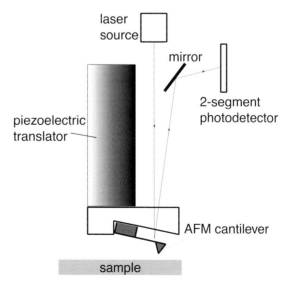

FIGURE 1. Schematic of AFM.

sitivity of the instrument. In our apparatus, the vertical movement of the cantilever is controlled by mounting it on a piezoelectric translator (Physik Instrumente, model P-821.10), which expands or contracts depending on the applied voltage. By changing the voltage across the piezo, the protein-coated cantilever and ligand can be brought into binding contact. The deflection of the cantilever is monitored by focusing a laser beam from a 3-mW diode laser (Oz Optics; em. 680 nm) on the upper surface of the cantilever. Changes in the reflection path of the laser are monitored by a two-segment photodiode (UDT Sensors; model SPOT-2D). The differential signal from the photodiode segments is then digitized by a data acquisition system equipped with an 18-bit optically isolated analog-to-digital converter (Instrutech Corp., Port Washington, New York). Control of the piezoelectric translator and timing of the measurements is done through custom software. Further reduction of mechanical vibration and temperature fluctuation is achieved by suspending the entire apparatus by bungee cords inside a large evacuated refrigerator. The detection limit of our AFM system is in the range of 20 piconewtons (pN). A piconewton-sensitive commercial AFM with necessary software is readily available from Asylum Research (Molecular Force Probe; Santa Barbara, California) and is specially designed to carry out force measurements similar to those carried out in our laboratory. A partial listing of AFM manufacturers is given below for the interested reader.

Manufacturers of AFM Instrumentation

The following is a partial list of AFM manufacturers.

Asylum Research, California
http://asylumresearch.com/

Burleigh Instruments Inc., New York
http://www.burleigh.com/Pages/surface.htm

Digital Instruments, California
http://www.di.com/

Molecular Imaging Corporation, Arizona
http://www.molec.com/index.html

OMICRON, GmbH, Germany
http://www.omicron-instruments.com/

ThermoMicroscopes/Park Scientific Instruments, California
http://www.thermomicro.com/

TopoMetrix, California
http://www.topometrix.com/

WITec GmbH, Germany
http://www.WITec.de/

Cantilever Calibration

To translate the deflection of the cantilever, x, to units of force, F, it is necessary to determine the spring constant of the cantilever, k_C (i.e., $F = k_C x$). There are several techniques for calibrating tips and theoretical methods that provide an approximation of k_C (Sader 1995). Determination of k_C using empirical methods involves taking measurements of cantilever deflection with application of a constant known force (Senden and Ducker 1994) or measurements of the cantilever's resonant frequency (Hutter and Bechhoefer 1993). The method we use for calibrating cantilevers is based on Hutter and Bechhoefer (1993) and is briefly outlined below.

For ligand–receptor force measurements, we use triangle-shaped, unsharpened, gold-coated silicon-nitride cantilever tips that have spring constants ranging from 10 mN/m to 50 mN/m. The cantilever tip can be treated as a simple harmonic oscillator whose power spectrum of thermal fluctuation can be used to derive the spring constant. In brief, the cantilever is raised several microns from the surface and its natural frequency of vibration (resonant frequency) is monitored for 2–3 seconds. Because each vibration mode of the cantilever receives the thermal energy commensurate with one degree of freedom, $k_B T/2$, the measured variance of the deflection $\langle x^2 \rangle$ can be used to calculate the spring constant, i.e., $\frac{1}{2} k_B T = \frac{1}{2} k_C \langle x^2 \rangle$, where k_B and T are Boltzmann's constant and temperature, respectively. To separate deflections belonging to the basic (and predominant) mode of vibration from other deflections or noise in the recording system, the power spectral density of the temperature-induced deflection is determined, and only the spectral component corresponding to the basal mode of vibration is used to estimate the spring constant. Using this approach, the spring constants of cantilevers can be calibrated in either air or solution. The calculated spring constant k_C can then be used to calculate rupture force, F, by $F = k_C C \Delta V$, where ΔV is the change in voltage detected by the photodiode just prior to and immediately after the rupture event and C is a calibration constant that relates deflection and photodiode voltage. C is determined from the deflection of the cantilever pressed against a rigid surface, such as the bottom of a plastic petri dish.

AFM APPLICATIONS FOR LIGAND–RECEPTOR INTERACTION STUDIES

The AFM can be a powerful tool to determine dynamic strength of ligand–receptor bonds. There are certain issues one must consider, however, when acquiring direct force measurements. First, appropriate measures must be taken to identify and/or eliminate any possible nonspecific binding forces. Second, the strength of a ligand–receptor bond is affected by the applied loading rate of the measurement. In other words, the binding properties can be influenced by the speed at which ligand and receptor are separated and the elasticity of the surfaces to which the proteins are anchored. In the following paragraphs, we go into greater detail about these two points and then follow with protocols for carrying out AFM force measurements on agarose beads and cells.

Because the nonspecific forces involved in the tip–sample interaction are typically electrostatic in origin, simple measures can be taken to reduce nonspecific binding. Minimization of these forces can be accomplished by carrying out experiments in ionic solutions, such as phosphate-buffered saline (PBS), that shield charges on the surfaces being brought into contact. Adsorbing receptor proteins directly on surfaces that are intrinsically charged (e.g., glass has a layer of negative charges on its surface) should also be avoided. Instead, uncharged substrates like agarose and dextran can be used to support the proteins. The addition of 0.1 mg/ml bovine serum albumin (BSA) to the sample reservoir may also help reduce nonspecific interactions.

Another concern when acquiring AFM force measurements is that the rate at which an external force is applied, i.e., the loading rate of the measurement, can influence the binding properties of the ligand–receptor interaction. At higher loading rates the rupture force is expected to be higher than at lower loading rates. Because loading rate is dependent on the speed at which the ligand and receptor are separated and the elasticity of the surfaces to which both ligand and receptor are attached, it is important to note the cantilever retract speed and support parameters (i.e., surface elasticity) when comparing the rupture force of different ligand–receptor pairs.

Recent experiments in affinity imaging and cell–cell adhesion point to promising future directions in the studies of ligand–receptor interaction by AFM. Affinity imaging combines the force measurements of ligand–receptor interaction with the imaging function of AFM. When the cantilever tip is coated with a specific receptor or ligand, AFM can provide an adhesion map detail-

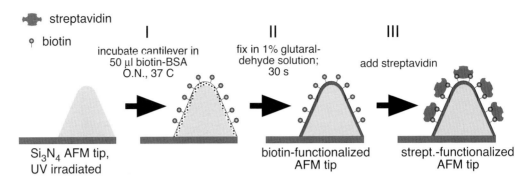

FIGURE 2. Schematics for the functionalization of AFM tips with streptavidin.

ing the density of the binding partner on a surface (Ludwig et al. 1997). Affinity images of antigen immobilized on a substrate were obtained using antibody-functionalized cantilevers (Willemsen et al. 1998; Raab et al. 1999). This technique has been extended to soft cellular surfaces. On the basis of the specific interaction between *Helix pomatia* lectin and Group A red blood cell plasma membrane proteins, Grandbois et al. (2000) were able to discriminate Group A red blood cells from Group O red blood cells.

AFM has also been adapted for studies in cell–cell adhesion (Razatos et al. 1998; Benoit et al. 2000). These experiments differed from those discussed earlier in that a cell is attached to the end of the cantilever and used as a probe in cell adhesion studies. This approach allowed both the ligand and receptor to be studied under conditions close to the native environment. Potential applications of this approach include the study of modulated adhesion following cell activation.

Although AFM is still being improved to enhance sensitivity (Viani et al. 2000), the major challenge for AFM research may stem from its restricted accessibility to a limited number of laboratories. With the completion of many of the genome projects, many protein sequences have been revealed; however, their biophysical properties and functions still need to be elucidated. AFM will be an important tool in this research.

The following sections present experimental protocols for carrying out AFM force measurements of ligand–receptor interactions. In brief, the experiments use an AFM cantilever tip coated with the ligand or receptor of interest while its binding partner is immobilized on another surface. The interaction between the two functionalized surfaces is acquired from the deflection of the cantilever during the approach and separation of the surfaces. Ligand–receptor interaction is derived from the adhesive force between the surfaces.

Protocol 1

Cantilever Functionalization

The ligand must be immobilized on the AFM tip to acquire direct force measurements of ligand–receptor pairs. The techniques commonly used involve either chemiadsorption (Moy et al. 1994) or covalent coupling of the ligand to the tip via an extended linker (Hinterdorfer et al. 1996). The linker between the tip and the ligand lends greater mobility and access to receptors on the surface being probed. The following outlines a method for functionalizing tips with streptavidin (Fig. 2, facing page). This method is advantageous because the streptavidin/biotin system has been well-characterized and has high affinity, and the initial layer of biotin-BSA may help to mask any electrical charges on the cantilever tip that could lead to nonspecific binding.

MATERIALS

CAUTION: See Appendix for appropriate handling of materials marked with <!>.

Buffers and Solutions

Acetone <!>
Biotin-BSA (Sigma A-6043) <!>
Phosphate-buffered saline (0.01 M, pH 7.4)
Sodium bicarbonate (0.1 M, pH 8.3)

Antibiotics

Streptavidin (Sigma)

Special Equipment

AFM cantilevers (MLCT-AUHW; Thermomicroscopes, Sunnyvale, California)

METHOD

1. Soak the cantilever for 5 minutes in acetone and then UV-irradiate <!> for 15 minutes.
2. Incubate the cantilever in a 50-µl drop of biotin-BSA (0.5 mg/ml in 0.1 M sodium bicarbonate, pH 8.3) overnight at 37°C in a humidified incubator.
3. Wash the cantilever three times in PBS (pH 7.4) to remove unbound protein.
   At this point the cantilevers can be stored in PBS for up to a week at 4°C.

4. Incubate the cantilever in a 50-μl drop of streptavidin (0.5 mg/ml in 0.01 M PBS, pH 7.4) for 10 minutes at room temperature.

> Cross-linking of surface-bound biotin-BSA with 1% glutaraldehyde before adding streptavidin helps stabilize the immobilized protein.

- It is important that biotin-BSA adsorption (step 2) takes place at pH 8.3 or higher, as the basic conditions seem to facilitate BSA adsorption to the cantilever.

- Kits for biotinylating the ligand of interest are available from Pierce Chemical Company.

5. Wash the cantilever three times in PBS before use.

6. Biotinylated ligand may be coupled to the streptavidin-functionalized tip at this step (e.g., biotinylated concanavalin A; 0.5 mg/ml in PBS, 10 minutes incubation at room temperature).

AFM Measurement of Ligand Immobilized on Agarose Beads

The following paragraphs discuss methods for acquiring force measurements of the strepta-vidin–biotin interaction. Numerous methods exist for immobilizing ligand to a variety of different substrates. Some commonly used ligands for affinity chromatography (e.g., biotin or D-man-nose) already attached to agarose beads can be purchased from vendors such as Sigma (St. Louis, Missouri) and Pierce (Rockford, Illinois). As a substrate for the force measurements, the agarose beads provide several attractive features. The agarose matrix has low affinity for most proteins, keeping nonspecific interaction at a minimum. Proteins can be readily coupled to activated agarose beads. Moreover, the elastic agarose substrate will conform to the shape of the cantilever tip, thus increasing the contact surface area and creating a higher probability for ligand–receptor interactions (Moy et al. 1994). Ligand–receptor binding is also greatly enhanced by attaching receptors to molecular tethers (e.g., dextran and polyethyleneglycol) that allow a wider range of lateral motion (Hinterdorfer et al. 1996; Rief et al. 1997a). In addition, tethers can provide lati-tude for proper reorientation of the molecule during stretching so that the external force being applied to the molecule is perpendicular to the surface.

The following is a protocol that we use for the preparation of biotinylated agarose beads for streptavidin–biotin rupture force measurements.

MATERIALS

CAUTION: See Appendix for appropriate handling of materials marked with <!>.

Buffers and Solutions

Bovine serum albumin (0.01%) in PBS
Phosphate-buffered saline (pH 7.4)
Sodium bicarbonate (pH 9.6)

Antibiotics

Streptavidin (Sigma)

Special Equipment

Biotinylated agarose beads (Sigma) <!>
Plastic petri dishes (35 mm)

METHOD

1. To prepare streptavidin-coated dishes for immobilizing biotinylated agarose beads:

   a. Place a 100-μl drop of streptavidin (0.05 mg/ml in sodium bicarbonate, pH 9.6) on the bottom of a 35-mm plastic petri dish.

   b. Incubate overnight at 37°C in a humidified incubator.

   c. Rinse the dish three times in PBS just before adding beads.

2. Add 100 µl of biotinylated agarose beads to 1.5 ml of PBS.

3. Centrifuge the beads at 10,000g for 10 seconds and remove the supernatant.

4. Repeat the wash two times.

5. After removing the supernatant from the last wash, resuspend the beads in 0.01% BSA in PBS.

6. Add 100 µl of washed beads to the streptavidin-coated plate. The beads should adhere to the dish almost immediately.

Figure 3 presents a representative AFM force measurement acquired using a streptavidin-functionalized AFM cantilever and a biotinylated agarose bead. The measurement consisted of an approach trace and a retract trace. During the approach trace, expansion of the piezoelectric translator lowered the cantilever onto the agarose bead and pressed the AFM tip into the elastic bead. Surface contact is registered by an upward deflection of the cantilever and allows for the formation of streptavidin–biotin complexes. The number of complexes formed depends on the area of surface contact, which can be estimated from the force exerted by the cantilever and the elasticity of the agarose bead. An applied force of one nanonewton frequently resulted in the formation of several complexes. The forced unbinding of these complexes is recorded in the retract trace. The sawtooth shape of the retract trace revealed that the unbinding of the complexes does not necessarily occur simultaneously. The retract trace showed multiple transitions in force that are attributed to the breakage of one or more streptavidin–biotin bonds. The interaction between the AFM tip and the substrate can be reduced to a single streptavidin–biotin linkage by the addition of either soluble biotin or streptavidin. Under these conditions, the force transition at surface separation corresponds to the rupture force of the single streptavidin–biotin complex. Control experiments to verify the specificity of the streptavidin–biotin interaction were carried out with the addition of either free streptavidin or free biotin.

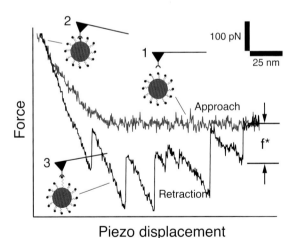

FIGURE 3. Force versus displacement curves of the interaction between a streptavidin-functionalized tip and a biotinylated agarose bead. The measurement recorded the force on the AFM cantilever on approach and retraction of the cantilever from the agarose bead. The inserts illustrate the deflection of the cantilever at different points during the measurement. f^* is the rupture force.

Protocol 3

AFM Measurements on Living Cells

The surface of a living cell is far more complicated than that of a bead or mica coated with protein. Furthermore, the low density of receptors on the cell surface and nonspecific interactions introduce additional challenges. Lehenkari and Horton (1999) were able to measure RGD (arginine-glycine-aspartic acid)-integrin binding on the surface of osteoblasts and osteoclasts, and recently, we measured the binding force between concanavalin A and its receptor on the surface of NIH-3T3 fibroblasts (Chen and Moy 2000). A brief generalizable protocol used to perform this work follows.

MATERIALS

CAUTION: See Appendix for appropriate handling of materials marked with <!>.

Buffers and Solutions

Biotinylated concanavalin A (ConA) (Sigma) <!>
Glutaraldehyde
NIH-3T3 fibroblasts
RPMI medium, glucose-free, supplemented with 0.01% bovine serum albumin (BSA) and
0.01 M magnesium chloride <!>

Special Equipment

Plastic tissue culture dishes (uncoated)

METHOD

1. Functionalize the AFM tip with ConA prior to the measurement (see Cantilever functionalization, p. 247).

2. Perform measurements at room temperature in glucose-free RPMI medium supplemented with 0.01% BSA and 0.01 mM $MnCl_2$.

   Eliminate glucose from the culture medium to prevent potential competitive binding with the ConA-functionalized tip and ConA receptors on the cell. Add BSA to reduce nonspecific binding and to promote adhesion of cultured NIH-3T3 fibroblasts to the bottom of an uncoated plastic tissue culture dish.

3. Carry out measurements on both unfixed cells and cells lightly fixed with glutaraldehyde to determine whether cross-linking of ConA receptors would have an effect on receptor unbinding strength.

   Compared to measurements on agarose beads and fixed cells, unfixed cells have much longer regions of stretch before final separation between the tip and membrane (compare Fig. 4A with Figs. 3 and 4B). Typical distances for unfixed cells spanned 500 nm. Thus, ConA receptors seemed to be anchored to cell membrane tethers that stretched as the receptor was pulled.

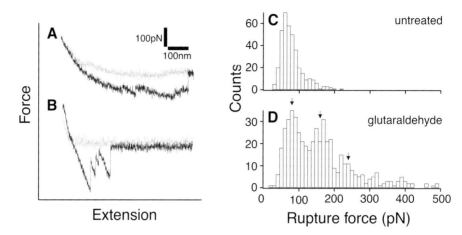

FIGURE 4. Force versus extension curves acquired from ConA-functionalized AFM tips interacting with ConA receptors on the surface of NIH-3T3 cells that were not fixed (A) or fixed with glutaraldehyde (B). Histograms of rupture force between ConA-functionalized AFM tips and ConA receptors on untreated cells (C) and glutaraldehyde-fixed cells (D). Arrows in D indicate quantized peaks at 80 pN, 160 pN, and 240 pN following fixation of cells in glutaraldehyde.

Rupture force measurements revealed a stronger rupture force for chemically fixed cells (173 ± 6.1 pN) compared to unfixed cells (86 ± 2.6 pN) (Fig. 4C,D). Moreover, differences in cell elasticity were readily apparent from the slope of the retract trace as the tip pulled on the surface of the cell. Force histograms revealed multiple quantal peaks that were absent in the unfixed cell histograms (Fig. 4D), suggesting that much of the increase in rupture force was due to a shift toward cooperative binding of cells.

Using this method, it is possible to acquire direct measurements of adhesion down to the level of single molecule pairs, unlike earlier cell adhesion assay studies and biochemical methods that relied on indirect measurements and could only access receptor group dynamics. Moreover, because the AFM is capable of both applying and detecting minute force changes on a cell or protein, simultaneous measurements of adhesion force and cell elasticity are possible at the cell surface.

ACKNOWLEDGMENTS

Work in our laboratory was supported by grants from the American Cancer Society and the National Institutes of Health (1R29 GM 55611-01) to V.T.M.

REFERENCES

Allen S., Chen X., Davies J., Davies M.C., Dawkes A.C., Edwards J.C., Roberts C.J., Sefton J., Tendler S.J.B., and Williams P.M. 1997. Detection of antibody-antigen binding events with the atomic force microscope. *Biochemistry* **36:** 7457–7463.

Ashkin A., Schutze K., Dziedzic J.M., Euteneuer U., and Schliwa M. 1990. Force generation of organelle transport measured in vivo by an infrared laser trap. *Nature* **348:** 346–348.

Baumgartner W., Hinterdorfer P., Ness W., Raab A., Vestweber D., Schindler H., and Drenckhahn D. 2000. Cadherin interaction probed by atomic force microscopy. *Proc. Natl. Acad. Sci.* **97:** 4005–4010.

Bell G.I. 1978. Models for the specific adhesion of cells to cells. *Science* **200:** 618–627.

Benoit M., Gabriel D., Gerisch G., and Gaub H.E. 2000. Discrete interactions in cell adhesion measured by single-molecule force spectroscopy. *Nat. Cell Biol.* **2:** 313–317.

Binnig G., Quate C.F., and Gerber C. 1986. Atomic force microscope. *Phys. Rev. Lett.* **56:** 930–933.

Burnham N.A. and Colton R.J. 1989. Measuring the nanomechanical properties and surface forces of materials using an atomic force microscope. *J. Vac. Sci. Technol.* **A7:** 2906–2913.

Butt H.J., Downing K.H., and Hansma P.K. 1990. Imaging the membrane protein bacteriorhodopsin with the atomic force microscope. *Biophys. J.* **58:** 1473–1480.

Chen A. and Moy V.T. 2000. Cross-linking of cell surface receptors enhances cooperativity of molecular adhesion. *Biophys J.* **78:** 2814–2820.

Dammer U., Popescu O., Wagner P., Anselmetti D., Güntherodt H.-J., and Misevic G.N. 1995. Binding strength between cell adhesion proteoglycans measured by atomic force microscopy. *Science* **267:** 1173–1175.

Engel A., Lyubchenko Y., and Müller D. 1999. Atomic force microscopy: A powerful tool to observe biomolecules at work. *Trends Cell Biol.* **9:** 77-80.

Evans E., Ritchie K., and Merkel R. 1995. Sensitive technique to probe molecular adhesion and structural linkages at biological interfaces. *Biophys. J.* **68:** 2580–2587.

Florin E.L., Moy V.T., and Gaub H.E. 1994. Adhesion forces between individual ligand-receptor pairs. *Science* **264:** 415–417.

Fritz J., Katopodis A.G., Kolbinger F., and Anselmetti D. 1998. Force-mediated kinetics of single P-selectin/ligand complexes observed by atomic force microscopy. *Proc. Natl. Acad. Sci.* **95:** 12283–12288.

Grandbois M., Dettmann W., Benoit M., and Gaub H.E. 2000. Affinity imaging of red blood cells using an atomic force microscope. *J. Histochem. Cytochem.* **48:** 719–724.

Hansma H.G., Vesenka J., Siegerist C., Kelderman G., Morrett H., Sinsheimer R.L., Elings V., Bustamante C., and Hansma P.K. 1992. Reproducible imaging and dissection of plasmid DNA under liquid with the atomic force microscope. *Science* **256:** 1180–1184.

Heinz W.F. and Hoh J.H. 1999. Spatially resolved force spectroscopy of biological surfaces using the atomic force microscope. *Trends Biotechnol.* **17:** 143–150.

Heymann B. and Grubmüller H. 1999. AN02/DNP-hapten unbinding force studied by molecular dynamics atomic force microscopy simulations. *Chem. Phys. Lett.* **303:** 1–9.

Hinterdorfer P., Baumgartner W., Gruber H.J., Schilcher K., and Schindler H. 1996. Detection and localization of individual antibody-antigen recognition events by atomic force microscopy. *Proc. Natl. Acad. Sci.* **93:** 3477–3481.

Hutter J.L. and Bechhoefer J. 1993. Calibration of atomic-force microscope tips. *Rev. Sci. Instrum.* **64:** 1868–1873.

Kishino A. and Yanagida T. 1988. Force measurements by micromanipulation of a single actin filament by glass needles. *Nature* **334:** 74–76.

Lee G.U., Chrisey L.A., and Colton R.J. 1994a. Direct measurement of the forces between complementary strands of DNA. *Science* **266:** 771–773.

Lee G.U., Kidwell D.A. and Colton R.J. 1994b. Sensing discrete streptavidin-biotin interactions with AFM. *Langmuir* **10:** 354–361.

Lehenkari P.P. and Horton M.A. 1999. Single integrin molecule adhesion forces in intact cells measured by atomic force microscopy. *Biochem. Biophys. Res. Commun.* **259:** 645–650.

Ludwig M., Dettmann W., and Gaub H.E. 1997. Atomic force microscope imaging contrast based on molecular recognition. *Biophys. J.* **72:** 445–448.

Marszalek P.E., Lu H., Li H., Carrion-Vazquez M., Oberhauser A.F., Schulten K., and Fernandez J.M. 1999. Mechanical unfolding intermediates in titin modules. *Nature* **402:** 100–103.

Merkel R., Nassoy P., Leung A., Ritchie K., and Evans E. 1999. Energy landscapes of receptor-ligand bonds explored with dynamic force spectroscopy. *Nature* **397:** 50–53.

Moy V.T., Florin E.-L., and Gaub H.E. 1994. Adhesive forces between ligand and receptor measured by AFM. *Colloids Surf.* **93:** 343–348.

Nakajima H., Kunioka Y., Nakano K., Shimizu K., Seto M., and Ando T. 1997. Scanning force microscopy of the interaction events between a single molecule of heavy meromyosin and actin. *Biochem. Biophys. Res. Commun.* **234:** 178–182.

Oesterhelt F., Oesterhelt D., Pfeiffer M., Engle A., Gaub H.E., and Müller D.J. 2000. Unfolding pathways of individual bacteriorhodopsins. *Science* **288:** 143–146.

Perkins T.T., Quake S.R., Smith D.E., and Chu S. 1994. Relaxation of a single DNA molecule observed by optical microscopy. *Science* **264:** 822–826.

Raab A., Han W., Badt D., Smith-Gill S.J., Lindsay S.M., Schindler H., and Hinterdorfer P. 1999. Antibody recognition imaging by force microscopy. *Nat. Biotechnol.* **17:** 901–905.

Razatos A., Ong Y.L., Sharma M.M., and Georgiou G. 1998. Molecular determinants of bacterial adhesion monitored by atomic force microscopy. *Proc. Natl. Acad. Sci.* **95:** 11059–11064.

Rief M., Oesterhelt F., Heymann B., and Gaub H.E. 1997a. Single molecule force spectroscopy on polysaccharides by atomic force microscopy. *Science* **275:** 1295–1297.

Rief M., Gautel M., Oesterhelt F., Fernandez J.M., and Gaub H.E. 1997b. Reversible unfolding of individual titin immunoglobulin domains by AFM. *Science* **276:** 1109–1112.

Sader J.E. 1995. Parallel beam approximation for V-shaped atomic force microscope cantilevers. *Rev. Sci. Instrum.* **66:** 4583–4587.

Schwesinger F., Ros R., Strunz T., Anselmetti D., Guntherodt H.J., Honegger A., Jermutus L., Tiefenauer L., and Pluckthun A. 2000. Unbinding forces of single antibody-antigen complexes correlate with their thermal dissociation rates. *Proc. Natl. Acad. Sci.* **97:** 9972–9977.

Senden T.J. and Ducker W.A. 1994. Experimental determination of spring constants in atomic force microscopy. *Langmuir* **10:** 1003–1004.

Smith S.B., Finzi L., and Bustamante C. 1992. Direct mechanical measurements of the elasticity of single DNA molecules by using magnetic beads. *Science* **258:** 1122–1126.

Strunz T., Oroszlan K., Schafer R., and Guntherodt H.J. 1999. Dynamic force spectroscopy of single DNA molecules. *Proc. Natl. Acad. Sci.* **96:** 11277–11282.

Svoboda K., Schmidt C.F., Schnapp B.J., and Block S.M. 1993. Direct observation of kinesin stepping by optical trapping interferometry. *Nature* **365:** 721–727.

Viani M.B., Pietrasanta L.I., Thompson J.B., Chand A., Gebeshuber I.C., Kindt J.H., Richter M., Hansma H.G., and Hansma P.K. 2000. Probing protein-protein interactions in real time. *Nat. Struct. Biol.* **7:** 644–647.

Willemsen O.H., Snel M.M., van der Werf K.O., de Grooth B.G., Greve J., Hinterdorfer P., Gruber H.J., Schindler H., van Kooyk Y., and Figdor C.G. 1998. Simultaneous height and adhesion imaging of antibody-antigen interactions by atomic force microscopy. *Biophys. J.* **75:** 2220–2228.

Wong S.S., Joselevich E., Woolley A.T., Cheung C.L., and Lieber C.M. 1998. Covalently functionalized nanotubes as nanometre-sized probes in chemistry and biology. *Nature* **394:** 52–55.

Yuan C., Chen A., Kolb P., and Moy V.T. 2000. Energy landscape of streptavidin-biotin complexes measured by atomic force microscopy. *Biochemistry* **39:** 10219–10223.

Zasadzinski J.A., Schneir J., Gurley J., Elings V., and Hansma P.K. 1988. Scanning tunneling microscopy of freeze-fracture replicas of biomembranes. *Science* **239:** 1013–1015.

14 Analysis of Interacting Proteins with Surface Plasmon Resonance Using Biacore

Maxine V. Medaglia and Robert J. Fisher
National Cancer Institute-FCRDC, Frederick, Maryland

INTRODUCTION

Both qualitative and quantitative techniques, including gel filtration, calorimetry, and ultracentrifugation, have been exploited in the study of interactions among macromolecules. Three complementary quantitative approaches have proven to be valuable in the study of protein-to-protein binding interactions: analytical ultracentrifugation, isothermal titration calorimetry, and surface plasmon resonance (SPR). Calorimetry and analytical ultracentrifugation require milligram amounts of pure protein, whereas SPR requires only microgram amounts of protein. SPR provides a means for collecting kinetic data (the microscopic rate constants for an interaction between macromolecules), whereas analytical ultracentrifugation provides the best evidence for the aggregation state of a protein, and isothermal titration calorimetry provides a complete ther-

The authors prepared this manuscript with the help of M. Fivash, C. Hixsonn, and L. Wilson (National Cancer Institute-FCRDC, Frederick, Maryland), and JoAnne Bruno (Biacore Inc., New York).
This chapter is modified from Chapter 18, Protocol 6, of Sambrook and Russell 2001, *Molecular Cloning*.

255

modynamic profile. including the equilibrium constant, reaction stoichiometry, enthalpy, and entropy. Ideally, these methods should be used in concert to characterize a molecular interaction completely.

SPR is an optical resonance effect in which the back side of a thin conductive mirror affects the angle at which a minimum of reflected light exists (Fig. 1). The SPR response is extremely sensitive to any change in the dielectric constant of the medium adjacent to the gold surface of a sensor chip. Changes in the SPR signal are monitored by recording the angular changes at which the reflected light is minimum. The introduction of commercial instruments capable of measuring SPR has simplified the study of macromolecular interactions by providing a format that may be used to measure molecular interactions in real time with small analytical amounts of material instruments (Jonsson et al. 1991). Various SPR instruments are available, ranging from very simple and inexpensive models to more integrated instruments with attendant robotics, optical interfaces, and software. The simplest device, called a Spreeta chip, was developed by Texas Instruments (www.ti.com/spreeta). Texas Instruments maintains an informative Web site with descriptions of the physical basis of SPR. More sophisticated instrumentation is available from Biacore, Inc. (www.biacore.com). Biacore instruments have been used to characterize several different aspects of macromolecular interactions, including protein–protein, protein–small molecule, protein–nucleic acid, and protein–lipid interactions (Celia et al. 1999; Raut et al. 1999; Saenko et al. 1999). The Biacore Web site maintains an extensive list of references covering the wide range of experimental designs that have used Biacore SPR technology, and provides information about the technology and applications of SPR and Biacore instrumentation. The company offers technical support as well as instruction in the use of Biacore instruments and BIAevaluation software. A full discussion of SPR can be found in the *Encyclopedia of Life Sciences* (www.els.net).

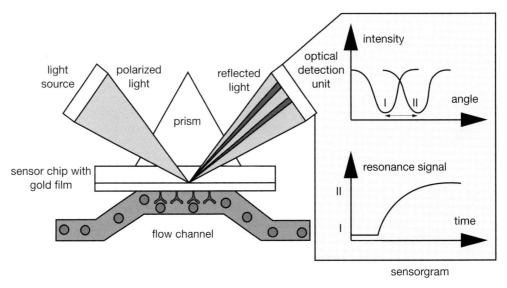

FIGURE 1. Schematic of SPR. BIAtechnology relies on the phenomenon of SPR, which occurs when surface plasmon waves are excited at a metal/liquid interface. Light is directed at, and reflected from, the side of the surface not in contact with sample, and SPR causes a reduction in the reflected light intensity at a specific combination of angle and wavelength. Biomolecular binding events cause changes in the refractive index at the surface layer, which are detected as changes in the SPR signal. (Redrawn, with permission, from Biacore and reprinted, with permission, from Sambrook and Russell 2001, *Molecular Cloning*.)

OUTLINE OF PROCEDURE

There are four major components of a Biacore instrument: (1) an optical unit to mount the sensor chip; (2) a microfluidics system (Sjolander and Urbaniczky 1991) to deliver analyte (molecule of interest) precisely to the ligand (molecule attached to the surface of the chip); (3) an autosampler to transfer samples from the sample rack to the microfluidics delivery system; and (4) data collection and data analysis software.

It is difficult to predict how long it will take to perform a Biacore experiment and analyze the data obtained. After preparing the appropriate sensor chip and optimizing conditions for the interaction between the ligand and analyte, the regeneration conditions must be determined. Once these preliminary studies are completed, the actual experiment can be run, and the data can be analyzed using BIAevaluation software. For some systems, such as epitope mapping, once the optimal conditions have been determined, the experimental design and analysis of the results are quite straightforward. However, experiments designed to elucidate the kinetics of an interaction may lead to the development of kinetic pathways that are more complex than the initial model. BIAsimulation software (available on the Web site) provides a means to predict what the results of the interaction would look like with different kinetic conditions.

The Sensor Chip

The chip technology introduced by Biacore consists of a 1-cm^2 glass slide coated with gold that is embedded in a plastic support for ease of handling. The standard chip is Sensor Chip CM5, but several varieties of chips are available (www.biacore.com) for specialized purposes. Sensor Chip CM5 has the equivalent of a 2–3% solution of carboxymethylated dextran covalently attached to the gold surface. A number of protocols have been developed for attaching ligand molecules to the dextran surface (O'Shannessy et al. 1992); these create active groups on the carboxymethyl dextran that are capable of reacting covalently with the ligand molecule. The choice of coupling chemistry to be used depends on characteristics of the ligand molecule (Fig. 2). Each chip has four different flow cells so that ligand density may be varied on three of the surfaces, and the fourth may be used for a control or blank surface.

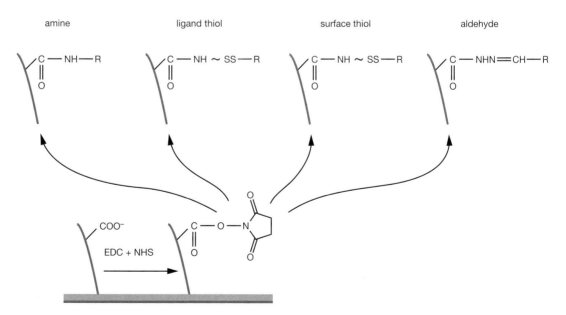

FIGURE 2. Ligand mobilization via native $-NH_2$, $-SH_2$, -CHO, and COOH. (Redrawn, with permission, from Biacore and reprinted, with permission, from Sambrook and Russell 2001, *Molecular Cloning*.)

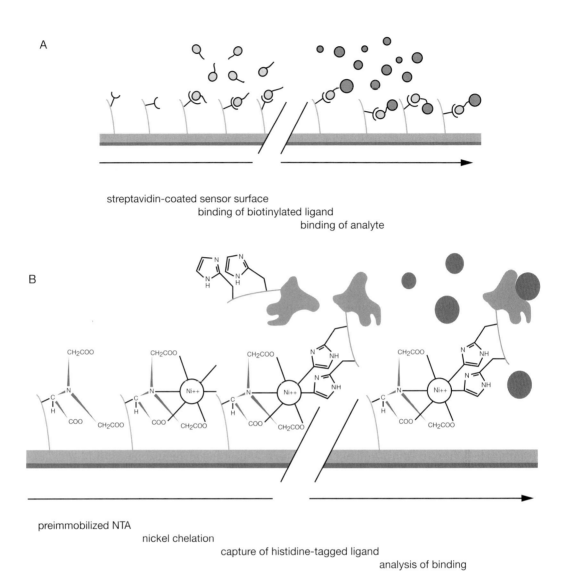

A

streptavidin-coated sensor surface
binding of biotinylated ligand
binding of analyte

B

CH₂COO
CH₂COO
CH₂COO
CH₂COO

preimmobilized NTA
nickel chelation
capture of histidine-tagged ligand
analysis of binding

FIGURE 3. Capture of target molecules. (*A*) Capture of biotinylated macromolecules. (*B*) Capture of histidine-tagged protein. (Redrawn, with permission, from Biacore, Inc. and reprinted, with permission, from Sambrook and Russell 2001, *Molecular Cloning.*)

Careful consideration should be given to the method used to immobilize the ligand to the Sensor Chip CM5. Whenever possible, the method of choice will be one that minimizes the heterogeneity and maximizes the functionality of the bound ligand. Techniques that allow capture of the ligand by another molecule covalently bound to the carboxymethyl dextran are useful because they can result in uniform orientation of the ligand for a single attachment point, and thus are preferred over direct coupling of the ligand to the chip surface. Examples of such capture surfaces are streptavidin (Morgan and Taylor 1992; O'Shannessy et al. 1992), nickle–chelate (NTA) (Nieba et al. 1997), rabbit anti-mouse Fc domain (RaM Fc), and anti-GST. Chips can be purchased with the streptavidin (Fig. 3A) or NTA (Fig. 3B) already coupled to the carboxymethyl dextran. Capture molecules such as antibodies and streptavidin can be coupled to the surface through primary amine groups using the Biacore Amine Coupling kit (as in the protocol below).

Other sensor chips are available that offer additional features. Certain chips promote the formation of a lipid monolayer (HPA) (Fig. 4) or bilayer (Pioneer Chip L1), whereas others have

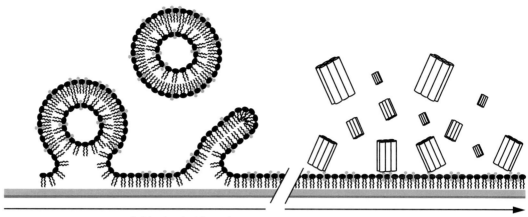

flat hydrophobic surface
coated with user-defined liposomes
hydrophilic surface-created
analyte binding

FIGURE 4. Membrane biochemistry and membrane-bound receptor studies. (Redrawn, with permission, from Biacore and reprinted, with permission, from Sambrook and Russell 2001, *Molecular Cloning*.)

shorter dextran molecules (Pioneer Chip F1), dextran molecules with a lower charge (Pioneer Chip B1), a carboxymethylated surface with no dextran matrix (Pioneer Chip C1), or a flat gold surface without any attached molecules (Pioneer Chip J1).

Regeneration of the Chip Surface

It is important to develop a robust regeneration system so that the ligand-immobilized surfaces may be reused. A wide variety of materials (acid, base, solvents, salt, detergents, or combinations of these reagents) may be used for regeneration (Andersson et al. 1999). An experiment should be designed to see how robust the regeneration is by repeated injections of analyte and regeneration solution to determine whether the ligand remains bound to the surface. With the correct regeneration conditions, it is possible to reuse a surface for up to 100 regeneration cycles.

Collection of Data

The instrumentation uses a flow system and a sensor chip to which a ligand has been attached (immobilized). The analyte solution is injected through the system and moves by diffusion and convection from the bulk solution to the sensor chip surface, where it interacts with the ligand. Figure 5 illustrates the progress of an interaction as monitored by a Biacore sensorgram. The interaction between the ligand and analyte molecules is followed in real time as a change of the SPR angle. Regeneration is accomplished by the removal of the non-covalently bound material from the surface in as gentle a manner as possible, allowing reuse of the ligand in subsequent binding/regeneration cycles.

The process of diffusion of the analyte is called material transport, and all binding systems are affected to some degree by this phenomenon. Part of the experimental design for kinetic studies involves determining a low enough surface density of ligand and a high enough flow rate to minimize transport effects. After satisfying these minimal criteria for an experiment, data may be collected for subsequent fitting to an appropriate binding model. We strongly recommend that three channels (also referred to as flow cells) of the sensor chip be modified to contain different levels of ligand and the fourth channel be used as an appropriately modified control surface. Another

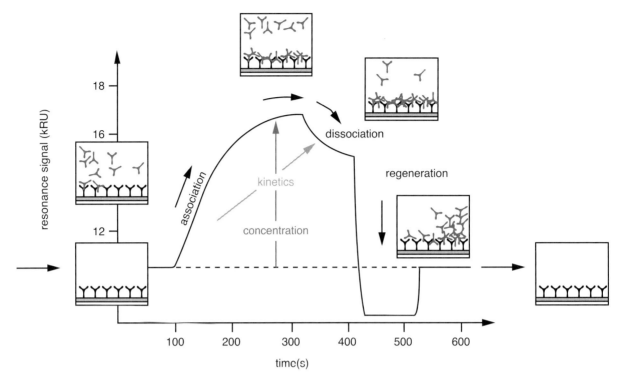

FIGURE 5. Progress of an interaction monitored as a sensorgram. Analyte binds to the surface-attached ligand during sample injection, resulting in an increase in signal. At the end of the injection, the sample is replaced by a continuous flow of buffer and the decrease in signal now reflects dissociation of the interactant from the surface-bound complex. A response of 1000 RU corresponds to a change in surface concentration of 1 ng/mm². (Redrawn, with permission, from Biacore and reprinted, with permission, from Sambrook and Russell 2001, *Molecular Cloning*.)

method has been suggested (Myszka 1999) in which a low-density ligand surface is corrected by double referencing. The signal from the analyte injection over a reference channel is subtracted from the specific binding signal of the analyte injection over the active surface. A further subtraction of a buffer injection from each of the injections of analyte is used to reduce machine effects in the data. Care should be taken when applying any of these techniques, especially when the signal noise is greater than the binding signal itself. If there is nonspecific binding to the reference channel (the control surface), this technique of subtracting the reference channel's signal from that of the active surface should be used with caution.

The resulting sensorgram or series of sensorgrams may be modeled using one of several binding models provided with the BIAevaluation software or by user-defined models. When analyzing data from the Biacore, it must be remembered that the sensitivity of the instrument returns a very low noise signal that is profoundly sensitive to machine effects, such as the opening and closing of valves, and bulk effects arising from temperature and refractive index (solute concentration) mismatch between the running buffer and the sample. The interpretation of the binding signal can be complicated by the presence of these frequently unavoidable effects. The basic training in the use of the Biacore (BIAuniversity) and the specialized training in BIAevaluation software (Kinetics I and Kinetics II) are excellent sources for learning instrument operation, experimental design, and data evaluation. Information pertaining to these courses is available on the Biacore Web site.

Kinetic Measurements

The early Biacore instruments, first introduced in 1990, had few applications and only a rudimentary method for analysis of the data. These limitations generated discussions as to whether data should be analyzed using nonlinear methods (O'Shannessy et al. 1993) or linear methods (Karlsson et al. 1991; Altschuh et al. 1992; Pellequer and Van Regenmortel 1993). The models now used are nonlinear, which utilize actual data rather than using replots as do the linear models. The modeling of kinetic data was much improved after Fisher and Fivash (1994) proposed a two-compartment model to describe the delivery of analyte from the flow buffer (compartment 1) to the binding region of the chip (compartment 2). This work also introduced a new method of analysis to be applied to Biacore data. In this method, known as global analysis, a series of sensorgrams scaled by different surface ligand densities and different analyte concentrations are solved with a single set of association and dissociation rate constants (Fisher and Fivash 1994; Fisher et al. 1994). The two-compartment model (Myszka et al. 1998) and global modeling (Roden and Myszka 1996; Karlsson and Falt 1997) were widely accepted in the field.

The consensus on appropriate modeling of Biacore data has led to improvements in the Biacore evaluation software that allows data grooming (baseline adjustment, time zeroing, and subtraction of a blank injection for a set of sensorgrams), and kinetic analysis using a set of models included in the software. This allows calculation of k_t (k transport—delivery of analyte to the chip surface), and the microscopic rate constants k_a (k association—association rate constant) and k_d (k dissociation—dissociation rate constant). The ratio of k_a to k_d can be used to determine K_D, the equilibrium constant. The possibility of obtaining these microscopic rate constants for a molecular interaction (sometimes referred to as the on- and off-rates) excites the imagination of researchers. However, one should regard the kinetic constants obtained using Biacore analysis as tentative until verified using other methods. (*Note:* A better description may be that the results are apparent rate and equilibrium constants. There is a large body of data to suggest that the results from SPR assays agree with all of the assays listed below and others.) The Molecular Interactions Research Group (MIRG) within the Association of Biomolecular Resource Facilities (ARBF) plans to address this issue by sending samples for analysis by SPR, analytical ultracentrifugation, and isothermal titration calorimetry. These results will provide, for the first time, a direct comparison of these different methodological approaches.

In addition, the SPR measurement is derived from an analyte that comes out of solution and binds to a surface matrix, which involves mixed phases (solid phase and solution phase). The relationship of measurements made using mixed phases to those made using a single phase has been and will continue to be a source of controversy (Ladbury et al. 1995; Nieba et al. 1996; Muller et al. 1998).

Concentration Measurement

Under material (mass) transport limited binding, the initial binding slope should be linearly related to concentration of analyte (Karlsson et al. 1993; Nieba et al. 1996; Christensen 1997). When the analyte and free ligand molecules are incubated before injection (to allow the system to reach equilibrium), the initial binding slope shows the amount of free (uncomplexed) analyte. This concept is particularly useful because it allows calculation of an equilibrium constant for a ligand–analyte interaction without any assumptions about a particular binding model. This technique also allows a calibrated system to be used to determine the concentration of an analyte in solution.

The equilibrium association constant K_A may be calculated, without resorting to modeling, in at least three ways. Finding the concentration of analyte in a competition experiment (see above) is one useful method. The constant found in this way is a solution constant. For a surface constant, one may inject sufficient analyte to show about half of the saturation signal. An alternative to this method is to circulate the analyte in the flow buffer over long periods of time to attain equilibrium (Myszka et al. 1998; Schuck et al. 1998).

Designing a Protocol

Several preliminary experiments to characterize the reagents are required to develop a well-behaved experimental protocol. The sequence of steps for analyzing a typical antibody–antigen system, including several details involved in the development of a Biacore experiment, are described here. The antibody and antigen used to optimize the parameters of the system under study are available commercially.

The protocol is divided into two parts. The first part describes the preparation of a RaM Fc capture surface, followed by tests for the binding of a specific antibody, anti-thyroid-stimulating hormone (anti-TSH), to the RaM Fc surface, and binding of the antigen thyroid-stimulating hormone (TSH) to the antibody (anti-TSH). The second part of the protocol presents a series of analysis programs for kinetic analysis; that is, determining the equilibrium constants for the interaction.

The first sequence of steps describes the method for capturing anti-TSH on a rabbit anti-mouse C domain (RaM Fc) surface, to prepare the anti-TSH surface for use in the kinetic analysis of the binding of TSH with anti-TSH. The RaM Fc is immobilized to the carboxymethyl dextran of the sensor chip by primary amine coupling, and when used as the capture molecule for the antibody, results in the bivalent antibody molecules being oriented in a manner that will allow binding of two antigen molecules to each antibody molecule. An alternative method of preparing a surface for kinetic analysis is to couple one of the reactants directly to the sensor chip via primary amine residues. If this method is used with an antibody, the heterogeneity of the resulting surface may result in less than two antigen molecules binding to each of the antibody molecules.

TSH binding to anti TSH is expected to have a 2:1 stoichiometry owing to the presence of two antigen-binding sites on the antibody. To determine the efficiency of the binding of the TSH (M_r = 50,000) to the anti-TSH (M_r = 150,000), calculate the maximum number of resonance units (RUs) of analyte that can bind to the captured antibody.

$$= (MWanalyte)/(MWligand) \times (RUs\ ligand) \times valency$$

$$= 2/3 \times RUs\ anti\text{-}TSH\ bound$$

In the case of 250 RUs of anti-TSH captured by the RaM Fc surface, 167 RUs of TSH are expected to bind if all of the antibody is active.

The second part of the protocol presents a series of method blocks or analysis programs that allow the determination of the equilibrium constants for the interaction under study. The program sequence is designed to allow capture of a known number of RUs of anti-TSH by the RaM Fc surface at the start of each cycle, followed by an injection of TSH. The RaM Fc surface is regenerated at the end of each cycle by the injection of 20 mM HCl. A total of six different concentrations of TSH (prepared by serial dilution) are injected over identically prepared anti-TSH surfaces. The high degree of reproducibility of the binding of antibody to the capture surface results in essentially the same anti-TSH surface being available for the interaction of the different concentration of TSH injected in each cycle.

The Interaction Wizard (available with Biacore Control Software version 3.0 and higher for Biacore 2000 and Biacore 3000) can be used to design this type of experiment under the Kinetic Analysis, Concentration Series, Binding Using Capture Molecule choices. If the Wizard is used, however, only one surface density of anti-TSH can be used at a time in the experiment. An advantage of writing a method directly is that different anti-TSH surface densities may be created that are then exposed to the same concentration series of TSH, thus satisfying the data requirements of global analysis (see Kinetic Measurements section, above). The Biacore *Instrument Handbook* gives a thorough discussion of writing methods using the Biacore Method Definition Language (MDL). The increased flexibility in experimental design realized by writing Biacore methods directly is worth the time spent becoming familiar with MDL. Alternatively, Custom Application Wizards in Biacore Control Software versions 3.0 and higher allow the user to program the same experiment.

Surface Plasmon Resonance Using Biacore

The Molecular Interactions Research Group (MIRG) within the Association of Biomolecular Resource Facilities (ARBF) (www.abrf.org) recommends that protein samples be prepared for Biacore analysis by buffer exchange dialysis (3 times in 500 volumes) against HEPES-buffered saline (HBS), then centrifuged at maximum speed in a refrigerated microcentrifuge, and finally, passed through a 0.2-μm filter. The resulting protein concentration is determined by absorbance at 280 nm using the calculated extinction coefficient for the particular protein or peptide.

MATERIALS

CAUTION: See Appendix for appropriate handling of materials marked with <!>.

Buffers and Solutions

Recipes for stock solutions of buffers and reagents can be found in Appendix 1 of Sambrook and Russell (2001). Dilute stock solutions to the appropriate working concentrations.

Ethanolamine (1 M ethanolamine hydrochloride, adjusted to pH 8.5 with NaOH) <!>
N-Ethyl-N′-(dimethylaminopropyl)-carbodiimide (0.2 M) (EDC) in H_2O <!>
> EDC, ethanolamine, and NHS are commercially available as part of the Amine-Coupling Kit from Biacore, Inc.

HBS buffer
> 10 mM HEPES (pH 7.4)
> 150 mM NaCl
> 3 mM EDTA
> 0.005% polysorbate 20 (Tween-20)
>> Commercially available from Biacore, Inc.

HCl (20 mM) <!>
N-Hydroxysuccinimide (0.05 M) (NHS) in H_2O <!>

Antibodies

Mouse anti-TSH monoclonal antibody (1 mg/ml in HBS) (available from Alexon-Trend, Inc., Ramsey, Minnesota)
> Dilute the antibody to 2 μg/ml and prepare 1000 μl or the volume required to achieve the target RUs as determined in steps 15–20.

Rabbit anti-mouse Fc domain (RaM Fc) (available from Biacore, Inc.)
> 30 μg/ml in 10 mM sodium acetate (pH 5.0) <!>

TSH (20 μM in HBS) (available from Sigma)
> For kinetic analysis, prepare serial dilutions of TSH for injection:
> 1. Prepare serial dilutions of 200, 100, 50, 25, 10, and 5 nM TSH, each in 280 μl of the running buffer used for the experiment.
> 2. Dispense 70 μl of each concentration in four plastic 7-mm vials and cap each vial.
> 3. Place the vials in the appropriate position in the Biacore rack, as directed by the sample loop parameters.

Special Equipment

The equipment needed here is available from Biacore, Inc. Information is available at www.biacore.com.

Biacore instrument (including Biacore Control Software and BIAcvaluation software)
Sensor Chip CM5
Sensor Chip CM5, coupled to RaM Fc (from step 14 for subsequent kinetic analysis)

METHOD

Preparation of the Capture Surface: Primary Amine Coupling of RaM Fc to Sensor Chip CM5 Surface

This procedure will take ~45 minutes. The commands can be accessed through the pull-down menus or by the icons of the toolbar of the Biacore control software.

1. Dock the Sensor Chip CM5 in the Biacore instrument.

2. Prime the equipment using filtered and degassed HBS.

3. Place tubes containing 100 µl each of NHS, EDC, ethanolamine, RaM Fc, and 20 mM HCl in appropriate positions in the autosampler rack in the Biacore instrument.

4. Place an empty tube in the autosampler rack in the Biacore instrument.

   The Immobilization Wizard, if available, can be used to perform this procedure. (The Wizards are standard with Biacore control software versions 3.0 and higher for Biacore 2000 and Biacore 3000.) If using the Wizard, dispense the volumes specified. The target number of RUs of bound RaM Fc is 13,000. An RU is arbitrarily defined as 1/10,000 of a degree in the Biacore instrument.

5. Start the instrument at a flow of 5 µl/minute over one flow cell.

6. Transfer 75 µl of NHS to the empty tube.

7. Transfer 75 µl of EDC to the same tube.

8. Mix 75 µl of the tube now containing the NHS and EDC.

9. Inject 35 µl of the NHS/EDC mixture to activate the surface.

10. Inject 35 µl of the RaM Fc to couple the antibody to the activated surface.

11. Inject 35 µl of ethanolamine to deactivate excess reactive groups.

12. Quickinject 10 µl of 20 mM HCl followed by Extraclean to remove non-covalently bound material.

13. Determine the level of RaM Fc bound by placing a baseline report point before the start of the RaM Fc injection and a second report point 2 minutes after the end of the 20 mM HCl injection.

    The final RUs of RaM Fc immobilized should be between 10,000 and 13,000; if a lower level is obtained, it could be due to a lower concentration of RaM Fc used, or using NHS and EDC solutions that have been stored for longer than 2 months. A surface with fewer RUs of RaM Fc is usable, but the volumes of anti-TSH used will need to be adjusted empirically; in general, more will be required to achieve the desired level of binding.

14. Stop the flow, close the command queue window, and save the result file.

> The Biacore can be left on Standby (Continue) at this point, or the surface can be directly in the next section. The RaM Fc surface is quite stable left in the instrument with continuous buffer flow. Alternately, the chip can be undocked and stored at 4°C in a 50-ml conical tube containing a small amount of H_2O; the surface will be stable for several weeks stored this way.

Testing Binding of the anti-TSH to the RaM Fc Surface

This procedure will take ~20 minutes. The commands can be accessed through the pull-down menus or by the icons of the toolbar of the Biacore control software.

15. Start the instrument at a flow of 10 μl/minute over the flow cell with the coupled RaM Fc.

16. Inject 10 μl of 2 μg/ml anti-TSH. (This will require a total of 40 μl of anti-TSH solution.)

> A total of 10 μl of the anti-TSH should result in an increase of ~250 RUs. If this level of binding to the RaM Fc surface is not obtained, regenerate (see step 17) and repeat the injection of anti-TSH using different volumes until the necessary injection volume is determined.

17. Quickinject 10 μl of 20 mM HCl followed by Extraclean to regenerate the RaM Fc surface.

18. Repeat the injection of the volume of anti-TSH required for 250 RUs bound anti-TSH to test the reproducibility of the binding to the RaM Fc surface.

19. Quickinject 10 μl of 20 mM HCl followed by Extraclean to regenerate the RaM Fc surface.

20. Stop the flow, close the command queue window, and save the result file.

Testing Binding of TSH to the Captured Anti-TSH Surface

This procedure will take ~20 minutes. The commands can be accessed through the pull-down menus or by the icons of the toolbar of the Biacore control software.

21. Start the instrument at a flow of 10 μl/minute over the flow cell with the coupled RaM Fc.

22. Inject 10 μl of 2 μg/ml anti-TSH (or the volume determined to give 250 RUs bound anti-TSH as determined in steps 15–20).

23. Inject 25 μl of 200 nM TSH specifying a 120-second dissociation time (this will require a total of 65 μl of TSH solution). The 120-second dissociation time will allow the rate at which the antibody–antigen complex dissociates to be gauged.

> After ~20 μl of 200 nM TSH has been injected, the curve should flatten (the slope should approach zero), representing a stochastic steady state of the interaction between the antibody and antigen molecules. If the curve does not flatten, regenerate the surface and repeat steps 21–23 using 400 nM TSH in step 23 (Figs. 5 and 6.)

24. Quickinject 10 μl of 20 mM HCl followed by Extraclean to regenerate the RaM Fc surface.

25. Stop the flow, close the command queue window, and save the report file.

Kinetic Analysis of the Interaction

The entire program sequence consists of three method blocks or sections: the MAIN, the APROG, and the LOOP. There are four different analysis programs (defined APROG blocks), one for each of the four APROGs referred to in the MAIN (bind1, bind2, bind3, and bind4). Each APROG

Discussion of Theory

The binding of the TSH to the anti-TSH is expected to have a 2:1 stoichiometry owing to the presence of two antigen-binding sites on the antibody. To determine the efficiency of the binding of the TSH (M_r = 28,000) to the anti-TSH (M_r = 150,000), calculate the maximum number of resonance units (RUs) of analyte that can bind to the captured antibody.

R_{max} = (MWanalyte)/MWligand) x (R_L) x valency

R_{max} is the theoretical maximum response if 100% of the ligand sites are (1) active and (2) bound by analyte, R_L is the amount of ligand immobilized in RU, and valency is the number of analyte molecules bound per ligand molecule. In the case of 250 RUs of anti-TSH captured by the RaM Fc surface, 93 RUs of TSH would be expected to bind if all of the antibody were active.

requires that a different volume of mouse anti-TSH monoclonal antibody be injected over the same RaM Fc surface. The injection of each volume of anti-TSH is followed by the injection of the series of TSH dilutions. For replicate injections of the TSH samples, the appropriate lines of the sample loops should be added. As the method is written, 1000 μl of the anti-TSH dilution is required: 400 μl for APROG1 (eight 20-μl injections, each requiring 50 μl), 320 μl for APROG2 (eight 10-μl injections, each requiring 40 μl), and 280 μl for APROG3 (eight 5-μl injections, each requiring 35 μl). If a different volume of the anti-TSH is needed to achieve the target RUs, the APROGS must be modified and the total volume of antibody required will also differ. It is important to remember that each INJECT command requires the specified volume plus an additional 30 μl for each injection. There are two buffer injections before the TSH series is injected over each anti-TSH level, to allow the binding level of antibody to stabilize.

When the Biacore software edits a method, the method commands will be written in capitals and the user-defined parameters will be in lowercase letters. The comments following an exclamation point (!) are seen by the instrument as text rather than command code, and are inserted to explain the significance of the preceding command.

FIGURE 6. Anti-TSH binding to RaM Fc. This plot shows the results when the same concentration series of TSH is injected over three levels of anti-TSH captured by a RaM Fc surface as well as over the RaM Fc surface itself. The data have been Y-transformed and the regeneration injection has been removed from each cycle. The high level of reproducibility of the anti-TSH binding to the RaM Fc is apparent, as is the stability of the interaction of these two molecules. The low level of interaction between the TSH and RaM Fc surface is also illustrated in this diagram. (Redrawn, with permission, from Biacore, Inc. and reprinted, with permission, from Sambrook and Russell 2001, *Molecular Cloning*.)

```
MAIN
    RACK                    1 thermo_a
    FLOWCELL 'X'

    LOOP sample1 STEP
        APROG          bind1 %p1 %conc
    ENDLOOP

    LOOP sample3 STEP
        APROG          bind3 %p3 %conc
    ENDLOOP

    LOOP sample2 STEP
        APROG          bind2 %p2 %conc
    ENDLOOP

    LOOP sample4 STEP
        APROG          bind4 %p4 %conc
    ENDLOOP

    APPEND                    continue

END
```

!Specifies instrument block to be used.
!Insert the flow cells to be used.

!anti-TSH; final RUs = 500–600

!anti-TSH; final RUs = 100–200

!anti-TSH; final RUs = 250–350

!Reference surface

!Keeps the running buffer flowing through
!instrument after the the end of the run.

```
DEFINE APROG bind1
    PARAM                   %p1 %conc
    KEYWORD CONC                 %conc
    FLOW 10
    * INJECT                    r2a1 20
    -0:05 RPOINT           Baseline -b
    1:55 RPOINT           Anti-TSH_Peak
    *KINJECT                %p1 25 240
    -0:05 RPOINT        Anti-TSH_bound -b
    2:25 RPOINT              TSH_Peak
    *QUICKINJECT              R2f3 10
    EXTRACLEAN
    -0:05 RPOINT            TSH_Washout
    3:00 RPOINT             Regeneration
    END
```

!anti-TSH; final RUs = 500–600
!sets a baseline report point
!sets a response report point
!TSH

!20 mM HCl to regenerate RaM Fc surface

```
DEFINE APROG bind2
    PARAM                   %p2 %conc
    KEYWORD CONC                 %conc
    FLOW 10
    *INJECT                     r2b1 10
    -0:05 RPOINT           Baseline -b
    0:55 RPOINT           Anti-TSH_Peak
    *KINJECT                %p2 25 240
    -0:05 RPOINT        Anti-TSH_bound -b
    2:25 RPOINT              TSH_Peak
    *QUICKINJECT              R2f3 10
    EXTRACLEAN
    -0:05 RPOINT            TSH_Washout
    3:00 RPOINT             Regeneration
    END
```

!anti-TSH; final RUs = 250–350

!20 mM HCl to regenerate RaM Fc surface

(Continued on next page.)

```
DEFINE APROG bind3
    PARAM                    %p3 %conc
    KEYWORD CONC                  %conc
    FLOW 10
    *INJECT                     r2c1  5
    -0:05 RPOINT              Baseline -b
    0:25 RPOINT             Anti-TSH_Peak
    *KINJECT                  %p3 25 240
    -0:05 RPOINT         Anti-TSH_bound -b
    2:25 RPOINT                 TSH_Peak
    *QUICKINJECT               R2f3  10
    EXTRACLEAN
    -0:05 RPOINT             TSH_Washout
    3:00 RPOINT              Regeneration
    END
```

!anti-TSH; final RUs = 100–200

!20 mM HCl to regenerate RaM Fc surface

```
DEFINE APROG bind4
    PARAM                    %p4 %conc
    KEYWORD CONC                  %conc
    FLOW 10
    *KINJECT                  %p4 25 240
    -0:05 RPOINT              Baseline -b
    2:25 RPOINT                 TSH_Peak
    *QUICKINJECT               R2f3  10
    EXTRACLEAN
    -0:05 RPOINT             TSH_Washout
    3:00 RPOINT              Regeneration
    END
```

!20 mM HCl to regenerate RaM Fc surface

```
DEFINE LOOP sample1
    LPARAM %p1 %conc
            r1a9    0n
            r1a10   0n
            r1a1    200n
            r1a6    5n
            r1a2    100n
            r1a5    10n
            r1a3    50n
            r1a4    25n
    END
```

!buffer
!buffer
!200 nM TSH

```
DEFINE LOOP sample2
    LPARAM %p2 %conc
            r1b9    0n
            r1b10   0n
            r1b1    200n
            r1b6    5n
            r1b2    100n
            r1b5    10n
            r1b3    50n
            r1b4    25n
    END
```

!buffer
!buffer

(Continued on next page.)

```
DEFINE LOOP sample3
    LPARAM %p3 %conc
            r1c9    0n
            r1c10   0n
            r1c1    200n
            r1c6    5n
            r1c2    100n
            r1c5    10n
            r1c3    50n
            r1c4    25n
    END
```

```
!buffer
!buffer
```

```
DEFINE LOOP sample4
    LPARAM %p4 %conc
            r1d9    0n
            r1d10   0n
            r1d1    200n
            r1d6    5n
            r1d2    100n
            r1d5    10n
            r1d3    50n
            r1d4    25n
    END
```

```
!buffer
!buffer
```

Data Analysis

After the run is completed, open the BIAevaluation software and import the cycles from the report file. For the run described here, the file should consist of 64 cycles. For a preliminary evaluation of the data, bring all the cycles (sensorgrams) into one plot and Y-transform the curves just before the start of the injection of the anti-TSH. This should show the results of the injections of the different concentrations of TSH over the three different surface densities of anti-TSH and the control surface (Fig. 6).

To perform a more detailed evaluation, select the cycles corresponding to each of the surface densities of anti-TSH into separate plots; there should be four plots corresponding to the three different levels of anti-TSH captured by the RaM Fc surface and the control surface. Working with one of these plots at a time, evaluate the completeness of the regeneration injection before removing it and any air spikes that may be present. It may also be helpful to remove the region corresponding to the binding of the anti-TSH, again after checking the reproducibility of the anti-TSH capturing step in each cycle. The sensorgrams (curves) should then be Y-transformed before the TSH injection start. Finally, the sensorgrams should be X-transformed so that each of the injections starts at the same time. When examining these plots, it is apparent that the level of binding of the TSH to the antibody surface is lower than anticipated from the stoichiometry and ratio of molecular weights of the molecules. For example, for the plot corresponding to the TSH concentration series injected over 340 RUs of captured anti-TSH (Fig. 7), the maximum response is around 140 RUs, although 127 RUs is expected (28,000 x 340 x 2/1 = 127). A comparison of the experimental and calculated maximum response indicates that 90% of the anti-TSH antibody is active.

Once these procedures are completed, the data can be reference-corrected by subtracting the sensorgram resulting from the injection of each concentration of TSH over the RaM Fc surface from the sensorgram resulting from the injection of the same concentration of TSH over the anti-TSH surfaces. This subtraction is valid if the TSH does not bind to RaM Fc control surface. Care must be exercised when using the in-line subtraction feature of Biacore 2000 and Biacore 3000, and the response of the individual flow cells must be examined before the subtracted data are analyzed further. For each of the three plots representing the anti-TSH interacting with TSH, a fur-

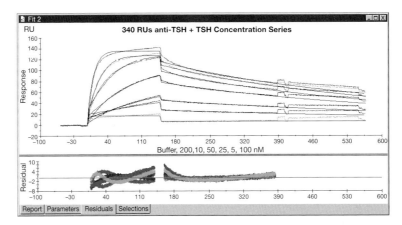

FIGURE 7. Langmuir binding model. This plot shows the results of fitting the sensorgrams resulting from the raw data to a 1:1 (Langmuir) binding model with a correction for mass transport and refractive index differences at both the start and end of the injection. The data in the analysis only included the first 380 seconds because of apparent machine effects resulting in higher RUs for the latter part of the dissociation part of the sensorgrams. Differences between the experimental and fitted data are apparent both in the curves and in the residual plot. Both show systemative deviations in the association and dissociation phases. (Reprinted, with permission, from Sambrook and Russell 2001, *Molecular Cloning.*)

ther correction can be made by subtracting the buffer injection sensorgram from those resulting from the TSH concentration series (Fig. 8). This subtraction will remove any changes that are common in all the curves, such as drift or a machine effect.

The resulting plots are now ready to fit to the kinetic model that best describes the reaction kinetics for the determination of the association and dissociation rate constants. The BIAevaluation software comes with a number of models and allows the user to make modifications to these or to add others. In the absence of addition SPR experiments or data from other biophysical techniques, the 1:1 (Langmuir) binding model should be used initially when performing the analysis of kinetic experiments. In the case of TSH as the analyte binding to the anti-

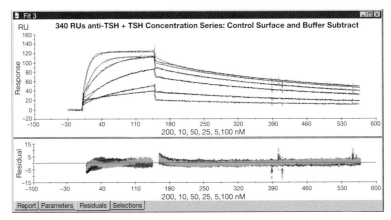

FIGURE 8. Corrected Langmuir model. The plot shows the results of fitting the double-reference subtracted data to the same model as used in Fig. 7 (1:1 [Langmuir] binding with a correction for mass transport and refractive index differences at both the start and end of the injection). The corrected data show better fitting to the curves and in the residual plot. (Reprinted, with permission, from Sambrook and Russell 2001, *Molecular Cloning.*)

TSH ligand captured by the RaM Fc surface, although each antibody is bivalent, the binding of each of the two TSH molecules is independent of the other. Therefore, the appropriate models to use are the 1:1 (Langmuir) binding models (A + B ↔ AB).

The model selected to illustrate the fitting was Langmuir binding with a correction for mass transport and refractive index differences at both the start and end of the injection. This model was used to fit data from both uncorrected (raw) data (Fig. 7) and double-referenced data (Fig. 8) using global settings for k_a and k_d. The only difference in the analysis of the raw and corrected data was that for the raw data the value of k_t was kept constant at 1×10^8 RU $M^{-1}s^{-1}$ (the default value); this ensured a reasonable value for this parameter. The actual value for k_t depends on the molecular mass and the flow rate; for TSH at a flow of 10 μl/min, the value is 5.4×10^8 RU $M^{-1}s^{-1}$. If k_t is not kept constant, the value obtained is greater than the maximum possible rate of diffusion. The model fits the corrected data more closely than it does the raw data, as the lines representing the fit data more closely follow those from the actual data (Fig. 7) than those of the raw data (Fig. 8). The residual plot shows the difference between the experimental and fitted data for each curve, and is another means of assessing the goodness of the fit. The residual plot for the corrected data (Fig. 8) shows less deviation than that of the raw data (Fig. 7), confirming the appropriateness of the double referencing.

For a test of the global fitting of the data, the curves resulting from the injections of 5, 50, and 100 nM TSH over the three anti-TSH surfaces were combined for analysis. The model fits the combined corrected sensorgrams similarly to the fit seen for the individual corrected surfaces.

In Table 1, the values obtained for the kinetic rate constants for all three levels of anti-TSH are compared for the data before and after the corrections were performed, as well as those obtained from the combined data. For the 340 RU surface, the values obtained for the constants for both sets of data are within 30% of each other, indicating that the model fits both sets of data similarly. When the values for the combined data are examined, the constants for the raw and corrected data are within 20% of each other. The high degree of consistency between the rate constants for the combined and individual anti-TSH surfaces reflects the appropriateness of the selected model.

TABLE 1. Kinetic Rate Constants for All Three Levels of Anti-TSH

Anti-TSH surface	Source	R_{max}	k_a $M^{-1}s^{-1}$	k_d s^{-1} RU $M^{-1}s^{-1}$	k_t s^{-1}	K_a M^{-1}	K_d M	χ^2
160 RUs	raw data*	62.8	4.42e^5	1.81e^{-3}	1.00e^8	2.44e^8	4.10e^{-9}	1.18
	corrected data	43.9	7.01e^5	3.80e^{-3}	3.37e^7	1.85e^8	5.41e^{-9}	1.08
340 RUs	raw data*	111	5.02e^5	2.25e^{-3}	1.00e^8	2.23e^8	4.48e^{-9}	2.71
	corrected data	89.9	6.23e^5	3.19e^{-3}	6.51e^7	1.95e^8	5.12e^{-9}	1.74
550 RUs	raw data*	166	4.90e^5	2.58e^{-3}	1.00e^8	1.90e^8	5.26e^{-9}	3.35
	corrected data	142	5.73e^5	3.06e^{-3}	8.08e^7	1.87e^8	5.34e^{-9}	2.28
Combined	raw data*		5.35e^5	2.34e^{-3}	1.00e^8	2.29e^8	4.37e^{-9}	2.06
	corrected data		5.72e^5	2.82e^{-3}	8.34e^7	2.03e^8	4.93e^{-9}	1.45

For fitting the raw data, the material transport value (k_t) was held constant at 1e^8.

Shown are the values for the kinetic constants and χ^2 for both the raw and corrected data for all three surface levels of anti-TSH as well as the combined values representing a global fit. For the raw data fits, the k_t was held constant at 1e^8, which provides a good estimate and eliminates numerical difficulties based on the sum-squared algorithm. The high degree of consistency between the rate constants for the combined and individual anti-TSH surfaces reflects the appropriateness of the selected model. Despite the improvement seen in the fit and in the residuals when the data was double-reference subtracted, the kinetic constants are quite close, again indicating that the selected model is appropriate. The χ^2 values are on the same order of magnitude as instrument noise (typically <2 RU), which is consistent with a good fit.

(Reprinted, with permission, from Sambrook and Russell 2001, *Molecular Cloning.*)

REFERENCES

Altschuh D., Dubs M.C., Weiss E., Zeder-Lutz G., and van Regenmortel M.H. 1992. Determination of kinetic constants for the interaction between a monoclonal antibody and peptides using surface plasmon resonance. *Biochemistry* **31:** 6298–6304.

Andersson K., Hamalainen M., and Malmqvist M. 1999. Identification and optimization of regeneration conditions for affinity-based biosensor assays. A multivariate cocktail approach. *Anal. Chem.* **71:** 2475–2481.

Celia H., Wilson-Kubalek E., Milligan R.A., and Teyton L. 1999. Structure and function of a membrane-bound murine MHC class I molecule. *Proc. Natl. Acad. Sci.* **96:** 5634–5639.

Christensen L.L. 1997. Theoretical analysis of protein concentration determination using biosensor technology under conditions of partial mass transport limitation. *Anal. Biochem.* **24:** 153–164.

Fisher R.J. and Fivash M. 1994. Surface plasmon resonance based methods for measuring the kinetics and binding affinities of biomolecular interactions. *Curr. Opin. Biotechnol.* **5:** 389–395.

Fisher R.J., Fivash M., Casas-Finet J., Erickson J.W., Bladen S.V., Fisher C., Watson D.K., and Papas T. 1994. Real-time DNA binding measurements of the ETS1 recombinant oncoproteins reveal significant kinetic differences between the p42 and p51 isoforms. *Protein Sci.* **3:** 257–266.

Jonsson U., Fagerstam L., Ivarsson B., Johnsson B., Karlsson R., Lundh K., Lofas S., Persson B., Roos H., Ronnberg I., et al. 1991. Real-time biospecific interaction analysis using surface plasmon resonance and a sensor chip technology. *Biotechniques* **11:** 620–627.

Karlsson R. and Falt A. 1997. Experimental design for kinetic analysis of protein-protein interactions with surface plasmon resonance biosensors. *J. Immunol. Methods* **200:** 121–133.

Karlsson R., Michaelsson A., and Mattsson L. 1991. Kinetic analysis of monoclonal antibody-antigen interactions with a new biosensor based analytical system. *J. Immunol. Methods* **145:** 229–240.

Karlsson R., Fagerstam L., Nilshans H., and Persson B. 1993. Analysis of active antibody concentration. Separation of affinity and concentration parameters. *J. Immunol. Methods* **166:** 75–84.

Ladbury J.E., Lemmon M.A., Zhou M., Green J., Botfielld M.C., and Schlessinger J. 1995. Measurement of the binding of tyrosyl phosphopeptides to SH2 domains: A reappraisal. *Proc. Natl. Acad. Sci.* **92:** 3199–3203.

Morgan H. and Taylor D.M. 1992. A surface plasmon resonance immunosensor based on the streptavidin-biotin complex. *Biosens. Bioelectron* **7:** 405–410.

Muller K.M., Arndt K.M., and Plückthun A. 1998. Model and simulation of multivalent binding to fixed ligands. *Anal. Biochem.* **261:** 149–158.

Myszka D.G. 1999. Improving biosensor analysis. *J. Mol. Recog.* **12:** 1–6.

Myszka D.G., Jonsen M.D., and Graves B.J. 1998. Equilibrium analysis of high affinity interactions using BIACORE. *Anal. Biochem.* **265:** 326–330.

Myszka D.G., He X., Dembo M., Morton T.A., and Goldstein B. 1998. Extending the range of rate constants available from BIACORE: Interpreting mass transport-influenced binding data. *Biophys. J.* **75:** 583–594.

Nieba L., Krebber A., and Plückthun A. 1996. Competition BIAcore for measuring true affinities: Large differences from values determined from binding kinetics. *Anal. Biochem.* **234:** 155–165.

Nieba L., Nieba-Axmann S.E., Persson A., Hamalainen M., Edebratt F., Hansson A., Lidholm J., Magnusson K., Karlsson A.F., and Plückthun A. 1997. BIACORE analysis of histidine-tagged proteins using a chelating NTA sensor chip. *Anal. Biochem.* **252:** 217–228.

O'Shannessy D.J., Brigham-Burke M., and Peck K. 1992. Immobilization chemistries suitable for use in the BIAcore surface plasmon resonance detector. *Anal. Biochem.* **205:** 132–136.

O'Shannessy D.J., Brigham-Burke M., Soneson K.K., Hensley P., and Brooks I. 1993. Determination of rate and equilibrium binding constants for macromolecular interactions using surface plasmon resonance: Use of nonlinear least squares analysis methods. *Anal. Biochem.* **212:** 457–468.

Pellequer J.L. and Van Regenmortel M.H. 1993. Measurement of kinetic binding constants of viral antibodies using a new biosensor technology. *J. Immunol. Methods* **166:** 133–143.

Raut S., Weller L., and Barrowcliffe T.W. 1999. Phospholipid binding of factor VIII in different therapeutic concentrates. *Br. J. Haematol.* **107:** 323–329.

Roden L.D. and Myszka D.G. 1996. Global analysis of a macromolecular interaction measured on BIAcore. *Biochem. Biophys. Res. Commun.* **225:** 1073–1077.

Saenko E., Sarafanov A., Greco N., Shima M., Loster K., Schwinn H., and Losic D. 1999. Use of surface plasmon resonance for studies of protein-protein and protein-phospholipid membrane interactions. Application to the binding of factor VIII to von Willebrand factor and to phosphatidylserine-containing membranes. *J. Chromatogr. A.* **852:** 59–71.

Schuck P., Millar D.B., and Kortt A.A. 1998. Determination of binding constants by equilibrium titration with circulating sample in a surface plasmon resonance biosensor. *Anal. Biochem.* **265:** 79–91.

Sjolander S. and Urbaniczky C. 1991. Integrated fluid handling system for biomolecular interaction analysis. *Anal. Chem.* **63:** 2338–2345.

15 Analysis of Protein Interactions Using a Quartz Crystal Microbalance Biosensor

Sabine Hauck,[1] Stephan Drost,[4] Elke Prohaska,[2] Hans Wolf, [2] and Stefan Dübel[3]

[1]Fraunhofer-Institute of Microelectronic Circuits and Systems, 80801 München, Germany; [2]Institute of Medical Microbiology and Hygiene, University of Regensburg, 93053 Regensburg, Germany; [3]Institute of Molecular Genetics, University of Heidelberg, 69120 Heidelberg, Germany

INTRODUCTION

ELISA, calorimetry, and fluorescence spectroscopy are well-established techniques for the investigation of protein–protein interactions. In addition, biosensor-based approaches have gained increasing importance due to the advantages of easy handling and high-throughput capability. Biosensors consist of a biologically active component and a transducer integrated into a single system for reagent-free measurements (Hall 1990). The most prominent biosensor systems for the study of protein–protein interactions are based on surface plasmon resonance (SPR) (Chapter 14). Recently, the quartz crystal microbalance (QCM) has been established as an alternative to SPR. A comparison of the two methods is given by Kösslinger et al. (1995, 1998).

The QCM is an acoustic sensor based on a piezoelectric crystal. Mass changes can be detected on the sensor surface in the nanogram range. The relation is expressed in the Sauerbrey formula (Eq. 1) (Sauerbrey 1959):

$$\Delta F = S \cdot \Delta m = \frac{-2F^2}{Zp} \cdot \Delta m \tag{1}$$

where ΔF, frequency shift [Hz]; Δm, change of mass per area [ng cm^{-2}]; S, mass sensitivity [Hz ng^{-1} cm^{-2}]; F, resonant frequency [Hz]; Zp, acoustical impedance [Hz ng cm^{-2}].

[4]Contact for instrument purchase requests.

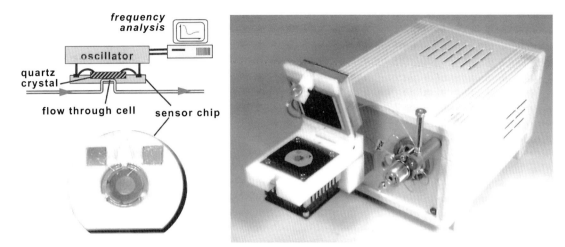

FIGURE 1. The QCM biosensor. (*A*) Scheme of the measurement setup. (*B*) The AFFCo2000 QCM system. The samples are transported to the oscillator-driven QCM sensor chip mounted in a flowthrough cell by a flow injection analysis (FIA) system. The frequency signal is detected by a frequency counter and processed electronically in a separate computer.

The sensor can be used for the direct, marker-free measurement of specific interactions between immobilized molecules and analytes in solution. Binding of a soluble analyte to the immobilized ligand causes a shift in the resonance frequency, and this signal can be recorded using a frequency counter with high resolution. This method, despite its existence for four decades, has only recently been developed for immunological measurements in a flowthrough system (Kösslinger et al. 1992), as shown in Figures 1 and 2. The resulting frequency versus time curve is called a "sensorgram." As real-time measurements are performed, the sensorgrams are capable of deducing not only the equilibrium binding constants (Sladal et al. 1994), but also the affinity rate constants. This methodology was recently applied to describe the binding kinetics of various phage-presented proteins (Hengerer et al. 1999a; Decker et al. 2000), including recombinant antibodies (Hengerer et al. 1999b).

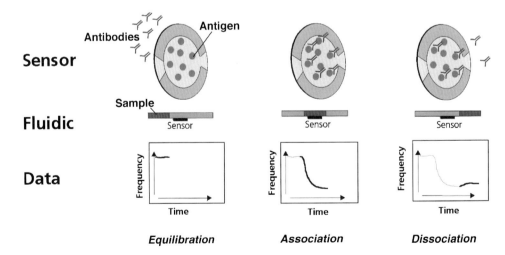

FIGURE 2. Scheme of the relation between action on sensor surface ("sensor"), status in FIA system ("fluidics"), and resulting frequency versus time curve ("data"), showing an antibody–antigen reaction as an example.

OUTLINE OF PROCEDURE

In QCM, the ligand is immobilized as a monolayer on a flat gold surface. Immobilization can be achieved using various methods. The easiest way is absorption of the ligand on the gold surface. This method, however, does not allow regeneration of the ligand layer (Uttenthaler et al. 1998). As the gold reacts with thiols, yielding a stable, semi-covalent bond, proteins can be immobilized by the thiol groups of their cysteine residues. Alternatively, the sensor surface can be activated by using a thiol-containing bifunctional linker. Reproducible results were obtained with dithiobis-succinimidyl-propionate (DSP) (Hermannson 1992). DSP forms disulfide bonds to the gold surface and provides N-hydroxysuccinimide (NHS) groups that can react with the free amino groups on the ligand. If streptavidin is immobilized using DSP, biotinylated ligands can be conveniently coated. The availability of in vivo (Weiss et al. 1994) or in vitro systems (Saviranta et al. 1998) for sequence-specific biotinylation should allow the generation of highly ordered ligand monolayers.

Samples are applied to the immobilized ligand on the sensor surface by a continuous constant flow. A constant analyte concentration at any part of the flowthrough cell is thus provided, and diffusion effects can be neglected. Moreover, an optimized cell design helps to avoid mass transport effects. Viscosity effects, however, can lead to frequency shifts interfering with the specific signal. Therefore, the biosensor should be calibrated with a viscous solution to determine the time in which the viscosity change generates a signal. This time interval should be excluded from the sensorgram. A sample sensorgram is given in Figure 3.

After a measurement, the sensor has to be regenerated to remove bound analyte from the sensor surface. For this regeneration step, elution buffers as used in affinity chromatography can be applied as long as they do not destroy the native structure of the ligand. A good overview regarding ligand stability is given in Harlow and Lane (1988).

It should be noted that the QCM method does not allow the measurement of true affinities. Because the ligand is immobilized, one degree of freedom is lost. Therefore, the measured affinities could be influenced by decreased mobility and sterical hindrance. Furthermore, avidity effects

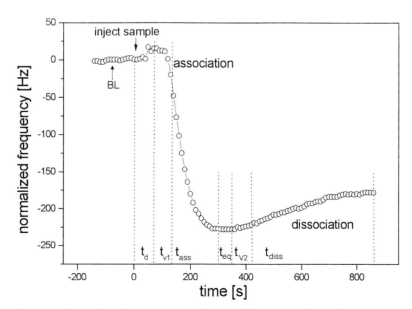

FIGURE 3. Evaluation scheme for the sensorgram. The sample is injected into the system at t_0. After the delay time t_d, the binding analyte generates a signal, which is interfered with by the viscosity effect during t_{v1}. The segment t_{ass} shows the association yielding a steady-state t_{eq}. When the sample buffer runs over the surface again, it induces a viscosity effect t_{v2} and the dissociation phase t_{diss}. (BL) Baseline.

resulting from multivalent binding might affect the apparent affinity. On the other hand, partial denaturation of the immobilized ligand may decrease the apparent binding affinity. In the case of phage-presented derivatives, as tested by Hengerer et al. (1999b), the affinity data obtained are only apparent, because the fraction of total phage carrying functional single-chain fragments of an antibody (scFv fragments) may not be 100%, and a fraction of surface-expressed scFv antibodies could be shed by proteolysis (Hengerer et al. 1999b). In some phage-display systems, soluble scFv fragments may be present in the phage supernatants due to the existence of an amber codon between scFv and the virus coat protein. Finally, more than one fragment could be presented per phage. Thus, for the determination of true affinities, an accurate determination of true molarities of scFv fragments would be required. Nevertheless, this method allows the ranking of a set of antibody clones from a phage-screening experiment according to their apparent affinities.

Quartz Crystal Microbalance Analysis

A QCM analysis in this protocol is broken down into the following steps: sensor cleaning, immobilization of the ligand on the sensor, sample preparation, system calibration, measurement procedure, optimization of regeneration buffer, and evaluation of sensorgram.

Sensor cleaning and immobilization of ligand are performed before the sensor chip is mounted onto the AFFCo 2000 device. Sensor cleaning should be performed directly before immobilization. The sensor chip with the immobilized ligand can be stored for several days at 4°C in a wet chamber (storage conditions depend on the immobilized protein).

MATERIALS

CAUTION: See Appendix for appropriate handling of materials marked with <!>.

Buffers and Solutions

Acetone <!>

Activation solution

0.4 mg/ml of dithiobissuccinimidyl-propionate (DSP) dissolved in water-free dimethylformamide (DMF)<!>

This solution must be prepared directly before use.

Blocking reagent

0.1 mg/ml bovine serum albumin (BSA) in PBS

Can be stored in aliquots of 0.5 ml for ~1 month at –20°C.

Calibration solution

2% (w/v) glucose

1 M hydrochloric acid (HCl) <!>

1 M sodium hydroxide (NaOH) <!>

PBS buffer

2.6 mM potassium chloride (KCl) (200 mg/liter if anhydrous salt is used)

138 mM sodium chloride (NaCl) (8 g/liter if anhydrous salt is used)

10 mM sodium hydrogen phosphate (Na_2HPO_4) (1.44 g/liter if anhydrous salt is used)

1.8 mM potassium hydrogen phosphate (KH_2PO_4) (240 mg/liter if anhydrous salt is used)

Adjust to pH 7.4. Can be stored for ~4 weeks at room temperature.

Piranha solution

3 parts (3 ml) of 98% sulfuric acid (H_2SO_4) <!>

1 part (1 ml) of 30% hydrogen peroxide (H_2O_2) <!>

This solution should be prepared directly before use. The solution heats up quickly upon mixing; therefore, wait ~2 minutes for it to cool down.

Regeneration buffer (glycine buffer)

100 mM glycine (0.7507 mg/100 ml if anhydrous salt is used)

100 mM NaCl (0.5855 mg/100 ml if anhydrous salt is used)

Adjust pH to 2.5 or 11.5 with 100 mM HCl <!> or 100 mM NaOH <!>. Can be stored in aliquots of 1 ml for ~1 year at –20°C.

Regeneration buffer (borate buffer)

borate buffer solution (pH 11) (Merck, Darmstadt, Germany)

Store at 4°C. Check pH before use. No stability data are available. Adjust pH to 11.5 or up to 12.5 with NaOH <!>

Regeneration buffer (sodium isothiocyanate)

4 M sodium isothiocyanate (NaSCN)<!>

Dissolve in distilled H_2O. Adjust pH to 11.0 with NaOH <!>, if necessary. Store at room temperature, and check pH before use. No stability data are available.

Streptavidin (1 mg/ml) in PBS

Can be stored in aliquots of 25 ml for ~1 year at –20°C.

Ligands

Ligands with free amino groups (for immobilization procedure 1) or biotinylated ligands (for immobilization procedure 2).

Negative Control

Nonbinding protein, nonspecific antibody, helper phage, or unspecific phage-displayed protein.

The control should be a molecule very similar to the sample, but showing no binding properties to the immobilized ligand. Use the control in a concentration similar to that of the highest sample concentration.

Biosensor Equipment

AFFCo 2000 Quartz Crystal Microbalance Device, developed at the Fraunhofer-Institute of Microelectronic Circuits and Systems, Munich, Germany (FhG-IMS)

A scheme of the device is depicted in Figure 1. The system consists of a mount for the exchangeable sensor chips (see below), an injection valve with, e.g., a 0.1-ml loop (range 0.050–0.250 ml) for the sample application, and a pump to maintain a continuous buffer flow of ~0.03 ml/minute (range 0.020–0.050 ml/minute). The sensors, produced at the FhG-IMS, are round-shaped 20-MHz quartz crystals with gold electrodes and are mounted on a flowthrough cell. The quick exchange mount for the chip provides temperature control of the buffer and sample solution. The measurement time depends on sample loop size and buffer flow rate. Optimal conditions should be determined by the user.

Additional Equipment

Gaseous N_2 or clean low-pressure air
Hamilton syringe (0.250 ml) for sample application
Wet chamber and dry chamber (blue silica gel)

METHOD

Sensor Cleaning

Procedure 1

1. Rinse the sensor with ~0.2 ml of acetone.
2. Dry the sensor carefully with N_2.
   If the gas pressure is too strong, the sensor can be pressed out of the chip.

Alternative: Procedure 2

1. Rinse the sensor with ~0.2 ml of piranha solution.
2. Rinse the sensor with ~0.2 ml of H_2O.
3. Rinse sensor with ~0.2 ml of ethanol.
4. Dry the sensor carefully with N_2.
   If the gas pressure is too strong, the sensor can be pressed out of the chip.

Alternative: Procedure 3

1. Rinse the sensor with 1 M NaOH (20 minutes of flowthrough, flow rate of 0.05 ml/minute).
2. Rinse the sensor with 1 M HCl (20 minutes, flow rate of 0.05 ml/minute).
3. Rinse the sensor with ~0.2 ml of ethanol.
4. Dry the sensor carefully with N_2.
   If the gas pressure is too strong, the sensor can be pressed out of the chip.
   Optimal results depend on properties of individual molecules. Flow rate is not critical except in steps where it is specifically mentioned.

Immobilization of the Ligand on the Sensor

Procedure 1: Ligands with Free Amino Groups

1. Rinse the sensor with ~0.2 ml of activation solution.
2. Activate the sensor surface for 20 minutes with 0.01 ml of activation solution at room temperature in a dry chamber.
3. Rinse the sensor with ~0.2 ml of PBS.
4. Incubate the sensor with 0.01 ml of ligand solution (1 mg/ml) for 16 hours at 4°C in a wet chamber.

Procedure 2: Alternative

1. Rinse the sensor with ~0.2 ml of activation solution.
2. Activate the sensor surface for 20 minutes with 0.01 ml of activation solution at room temperature in a dry chamber.

3. Rinse the sensor with ~0.2 ml of PBS.

4. Incubate the sensor with 0.01 ml of streptavidin solution (1 mg/ml) for 16 hours at 4°C in a wet chamber.

5. Rinse the sensor with ~0.2 ml of PBS.

6. Incubate the sensor with 0.01 ml of biotinylated ligand solution (1 mg/ml) for at least 1 hour at room temperature in a wet chamber.

Sample Preparation

1. Dilute the sample (containing the analyte) in PBS. The resulting concentrations should be as follows:

Protein <100 kD	100–10.000 ng/ml
Antibody	10 to 1000 ng/ml
Phage	10^{11} to 10^{13} pfu/ml

2. Approximately five different concentrations should be prepared and tested. One sample, consisting only of diluent, should be prepared as a negative control. At least 0.5 ml of every concentration (corresponding to two sample volumes) is required.

   All sample dilutions should be done in flow buffer to avoid matrix effects.

System Calibration

1. Inject the calibration solution.

2. Evaluate the sensorgram according to the Evaluation of Sensorgram protocol (see p. 281) to determine the time of a measurement, the delay time t_d, and time of viscosity effects t_{v1} and t_{v2}.

Measurement Procedure

1. Rinse the sensor with ~0.2 ml of PBS.

2. Dry the sensor carefully with gaseous N_2.

3. Insert the sensor in the exchange mount of the AFFCo 2000 device.

4. Start the pump at a rate of 0.030 ml/minute.

5. Wait until the baseline drift is less than 2 Hz/10 minutes and the baseline noise is less than 5 Hz (BL).

6. Rinse and fill the loop with 0.25 ml of blocking reagent.

7. Inject the blocking reagent into the buffer stream. Note the time.

8. A starting "peak" should be seen after ~2 minutes.

9. Wait until the new stable baseline is reached.

10. If the resulting frequency shift is more than 30 Hz when compared to the baseline before blocking reagent injection, repeat blocking.

    This may indicate incomplete immobilization.

11. Wait until a new stable baseline is reached.

12. Rinse and fill loop with 0.25 ml of negative control.

13. Inject the negative control into the buffer stream. Note the time.

14. No decrease of the baseline should be seen after ~2 minutes.

15. Wait until the measurement time, as determined in system calibration, has passed.

16. Rinse and fill the loop with 0.25 ml of sample.

17. Inject the sample into the buffer stream. Note the time.

18. An initial drop should be seen after ~2 minutes.

19. With continuous flow, a decreasing frequency, which indicates the association phase, is followed by an increasing frequency, which indicates the dissociation of the analyte after the sample has completely passed through the cell.

20. Rinse and fill the loop with 0.25 ml of the appropriate regeneration buffer.

21. Inject the regeneration buffer into the buffer stream. Note the time.

22. An increase or decrease of the frequency should be seen (viscosity change) after ~2 minutes. After regeneration, the frequency should have adjusted to the same level observed before the sample measurement.

23. Wait until a new stable baseline is reached.

24. Inject the next sample and repeat the procedure from steps 16 to 23.

    Start measurements with a negative control sample and continue from low to higher concentrations.

Optimization of Regeneration

1. Note the frequency before the first sample injection.

2. Inject the regeneration buffer. Start with a mild buffer (for example, glycine at pH 2.5).

3. Note the frequency of the new stable baseline after regeneration. It should approximately equal the baseline before sample injection.

4. If the frequency is higher than before sample injection (more than 50 Hz), the ligand layer is partially lost. In this case, use less stringent conditions (more neutral pH or other regeneration buffer).

5. If the frequency is lower than before sample injection, bound analyte is not completely removed. In this case, use more stringent conditions (lower pH down to 2.0, try alkaline pH [9–11], or 1–3 M NaSCN).

6. Inject the first sample again. The frequency shift should be similar to the first measurement. If this is not the case, try another regeneration buffer. The amount of deviation may be up to 20 –30%; the deviation limit should be determined by the required accuracy of the individual experiment.

Evaluation of Sensorgram

1. Analyze sensorgram according to the following scheme. An example curve is given in Figure 3.

   a. BL, stable frequency before injection (drift less than 2 Hz/10 minutes).

   b. t_d, delay time. Time between injection and start of signal.

c. t_{v1}, viscosity effect. During this time, the viscosity effects overlay the association effect.

d. t_{ass}, association. Time during which the sample flows over the sensor and association can take place.

e. t_{eq}, steady state is reached.

f. t_{v2}, viscosity effect. Buffer reaches the sensor again. During this time, the viscosity effects overlay the dissociation effect.

g. t_{diss}, dissociation. Time during which buffer flows over the sensor and dissociation takes place. This curve segment should contain at least 50 data points.

2. Fit curve segment of t_{diss} according to Equation 2. This curve segment should consist of at least 50 data points.

$$F_{(t)} = F_0 \cdot e^{k_{diss}(t-t_0)}$$ (2)

Where $F_{(t)}$, Frequency at time t [Hz]; t_0, start of dissociation; F_0, frequency at t_0 [Hz]; k_{diss}, dissociation rate constant [s^{-1}].

3. F_0 is the frequency at the start of t_{diss}, t_0 is the time at the start of t_{diss}. Insert F_0 in Hz and t_0 in s. The dissociation rate constant k_{diss} [s^{-1}] is obtained.

4. Fit curve segment t_{ass} according to Equation 3. This curve segment should contain at least 50 data points.

$$F_{(t)} = \frac{k_{ass} \cdot C \cdot \Delta F_{max}}{k_{ass} \cdot C + k_{diss}} \cdot \left(e^{-(k_{ass} \cdot C + k_{diss}) \cdot (t-t_0)} - 1 \right) + F_0$$ (3)

where k_{ass}, association rate constant [M^{-1}s^{-1}]; C, concentration of analyte [M]; ΔF_{max}, frequency obtained for a complete analyte monolayer [Hz].

5. F_{max} is the frequency shift for a completely occupied surface, C is the concentration of the analyte, t_0 is time at start of t_{ass}, k_{diss} is determined from step 3. Insert assumed or experimentally determined values for F_{max} in Hz, C in mol/liter, t_0 in s, and k_{diss} in s^{-1}. The association rate constant k_{ass} is obtained.

6. Calculate the dissociation constant K_D using Equation 4.

$$K_D = \frac{k_{diss}}{k_{ass}}$$ (4)

where K, dissociation constant [M^{-1}].

7. Repeat for each sample concentration.

REFERENCES

Decker J., Weinberger K., Prohaska E., Hauck S., Kösslinger C., Wolf H., and Hengerer A. 2000. Characterization of a human pancreatic secretory trypsin inhibitor mutant binding to *Legionella pneumophila* as determined by a quartz crystal microbalance. *J. Immunol. Methods* **233**: 159–165.

Hall E. 1990. *Biosensoren*. Springer Verlag, Berlin. p. 4,

Harlow E. and Lane D. 1988: *Antibodies: A laboratory manual*, p. 549. Cold Spring Harbor Laboratory, Cold Spring Harbor, New York.

Hengerer A., Decker J., Prohaska E., Hauck S., Kösslinger C., and Wolf H. 1999a: Quartz crystal microbalance (QCM) as a device for the screening of phage libraries. *Biosens. Bioelectron.* **14**: 139–144.

Hengerer A., Kösslinger C., Decker J., Hauck S., Queitsch I., Wolf H., and Dübel S. 1999b. Determination of phage antibody affinities to antigen by a microbalance sensor system. *Biotechniques* **26**: 956–960.

Hermannson, G.T. 1992: *Immobilized affinity ligand techniques*. Academic Press, San Diego.

Kösslinger C., Uttenthaler E., Abel T., Hauck S., and Drost S. 1998. Comparison of the determination of affinity constants with surface plasmon resonance and quartz crystal microbalance. *Proceedings of Eurosensors XII*, p. 845. IOP Publishing.

Kösslinger C., Drost S., Aberl F., Wolf H., Koch S., and Woias P. 1992. A quartz crystal biosensor for measurement in liquids. *Biosens. Bioloectron.* **7**: 397–404.

Kösslinger C., Uttenthaler E., Drost S., Aberl F., Wolf H., Brink G., Stanglmaier A., and Sackmann E. 1995. Comparison of the QCM and the SPR method for surface studies and immunological applications. *Sensors and Actuators B* 24–25, 107–112.

Sauerbrey G. 1959. Verwendung von Schwnigquarzen zur Wägung dünner Schichten und zur Mikrowägung. *Z. Phys.* **155**: 206.

Saviranta P., Haavisto T., Rappu P., Karp M., and Lovgren T. 1998. In vitro enzymatic biotinylation of recombinant fab fragments through a peptide acceptor tail. *Bioconjug. Chem.* **9**:725–735.

Skladal P., Minunni M., Mascini M., Kolar V., and Franek M. 1994. Characterization of monoclonal antibodies to 2,4-dichlorophenoxyacetic acid using a piezoelectric quartz crystal microbalance in solution. *J. Immunol. Methods* **176**: 117–125.

Uttenthaler E., Kösslinger C., and Drost S. 1998. Characterization of immobilization methods for African swine fever virus protein and antibodies with a piezoelectric immunosensor. *Biosens. Bioelectron.* **13**: 1279–1286.

Weiss E., Chatellier J., and Orfanoudakis G. 1994. *In vivo* biotinylated recombinant antibodies: construction, characterization, and application of a bifunctional Fab-BCCP fusion protein produced in *Escherichia coli*. *Prot. Expr. Purif.* **5**: 509–517.

16 Protease Footprinting

Rod Hori[1] and Noel Baichoo[2]

[1]*Department of Molecular Sciences, College of Medicine, University of Tennessee, Memphis, Tennessee 38163;* [2]*Department of Microbiology, Cornell University, Ithaca, New York 14853*

INTRODUCTION

In this chapter, we focus on the use of chemical and enzymatic proteases to determine the regions of a protein that become buried or exposed upon assembly into a macromolecular complex in vitro (for review, see Hori and Carey 1998; Loizos and Darst 1998; Datwyler and Meares 2000; Heyduk et al. 2001). The primary method described is protease (or protein) footprinting. Protease footprinting is analogous to the commonly used DNase I footprinting technique except that, rather than using DNase I cleavage of end-labeled DNA to identify where proteins bind, pro-

285

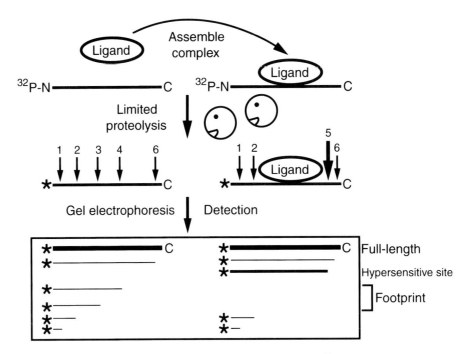

FIGURE 1. Principle of protease footprinting. A protein end-labeled with ^{32}P on its amino terminus (N) is assembled into complexes with its ligand. In parallel reactions, the free and complexed proteins are then immediately subjected to limited proteolysis. Cleavage sites are represented by arrows and are numbered. At this point, the end-labeled amino terminus is represented by an asterisk. The cleavage products are separated on a denaturing protein gel. Upon detection, a protected region (sites 3 and 4 are occluded) and a hypersensitive site (site 5 is cleaved only upon binding of the ligand) are identified.

teases are employed to identify the sites of contact and/or induced conformational changes made by an end-labeled protein upon complex formation (Fig. 1). One major advance in the development of protease footprinting has been the ability to generate an end label easily by engineering the recognition site for protein kinase A (PKA) at either the amino or carboxyl terminus of a protein of interest, allowing isotopic labeling with [γ-^{32}P]ATP or [γ-^{33}P]ATP (Li et al. 1989; Heyduk et al. 1996). The end-labeled protein, both in its free state and within a macromolecular complex, can then be subjected to partial cleavage, generating a nested set of digestion products that initiate at the end label and terminate at the point of scission (see Fig. 1 in Chapter 11). Both enzymatic and chemical methods have been used to perform the proteolytic reaction and are described elsewhere in this chapter. The cleavage products are then resolved on a denaturing gel adjacent to the appropriate markers and detected using the end label. When the digestion pattern of the complexed protein is compared to that of the free protein, some sites show decreased proteolysis (protections), whereas others show increased proteolysis (hypersensitivity).

The protections are the result of either direct occlusion in the complex or a conformational change that inhibits the ability of the protease to access those regions of the protein. Hypersensitive sites are often caused by a conformational change that allows more efficient cleavage at those residues. The positions of the cleavage sites are assigned by creating "calibration ladders" (described in Identification of Cleavage Sites, p. 296) that serve as mobility standards to interpret amino acid position from migration. A relatively precise assignment of the affected cleavage sites (usually within 5 residues) can usually be made using these ladders. Proteolysis has historically been used for examining protein structure and mapping domains. By introducing an end label and the principles behind DNase footprinting, protease footprinting increases the resolution of the mapping and the amount of information that can be obtained using proteolysis, thus making it an attractive alternative (or supplement) to other methods.

Advantages and Limitations

Although there are many techniques for identifying interactions between a protein and another molecule, only a few can be used to specify the precise regions involved. Mutational analysis, chemical cross-linking, and structure determination by X-ray crystallography or nuclear magnetic resonance (NMR) are the most common methods for examining the structure/function relationship of a protein. Protease footprinting provides a direct biochemical method readily amenable to most laboratories for identifying the domains involved in complex assembly. Protease footprinting is an in vitro experiment executed using native proteins in solution and can be used to simultaneously examine multiple regions of a protein. In comparison, in vivo genetic analysis is often only inferential and requires the time-consuming construction of numerous mutations to study multiple domains. Structural determination under aqueous conditions of a single-chain protein by NMR has been limited to proteins that are 30 kD or less (Clore and Gronenborn 1998). Structural determination of proteins by X-ray crystallography requires non-physiological conditions to nucleate crystal formation and often requires complementary experiments such as a genetic analysis or protease footprinting for verification of their significance.

The limitations or difficulties of other protein-mapping techniques underscore the advantages of protease footprinting.

1. The only protein engineering required is the addition of the phosphorylation site to generate the end label. The termini of most proteins are exposed, thus an end label can usually be tolerated at those positions without disrupting the structure. If direct methods of end labeling perturb the structure, visualization by an indirect end-labeling procedure such as immunoblots using antibodies specific to one terminus can be used. In contrast, mutagenesis studies require amino acid changes throughout the coding region, which have an increased probability of causing structural alterations.

2. Protease footprinting can potentially identify sequential conformational changes that occur during the assembly of a complex. X-ray crystal structures can also be used to define these structural changes, but this requires solving and comparing the free (unbound) protein and each of the higher-order species. Mutational studies are usually not as straightforward in defining allosteric transitions, especially in the absence of structural data.

3. Protease footprinting experiments require only a small fraction of the amount of protein required for NMR or X-ray crystallography, and the necessary reagents and equipment are accessible to most laboratories.

4. The size of the protein makes little difference in executing a protease footprinting experiment. In contrast, large proteins are typically more difficult to characterize using NMR and X-ray crystallography because more data points are required to obtain the same resolution relative to a smaller protein. For protease footprinting experiments, additional footprinting reactions are typically not required for a larger protein, although multiple gels of varying polyacrylamide concentrations may be necessary to examine the entire length of the protein.

Protease footprinting has several limitations.

1. The affinity of the end-labeled molecule for its ligand(s) must be strong enough to assemble the complex quantitatively. Alternatively, there must be a method for purifying the bound protein from the free protein. One potential issue with purifying the bound complex is the presence of misfolded proteins among the free protein. If the initial solution of free protein contains a significant percentage of misfolded protein, it is possible to obtain a false footprint that reflects differences in the conformation between the active and inactive protein.

2. The protein should not contain any endogenous phosphorylation sites that preclude the introduction of a specific end label. However, the presence of an endogenous PKA consensus

sequence does not necessarily mean it will be accessible to phosphorylation in vitro; it is best to check this before proceeding to remove the site by mutagenesis.

3. The protein should be purified homogeneously as full length. Degradation products (containing the phosphorylation site) will also be labeled and lead to background fragments that obscure the signal. Methods for eliminating this problem are described in the End-labeling section, p. 290.

Although there are some limitations, protease footprinting is a relatively simple method of directly mapping regions of proteins involved in interactions with other molecules. Table 1 includes a representative list of studies that have used protease footprinting and classifies them by the method of proteolysis employed and type of complex(es) characterized. A list of the studies that have comparisons to known three-dimensional structures and the types of detection methods used is summarized at the bottom of Table 1. This table helps to illustrate the flexibility this procedure has to examine nearly any type of macromolecular assembly.

Types of Proteolysis

The two most common reagents used for digestion of proteins are small chemicals (primarily metal–chelate complexes that produce hydroxyl radicals) and enzymes (e.g., proteinase K). Both methods have been used extensively (see Table 1). Chemicals such as cyanogen bromide and BNPS (3-bromo-3-methyl-2-(2 nitrophenylmercapto)-3H-indole skatole) cleave proteins at specific residues (methionine and tryptophan, respectively) and have been used for decades to map protein structures and interactions successfully. However, most of these reagents require extreme

TABLE 1. Examples of Complexes Characterized by Protease Footprinting

Type of proteolysis used	
enzymatic	chemical
Binary complexes:	
Protein: protein	
1. σ^{70}: phage T4 AsiA	4. α, β, and β': other subunits of RNA polymerase
2. CRM1 (Exportin 1): Rev	5. gp55 (gene 55 protein: core RNA polymerase
3. Transcription factor IIB: VP16 or TAF40	6. ADP ribosylation factor 1: Sec7 domain
Protein: nucleic acid	
7. TFIIIA: DNA or RNA	10. HMGI/C: DNA
8. cAMP receptor protein/cAMP:DNA	11. HMGI/Y: DNA
9. Rev: RNA	12. cAMP receptor protein/cAMP: DNA
Protein: small ligand	
13. DNA topoisomerase II: ATP	16. BirA: biotin
14. Myoglobin: heme	17. cAMP-dependent protein kinase: Mg-ATP
15. Iron regulatory protein-1: Fe-S cluster	18. I-*Por*I and I-*Dmo*I: Fe^{++}
Ternary complexes and higher	
19. TFIIA: TBP: DNA	22. B'': TATA-box binding protein: B-related factor: DNA
20. gp61 (gene 61 protein): gp41: DNA	23. GreB: RNA polymerase: DNA: RNA
21. σ^{70}: core RNA polymerase: DNA	24. α: β, β', and σ (sequential assembly)

The first protein listed for each entry is the one that was directly or indirectly end-labeled.
(5, 6, 12, 14, 15, 16, 17, 19, 20, 21, 23) Results were mapped on a known or homologous three-dimensional structure.
(4, 7, 12, 17) Used an antibody against an endogenous terminus.
(13, 15) Used an antibody against an engineered epitope tag.
(8) His$_6$-CRP was bound to Ni-NTA affinity after digestion to purify the nested set initiating at the His$_6$ tag.
(21) Used biotinylated nitroloacetic acid to detect a hexahistidine tag as an end label.
(All others) Used either ^{32}P or ^{33}P.

conditions that make it difficult to use them with native proteins and generate sparse data because they cleave infrequently due to their specificity. In contrast, metal–chelate complexes, such as iron–EDTA and phenanthroline–copper (Baichoo and Heyduk 1999), generate a more comprehensive pattern of cleavage. Initially, these reagents were instrumental in producing DNA footprints with a uniform laddering pattern. Similar benefits are provided to those performing protease footprinting. Under mild conditions, iron–EDTA generates hydroxyl radicals that cleave proteins and nucleic acids nonspecifically. The fact that these radicals are comparable in size to a water molecule allows the entire solvent-exposed surface of a protein to be available for cleavage, resulting in more detailed and precise delineation of regions experiencing contact with other molecules, or conformational changes.

There are two major disadvantages to using Fe–EDTA chelates to digest proteins. First, their cleavage sites are not defined by sequence, so assigning cleavage positions requires detailed analysis of their relationship to the calibration markers. Second, extensive data analysis (described in Quantitative Analysis, p. 297) and averaging of several experiments is necessary to identify regions within proteins where only subtle changes in sensitivity to cleavage occur. A phosphor screen reader or exposure to film followed by densitometry or a fluorescence image reader, depending on the type of end label used, is necessary to collect the data required for this type of analysis.

Enzymatic proteases, on the other hand, are more familiar to most researchers and have several advantages. First, they are simpler to manipulate. Protease can easily be titrated over several orders of magnitude. Second, a relatively small number of digestion products are generated, usually making data simpler to interpret.

The main disadvantages of using enzymatic proteases are (1) the sequence specificity of the cleavage reaction and (2) their large size, which further limits their accessibility to potential cleavage sites. Thus, most proteins will have regions that contain only a few cleavage sites or none at all. Enzymatic proteases will be unable to characterize these regions fully.

Alternative Methods

Another way of employing chemical proteolysis to map macromolecular interactions is the use of tethered metal–chelate complexes. In this method, the tethered metal–chelate is covalently attached to either a cysteine or lysine residue on an unlabeled protein, which is allowed to form a macromolecular complex containing the end-labeled protein(s) of interest. Subsequently, cleavage is initiated by adding hydrogen peroxide and ascorbate to generate hydroxyl radicals (for review, see Datwyler and Meares 2000). If the cysteine or lysine residue is unique, this allows the specific cleavage and mapping of regions within other proteins that are close to that particular residue. Alternatively, the metal–chelate complex can be tethered to multiple lysine or cysteine residues to produce a more general characterization of interactions within the complex (Traviglia et al. 1999). When the cleavage products are resolved on a denaturing protein gel, residues within the end-labeled protein, juxtaposed within the complex near the site conjugated to the metal–chelate complex, are defined by identifying the positions of the cleavage sites (as described in Quantitative Analysis, p. 297).

One advantage of using a tethered protease is that it identifies only residues that are within its vicinity. Diffusable proteases do not allow the mapping of proximity to a specific residue. Another advantage is that results are easier to interpret than when diffusible metal–chelates are used, because relatively few fragments are produced. Additionally, this technique can be more sensitive because interactions generate the appearance of cleavage products against a null background rather than the decrease of a signal, and this is better for characterizing low-affinity associations. The disadvantages are the limitation of using only lysine or cysteine residues to attach the metal–chelate group, the identification of residues/regions within the distance of the linker arm used as a tether rather than the direct contact site, and the inability to define induced conformational changes.

Another protein footprinting method that can yield high-resolution data is based on chemical modification rather than direct cutting (Hanai and Wang 1994). In addition, this technique can be especially useful for examining proteins that denature upon even slight cleavage, but remain stable in response to the chemicals being used. In this method, free (unbound) protein and protein assembled into complexes are treated in parallel with a limiting amount of citraconic anhydride that reversibly modifies exposed lysine residues. Subsequently, the protein is denatured and all of the lysines that were not previously citraconylated are acetylated. Next, the citraconyl groups are removed by incubating the modified protein in acetic acid. The final product is recovered and the deblocked lysine residues are then digested to completion. Exposed lysine residues that become buried upon assembly of the complex exhibit a decreased rate of digestion (i.e., protection). The procedure has the advantages that a small chemical can be used that can access nearly all exposed regions, and the assembled complex is not subjected to proteolysis that could disrupt its integrity. The disadvantage is that only regions containing lysine residues can be characterized. An analogous method to examine domains containing cysteine residues has also been developed (Tu and Wang 1999) and can be used to gain additional information.

Protease footprinting is a method within the capabilities of the average molecular biology laboratory. A flowchart in Figure 2 outlines the basic steps that are common to the different versions of this technique. This method provides a powerful way to gain structural information about specific macromolecular interactions. With many genomes fully sequenced, the question remains—What is the function or role of these newly sequenced proteins? Direct structural studies of the macromolecular complexes that perform key enzymatic and regulatory functions in the cell are a critical adjunct to the genetic, computational, and other approaches of modern biology.

OUTLINE OF PROCEDURE

End Labeling

A specific end label is crucial to the success of protease footprinting, and protein kinase A provides a simple and reproducible method for achieving this goal (Pestka et al. 1999). Proteins containing a PKA recognition site (X-Arg-Arg-X-Ser-Y, where X and Y can be most amino acids, although arginine and hydrophobic residues, respectively, are optimal) engineered onto either the amino or carboxyl termini, respectively, have been used in protease footprinting experiments (Table 1). For greater detail regarding the recognition site, see Songyang et al. (1994). The most complete set of data is obtained when both amino- and carboxy-terminally labeled proteins are used for analysis. In the presence of either $[\gamma\text{-}^{32}P]ATP$ or $[\gamma\text{-}^{33}P]ATP$ and the catalytic subunit of PKA, the serine residue becomes phosphorylated and creates an end label. Both ^{32}P and ^{33}P have been used for end-labeling proteins in protease footprinting experiments.

Several properties need to be examined prior to using this methodology for end-labeling a protein. First, the introduced PKA recognition site should be unique, or the other site(s) should be buried and inaccessible. This is important because internal or multiple sites will obscure the nested set of fragments containing the phosphorylation at the amino or carboxyl terminus. The sequence of the wild-type protein should be examined for the presence of endogenous PKA recognition sequences. If any are present, the phosphorylation efficiency of that wild-type protein can be compared in a trial reaction against a protein known to be phosphorylated by PKA. If the protein of interest is phosphorylated efficiently, these sites can be eliminated by site-directed mutagenesis, assuming the mutations do not affect the structure or function of the protein. (Alternatively, the end label can be generated using casein kinase II and a terminus engineered to contain its recognition site rather than that of PKA [Schwanbeck et al. 2000].) If the protein lacks endogenous PKA recognition sites, one is engineered onto a terminus of the protein. In addition, the presence of cryptic phosphorylation sites should be checked by incubating equal amounts of

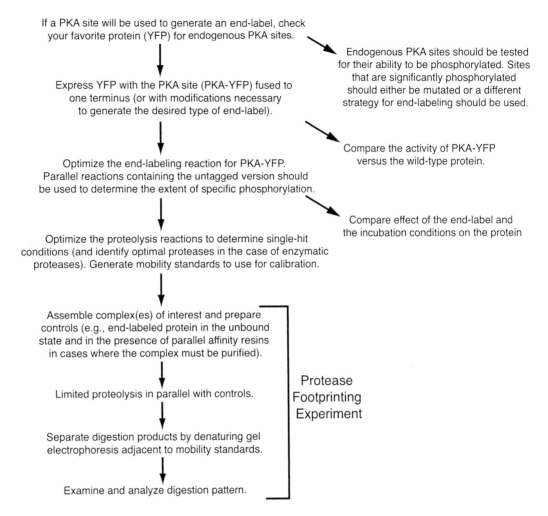

FIGURE 2. Flowchart of steps involved in protease footprinting.

the protein containing and lacking the engineered PKA–tag in a phosphorylation reaction and examining small aliquots of these reactions on an SDS-polyacrylamide protein gel. The percentage incorporation and the ratio of specific phosphorylation to fortuitous background phosphorylation should be determined using a phosphor screen reader or autoradiography followed by densitometry. For use in protease footprinting, this ratio should be at least 10:1. A time course of the reaction and a titration of enzyme should be performed to optimize the extent of labeling and the specific labeling.

The amount of protein in the end-labeling reaction depends on the efficiency of the phosphorylation reaction and complex assembly. For the highest specific-activity end-labeled protein, the optimal PKA site (Songyang et al. 1994) and a reaction containing 10–30 pmoles of protein and a three- to tenfold excess of $[\gamma\text{-}^{32}P]ATP$ or $[\gamma\text{-}^{33}P]ATP$ should be used. The use of optimal labeling conditions is especially important when (1) only low incorporation rates can be obtained or (2) only a fraction of the free protein can be driven into complexes, due to low-affinity interactions, and purifying the bound protein results in significant loss of material (see Assembly of Complexes, p. 293). Alternatively, lower specific activity, achieved by increasing the protein or lowering the ATP concentration in the reaction, can be tolerated if the signal is sufficient to observe the digestion pattern and may provide an advantage if larger amounts of protein facilitate the reactions used to assemble complexes. Phosphorylation reactions involving human

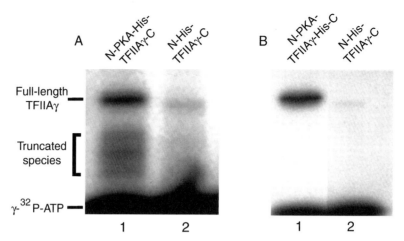

FIGURE 3. End-labeling of the γ subunit within human TFIIA. (*A*) End-labeling of TFIIAγ with both the PKA site and His$_6$ tag on the amino terminus (lane *1*) versus one with only a His$_6$ tag on the amino terminus (lane *2*). (*B*) End-labeling of TFIIAγ with the PKA site and the His$_6$ tag on opposite termini (lane *1*) versus one with only a His$_6$ tag on the amino terminus (lane *2*).

TFIIA, a general transcription factor, are illustrated in Figure 3. TFIIA can be reconstituted using two subunits named αβ and γ. Reactions using molecules of TFIIA, in which the γ subunit either contains (lane 1) or lacks (lane 2) the PKA recognition site are shown in Figure 3B. The ratio of specific to nonspecific labeling is approximately 20:1.

Second, it is important to compare the specific activity of the phosphorylated protein to that of the wild-type version to ensure that the phosphorylation or the incubation conditions have not led to a significant loss in activity. As discussed above, the presence of inactive/misfolded protein can lead to a "false" footprint. A test phosphorylation reaction should be performed with and without [γ-^{32}P]ATP with a known amount of protein. If phosphorylation results in loss of activity, placing the PKA recognition site on the opposite terminus may be enough to eliminate this effect. If inactivity results from the incubation alone, the phosphorylation reaction can be performed at 4°C overnight, or alternative reaction conditions can be screened.

Third, the integrity of the protein to be labeled is also an important issue. Any degradation products that contain the PKA recognition site will be labeled and generate background bands upon detection. There are two methods for eliminating potential problems due to degradation. In the first, a sequence, such as a hexahistidine (His$_6$) tag, that can be used to purify the protein can be engineered on the terminus opposite the PKA site. Because the purification tag and the PKA site are on opposite termini, only full-length protein can be both purified and labeled. TFIIA, reconstituted using a γ subunit containing the His$_6$ tag and the PKA recognition site on the same terminus, yields background phosphorylated products due to degradation (see Fig. 3A, lane 1). The use of a His$_6$ tag on the opposite terminus to purify the same protein eliminated the background labeling (cf. Fig. 3A, lane 1, and 3B, lane 1). Alternatively, data analysis can be used to correct for the presence of degradation products. For a given preparation, there is a definable relationship between the full-length, end-labeled protein and the truncated species. These values can be used to normalize the pattern detected in each lane. A potential issue with this analysis is the increased possibility of misfolded forms among the truncated versions. In some cases, this may expose (or occlude) regions of the protein and generate (or diminish) those digestion fragments relative to the properly folded species.

Upon completing the phosphorylation reaction, the labeled protein can be purified from the unincorporated [γ-^{32}P]ATP using one of several methods, including a spin column or dialysis. Alternatively, an excess of nonradioactive ATP can be added to the reaction to prevent significant labeling of other proteins in subsequent experiments. This mixture can be used without further purification as long as the components do not inhibit any subsequent steps. Because ATP is small, it migrates at a position that obscures fragments of less than 10 residues. For example, in Figure 5, the solution of end-labeled TFIIB also contains unincorporated [γ-^{32}P]ATP, but it is not observed although fragments of ~5 residues generated by clostripain or trypsin (lanes 2–11) can be distinguished.

If phosphorylation is not an effective method for end-labeling a protein, there are other means available to achieve this goal. A direct end label can also be generated by specifically attaching a fluorescent tag at one terminus using either a lysine or cysteine residue. In one example, the Cy5 fluorescent dye was modified to allow it to be attached to a unique cysteine residue (Callaci et al. 1998). Alternatively, an indirect end label can be generated using immunoblots. The protein can be engineered to contain an epitope, such as the FLAG, myc, or His$_6$ tag, to which there are commercially available high-affinity antibodies (Lindsley and Wang 1993).

Some of these antibodies do not work when the epitope is at one of the termini, thus their ability to work should be carefully considered. As another option, instead of an antibody, biotinylated nitriloacetic acid can be used to probe a membrane containing a His$_6$-tagged protein (McMahan and Burgess 1999). If the protein does not tolerate the addition or alteration of residues on the termini, antibodies that recognize one of the endogenous termini can be used (Heyduk and Heyduk 1994; Cheng et al. 1998). In these cases, multiple proteins in a complex can be examined within the same experiment by using multiple fluorescent tags or antibodies (Greiner et al. 1996; Owens et al. 1998) to detect different proteins. Experiments to test the effect of any engineered residues on the activity of the protein and to measure the background generated from these methods must be performed.

Once a method for end-labeling the protein of interest has been established, the electrophoresis standards that will be used to determine which cleavage site generates each proteolytic fragment should be generated. These markers are described in more detail in Identification of Cleavage Sites (p. 296) and include those made by site-specific proteolysis of the end-labeled protein and engineered deletion mutants that contain the same end label, but are truncated from the opposite terminus.

Assembly of Complexes

In protease footprinting experiments, the end-labeled protein must be assembled quantitatively into the desired complexes because background signal that obscures the protections will be generated when the uncomplexed protein is digested at those sites. If the affinity is high, this can be accomplished by simply adding an excess of the ligand. A sufficient amount should be used so that the resulting digestion pattern can be observed, and that value depends on the specific activity of the end-labeled protein. A typical concentration to use in each digest is an amount of protein containing 10,000–30,000 Cerenkov counts of incorporated label. Higher concentrations of protein will be necessary to observe results when it has been less efficiently labeled. In other cases, the affinities are not sufficient and other measures must be employed to separate the complexed protein from the uncomplexed protein. Several of these methods are covered elsewhere in this manual and are only discussed briefly here. In the case of protein–protein complexes, the end-labeled protein can be bound on an affinity resin that displays the ligand protein. A simple procedure for generating such a resin is to express the ligand protein fused to the protein glutathione-*S*-transferase (GST) and to purify this fusion protein using glutathione–agarose. The fusion protein should be tested for activity as described in Chapter 4 and, if active, used subsequently to assemble complexes. Parallel affinity resins displaying the same fusion protein containing mutations that are known to abolish complex formation specifically, and/or GST alone, should be used as controls to establish conditions for specific complex assembly.

Protein–DNA and protein–RNA complexes can be purified by using biotinylated nucleic acids attached to a matrix containing streptavidin. Affinity resins displaying oligonucleotides containing crucial mutations, or the removal of a required protein from the assembly reaction, should be used as controls to establish conditions for specific complex formation and the subsequent proteolysis reactions. When the affinity is sufficiently low to require one of these (or some other) methods to purify the complex of interest, a significant fraction of the material is not recovered. This must be taken into account when designing the assembly. Typically, retaining >10,000 Cerenkov counts of end-labeled protein per digest (i.e., lane) is sufficient. The actual amount of end-labeled protein that must be incubated with the affinity matrix will depend on its specific activity and the efficiency of complex formation.

Proteolysis

As in the case of DNase footprinting, it is essential that each protein molecule be cut a maximum of once along its length. To achieve these "single-cut conditions," the proteolysis reactions should be manipulated such that >50% of the protein is uncut (Brenowitz et al. 1986). This can be determined by comparing the intensity of the uncut band in a lane to the total intensity within the lane. The time of proteolysis necessary for single-cut conditions is determined empirically and should take into account the stability of the complexes being characterized. Stable complexes such as that of an antibody with myoglobin (Zhong et al. 1995) can survive longer incubations with lower concentrations of proteases. However, it is often preferable to minimize the time of digestion because the initial cleavage may lead to some protein unfolding. Increasing the length of the digestion reaction provides time for cleavage at secondary sites that have become exposed. When the complexes are less stable, shorter incubations with higher concentrations of proteases should be used. In either case, optimization using a range of concentrations and times is necessary to balance single-hit conditions with complex stability.

Parallel digestion reactions of the complex containing the end-labeled protein and, as a control, the free (unbound) end-labeled protein should always be performed. In addition, other controls may be necessary, depending on the method used to assemble (and purify) the complex being studied. If the complex has been assembled on an affinity matrix, the end-labeled protein should be digested in the presence of control affinity resin(s). For example, when a protein–protein complex is assembled and purified on a GST-based affinity matrix, the end-labeled protein should be examined after incubation with both GST–ligand and GST alone.

The GST resin will retain only a small fraction of end-labeled protein under conditions required for specific complex formation. In this case, it may be necessary to subject this assembly reaction to one less cycle of "washing" to retain sufficient end-labeled protein for detection. In another example, when ternary complexes are being examined, to determine the contribution of each component, the controls should include both the free (unbound) protein and binary species. In general, the controls required for protease footprinting will be similar to those required to establish the specificity of complex assembly.

Hydroxyl radical cleaving experiments require additional considerations. The time between preparation of reagents and their addition to proteins is crucially important. An iron–EDTA complex is generated by mixing solutions of ferrous ammonium sulfate and EDTA (Dixon et al. 1991; Heyduk and Heyduk 1994), or it may be prepared from preformed iron–EDTA (Greiner et al. 1996; Kumar et al. 1997). These solutions should be made just prior to use. A typical titration to determine the optimal concentrations uses 1 mM EDTA and 0.5 mM Fe as the minimal concentrations, with constant concentrations of ascorbate and hydrogen peroxide. Figure 4 illustrates an example of an experiment examining cAMP receptor protein (CRP)-cAMP bound to DNA with proteolysis by Fe–EDTA. The regions of CRP that become buried upon binding to DNA will produce a decrease in the signal intensity of fragments derived from that domain, whereas the ones that become more exposed will generate an increase in signal intensity. The sites that are significantly affected when CRP binds DNA are identified using a difference plot to determine the rela-

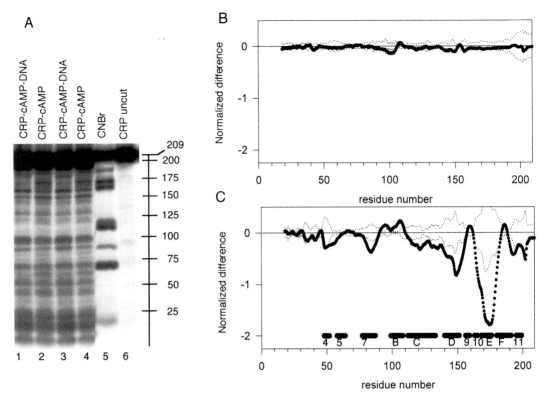

FIGURE 4. Fe-EDTA cleavage of CRP–cAMP and its complex with DNA. (*A*) Autoradiogram of a protease footprinting experiment for CRP–cAMP and its complex bound to DNA containing the consensus site. CRP indicates end-labeled CRP. Lanes *1* and *3* contain two parallel trials of CRP bound to cAMP and DNA containing the consensus binding site treated with Fe-EDTA. Lanes *2* and *4* contain two parallel trials of CRP bound to cAMP treated with Fe-EDTA. Lane *5* contains CRP digested with CNBr to produce a calibration ladder. Lane *6* contains CRP untreated by Fe-EDTA. (*B*) Averaged difference plot of CRP–cAMP versus CRP–cAMP complexes to determine the range of experimental variation. The dotted line indicates 2x standard error and the heavy line is the average normalized difference. The difference is calculated as (signal intensity of a given fragment within the bound protein – signal intensity of the same fragment within the free protein)/(signal intensity of that fragment within the bound protein). (*C*) Averaged difference plot for the CRP–cAMP-DNA ternary complex versus CRP–cAMP binary complex to determine the effect of binding to DNA. The dotted line indicates 2x standard error and the heavy line is the average normalized difference. Letters and numbers below the graph refer to the α helices and β strands, respectively.

tive change at any given site and are calculated as described in the figure legend. Negative values correspond to more buried (protected) regions, and positive values correspond to more exposed (hypersensitive) regions. Changes larger than two times the standard error (the thin lines) calculated from multiple trials (lanes 1 and 3) are considered to be significant. Note that the protection in residues 161–183 is evident from visual inspection of the gel, whereas the protections in residues 48–58 and in residues 79–88 require analysis to be discerned.

Because iron–chelate complexes must generate hydroxyl radicals to work in this procedure (Croft et al. 1992), it is necessary to perform reactions in buffers free from radical scavengers such as glycerol and thiols. If a protein is stored in buffers containing scavengers, dialysis or desalting columns can be used to exchange the buffer.

Many different enzymatic proteases have been used in protease footprinting experiments (see Table 1). They will cleave at a more limited set of positions than the chemical proteases, and the ones that generate the best signal for each situation must be determined empirically. When deter-

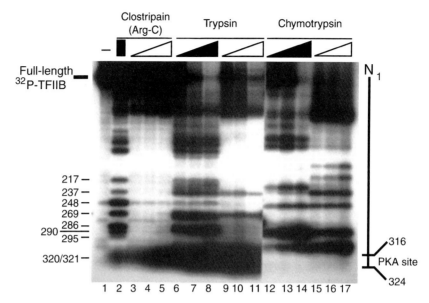

FIGURE 5. Test proteolysis reactions of TFIIB. Human TFIIB end-labeled on a carboxy-terminal PKA recognition site was digested for 2 minutes with increasing amounts (denoted by *triangles*) of clostripain (Endoproteinase Arg-C; lanes *2–5*), trypsin (lanes *6–11*), and chymotrypsin (lanes *12–17*). Undigested end-labeled TFIIB is in lane *1*. The open and closed triangles represent digestions using native and denatured TFIIB, respectively. The positions of the amino terminus and the PKA site on the 324-residue protein are shown on the right-hand side. The cleavage products corresponding to the nine arginine residues in the carboxy-terminal third of the protein identified when denatured TFIIB is digested with clostripain (lane *2*) are identified on the left side.

mining the digestion conditions to use, the amount of protease used and the reaction time can be varied. The optimal protease concentrations can be defined by making serial dilutions and examining the resulting proteolysis products over a wide range of concentrations. A typical range of protease is 20–2000 ng. Figure 5 illustrates a set of reactions for establishing the conditions for digesting human TFIIB.

The optimal amount of protease is one in which the full-length protein is 30–50% digested. Under these conditions, the signal of the proteolytic fragments will be strongest while staying within the range that results in single-hit digestion. Different proteases digest different portions of the protein, and the ones that cleave in the regions of interest should be used in the experiments. Usually, a larger amount of protease is required to digest the assembled complex relative to the free protein. Even after an optimal concentration has been identified within one trial, it is often useful to use three concentrations centered around that amount in subsequent experiments to obtain the best results. To obtain the most consistent results, the proteases should be stored frozen in single-use aliquots at a convenient stock concentration, typically 1 mg/ml. Multiple use of aliquots results in increased variability due to either autoproteolysis and/or denaturation during the freeze/thaw cycle (R. Hori, unpubl.).

Identification of Cleavage Sites

The better the resolution of the system used for separating the digestion products, the more precise the assignment of cleavage sites and therefore the assignment of residues involved in protections and hypersensitivities. The use of tricine instead of glycine as a carrier ion allows higher resolution of smaller fragments (Schagger and von Jagow 1987) and, although it can be used to analyze fragments as large as 100 kD, its biggest advantage is the increased resolution of fragments from 20 kD down to 1 kD. In addition, varying the percentage of acrylamide in the gel allows different ranges of cleavage products to be targeted and analyzed. Tricine- and glycine-based systems have been used in conjunction to analyze fragments from large proteins (Casaz and Buck 1999). Glycine gels are effective for resolving fragments from 15 to 100 kD. Larger resolving gels also increase resolution. Other potential candidates for separating fragments are capillary electrophoresis and mass spectrometry (Cohen et al. 1995; Kriwacki et al. 1996). The high sensitivity of mass spectrometry allows it to resolve fragments very close in size. However, results from cleavage by metal–chelate complexes can be difficult to analyze using this method. Hydroxyl radicals can attack amino acid side chains as well as peptide bonds; cleavage generates a number of different-sized fragments each derived from the same portion of the protein. Recently, *matrix assisted laser desorption ionization* (MALDI) was used to map modifications of side chains by hydroxyl radicals, a promising sign for further use of this method (Maleknia et al. 1999).

The method used for detecting the cleavage fragments depends on the type of end label and separation system used. In the case of a denaturing polyacrylamide gel, a phosphor screen reader or autoradiography can be used for a radioactive end label, and a fluorescence image reader for a fluorescent one. An immunoblot will be used in the case of using an antibody directed against the endogenous or engineered terminus. Regardless of how the cleavage products are detected, they need to be mapped to identify the sites of interaction and/or conformational changes. This is accomplished by comparing the mobility of the fragments generated proteolytically to that of known standards, including commercial prestained molecular-weight markers, deletion mutants, and site-specific cleavage products of the end-labeled protein. Because the molecular weights of discrete peptide fragments are not likely to retain a precise logarithmic relationship to their mobility, the prestained markers provide only a first approximation of the cleavage positions. The precise identification of sites (within about 5 residues) requires the use of the markers described next. To generate mobility standards by site-specific cleavage, the end-labeled protein is unfolded and then digested by either chemical or enzymatic proteases that target specific amino acid residues such as cyanogen bromide (which targets methionine residues) and endoproteinase Lys-C (which targets lysine residues), respectively. Cleavage sites are assigned by comparing fragments generated to the number and sizes of expected fragments predicted from the protein's primary sequence. In practice, however, some sites are cut more than others due to their environment or residual protein structure. To improve the confidence in assignment of fragments, one set of markers produced by a particular method can be used to assign another set produced by a different method. A set of deletion mutants can be engineered using standard molecular cloning techniques and must contain the same end label as the full-length molecule. These are particularly useful for delineating crucial regions that are affected within a complex but are not mapped by the site-specific cleavage markers. Using a combination of chemical and enzymatically produced markers along with deletion mutants, accurate assignment of cleavage sites is possible.

Quantitative Analysis

When using metal–chelate complexes, the large number of cleavage sites makes it difficult to see subtle, but significant, changes in cleavage susceptibility or to be certain that differences in the intensity of bands correspond to an effect due to complex formation. Therefore, it is necessary to use statistical or graphical methods to identify regions of protein involved in interactions. The first step for quantitative analysis is to obtain a digital image using either a densitometer, phosphor screen reader, or fluorescence image reader. Each lane is quantified by full-lane-width scanning along its length, producing a graph of intensity as a function of migration. This can be done using software associated with the image reader or by exporting the data points to a graphing or spread sheet program. Analogous bands in different lanes are then aligned to correct for minor variations in migration. This can be done by manually adjusting the rows in a spread sheet or drawing program or using specialized software such as ALIGN (written in BASIC; available from T. Heyduk). Aligned scans are then corrected for differences in gel loading and cleavage efficiency, converted from mobility to amino acid cleavage site, and compared to each other as difference plots. In some cases, averaging of multiple experiments using statistical analysis will be necessary to observe differences in the cleavage patterns and can be accomplished using standard spread sheet software. Alternatively, specialized software such as DIFPLOT (written in SigmaPlot transform language; available from T. Heyduk) to consolidate normalization and averaging can be used (Baichoo and Heyduk 1999; Loizos and Darst 1999). These plots graph normalized differences between free and bound protein as a function of amino acid residue number. An example of this analysis can be observed in Figure 4 (described above). To eliminate artifacts and emphasize real differences, averaging of difference plots from multiple experiments is done and statistical methods such as standard error (the thin line in Figure 4) and Student's t-test can be applied. After this extensive analysis, regions of protein where changes occur can be assigned confidently.

End-labeling a Protein Using Protein Kinase A

Protein kinase A provides a simple and reproducible method for end-labeling.

MATERIALS

CAUTION: See Appendix for appropriate handling of materials marked with <!>.

Buffers and Solutions

5x Phosphorylation buffer
100 mM Tris (pH 7.6)
500 mM NaCl
60 mM magnesium chloride (MgCl$_2$) <!>

> If the protein solution contributes significant NaCl or MgCl$_2$, adjust the 5x phosphorylation buffer such that their concentrations in the reaction are 100 mM and 12 mM, respectively.

2x SDS gel-loading buffer
100 mM Tris-HCl (pH 6.8)
200 mM dithiothreitol (DTT) <!>
4% SDS (electrophoresis grade) <!>
0.2% bromophenol blue <!>
20% glycerol
Your Favorite Protein (YFP) containing a terminal PKA recognition site

> For high-specific-activity end-labeling, 10–30 pmoles is typically used per labeling reaction. For proteins between 10 and 100 kD, this is 0.1–3 µg.

Nontagged version of YFP (i.e., typically wild-type)

Enzymes and Buffers

Catalytic subunit (of PKA) from Bovine Heart (Sigma P 2465)

> Resuspend the entire contents of the bottle in 6 mg/ml DTT, as specified by the manufacturer, at a concentration of 5 units/µl. Divide into single-use aliquots, and then freeze on dry ice and store at –70°C. No significant loss of enzyme activity has been observed with a single freeze/thaw cycle (R. Hori, unpubl.).

Gels

Gel filtration columns
SDS-polyacrylamide protein gel (4% stacking; 8–15% separating, depending on the size of the protein being labeled)

> For further details on SDS-polyacrylamide gels, see Sambrook and Russell 2001, *Molecular Cloning.*

Additional Reagents

Bovine serum albumin
Milli Q (Millipore)
Trichloroacetic acid <!>

Labeled Compounds

[γ-^{32}P]ATP (>5000 Ci/mmole) or [γ-^{33}P]ATP (1000–3000 Ci/mmole) <!>

METHOD

1. Combine the following:

5x phosphorylation buffer	20 µl
catalytic subunit of PKA (30 units)	6 µl
[γ-^{32}P]ATP (>5000 Ci/mmole; ~100 pmole)	500 µCi
or	
[γ-^{33}P]ATP (1000–3000 Ci/mmole)	100–300 µCi
wild-type and PKA-tagged YFP, respectively	10–30 pmole

   For high specific activity, use a three- to tenfold molar excess of labeled ATP. The optimal amount can be determined empirically.

   Add Milli-Q H$_2$O to 100 µl.

2. Incubate for 45 minutes at 25°C.

3. Add 1 µl of 100 mM ATP to quench the reaction, and place on ice.

   Alternatively, the end-labeled protein can be purified from the unincorporated ATP using either a gel filtration column (Kumar et al. 1997) or dialysis.

4. Add 2 µl of each reaction to 10 µl of 2x SDS gel-loading dye and heat 3 minutes at 90°C.

5. Load these samples onto an SDS-polyacrylamide protein gel and electrophorese until the bromophenol blue dye has migrated approximately two-thirds of the way to the bottom. Transfer the gel to an old piece of film and wrap thoroughly with plastic wrap. Use autoradiography and densitometry (or a phosphor screen reader after drying the gel under a vacuum for 1 hour at 70°C) to determine the efficiency of radioactive incorporation and the ratio of specific to nonspecific labeling.

   Alternatively, the reactions can be analyzed using gel filtration columns (Kumar et al. 1997) or by precipitation:

   a. Incubate 3 µl of the labeling reaction, 0.5 ml of 0.1 mg/ml bovine serum albumin (as carrier), and 0.5 ml of ice-cold 20% trichloroacetic acid on ice for 30 minutes.

   b. Centrifuge for 5 minutes at 10,000g, and then wash the pellet with 1 ml of ice-cold acetone.

   c. Count the precipitated material and 3 µl of the initial labeling reaction directly (Cerenkov) in a scintillation counter. Determine the incorporation rate by comparing the values.

6. Divide the protein into five 20-µl aliquots to minimize freeze/thaw cycles (typically each one is sufficient for one experiment), freeze on dry ice, and store at –70°C.

Fe–EDTA Protease Footprinting Reactions

This protocol was adapted from Heyduk and Heyduk (1994). In the following protocol, the end-labeled protein is assembled quantitatively into the desired complexes.

MATERIALS

CAUTION: See Appendix for appropriate handling of materials marked with <!>.

Buffers and Solutions

Ammonium iron(II) sulfate (40 mM), hexahydrate <!>
EDTA (80 mM) (diluted in reaction buffer from a 0.5 M EDTA, pH 8.0 stock solution)
Hydrogen peroxide (10 mM) <!>
Reaction buffer
  10 mM MOPS (3-[*n*-morpholino]propane sulfonic acid) (pH 7.2) <!>
  250 mM NaCl
  10 mM MgCl$_2$ <!>
3x Sample buffer
  150 mM Tris-HCl (pH 6.8)
  36% glycerol
  12% SDS <!>
  6% β-mercaptoethanol <!>
  0.01% bromophenol blue <!>
Sodium ascorbate (0.2 M) in reaction buffer <!>

Gels

SDS-polyacrylamide protein gels (glycine or tricine-based)
  For further details on SDS-polyacrylamide protein gels, see Sambrook and Russell 2001, *Molecular Cloning* (glycine-based) and this section (tricine-based).

Additional Reagents

Complexes from Protocol 1 and controls

METHOD

1. Prepare the Fe–EDTA complex just prior to use by combining 25 μl of 40 mM ammonium iron(II) sulfate with 25 μl of 80 mM EDTA and then adding 50 μl of reaction buffer. Alternatively, preformed Fe–EDTA can be used (Greiner et al. 1996; Kumar et al. 1997).

2. Assemble complexes for the proteolysis reactions. Controls for the digestion reaction should be prepared in parallel, and among others, typically include end-labeled protein alone (i.e., uncomplexed), in the presence of control affinity matrices that are unable to bind specifically, or in binary species (when a ternary complex is being characterized).

3. Add in the following order: Fe–EDTA, hydrogen peroxide, and ascorbate to assembled complexes and controls.

   If there are numerous samples to process, it may simplify this step to add the reagents by placing the appropriate amounts of each reagent at different positions on the underside of the microcentrifuge tube cap, then closing it and centrifuging to simultaneously initiate the reaction.

   For cAMP Receptor Protein (CRP), the following amounts were used:

   1 μl of Fe–EDTA complex
   1 μl of 0.2 M ascorbate
   1 μl of 10 mM hydrogen peroxide

   Bring the reaction volume to 10 μl with protein complex (6 μM CRP) and reaction buffer.

   Figure 4 is an example of a footprinting experiment comparing CRP in the presence and absence of its DNA-binding site.

4. Digest for the appropriate time. (Reaction times of 1–30 minutes have been used in various studies.)

5. Stop cleavage by adding 5 μl of 3x sample buffer. Store the samples at –70°C until separated by electrophoresis.

6. Separate samples on a SDS-polyacrylamide protein gel (glycine or tricine, depending on the molecular weight of the end-labeled protein and the region[s] of interest). Controls that should be loaded onto the same gels include the undigested end-labeled protein, calibration ladders (ones made by site-specific cleavage and/or engineered truncations), and prestained molecular-weight markers. Heat each sample for 3 minutes at 90°C. Load 14 μl of each sample.

7. Dry the gel under vacuum for 1.5 hours at 70°C. Expose the gel to either a phosphor screen reader or film.

Protocol 3

Proteolysis Using Enzymatic Proteases

In this protocol, each protein molecule is cut a maximum of once along its length.

MATERIALS

CAUTION: See Appendix for appropriate handling of materials marked with <!>.

Buffers and solutions

Bovine serum albumin (10 mg/ml) (BSA)

Protease inhibitors (e.g., 100 mM phenylmethylsulfonyl fluoride [PMSF] <!>, 1 mg/ml leu-peptin)

2x SDS gel-loading buffer
  100 mM Tris-HCl (pH 6.8)
  200 mM dithiothreitol <!>
  4% SDS (electrophoresis grade) <!>
  0.2% bromophenol blue <!>
  20% glycerol

Complex assembly buffer (varies according to the particular system)
  For the experiment in Figure 4, the buffer contained
    10 mM MOPS (pH 7.2)
    250 mM NaCl
    10 mM $MgCl_2$

Enzymes and Buffers

Various proteases, e.g., alkaline protease (subtilisin), trypsin, chymotrypsin, typically resuspend-ed at 1 mg/ml as specified by the manufacturer, stored at –70°C in single-use aliquots

Gels

SDS-polyacrylamide protein gels (glycine- or tricine-based)

For further details on SDS-PAGE, see Sambrook and Russell 2001, *Molecular Cloning* (glycine) and this section (tricine).

METHOD

Test Reaction to Optimize Proteolysis

1. On ice, make serial threefold dilutions, from 1:3 to 1:729 of each protease(s) in complex assembly buffer + 0.1 mg/ml BSA.

2. At room temperature, combine 1 pmole of end-labeled protein (~3% of the labeling reaction for high specific activity). Bring the reaction volume to 140 μl with complex assembly buffer. Divide this solution into 20-μl aliquots.

3. Add 2 μl of each protease dilution at 0, 12, 24, 36, 48, and 60 seconds.

4. Incubate each reaction for 2 minutes at room temperature. Add 2 μl of the appropriate protease inhibitor to each reaction and freeze immediately on dry ice.

5. Separate samples on SDS-polyacrylamide protein gel (glycine or tricine, depending on the molecular weight of the end-labeled protein and the region[s] of interest). Add 24 μl of 2x SDS gel-loading buffer. Heat each sample for 3 minutes at 90°C. Load 16 μl of each sample. Electrophorese until the bromophenol blue is ~75% of the distance to the bottom of the gel.

6. Dry the gel and expose to either film or a phosphor storage screen to detect the cleavage pattern.

   The conditions established from this experiment can be used to perform the protease footprinting experiments. Substitute the following for step 2 above. The complexes are assembled along with the appropriate controls (e.g., uncomplexed protein or protein in the presence of control affinity matrices). After assembly of the complex, it is immediately divided into four aliquots, and three are subjected to limited proteolysis using varying amounts of protease and, as a control, one is not digested. The assembled complexes will generally require two- to five-fold more protease than the free protein.

Tricine SDS-polyacrylamide Protein Gels

This protocol was adapted from Schagger and von Jagow (1987). This method for detecting cleavage fragments depends on the type of end label and separation system used. It is for a single 16 × 14 × 0.75-cm 16.5% gel. For increased resolution, the length of these gels can be increased.

MATERIALS

CAUTION: See Appendix for appropriate handling of materials marked with <!>.

Buffers and Solutions

Acrylamide (50%) (48.5% w/v acrylamide, 1.5% w/v bisacrylamide) <!>
Ammonium persulfate (10%) <!>
Anode buffer (used in the bottom buffer tank)
  0.2 M Tris-HCl (pH 8.9)
Cathode buffer (used in the top buffer tank)
  0.1 M Tris base
  0.1 M tricine
  0.1% SDS <!>
    The pH of this solution is correct without adjusting it.
Gel buffer
  3 M Tris-HCl (pH 8.45)
  0.3% SDS <!>
Glycerol (40%) (v/v)
H_2O (deionized)
TEMED (N',N',N',N'-tetramethylethylenediamine) <!>

METHOD

1. For the separating gel, combine 7 ml each of 50% acrylamide, gel buffer, and 40% glycerol. Filter, then add 70 µl of 10% ammonium persulfate and 7 µl of TEMED and pour into a vertical gel mold. Overlay with deionized H_2O (or 20% ethanol) until polymerized.

   In some cases, eliminating or reducing the glycerol in the gel has no detrimental effects (R. Hori, unpubl.).

2. Pour off H_2O from the separating gel.

3. Overlay with a stacking gel and then insert a gel comb. For the stacking gel, combine 1 ml of 50% acrylamide, 3.1 ml of gel buffer, 12.5 ml of deionized H_2O; filter, then add 100 µl of 10% ammonium persulfate and 10 µl of TEMED.

4. Inject the samples into the wells at 14 mAmp and run the gels overnight at 24 mAmp.

5. Dry the gels under a vacuum for 1.5 hours at 70°C and expose to a phosphor storage screen or film.

   If desired, the gels can be fixed using a solution of 45% methanol, 10% acetic acid. One problem associated with fixing the gels is an increased tendency to crack during drying.

Calibration Ladder Generated by Cleavage of Proteins by CNBr

This method, adapted from Gross (1967), targets methionine residues to generate mobility standards.

MATERIALS

CAUTION: See Appendix for appropriate handling of materials marked with <!>.

Buffers and Solutions

Acetonitrile <!>
Crystalline cyanogen bromide (CNBr) (stored in a can at 4°C) (very toxic) <!>
H₂O (deionized)
HCl (1 N) <!>
3x Sample buffer (see Protocol 2)
SDS (10%) <!>

Gels

SDS-polyacrylamide protein gels (glycine- or tricine-based)
> For further details on SDS-PAGE, see Sambrook and Russell 2001, *Molecular Cloning*, and this section.

Special Equipment

Chemical fume hood
Phosphor storage screen
SpeedVac
Ultrafreezer

METHOD

1. Prefreeze a plastic rack in a –70°C ultrafreezer.

2. Allow CNBr (within its container) to warm up under a chemical fume hood to room temperature.

3. Tare a balance with an empty microfuge tube.

4. Under the fume hood, add a small amount (a few crystals) of CNBr to microfuge tube (as quickly as possible to reduce inhalation).

5. Weigh CNBr in the closed microfuge tube.

6. Under the fume hood, add appropriate amount of acetonitrile to obtain a 106 mg/ml stock of CNBr.

7. Assemble the following reaction mix:

10% SDS	0.75 µl
1 N HCl	0.4 µl
CNBr solution	10 µl

   Add CNBr last to initiate the reaction. Bring the final volume to 25 µl with end-labeled protein and deionized H_2O.

8. Incubate at room temperature. Take the reactions to the –70°C ultrafreezer. Remove the samples (5 µl) at different time points (e.g., typically every 5 or 10 minutes) and place in the rack in the –70°C ultrafreezer.

9. At the conclusion of the reactions, dry the frozen reactions in a SpeedVac.

10. Resuspend the dried pellet in 3x sample buffer (50 µl). Run the samples (typically, 5 µl, but it depends on the specific activity of the end-labeled protein) from different time points on a denaturing SDS-polyacrylamide gel to determine optimal digestion time.

11. Dry the gel under vacuum for 1.5 hours at 70°C and expose to film or a phosphor storage screen to detect the cleavage pattern. This pattern should be compared with the position of methionine residues in the protein's primary sequence.

12. Add H_2O to stock CNBr for disposal. Soak the spatula that was used to weigh the CNBr crystals overnight in H_2O.

Calibration Ladder Generated by Enzymatic Cleavage

Here, mobility standards are generated by targeting arginines (Clostripain) or lysines (Endoproteinase Lys-C).

MATERIALS

CAUTION: See Appendix for appropriate handling of materials marked with <!>.

Buffers and Solutions

Bovine serum albumin (10 mg/ml) (BSA)
Denaturation buffer
  20 mM HEPES (pH 7.9)
  100 mM KCl <!>
  2 M urea <!>
  0.2% SDS <!>
  0.5 M EDTA
EDTA (0.5 M)
2x SDS gel-loading buffer
  100 mM Tris-HCl (pH 6.8)
  200 mM dithiothreitol <!>
  4% SDS (electrophoresis grade) <!>
  0.2% bromophenol blue <!>
  20% glycerol

Enzymes and Buffers

Site-specific proteases (sequencing grade or the best available), e.g., Clostripain (which cleaves at arginines) or Endoproteinase Lys-C (which cleaves at lysines). Resuspend as specified by the manufacturer.

Additional Reagents

End-labeled protein of interest

Special Equipment

Autoradiography or phosphor image reader

METHOD

1. Add 1 pmole (~3% of a labeling reaction for high specific activity) of the end-labeled protein to 50 μl of denaturation buffer.

2. Heat for 30 minutes at 65°C.

3. Transfer the reaction mixture to 37°C.

4. Add 2 μg of Clostripain (or other site-specific protease such as Endoproteinase Lys-C).

5. Remove 10 μl at 2, 6, 18, and 54 minutes. Terminate the reaction by adding 1 μl of 0.5 M EDTA and 1 μl of 10 mg/ml BSA and freezing on dry ice. Store at –70°C.

6. Separate the samples on a tricine-based denaturing SDS-polyacrylamide gel to examine the proteolysis pattern. A standard SDS-polyacrylamide gel can be used if the end-labeled protein is very large or high resolution of small fragments is not necessary. Add 12 μl of 2x SDS gel-loading buffer. Heat each sample for 3 minutes at 90°C. Load 6 μl of each sample. Electrophorese until the bromophenol blue is ~75% of the distance to the bottom of the gel.

7. Dry the gel under vacuum for 1.5 hours at 70°C.

8. Visualize the cleavage pattern by autoradiography or a phosphor image reader.

REFERENCES

Baichoo N. and Heyduk T. 1999. DNA-induced conformational changes in cyclic AMP receptor protein: Detection and mapping by a protein footprinting technique using multiple chemical proteases. *J. Mol. Biol.* **290:** 37–48.

Brenowitz M., Senear D.F., Shea M.A., and Ackers G.K. 1986. Quantitative DNase footprint titration: A method for studying protein-DNA interactions. *Methods Enzymol.* **130:** 132–181.

Callaci S., Heyduk E., and Heyduk T. 1998. Conformational changes of *Escherichia coli* RNA polymerase σ^{70} factor induced by binding to the core enzyme. *J. Biol. Chem.* **273:** 32995–33001.

Casaz P. and Buck M. 1999. Region I modifies DNA-binding domain conformation of sigma 54 within the holoenzyme. *J. Mol. Biol.*. **285:** 507–514.

Cheng X., Shaltiel S., and Taylor S.S. 1998. Mapping substrate-induced conformational changes in cAMP-dependent protein kinase by protein footprinting. *Biochemistry* **37:** 14005–14013.

Clore G.M. and Gronenborn A.M. 1998. NMR structure determination of proteins and protein complexes larger than 20 kDa. *Curr. Opin. Chem. Biol.* **2:** 564–570.

Cohen S.L., Ferre-D'Amare A.R., Burley S.K., and Chait B.T. 1995. Probing the solution structure of the DNA-binding protein Max by a combination of proteolysis and mass spectrometry. *Protein Sci.* **4:** 1088–1099.

Croft S., Gilbert B.C., Smith J.R., and Whitwood A.C. 1992. An E.S.R. investigation of the reactive intermediate generated in the reaction between FeII and H_2O_2 in aqueous solution. Direct evidence for the formation of the hydroxyl radical. *Free Radic. Res. Commun.* **17:** 21–39.

Datwyler S.A. and Meares C.F. 2000. Protein-protein interactions mapped by artificial proteases: Where sigma factors bind to RNA polymerase. *Trends Biochem. Sci.* **25:** 408–414.

Dixon W.J., Hayes J.J., Levin J.R., Weidner M.F., Dombroski B.A., and Tullius T.D. 1991. Hydroxyl radical footprinting. *Methods Enzymol.* **208:** 380–413.

Greiner D.P., Hughes K.A., Gunasekera A.H., and Meares C.F. 1996. Binding of the sigma 70 protein to the core subunits of *Escherichia coli* RNA polymerase, studied by iron-EDTA protein footprinting. *Proc. Natl. Acad. Sci.* **93:** 71–75.

Gross E. 1967. The cyanogen bromide reaction. *Methods Enzymol.* **11:** 238–255.

Hanai R. and Wang J.C. 1994. Protein footprinting by the combined use of reversible and irreversible lysine modifications. *Proc. Natl. Acad. Sci.* **91:** 11904–11908.

Heyduk E. and Heyduk T. 1994. Mapping protein domains involved in macromolecular interactions: A novel protein footprinting approach [published erratum appears in *Biochemistry* 1995 Nov 21; **34(46):** 15388]. *Biochemistry* **33:** 9643–9650.

Heyduk T., Baichoo N., and Heyduk E. 2001. Probing of proteins by metal ions and their low molecular weigh complexes. In *Metal ions in biological systems* (ed. A. Sigel and H. Sigel), vol. 38, pp. 255–287. M. Dekker, Inc., New York.

Heyduk T., Heyduk E., Severinov K., Tang H., and Ebright R.H. 1996. Determinants of RNA polymerase alpha subunit for interact ion with beta, beta′, and sigma subunits: Hydroxyl-radical protein footprinting. *Proc. Natl. Acad. Sci.* **93:** 10162–10166.

Hori R. and Carey M. 1998. Protease footprinting analysis of protein-protein and protein-DNA interactions. In *Human genome methods* (ed. K. Adolph), pp. 73–90. CRC Press, Boca Raton.

Kriwacki R.W., Hengst L., Tennant L., Reed S.I., and Wright P.E. 1996. Structural studies of p21Waf1/Cip1/Sdi1 in the free and Cdk2-bound state: Conformational disorder mediates binding diversity. *Proc. Natl. Acad. Sci.* **93:** 11504–11509.

Kumar A., Kassavetis G.A., Geiduschek E.P., Hambalko M., and Brent C.J. 1997. Functional dissection of the B″ component of RNA polymerase III transcription factor IIIB: A scaffolding protein with multiple roles in assembly and initiation of transcription. *Mol. Cell. Biol.* **17:** 1868–1880.

Li B.L., Langer J.A., Schwartz B., and Pestka S. 1989. Creation of phosphorylation sites in proteins: Construction of a phosphorylatable human interferon alpha. *Proc. Natl. Acad. Sci.* **86:** 558–562.

Lindsley J.E. and Wang J.C. 1993. Study of allosteric communication between protomers by immunotagging. *Nature* **361:** 749–750.

Loizos N. and Darst S.A. 1998. Mapping protein-ligand interactions by footprinting, a radical idea. *Structure* **6:** 691–695.

———. 1999. Mapping interactions of *Escherichia coli* GreB with RNA polymerase and ternary elongation complexes. *J. Biol. Chem.* **274:** 23378–23386.

Maleknia S.D., Brenowitz M., and Chance M.R. 1999. Millisecond radiolytic modification of peptides by synchrotron X-rays identified by mass spectrometry. *Anal. Chem.* **71:** 3965–3973.

McMahan S.A. and Burgess R.R. 1999. Mapping protease susceptibility sites on the *Escherichia coli* transcription factor sigma70. *Biochemistry* **38:** 12424–12431.

Owens J.T., Chmura A.J., Murakami K., Fujita N., Ishihama A., and Meares C.F. 1998. Mapping the promoter DNA sites proximal to conserved regions of sigma 70 in an *Escherichia coli* RNA polymerase-lacUV5 open promoter complex. *Biochemistry* **37**: 7670–7675.

Pestka S., Lin L., Wu W., and Izotova L. 1999. Introduction of protein kinase recognition sites into proteins: A review of their preparation, advantages, and applications. *Protein Exp. Purif.* **17**: 203–214.

Schagger H. and von Jagow G. 1987. Tricine-sodium dodecyl sulfate-polyacrylamide gel electrophoresis for the separation of proteins in the range from 1 to 100 kDa. *Anal. Biochem.* **166**: 368–379.

Schwanbeck R., Manfioletti G., andWisniewski J.R. 2000. Architecture of high mobility group protein I-C-DNA complex and its perturbation upon phosphorylation by Cdc2 kinase. *J. Biol. Chem.* **275**: 1793–1801.

Songyang Z., Blechner S., Hoagland N., Hoekstra M.F., Piwnica-Worms H., and Cantley L.C. 1994. Use of an oriented peptide library to determine the optimal substrates of protein kinases. *Curr. Biol.* **4**: 973–982.

Traviglia S.L., Datwyler S.A., and Meares C.F. 1999. Mapping protein-protein interactions with a library of tethered cutting reagents: The binding site of sigma 70 on *Escherichia coli* RNA polymerase. *Biochemistry* **38**: 4259–4265.

Tu B.P. and Wang J.C. 1999. Protein footprinting at cysteines: Probing ATP-modulated contacts in cysteine-substitution mutants of yeast DNA topoisomerase II. *Proc. Natl. Acad. Sci.* **96**: 4862–4867.

Zhong M., Lin L., and Kallenbach N.R. 1995. A method for probing the topography and interactions of proteins: Footprinting of myoglobin. *Proc. Natl. Acad. Sci..* **92**: 2111–2115.

17 Tandem Affinity Purification to Enhance Interacting Protein Identification

Bertrand Séraphin,[1,2] Oscar Puig,[1] Emmanuelle Bouveret,[1]
Berthold Rutz,[1] and Friederike Caspary[1]

[1]EMBL, D-69117 Heidelberg, Germany; [2]CGM, CNRS, F-91198 Gif sur Yvette, France

INTRODUCTION

With the number of complete genome sequences increasing at an ever-faster pace, we are getting sequence data for the full complement of proteins or "proteome" (Wasinger et al. 1995; Wilkins et al. 1996), encoded by various organisms. To understand how the proteome confers specific properties to a host cell, one needs to obtain information regarding several parameters, including the level of expression of all proteins, their posttranslational modifications, their cellular localizations, and their interactions with different partners (proteins, nucleic acids, etc.). This task is often complicated by the wide variety of expression levels of different proteins in a given cell: The steady-state level of cellular proteins may differ by several orders of magnitude even though they are encoded by genes approximately equally represented in the genome (Gygi et al. 2000). In recent years, mass spectrometry has become a powerful tool for protein identification at an

313

unequaled level of sensitivity (Wilm et al. 1996). This strength has been exploited to analyze complex protein mixtures fractionated on two-dimensional gels (Shevchenko et al. 1996). However, it has become clear that only a fraction of the total cellular proteins, representing the most highly expressed polypeptides, can be detected using this strategy (Gygi et al. 2000).

Given its high sensitivity and the possibility of processing numerous samples, mass spectrometry (see Chapter 12) is also well adapted for the identification of subunits of protein complexes or interacting protein partners (Lamond and Mann 1997). The limiting step for this application of mass spectrometry resides in the difficulties encountered to obtain sufficient amounts of purified complexes. Several strategies have been successful for the purification of such protein complexes. Standard biochemical methods can be used and will remain essential for some applications (Deutscher 1990). They are, however, often cumbersome and time-consuming because optimal purification steps have to be determined empirically for each complex. Furthermore, a specific, reliable and, if possible, rapid assay needs to be designed for each new purification (activity assay, western blotting, ELISA, etc.). Designing such assays is not always easy, especially if only the nucleotide sequence coding for the target protein has been obtained from a sequencing project and no known activity is associated with the predicted polypeptide.

Interacting protein partners may also be recovered by passing cellular extracts or cellular fractions through columns containing the target protein as a bait. A prerequisite for this strategy is the recovery of a substantial amount of (recombinant) bait protein to build the affinity column. This strategy will not be successful, however, if the endogenous bait proteins form tight complexes with interacting partners, preventing their retention on the column. Additional problems may be encountered if the bait protein needs to be modified posttranslationally for successful association, if interacting partners are present in minute amounts, and/or if abundant cellular proteins interact nonspecifically with the bait protein, masking the specific interacting partners present at low level.

Coimmunoprecipitation (see Chapter 5 and Sambrook and Russell 2001, *Molecular Cloning*, pp. 18.60–18.68) is a very useful method to copurify interacting protein partners. The success of this strategy depends on the availability of good-quality antibodies directed against the target protein. This often requires the production of recombinant target protein or a fragment thereof to generate monoclonal or polyclonal antibodies of high affinity and specificity. Indeed, immunocopurification of target protein and interacting partners is most often performed in a single step. Thus, the efficiency of the method and the purity of the recovered material will be totally dependent on the quality of the antibodies used. In addition, because this procedure involves a single purification step, this limits the use of this strategy for protein present in very low abundance. Furthermore, the purified material is often contaminated by immunoglobulins used for the purification that may mask unknown protein partners. A strength of the coimmunoprecipitation strategy is, however, that proteins associated in vivo are copurified in a rapid and simple manner. Therefore, coimmunoprecipitation remains a rigorous method to validate the physiological significance of protein interaction.

Recently, we have developed a new strategy for protein purification that is particularly well suited for protein complex purification and for the identification of interacting protein partners. This strategy, dubbed TAP (for *t*andem *a*ffinity *p*urification; Rigaut et al. 1999), involves the fusion of a protein tag to the target protein and its expression in a natural host cell or organism. Extracts prepared from these cells or organisms are then used to recover the target protein and associated interacting partners in a standard two-step process. The purified material can then be fractionated on a gel, and copurified proteins can be identified by mass spectrometry or Edman degradation. The TAP strategy combines several advantages of the standard biochemical purification and the coimmunoprecipitation strategies. As for classic biochemical purification, the successive purification steps of the TAP strategy allow the recovery of highly purified protein complexes present in low amounts in the starting extract. However, the use of a constant peptide tag and standard purification conditions avoids the problem of designing new purification schemes for each new target that is associated with classic biochemical purification. Like coimmunoprecipitation, the TAP strategy relies on specific protein–protein (affinity matrix-tag) interaction of high affinity allowing the rapid, selective, and efficient recovery of the in-vivo-associated target complex from extracts. However, in contrast to coimmunoprecipitation, the various steps of the TAP method reduce background contamination by abundant cellular proteins or material leaking from the affinity column. This allows the recovery of highly purified complexes present at very low concentration. Like any method, the TAP strategy has some limitations (discussed below). Nevertheless, it appears applicable to a wide variety of proteins. Indeed, the TAP strategy has proven extremely useful in our hands and in other laboratories as a generic method to quickly purify protein complexes and associated proteins, overcoming a limiting step in the analysis of protein interaction by mass spectrometry.

Tag Selection and Design

Optimal tags for protein complex purification and the analysis of protein interaction should have the following characteristics:

1. High affinity for the cognate matrix for quantitative recovery of low-abundance target proteins in dilute solutions.

2. Highly specific binding to increase the ratio of specifically to nonspecifically bound material to the affinity material.

3. Efficient and specific elution allowing high-level and specific recovery of the target protein.

4. Mild conditions of elution to preserve protein interactions and protein complex structure.

These characteristics are obviously somewhat contradictory, because efficient elution and/or elution under mild conditions is often difficult if a high-affinity interaction is involved. Similarly, efficient elution often requires harsh conditions that are not compatible with the preservation of protein–protein interaction. In addition, optimal tags should be as neutral as possible for the activity of the passenger protein and should work under a wide variety of salt and detergent conditions to allow purification.

To select an efficient tag for protein purification, we screened various available tags such as FLAG (Brizzard et al. 1994; Sambrook and Russell 2001, *Molecular Cloning*, pp. 15.4–15.6), His (Hochuli et al. 1988), Strep (Schmidt and Skerra 1993; Sambrook and Russell 2001, *Molecular Cloning*, pp. 11.118–11.119), two immunoglobulin G (IgG) binding units of *Staphylococcus aureus* Protein A (ProtA, Lohman et al. 1989), calmodulin-binding peptide (CBP, Stofko-Hahn et al. 1992), and chitin-binding domain (CBD, Chong et al. 1997) for optimal characteristics. Efficiency of recovery of a low-abundance passenger protein from yeast and tag selectivity (level of nonspecific background binding to the affinity support) was assayed semiquantitatively. Complementation analysis indicated that none of these tags impaired function of the Lsm3 target protein. In our hands, ProtA and CBP turned out to be the most efficient tags, with approximately 80% and 50% recovery of the target complex, respectively (Rigaut et al. 1999).

Binding of CBP to calmodulin in the presence of calcium is efficient and highly specific. Furthermore, this interaction can be easily reversed by the removal of calcium (by addition of EGTA; Stofko-Hahn et al. 1992). This represents a mild condition that does not appear to affect protein–protein interaction in many complexes. The binding of ProtA to rabbit IgG is also very efficient and highly specific (Lohman et al. 1989). However, elution of the bound material usually requires harsh conditions (e.g., very low pH) that would not preserve protein complex integrity and would also cause elution of material nonspecifically bound to the affinity matrix. Therefore, we decided to include a protease cleavage site between the passenger protein and the ProtA tag to allow specific and efficient elution under mild conditions. The TEV protease was selected for this purpose because of its high specificity and commercial availability (see Materials)

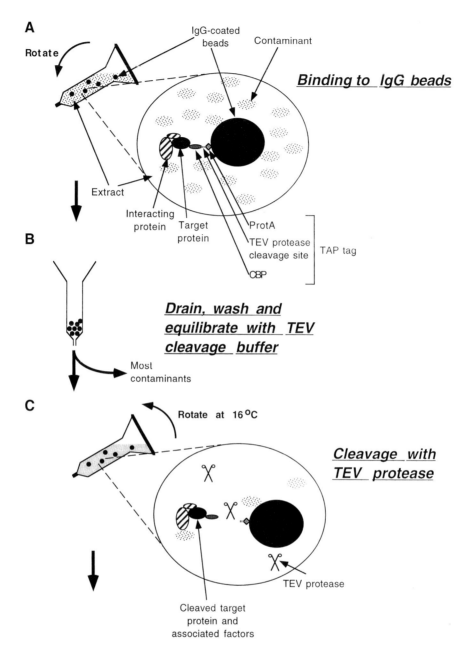

FIGURE 1. (*Continues on facing page.*) Theoretical and practical overview of the TAP method. The various steps of the TAP method are shown. Technical conditions are indicated for each step. When relevant, a detailed description of the reaction occurring at a specific step is provided in an enlarged view. The names of the various proteins, enzymes, and affinity media are indicated.

and because relatively mild conditions are required for efficient cleavage activity (Dougherty et al. 1989). Preliminary experiments demonstrated that single-step purifications were not sufficient to recover highly purified proteins from total cellular extracts. Therefore, we combined the CBP tag and ProtA flanked by a TEV protease cleavage site to create the TAP tag (Rigaut et al. 1999). Purification of a TAP-tagged protein present in a crude extract involves four main steps (Fig. 1):

1. Binding of the target protein to the IgG beads (Fig. 1A).

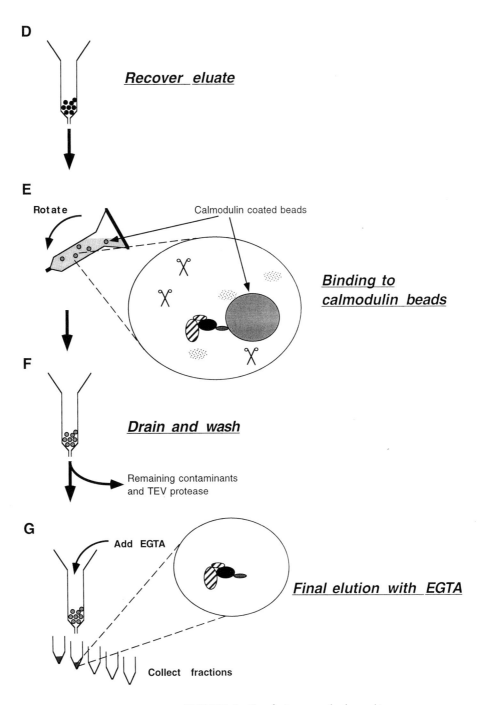

FIGURE 1. (*See facing page for legend.*)

2. After extensive washing (Fig. 1B), elution of the bound material with the help of the TEV protease (Fig. 1C).

3. Binding of the protein present in the first eluate (Fig. 1D) to calmodulin-coated beads in the presence of calcium (Fig. 1E).

4. After extensive washing (Fig. 1F), elution of the bound material by addition of EGTA (Fig. 1G).

It is noteworthy that the four different binding and elution steps of this procedure are highly specific, thereby increasing the level of purification of the target protein complex. Although all of these steps are not necessary for abundant proteins, they are required when the target is present at a low level. Theoretically, binding to IgG or to calmodulin could be used indifferently as the first purification step. However, because elution from IgG requires the addition of the TEV protease that will contaminate the material eluted from this affinity material, it is preferable to perform affinity purification on an IgG matrix first. The second affinity step then allows the removal of contaminating TEV protease in addition to nonspecific contaminants. As a consequence, TEV protease cleavage should remove only the protein A (ProtA) part of the TAP tag, leaving the CBP fused to the target protein behind. ProtA should therefore preferably be located at the extreme amino or carboxyl terminus of the passenger protein. Due to this constraint, amino- and carboxy-terminal TAP tags containing the CBP, TEV protease cleavage site, and ProtA moiety of the TAP tag in reverse order have been constructed (Puig et al. 2001). Because of the modularity of the TAP tag, the TAP method is highly flexible. Indeed, on some occasions, one may prefer to use a "split TAP tag" by fusing the CBP and TEV-ProtA fragment cassettes to two different passenger proteins, allowing the purification of subcomplexes containing the two target proteins simultaneously (Caspary et al. 1999). Alternatively, it is possible in some cases to subtract undesired proteins by fusing them to a ProtA cassette lacking a TEV protease cleavage site (Bouveret et al. 2000).

Vectors encoding amino- and carboxy-terminal TAP tag cassettes have been built (Rigaut et al. 1999; Puig et al. 2001). The structures of these cassettes, including the location of useful restriction sites, are presented in Figure 2. It is noteworthy that, in the case of the amino-terminal TAP tag, an additional cleavage site for the enterokinase protease has been added, allowing the (near) complete removal of the TAP tag from the target protein following the second purification step.

OUTLINE OF PROCEDURE

The procedure involves construction of cells or organisms expressing the TAP-tagged protein, preparation of extracts, purification of the target protein according to the TAP procedure, and analysis of the purified complex as originally reported by Rigaut et al. (1999).

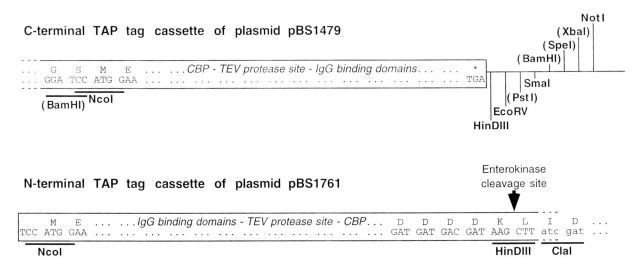

FIGURE 2. Structures of the carboxy- and amino-terminal TAP tags. The tags are shown as open boxes, and nucleotides and amino acids relevant for in-frame fusion are shown. Useful restriction sites are also indicated. The enzymes in brackets are not unique in the corresponding vectors, but they may still be used for the fusion of the TAP tag to the target open reading frame.

Construction of Cells or Organisms Expressing a TAP-tagged Protein

When constructing a protein fusion, one should also consider how the vector carrying the tagged fusion will be introduced in the recipient cell or organism and how expression of the fusion protein will be controlled. Because the goal of a TAP purification experiment is often to identify a physiological partner(s), one should express the tagged protein in cells or organisms naturally expressing the target protein or in the closest possible system available. This feature may become critically important in multicellular organisms because the target protein may interact with different partners in different tissues or at different developmental stages. It is also important to avoid high overexpression of the target protein. Although this strategy is often used to out-compete binding of the endogenous target protein with natural partners, exaggerated overexpression of the target protein most often results in the formation of a pool of free protein because natural interacting partners are not themselves overexpressed. Target proteins present in this free pool frequently interact with nonnatural partners such as heat-shock factors, the proteasome, or abundant cellular proteins (e.g., translation factors or glycolytic enzymes) (Swaffield et al. 1995, 1996). If the free pool of target protein is in large excess over interacting partners, it is possible that the nonnatural partners will be in excess or even mask natural interacting partners in the purified fraction. Optimally, one should therefore replace the gene(s) coding for the target protein by a TAP-tagged version. However, this is not always possible. Nevertheless, these considerations should be taken into account in building the protein fusion.

Usually, a standard cloning technique can be used for constructing the appropriate vector carrying the TAP tag fused to the target protein (see Sambrook and Russell 2001, *Molecular Cloning*, Chapters 1–4 for vector cloning methods). This vector is then introduced into recipient cells or organisms following the best procedure available for this specific system. However, in some cases, strategies bypassing the cloning step are available. A striking example is provided by the yeast *Saccharomyces cerevisiae*, where the endogenous target gene can be replaced by a TAP-tagged version through the simple transformation of a wild-type strain with a PCR fragment (Baudin et al. 1993; Puig et al. 1998).

Preparation of Extracts

Even though progress in mass spectrometry has significantly increased the sensitivity of protein identification methods, a significant amount of cells will be required for protein purification. Using the TAP method, this quantity can be determined empirically by comparing, using western blotting, the level of expression of the target protein to the level of expression of a previously purified protein carrying the TAP tag (e.g., by using either the Mud13p, Snu71p, or Lsm9p protein [Rigaut et al. 1999] as control). Extracts can be prepared using a variety of conditions. They can be prepared from cell culture, tissues (from transgenic organisms), or even whole organisms. Obviously, any step increasing the number of cells expressing the target protein relative to other proteins will facilitate the purification. This could involve the use of tissues or homogeneous cell cultures instead of whole organisms or the use of cellular fractions instead of whole cellular extracts. Extracts should be prepared in conditions that optimize target protein yield and solubility without interfering with protein–protein interaction. In some cases, these conditions may be nearly impossible to satisfy, because interaction of the target protein with some structure may be more difficult to break up than interaction of the target protein with another protein. (For example, solubilization of a membrane protein may require conditions that destroy interaction of the target protein with specific interacting partners.)

To establish a basis for the design of the extract preparation procedure, it is usually useful to explore the literature to define conditions that have been used for preparation of extracts using the same cell system or for purifying related complexes. Pilot experiments can be used to determine optimal conditions for protein extraction. However, one should remember that stronger conditions might increase the target protein yield while concomitantly reducing the level of the interacting partners. It is also possible that the target protein will interact with various partners and that these various complexes will be extracted with very different efficiencies. Generally, increasing salt and/or detergent concentrations increases protein release but tends to disrupt natural protein interactions, whereas low salt and detergent concentrations reduce protein release while favoring nonspecific interactions. If possible, we suggest using conditions that mimic the intracellular environment of the target protein and that should therefore preserve its solubility and interaction potential. One should remember, however, that any change in conditions during extract preparation may irreversibly disrupt a protein interaction: Even if proper conditions are restored afterward, the interacting partner may be unfolded, modified, or present at too low a concentration to allow reassociation.

It is also noteworthy that, during extract preparation, components of different cellular organelles or of different cell types are being mixed together, creating opportunities for proteins to form nonnatural associations. This situation also favors protein degradation and denaturation. This should be prevented as much as possible by reducing incubation time, lowering the temperature, and/or including protease inhibitors. Therefore, optimal conditions for protein extraction will vary from protein to protein and from one cell type to another. They should be determined empirically following the guidelines given above. The possibility of detecting a small amount of the TAP-tagged target protein by western blotting provides a useful tool for testing various extraction conditions at reduced scale in pilot experiments. Extracts may be used directly for purification or may be stored frozen after addition of glycerol or after dialysis against a glycerol-containing buffer. In some cases, it may be helpful to remove insoluble material from the extract by centrifugation before using it for purification. (An example of extract preparation conditions is given for the yeast Lsm3 protein on p. 322.)

Tandem Affinity Purification

The various steps of the TAP procedure are depicted in Figure 1. The first affinity selection is performed in a batch format by mixing the extract with IgG beads (Fig. 1A). Unbound material is removed by emptying the column by gravity, and nonspecifically bound material is removed by extensive washing with the binding buffer (Fig. 1B). Beads are then washed with TEV cleavage buffer (see Materials, pp. 324–325) before performing the TEV cleavage reaction (Fig. 1C). After cleavage, the eluate (containing the TEV protease) is recovered (Fig. 1D), diluted in the appropriate buffer, and incubated in batch with calmodulin-coated beads (Fig. 1E). Contaminants and the TEV protease are removed by extensive washing (Fig. 1F) before elution of the purified complex with EGTA (Fig. 1G). From some test experiments, we have estimated the overall yield of the method to be roughly 20–30% recovery of the starting material. The purified material can be kept frozen before analysis. However, if functional tests are planned, the purified material may have to be stored in a buffer containing glycerol or another component to preserve its activity. This may be achieved through dialysis or direct addition of requested chemicals (e.g., glycerol) to the eluate. In this vein, it is noteworthy that even if the last elution step of the TAP procedure disrupts the target complex (e.g., if stability of this complex requires calcium), this will not affect the analysis of the composition of the complex, although this is likely to disrupt its activity.

The various reagents used for the TAP purification are described in the Materials section (p. 324), and a detailed step-by-step protocol is given in the Methods section. As for extract preparation, conditions may be adapted for different target proteins (salt concentration, detergents, etc.; see above). Similarly, reducing the temperature and the handling time as well as including protease inhibitors may help to preserve the integrity of proteins and protein complexes. Extensive washing steps should be included between the two binding and elution steps to prevent contamination of the purified fraction with background material. Similarly, important changes in buffer conditions should be avoided because they tend to provoke the release of material bound non-specifically to the affinity matrix and because they could affect the stability of the target protein complex. A potential concern with the TAP method is that calmodulin from the host cell might interact with the CBP moiety of the TAP tag and prevent its association with the affinity matrix during the second step of the purification. Although this is theoretically possible, we have not yet encountered such a situation. Yet, if this becomes a problem, inclusion of EGTA during extract preparation and/or the first purification step on IgG-coated beads should remove endogenous calmodulin, exposing CBP for the second affinity step.

Analysis of the Purified Material

Material recovered from a TAP purification may be analyzed in various ways, including testing its activity in vitro, analyzing the structural shape of the complex by electron microscopy, and assaying for the presence of candidate protein(s) by western blotting (Rigaut et al. 1999; O. Puig et al., unpubl.). In the analysis of interacting proteins, mass spectrometry appears to be the method of choice for protein identification. We routinely fractionate the purified material on monodimensional denaturing protein gels. Gels are poured with an exponential gradient of acrylamide concentration to favor resolution of proteins in a wide variety of molecular-weight ranges. Great care should be taken in handling such gels to avoid introduction of human keratin that could compromise analysis of the purified products by mass spectrometry. Gels are stained with Coomassie Blue or with silver nitrate. A scan of the gel is performed before cutting out the desired band for analysis by mass spectrometry. An advantage of protein fractionation before analysis by mass spectrometry is that one can get a rough estimate of the complex abundance, the number of proteins present, and their relative stoichiometric levels. However, it is possible that some material is lost during this process and that direct analysis of the eluted material by mass spectrometry (Link et al. 1999) could be more sensitive.

Controls and Verification

To control for the specificity of the copurification, one can perform a purification with extracts prepared from cells that do not express a TAP-tagged protein. Although this is an important control, the mere fact that a protein is recovered in the purified fraction with a given TAP-tagged protein but is not recovered in a purification with material obtained from a nontagged strain does not demonstrate that this interaction is specific. Indeed, interaction with the target protein may represent interaction between partly denatured proteins or the formation of aggregates. The comparison of several purifications using unrelated target proteins is often instructive. Indeed, if the same proteins are repeatedly recovered in the purified fractions, they are likely to represent contaminants, particularly if they represent abundant cellular proteins (e.g., translation factors) or proteins known to interact with unfolded polypeptides (e.g., heat shock proteins, proteasome). In any case, it is important to verify that a protein identified in a TAP-purified fraction interacts physiologically with the target protein using an independent method. Coimmunoprecipitation (see Chapter 5) appears as a method of choice for such verification. In the absence of antibody, the ProtA and CBP tags can be used to test such coprecipitations. First, one of the target proteins is tagged with CBP (or a full TAP tag) while the putative interacting partner is tagged with ProtA (or the fusion TEV protease cleavage site-ProtA). Cells expressing both fusion proteins are lysed and incubated with calmodulin-coated beads in the presence of calcium. Samples from the input extracts, supernatant fraction, and purified material are then fractionated by gel electrophoresis before detecting the ProtA-protein fusion(s) by western blotting (Salgado-Garrido et al. 1999; Bouveret et al. 2000).

Preparation of Extracts to Purify the Yeast Lsm3 Protein

Using this protocol, more than 20 proteins ranging from 8 to 250 kD were visible by Coomassie staining in the purified fraction and were identified by mass spectrometry (Bouveret et al. 2000). Quantification of the associated U6 snRNA by primer extension indicated that about 25–40% of the starting extract material was recovered in the final fraction. The Lsm3 protein belongs to the Sm protein family (Séraphin 1995) and is used here as a representative sample.

MATERIALS

CAUTION: See Appendix for appropriate handling of materials marked with <!>.

Buffers and Solutions

Buffer A
  10 mM K-HEPES (pH 7)
  10 mM potassium chloride (KCl) <!>
  1.5 mM magnesium chloride ($MgCl_2$) <!>
  0.5 mM dithiothreitol (DTT) <!>
  0.5 mM phenylmethylfulfonyl fluoride (PMSF) <!>
  2 mM benzamidine
  0.5 μg/ml (1 μM) leupeptin <!>
  1.4 μg/ml (2 μM) pepstatin A <!>
  2.4 μg/ml (4 μM) cymostatin
  17 μg/ml (2.6 μM) aprotinin <!>
Buffer D
  20 mM K-HEPES (pH 7.9)
  50 mM KCl <!>
  0.2 mM EDTA (pH 8.0)
  0.5 mM DTT <!>
  20% glycerol
  0.5 mM PMSF <!>
  2 mM benzamidine

Additional Reagents

LSM3
Tag for protein purification

Special Equipment

French press

METHOD

1. To identify interacting factors in yeast (or another protein of interest), fuse a cassette encoding the TAP tag to the carboxyl terminus of the *LSM3* reading frame using standard cloning techniques on a low-copy centromeric plasmid. The resulting construct should complement a disruption of the essential *LSM3* gene to indicate that it is fully functional.

2. Grow up 2 liters of culture of the resulting strain to saturation.

3. Recover cells by centrifugation, wash with H_2O, pellet again, and freeze at $-80°C$ before extract preparation.

4. Resuspend the cells in one pellet volume of Buffer A. Break by passing two times through a French press at 8.27 MPa (1200 psi) at 4°C.

5. Adjust this crude extract to 200 mM KCl using a 2 M stock.

6. Remove cell debris by two consecutive centrifugation steps at 4°C (50,000g for 30 minutes followed by 130,000g for 85 minutes).

7. Dialyze the cleared extracts for 3 hours against Buffer D at 4°C and store frozen at $-80°C$ before using for TAP purification using the protocol described below. The final volume of extract from 2 liters of starting culture should be approximately 10 ml.

Protocol 2

Dissecting Protein Interactions with the TAP Method

The TAP method can be used to identify interacting partners, and it can also be used to characterize protein interactions occurring in a complex in more detail. For example, purification of a mutant or a truncated protein fused to the TAP tag can be used to map interaction domains. Similarly, complexes may be purified from cells carrying a mutation or even a deletion of one of the complex subunits that is different from the target protein. This type of analysis can give some insight into interactions occurring in the complex or on the assembly pathway of this complex. Repeating the TAP purification of a protein complex under various conditions (e.g., increased salt or detergent concentration) may also be used to differentiate protein(s) interacting directly with the target protein from more peripheral subunits. Phylogenetic comparison of complexes may also be instructive because different subcomplexes may exist in various species, giving information about protein interaction order. Furthermore, phylogeny may be used to identify protein partners. For example, if it is not possible to purify a protein from a given species, one may consider purifying its homolog from a species where the TAP method is easy to implement. This strategy has proven useful in the analysis of an RNA helicase involved in mRNA transport (Schmitt et al. 1999).

MATERIALS

CAUTION: See Appendix for appropriate handling of materials marked with <!>.

Buffers and Solutions

Unless otherwise stated, all buffers are cooled to the working temperature of 4°C before use.

IPP150 buffer
    10 mM Tris-Cl (pH 8.0)
    150 mM NaCl
    0.1% Nonidet P-40 (NP-40)
IPP150 calmodulin-binding buffer
    10 mM β-mercaptoethanol (dilute and add immediately before use) <!>
    10 mM Tris-Cl (pH 8.0)
    150 mM NaCl
    1 mM magnesium acetate <!>
    1 mM imidazole <!>
    2 mM $CaCl_2$
    0.1% NP-40
IPP150 calmodulin elution buffer
    10 mM β-mercaptoethanol (dilute immediately before use) <!>
    10 mM Tris-Cl (pH 8.0)
    150 mM NaCl
    1 mM magnesium acetate <!>
    1 mM imidazole <!>
    2 mM EGTA
    0.1% NP-40

324

Enzymes and Buffers

TEV cleavage buffer
   10 mM Tris-Cl (pH 8.0)
   150 mM NaCl
   0.1% NP-40
   0.5 mM EDTA
   1 mM DTT (add immediately before use from 1 M stock) <!>
TEV protease (GIBCO 10127-017)

Affinity Media

Calmodulin affinity resin (Stratagene 214303)
IgG agarose (Sigma A 2909) or IgG Sepharose (Pharmacia 17-0969-01)

Extracts

Appropriate cell lines or strains expressing a TAP-tagged protein to prepare extracts. As a guideline, we store yeast extracts before purification in 20 mM HEPES (pH 7.9) (adjusted with KOH), 50 mM KCl, 0.2 mM EDTA, 0.5 mM DTT <!>, 20% glycerol, 0.5 mM PMSF <!>, 2 mM benzamidine.

Special Equipment

Cold room space
Incubator (16°C) with rotating wheel
Poly-Prep Chromatography Column (10-ml; Bio-Rad 731-1550)
Rotating wheel

METHOD

1. Wash 200 µl of IgG agarose or IgG Sepharose beads with 5 ml of IPP150 buffer in a closed purification column. Rotate the beads and wash buffer in the cold room for 5 minutes. Remove the IPP150 buffer by letting the column drain by gravity.

2. Adjust the extract buffer concentration by adding 50 µl of 2 M Tris-Cl (pH 8.0), 200 µl of 5 M NaCl, and 100 µl of 10% NP-40 per 10 ml of extract.

3. Add equilibrated extract (10 ml) to the washed beads. Close the column (top and bottom). Rotate for 2 hours at 4°C.

4. Remove the unbound material by opening the column (top first) and letting it drain by gravity. This fraction may be saved for control analysis (e.g., estimation of the binding efficiency by western blotting).

5. Wash the column with 3 × 10 ml of IPP150 buffer.
   Careful cleaning of the column wall and column can help in reducing the level of contaminants.

6. Equilibrate the column for the TEV cleavage reaction by washing the column with 10 ml of TEV cleavage buffer at 16°C.

7. Close the bottom of the column. Add 1 ml of TEV cleavage buffer and 100 units of TEV protease. Close the top of the column. Rotate for 2 hours at 16°C.

8. Remove first the top and then the bottom plug of the column. Recover the eluate by gravity flow. The solution remaining in the column dead-volume and on the column walls may be eluted with an additional 200 µl of TEV cleavage buffer.

   Material remaining on the column may be eluted with a buffer containing SDS, such as gel loading buffer or buffer 1× PK (100 mM Tris-Cl, pH 7.5, 12.5 mM EDTA, pH 8.0, 150 mM NaCl, 1% SDS) and saved for control analysis.

9. Wash 200 µl of calmodulin affinity resin with 5 ml of IPP150 calmodulin-binding buffer in a new purification column. Rotate the beads and the IPP150 calmodulin-binding buffer in the closed column for 5 minutes in the cold room. Remove the IPP150 calmodulin-binding buffer by letting the column drain by gravity.

10. To the previous eluate add 3 volumes of IPP150 calmodulin-binding buffer and 1 µl of 1 M CaCl$_2$ per milliliter of eluate to titrate the EDTA present in the TEV cleavage buffer.

11. Transfer the equilibrated solution to the column containing the washed calmodulin affinity resin. Close the column (top and bottom). Rotate for 2 hours at 4°C.

12. Remove the unbound material by opening the column (top first) and letting it drain by gravity. This fraction may be saved for control analysis.

13. Wash the column with 3 × 10 ml of IPP150 calmodulin-binding buffer.

    Careful cleaning of the column wall may reduce the level of contaminants.

14. Elute 5 fractions of 200 µl with IPP150 calmodulin elution buffer.

Proteins present in these fractions may be dialyzed against the appropriate buffer for functional analysis or concentrated (e.g., by TCA or methanol-chloroform precipitation; Wessel and Flugge 1984; Ozols 1990) before fractionation on a denaturing gel.

TROUBLESHOOTING AND LIMITATIONS

It is often valuable to follow quantitatively the purification of the target protein by western blotting. This allows the determination of problematic steps and the design of potential solutions. It is also valuable to obtain cells expressing a known TAP-tagged fusion protein as a positive control for the first purification to check that the various buffers and affinity material are fully functional.

Although the TAP method is broadly applicable, there are some limitations. First, a functional TAP-tagged protein must be produced. For some proteins, tagging at both the amino and the carboxyl termini of the protein may affect its activity. In our experience, however, this is uncommon. Even if the protein is functional, a similar problem may occur if the TAP tag is inaccessible in the native protein complex. In this case, we suggest testing a construct expressing the target protein fused to the TAP tag at its other extremity. Although some proteins also contain an endogenous TEV protease cleavage site that interferes with the purification, this is not likely to be a very common problem. Indeed, as an example, only a dozen polypeptides among the 6000 yeast proteins are predicted to contain a TEV protease cleavage site (consensus sequence ENLYFQG, Dougherty et al. 1989). One should remember, however, that the TEV protease is not a restriction enzyme and that degenerate sites may be cleaved if well exposed, whereas perfect sites buried in the interior of the protein will not be accessible to the protease. The presence of a bona fide TEV protease cleavage site in the target protein remains to be determined by an experimental approach.

REFERENCES

Baudin A., Ozier K.O., Denouel A., Lacroute F., and Cullin C. 1993. A simple and efficient method for direct gene deletion in *Saccharomyces cerevisiae*. *Nucleic Acids Res.* **21**: 3329–3330.

Bouveret E., Rigaut G., Shevchenko A., Wilm M., and Seraphin B. 2000. A Sm-like protein complex that participates in mRNA degradation. *EMBO J.* **19**: 1661–1671.

Brizzard B.L., Chubet R.G., and Vizard D.L. 1994. Immunoaffinity purification of FLAG epitope-tagged bacterial alkaline phosphatase using a novel monoclonal antibody and peptide elution. *BioTechniques* **16**: 730–735.

Caspary F., Shevchenko A., Wilm M., and Séraphin B. 1999. Partial purification of the yeast U2 snRNP reveals a novel yeast pre-mRNA splicing factor required for pre-spliceosome assembly. *EMBO J.* **18**: 3463–3474.

Chong S., Mersha F.B., Comb D.G., Scott M.E., Landry D., Vence L.M., Perler F.B., Benner J., Kucera R.B., Hirvonen C.A., Pelletier J.J., Paulus H., and Xu M.Q. 1997. Single-column purification of free recombinant proteins using a self-cleavable affinity tag derived from a protein splicing element. *Gene* **192**: 271–281.

Deutscher M.P. 1990. *Guide to protein purification*. Academic Press, San Diego.

Dougherty W.G., Cary S.M., and Parks T.D. 1989. Molecular genetic analysis of a plant virus polyprotein cleavage site: A model. *Virology* **171**: 356–364.

Gygi S.P., Corthals G.L., Zhang Y., Rochon Y., and Aebersold R. 2000. Evaluation of two-dimensional gel electrophoresis-based proteome analysis technology. *Proc. Natl. Acad. Sci.* **97**: 9390–9395.

Hochuli E., Bannwarth W., Doebeli H., Gentz R., and Stueber D. 1988. Genetic approach to facilitate purification of recombinant proteins with a novel metal chelate adjustment. *Bio/Technology* **6**: 1321–1325.

Lamond A.I. and Mann M. 1997. Cell biology and the genome projects—A concerted strategy for characterizing multi-protein complexes using mass-spectrometry. *Trends Cell Biol.* **7**: 139–142.

Link A.J., Eng J., Schieltz D.M., Carmack E., Mize G.J., Morris D.R., Garvik B.M., and Yates J.R., III. 1999. Direct analysis of protein complexes using mass spectrometry. *Nat. Biotechnol.* **17**: 676–682.

Lohman T.M., Chao K., Green J.M., Sage S., and Runyon G.T. 1989. Large-scale purification and characterization of the *Escherichia coli* rep gene product. *J. Biol. Chem.* **264**: 10139–10147.

Ozols J. 1990. Amino acid analysis. *Methods Enzymol.* **182**: 587–601.

Puig O., Rutz B., Luukkonen B.G., Kandels L.S., Bragado N.E., and Séraphin B. 1998. New constructs and strategies for efficient PCR-based gene manipulations in yeast. *Yeast* **14**: 1139–1146.

Puig O., Caspary F., Rigaut G., Rutz B., Bouveret E., Bragado-Nilsson E., Wilm M., and Séraphin B. 2001. The Tandem Affinity Purification (TAP) method: A general procedure of protein complex purification. *Methods* **24**: 218–229.

Rigaut G., Shevchenko A., Rutz B., Wilm M., Mann M., and Séraphin B. 1999. A generic protein purification method for protein complex characterization and proteome exploration. *Nat. Biotechnol.* **17**: 1030–1032.

Salgado-Garrido J., Bragado-Nilsson E., Kandels-Lewis S., and Séraphin B. 1999. Sm and Sm-like proteins assemble in two related complexes of deep evolutionary origin. *EMBO J.* **18**: 3451–3462.

Sambrook J. and Russell D.W. 2001. *Molecular cloning: A laboratory manual*, 3rd edition. Cold Spring Harbor Laboratory Press, Cold Spring Harbor, New York.

Schmidt T.G. and Skerra A. 1993. The random peptide library-assisted engineering of a C-terminal affinity peptide, useful for the detection and purification of a functional Ig Fv fragment. *Protein Eng.* **6**: 109–122.

Schmitt C., von Kobbe C., Bachi A., Pante N., Rodrigues J.P., Boscheron C., Rigaut G., Wilm M., Séraphin B., Carmo-Fonseca M., and Izaurralde E. 1999. Dbp5, a DEAD-box protein required for mRNA export, is recruited to the cytoplasmic fibrils of nuclear pore complex via a conserved interaction with CAN/Nup159p. *EMBO J.* **18**: 4332–4347.

Séraphin B. 1995. Sm and Sm-like proteins belong to a large family: Identification of proteins of the U6 as well as the U1, U2, U4 and U5 snRNPs. *EMBO J.* **14**: 2089–2098.

Shevchenko A., Jensen O.N., Podtelejnikov A.V., Sagliocco F., Wilm M., Vorm O., Mortensen P., Shevchenko A., Boucherie H., and Mann M. 1996. Linking genome and proteome by mass spectrometry: Large-scale identification of yeast proteins from two dimensional gels. *Proc. Natl. Acad. Sci.* **93**: 14440–14445.

Stofko-Hahn R.E., Carr D.W., and Scott J.D. 1992. A single step purification for recombinant proteins. Characterization of a microtubule associated protein (MAP 2) fragment which associates with the type II cAMP-dependent protein kinase. *FEBS Lett.* **302**: 274–278.

Swaffield J.C., Melcher K., and Johnston S.A. 1995. A highly conserved ATPase protein as a mediator between acidic activation domains and the TATA-binding protein. *Nature* **374**: 88–91.

—— 1996. A highly conserved ATPase protein as a mediator between acidic activation domains and the

TATA-binding protein (Correction). *Nature* **379:** 658.

Wasinger V.C., Cordwell S.J., Cerpa-Poljak A., Yan J.X., Gooley A.A., Wilkins M.R., Duncan M.W., Harris R., Williams K.L., and Humphery-Smith I. 1995. Progress with gene-product mapping of the Mollicutes: *Mycoplasma genitalium. Electrophoresis* **16:** 1090–1094.

Wessel D. and Flugge U.I. 1984. A method for the quantitative recovery of protein in dilute solution in the presence of detergents and lipids. *Anal. Biochem.* **138:** 141–143.

Wilkins M.R., Pasquali C., Appel R.D., Ou K., Golaz O., Sanchez J.C., Yan J.X., Gooley A.A., Hughes G., Humphery-Smith I., Williams K.L., and Hochstrasser D.F. 1996. From proteins to proteomes: Large scale protein identification by two-dimensional electrophoresis and amino acid analysis. *Bio/Technology* **14:** 61–65.

Wilm M., Shevchenko A., Houthaeve T., Breit S., Schweigerer L., Fotsis T., and Mann M. 1996. Femtomole sequencing of proteins from polyacrylamide gels by nano-electrospray mass spectrometry. *Nature* **379:** 466–469.

18 Protein Purification by Inverse Transition Cycling

Dan E. Meyer and Ashutosh Chilkoti

Department of Biomedical Engineering, Duke University, Durham, North Carolina 27708

INTRODUCTION

Elastin-like polypeptides (ELPs) are environmentally responsive biopolymers based on the elastin-derived pentapeptide repeat Val-Pro-Gly-Val-Gly. ELPs undergo a reversible phase transition termed an "inverse temperature transition" (Urry 1992, 1997). Below their *transition temperature* (T_t), the polypeptides are highly soluble in aqueous solutions. However, when the temperature is raised above T_t, the hydrated polypeptide chains hydrophobically collapse and aggregate, forming a separate, ELP-rich phase. The T_t of an ELP can be conveniently controlled at several different levels. For example, the T_t can be adjusted over a wide range of temperatures through control of the amino acid sequence. The transition can also be triggered isothermally by modulation of environmental conditions, in particular by the type and concentration of added co-solutes.

Significantly, ELP fusion proteins, which are produced by joining the gene encoding a protein of interest with an ELP gene segment, can also undergo a reversible phase transition similar to that of the free ELP (Meyer and Chilkoti 1999). Thus, the environmental responsiveness of ELPs

329

TABLE 1. Uses of ELP Fusion Proteins and ITC

Mode	Use
ELP fusion proteins per se	postexpression purification of recombinant proteins (scale-up: industrial quantities; scale-down: parallel, high-throughput purification of expressed gene libraries) separation of reactants (e.g., after protein labeling) enzyme recycling buffer exchange and concentration of the fusion protein
Non-covalent capture using ELP fusions	characterization of unknown binding partners (in serial or high-throughput format) determination of equilibrium binding and rate constants competitive immunoassays

can be easily imparted by genetic fusion to a protein of interest. This is useful because, when the transition is triggered, the fusion protein aggregates and can be collected by centrifugation or filtration. The transition is reversible, and therefore the pelleted ELP fusion protein can be re-solubilized after removal of the supernatant by dissolution in aqueous solution at a temperature less than the T_t. It is important to note that the fused target protein does not denature and precipitate when the phase transition is induced, but rather, it is the ELP that aggregates. For free ELPs, the aggregated phase contains up to ~60% water by weight (Urry et al. 1985), and the target protein apparently remains sufficiently hydrated to prevent its denaturation. Thus, an ELP tag genetically incorporated into a protein of interest provides a "molecular handle" to manipulate the protein, which allows easy and rapid purification, concentration, and buffer exchange. Potential applications of inverse transition cycling (ITC) are summarized in Table 1.

In an extension of this concept, ELP fusion proteins can also be used for non-covalent capture of affinity-binding partners from solution (D.E. Meyer and A. Chilkoti, in prep.). After binding, the non-covalent complex can be separated from solution upon triggering the ELP phase transition. This concept has many potential applications. First, non-covalent capture of molecules by an ELP fusion protein is a potentially useful tool to study protein–ligand interactions. For example, a protein of interest (antibody or antibody fragment, antigen, receptor, ligand, etc.) could be fused to an ELP and used to "fish out" an unknown binding partner from a solution-phase combinatorial library. After capture and purification, the complex can be interrogated by mass spectrometry or other spectroscopic techniques to identify the bound ligand. Second, capture of a labeled affinity partner from solution can also be used to determine equilibrium binding and rate constants, based on the quantification of the free and bound (i.e., captured) binding partner as a function of concentration and time. Third, non-covalent binding of a target analyte by an ELP fusion protein can also be used as the basis for competitive immunoassays for the quantification of analyte concentration in complex mixtures.

For each of these applications, the principle of ITC is used to selectively remove the ELP fusion protein or its non-covalent complex with a ligand from solution. Figure 1A shows a schematic of ITC, and Figure 1B shows typical SDS-PAGE results of protein purification by ITC from cell lysate. Initially, the ELP fusion protein is below its T_t and is soluble. The inverse temperature transition is triggered by increasing the solution temperature to above the T_t or by adding salt to depress isothermally the T_t to below the solution temperature, or by a combination of both. Upon triggering the transition, the fusion protein aggregates and can be separated from other molecules present in solution by centrifugation or filtration. The remaining soluble molecules are removed by decanting or pipetting off the supernatant. The purified fusion protein is then resolubilized at a temperature below its T_t in the buffer and volume of choice. The resolubilized fusion protein is often centrifuged (or filtered) a final time at a temperature below its T_t to remove any remaining insoluble matter that may have been trapped in the pellet along with the aggregated ELP fusion protein. Additional rounds of ITC can be undertaken until the fusion protein is purified to the desired degree.

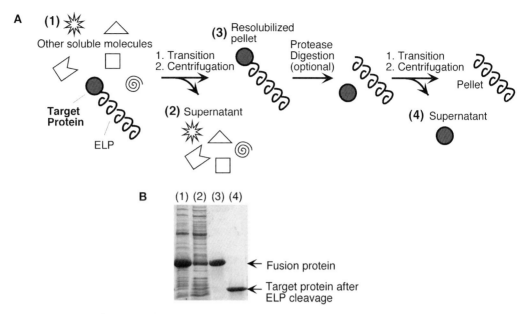

FIGURE 1. (*A*) Schematic of protein separation by inverse transition cycling (ITC). The ELP–target fusion protein is initially present in a complex mixture (e.g., cellular lysate, immunoassay sample solution). The ELP inverse transition is triggered by increasing the solution temperature and/or by adding NaCl, resulting in aggregation of the fusion protein. After centrifugation (or filtration), the supernatant (or filtrate) containing the contaminating soluble molecules is discarded, and the fusion protein is recovered by resolubilizing the pellet (or washing the filter) with buffer at a temperature below the T_t. If necessary, the target protein can be separated from the ELP by cleavage with a site-specific protease at a recognition site between the target protein and ELP, followed by a final round of ITC. (*B*) Typical SDS-PAGE results illustrating the purification of an ELP fusion protein from the lysate. Lane numbers correspond to the steps as labeled in *A*. Lane *1* is the soluble cell lysate with the overexpressed *E. coli* thioredoxin/ELP[V-20] fusion protein and contaminating *E. coli* proteins. Lanes *2* and *3* are the supernatant and pellet, respectively, after one round of ITC. Purified fusion protein is obtained in the pellet, while the contaminating *E. coli* proteins are discarded in the supernatant. Lane *4* shows purified thioredoxin after cleavage with thrombin and purification by ITC from the cleaved ELP tag.

ITC is technically simple, fast, economical, and requires no specialized equipment or reagents. Separations can be completed in minutes using standard laboratory centrifuges and NaCl. Furthermore, these advantages can be realized over a wide range of purification scales. For microliter-scale expression and purification, ITC can be employed in parallel for high-throughput applications. At the other extreme, because it avoids the expense associated with traditional chromatographic separations, ITC is also likely to be useful for the purification of single proteins at the gram-to-kilogram scale in industrial bioprocessing.

OUTLINE OF PROCEDURE

General

ITC purification is a convenient method for a variety of separations. Common to all these applications is the concept that genetic fusion of a protein of interest to an ELP imparts the environmental sensitivity of the ELP to the fusion protein. Triggering the ELP inverse temperature transition by changing environmental parameters, which are typically the solution temperature and ionic composition, allows the target protein to be removed selectively from solution.

A simple but important concept to remember when working with ELPs and their fusion proteins is that there is no unique and fixed T_t associated with a given construct. Rather, the T_t (i.e.,

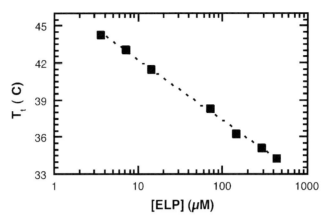

FIGURE 2. Effect of ELP concentration on the T_t. The T_t was determined for ELP[$V_5A_2G_3$-180] at each concentration in PBS as follows. The solutions were heated at a rate of 1°C per minute in a temperature-controlled spectrophotometer, and the optical density at 350 nm was measured as a function of temperature. The T_t was defined at the point of 50% maximal turbidity. The data show that, for a given ELP sequence and molecular weight, the T_t decreases logarithmically (*dashed line*) with increasing ELP concentration. For ITC applications, this effect is of practical importance for ELP fusion protein concentrations lower than ~100 μM, which are typical for ELP fusion proteins in the cellular lysate.

the temperature at which the transition occurs for given environmental conditions) is dependent on a number of factors. Note that, for the purpose of this discussion, reported T_t values are at an ELP fusion protein concentration of 25 μM in phosphate-buffered saline (PBS) unless otherwise specified.

For separation applications using ITC, the two most relevant factors affecting T_t are (1) the concentration of the ELP fusion protein and (2) the solution composition (in particular, types of ions and their concentrations). Examples of the effects of these two factors on the T_t for a given ELP construct are shown in Figures 2 and 3. Because the T_t decreases logarithmically with increasing concentration (Fig. 2), the effect of ELP concentration is of greatest practical importance at concentrations lower than ~100 μM. Increasing the NaCl concentration also decreases the T_t, and, as a rule of thumb, the T_t is depressed by ~15°C for each 1 M increase in NaCl concentration (although this effect could vary for ELPs with different fractions of charged residues, for example). The effect of ions on the T_t of ELPs in solution appears to follow the Hofmeister series (Cacace et al. 1997), and as shown in Figure 3, ions such as guanidinium can be used to increase the T_t, as well.

The T_t of an ELP also depends on its amino acid sequence and its molecular weight, as illustrated in Figure 4. The sequence dependence of the ELP T_t has been studied extensively by Urry and coworkers, who have found that substitution at the fourth residue of the Val-Pro-Gly-Val-Gly pentapeptide, termed the "guest residue," with residues more hydrophilic than Val increases the T_t from that of the native sequence, whereas substitution with more hydrophobic residues reduces the T_t (Urry et al. 1991). These data allow the rational design of ELPs exhibiting a specific target T_t through selection of the type of guest residues and their relative frequency of substitution in the ELP sequence (Table 2). The effect of molecular weight is of greatest practical importance for short ELPs (e.g., less than ~50 pentapeptides in length) and becomes less important for higher molecular weights (Fig. 4). Here, the different ELP constructs are distinguished using the notation ELP[X_iY_j-n], where the bracketed capital letters are the single-letter amino acid codes and the corresponding subscripts designate the frequency of each guest residue in the repeat unit, and n describes the total length of the ELP in number of pentapeptides. For example, ELP[$V_5A_2G_3$-180] is an ELP that has a repeating sequence of 10 pentapeptides with the guest residues Val, Ala, and Gly in a 5:2:3 ratio, respectively, and is 180 pentapeptides in length.

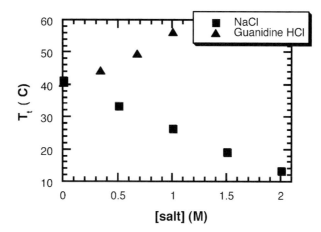

FIGURE 3. Effect of ions on the T_t. The T_t was determined for ELP[$V_5A_2G_3$-180] at 25 μM in PBS supplemented with increasing concentrations of NaCl or guanidine HCl. (The T_t was defined as described in Fig. 2.) The data show that the T_t can be modulated by the ionic composition of the solution for a given ELP sequence, molecular weight, and concentration. This allows "fine tuning" of the T_t to a desired temperature during ITC.

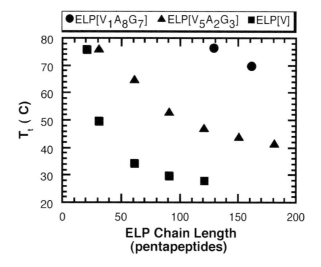

FIGURE 4. Effect of ELP sequence and length on T_t. The T_t was determined (as described in Fig. 2) in PBS for 25 μM solutions of ELP with varied sequences and chain lengths. The T_t increases for a given molecular weight with increasing hydrophilicity of the guest residue in the ELP sequence (see text for sequence notation), and increases with decreasing molecular weight for a given sequence. The molecular-weight effect is most dramatically observed at lower chain lengths (e.g., less than ~50 pentapeptides). However, even ELPs with short chain lengths can be useful in ELP fusion protein applications because their relatively high T_t values can be functionally depressed by the addition of NaCl or by increasing the ELP concentration (9).

TABLE 2. Examples of ELP Sequences

Construct	Monomer sequence[a]	T_t (°C)[b]
ELP[V]	(VPGVG)$_5$	30
ELP[$V_5A_2G_3$]	(VPGVG)$_2$VPGAG VPGGG(VPGVG)$_3$VPGGG VPGAG VPGGG	40
ELP[$V_1A_8G_7$]	(VPGVG)VPGAG(VPGGG VPGAG)$_7$	70

[a]Polypeptide sequence encoded by the monomer gene.

[b]Approximate transition temperature for high-molecular-weight ELPs (i.e., greater than 100 pentapeptides) at 25 μM concentration in PBS.

The T_t of an ELP with a particular sequence and molecular weight can also be affected by the target protein to which it is fused. Proteins with a large fraction of solvent-exposed hydrophobic surface area can depress the T_t of a fused ELP. In one example, the protein tendamistat, with a calculated exposed hydrophobic surface area of 53%, caused a ~26°C reduction in T_t (Meyer and Chilkoti 1999). Most soluble proteins (>90% in a survey of 100 proteins from the Protein Data Bank), however, have a calculated exposed hydrophobic surface area of less than 30%. Based on experimental observations, fusion to proteins with this degree of exposed hydrophobic surface area should have an insignificant effect on the ELP T_t (less than a 5°C change relative to free ELP).

Fusion Protein Design Considerations

A gene for a typical ELP fusion protein includes three components: (1) the gene of the target protein of interest, (2) an intervening peptide linker sequence, and (3) the ELP gene. Typically, the gene of the fusion is constructed in three sequential steps. If the gene of interest (encoding the target protein) is already available in an expression vector, this vector can be modified using a synthetic oligonucleotide cassette to include the endonuclease recognition sequence for ELP insertion and codons for a peptide linker sequence. An ELP gene segment can then be cloned into the ELP insertion site as described below in the Protocols section, completing the assembly of the gene for the ELP fusion protein. This last step can be completed in parallel, if desired, to create a library of different ELPs fused to the same target protein.

Alternatively, if a number of different target proteins are to be studied, it may be preferable to synthesize a generic expression vector containing an ELP tag into which the genes of the various target proteins can be cloned. This can be achieved by first modifying an expression vector, adjacent to a multiple cloning site, to introduce the ELP insertion site and to encode for the linker sequence. Next, the selected ELP gene segment is cloned into the ELP insertion site. The vector is then ready for the parallel cloning of a library of target genes into the multiple cloning site adjacent to the ELP.

The peptide linker sequence can be designed to incorporate several useful features. First, it can provide a spacer between the target protein and ELP. This may help to retain functional activity of the target protein if the ELP is fused near the active site of the folded protein. Second, the linker sequence can encode for a protease recognition site so that the ELP can be cleaved from the target protein after ITC purification. Thrombin, with a Leu-Val-Pro-Arg|Gly-Ser recognition sequence, has been used successfully with ELP fusion proteins (Meyer and Chilkoti 1999). Factor Xa protease has also been used with ELP fusion proteins (McPherson et al. 1992), and other proteases may also be useful. Finally, a secondary purification tag (e.g., a His$_6$ tag) can be included between the protease recognition site and the target protein for purification after cleavage from the ELP.

There is no clear rationale for the choice of constructing amino- or carboxy-terminal ELP fusions, and this decision may depend on the properties of the target protein. If the target protein is known to express at a high level, it may be preferable to fuse the ELP to the carboxyl terminus. (ELP incorporated at the amino terminus could interfere with proper folding of the target protein, in which case a carboxy-terminal ELP fusion may be preferable because it is translated after synthesis of the target protein is completed.) On the other hand, an amino-terminal ELP tag has been observed to increase the soluble expression of a poorly expressed target protein (Meyer and Chilkoti 1999). This may be because an amino-terminal ELP aids folding of the target protein by hydrophobic interactions with the nascent target protein chain or by sterically preventing intermolecular aggregation of folding intermediates of the target protein. Given all of these considerations, it may be best to try both configurations.

Selection of the ELP Tag and ITC Conditions

The implementation of ITC can be optimized in terms of fusion protein expression and separation efficiency for a given target protein by control of ELP amino acid sequence, ELP molecular weight, and solution conditions (temperature and NaCl concentration). This may be important for some applications, such as large-scale production of a target protein where maximization of yield is an important goal. Alternatively, a generic protocol has been developed that may not be optimal for all proteins, but should provide acceptable separations for a wide variety of proteins. This approach would be more useful, for example, in high-throughput expression and purification of a large number of different proteins with unknown properties.

For general implementation, an ELP of moderate length (~36 kD) with a T_t of ~54°C (e.g., ELP[$V_5A_2G_3$-90]) has been found to be useful for a range of target proteins. The T_t of this ELP is high enough to withstand a downward shift induced by fusion to a hydrophobic target protein, yet low enough so that the T_t can be reduced to 30–35°C with the addition of only moderate amounts of NaCl (1–2 M). Use of this (or a similar) ELP tag is recommended for initial, pilot ITC purification of a new target protein. For poorly expressed target proteins, for which the T_t is too high or for which no phase transition is observed due to low concentration, the soluble lysate can be spiked with free ELP prior to purification by ITC. The addition of free ELP reduces the T_t by the concentration effect and helps to ensure efficient capture of the aggregated proteins by centrifugation.

Complete optimization of ITC may require empirical testing of different ELP tags for a given target protein. Because the expression yields of a target protein, *E. coli* thioredoxin, have been found to increase with decreasing ELP tag length (Meyer and Chilkoti 1999), the goal is to determine the shortest ELP tag that will still allow efficient purification at moderate temperatures and NaCl concentrations. This can be achieved using test expression cultures of the target protein fused to a small library of ELPs (e.g., three or four constructs with the ELP tag ranging in size from ~10 kD to ~35 kD). One ELP at the low end of this range, the ~9 kD ELP[V-20] tag, has been successfully used for purification of thioredoxin by ITC, and provided a fourfold increase in target protein yield versus a control 35-kD ELP tag (Meyer et al. 2001). However, the inverse transition of the fusion protein with the 9-kD tag had a more complex dependence on temperature (Meyer et al. 2001), and furthermore, some target proteins failed to exhibit an inverse temperature transition when fused to tags of this size (D.E. Meyer and A. Chilkoti, unpubl.). From a design perspective, because the T_t increases with decreasing molecular weight, smaller tags require the incorporation of a greater fraction of hydrophobic guest residues to keep the T_t low enough such that ITC can be performed at conditions that do not denature the target protein. At the other end of the ELP molecular-weight range (e.g., the 35-kD ELP[$V_5A_2G_3$-90]), all fusion proteins tested to date, albeit limited in number, have been successfully purified by ITC. Larger ELP tags may also protect the reversibility of the temperature-induced aggregation by limiting intermolecular contact of the target protein in the aggregated state.

For postexpression purification, a target T_t above 37°C is typically selected so that the fusion protein remains soluble in vivo during culture, while allowing the inverse temperature transition to be triggered by only a small increase in temperature or NaCl concentration in the soluble cell lysate. This selection of the T_t is based on the assumption that expression of the ELP in its stable state is likely to result in a higher yield of properly folded target protein. However, this assumption has not yet been tested in detail, and it is possible that a T_t below the culture temperature may be preferable because formation of an aggregated ELP phase within the cell may protect the target protein from degradation by intracellular proteases.

For ITC applications involving the noncovalent binding of a target molecule to an ELP fusion protein, the effect of ITC conditions on binding affinity must also be considered. A T_t slightly above (e.g, ~5°C or less) the incubation temperature of the reaction allows separation of the complex from solution by only a modest increase in temperature (0–5°C) and/or a modest increase in NaCl concentration (0–300 mM NaCl). Any decrease in affinity due to ITC performed under these conditions is likely to be negligible. On the other hand, larger swings in temperature or ionic strength could lead to undesired shifts in binding equilibrium, resulting in release of the target molecule prior to capture by ELP aggregation.

Fusion of an ELP Moiety to a Protein of Interest

Genes for ELPs and for ELP fusion proteins can be readily synthesized using standard methods of molecular biology. A number of approaches for synthesizing ELP genes have been described in the literature (McPherson et al. 1996; McMillan et al. 1999; Meyer and Chilkoti 1999). A protocol for producing ELP fusion proteins using genes generated by one of these methods is presented below. At the time of writing, genes encoding ELPs as described in this protocol are available free of charge from the authors for noncommercial research purposes. Note that this protocol serves only as an example, and the DNA amounts, fluid volumes, etc., can be successfully varied from those shown here.

MATERIALS

Buffers and Reagents

Calf intestinal alkaline phosphatase (CIAP)
Enzyme buffers
10x Ligase buffer
Restriction enzymes
T4 DNA ligase

Vectors

Expression vector containing the target protein gene that has been modified to include a unique restriction endonuclease recognition site for ELP insertion
Vector containing an ELP gene, which is flanked by *Pfl*MI (at the 5′ end of the gene) and *Bgl*I (at the 3′ end) recognition sequences, which are both cleaved to form 3′ single-stranded GGC extensions

Special Equipment

Gel Electrophoresis and PCR
Agarose gel electrophoresis equipment
Agarose gel extraction kit
PCR purification kit

METHOD

Preparation of ELP Insert

Due to their repetitive nature, larger ELP genes produced by gene oligomerization are not effectively amplified using PCR.

1. Digest ~4 µg of the ELP plasmid DNA to completion with *Pfl*MI and *Bgl*I.

2. Purify the ELP fragment by agarose gel extraction (commercial kits are available for this purpose). The ELP gene fragment size in base pairs is the number of pentapeptides encoded mul-

tiplied by 15 (3 bases per amino acid × 5 amino acid residues per pentapeptide). Elute in 30 μl of H$_2$O and store frozen at –20°C until further use.

> Avoid heating or vortexing the sample during gel extraction.

Vector Preparation

3. Digest ~1 μg of the expression vector plasmid DNA with the appropriate restriction endonuclease that specifically cleaves at the ELP insertion site.

> The ELP insertion site is a recognition site for a restriction endonuclease that (1) is unique to the expression vector and (2) creates a nonpalindromic overhang compatible with the ELP insert upon cleavage (e.g., 5′-GGC-3′). The insertion site must be positioned such that it is in frame with the coding sequence of the fusion protein, and stop codons must be included downstream from the insertion site. Examples of restriction enzymes that are compatible with this particular system include: *Alw*NI, *Bgl*I, *Bsl*I, *Bst*API, *Dra*III, *Mwo*I, *Pfl*MI, and *Sfi*I. Introduction of the ELP insertion site can be achieved by cassette mutagenesis of the expression vector that already contains the gene of the target protein.

   Confirm the plasmid digestion by agarose gel electrophoresis.

> It is critical that the target protein vector be fully digested. Incomplete digestion can produce an overwhelming background of wild-type vector upon transformation.

> If digestion of the plasmid DNA is incomplete, add more enzyme and/or allow the reaction to proceed longer until the vector is fully linearized. If applicable, heat-inactivate the restriction enzyme for 20 minutes at 65°C, or as suggested by the supplier.

4. Dephosphorylate the linearized vector (e.g., using 1 unit of CIAP per pmole of vector for 1 hour at 50°C). Heat-inactivate the CIAP for 10 minutes at 75°C, or as suggested by the supplier.

5. Purify the linearized, dephosphorylated vector (e.g., using a commercial PCR purification kit). Elute in 30 μl of sterile H$_2$O and store frozen at –20°C.

Ligation

6. Set up the ligation reaction by adding 5 μl of vector (~150 ng); 10 μl of the purified insert (molar insert to vector ratio of ~5–10:1); 2 μl of 10× ligase buffer, 1 μl of T4 DNA ligase (1 unit/μl), and 2 μl of sterilized H$_2$O (20 μl total reaction volume). Incubate for 2–4 hours at 16°C.

7. Transform into an *E. coli* strain (or other host). A noninducible, *recA*⁻ *E. coli* strain is recommended.

8. Spread on agar plates with antibiotic selection, and allow colonies to grow overnight.

9. Screen colonies using colony PCR for smaller ELP genes (note that due to their repetitive nature, larger ELP genes produced by gene oligomerization are *not* effectively amplified using PCR) or plasmid minipreps followed by diagnostic digests for larger ELP genes. Expect 50–90% of colonies to harbor the fused ELP gene.

> A high wild-type background can be due to incomplete digestion of the vector and/or to incomplete dephosphorylation. Direct transformation of the vector without ligation can be used to test the completeness of the restriction digest. If this test produces significantly fewer colonies (e.g., tenfold less) than does ligation with the ELP insert, then incomplete digestion of the vector is not the problem. In this case, ligation and transformation of the vector without adding the ELP insert can be used to determine the extent of self-ligation allowed by incomplete dephosphorylation of the vector. Again, the number of colonies without insert should be much less than for ligation with the ELP insert.

> Also expect some dimer or trimer inserts for small ELP gene inserts (e.g., less than ~300 bp), typically in less than 10% of the colonies.

10. Transform the expression vector encoding the ELP–target protein fusion into a suitable expression host (*recA*⁻ recommended for *E. coli* expression).

Purification of ELP Fusion Proteins from Cell Lysate

This protocol is useful for the purification of ELP–target fusion proteins after expression. Briefly, the cells are lysed, DNA is precipitated, and insoluble cellular debris is removed to yield the soluble cell lysate. The ELP–target fusion protein is then selectively aggregated and removed from solution by centrifugation at a temperature above T_t. After dissolution of the pelleted fusion protein in the buffer and volume of choice, any remaining insoluble material is removed by a final round of centrifugation at a temperature below the ELP T_t.

MATERIALS

Buffers and Reagents

Additional buffer
NaCl
Expression culture harvested and resuspended in low-salt buffer (e.g., Tris or phosphate buffered at pH 7–7.5 with 20–150 mM NaCl) in ~1/20 of the culture volume.
Polyethylenimine, 10% aqueous solution (Optional, see box below step 3).

METHOD

1. Lyse the cells. Any lysis method should work, although ultrasonic disruption may help to break apart and resolubilize any ELP aggregates that have formed during culture. If sonication is used, lyse on ice to ensure that the solution temperature remains below the ELP T_t (e.g., ~5°C or more lower than the T_t). One protocol for preparation of bacterial lysates is provided in Chapter 4.

2. Centrifuge the lysed cell suspension at ~20,000g for 15–20 minutes at ~4°C. Remove the supernatant to a new centrifuge tube and discard the pellet, which contains insoluble cellular debris.

3. *Optional:* Add polyethylenimine to a final concentration of 0.5% and mix gently. The solution should turn white due to coprecipitation of the polyethylenimine and the DNA present in the suspension. Incubate for 10–20 minutes on ice, occasionally mixing the suspension gently. Centrifuge at ~20,000g for 15–20 minutes at 4°C. Remove the supernatant to a new centrifuge tube and discard the pellet, which contains precipitated DNA and insoluble cellular debris.

Precipitation of nucleic acids with polyethylenimine prior to induction of the ELP inverse transition is recommended because some nucleic acids present in the soluble lysate can pellet with the ELP in the centrifugation steps of ITC, particularly when higher salt concentrations are used (e.g., greater than ~1.5 M NaCl). If the polyethylenimine precipitation step is omitted, separation from nucleic acids can still be achieved using additional rounds of ITC.

4. Increase the solution temperature and/or add NaCl to trigger the inverse transition.

The solution temperature and NaCl concentration required to trigger the transition depend on the ELP sequence and molecular weight, the fusion partner, and ELP concentration. The goal is to lower the ELP T_t at least 3–5°C below the solution temperature during centrifugation. An effective T_t of 25–30°C (achieved using NaCl concentrations in the 0.1–2 M range) and centrifugation at 35–40°C typically works well, although higher temperatures may be used if the target protein is thermally tolerant. However, fusion to the ELP can reduce the thermal denaturation temperature of the target protein, and, thus, care must be taken not to exceed this temperature during ITC. If irreversible precipitation of the fusion protein is observed during ITC, consider using a different combination of temperature and NaCl concentration to prevent denaturation of the target protein, which is a separate event from the ELP transition. When working with a new construct for which the precise T_t is unknown, it is helpful to increase the temperature and/or NaCl (e.g., in 5°C or 500 mM increments) slowly until the transition is observed in the cell lysate. If solid NaCl crystals are used, dissolve them gently by inverting the tube. Do not vortex vigorously, because the ELP fusion protein can aggregate at the air–water interface of bubbles and be subsequently lost when the supernatant is decanted.

The solution should become turbid because of aggregation of the ELP.

Before proceeding to the centrifugation (step 5), it is essential that (1) the ELP fusion protein has undergone its transition and (2) the solution temperature is equal to or less than the centrifugation temperature. If both of these conditions are met, the ELP fusion protein will remain aggregated during the centrifugation and therefore will be captured. If, on the other hand, the centrifugation temperature is lower than the precentrifugation solution temperature, some or all of the ELP fusion protein may resolubilize and remain in the supernatant.

If no turbidity is observed after increasing the temperature and/or adding salt, confirm expression levels by SDS-PAGE. Although not generally required, spiking of additional free ELP (e.g., to a concentration of ~10–25 µM) with a matched T_t may be necessary for poorly expressed fusion proteins. If the protein is highly overexpressed but turbidity is still not observed, it may be necessary to use an ELP tag either with a lower T_t or one with a greater molecular weight (for examples, see Table 2 and Fig. 4).

5. Centrifuge at a temperature above the T_t at ~15,000g for 5–10 minutes to pellet the aggregated ELP fusion protein.

Typically this centrifugation is completed at a temperature in the range of ~25–40°C.

6. Discard the supernatant, which should be clear, and resuspend the pellet in a buffer of choice that is at a temperature below T_t.

The pellet is often brown or off-white during the first round of purification from cell lysate but is typically translucent in later rounds of ITC. The pellet can be resuspended by working aggressively with a pipettor (a 1000-µl pipettor with disposable tips is recommended). Place the pipette tip near the pellet, and repeatedly withdraw and expel the solution over the pellet. The aggregated ELP fusion protein is also typically adherent to the centrifuge tube, and therefore the tube walls should be washed thoroughly. A period of soaking (again, at a temperature below T_t) can help in the resolubilization. Mechanical scraping of the pellet and tube walls may also be helpful. Ensure that the pellet has been entirely resuspended. During the first round of ITC from the cell lysate, the solution typically remains turbid after resuspension of the pellet due to insoluble contaminants that will be removed in step 7. However, for later rounds of the ITC, the solution is generally clear after dissolution of the pellet.

The pellet can be resuspended in any desired volume (~1/50 to 1/10 of the lysate volume is recommended, depending on the expression level of the target protein).

7. Centrifuge at ~20,000g for 10–15 minutes at 4°C to remove any remaining insoluble matter. Retain the supernatant, and discard any pellet that is formed.

For the first round of purification from cell lysate, the pellet from this cold centrifugation step can be relatively large. However, it typically does not contain significant amounts of the ELP fusion protein.

8. Repeat steps 4–7 at least one additional time.

> Typically, two rounds of transition cycling provide a high level of purity, although additional rounds may be undertaken if desired. Also note that a lower salt concentration can generally be used for the second round because the transition temperature is lowered by the significantly increased ELP fusion protein concentration achieved in the first round.

9. *Optional:* If a protease recognition site has been incorporated in the primary amino acid sequence between the ELP and the target protein, free target protein can be obtained by incubating the purified ELP fusion protein with the protease to cleave the ELP tag.

> The solubility of the ELP fusion protein can be significantly greater than the solubility of the free target protein. Therefore, ensure that the protease cleavage step is performed at a concentration within the solubility limits of the free target protein. If the protein concentration is too high, addition of the protease and subsequent cleavage of the ELP can cause irreversible precipitation of the liberated target protein. Also note that, unless a secondary purification is performed, the protease remains with the purified target protein. Although not currently commercially available, the use of ELP-tagged proteases could solve this problem by enabling the simultaneous removal of both the protease and the cleaved ELP tag from solution by ITC, leaving only the purified target protein in the retained supernatant.

> A final round of ITC can then be used to remove the free ELP, retaining the purified, free target protein in the supernatant.

10. Purified ELP fusion proteins can often be stored frozen in water at –80°C until use. Use a small aliquot to test for denaturation of the target protein after freezing; if the protein denatures, storage at –20°C in a 50% glycerol solution may be necessary.

General Transition Cycling of ELP Fusion Proteins

This protocol is useful for buffer exchange, increasing the concentration of an ELP fusion protein in solution, and for general separation applications (e.g., purification from coreactants after chemical reaction or labeling, recovery of ELP–enzyme fusions after catalysis). It is also useful for the study of protein interactions through the capture of affinity-binding partners.

MATERIALS

Buffers and Reagents

Buffer
Purified ELP fusion proteins (typically in the micromolar concentration range)
NaCl

Additional Equipment

Centrifuge and centrifuge tubes
Pipettor (1000 ml) with disposable tips

METHOD

1. Add NaCl and/or increase the solution temperature to induce ELP aggregation.

    The solution temperature and NaCl concentration required to trigger the transition depend on the ELP sequence and molecular weight, the fusion partner, and ELP concentration. The goal is to lower the ELP T_t at least 3–5°C below the solution temperature during centrifugation. An effective T_t of 25–30°C (achieved using NaCl concentrations in the 0.1–2 M range) and centrifugation at 35–40°C typically work well, although higher temperatures may be used if the target protein is thermally tolerant. However, fusion to the ELP can reduce the thermal denaturation temperature of the target protein, and thus, care must be taken not to exceed this temperature during ITC. If irreversible precipitation of the fusion protein is observed during ITC, consider using a different combination of temperature and NaCl concentration to prevent denaturation of the target protein, which is a separate event from the ELP transition. If solid NaCl crystals are used, dissolve them gently by inverting the tube. Do not vortex vigorously, because the ELP fusion protein can aggregate at the air–water interface of bubbles and be subsequently lost when the supernatant is decanted.

    Before proceeding to the centrifugation (step 2), it is essential that (1) the ELP fusion protein has undergone its transition and (2) the solution temperature is equal to or less than the centrifugation temperature. If both of these conditions are met, the ELP fusion protein will remain aggregated during the centrifugation and therefore will be captured. If, on the other hand, the centrifugation temperature is lower than the pre-centrifugation solution temperature, some or all of the ELP fusion protein may resolubilize and remain in the supernatant.

The solution will become visibly turbid for concentrations of ~1–5 µM and greater. Because ITC may not be effective for concentrations lower than this, spiking the solution with additional free ELP (e.g., to a concentration of ~10–25 µM) with a matched T_t may be necessary.

2. Centrifuge at a temperature above the transition temperature at ~15,000g for 5–10 minutes.

3. Discard the supernatant, and resuspend the pellet in cold buffer of choice.

The pellet can be resuspended by working aggressively with a pipettor (a 1000-µl pipettor with disposable tips is recommended). Place the pipette tip near the pellet, and repeatedly withdraw and expel the solution over the pellet. The aggregated ELP fusion protein is also typically adherent to the centrifuge tube, and therefore the tube walls should be washed thoroughly. A period of soaking (again, at a temperature below T_t) can help in the resolubilization. Mechanical scraping of the pellet and tube walls may also be helpful. Ensure that the pellet has been entirely resuspended. The solution is generally clear after dissolution of the pellet.

The pellet can be resuspended in a volume of choice to achieve a desired concentration.

4. Centrifuge cold (4°C, ~15,000g, 5–10 minutes) to remove any remaining insoluble matter. Keep the supernatant and discard any pellet. In many instances, no pellet may be observed; however, it is good practice to always perform this cold centrifugation step after resuspending the pellet from the warm centrifugation step.

5. For additional rounds of ITC, repeat steps 1–4 as desired.

CONCLUSION

Purification of recombinant proteins using ELP fusion tags and ITC is a simple and powerful method for the purification of recombinant target proteins from cell lysate after expression. By eliminating traditional chromatographic purification methods, ITC provides substantial savings in time and expense. Furthermore, it is also useful for wide-ranging bioseparation applications following initial purification of a recombinant protein. Because it can be used to capture non-covalently a ligand specific to the target protein from solution, ITC is also useful for the study of protein–ligand interactions and forms the basis for competitive solution immunoassays. For all of these applications, ITC is flexible enough to allow optimization for a specific protein of interest using different ELP tags and solution conditions, yet is general enough to be useful for high-throughput applications involving different proteins with varied physicochemical properties. In conclusion, ITC is a new bioseparation technology that should have a significant impact on biotechnology and the biological sciences.

REFERENCES

Cacace M.G., Landau E.M., and Ramsden J.J. 1997. The Hofmeister series: Salt and solvent effects on interfacial phenomena. *Q. Rev. Biophys.* **30:** 241–277.

McMillan R.A., Lee T.A.T., and Conticello V.P. 1999. Rapid assembly of synthetic genes encoding protein polymers. *Macromolecules* **32:** 3643–3648.

McPherson D.T., Xu J., and Urry D.W. 1996. Product purification to reversible phase transition following *Escherichia coli* expression of genes encoding up to 251 repeats of the elastomeric pentapeptide GVGVP. *Protein Exp. Purif.* **7:** 51–57.

McPherson D.T., Morrow C., Minehan D.S., Wu J., Hunter E., and Urry D.W. 1992. Production and purification of recombinant elastomeric polypeptide, G-(VPGVG)$_{19}$-VPGV from *Escherichia coli. Biotechnol. Prog.* **8:** 347–352.

Meyer D.E. and Chilkoti A. 1999. Purification of recombinant proteins by fusion with thermally responsive polypeptides. *Nat. Biotechnol.* **17:** 1112–1115.

Meyer D.E., Trabbic-Carlson K., and Chilkoti A. 2001. Protein purification by fusion with an environmentally responsive elastin-like polypeptide: Effect of polypeptide length on the purification of thioredoxin. *Biotechnol. Prog.* **17:** 720–728.

Urry D.W. 1992. Free energy transduction in polypeptides and proteins based on inverse temperature transitions. *Prog. Biophys. Mol. Biol.* **57:** 23–57.

———. 1997. Physical chemistry of biological free-energy transduction as demonstrated by elastic protein-based polymers. *J. Phys. Chem.* **101:** 11007–11028.

Urry D.W., Trapane T.L., and Prasak K.U. 1985. Phase-structure transitions of the elastin polypentapeptide-water system within the framework of composition-temperature studies. *Biopolymers* **24:** 2345–2356.

Urry D.W., Luan C.-H., Parker T.M., Gowda D.C., Parasad K.U., Reid M.C., and Safavy A. 1991. Temperature of polypeptide inverse temperature transition depends on mean residue hydrophobicity. *J. Am. Chem. Soc.* **113:** 4346–4348.

19 Using In Vitro Expression Cloning to Identify Interacting Proteins

Fumihiko Kanai,[1] Michael B. Yaffe,[1] and P. Todd Stukenberg[2]

[1]*Center for Cancer Research, Massachusetts Institute of Technology, Cambridge, Massachusetts 02139;*
[2]*Department of Biochemistry and Molecular Genetics, University of Virginia Medical School,*
Charlottesville, Virginia 22908-0733

INTRODUCTION

An important step in the characterization of any protein is to determine whether it exists in a complex with other proteins and, if it does, to identify its partners. This chapter describes a new technique to identify interacting proteins. It is a based on in vitro expression cloning (IVEC), which was recently developed to screen cDNA libraries rapidly and systematically according to the function of the protein encoded on each gene. This variation on traditional expression cloning has been successfully used to identify the substrates of kinases (Stukenberg et al. 1997), substrates of proteases (Cryns et al. 1997; Kothakota et al. 1997; Li 1998; McGarry and Kirschner 1998), DNA-binding activities (Mead et al. 1998), and components of signaling pathways (Andresson and Ruderman 1998). To date, IVEC has not been used to identify proteins that interact in complexes. This chapter describes a modification of the technique that successfully identified five binding partners of the specific phosphoserine/phosphothreonine-binding protein 14-3-3 (Kanai et al. 2000). Thus, IVEC can also complement the two-hybrid technique and biochemical purification to identify interacting proteins systematically.

IVEC was originally reported in 1997 (Lustig et al. 1997; Stukenberg et al. 1997). The technique combines the power of expression cloning with the ability to manipulate a reaction exper-

345

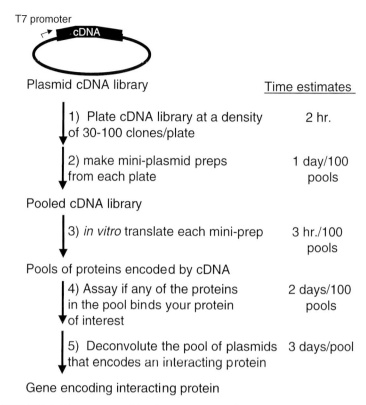

T7 promoter

cDNA

Plasmid cDNA library

Time estimates

1) Plate cDNA library at a density of 30-100 clones/plate — 2 hr.

2) make mini-plasmid preps from each plate — 1 day/100 pools

Pooled cDNA library

3) *in vitro* translate each mini-prep — 3 hr./100 pools

Pools of proteins encoded by cDNA

4) Assay if any of the proteins in the pool binds your protein of interest — 2 days/100 pools

5) Deconvolute the pool of plasmids that encodes an interacting protein — 3 days/pool

Gene encoding interacting protein

FIGURE 1. The strategy of in vitro expression cloning. See text for details.

imentally in vitro. An overview of this strategy is shown in Figure 1. The first step involves constructing or purchasing a cDNA plasmid library in which the cDNAs have been directionally cloned downstream from a T7, SP6, or T3 promoter and transformed into bacteria. (This chapter focuses on the T7 promoter, although IVEC has also been succesfully employed using SP6-based vectors.) Bacteria containing the library plasmids are plated at a density of about 30–100 clones per plate, and plasmid DNA is pooled by scraping colonies from each plate and performing small-scale plasmid purification ("minipreps"). Each plasmid pool, containing 30–100 distinct cDNAs, is then transcribed and translated in vitro in the presence of [^{35}S]methionine. The resulting labeled pools of proteins are assayed for a particular activity such as protein or ligand binding, phosphorylation by a particular kinase, or cleavage by a particular protease. Once a pool that contains the gene expressing the candidate activity has been identified, the plasmids encoding the cDNAs within the pool are subdivided and individually retested until the single cDNA encoding the protein of interest is isolated.

We describe here the general methods used to prepare library pools of cDNAs and protein, and the sib-selection techniques used to subdivide a pool once it is found to contain a candidate activity. These methods were developed in traditional academic research settings, but they could be easily modified to allow high-throughput using robotics. As an example, we describe how protein pools can be screened in vitro for binding to the specific phosphoserine/phosphothreonine-binding molecule, 14-3-3. We conclude with a discussion about the practicalities of performing IVEC and the situations where IVEC may be better than two-hybrid techniques to screen for interacting proteins.

In Vitro Expression Cloning

In this protocol, we use the example of phosphoserine/phosphothreonine-binding molecule 14.3.3 to describe how protein pools can be screened for binding using IVEC.

MATERIALS

CAUTION: See Appendix for appropriate handling of materials marked with <!>.

Buffers and Reagents

E. coli, DH5α or XL2-Blue Strain
Glutathione-agarose (Sigma)
Glycerol (40%)
GST–14-3-3 fusion protein
LB-agar plates (10 cm) containing antibiotic
LB media
[^{35}S]Methionine
Plasmid cDNA library (Discovery Line Premade cDNA Library, Invitrogen)
Recombinant RNasin (ribonuclease inhibitor) (40 units/µl) (Promega)
TNT T7 coupled reticulocyte lysate system (Promega)
Tris-HCl (10 mM, pH 8.0)
Unamplified library and bacteria

Special Equipment

Eppendorf tubes (1.5 ml)
Microfuge
Molecular Imager System (Model GX-525; Bio-Rad)
Pipettor
Plates (96-well)
QIAprep Spin Miniprep Kit (Qiagen)
Rubber policeman
SDS-PAGE equipment

METHOD

Stage 1: Preparation of cDNA Pools

To obtain detectable amounts of protein by in vitro translation, it is critical to produce plasmid pools that encode a limited number of individual cDNAs (30–100 clones/pool). Although this step constitutes the major time and expense of IVEC, this single investment can be used for ~50 IVEC screens. In fact, at least three laboratories that constructed pooled libraries have used them to screen for multiple activities and have given the libraries out as collaborations with other laboratories as well.

Preparation of cDNA pools and in vitro translation reactions are performed essentially as described previously (Lustig et al. 1997). We used a commercially available plasmid cDNA library of a human HeLa cell line from Invitrogen. This cDNA library is primed with the oligo(dT) (*Not*I) primer, and double-stranded cDNA is subcloned unidirectionally downstream from the CMV/T7 promoter of pcDNA3.1 (Invitrogen). Although this cDNA library was originally made for in vivo mammalian expression cloning, it functions well for in vitro expression screening. Furthermore, after identifying the cDNA having the desired interaction, the same construct can then be used for gene expression in mammalian cells without the necessity of recloning. For other applications, it may be advantageous to purchase or prepare cDNA libraries from a particular tissue source or organism. In each case, however, it is necessary to ensure that the library is of good quality by examining the average insert size and the percentage of empty vectors. In the human HeLa library available from Invitrogen, we found that 11 of 12 randomly chosen colonies contained plasmids with cDNA inserts, and the average insert size was ~1.2 kb.

1. To prepare pools of library cDNA that contain ~100 cDNA clones per pool, thaw an aliquot of the frozen unamplified library in bacteria, dilute with LB medium, and plate on LB-agar plates (10-cm diameter) containing appropriate antibiotics (ampicillin in the case of the Invitrogen HeLa cell library).

    It is crucial to know the titer of the original library before plating. For example, if the titer of the original library is 1×10^7 colonies/µl, then a $1:10^6$ dilution is performed and 10 µl of the final dilution is plated to get ~100 colonies/plate. The number of plates that must be prepared is calculated based on the expected number of distinct cDNAs expected to be present in the library and the depth of coverage desired during the screening. For example, we prepared 680 plates of cDNA pools that we estimated contained more than 20,000 total proteins (a density of 100 clones per pool gives ~30 proteins when translated.

In the literature, most cases have used pool sizes of ~100 clones/pool. However, we have found that translation of these pools only gives about 30 bands, suggesting that there is competition among the plasmids, and only the transcripts with the best Kozak sequences are translated in sufficient quantities to be detected. A titration of pool size suggested that the ideal pool size is closer to 30 proteins (P.T. Stukenberg and M. Kirschner, data not shown). At this lowered pool size, there are about 25 bands/pool, suggesting that most of the cDNAs in the pool are translated to a high enough concentration to be detected (Stukenberg 1997). Thus, by lowering the pool size, one can increase the coverage of the cDNA library almost threefold. To assay more than 30 proteins per assay, one can easily combine pools after the in vitro translation step (P.T. Stukenberg and M. Kirschner, data not shown).

2. Incubate the plates at 37°C until the size of bacterial colonies reaches ~1.5 mm in diameter (this typically takes overnight to 24 hours). Wrap the plates with Parafilm and store at 4°C.

    To keep the number of plates and plasmid pools manageable, the following procedures are performed using only 100 plates at a time, and subsequently repeated until all the plates have been processed. In our experience, the procedure seems to work best if the work is shared by two people working simultaneously.

3. Add 1 ml of LB medium to each of the 100 plates, and collect the bacterial colonies into microcentrifuge tubes by scraping and resuspending them using a rubber policeman. Store a small aliquot of the pooled bacteria as a 20% (v/v) glycerol stock at –80°C in either 1.5-ml Eppendorf tubes or 96-well plates.

4. Centrifuge the remainder of the pooled bacteria briefly in a microfuge, and isolate the plasmid DNA from the 100 bacterial pellets using commercially available miniprep kits. Typically, we use the miniprep kit from Qiagen (QIAprep Spin) and have consistently obtained good results for in vitro translation. The plasmid yield from one 10-cm plate with the Qiagen miniprep kit is ~5–30 µg, representing 300 ng/starting colony. The purified plasmid DNA is stored at –20°C.

Some commercially available kits for plasmid purification do not provide sufficiently good quality plasmid DNA for in vitro translation. Best results were obtained with Qiagen QIAprep Spin, and both Bio101 and Promega Wizard preps have been used successfully.

Stage 2: Preparation of In-vitro-labeled Protein Pools

1. Prepare the protein pools directly from cDNA pools using a coupled transcription/ translation system.

   > We prepare protein pools using the TNT-coupled reticulocyte lysate system (Promega) according to the manufacturer's instructions, although the conditions are rescaled to use 0.1–0.6 μg of pooled cDNA per reaction in a 10-μl reaction volume. To prepare radioactively labeled protein pools, each 10-μl translation reaction contains 8 μCi of [^{35}S]methionine (10 mCi/ml) (New England Nuclear) and a mixture of amino acids lacking methionine that is supplied with the TNT kit.

   Amounts of each reagent for a single translation reaction are tabulated below. It is generally advantageous to prepare a master mix containing enough of each reagent EXCEPT THE DNA for 100 reactions.

	Per reaction	Master mix -100 reactions
TNT lysate	5.0 μl	500 μl
TNT reaction buffer	0.4 μl	40 μl
T7 RNA polymerase	0.2 μl	20 μl
Amino acid mix (–methionine)	0.2 μl	20 μl
[^{35}S]Methionine (10 mCi/ml)	0.8 μl	80 μl
Recombinant RNasin (40 units/μl)	0.2 μl	20 μl
ddH$_2$O	2.2 μl	220 μl
DNA (0.1~0.6 μg/μl)	1.0 μl	—
Total	10.0 μl	900 μl

   Perform the in-vitro-coupled transcription/translation reactions either in 1.5-ml microcentrifuge tubes or in 96-well plates. If the master mix is prepared, place 1 μl of DNA from each pool into each tube or well on the 96-well plate; then add 9 μl of the master mix to each tube/well. Incubate the reactions for 90 minutes at 30°C either in a water bath or hot block

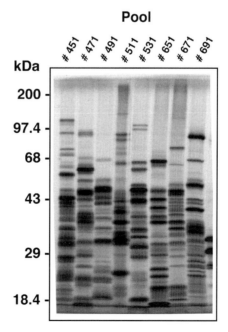

FIGURE 2. Example of in-vitro-translated pools. A total of 2 μl of each in-vitro-translated pool was analyzed by 10% SDS-PAGE and autoradiography. The mobility of molecular mass standards in kilodaltons is indicated to the left; pool numbers are shown above lanes.

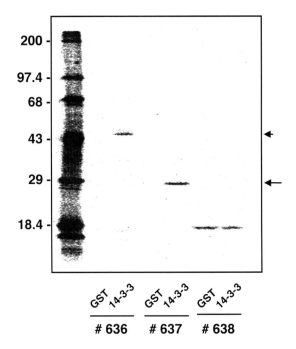

FIGURE 3. Identification of 14-3-3-binding proteins. ^{35}S-Labeled proteins from cDNA pools #636–638 were incubated with Sepharose beads containing GST or GST-14-3-3 (indicated as 14-3-3) for 2 hours at 4°C. Bound proteins were analyzed by 10% SDS-PAGE and autoradiography. The arrow and arrowhead indicate specific 14-3-3-interacting proteins later identified as a partial fragment of human TAZ (Pool #636, *arrow*) and 14-3-3 beta (Pool #637, *arrowhead*). (14-3-3 proteins are known to form homo- and heterodimers.) There is also a single band in Pool #638 that binds to both GST and GST-14-3-3, suggesting an interaction with GST rather than 14-3-3. The mobility of molecular mass standards in kilodaltons is indicated to the left.

(for microcentrifuge tubes) or in a 30°C hybridization oven (for 96-well plates). Store the radiolabeled protein pools on ice for up to 2 hours or at –80°C for 1 week.

2. Check the efficiency of translation of each pool by SDS-PAGE using 10% polyacrylamide gels followed by standard autoradiography using an overnight exposure on Model GS-525 Molecular Imager System (Bio-Rad) or a 2- to 5-day exposure on conventional X-ray film (Fig. 2).

AN EXAMPLE OF A BINDING REACTION: USING IVEC TO IDENTIFY NOVEL 14-3-3-BINDING PROTEINS

To search for novel 14-3-3 binding proteins:

1. Preclear each ^{35}S-labeled pool by incubation with 10 pmole of GST beads for 2 hours at 4°C to reduce background binding. Incubate with 10 pmole of bead-immobilized 14-3-3–GST fusion proteins for 2 hours at 4°C with gentle agitation.

2. Centrifuge briefly, then wash the beads four times. Resuspend in Laemmli sample buffer.

3. Analyze the aliquots by SDS-PAGE and detect the radioactive protein bands using a Model GS-525 Molecular Imager System.

   In addition, it is also worthwhile to expose the gels on X-ray film because long exposures (6 days) can often uncover bands that are not detectable by phosphorimager. Using this approach, five distinct 14-3-3-binding proteins were identified (Fig. 3) (Kanai et al. 2000).

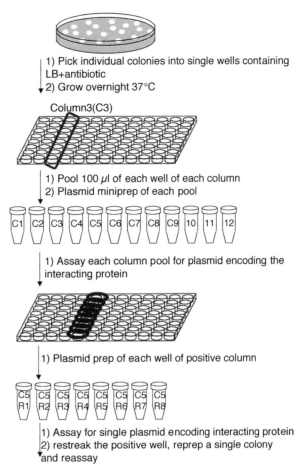

FIGURE 4. Strategy to isolate a single plasmid encoding the 14-3-3 interacting proteins from a positive pool of 100 plasmids. See text for details.

Stage 3: Sib Selection of Positive Pools

Once a cDNA pool containing a desired activity has been identified, it is progressively subdivided until a single cDNA clone containing that activity has been isolated (Fig. 4). The initial step is to retransform or replate the bacteria containing the cDNA pool to provide >100 colonies. Individual bacteria are grown in a 96-well plate format, and wells from individual columns and rows in the plate are pooled and re-screened to identify the isolated cDNA clone unambiguously.

1. To subdivide the pooled cDNAs, perform a 1:100–1:1000 dilution of the original miniprep DNA into sterile H_2O or a 10 mM Tris-HCl (pH 8.0) solution. Use 1 μl of this dilution to transform 50 μl of competent DH5α or XL2-Blue strains of *E. coli* by electroporation. The exact dilution of the library that is used depends on the transformation efficiency of the bacterial strain used.

2. Following electroporation, add 1 ml of LB medium and allow the transformed bacteria to recover at 37°C for 1 hour with shaking. Pellet the bacteria briefly in a microfuge, resuspend in ~100 μl of LB, and plate the entire suspension onto a 10-cm LB-agar plate with the appropriate antibiotic. In our experience, this typically yields 200–1000 colonies on the plate.

   Alternatively, it is acceptable to plate a small aliquot of the corresponding glycerol bacterial stock onto LB-agar plates containing the appropriate antibiotic.

3. After the plates are incubated overnight at 37°C, use an autoclaved toothpick to transfer a small sample of each colony to a single well of a 96-well flat-bottom plate containing 150 µl of LB medium with antibiotic. Avoid cross-contamination between wells.

4. Incubate the 96-well plate overnight at 37°C without shaking. Mix the cultures by repipetting, and remove 100-µl samples from each well in a column of the 96-well plate and pool in a separate bacterial culture tube. This yields 12 samples corresponding to each of the 12 columns on the 96-well plate, and each sample contains bacteria from eight different wells. Add 2 ml of LB containing antibiotic to each culture tube and shake the culture for several hours to get enough bacteria for plasmid purification. Then supplement the original cultures remaining in the 96-well plate (50 µl should be left in each well) with 50 µl of LB medium containing 40% glycerol and the appropriate antibiotic, mix by pipetting, and freeze at –80°C to provide glycerol stocks of the individual cDNA clones. The final glycerol concentration is 20%.

5. Isolate plasmid DNA from each of the 12 column pools, synthesize the proteins, and assay as described above. If no positives are detected, pick 96 more colonies and repeat the same procedure.

6. Further subdivide the positive column pools by individually preparing and testing the cDNA in each of the eight wells using the glycerol stock of the 96-well plate described above. Cross-contamination between wells can occur; therefore, it is critical to verify that the final observed activity is a consequence of the action of a single cDNA by restreaking the final clone on a fresh LB plate, picking a single clone, and verifying that it maintains the desired activity. Sequence the isolated cDNA using a T7 oligonucleotide primer to identify the protein of interest. Alternatively, a two-dimensional matrix can also be used to find a single active cDNA in one step (e.g., 12 column pools by 8 row pools to make 20 pools). However, because one can pick a positive clone more than one time (which confuses interpretation of a two-dimensional matrix), the outlined two-step procedure is often less work.

WHEN TO USE IVEC

IVEC has been most successful when used to isolate cDNAs for simple biochemical activities that were not possible by traditional methods. For example, before IVEC, effective genetic screens to identify kinase or protease substrates were elusive. A few practical considerations may help you decide whether IVEC is the correct approach for your application.

1. Because small pools of cDNAs are being expressed, it is unlikely that IVEC could identify activities that require more than one subunit being encoded by the library. To circumvent this limitation, other components or crude extracts to complement missing activities can be added.

2. IVEC may not be the best technique to screen for rare messages. In most of the screens performed to date, the rate-limiting step has been the running of SDS-PAGE gels, and it is difficult to electrophorese more than 50–100 pools of clones (1500–3000 proteins) per day. Thus, until good techniques are developed to normalize libraries, saturating screens are not practical. For smaller genomes (i.e., viruses, yeast, and worms) it would be ideal to prearray clones where each plasmid encoding cDNA is located in individual wells, which are then later pooled to simplify the translation. Once such a library is made, it allows rapid saturating IVEC screens.

3. There are presently at least three complementary techniques to identify interacting proteins, and all of them have distinct advantages and disadvantages. Protein purification is a powerful tool for identifying interacting proteins. Some advantages of protein purification are that it makes no assumptions about the binding conditions, and a whole native complex can be isolated. Biochemical purifications are technically difficult, and isolating the genes that encode these proteins is often cumbersome, requiring multiple steps of purification, peptide sequencing, and cDNA isolation. However, it is important to note that

direct complex purification has become more attractive, because each of these steps has been significantly simplified by recent technical advances such as epitope tagging proteins, mass spectrometry, and genome/expressed sequence tag (EST) sequencing projects.

The two-hybrid approach revolutionized the identification of protein–protein interactions by directly identifying the gene of an interacting protein. A second reason that two-hybrid systems are so widely used is that the techniques are well established and can be performed in most laboratory settings. IVEC also has both of these advantages. The two techniques are actually complementary because some of the limitations of two-hybrid systems are the strengths of IVEC. First, some important interactions may not occur in yeast nuclei because of cellular compartmentalization. IVEC interactions occur in cytoplasmic lysates that are cleared of compartmentalizing membranes. Two-hybrid screens cannot identify interactions that require more than two metazoan proteins. As discussed earlier, the experimental malleability of an in vitro reaction allows the easy addition of protein or small-molecule components. Third, two-hybrid interactions are prone to false positives, which must be excluded using the kind of interaction assay that is the basis of IVEC screening. Fourth, metazoan proteins may not be properly modified posttranslationally in budding yeast nuclei. Reticulocyte lysates most often generate well-folded proteins that are modified posttranslationally in a manner analogous to metazoan interphase cytoplasm. Thus, for metazoan systems the binding interactions using IVEC may actually occur at more physiological concentrations and conditions than in two-hybrid reactions.

14-3-3 binds phosphorylated proteins. Thus, the identification of 14-3-3 interacting proteins is an example of a screen that may have been difficult using typical two-hybrid techniques, because the "bait" proteins must be properly modified posttranslationally. In the screen presented here, we depended on the kinases in the reticulocyte lysate to phosphorylate the translated protein pools properly.

REFERENCES

Andresson T. and Ruderman J.V. 1998. The kinase Eg2 is a component of the *Xenopus* oocyte progesterone-activated signaling pathway. *EMBO J.* **17:** 5627–5637.

Cryns V.L., Byun Y., Rana A., Mellor H., Lustig K.D., Ghanem L., Parker P.J., Kirschner M.W., and Yuan J. 1997. Specific proteolysis of the kinase protein kinase C-related kinase 2 by caspase-3 during apoptosis. Identification by a novel, small pool expression cloning strategy. *J. Biol. Chem.* **272:** 29449–29453.

Kanai F., Marignani P.A., Sarbassova D., Yagi R., Hall R.A., Donowitz M., Hisaminato A., Fujiwara T., Ito Y., Cantley L.C., and Yaffe M.B. 2000. TAZ: A novel transcriptional co-activator regulated by interactions with 14-3-3 and PDZ domain proteins. *EMBO J.* **19:** 6778–6791.

Kothakota S., Azuma T., Reinhard C., Klippel A., Tang J., Chu K., McGarry T.J., Kirschner M.W., Koths K., Kwiatkowski D.J., and Williams L.T. 1997. Caspase-3-generated fragment of gelsolin: Effector of morphological change in apoptosis. *Science* **278:** 294–298.

Li H., Zhu H., Xu C.-J., and Yuan Y. 1998. Cleavage of BID by caspase-8 mediates the mitochondrial damage in the Fas pathway of apoptosis. *Cell* **94:** 491–501.

Lustig K.D., Stukenberg P.T., McGarry T.J., King R.W., Cryns V.L., Mead P.E., Zon L.I., Yuan J., and Kirschner M.W. 1997. Small pool expression screening: Identification of genes involved in cell cycle control, apoptosis, and early development. *Methods Enzymol.* **283:** 83–99.

McGarry T.J. and Kirschner M.W. 1998. Geminin, an inhibitor of DNA replication, is degraded during mitosis. *Cell* **93:** 1043–1053.

Mead P.E., Zhou Y., Lustig K.D., Huber T.L., Kirschner M.W., and Zon L.I. 1998. Cloning of mix-related homeodomain proteins using fast retrieval of gel shift activities (FROGS), a technique for the isolation of DNA-binding proteins. *Proc. Natl. Acad. Sci.* **95:** 11251–11256.

Stukenberg P.T., Lustig K.D., McGarry T.J., King R.W., Kuang J., and Kirschner M.W. 1997. Systemic identification of mitotic phosphoproteins. *Curr. Biol.* **7:** 338–348.

20 Using *Xenopus* Egg Extracts to Modify Recombinant Proteins

Boris Spodik, Steven H. Seeholzer, and Thomas R. Coleman

Division of Molecular Oncology, Institute for Cancer Research, Fox Chase Cancer Center, Philadelphia, Pennsylvania 19111

INTRODUCTION

General Considerations

Nearly every protein undergoes posttranslational modifications that can dramatically affect many of the protein's basic functional characteristics. Broadly speaking, posttranslational modifications fall into three categories: processing, chemical addition, and formation of complexes with binding partners. Processing encompasses proteolytic cleavage wherein a polypeptide is synthesized as a longer inactive precursor (e.g., prohormone) that is consequently cleaved to generate the mature active molecule. Chemical modifications can be quite pleiotropic, altering protein activi-

355

ty (disulfide bond formation, phosphorylation), localization (fatty acid acylation, glycosylation, phosphorylation), and/or stability (acetylation, phosphorylation). Similarly, complexes formed between two or more protein-binding partners can dramatically affect protein activity (e.g., cyclin potentiation of *cyclin dependent kinase* [cdk] activity), localization (e.g., piggyback nuclear transport), or stability (e.g., ubiquitin-mediated degradation).

Of these posttranslational modifications, protein–protein interactions are extremely important. That is, the protein-binding partner(s) with which a given protein interacts frequently governs its ultimate biology. For example, phosphorylation of the Cdc25 mitotic phosphatase by the checkpoint-activated kinase, Chk1, ultimately changes its subcellular localization, but it does so by altering the ability of Cdc25 to interact with multiple binding partners. Briefly, the Chk1 kinase phosphorylates Cdc25 on a critical serine residue, creating a binding site for 14-3-3 protein family members (Peng et al. 1997; Kumagai et al. 1998b). Binding by 14-3-3 is critical for maintaining the checkpoint-induced G_2 arrest but does not alter in vitro Cdc25 phosphatase activity. Binding of 14-3-3, however, does inhibit the association of Cdc25 with importin-α, markedly reducing the nuclear importation of Cdc25 (Kumagai and Dunphy 1999; Yang et al. 1999). Thus, chemically modifying Cdc25 (phosphorylation) ultimately down-regulates Cdc25 activity, but it does so by shifting the equilibrium from a nuclear to a cytoplasmic localization, through the action of multiple Cdc25-binding partners.

Using Cellular Extracts as Model Systems

Recombinant bacterially derived or insect cell (Sf9)-derived proteins frequently lack many of the characteristics of the native molecule, presumably due to the absence of one or more functionally significant posttranslational modifications. Because of the low cost and ease of working with bacterially expressed proteins, it is desirable to identify convenient means of conferring suitable eukaryotic modifications on such recombinant proteins. To this end, incubating bacterially derived recombinant proteins in crude cellular extracts has been a productive means through which to confer eukaryotic modifications on such recombinant proteins. For example, physiologically relevant phosphorylation and protein-binding partners can frequently be detected in vitro using crude cellular extracts. Recombinant cyclin D and cdk4 subunits produced in bacteria or insect cells do not assemble efficiently when combined in vitro but can be activated in the presence of lysates derived from proliferating mammalian cells (Kato et al. 1994). The mammalian cell lysates appear to provide two components: an assembly factor, which stabilizes the cyclin/cdk complex, and *cdk activating kinase* (CAK), which activates the complex by phosphorylating a conserved threonine residue on cdk4 (Kato et al. 1994). The crystal structure of human cdk2 reveals that the analogous threonine residue resides on a structural loop that occludes the substrate-binding cleft (De Bondt et al. 1993). Thus, phosphorylation on this residue likely alters the geometry of this inhibitory loop and stabilizes an active conformation, enabling important protein–protein interactions. Similarly, crude cellular extracts have also been used to identify physiologically relevant protein-binding partners of recombinant proteins. For example, the cyclin/cdk inhibitor p27[Kip1] can be purified by cyclin E/cdk2 affinity chromatography from growth-arrested cell lysates, but not from proliferating cell lysates (Polyak et al. 1994). This differential binding pattern is due to high levels of cyclin D in proliferating cells, wherein p27[Kip1] is sequestered in cyclin D/cdk4,6 ternary complexes and is, therefore, unable to interact with cyclin E/cdk2 (Polyak et al. 1994; Reynisdottir et al. 1995).

To be of general utility, a recombinant protein salvage/modifying system must function on a wide variety of proteins from diverse cellular pathways. With the emerging results of the genome sequencing projects, it is clear that significant evolutionary conservation of protein structure and enzymatic pathways exists among all vertebrates. We have been using *Xenopus* eggs as a starting material for our salvage/modifying system. *Xenopus* oocytes and eggs have been used to study a remarkable array of pathways conserved among frogs, mice, and humans. For example, the fact that *Xenopus* oocytes injected with mammalian RNAs express, assemble, and insert functional mammalian ion channels and neurotransmitters into *Xenopus* membranes strongly argues for the conservation of protein–protein interactions among these species (Goldin 1991). Similarly, extracts from *Xenopus* eggs have been used to study a wide variety of biological problems, including DNA metabolism, protein processing and transport, and cellular proliferation and death (Table 1). All of these pathways are rich in conserved protein–protein interactions. Thus, the *Xenopus* egg represents an excellent starting material for a recombinant protein salvage/modifying system. Frog egg extracts offer a major advantage over mammalian cell lysates because they are cost-effective, easy to prepare, highly concentrated, and have a proven track record (Table 1). Finally, in addition to their general utility as a modifying system, egg extracts are uniquely suited for cell cycle analyses (see below).

Metazoan oocyte/egg extracts have also proven invaluable for identifying cell-cycle-dependent phosphorylation and binding partners. For example, one of the first purifications of *maturation promoting factor* (MPF) was accomplished using fission yeast p13[suc1] affinity chromatography of starfish oocyte extracts (Labbe et al. 1989). Similarly, starfish CAK was detected and purified on the basis of its ability to activate *Xenopus* cdc2 in a cyclin-dependent manner (Fesquet et al. 1993). More recently, activated *anaphase promoting complex* (APC) or cyclosome has been purified from clam oocytes by affinity chromatography on fission yeast p13[suc1] (Sudakin et al. 1997). Finally, as detailed below, *Xenopus* egg extracts have been used to reveal physiologically relevant binding partners and/or specific phosphorylation of a large number of recombinant proteins. Indeed, the fact that specific phosphorylation events can be preserved in such crude cellular extracts, coupled with the ubiquitous nature of phosphorylation, speaks to the strength of using cell lysates to phosphorylate recombinant proteins. Moreover, because many proteins are phosphorylated in a cell-cycle-dependent manner, having a means to assess this modification in vitro in the context of the cell cycle is of enormous utility.

The *Xenopus* Egg Extract Model System

Xenopus egg extracts are a premiere system in which to evaluate cell-cycle-dependent phosphorylation of recombinant proteins. *Xenopus* egg extracts arrested specifically in mitosis or interphase can be generated, thereby facilitating study of cell-cycle-dependent modifications. Because of the conserved nature of the cell cycle machinery, physiologically meaningful modifications of cell-cycle-regulated proteins derived from any eukaryotic host can be studied in *Xenopus*. For example, cyclins from sea urchins (Murray and Kirschner 1989) and clams (Swenson et al. 1986), p13[suc1] from fission yeast (Dunphy and Newport 1989), and Cdc25 from *Drosophila* (Kumagai and Dunphy 1991) all function in a physiologically relevant manner in *Xenopus* egg extracts.

TABLE 1. Conserved Metazoan Pathways Studied in *Xenopus* Egg Extracts

Pathways	Comments
Cell cycle[a]	Cdk/cyclin-driven cell cycle regulated by known checkpoints (e.g., DNA replication/repair, mitotic spindle, chromosome cohesion)
Apoptosis[b]	Nuclear events typical of apoptosis, including Bcl2-mediated fragmentation of nuclei with resultant "laddering" of DNA
Chromatin assembly[c]	Nucleoplasmin-dependent assembly of nucleosomes around exogenous DNA template
Nuclear import/assembly[d]	Cell-cycle-regulated assembly of double-bilayer membrane, nuclear lamina, and functional nuclear pores
DNA replication[e]	Replication checkpoint-dependent single round of semiconservative DNA replication in ~1 hour
DNA recombination/repair[f]	Homologous and nonhomologous recombination, repair of mismatches, abasic sites, UV and X-ray lesions
Translocation/processing[g]	Efficient posttranslational modification (e.g., translocation, glycosylation, and signal sequence cleavage) of exogenous proteins

[a]Murray (1991); [b]Kornbluth (1997); [c]Wolffe and Schild (1991); [d]Newmeyer and Wilson (1991); [e]Madine and Coverley (1997); [f]Carroll and Lehman (1991); [g]Zhou et al. (2000).

OUTLINE OF PROCEDURE

The following protocol makes use of the *Xenopus* egg extract to phosphorylate a recombinant protein with the ultimate goal of identifying the specific phosphorylated residues. A similar approach can be used to isolate binding partners or otherwise modify a recombinant protein. Whether the goal is to identify a chemical modification or binding partner, or to attempt to activate a recombinant protein, the protocol is the same. The basic technique is to modify the recombinant protein by incubation in either a mitotic or interphase *Xenopus* egg extract, reisolate the protein by affinity chromatography, and either analyze the modification or use the modified protein as a probe for further studies (Fig. 1). An example of the latter interaction studies would be to use the unmodified and modified protein to detect differential binding in a far western (Chapter 4), co-immunoprecipitation (Chapter 5), GST-fusion pull-down (Chapter 4), or other biophysical approach (e.g., Biacore, Chapter 14).

The remainder of this protocol serves as a guide for optimizing and troubleshooting the use of *Xenopus* egg extracts to modify recombinant proteins. As an illustration of the protocol, we specifically describe analysis of the cell-cycle-dependent modifications of the *Xenopus* Cdc6 (Xcdc6) protein. The Cdc6 protein is a key regulator of the initiation of DNA replication in all eukaryotes (Diffley 1996). Cdc6 homologs bind to chromosomes in an *origin recognition complex* (ORC)-dependent manner and mediate the binding of other essential DNA replication machinery to DNA (Coleman et al. 1996; Tanaka et al. 1997). Recent genetic and biochemical evidence suggests that Cdc6 homologs function to load the ring-shaped *minichromosome maintenance* (MCM) complex on the DNA (Donovan et al. 1997; Perkins and Diffley 1998; Weinreich et al. 1999). MCM, in turn, appears to be one of the essential helicases involved in initiating DNA

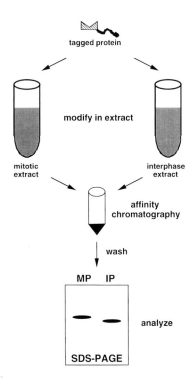

FIGURE 1. Outline showing the use of *Xenopus* egg extracts to modify proteins. Recombinant protein (which may be tagged with a fused peptide for convenient reisolation) is incubated in either an M-phase (MP) or interphase (IP) *Xenopus* egg extract. The recombinant protein is then reisolated by affinity chromatography using either the fused tag or specific antibodies, and nonspecific interactors are removed by repeated washes. Finally, the modified protein is released, and the modification is analyzed.

replication (Kelman et al. 1999; Chong et al. 2000; Shechter et al. 2000; Tye and Sawyer 2000). Both Cdc6 and the MCM family are involved in limiting DNA replication to one round per S phase. This regulatory process, called the block to rereplication, likely involves a positively acting license that can be utilized only once each S phase (Blow and Laskey 1988). In yeast, phosphorylation of Cdc6 homologs has been shown to play a role in its destruction (Jallepalli et al. 1997; Elsasser et al. 1999), whereas in metazoans phosphorylation of Cdc6 appears to cause its redistribution from the nucleus to the cytoplasm (Saha et al. 1998; Jiang et al. 1999; Petersen et al. 1999; Pelizon et al. 2000). Thus, the regulation of the Cdc6 protein and its cell-cycle-dependent phosphorylation is of general interest and also represents an excellent example wherein phosphorylation regulates the interactions and activity of a protein.

Production of *Xenopus* Egg Extracts

Xenopus egg extracts provide a biochemically tractable system that can be generated in multiple forms (Murray 1991). For the purposes of modifying recombinant proteins, we primarily use "CSF-arrested" egg extracts. The CSF-arrested extract is stalled in metaphase by the action of *cytostatic factor* (CSF). CSF-arrested extracts can be induced to progress through one cell cycle following the addition of calcium. Each in vitro cell cycle consists of chromosome decondensation, nuclear envelope formation, semiconservative DNA replication, chromosome condensation, nuclear envelope breakdown, and chromosome separation. Because CSF-arrested extracts are stalled in metaphase of meiosis II, they are not in a true mitotic state. For most purposes, however, this meiotic metaphase is identical to mitosis, and this subtlety does not affect the modification of recombinant proteins. Thus, CSF-arrested extracts and CSF-arrested extracts after calcium addition are regularly used to simulate mitotic and interphase states. To ensure that interphase extracts do not return to mitosis, the protein translation inhibitor cycloheximide is added before calcium addition because in the absence of cyclin B synthesis, these extracts cannot enter mitosis. For the purposes of familiarity and brevity, we refer to CSF-arrested extracts as "M-phase" and CSF-arrested extracts after calcium addition as "interphase" throughout this procedure.

Purified recombinant protein also needs to be prepared, and this can be done using any number of schemes. Because the production of recombinant proteins is covered elsewhere in this volume (Chapter 4) and must be tailored to individual needs, we do not address this issue here. For the purposes of these experiments, it is convenient to express the protein of interest with a fused tag, for example, hexahistidine, glutathione-*S*-transferase (GST), or maltose-binding protein, to facilitate purification and reisolation. Alternatively, if specific antibodies are available, one can recover the protein by immunoprecipitation (Chapter 5). In the specific example outlined below, we expressed recombinant Xcdc6 with an amino-terminal hexahistidine tag in baculovirus-infected Sf9 insect cells and purified it by Ni-agarose chromatography (Coleman et al. 1996). Similar binding and phosphorylation studies have been performed using GST-tagged proteins expressed in bacteria (data not shown).

Modification and Reisolation of Recombinant Protein

A detailed modification protocol is provided at the end of this chapter. Briefly, recombinant protein is modified by incubating it in the *Xenopus* egg extract for 30 minutes at 23°C. Typically, we add one volume of purified recombinant protein (0.5–1 mg/ml) to nine volumes of extract. In some instances, the recombinant protein can be modified while it is bound to affinity beads, although the beads can sterically inhibit modifications in some circumstances. Thus, the utility of leaving the protein bound to beads needs to be determined empirically. We next dilute the extract containing the purified recombinant protein with four volumes of buffer, clarify it by brief centrifugation, and incubate the supernatant with affinity matrix for 30 minutes at 4°C. Finally, the matrix-bound recombinant protein is isolated by centrifugation, washed, and analyzed.

As a specific example, we observed that the endogenous Xcdc6 protein displayed a slower electrophoretic mobility during M-phase than during interphase (Fig. 2A). To characterize this modification, we wished to modify a sufficient quantity of Xcdc6 for subsequent analysis. For this purpose, full-length Xcdc6 was overproduced in Sf9 insect cells as a hexahistidine-tagged fusion protein and purified to near homogeneity by Ni-agarose chromatography (Coleman et al. 1996). To modify the eluted recombinant protein, we incubated it in either M phase or interphase egg extracts at 23°C. The histidine-tagged protein was then reisolated on Ni-agarose beads, washed, and analyzed by SDS-polyacrylamide gel electrophoresis (SDS-PAGE) (Fig. 2B). In this example, the purified recombinant Xcdc6 protein migrated with the anticipated molecular mass of 61 kD (Fig. 2B, lane 3). Consistent with the endogenous protein, which displayed a retarded relative mobility when modified during M phase (Fig. 2A, lane 2), recombinant protein modified in an M-phase extract displayed a similar slower mobility (Fig. 2B, lane 4). As discussed in detail below, this M-phase modification can be attributed entirely to phosphorylation. In control experiments, the relative mobility of recombinant Xcdc6 incubated in an interphase extract was not altered significantly (Fig. 2B, lane 5). Other control experiments that did not include recombinant protein revealed a prominent Ni-agarose interactor of ~51 kD in either M-phase or interphase extracts (Fig. 2B, lanes 1,2). The identity of this 51-kD Ni-agarose interactor remains unknown (see Fig. 2B).

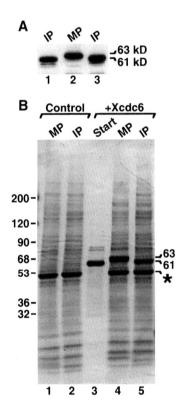

FIGURE 2. *Xenopus* Cdc6 is differentially modified during the cell cycle. (*A*) Endogenous Xcdc6 has a slower relative mobility in M phase (MP, lane *2*) than interphase (IP, lanes *1* and *3*), as detected by immunoblotting with anti-Xcdc6 antibodies. The apparent molecular weight of the M-phase and interphase forms of Xcdc6 are indicated (63 kD and 61 kD, respectively). (*B*) The M-phase modification can be recapitulated with a recombinant hexahistidine-tagged Xcdc6 (starting material, lane *3*) incubated in either an M-phase (MP, lane *4*) or interphase (IP, lane *5*) egg extract. Note: Control preparations of Ni-agarose beads treated with either M-phase (MP) or interphase (IP) extracts showed a prominent *Xenopus* protein of ~51 kD (*asterisk*) which binds Ni-agarose (lanes *1* and *2*), but is nonspecific. Following incubation in egg extracts, samples were recovered on Ni-agarose, washed, and analyzed by SDS-PAGE and Coomassie Blue staining.

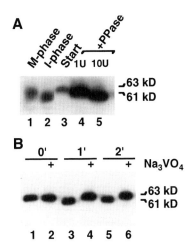

FIGURE 3. Xcdc6 is phosphorylated in a mitotic egg extract. (*A*) M-phase-modified recombinant Xcdc6 was treated with either control buffer (Start, lane *3*) or alkaline phosphatase (1 or 10 units, lanes *4* and *5*, respectively). Control lanes show the migration of endogenous M-phase (lane *1*) and interphase (I-phase, lane *2*) Xcdc6. (*B*) M-phase-modified recombinant Xcdc6 was treated with λ phosphatase for the indicated times in the absence (odd-numbered lanes) or presence (even-numbered lanes) of the phosphatase inhibitor sodium orthovanadate. All samples were processed for immunoblotting with anti-Xcdc6 antibodies.

An obvious explanation for the M-phase modification is that the Xcdc6 protein is phosphorylated in a cell-cycle-dependent manner. To confirm this, we modified recombinant Xcdc6 protein in an M-phase *Xenopus* egg extract and treated it with either alkaline (Fig. 3A) or λ (Fig. 3B) phosphatase. In either case, the electrophoretic mobility of the hyperphosphorylated form was increased upon phosphatase treatment. Significantly, the relative mobility of the phosphatase-treated protein returned to that of the endogenous interphase protein (Fig. 3A). Moreover, this increase in electrophoretic mobility was sensitive to the phosphatase inhibitor vanadate (Fig. 3B). Taken together, these results suggest that the cell-cycle-dependent shift in the molecular weight of Xcdc6 can be attributed entirely to phosphorylation during M phase and that the recombinant protein can be phosphorylated in an M-phase egg extract in a manner consistent with that seen in vivo.

Evaluation of Modification and Utilization of Modified Protein

As a validation step to ensure that a recombinant functional consequence of protein has been modified in a physiologically meaningful manner, it is important to analyze the modification. For example, Hunt and coworkers found that the catalytic subunit of CAK was inactive when isolated from bacteria, but it could be activated by "marinating" it in a *Xenopus* egg extract (Poon et al. 1993). Thus, assessing this modification required comparing the activity of the protein before and after incubation in the extract. Having established that the recombinant protein is modified in a physiologically meaningful manner, one may next define the modification. The detection of a modification depends in large part on the technique used. Detection techniques can range from assays that assess the biochemical activity of a protein to simply performing SDS-PAGE on the protein bound to washed beads. In the above example, subsequent studies revealed that the *Xenopus*-extract-mediated activation of CAK was due to phosphorylation (Poon et al. 1994).

Alternatively, if one is looking for a specific binding partner, components of protein complexes purified by affinity chromatography can be analyzed by separation on SDS-PAGE and subsequent staining. The presence of protein bands that migrate differentially from the recombinant protein of interest would indicate that a protein complex has formed. *Xenopus* crude extracts have facilitated the identification of many such binding partners using a wide variety of "baits" in the pull-down experiments, including fission yeast p13[suc1] (Dunphy and Newport 1989), Cdc25 phosphatase (Crenshaw et al. 1998; Kumagai et al. 1998a), CRM1 nuclear export factor (Yang et al. 1998), importin (Yang et al. 1999), Pin1 propyl isomerase (Crenshaw et al. 1998), cyclin B (Yang et al. 1998), and cyclin E (Yang et al. 1999). One can increase the sensitivity for detecting a binding partner by several means. The most common is immunoblotting with an antibody directed against a known (or suspected) interactor. A more general approach involves biotinylating all proteins bound to the bait (e.g., using an Amersham biotinylation module). After transfer to a membrane, the biotinylated proteins can be detected by staining with horseradish peroxidase (HRP)-streptavidin. This approach has been used successfully to identify the *Xenopus* apoptotic regulator, Scythe, using the *Drosophila* reaper protein as bait (Thress et al. 1998). Having identified Scythe by biotinylating the proteins bound to reaper conjugated beads, these researchers expanded the purification scheme to obtain Coomassie Blue-detectable levels of Scythe for microsequencing. This approach underscores another advantage of the *Xenopus* system, in that large amounts (gram) of extract can easily be generated from a modest number of frogs.

Two approaches are generally used when analyzing phosphorylation modifications. First, one can incubate the protein in an extract in the presence of radioactive ATP and perform autoradiography to assess the extent of phosphorylation. Because one is detecting only the phosphorylated protein, this method does not require a shift in electrophoretic mobility for detection. The extract contains millimolar concentrations of endogenous ATP, and therefore this method may not produce a readily detectable product due to the limited incorporation of radioactive label. A second approach, which relies on an electrophoretic mobility shift, makes use of ^{35}S. By incubating ^{35}S-labeled recombinant protein (e.g., translated in a reticulocyte system charged with [^{35}S]methionine) in the extract, one can assess the phosphorylation (shift) over time. For example, this method allowed the rapid mapping of phosphorylated domains of Cdc25 (Kumagai and Dunphy 1992) and fission yeast Wee1 (Tang et al. 1993).

In the case of the Xcdc6 protein, we could detect the M-phase-specific modification (phosphorylation) by SDS-PAGE. To pinpoint which residues were phosphorylated during M phase, we used *m*atrix-*a*ssisted *l*aser *d*esorption/*i*onization *m*ass *s*pectrometry (MALDI-TOF-MS). This technique offers the advantage that one can obtain rapid and accurate molecular weight information on low picomole amounts of peptides present in complex mixtures, such as those that result from a proteolytic digest. Moreover, phosphorylated peptides are readily detected by MALDI-TOF-MS due to a mass shift of 80 D, or multiples of 80 D (the molecular mass of HPO_3), when comparing the phosphorylated and nonphosphorylated species.

MALDI-TOF-MS techniques are covered in more detail in Chapter 12. Briefly, following SDS-PAGE and staining, the mitotic (63 kD) and interphase (61 kD) Xcdc6 protein bands were excised, destained, and dried. Next, the gel slices containing derivatized Xcdc6 were reduced, alkylated, and digested with trypsin, and the resultant peptides were extracted and subjected to MALDI-TOF-MS analysis (Shevchenko et al. 1996). All MALDI-TOF-MS spectra were obtained on a Voyager DE time-of-flight MS (PerSpective Biosystems) operating in linear mode with delayed ion extraction. Data were analyzed using the *Protein Analysis Worksheet* (PAWS) program to match the observed with theoretical masses. We then assigned many of the observed mass values to the anticipated mass values, assuming first unphosphorylated and then monophosphorylated peptides. To ascertain which sites were candidate phosphorylation sites, we needed to establish stringent criteria to test our approach. First, we identified the mass peaks in the peptides derived from M-phase-treated Xcdc6 whose sizes were consistent with being monophosphorylated. Next, we verified that only the phosphorylated form of these peptides was present in M-phase-treated Xcdc6, whereas only the nonphosphorylated form was found in the peptide mix derived from interphase-treated Xcdc6. Only those shifted peaks that satisfied these stringent criteria are shown (Fig. 4). By imposing these stringent criteria, the peptides identified are likely phosphorylated stoichiometrically during mitosis and completely unphosphorylated during interphase.

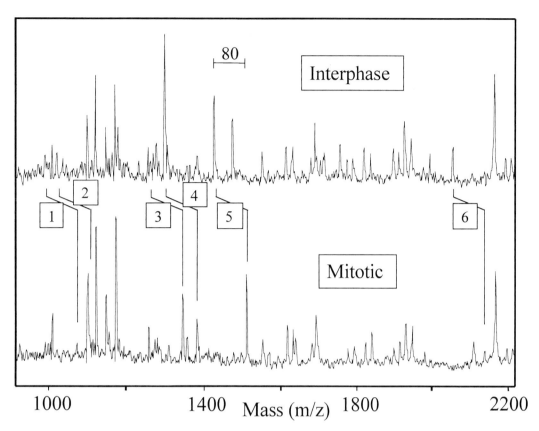

FIGURE 4. MALDI-TOF-MS peptide mass mapping of Xcdc6 phosphorylation sites. Following SDS-PAGE and staining, the M-phase- and interphase-modified Xcdc6 protein bands were excised, destained, and digested with trypsin. The extracted peptides were subjected to MALDI-TOF-MS analysis. We identified the mass peaks in the peptides derived from M-phase-treated Xcdc6 whose sizes were consistent with being monophosphorylated. Next, we verified that only the nonphosphorylated form, but not the phosphorylated form, of these peptides was found in the peptide mix derived from interphase-treated Xcdc6.

Note the presence of several peaks in the mitotic spectrum that have been displaced by 80.0 D (±0.3) relative to those of the interphase spectrum (Fig. 4, peaks 1–6). Because MALDI-TOF-MS analysis is inherently nonquantitative, it is not surprising that the relative peak heights of these interphase and mitotic peptides are not equivalent. Importantly, all of the tryptic peptides shown as M-phase-shifted peaks contain one or more serine or threonine residues (Table 2). In fact, four of the putative phosphorylation sites contain S/P or T/P residues, which are the preferred consensus recognition sites of several mitotic kinases (Table 2, peptides 4a, 4b, 5, and 6). Moreover, several of these identified phosphopeptide sequences were observed in multiple overlapping peptides (Table 2, peptides 2 and 3; 4b and 5), underscoring their strong candidacy as legitimate phosphorylation sites, a point confirmed separately by liquid chromatography electrospray ionization tandem mass spectrometry (LC-MSMS) (S. Seeholzer and T.R.Coleman, in prep.).

In summary, the *Xenopus* egg extract represents a powerful system in which to modify recombinant proteins. Most importantly, the numerous protein modifications revealed with this system are physiologically relevant. As outlined above, these modifications include phosphorylation and association with appropriate protein partners. By using recombinant proteins in such an assay, one can maximize the quantity of modified protein obtained, which can then be analyzed conveniently for biological activity, association with a particular binding partner, or presence of chemical modification. Moreover, in the specific example cited above, it is likely that the *Xenopus* egg extract phosphorylates proteins in a stoichiometric manner, greatly aiding analysis. Finally, the

TABLE 2. Potential Xcdc6 Phosphopeptide Mass Spectrometry Identification

Peptide number	Interphase MH+		Mitotic MH+		Peptide site	Peptide
	observed	calculated	observed	calculated		
1	991.9	992.2	1072.4	1072.2	137–145	NSVGVQLFK
2	1021.7	1022.1	1100.9	1102.1	8–16	SQSSIQFPK
3	1264.0	1265.4	1345.4	1345.4	6–16	SRSQSSIQFPK
4a	1300.9	1300.4	1381.6	1380.4	81–92	KETGQPTTPKGR
4b	1300.9	1301.5	1381.6	1381.5	113–123	LQDPYLLSPVR
5	1429.0	1429.7	1509.7	1509.7	(112)–123 or 113–(124)	(K)LQKPYLLSPVR(K)
6	2056.6	2056.4	2137.0	2136.4	93–111	RLLFDENQAAAATPLSPK

production of egg extracts is well established, and the entire procedure takes only a half-day from production of extract to modification of protein.

Controls

The *Xenopus* egg extract is ideally suited to address cell-cycle-dependent modifications. By comparing the ability of an M-phase versus an interphase extract to modify a recombinant protein, one immediately acquires specificity. That is, the *Xenopus* egg extract system frequently has built-in internal controls for cell cycle modifications, in that parallel incubations of a desired protein can be performed in both M-phase and interphase extracts. For example, Xcdc6 is hyperphosphorylated, as revealed by a shift in relative electrophoretic mobility, when incubated in M-phase, but not interphase, egg extracts (Fig. 2B). Conversely, Cdc25 associates preferentially with 14-3-3 proteins when incubated in an interphase, but not an M-phase, egg extract (Kumagai et al. 1998a). Essential controls include treating an extract with affinity beads alone to assess the background levels of copurifying proteins. For example, in the case of Ni-agarose, we noticed a prominent 51-kD protein that bound in the presence and absence of Xcdc6 (Fig. 2B, see legend). Similarly, if one were reisolating recombinant protein from extracts via antibody immunoprecipitation, it would be critical to perform a parallel immunoprecipitation using a nonspecific control antibody.

Depending on the reagents used, the number and intensity of copurifying proteins that bind to parallel control bead preparations can confound subsequent analysis, but these difficulties can be surmounted by a number of means. A primary consideration is to limit the amount of affinity beads used in any binding reaction. That is, the binding capacity of these resins is typically quite high, and, by using a small amount (<10 µl), one can frequently favor the high-affinity interactions (i.e., tagged recombinant protein) at the expense of lower-affinity nonspecific endogenous protein interactions. One can also pretreat the extract with control beads (0.5 hour, 4°C) before actually incubating with recombinant protein. In this manner, one can clarify the extract of endogenous matrix-binding proteins before performing any modification. Similarly, one can pretreat the affinity beads with a concentrated protein mixture (e.g., fetal bovine serum; 0.5 hour, 4°C) to block nonspecific interactions (Yang et al. 1999). Finally, one can tailor specific beads to individual applications. For example, in the case of Ni-agarose, we found that it was critical to reisolate recombinant Xcdc6 in the presence of imidazole. Most hexahistidine-tagged proteins remain bound to Ni-agarose in concentrations <60 mM imidazole. We found that reisolating Xcdc6 in the presence of 50 mM imidazole significantly reduced the number of endogenous frog proteins that bound to Ni-agarose without diminishing recombinant Xcdc6 binding. Similar approaches have used low concentrations (0–10 mM) of EGTA in the binding buffer to reduce nonspecific binding without chelating nickel from the column (Kumagai and Dunphy 1995). Because hexahistidine binds Ni-agarose in the presence of denaturants (e.g., 4–8 M urea), one can also wash the Ni-agarose with these agents and decrease nonspecific endogenous frog protein binding. Of course, this treatment can only be used in those cases where one is isolating either a very stable complex or a covalently modified fusion protein. Another elegant approach is to combine several fusion tags on one bait. For example, Claspin was recently identified as an important regulator of the Chk1 kinase using recombinant Chk1 protein fused to both GST and hexahistidine tags (Kumagai and Dunphy 2000). In this two-step approach, Chk1-binding proteins were isolated from interphase extracts on Ni-agarose beads containing bound Chk1-GST-His6 protein. These proteins were then released with imidazole and bound to glutathionine-agarose followed by SDS-PAGE. The use of two independent affinity purification steps reduces nonspecific binding. Unfortunately, no panacea exists, but rather, it is necessary to empirically determine what precautions are necessary to reduce nonspecific interactions for each application.

Egg Extracts

With this protocol, CSF-arrested M-phase and interphase egg extracts are prepared.

MATERIALS

CAUTION: See Appendix for appropriate handling of materials marked with <!>.

Buffers and Solutions

Chorionic gonadotropin (CG-10, Sigma) <!>

Dissolve 10,000 IU in 10 ml of sterile H_2O; store at 4°C.

CSF-XB

100 mM potassium chloride (KCl) <!>

0.1 mM calcium chloride ($CaCl_2$) <!>

2 mM magnesium chloride ($MgCl_2$) <!>

10 mM HEPES (pH 7.7)

50 mM sucrose

5 mM EGTA (KOH)

Can be made as a 10x stock; store at –20°C; pH at 1x concentration (KOH) before use.

Cycloheximide (Sigma) <!>

Add 10 mg/ml in H_2O; store in small aliquots at –20°C.

Cysteine (2%) (Sigma) titrated to pH 7.8 with NaOH

Make up immediately before use.

Cytochalasin B (Sigma)

Dissolve in DMSO <!> (10 mg/ml); store in small aliquots at –20°C.

Dejelly solution

20x Energy mix

150 mM creatine phosphate (Sigma)

20 mM ATP

2 mM EGTA

20 mM $MgCl_2$ <!>

Store in small aliquots at –70°C.

MMR

5 mM HEPES (pH 7.8)

100 mM NaCl

2 mM KCl <!>

1 mM $MgCl_2$ <!>

2 mM $CaCl_2$ <!>

0.1 mM EDTA

Make as a 10x stock, autoclave, and store at room temperature.

Pregnant mare serum gonadotropin (Calbiochem) <!>

Dissolve 5000 IU in 1 ml of sterile H_2O; store at –20°C.

Protease inhibitor cocktail
>   pepstatin A (Boehringer Mannheim)
>   chymostatin (Boehringer Mannheim)
>   leupeptin (Sigma)
>
>> Add 10 mg/ml each in DMSO <!>; store in small aliquots at –20°C.

Special Equipment

>   Beckman TL ultracentrifuge (or equivalent) and swinging bucket rotor (TLS-55)
>
>> Although any centrifuge/swinging bucket rotor combination capable of achieving 12,000g is acceptable, we find that the capacity (~2 ml) of this rotor is ideal for extract production.
>
>   Clinical centrifuge
>   Microfuge equipped with horizontal or swinging bucket rotor (e.g., Beckman microfuge E) able to hold elongated (8 × 28 mm, 0.5 ml) microfuge tubes (Beckman)

METHOD

1. Female frogs can be obtained from a number of supply houses (e.g., Nasco). *Xenopus* is well suited to life in the laboratory. (For an excellent review of *Xenopus* husbandry, see Wu and Gerhart 1991.) Typically, maintain frogs in tap water with 30 mM NaCl at 16–22°C and feed trout chow (Purina) or *Xenopus* pellets (Nasco).

2. Prime female frogs for ovulation 3–14 days prior to use. Use 75 IU of pregnant mare serum gonadotropin per frog in 0.5 ml of sterile H_2O. Make all injections into the dorsal lymph sac using a 23-gauge needle.

3. At 12–14 hours before extract preparation, induce ovulation by injecting each frog with 800 IU of chorionic gonadotropin in 0.8 ml of sterile H_2O. Following injection, maintain each frog in a separate container containing 5 liters of H_2O with 25 g of NaCl. Typically, we inject 2–3 frogs and use the eggs of the highest quality derived from one frog.

   > Unfortunately, high-quality eggs have no rigorous definition but are uniform in appearance and size and have a single white spot (the female pronucleus, or germinal vesicle) in the dark animal pole when viewed from above. Normally, >95% of the eggs from a frog are uniform, and the few substandard eggs are less dense and can be easily removed with a transfer or Pasteur pipette. Large white "snowball" eggs need to be removed. Occasionally a frog will lay "stringy" eggs, which are attached end to end and frequently of poor quality. One frog produces sufficient eggs (20–50 ml with jelly) for 1–2 ml of extract. We always use fresh extracts for our experiments because they only require about 1 hour to make, and they lose activity upon freezing.

4. Wash eggs two to three times in ~200 ml of MMR each time. Remove any debris or substandard eggs during these washes using a transfer pipette.

5. Make up ~200 ml of 2% cysteine (pH 7.8). Decant MMR and add ~50 ml of cysteine solution, swirl, decant, add ~100 ml of fresh cysteine solution, and swirl occasionally. The jelly coat is nearly invisible and roughly doubles the volume of an egg. Eggs are dejellied when they closely pack due to absence of jelly coat. Decant the cysteine solution as soon as eggs are closely packed (<5 minutes). Wash eggs with the final ~50 ml of cysteine solution and decant.

6. Wash eggs two or three times with ~200 ml of MMR each time.

7. Wash eggs four times with ~200 ml of CSF-XB each time.

8. Break off the point of a Pasteur pipette and insert a bulb over this end to produce a blunt pipette. Using this instrument, which minimizes physical damage, transfer the eggs to swing-

ing bucket elongated microfuge tubes. To minimize the transfer of buffer with eggs, tip the beaker of eggs and plunge the blunt pipette into the eggs. Fill the pipette.

9. Remove all CSF-XB from above eggs with a Pasteur pipette.

10. To each tube of eggs, add 0.5 ml of CSF-XB containing 6.6 µl of cytochalasin B stock and 2.2 µl of protease inhibitor stock.

11. Cover the top of the tube with Parafilm and gently invert several times to mix.

12. Transfer the tubes to a homemade apparatus consisting of two microfuge tubes with caps removed, one on top of the other, inside a snap-cap, 14-ml tube (e.g., Falcon 2059). The microfuge tubes form a pedestal on which the egg-containing tube conveniently rests.

13. Pack the eggs into a clinical centrifuge in three stages:

    setting 3 (~1300 rpm), 15 seconds
    setting 5 (~2000 rpm), 60 seconds
    setting 7 (~3000 rpm), 30 seconds

14. Remove all buffer from above the packed eggs with a Pasteur pipette. At this point, the eggs have not lysed.

15. Crush the eggs by centrifugation at 12,000g for 10 minutes at 16°C. Crushing the eggs by centrifugation, rather than homogenization, produces a more concentrated cytoplasmic extract and minimizes yolk and pigment contaminants.

16. Following centrifugation, the tubes will contain three layers: top, yellow lipid; middle, cytoplasm; bottom, yolk and packed pigment granules. Using a 21-gauge needle attached to a 3-ml syringe, slowly remove the middle (cytoplasmic) layer. Insert the needle at the interface of the cytoplasm with the yolk layer. Note the "buffy" sublayer at this interface. Because this sublayer contains many of the membranes to make nuclei, it is important to "vacuum" this sublayer with the cytoplasm.

17. Transfer the extracted layer to elongated 0.5-ml tubes and centrifuge at 14,000g for 2 minutes at 4°C in a swinging bucket rotor outfitted with adapters. This spin removes any contaminating yolk and lipid.

18. Using a single-edge razor, cut off top of tube just below lipid layer. In this manner, one can remove any contaminating lipid. Using a Pasteur pipette, remove the extract, taking care to avoid any yolk.

19. Weigh the cytoplasm and add energy mix (1/19 volume of 20X stock). This is the cytostatic factor (CSF)-arrested or M-phase extract.

20. Add cycloheximide (100 µg/ml) to M-phase extract. Leave on ice until ready to use (up to 6 hours).

21. Prepare the interphase extract (0.2–0.5 ml, depending on number of assays) by making 0.4 mM in $CaCl_2$ and incubating for 15 minutes at 23°C.

Protocol 2

Modification, Reisolation, and Evaluation

The recombinant protein is modified by incubation in the *Xenopus* egg extract. After binding to the matrix, the recombinant protein is isolated by centrifugation, washed, and analyzed.

MATERIALS

CAUTION: See Appendix for appropriate handling of materials marked with <!>.

Buffers and Solutions

Gel destain I
 50% ethanol <!>
 7% HOAc
Gel destain II
 10% ethanol
 5% HOAc
Gel stain
 50% methanol <!>
 5% HOAc
 2.5% Coomassie Blue R250
Gel storage buffer
 3% glycerol
 5% HOAc
HBS
 10 mM HEPES (pH 7.4)
 150 mM NaCl

Chromatography Equipment

Antibody–protein A agarose (Sigma)
> In our hands, we see lower nonspecific binding when we use freshly hydrated and washed protein A agarose beads.

Glutathione beads (Pharmacia)
Ni-agarose beads
> We make Ni-agarose using iminodiacetic acid-agarose (Sigma). The resin is washed with H_2O in a sintered glass funnel and then incubated with 100 mM $NiCl_2$ for 15 minutes, followed by washing with H_2O and PBS with 0.02% sodium azide <!>.

Gels

Discontinuous SDS-PAGE gradient gel (Bio-Rad)
> When performing mass spectrometry analysis, precast 4–15% gradient minigels offer two major advantages over freshly poured gels. First, they have minimal unpolymerized acrylamide, which can covalently modify cysteine residues and complicate any MS data. Second, minigels are preferable to full-sized gels owing to the reduced acrylamide:protein ratio within a protein band.

369

METHOD

1. Prepare and purify recombinant fusion protein of interest (see Chapter 5).

2. Prepare *Xenopus* CSF-arrested "M-phase" and interphase egg extract (see Egg Extracts protocol, p. 366).

3. Incubate recombinant protein in M-phase or interphase egg extract (typically 20 μl of protein in 180 μl of extract) for 30 minutes at 23°C.

4. Dilute egg extracts with 5 volumes of HBS.

5. Centrifuge at full speed (~3000 rpm) in clinical centrifuge for 2 minutes.

6. Bind recombinant protein to affinity beads (5–10 μl) for 1 hour at 4°C with rotation.

   In the case of hexahistidine-tagged recombinant protein, we found the addition of imidazole (<50 mM) considerably reduced nonspecific frog proteins from binding to the Ni-agarose (see text).

7. Wash the bound protein mixture by resuspending in 400 μl of HBS.

8. Centrifuge at full speed in clinical centrifuge for 15 seconds.

9. Repeat this wash three more times.

10. Remove as much liquid as possible, taking care to leave the bead pellet intact.

11. Add SDS loading buffer (15 μl) and boil for 4 minutes.

12. Load the sample onto a discontinuous SDS-PAGE gradient gel.

13. Stain with freshly prepared Coomassie Blue R250.

    We found that using freshly prepared stain was critical for MS analysis because it assured a more complete destain. Similarly, using a stain with high dye content (e.g., Bio-Rad) improved the subsequent MS analysis.

14. Destain with gel destain I for 1 hour at room temperature.

15. Decant and repeat step 14.

16. Destain with gel destain II overnight at room temperature. Store in gel storage buffer.

Protocol 3

Mass Spectrometry

Processing of samples for mass spectrometry analysis varies greatly depending on the individual running the machine. For a review of methods, see Chapter 12.

MATERIALS

CAUTION: See Appendix for appropriate handling of materials marked with <!>.

Buffers and Solutions

Acetonitrile (Burdick and Jackson) <!>
Alkylating buffer
    50 mM iodoacetamide <!>
    50 mM ammonium bicarbonate <!>
Ammonium bicarbonate (100 mM) <!>
Destain (50% methanol <!>, 5% formic acid <!>)
Extraction solution
    5% formic acid <!>
Reducing buffer
    10 mM dithiothreitol (DTT) <!>
    50 mM ammonium bicarbonate <!>

Biological Molecules

Trypsin solution
    20 µg/ml sequencing grade modified porcine trypsin (Promega) in 25 mM ammonium bicarbonate <!>.

Special Equipment

Mass spectrometer
Microcentrifuge
SpeedVac

METHODS

1. Closely excise the bands of interest from the gel and divide into smaller pieces (~0.5 mm^3).

2. Destain bands in 0.5 ml of destain overnight (or longer) at room temperature.

3. Discard destain and replace with 0.5 ml of destain for 2–3 hours at room temperature.

4. Discard destain and dehydrate gel slices in 0.2 ml of acetonitrile for about 5 minutes. Discard acetonitrile.

5. Repeat step 4.

6. Dehydrate the gel piece in a SpeedVac until dry.

7. Incubate the gel pieces in 100 µl of reducing buffer for 30 minutes at room temperature.

8. Centrifuge briefly to pellet the gel pieces and remove the DTT solution.

9. Incubate the gel pieces in 200 µl of alkylation buffer for 30 minutes at room temperature.

10. Centrifuge briefly to pellet the gel pieces and remove iodoacetamide solution.

11. Dehydrate gel slices in 0.2 ml of acetonitrile as in steps 4 and 5.

12. Rehydrate gel pieces in 0.2 ml of 50 mM ammonium bicarbonate.

13. Centrifuge briefly to pellet the gel pieces and remove ammonium bicarbonate.

14. Dehydrate with acetonitrile as above.

15. Dry the gel pieces in a SpeedVac.

16. Freshly prepare 20 µg of Promega trypsin in 1 ml of ice-cold 25 mM ammonium bicarbonate.

17. Add a minimal volume of trypsin (10–30 µl, depending on gel band volume) to rehydrate the dried gel pieces. Keep on ice during rehydration.

18. Centrifuge briefly to pellet the gel pieces and remove any excess trypsin solution and add a small volume of ammonium bicarbonate solution to assure complete coverage of the gel pieces.

19. Incubate overnight at 30–37°C.

20. Extract peptides by adding 1 volume of extraction solution. Collect the supernatant in a separate siliconized tube.

21. Repeat the extraction with a second volume of extraction solution. Combine the supernatants. Recovery of hydrophobic peptides may be improved by including 50% methanol or isopropanol in this second extraction.

ACKNOWLEDGMENTS

We thank the members of the Coleman laboratory as well as Drs. Eric Moss and Sarah Fashena for helpful discussions and critical readings of the manuscript. This work has been supported by an appropriation from the Commonwealth of Pennsylvania, an award from the ACS IRG and NCI CA-06927, and a grant from the National Institutes of Health (GM-58924). We also gratefully acknowledge The V Foundation for support and the Fannie E. Rippel Foundation for support of the Biotechnology Core Facility at Fox Chase Cancer Center directed by Dr. Anthony T. Yeung.

REFERENCES

Blow J.J. and Laskey R.A. 1988. A role for the nuclear envelope in controlling DNA replication within the cell cycle. *Nature* **332:** 546–548.

Carroll D. and Lehman C.W. 1991. DNA recombination and repair in oocytes, eggs, and extracts. *Methods Cell Biol.* **36:** 467–486.

Chong J.P., Hayashi M.K., Simon M.N., Xu R.M., and Stillman B. 2000. A double-hexamer archaeal minichromosome maintenance protein is an ATP-dependent DNA helicase. *Proc. Natl. Acad. Sci.* **97:** 1530–1535.

Coleman T.R., Carpenter P.B., and Dunphy W.G. 1996. The *Xenopus* cdc6 protein is essential for the initiation of a single round of DNA replication in cell-free extracts. *Cell* **87:** 53–63.

Crenshaw D.G., Yang J., Means A.R., and Kornbluth S. 1998. The mitotic peptidyl-prolyl isomerase, Pin1, interacts with Cdc25 and Plx1. *EMBO J.* **17:** 1315–1327.

De Bondt H.L., Rosenblatt J., Jancarik J., Jones H.D., Morgan D.O., and Kim S.H. 1993. Crystal structure of cyclin-dependent kinase 2. *Nature* **363:** 595–602.

Diffley J.F. 1996. Once and only once upon a time: Specifying and regulating origins of DNA replication in eukaryotic cells. *Genes Dev.* **10:** 2819–2830.

Donovan S., Harwood J., Drury L.S., and Diffley J.F.X. 1997. Cdc6p-dependent loading of Mcm proteins onto pre-replicative chromatin in budding yeast. *Proc. Natl. Acad. Sci.* **94:** 5611–5616.

Dunphy W.G. and Newport J.W. 1989. Fission yeast p13 blocks mitotic activation and tyrosine dephosphorylation of the *Xenopus* cdc2 protein kinase. *Cell* **58:** 181–191.

Elsasser S., Chi Y., Yang P., and Campbell. J.L. 1999. Phosphorylation controls timing of Cdc6p destruction: A biochemical analysis. *Mol. Biol. Cell.* **10:** 3263–3277.

Fesquet D., Labbe J.C., Derancourt J., Capony J.P., Galas S., Girard F., Lorca T., Shuttleworth J., Doree M., and Cavadore J.C. 1993. The MO15 gene encodes the catalytic subunit of a protein kinase that activates cdc2 and other cyclin-dependent kinases (CDKs) through phosphorylation of Thr161 and its homologues. *EMBO J.* **12:** 3111–3121.

Goldin A.L. 1991. Expression of ion channels by injection of mRNA into *Xenopus* oocytes. *Methods Cell Biol.* **36:** 487–509.

Jallepalli P.V., Brown G.W., Muzi-Falconi M., Tien D., and Kelly T.J. 1997. Regulation of the replication initiator protein p65[cdc18] by CDK phosphorylation. *Genes Dev.* **11:** 2767–2779.

Jiang W., Wells N.J., and Hunter T. 1999. Multistep regulation of DNA replication by Cdk phosphorylation of HsCdc6. *Proc. Natl. Acad. Sci.* **96:** 6193–6198.

Kato J.Y., Matsuoka M., Strom D.K., and Sherr C.J. 1994. Regulation of cyclin D-dependent kinase 4 (cdk4) by cdk4-activating kinase. *Mol. Cell. Biol.* **14:** 2713–2721.

Kelman Z., Lee J.K., and Hurwitz J. 1999. The single minichromosome maintenance protein of *Methanobacterium thermoautotrophicum* DeltaH contains DNA helicase activity. *Proc. Natl. Acad. Sci.* **96:** 14783–14788.

Kornbluth S. 1997. Apoptosis in *Xenopus* egg extracts. *Methods Enzymol.* **283:** 600–614.

Kumagai A. and Dunphy W.G. 1991. The cdc25 protein controls tyrosine dephosphorylation of the cdc2 in a cell-free system. *Cell* **64:** 903–914.

———. 1992. Regulation of the cdc25 protein during the cell cycle in *Xenopus* extracts. *Cell* **70:** 139–151.

———. 1995. Control of the cdc2/cyclin B complex in *Xenopus* egg extracts arrested at a G2/M checkpoint with DNA synthesis inhibitors. *Mol. Biol. Cell* **6:** 199–213.

———. 1999. Binding of 14-3-3 proteins and nuclear export control the intracellular localization of the mitotic inducer Cdc25. *Genes Dev.* **13:** 1067–1072.

———. 2000. Claspin, a novel protein required for the activation of Chk1 during a DNA replication checkpoint response in *Xenopus* egg extracts. *Mol. Cell* **6:** 839–849.

Kumagai A., Yakowec P.S., and Dunphy W.G. 1998a. 14-3-3 proteins act as negative regulators of the mitotic inducer Cdc25 in *Xenopus* egg extracts. *Mol. Biol. Cell* **9:** 345–354.

Kumagai A., Guo Z., Emami K.H., Wang S.X., and Dunphy W.G. 1998b. The *Xenopus* Chk1 protein kinase mediates a caffeine-sensitive pathway of checkpoint control in cell-free extracts. *J. Cell. Biol.* **142:** 1559–1569.

Labbe J.C., Capony J.P., Caput D., Cavadore J.C., Derancourt J., Kaghad M., Lelias J.M., Picard A., and Doree M. 1989. MPF from starfish oocytes at first meiotic metaphase is a heterodimer containing one molecule of cdc2 and one molecule of cyclin B. *EMBO J.* **8:** 3053–3058.

Madine M.A. and Coverley D. 1997. *Xenopus* replication assays. *Methods Enzymol.* **283:** 535–549.

Murray A.W. 1991. Cell-cycle extracts. *Methods Cell Biol.* **36:** 581–605.

Murray A.W. and Kirschner M.W. 1989. Cyclin synthesis drives the early embryonic cell cycle. *Nature* **339:** 275–280.

Newmeyer D.D. and Wilson K.L. 1991. Egg extracts for nuclear import and nuclear assembly reactions. *Methods Cell Biol.* **36:** 607–634.

Pelizon C., Madine M.A., Romanowski P., and Laskey R.A. 2000. Unphosphorylatable mutants of Cdc6 disrupt its nuclear export but still support DNA replication once per cell cycle. *Genes Dev.* **14:** 2526–2533.

Peng C.Y., Graves P.R., Thoma R.S., Wu Z., Shaw A.S., and Piwnica-Worms H. 1997. Mitotic and G2 checkpoint control: Regulation of 14-3-3 protein binding by phosphorylation of Cdc25C on serine-216 (see comments). *Science* **277:** 1501–1505.

Perkins G. and Diffley J.F.X. 1998. Nucleotide-dependent prereplicative complex assembly by Cdc6p, a homolog of eukaryotic and prokaryotic clamp-loaders. *Mol. Cell* **2:** 23–32.

Petersen B.O., Lukas J., Sorensen C.S., Bartek J., and Helin K. 1999. Phosphorylation of mammalian CDC6 by cyclin A/CDK2 regulates its subcellular localization. *EMBO J.* **18:** 396–410.

Polyak K., Kato J.Y., Solomon M.J., Sherr C.J., Massague J., Roberts J.M., and Koff A. 1994. p27Kip1, a cyclin-Cdk inhibitor, links transforming growth factor-beta and contact inhibition to cell cycle arrest. *Genes Dev.* **8:** 9–22.

Poon R.Y., Yamashita K., Adamczewski J.P., Hunt T., and Shuttleworth J. 1993. The cdc2-related protein p40MO15 is the catalytic subunit of a protein kinase that can activate p33cdk2 and p34cdc2. *EMBO J.* **12:** 3123–3132.

Poon R.Y., Yamashita K., Howell M., Ershler M.A., Belyavsky A., and Hunt T. 1994. Cell cycle regulation of the p34cdc2/p33cdk2-activating kinase p40MO15. *J. Cell. Sci.* **107:** 2789–2799.

Reynisdottir I., Polyak K., Iavarone A., and Massague J. 1995. Kip/Cip and Ink4 Cdk inhibitors cooperate to induce cell cycle arrest in response to TGF-beta. *Genes Dev.* **9:** 1831–1845.

Saha P., Chen J., Thome K.C., Lawlis S.J., Hou Z.H., Hendricks M., Parvin J.D., and Dutta A. 1998. Human CDC6/Cdc18 associates with Orc1 and cyclin-cdk and is selectively eliminated from the nucleus at the onset of S phase. *Mol. Cell. Biol.* **18:** 2758–2767.

Shechter D.F., Ying C.Y., and Gautier J. 2000. The intrinsic DNA helicase activity of *Methanobacterium thermoautotrophicum* delta H minichromosome maintenance protein. *J. Biol. Chem.* **275:** 15049–15059.

Shevchenko A., Wilm M., Vorm O., and Mann M. 1996. Mass spectrometric sequencing of proteins silver-stained polyacrylamide gels. *Anal. Chem.* **68:** 850–858.

Sudakin V., Shteinberg M., Ganoth D., Hershko J., and Hershko A. 1997. Binding of activated cyclosome to p13(suc1). Use for affinity purification. *J. Biol. Chem.* **272:** 18051–18059.

Swenson K.I., Farrell K.M., and Ruderman J.V. 1986. The clam embryo protein cyclin A induces entry into M phase and the resumption of meiosis in *Xenopus* oocytes. *Cell* **47:** 861–870.

Tanaka T., Knapp D., and Nasmyth K. 1997. Loading of an Mcm protein onto DNA replication origins is regulated by Cdc6p and CDKs. *Cell* **90:** 649–660.

Tang Z., Coleman T.R., and Dunphy W.G. 1993. Two distinct mechanisms for negative regulation of the wee1 protein kinase. *EMBO J.* **12:** 3427–3436.

Thress K., Henzel W., Shillinglaw W., and Kornbluth S. 1998. Scythe: A novel reaper-binding apoptotic regulator. *EMBO J.* **17:** 6135–6143.

Tye B.K. and Sawyer S.L. 2000. The hexameric eukaryotic MCM helicase: Building symmetry from non-identical parts. *J. Biol. Chem.* **275:** 34833–34836.

Weinreich M., Liang C., and Stillman B. 1999. The Cdc6p nucleotide-binding motif is required for loading mcm proteins onto chromatin. *Proc. Natl. Acad. Sci.* **96:** 441–446.

Wolffe A.P. and Schild C. 1991. Chromatin assembly. *Methods Cell Biol.* **36:** 541–559.

Wu M. and Gerhart J. 1991. Raising *Xenopus* in the laboratory. *Methods Cell Biol.* **36:** 3–18.

Yang J., Winkler K., Yoshida M., and Kornbluth S. 1999. Maintenance of G2 arrest in the *Xenopus* oocyte: A role for 14-3-3-mediated inhibition of Cdc25 nuclear import. *EMBO J.* **18:** 2174–2183.

Yang J., Bardes E.S., Moore J.D., Brennan J., Powers M.A., and Kornbluth S. 1998. Control of cyclin B1 localization through regulated binding of the nuclear export factor CRM1. *Genes Dev.* **12:** 2131–2143.

Zhou X., Tsuda S., Bala N., and Arakaki R.F. 2000. Efficient translocation and processing with *Xenopus* egg extracts of proteins synthesized in rabbit reticulocyte lysate. *In Vitro Cell Dev. Biol. Anim.* **36:** 293–298.

Using λ Repressor Fusions to Isolate and Characterize Self-assembling Domains

Leonardo Mariño-Ramírez and James C. Hu

Department of Biochemistry and Biophysics, and Center for Advanced Biomolecular Research, Texas A&M University, College Station, Texas 77842-3017

INTRODUCTION

Although there are a wide variety of methods available to study protein–protein interactions using biochemical approaches, the yeast two-hybrid system (Fields and Song 1989) has been the method of choice for genetic analysis of pairs of proteins that physically interact. The assay is simple and can be scaled to study a large number of interactions (for review, see Uetz and Hughes 2000). However, there are a number of reasons why similar genetic assays in *Escherichia coli* should be useful (Hu et al. 2000). First, *E. coli* has higher transformation efficiencies than *Saccharomyces cerevisiae*, allowing the construction of more complex libraries. Second, the endogenous proteins present in the cell can affect the assay. This is less likely to occur in a different cellular background like *E. coli*. Third, the yeast two-hybrid system requires nuclear localization of both prey and bait whereas a bacterial system would not, as bacteria lack nuclei. Until recently, bacterial two-hybrid systems that exploit these advantages were not available. During the last few years, however, several labs have described bacterial two-hybrid systems. Chapters 25, 26, and 30 describe two of the new bacterial systems. Others have been reviewed elsewhere (Hu et al. 2000).

This chapter describes an established system based on fusions to the amino-terminal DNA-binding domain of bacteriophage λ repressor, which has been used to examine a more specialized kind of protein–protein interaction: self-assembly into homotypic oligomers. Many proteins self-assemble into oligomers, and detecting and mapping oligomerization domains has become a standard part of structure–function analysis in the study of single-gene products. Genetic methods based on fusion proteins provide a convenient way to map and characterize oligomerization domains in the absence of an assay for the protein's normal function. Although some homotypic interactions are clearly detected by yeast two-hybrid systems, across the proteome yeast two-hybrid screens seem to underrepresent them (Hu 2000; Newman et al. 2000).

The activity of λ repressor depends on the presence of both the amino-terminal DNA-binding domain and the carboxy-terminal oligomerization domain (Pabo et al. 1979). Removing the carboxy-terminal domain reduces DNA-binding activity, inactivating the repressor (Fig. 1). However, repressor activity can be reconstituted when a heterologous oligomerization domain is fused to the amino-terminal domain (Hu et al. 1990). Since its introduction in 1990, the repressor fusion system has proven to be an easy-to-use genetic tool to study protein oligomerization in vivo. Bacteriophage λ repressor fusions have been used extensively to study the oligomerization properties of proteins from various organisms in *E. coli*. Repressor fusions have been used successfully to characterize oligomerization domains from bacterial proteins (Amster-Choder and Wright 1992; Turner et al. 1997; Kennedy and Traxler 1999; Jakimowicz et al. 2000), fungal proteins (Hu et al. 1990; Strauss et al. 1998; Nikolaev et al. 1999), plant proteins (Edgerton and Jones 1992; Gonzalez et al. 1997; Palena et al. 1997), insect proteins (Gigliani et al. 1996), and mammalian proteins (Lee et al. 1992; Romano et al. 1998; Tan et al. 1998). One feature of the repressor fusion system is its ability to distinguish between dimers and higher-order oligomers (Zeng and Hu 1997). This property of the system could be used to eliminate protein aggregates from the screen.

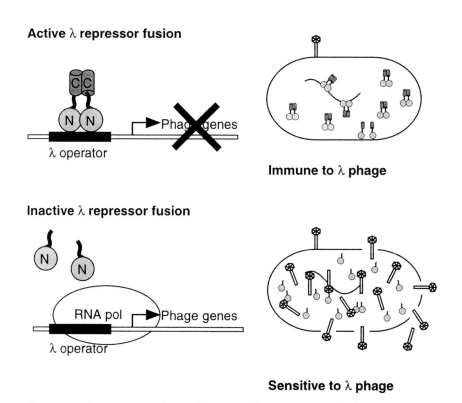

FIGURE 1. λ repressor fusions as a tool to study protein oligomerization. The λ repressor system is used to identify self-interacting domains. A positive interaction reconstitutes the activity of the repressor and the clone becomes immune to λ phage.

The use of repressor fusions has gone beyond the characterization of a single protein. Other applications include selections to disrupt interactions (Park and Raines 2000), design of peptide inhibitors (Jappelli and Brenner 1996), and characterization of proteins that require cofactors, posttranslational modifications, or molecular signals for oligomerization (Fiedler and Weiss 1995; Qin et al. 2000; Rashkova et al. 2000). Additionally, transmembrane segment interactions have been detected in vivo (Leeds and Beckwith 1998), opening new avenues for the study of interactions in nonsoluble proteins. Finally, library approaches include selections for high-affinity peptide ligands (Zhang et al. 2000) and selections to find novel self-assembly peptides encoded in *S. cerevisiae* and *E. coli* (Jappelli and Brenner 1999; Zhang et al. 1999).

Repressor fusions as well as other genetic methods to study protein–protein interactions have limitations; a genetic test can only suggest a physical interaction. Biochemical methods are then used to confirm the putative interaction. Like any genetic method, repressor fusions will miss certain interactions such as those that require posttranslational modifications for proper assembly. Therefore, repressor fusions should be thought of as providing complementary information to, rather than replacing, the ongoing large-scale two-hybrid screens (Fromont-Racine et al. 1997; Ito et al. 2000; Uetz et al. 2000) and other biochemical and genetic approaches (Phizicky and Fields 1995).

OUTLINE OF PROCEDURE

Whether the goal is to map a specific homotypic interaction from a known protein or to identify novel self-assembling domains, the repressor fusion system can be used in basically the same way. Because the characterization of self-assembling domains in known proteins has been the main application of repressor fusions, we focus below on how to find novel interacting domains from large-scale screens, such as a screen we have started in our laboratory for self-interaction domains from the yeast genome. However, the general approach for both small-scale and large-scale studies consists of the following steps (Fig. 2). First, individual clones or libraries are constructed from fragments of the target gene or genome fused to the amino-terminal DNA-binding domain of λ repressor. Second, selection and screening for repressor activity is carried out to identify fragments encoding self-assembling domains. Third, positive clones are identified by DNA sequencing and database searches to find the corresponding open reading frame (ORF) in the genome. Fourth, additional characterization can be done to examine the properties of any positive fusions.

Library Construction

Vectors

The repressor fusion vectors used for the identification of self-assembling domains are multicopy plasmids that allow cloning of the desired insert downstream and in frame with the DNA-binding domain of λ repressor. The identification of self-assembling domains requires low constitutive expression levels of the fusion proteins. This is because the amino-terminal DNA-binding domain when overexpressed is able to dimerize and thus confer immunity to phage infection even in the absence of its carboxy-terminal domain (Sauer et al. 1990).

There are three generations of repressor fusion vectors available in our lab. The differences are diagrammed in Figure 3. All of the ampicillin-resistant plasmids we use are descendants of pZ150 (Zagursky and Berman 1984), a pBR322 derivative with an M13 single-stranded (ss)DNA origin to facilitate M13-mediated transduction (Vershon et al. 1986). The region encoding the repressor fusion is placed between the *Eco*RI and *Eco*RV sites on the pBR322 map. The fusion protein is transcribed in the clockwise direction; a transcription terminator from the phage T7 φ10 gene is downstream of the cloning site. The plasmids also contain a deletion between *Bam*HI and *Sal*I sites in the *tet* gene that removes some restriction sites.

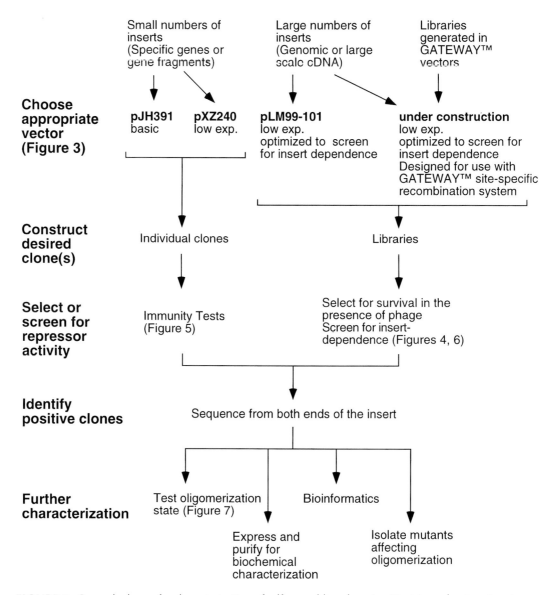

FIGURE 2. General scheme for characterization of self-assembling domains. Decisions about vectors to use and whether to use selections or screens are dictated by the source of the DNA to be tested. The figure shows different paths leading to the generation of active repressor fusions and their characterization.

In the original vector, pJH391 (Fig. 3B) (GenBank Acc. No. AF316554), the fusion protein was expressed from a minimal *lac*UV5 promoter. The 654 bp between the *Eco*RI site and the start of the repressor coding sequence contains the promoter region, a single *lac* operator, and the Shine–Dalgarno sequence from *lacZ*. There is no CAP-binding site in the promoter region. Constructs made using pJH391 must be used in a strain with the *lacI*q allele, which overexpresses the *lac* repressor. The basal level of leaky expression is more than enough to give an immune (repressor-positive) phenotype.

In pJH391, the desired inserts can be cloned between a *Sal*I site and a *Bam*HI site. Because the plasmid was designed to accommodate a specific artificial gene cassette encoding the *GCN4* leucine zipper, polylinkers were not included. pJH391 contains a "stuffer" fragment between the *Sal*I and *Bam*HI sites that allows easier purification of backbone DNA cut by these enzymes. Note

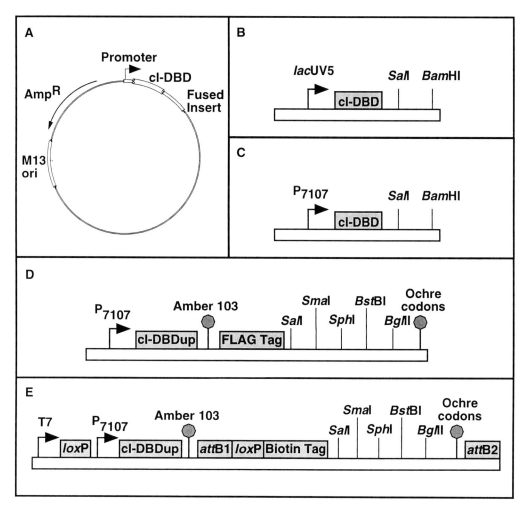

FIGURE 3. Important features in repressor fusion vectors. (*A*) Circular map indicating relevant features present in our repressor fusion vectors. Repressor fusions are constructed in a plasmid that contains an M13 origin of replication and an ampicillin-resistant marker. (*B*) First generation. A minimal *lac*UV5 promoter drives the expression of the fusions; inserts can be cloned using the *Sal*I and *Bam*HI sites. (*C*) Second generation. Repressor fusions are expressed at lower levels; P_{7107} replaces *lac*UV5. (*D*) Third generation. An amber mutation is introduced at position 103; additional unique restriction sites facilitate cloning. A translational terminator is introduced after the multiple cloning site (see text for details). (*E*) Fourth generation. Site-specific recombination will facilitate subcloning and overexpression of oligomerization domains. Segments are not drawn to scale.

that stop codons must be designed into the downstream end of the insert; readthrough into the vector can lead to the addition of a substantial amount of extra polypeptide, depending on the reading frame.

A kit based on pJH391 has been widely distributed and has been used successfully to study many different oligomerization domains. In 1997, we introduced a second generation of vectors (Fig. 3C). The main difference was the replacement of *lac*UV5 by a weak constitutive promoter called P_{7107} (Zeng et al. 1997). pXZ240 (GenBank Acc. No. AF316555) is identical to pJH391 except for the promoter replacement, which also introduces a unique *Xba*I site upstream of the cI start point. Operationally, P_{7107} seems to express the fusion protein at even lower levels than the basal level provided by *lac*UV5, even in the absence of isopropyl-β-D-galactopyranoside (IPTG); the actual differences in promoter strength have not been quantified. This lower level was important to set up an in vivo assay for dimerization specificity based on negative dominance (Zeng et al. 1997). However, pJH391 is more permissive for weak oligomers and is still often the vector of choice for dissecting the oligomerization unit of a specific gene product.

To characterize oligomerization domains from a specific gene, one usually makes a series of constructs with different end points and then tests them by screening for repressor activity. When active fusions are isolated by selection, it is important to determine that positive candidates are displaying repressor activity due to self-assembly, rather than due to a mutation in the host or the vector (e.g., up-promoter or plasmid copy number mutations) that simply increase the expression level of the fusion protein. For our first- and second-generation vectors, this is done by recloning the inserts isolated by selection. However, recloning can become cumbersome for large numbers of clones. We are currently testing a third generation of vectors that allow a rapid assay for insert dependence (Figs. 3D and 4). These vectors contain an amber mutation at position 103 between the DNA-binding domain and the DNA insert. The resultant amber fragment, comprising residues 1–102 of λ repressor, has been shown to be stable in *E. coli* (Parsell et al. 1990). As illustrated in Figure 4, if the repressor-positive phenotype is dependent on self-assembly of the insert-encoded domain, it should be dependent on suppression of the amber mutation. This can be determined by M13-mediated transduction (Vershon et al. 1986) to suppressor (*supF*; LM58) and nonsuppressor (*sup⁰*; LM59) strains (see below).

Our third-generation plasmids have multiple cloning sites to accommodate different kinds of DNA fragments. Our cloning site cassette contains recognition sites for *SalI*, *SmaI*, *SphI*, *BstBI*, and *BglII*. In addition to DNA cut with the same enzymes, these sites can accommodate a wide variety of overhanging and blunt ends. Plasmids pLM99 (GenBank Acc. No. AF308739), pLM100 (GenBank Acc. No. AF308740), and pLM101 (GenBank Acc. No. AF308741) are third-generation repressor fusion plasmids with this cloning site cassette in different reading frames.

When using designed PCR fragments or synthetic cassettes to create the inserts, it is easy to control the reading frame and to make sure that there are stop codons at the end of the insert.

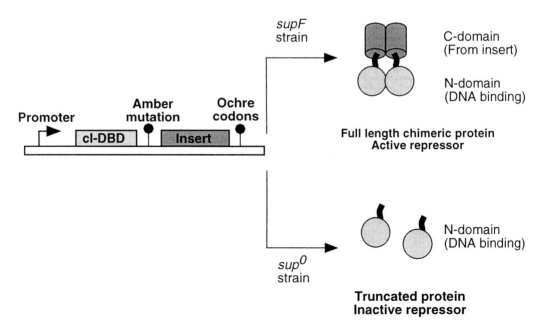

FIGURE 4. Nonsense suppression as a test for insert dependence. The presence of an amber mutation at position 103 allows a rapid screening for insert dependence using a reporter gene for the activity of the repressor fusion. In this particular case, an insert encoding a dimerization domain will display distinct phenotypes in suppressor (*supF*) and nonsuppressor (*sup⁰*) strains.

When inserting random DNA fragments, it is important to prevent translation from vector sequences, which may contain fortuitous interacting sequences. The multiple cloning site is therefore followed by a series of stop codons in all three reading frames.

The third-generation plasmids also contain mutations in the DNA-binding domain that increase its ability to activate the P_{RM} promoter (Bushman et al. 1989) and a FLAG epitope tag in the linker between the repressor domain and the insert. The third-generation plasmids express the fusion proteins from the P_{7107} promoter. However, an additional deletion in the backbone increases the copy number of these plasmids; thus, expression levels are probably somewhat higher than in the second-generation plasmids.

Currently, we are developing fourth-generation vectors that will contain several other features to facilitate working with any self-assembling domains identified through large-scale library screening. In particular, we plan to build overexpression and purification features into the fusion vectors to allow biochemical characterization. We are also including site-specific recombination sequences to facilitate insert transfer to other kinds of vectors (Walhout et al. 2000). For example, this might be used to transfer an assembly domain identified in a repressor screen to an appropriate vector for expression in a eukaryotic cell, where it might generate a dominant negative phenotype (Herskowitz 1987).

The choice of which vector to use depends on the problem to be addressed. The higher expression level from *lac*UV5 can be useful for finding weaker oligomerization domains. For example, we have found that whereas the *GCN4* leucine zipper, which has a nanomolar dissociation constant, gives repressor activity when transcribed from both *lac*UV5 and P_{7107}, a construct containing the C/EBP leucine zipper only works when expressed from the *lac*UV5 promoter. However, the *lac*UV5 promoter is too active to use in the negative dominance assay, and using the *lac* promoter when selecting for active clones in a library can lead to false positives due to mutations in *lacI*.

Inserts

To find oligomerization domains in specific genes, restriction fragments or PCR products are generally used. For very small domains, such as leucine zippers, it is often practical to generate the insert entirely from synthetic DNA. When synthetic DNA is being used, convenient restriction sites can be designed into the insert to facilitate future mutagenesis studies. Several computer programs are available to identify places in a sequence where restriction sites can be introduced without changing the coding sequence. We have implemented a Web-based version of the program Seqsearcher (Reidhaar-Olson et al. 1991) at http://tofu.tamu.edu/seqsearch/.

For our studies on whole genomes, we have only used partially digested genomic DNA from organisms that have few or no introns. However, in principle, PCR products or cDNA can also be used as a source of inserts. For genome-wide studies, the construction of high-quality plasmid libraries is critical for the success of the screen. The domain responsible for oligomerization may be located anywhere within a protein, and many fusions may be nonfunctional due to frameshifts or insert orientation. Therefore, a highly representative library must not only represent each gene, but should also provide as many different fusion end points within each gene as possible. We fragment genomic DNA by partial digestion with the restriction enzyme *Cvi*TI*, which has the recognition sequence ~NGCN. This enzyme gives a quasi-random distribution of DNA fragments that can be used for shotgun sequencing (Fitzgerald et al. 1992; Gingrich et al. 1996). *Cvi*TI* generates blunt ends that can be cloned into the *Sma*I site of the repression fusion vectors.

E. coli *Strains*

In principle, a wide variety of *E. coli* strains can be used as hosts for the repressor fusion vectors. For the first-generation vectors, a strain with the *lacI*q allele must be used. If M13-mediated transduction is going to be used, then strains must be F⁺. In practice, we have found subtle differences in the behavior of control plasmids in different strains. For example, the introduction of reporter prophages seems to decrease repressor activity slightly; some constructs that are immune in an isogenic non-lysogen become sensitive in a lysogen. In contrast, some commercially available strains, including some XL-1 Blue derivatives, allow some of our negative controls, which lack a dimerization domain, to behave as if they are immune. This suggests that there may be differences in plasmid copy number or supercoiling among different laboratory strains. In addition, some strains contain unannotated mutations that affect sensitivity to λ or φ80. We recommend the use of AG1688 when the first and second generations of vectors are used. The third and fourth generations require an amber suppressor to obtain a full-length fusion; in this case we recommend the use of JH787 as the host for these repressor fusion vectors. Note, however, there are two disadvantages for these strains. First, neither is available commercially as competent cells. Second, AG1688 is *endA⁺*, which means that plasmid preps from these strains should be deproteinized to prevent DNA loss due to endonuclease activity.

Selection and Screening

There are two classes of assays for repressor activity in *E. coli*: phage immunity tests and reporter constructs (Fig. 5). The ability of a strain expressing a repressor fusion to grow in the presence of phage can be used as either a selection or a screen. When testing a small number of specific constructs, screening is usually adequate. There are several simple assays for screening candidates. Cross-streak assays are done by placing a "line of death" on a plate by streaking a small amount of phage λ in a narrow line across a plate. Candidates are then streaked in single lines that cross the phage line at right angles. The plate is then incubated for anywhere from 5–6 hours to overnight. All candidates will grow up to the line of phage; only those restoring repressor oligomerization will grow through the line of death, whereas growth of sensitive cells will stop when they reach the phage. Individual clones can also be tested by examining the plating efficiency of phage λ. This can be done in several ways. The most rigorous way is to carefully titer a phage stock on the strain

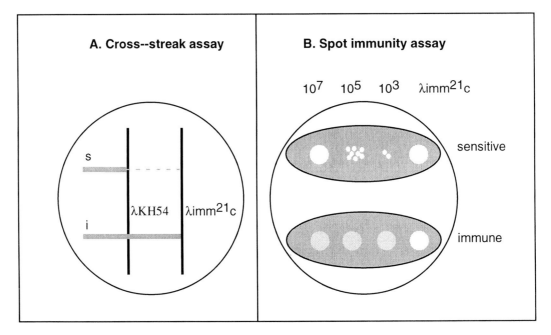

FIGURE 5. Simple immunity tests for repressor function. (*A*) Cross-streak assay. (*B*) Spot immunity test. See protocols for details.

of interest. An easier way to examine several candidates is to spot phage dilutions on preformed cell lawns in top agar. Selection for phage immunity is done by spreading cells on plates seeded with λKH54 and λKH54h80 phages. The KH54 mutation is a deletion in cI, which prevents the phage from forming lysogens. (A lysogen is a bacterial cell that has a prophage integrated in its chromosome; because λ lysogens express λ repressor and are immune to superinfecting phage, lysogens would show up as false positives.) The h80 phage uses the TonB protein as a surface receptor, whereas λ normally recognizes the LamB porin. A false positive could arise from mutations in recipient bacteria causing a loss of the surface receptor for phage attachment. Thus, by using a mixture of phages with two different receptor specificities, the background of survivors due to host mutations in receptor genes is reduced to a minimum.

Repressor activity can also be examined using a variety of reporter constructs that place a screenable or selectable marker under the control of λ operators. Operon (O) fusions that place the *lac* genes under the control of λ repressor were used to elucidate the regulation of the λP_R and P_{RM} promoters (Maurer et al. 1980; Meyer et al. 1980; Meyer and Ptashne 1980). For example, λ202 is a phage containing *lacZ* driven by λ$P_R O_R$. In this phage, O_R contains an $O_R 2^-$ mutation to eliminate cooperative binding by the wt repressor to the adjacent $O_R 1$ and $O_R 2$ operator sites. We have also constructed operon fusions where a *cat-lacZY* fusion is controlled by the λP_L promoter. In strains carrying this reporter, loss of repressor activity can be selected or screened using chloramphenicol (*cat*) resistance. Repression of this reporter is used in the screening strategy described below, to confirm the requirement for full-length fusions. Additionally, the β-galactosidase levels in the strains that contain these reporters can be determined by enzyme assays. However, whereas the difference between the repressed and unregulated levels is about 10-fold, the amount of enzyme made in the presence of an active repressor is still well above the level that gives a blue color to a colony on medium containing X-gal.

Selections based on negative control of a reporter, like the ones described above, are problematic. When a plasmid expressing a repressor fusion is introduced into a cell, the cell is already expressing the derepressed level of the reporter gene. If the reporter gene product is stable, its level is decreased only by dilution due to cell growth and division. This phenotypic lag means that libraries would have to be amplified before selection was applied. Selections based on activation of a reporter would not have this problem, because a small increase over the basal activity is easier to select, although it still takes several generations to reach steady state. In addition to its function as a repressor, λ repressor can act as an activator at the P_{RM} promoter (Meyer and Ptashne 1980). Although this activity should be useful, it has not been extensively exploited to examine repressor fusions. Note, however, that the phenotypic lag problem does not apply to phage selection because the phage promoters are not present before the repressor fusions are expressed.

Figure 6 shows a flowchart for the selection and screening of candidates from a genomic library using the third generation of vectors, illustrating the combination of phage selections and reporter constructs. The initial selection of candidates is for immunity to phage infection, as described above. Positive candidates are picked into wells in microtiter plates and grown in broth, which are used to generate M13 transducing stocks. The M13 stocks are used to do parallel transductions in microtiter plates into suppressor (*supF*) (LM58) and nonsuppressor (*sup⁰*) (LM59) strains carrying the λP_L-*cat-lacZ* reporter phage. Transductants are patched onto plates with chloramphenicol and screened for drug sensitivity, which indicates that *cat* expression has been repressed. A desired candidate, where oligomerization is dependent on the insert, will be chloramphenicol-resistant in the nonsuppressing strain and chloramphenicol-sensitive in the suppressor strain.

Requiring a candidate to repress both phage infection and reporters is a stringent test. Although many repressor fusions are active in both kinds of screens, others act as repressors only in phage infection or only with reporters. The reasons for these differences are unclear. However, this may reflect the role of cooperative binding to multiple operators during phage infection, while the reporters have been chosen to reflect occupancy of single operator sites.

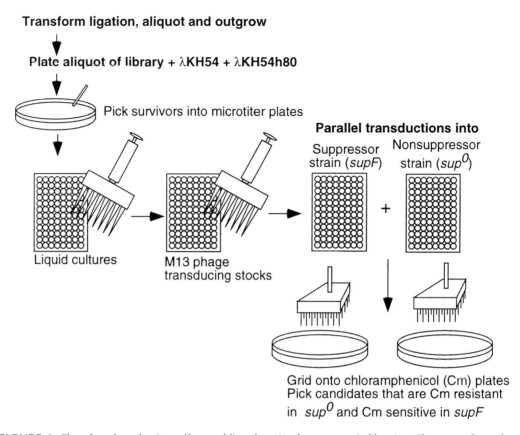

Transform ligation, aliquot and outgrow

Plate aliquot of library + λKH54 + λKH54h80

Pick survivors into microtiter plates

Liquid cultures

M13 phage
transducing stocks

Parallel transductions into

Suppressor
strain (*supF*)

Nonsuppressor
strain (*sup⁰*)

Grid onto chloramphenicol (Cm) plates
Pick candidates that are Cm resistant
in *sup⁰* and Cm sensitive in *supF*

FIGURE 6. Flowchart for selecting self-assembling domains from genomic libraries. Clones are first selected on plates seeded with λ phage. Survivors are then transduced to *supF* and *sup⁰* strains, where clones expressing a repressor fusion encoding a putative self-assembling domain are identified. See protocol for details.

Self-assembling Domain Identification

Self-assembling domains isolated from libraries can be identified by DNA sequencing and database searches to assign the corresponding ORF in the genome. Plasmid DNA is extracted by a high-throughput method (Marra et al. 1999) and DNA sequence is obtained from both ends. We use two primers: the cI primer (5′-AGGGATGTTCTCACCTAAGCT-3′) reads clockwise starting in the linker region between the repressor DNA-binding domain and the insert and T-φ (5′-CTCAGCGGTGGCAGCAGCCAA-3′), which reads counterclockwise starting in the T7 gene 10 transcription terminator. Sequencing from both ends is important for the identification of chimeric inserts or peptides outside of the protein space. We use the BLAST (Altschul et al. 1997) programs to assign the positive clones to specific open reading frames (ORF), and the WebBLAST (Ferlanti et al. 1999) package to organize the reports generated by BLAST searches.

Testing Oligomerization States

Repressor fusions can also be used in a genetic assay to distinguish between dimers and higher-order oligomers (Fig. 7) (Zeng and Hu 1997). This assay is performed by comparing two strains. The first strain (JH607) contains two λ repressor operator sequences with different binding affinities: a proximal low-binding-affinity operator labeled "weak" and a distal high-binding-affinity operator labeled "strong." The second strain (XZ970) contains only the proximal low-binding-

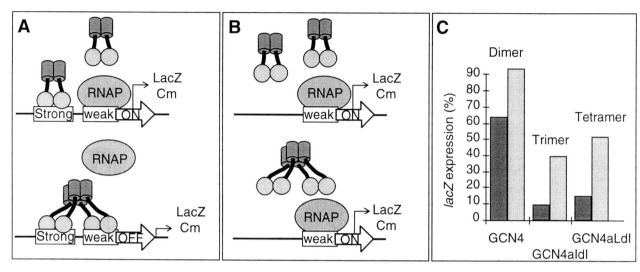

FIGURE 7. Genetic test to distinguish between dimers and higher-order oligomers. (*A*) Dimeric fusions will not fully repress reporter genes in the strain with two operators, whereas tetrameric fusions will bind to the two operators and fully repress the reporter genes. (*B*) In a control strain where only a single operator controls the reporter genes, both dimers and tetramers give comparable repression. (*C*) β-Galactosidase assays distinguish between dimers and higher-order oligomers. (*Light bars*) Single operator reporter; (*dark bars*) dual-operator reporter. (Adapted, with permission, from Zeng et al. 1997, © National Academy of Sciences, U.S.A.)

affinity operator. Reporter gene repression is achieved only when the proximal operator is occupied. Figure 7C illustrates how β-galactosidase assays can be used to quantify the difference in expression of *lacZ* for model leucine zipper dimers, trimers, and tetramers. Only trimers and tetramers are able to repress efficiently the *lacZ* reporter in the strain containing two operators due to cooperative DNA binding. This extension of the repressor system has also been used to examine other bacterial proteins besides a variety of λ repressor mutants and leucine zippers (Liu et al. 1998; Xia and Uhlin 1999), suggesting that other oligomers behave similarly in the system.

λ Repressor Fusions

Preparation of individual clones or libraries in pJH391, pXZ240, or pLM99-101 should be done using standard molecular biology methods. Similarly, processing plasmid-containing strains for DNA sequencing can be done by a variety of standard methods (see Sambrook and Russell 2001, *Molecular Cloning*). Detailed protocols provided below are for methods that are either specific to repressor fusions or likely to be unfamiliar to labs not working with phage.

MATERIALS

CAUTION: See Appendix for appropriate handling of materials marked with <!>.

Buffers and Solutions

Chloroform <!>
LB broth and agar (see Sambrook and Russell 2001, *Molecular Cloning*)
> Premixed LB broth and agar are prepared according to the vendor's instructions.

Sodium citrate (1 M) (sterile)
TM buffer
   10 mM Tris-HCl (pH 8.0)
   10 mM MgSO$_4$ <!>
> Autoclave.

Tryptone agar
> Add 13 g of Bacto-agar per liter of tryptone broth before autoclaving.

Tryptone broth (per liter)

tryptone	10 g
NaCl	5 g

> Add distilled H$_2$O to 1 liter. Dissolve, aliquot, and sterilize by autoclaving.

Tryptone top agar
   0.7 g of Bacto-agar per 100 ml
> We make this in bottles ahead of time and melt it in a microwave oven before use.

2x YT broth (per liter) (see Sambrook and Russell 2001, *Molecular Cloning*)

tryptone	16 g
yeast extract	10 g
NaCl	10 g

> Dissolve in 1 liter of distilled H$_2$O. Divide into 100-ml aliquots. Autoclave.

E. coli Strains, Phage, and Plasmids

Strain	Genotype or description	Use	Reference
MC1061F⁻	*araD139*, Δ(*ara-leu*)7697, Δ(*lac*)X74, *galE15*, *galK16*, *rpsL*(Str^R), *hsdR2*, *mcrA*, mcrB1	λ phage propagation	(Casadaban and Cohen 1980)
AG1688	MC1061 F'128 *lacI^q lacZ::Tn5*	host for first- and second-generation vectors; M13 transduction	(Hu et al. 1993)
JH787	AG1688 (φ80 Su-3)	host for third- and fourth-generation vectors; M13 transduction	(J.C. Hu, unpubl.)
LM58	JH787 (λLM58)	suppressor strain used for library screening of oligomerization domains	(L. Mariño-Ramírez, unpubl.)
LM59	AG1688 (λLM58)	nonsuppressor strain used for library screening of oligomerization domains	(L. Mariño-Ramírez, unpubl.)
JH607	AG1688 (λ112O$_s$P$_s$)	host used for testing oligomerization states	(Zeng and Hu 1997)
XZ970	AG1688 (λXZ970)	host used for testing oligomerization states	(Zeng and Hu 1997)

Phage	Genotype or description	Use	Reference
λKH54	λΔcI	λ phage used in selections and screens	(K.-C. Luk and W. Szybalski, unpubl. 1983)
λKH54h80	λ–φ80 hybrid, has the host range of φ80	λ phage used in selections and screens	
λimm²¹c or λimm⁴³⁴c	heteroimmune controls	used to test whether cells are sensitive to λ	
M13 rv-1	M13 helper phage	M13 transduction	(Zagursky and Berman 1984)
λLM58	λLM58 is λimm²¹ P$_L$-*cat-lac*; constructed by recombination between λXZ1 (Zeng and Hu 1997) and Plasmid pLM53 (GenBank Acc. No. AF179893)	contains the *cat* reporter used in library screens	(L. Mariño-Ramírez, unpubl.)

Plasmid	Description (see Fig. 3)	GenBank Acc. No.
pJH391	Amp^R, P$_{lac}$UV5-λcI[1–132]-*Sal*I-stuffer-*Bam*HI	AF316554
pXZ240	Amp^R, P$_{7107}$-λcI[1-132]-*Sal*I-stuffer-*Bam*HI	AF316555
pLM99	Amp^R, P$_{7107}$-λcI[1-116]-am103-FLAG-multiple cloning site cassette, frame 1	AF308739
pLM100	Amp^R, P$_{7107}$-λcI[1-116]-am103-FLAG-multiple cloning site cassette, frame 2	AF308740
pLM101	Amp^R, P$_{7107}$-λcI[1-116]-am103-FLAG-multiple cloning site cassette, frame 3	AF308741

Antibiotics

Ampicillin (200 mg/ml) in H$_2$O
 1000x stock; use at a final concentration of 200 μg/ml.
Kanamycin (20 mg/ml) in H$_2$O
 1000x stock; use at a final concentration of 20 μg/ml.

Chloramphenicol (25 mg/ml) <!> in 100% ethanol <!>
1000x stock; use at a final concentration of 25 µg/ml.

Gentamicin (15 mg/ml) (gentamicin sulfate) in ddH$_2$O
1000x stock; use at a final concentration of 15 µg/ml.

Special Equipment

Centrifuge (Sorvall Model RC-5B, Rotor SH-3000 or similar) for 96-well microplates
Heat block (Baxter Scientific Model H2025-1A)
Incubator (37°C)
Microplate replicator (96 pin; Boekel Model 140500)
Sterile toothpicks
Water bath (50°C)

METHODS

Preparation of λ Phage Stocks

λKH54 and λKH54h80 are used in both selections and screens for active repressor fusions. In this protocol, phages are plaque-purified before being amplified to generate working stocks.

On day 1:

1. Grow a 3–5-ml overnight culture of MC1061 in tryptone broth.

    Many other strains can be used to grow bacteriophage λ, but we prefer to use an F⁻ strain to grow our stocks to avoid the possibility of M13 contamination. Although fresh overnights are best, the same culture can be kept at room temperature and used for several days.

On day 2:

2. Melt and preincubate tryptone top agar at 50°C. It is important to let the agar cool to ~50°C before adding it to cells. If it is too hot, the cells will be killed. Prewarm three 100-mm-diameter tryptone agar plates to room temperature. Tryptone agar and top agar give better results than LB.

3. Make a set of serial tenfold dilutions of an existing phage stock (preferably an archival stock, not a repeated serial passage of a working stock) in TM buffer, such that the final dilutions have titers of ~100–500 pfu/ml. Prepare three 13 x 100-mm test tubes with 100 µl of MC1061 overnight and 100 µl of diluted phage from the last three dilutions. Incubate for 20 minutes at room temperature to allow the phage to adsorb to the cells.

4. Using a disposable 5-ml pipette, quickly add 3 ml of melted top agar to each of the tubes. It is not necessary to do anything to mix the agar with the cells and phage. Vortexing is counterproductive because it introduces bubbles into the top agar.

5. Immediately pour the top agar onto a plate and tip it back and forth to allow the top agar to spread over the surface. Do not overdo this; the top agar will solidify quickly.

6. Incubate overnight at 37°C.

On day 3:

7. Prewarm tryptone agar and tryptone top agar as before. Examine the plates from the incubator to find one with well-isolated plaques. Pick a single isolated plaque by stabbing it with a sterile Pasteur pipette. Move the pipette sideways to break the seal between the agar and the

bottom of the plate and withdraw the plaque as a plug of agar. Eject it into 0.5 ml of TM buffer. Add one or two drops of chloroform, vortex, and place on ice for a few minutes to allow the phage to diffuse out of the plug.

8. Transfer 0.2 ml of the "pickate" to each of two new test tubes. Be careful to avoid the chloroform. Vortex briefly to allow any residual chloroform to evaporate. Make a third tube with 0.2 ml of TM buffer as a control.

9. Add 0.1 ml of the MC1061 bacteria overnight to each tube. Preincubate, add top agar, and plate as before.

10. Incubate at 37°C until the plates with phage are covered with confluent plaques.

    Very small areas of the bacterial lawn should be seen where clusters of plaques converge. It may help to compare the phage plates to the control to follow the development of the uninfected lawn. This should take anywhere from 5 to 8 hours. Do not let the incubation go overnight; the titer will be decreased.

11. Add 5 ml of TM buffer to each confluent lysis plate. Leave overnight at 4°C in the dark to allow the phage to elute.

On day 4:

12. Transfer the liquid from the plates to a sterile screw-cap tube. A little less than 5 ml per plate will be recovered. Add a few drops of chloroform to kill any surviving cells.

13. Titer the phage stock using the same methods. Expect a titer of 10^9–10^{10} pfu/ml.

Screening by Cross-streak Assays

Cross-streak assays are a very fast and simple way to determine the immunity status of a particular fusion. This is usually the first test that is done once a fusion is constructed.

14. Using a marker, draw a pair of straight lines, about 1 cm apart, on the back of a LB ampicillin plate. Label one KH54 and the other imm21c (or imm434c, as appropriate).

15. Using a 0.1-ml pipette, a sterile stick, or an inoculating loop, paint about 5–10 µl of phage along the appropriate line.

16. Pick a fresh colony of a clone of interest with a sterile toothpick. Starting about 2 cm from the line of KH54, draw a single line of cells across both lines of phage.

17. Incubate for 6 hours to overnight at 37°C. Sensitive cells will die at the first line. Immune cells will cross the first line but not the second. Anything that crosses both lines has lost the λ receptor, or may be a non-*E. coli* contaminant.

    Results will vary depending on how heavy-handed you are when you pick the colony. It is better to examine the plates at earlier times, because at later times resistant cells can take over the streak. We recommend AG1688 or JH787 as the hosts for immunity assays.

Screening by Spot Immunity Tests

Plating efficiency can be measured by titering phage in parallel on a candidate strain and a strain with no repressor fusion. Spot titers are a rapid way to test large numbers of candidates; they can also be done in a way that uses fewer plates.

18. For each candidate, inoculate a fresh single colony into 2 ml of LB ampicillin. Grow for 8 hours at 37°C. For the best reproducibility, do not vary this outgrowth time.

19. Mix 50 µl of the culture with 2 ml of tryptone top agar. Use a disposable pipette to paint a sector of top agar onto a prewarmed LB ampicillin plate. We routinely use three or four different candidates per 100-mm plate.

20. Spot 2–3 µl of λKH54 diluted in TM onto the lawn. Use dilutions with 10^7, 10^5, and 10^3 pfu/ml. We use a multichannel pipettor to do the spots when there are a large number of candidates to process.

21. Incubate overnight at 37°C. Strains that do not show any plaques are considered immune. Strains with clear or turbid plaques are considered sensitive. We find that some strains give day-to-day variation in this assay.

M13-mediated Transduction

M13-mediated transduction is a simple, rapid, and inexpensive way to transfer plasmids containing the M13 ssDNA origin between F^+ cells. This can be used to test insert dependence (see below) or to transfer plasmids to reporter strains, such as those used to distinguish dimers from higher-order oligomers (Zeng and Hu 1997).

22. Use a single colony of cells containing each cI^+ fusion plasmid to inoculate overnight cultures of LB ampicillin. Prepare an overnight culture of the appropriate recipient strain by using a single colony of cells to inoculate overnight cultures in LB supplemented with appropriate antibiotics.

23. Mix 20 µl of the overnight culture with 2×10^{11} pfu phage M13 in 0.1 ml.

24. Incubate for 10 minutes at 37°C. Add 2 ml of 2x YT broth to the mixture and allow the cells to grow with aeration for another 8 hours or overnight.

25. Transfer part of the culture to a microfuge tube and spin down the cells for 10 minutes.

26. Transfer the supernatant containing the M13-transducing phage to a new tube and incubate for 15 minutes at 60°C to kill any remaining cells.

    The pasteurized phage stock can be stored for months to years at 4°C.

27. Mix 50 µl of the overnight culture of the recipient with 5 µl of each cI^+ fusion M13 transducing stock.

28. Incubate for 30 minutes at 37°C.

29. Spot 5 µl of each transduction onto a LB ampicillin plate. After the spot dries, use a toothpick to streak out from it for single colonies. The single colonies that grow should be struck for singles again the next day.

Selection and Screening of Candidates from Libraries Made with pLM99-101

Repressor fusions can be used to explore genomes for proteins that contain oligomerization domains (Jappelli and Brenner 1999; Zhang et al. 1999). The procedure below has significant improvements that allow the recovery of a wide variety of oligomerization domains with a low background (L. Mariño-Ramírez and J.C. Hu, unpubl.). Libraries are constructed using standard molecular biology methods (see Sambrook and Russell 2001, *Molecular Cloning*). The complexity needed for the isolation of oligomerization domains should be at least 10^6 independent clones. We recommend the use of JH787 as the host for selections.

1. For selection of immune clones: Plate ~10^8 JH787 cells containing fusion libraries on LB-ampicillin-kanamycin plates seeded with 10^8 pfu/plate λKH54 and λKH54h80. The libraries can be used for selection without amplification; this will increase the number of independent clones. However, the plating efficiency of immune clones will decrease because of the lag between transformation and achieving the steady-state level of a repressor fusion protein. Incubate overnight at 37°C.

2. Prepare three or more 96-well microplates (conical "V" bottom) for each library by adding 0.15 ml of 2x YT-ampicillin-kanamycin supplemented with 25 mM sodium citrate. The sodium citrate chelates divalent cations that would allow λ phage carried over from the selection plate to grow. Inoculate the wells with immune clones. Grow for 16 hours at 37°C.

   Keep using 96-well microplates for the rest of the protocol to facilitate manipulation.

3. Mix 5 μl of M13 rv-1 helper phage (~10^8 pfu) and 5 μl of overnight culture from immune clones. Incubate for 10 minutes at 37°C. Add 0.15 ml of 2x YT. Grow for 6 hours at 37°C.

4. Kill cells by heating for 20 minutes at 65°C. Centrifuge the plates at 1000g for 15 minutes. Store the plate, which contains 96 M13-transducing phage stocks, at 4°C.

5. For M13 transductions on LM58 and LM59 bacteria, mix 5 μl of M13-transducing phage and 50 μl of overnight culture from LM58 and LM59 (use different plates for each strain). Incubate for 30 minutes at 37°C. Use the microplate replicator to transfer the transductions to large 15 × 150-mm LB-ampicillin plates. Incubate overnight at 37°C.

6. From the LB-ampicillin plates, replicate onto LB-ampicillin-chloramphenicol plates using the 96-prong replicator. Incubate overnight at 37°C.

   A positive clone will be chloramphenicol-sensitive in LM58 but chloramphenicol-resistant in LM59. Expect about 50% clones that show a positive phenotype dependent on insert.

7. Prepare glycerol stocks from the clones transduced on LM58 by growing overnight cultures on LB-ampicillin and adding glycerol to 12%. Freeze at –70°C.

ACKNOWLEDGMENTS

The authors thank the members of the Hu lab and Debby Siegele for critical comments on the manuscript. Work related to this review was supported by funding from the National Science Foundation (MCB-9808474), the Robert A. Welch Foundation (A-1354), and the Advanced Research Program of the Texas Higher Education Coordinating Board (Award 999902-116). L.M. was supported by a fellowship from Fulbright/Colciencias/IIE.

REFERENCES

Altschul S.F., Madden T.L., Schaffer A.A., Zhang J., Zhang Z., Miller W., and Lipman D.J. 1997. Gapped BLAST and PSI-BLAST: A new generation of protein database search programs. *Nucleic Acids Res.* **25:** 3389–3402.

Amster-Choder O. and Wright A. 1992. Modulation of the dimerization of a transcriptional antiterminator protein by phosphorylation. *Science* **257:** 1395–1398.

Bushman F.D., Shang C., and Ptashne M. 1989. A single glutamic acid residue plays a key role in the transcriptional activation function of lambda repressor. *Cell* **58:** 1163–1171.

Casadaban M. and Cohen S. 1980. Analysis of gene control signals by DNA fusion and cloning in *Escherichia coli. J. Mol. Biol.* **138:** 179–207.

Edgerton M.D. and Jones A.M. 1992. Localization of protein-protein interactions between subunits of phytochrome. *Plant Cell* **4:** 161–171.

Ferlanti E.S., Ryan J.F., Makalowska I., and Baxevanis A.D. 1999. WebBLAST 2.0: An integrated solution for organizing and analyzing sequence data. *Bioinformatics* **15:** 422–423.

Fiedler U. and Weiss V. 1995. A common switch in activation of the response regulators NtrC and PhoB: Phosphorylation induces dimerization of the receiver modules. *EMBO J.* **14:** 3696–3705.

Fields S. and Song O. 1989. A novel genetic system to detect protein-protein interactions. *Nature* **340:** 245–246.

Fitzgerald M.C., Skowron P., Van Etten J.L., Smith L.M., and Mead D.A. 1992. Rapid shotgun cloning utilizing the two base recognition endonuclease *Cvi*JI. *Nucleic Acids Res.* **20:** 3753–3762.

Fromont-Racine M., Rain J.C., and Legrain P. 1997. Toward a functional analysis of the yeast genome through exhaustive two-hybrid screens. *Nat. Genet.* **16:** 277–282.

Gigliani F., Longo F., Gaddini L., and Battaglia P. 1996. Interactions among the bHLH domains of the proteins encoded by the Enhancer of split and achaete-scute gene complexes of *Drosophila*. *Mol. Gen. Genet.* **251:** 628–634.

Gingrich J.C., Boehrer D.M., and Basu S.B. 1996. Partial *Cvi*JI digestion as an alternative approach to generate cosmid sublibraries for large-scale sequencing projects. *BioTechniques* **21:** 99–104.

Gonzalez D.H., Valle E.M., and Chan G.G. 1997. Interaction between proteins containing homeodomains associated to leucine zippers from sunflower. *Biochim. Biophys. Acta* **1351:** 137–149.

Herskowitz I. 1987. Functional inactivation of genes by dominant negative mutations. *Nature* **329:** 219–222.

Hu J.C. 2000. A guided tour in protein interaction space: Coiled coils from the yeast proteome. *Proc. Natl. Acad. Sci.* **97:** 12935–12936.

Hu J.C., Kornacker M.G., and Hochschild A. 2000. *Escherichia coli* one- and two-hybrid systems for the analysis and identification of protein-protein interactions. *Methods* **20:** 80–94.

Hu J., Newell N., Tidor B., and Sauer R. 1993. Probing the roles of residues at the **e** and **g** positions of the GCN4 leucine zipper by combinatorial mutagenesis. *Protein Sci.* **2:** 1072–1084.

Hu J.C., O'Shea E.K., Kim P.S., and Sauer R.T. 1990. Sequence requirements for coiled-coils: Analysis with lambda repressor-GCN4 leucine zipper fusions. *Science* **250:** 1400–1403.

Ito T., Tashiro K., Muta S., Ozawa R., Chiba T., Nishizawa M., Yamamoto K., Kuhara S., and Sakaki Y. 2000. Toward a protein-protein interaction map of the budding yeast: A comprehensive system to examine two-hybrid interactions in all possible combinations between the yeast proteins. *Proc. Natl. Acad. Sci.* **97:** 1143–1147.

Jakimowicz D., Majkadagger J., Konopa G., Wegrzyn G., Messer W., Schrempf H., and Zakrzewska-Czerwinska J. 2000. Architecture of the *Streptomyces lividans* DnaA protein-replication origin complexes. *J. Mol. Biol.* **298:** 351–364.

Jappelli R. and Brenner S. 1996. Interaction between cAMP-dependent protein kinase catalytic subunit and peptide inhibitors analyzed with λ repressor fusions. *J. Mol. Biol.* **259:** 575–578.

———. 1999. A genetic screen to identify sequences that mediate protein oligomerization in *Escherichia coli*. *Biochem. Biophys. Res. Commun.* **266:** 243–247.

Kennedy K.A. and Traxler B. 1999. MalK forms a dimer independent of its assembly into the MalFGK2 ATP-binding cassette transporter of *Escherichia coli*. *J. Biol. Chem.* **274:** 6259–6264.

Lee J.W., Gulick T., and Moore D.D. 1992. Thyroid hormone receptor dimerization function maps to a conserved subregion of the ligand binding domain. *Mol. Endocrinol.* **6:** 1867–1873.

Leeds J.A. and Beckwith J. 1998. Lambda repressor N-terminal DNA-binding domain as an assay for protein transmembrane segment interactions *in vivo*. *J. Mol. Biol.* **280:** 799–810.

Liu T., Renberg S.K., and Haggard-Ljungquist E. 1998. The E protein of satellite phage P4 acts as an anti-repressor by binding to the C protein of helper phage P2. *Mol. Microbiol.* **30:** 1041–1050.

Marra M.A., Kucaba T.A., Hillier L.W., and Waterston R.H. 1999. High-throughput plasmid DNA purification for 3 cents per sample. *Nucleic Acids Res.* **27:** e37.

Maurer R., Meyer B., and Ptashne M. 1980. Gene regulation at the right operator (OR) bacteriophage lambda. I. OR3 and autogenous negative control by repressor. *J. Mol. Biol.* **139:** 147–161.

Meyer B.J. and Ptashne M. 1980. Gene regulation at the right operator (OR) of bacteriophage lambda. III. lambda repressor directly activates gene transcription. *J. Mol. Biol.* **139:** 195–205.

Meyer B.J., Maurer R., and Ptashne M. 1980. Gene regulation at the right operator (OR) of bacteriophage lambda. II. OR1, OR2, and OR3: Their roles in mediating the effects of repressor and cro. *J. Mol. Biol.* **139:** 163–194.

Newman J.R., Wolf E., and Kim P.S. 2000. A computationally directed screen identifying interacting coiled coils from *Saccharomyces cerevisiae*. *Proc. Natl. Acad. Sci.* **97:** 13203–13208.

Nikolaev I., Lenouvel F., and Felenbok B. 1999. Unique DNA binding specificity of the binuclear zinc AlcR activator of the ethanol utilization pathway in *Aspergillus nidulans*. *J. Biol. Chem.* **274:** 9795–9802.

Pabo C.O., Sauer R.T., Sturtevant J.M., and Ptashne M. 1979. The lambda repressor contains two domains. *Proc. Natl. Acad. Sci.* **76:** 1608–1612.

Palena C., Chan R., and Gonzalez D. 1997. A novel type of dimerization motif, related to leucine zippers, is present in plant homeodomain proteins. *Biochim. Biophys. Acta* **1352:** 203–212.

Park S.H. and Raines R.T. 2000. Genetic selection for dissociative inhibitors of designated protein-protein interactions. *Nat. Biotechnol.* **18:** 847–851.

Parsell D.A., Silber K.R., and Sauer R.T. 1990. Carboxy-terminal determinants of intracellular protein degradation. *Genes Dev.* **4:** 277–286.

Phizicky E.M. and Fields S. 1995. Protein-protein interactions: Methods for detection and analysis. *Microbiol. Rev.* **59:** 94–123.

Qin Y., Luo Z.Q., Smyth A.J., Gao P., Beck von Bodman S., and Farrand S.K. 2000. Quorum-sensing signal binding results in dimerization of TraR and its release from membranes into the cytoplasm. *EMBO J.* **19:** 5212–5221.

Rashkova S., Zhou X.R., Chen J., and Christie P.J. 2000. Self-assembly of the *Agrobacterium tumefaciens* VirB11 traffic ATPase. *J. Bacteriol.* **182:** 4137–4145.

Reidhaar-Olson J., Bowie J.U., Breyer R.M., Hu J.C., Knight K.L., Lim W.A., Mossing M.C., Parsell D.A., Shoemaker K.R., and Sauer R.T. 1991. Random mutagenesis of protein sequences using oligonucleotide cassettes. *Methods Enzymol.* **208:** 564–586.

Romano P.R., Zhang F., Tan S.L., Garcia-Barrio M.T., Katze M.G., Dever T.E., and Hinnebusch A.G. 1998. Inhibition of double-stranded RNA-dependent protein kinase PKR by vaccinia virus E3: Role of complex formation and the E3 N-terminal domain. *Mol. Cell. Biol.* **18:** 7304–7316.

Sambrook J. and Russell D. 2001. *Molecular cloning: A laboratory manual,* 3rd edition. Cold Spring Harbor Laboratory Press, Cold Spring Harbor, New York.

Sauer R.T., Jordan S.R., and Pabo C.O. 1990. Lambda repressor: A model system for understanding protein-DNA interactions and protein stability. *Adv. Protein Chem.* **40:** 1–61.

Strauss J., Muro-Pastor M.I., and Scazzocchio C. 1998. The regulator of nitrate assimilation in ascomycetes is a dimer which binds a nonrepeated, asymmetrical sequence. *Mol. Cell. Biol.* **18:** 1339–1348.

Tan S.L., Gale Jr., M.J., and Katze M.G. 1998. Double-stranded RNA-independent dimerization of interferon-induced protein kinase PKR and inhibition of dimerization by the cellular P58IPK inhibitor. *Mol. Cell. Biol.* **18:** 2431–2443.

Turner L.R., Olson J.W., and Lory S. 1997. The XcpR protein of *Pseudomonas aeruginosa* dimerizes via its N-terminus. *Mol. Microbiol.* **26:** 877–887.

Uetz P. and Hughes R.E. 2000. Systematic and large-scale two-hybrid screens. *Curr. Opin. Microbiol.* **3:** 303–308.

Uetz P., Giot L., Cagney G., Mansfield T.A., Judson R.S., Knight J.R., Lockshon D., Narayan V., Srinivasan M., Pochart P., Qureshi-Emili A., Li Y., Godwin B., Conover D., Kalbfleisch T., Vijayadamodar G., Yang M., Johnston M., Fields S., and Rothberg J.M. 2000. A comprehensive analysis of protein-protein interactions in *Saccharomyces cerevisiae. Nature* **403:** 623–627.

Vershon A.K., Bowie J.U., Karplus T.M., and Sauer R.T. 1986. Isolation and analysis of Arc repressor mutants: Evidence for an unusual mechanism of DNA binding. *Proteins Struct. Funct. Genet.* **1:** 302–311.

Walhout A.J., Temple G.F., Brasch M.A., Hartley J.L., Lorson M.A., van den Heuvel S., and Vidal M. 2000. GATEWAY recombinational cloning: Application to the cloning of large numbers of open reading frames or ORFeomes. *Methods Enzymol.* **328:** 575–592.

Xia Y. and Uhlin B.E. 1999. Mutational analysis of the PapB transcriptional regulator in *Escherichia coli.* Regions important for DNA binding and oligomerization. *J. Biol. Chem.* **274:** 19723–19730.

Zagursky R.J. and Berman M.L. 1984. Cloning vectors that yield high levels of single-stranded DNA for rapid DNA sequencing. *Gene* **27:** 183–191.

Zeng X. and Hu J.C. 1997. Detection of tetramerization domains *in vivo* by cooperative DNA binding to tandem lambda operator sites. *Gene* **185:** 245–249.

Zeng X., Herndon A.M., and Hu J.C. 1997. Buried asparagines determine the dimerization specificities of leucine zipper mutants. *Proc. Natl. Acad. Sci.* **94:** 3673–3678.

Zhang Z., Zhu W., and Kodadek T. 2000. Selection and application of peptide-binding peptides. *Nat. Biotechnol.* **18:** 71–74.

Zhang Z., Murphy A., Hu J.C., and Kodadek T. 1999. Genetic selection of short peptides that support protein oligomerization *in vivo. Curr. Biol.* **9:** 417–420.

22 The Membrane-based Yeast Two-hybrid System

Safia Thaminy[1] and Igor Stagljar[1,2]

[1]Institute of Veterinary Biochemistry, University of Zurich-Irchel, CH-8057 Zurich, Switzerland;
[2]DUALSYSTEMS BIOTECH, CH-8057 Zurich, Switzerland

INTRODUCTION

Biochemical and genetic approaches have been developed to identify and study interacting proteins, but beyond any doubt, the yeast two-hybrid system has been shown to be a revolutionary method for detection and analysis of protein–protein interaction (Fields and Song 1989). The two-hybrid system is an in vivo assay designed to detect protein–protein interactions in their native conformation (Chien et al. 1991). Generally in this assay, the DNA-binding domain (BD) of a transcription factor is fused to a target protein X, whereas its transcriptional activation domain (AD) is fused to a protein Y. These two hybrid proteins BD-X and AD-Y are coexpressed in a yeast strain containing reporter genes that require the activity of a transcriptionally complete activator for expression. If X and Y associate, a functional transcription factor is reconstituted, leading to expression of the reporter genes. The detection of the activity of the transcription factor via selection of the reporter gene expression is the basis for the identification of interacting proteins.

LIMITATIONS OF THE YEAST-TWO HYBRID SYSTEM

Despite the fact that the two-hybrid system is a powerful and versatile technique to identify new partners, this method is limited because it cannot be applied to all proteins: (1) For example, it is difficult to use this system for transcriptional activators, because these proteins, when used as a bait, often constitutively activate transcription of the reporter gene; (2) the expression of the bait and the prey as fusion proteins could sometimes result in loss of function because of steric obstruction or partial unfolding induced by the chimeric domain; (3) in higher eukaryotic organisms, a broad range of proteins undergoes extensive posttranslational modifications that are essential to their function. The interactions of these proteins depend on posttranslational modifications such as glycosylation, disulfide bond formation, and phosphorylation, which may not occur in the nucleus; (4) the fusion proteins used in the two-hybrid system have to be transported to and properly folded in the nucleus. This may cause some problems, because for a large variety of proteins, such as membrane-anchored proteins or proteins with a strong organelle targeting sequence, the nucleus does not represent the appropriate organelle for the folding, stability, and interactions with other partners.

NON-TRADITIONAL YEAST TWO-HYBRID SYSTEMS AS TOOLS TO STUDY MEMBRANE PROTEIN INTERACTIONS

It has been proposed that approximately 40% of all proteins (including many important drug receptors) are anchored in the lipid bilayer and are unlikely to enter the nucleus (Goffeau et al. 1996). Because membrane protein interactions take an important place in cellular functions and biological responses, the understanding of these processes requires the identification and the study of new partners interacting with the protein of interest. For the reasons mentioned above, the conventional yeast two-hybrid system is not optimal for the study of these proteins, although in some limited cases, it has been successfully applied. For example, despite the fact that some of the posttranslational modifications important in protein interactions in mammalian cells (e.g., tyrosine phosphorylation, glycosylation, and disulfide bond formation) are underrepresented in *Saccharomyces cerevisiae* (Chervitz et al. 1998), some membrane proteins have been successfully expressed as a partial extracellular or intracellular domain and shown to interact with their specific ligand. Appropriate extracellular receptor–ligand interactions have been shown for growth hormone and prolactin (Ozenberger and Young 1995). In this case, the system has been proved to work because the extracellular domain used as a bait contains whole critical ligand-binding determinants. Using the cytoplasmic domain of the platelet-derived growth factor receptor as a bait, Keegan and Cooper found that it interacts and phosphorylates SHPTP2, a ubiquitously expressed SH2-containing tyrosine phosphatase, allowing the interaction of the phosphorylated SHPTP2 with the signaling protein Grb7 (Keegan and Cooper 1996). More recently, the traditional yeast two-hybrid procedure has been used to identify new proteins interacting with the ErbB2 receptor (Borg et al. 2000). Using only the nine carboxy-terminal residues of the intracellular domain of ErbB2 as a bait, a new PDZ protein ERBIN (ErbB2 interacting protein) that acts as an adapter for the receptor in epithelia has been identified.

As mentioned previously, the traditional yeast two-hybrid system has been in some cases successfully applied for study of membrane proteins, but these examples are limited and cannot be used for all transmembrane proteins such as membrane proteins that undergo posttranslational modifications (e.g., glycosylation, phosphorylation), receptors with intramembranous ligand-binding pockets, or transmembrane proteins that require oligomerization via interactions between their respective transmembrane domains. To overcome these problems, several strategies have been designed to study protein interactions outside the yeast nucleus (Table 1).

TABLE 1. New Systems to Monitor Membrane Protein Interactions in Yeast

System	Sensor to monitor protein–protein interaction	Reconstitution of known protein–protein interactions	Identification of new protein–protein interaction	Reference
Membrane-based yeast two-hybrid system	ubiquitin	Wbp1 and Ost1p	in progress	Stagljar et al. (1998)
Modified split-ubiquitin system	ubiquitin	Gal4p and Gal80p Tup1p and Ssn6p	Gal4p and Nhp6B Tup1p and Nhp6B	Laser et al. (2000)
Sos recruitment system RRS	Sos	c-Jun and c-Fos PI-3 kinase subunit p110 and p85	c-Jun and JDP2 c-Jun and JDP1	Aronheim et al. (1997)
Ras recruitment system RRS	Ras	PI-3 kinase subunit p110 and p85 c-Jun and c-Fos hSos and Grb2 Pak65 and Src Pak65 and Rac1	JDP2 and C/EBPγ Pak65 and G protein	Broder et al. (1998)
Reverse RRS	Ras	Chp and Pak65	uncharacterized proteins	Hubsman et al. (2001)
G-protein-based system	Protein G	FGR3 and SNT-1 Syntaxin 1a and nSec1	nSec1 mutant (drug design)	Ehrhard et al. (2000)

The Sos recruitment system (SRS) developed by Aronheim et al. (Fig. 1) is based on the translocation of active human Sos to its site of action at the inner leaflet of the plasma membrane (Aronheim et al. 1997). The bait protein is expressed as a fusion protein with hSos, and the prey is targeted to the membrane via a myristoylation signal sequence. Interacting fusion proteins recruit Sos to the membrane where it stimulates guanyl nucleotide exchange on yeast Ras, thus rescuing a temperature-sensitive mutant. Although this system has been used to identify new regulators of c-Jun function (Table 1) (Aronheim et al. 1997), the strategy has been improved with the advent of the Ras recruitment system (RRS) (Broder et al. 1998) and the reverse RRS (Hubsman et al. 2001).

More recently, a new method using the G-protein signaling pathway has been described to monitor the binding properties of proteins expressed in their native form and targeted to the membrane (Fig. 2) (Ehrhard et al. 2000). One of the proteins to be tested in this system is fused to the G-protein βγ subunit, and the other is targeted to the membrane. The interaction between

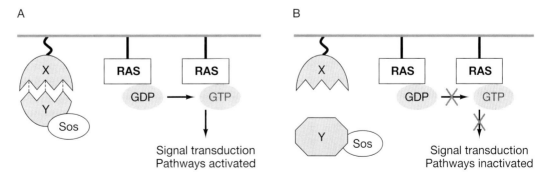

FIGURE 1. The Sos recruitment system (SRS). A protein X is targeted to the inner surface of the plasma membrane via a fused myristoylation signal sequence, and the putative interacting partner Y is expressed as a fusion protein to the human Sos protein. (A) Interaction between protein X and Y brings the Sos protein to the membrane, stimulating the conversion of the inactive Ras-GDP to an active Ras-GTP and subsequently allowing cell growth. (B) If protein X does not interact with protein Y, neither guanyl nucleotide exchange on Ras or signal transduction cascade is stimulated.

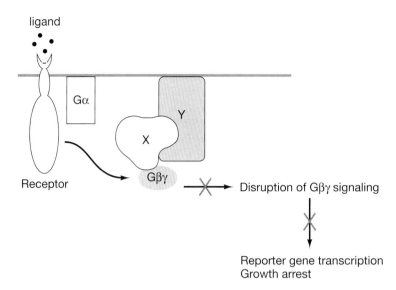

FIGURE 2. The G-protein-based system. Activation of G-protein coupled receptor by pheromones induces a conformational change in the Gα protein subunit and its dissociation from Gβγ subunits. Gβγ is then able to activate the signaling cascade. A protein X is fused to the Gβγ subunits of G-protein. The protein Y is targeted to the membrane. If protein X interacts with protein Y, the complex formation that occurs at the membrane sequesters Gβγ subunits and disrupts G-protein signaling. Gβγ is then not able to activate new gene transcription and growth arrest.

the two proteins leads to the sequestration of Gβγ subunits and disruption of the G-protein signaling. This system has been successfully applied for the reconstitution of the interaction between syntaxin 1, and fibroblast growth factor receptor and neuronal Sec 1, respectively (Table 1).

Recently, we have developed a new genetic method for the in vivo detection of membrane protein interactions in *Saccharomyces cerevisiae* (Stagljar et al. 1998; for basic reagents and method, see Stagljar and te Heesen 2000). Compared to the other systems mentioned above, our system is flexible and can be used to study interactions between two membrane proteins and/or one membrane and one cytosolic protein. The strategy (Fig. 3) uses the split-ubiquitin approach based on the detection of the in vivo processing of a reconstituted split ubiquitin (Johnsson and Varshavsky 1994a,b). Ubiquitin is a small, conserved protein of 76 amino acids involved in protein degradation (Hershko and Ciechanover 1992). Attachment of a ubiquitin molecule to a cellular protein represents a signal for the degradation of the target protein. Ubiquitin-specific proteases (UBPs) recognize specifically the folded conformation of the ubiquitin part of this target protein and cleave the bound ubiquitin–protein at the junction between the carboxy-terminal (Gly-76) residue of ubiquitin and the first amino group in the acceptor protein. The protein released is then degraded by the 26S proteasome, an ATP-dependent multisubunit protease. When ubiquitin is experimentally cleaved into carboxy-terminal (Cub) and amino-terminal (Nub) domains, neither domain can mediate ubiquitin function, whereas simultaneous expression of both domains reconstitutes in *trans* active ubiquitin. However, a particular allele of the amino-terminal domain (NubG) is unable to mediate this functional interaction and, because of reduced affinity between NubG and Cub, these two domains fail to reconstitute active ubiquitin (Johnsson and Varshavsky 1994a). We note, in eukaryotes, ubiquitin fusions are rapidly cleaved by UBPs after the last residue of ubiquitin at the ubiquitin–polypeptide junction. To prevent the degradation of the cleaved protein by the proteosome, the presence of destabilizing residues such as lysine or arginine at the amino-terminal end of the acceptor protein is avoided when adapting a ubiquitin system for protein-interaction studies (Varshavsky 1996).

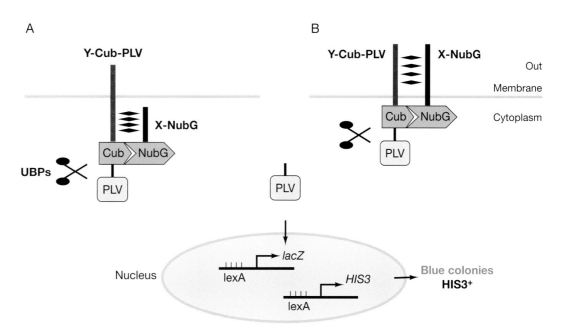

FIGURE 3. Outline of the membrane-based yeast two-hybrid system. A hybrid protein is generated that includes a transmembrane bait protein Y (*red line*) fused to Cub, followed by the artificial transcription factor ProteinA-LexA-VP16 (PLV), thus generating a Y-Cub-PLV chimera. Another hybrid prey protein X (*black line*) is generated as the fusion to the NubG domain. Note that X can also be a cytosolic protein. Interaction between Y and X results in a local increase of Cub and NubG concentration, leading to the formation of a split ubiquitin heterodimer. The heterodimer is recognized and cleaved by the UBPs (*open scissors*), liberating PLV, which enters the nucleus by diffusion and binds to lexA-binding sites upstream of the *lacZ* and *HIS3* reporter genes, resulting in β-galactosidase activity. Yeast cells are blue in the presence of X-gal and grow on agar plates lacking histidine.

The membrane-based yeast two-hybrid system takes advantage of these observations to detect and study membrane–protein interactions. The strategy described in Figure 3 consists of fusing the membrane bait protein of interest (Y) to Cub followed by the artificial transcription factor PLV (Protein A, LexA, and VP16), and the potential interacting partner (X) to NubG: Note, the inclusion of Protein A provides for an easy immunological detection of the fusion protein using the immunoglobulin G (IgG) antibody. If interaction between the bait Y and the prey X occurs, a functional split-ubiquitin molecule is reconstituted, leading to the proteolytic cleavage and the release of the transcription factor PLV, which subsequently activates a reporter gene (Stagljar and te Heesen 2000).

As a model system, yeast membrane proteins of the endoplasmic reticulum have been used. Wbp1 and Ost1p represent both subunits of the oligosaccharyl transferase membrane protein complex. The Alg5 protein also localizes to the membrane of the endoplasmic reticulum, but does not interact with the oligosaccharyltransferase. Specific in vivo interactions were detected between Wbp1 and Ost1p but not between Wbp1 and Alg5 (Stagljar et al. 1998). Beside this application, the membrane-based split ubiquitin system was successfully used to demonstrate an interaction between the yeast endoplasmic reticulum α,1,2-mannosidase and the Rer1p (M. Massaad, S. te Heesen, and A. Herscovics, unpubl.). In addition, our recent collaborative work has shown that this system can also be applied to heterologous transmembrane proteins such as plant transporter proteins. For example, we found physical interactions between plant SUT proteins and their subdomains (A. Reinders et al., in prep.).

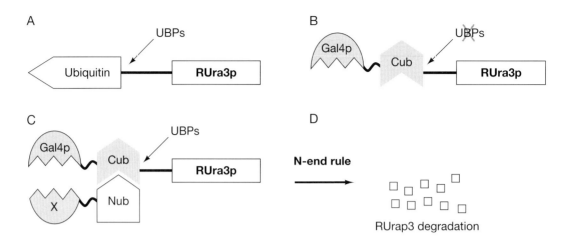

FIGURE 4. The modified split ubiquitin system. (A) The ubiquitin protein is fused to the amino-terminal part of Rura3p. Rura3p is recognized by UBPs because it contains an arginine at the first amino acid. (B) When the Gal4p is fused to the carboxy-terminal part of ubiquitin (Cub), Rura3p is not recognized by UBPs because ubiquitin is not functional. (C,D) A protein X is fused to the amino-terminal part of ubiquitin (Nub). If the protein X interacts with the Gal4p, a functional ubiquitin is reconstituted and the cleaved Rura3p is then degraded by the N-end rule pathway, leading to the cells growing on uracil and resistant to 5-FOA.

Very recently, Laser et al. reported a similar assay based on the split-ubiquitin approach for identifying proteins interacting with the transcriptional regulators Gal4p and Tup1p (Table 1) (Laser et al. 2000). Their technique has been developed to screen for interacting proteins localized in the cytosol of the cell. In this system (Fig. 4), the Gal4 protein is fused to the Cub domain, followed by the reporter protein RUra3p (with an arginine as its first amino acid recognized by the UBPs). The prey is fused to the Nub domain. The reconstitution of an interaction between the Gal4p-fusion and the prey leads to the cleavage of RUra3p. This free RUra3p is then rapidly degraded by the N-end rule pathway of protein degradation, resulting in uracil auxotrophy and resistance to the drug 5-fluoroorotic acid (5-FOAR). The authors showed that their system can monitor the interaction between transcription factors by in vivo reconstitution of the Gal4p/Gal80p and Ssn6p/Tup1 complexes.

PROSPECTS FOR FUTURE APPLICATIONS OF THE MEMBRANE-BASED YEAST TWO-HYBRID SYSTEM

The membrane-based two-hybrid system has several novel applications, including (1) testing known membrane proteins for interaction; (2) screening libraries for proteins that bind a defined target membrane protein; (3) use on a genome-wide basis to generate a comprehensive map of membrane protein interactions of a certain organism; (4) defining amino acids critical for an interaction; and (5) identification of peptides or small molecules that inhibit certain membrane protein interaction. Here we describe some of the potential applications.

Screening of NubG-fused cDNA Libraries

Perhaps the most powerful application of the membrane-based two-hybrid system is in identifying proteins that interact with a given membrane target protein fused to the Cub-PLV domain. Typically, libraries are constructed in which total cDNA from an organism or tissue is fused either amino- or carboxy-terminally to the NubG domain. Both oligo(dT) and random-primed cDNA libraries from yeast, fly, worm, mouse, and human sources are currently being constructed, as are size-selected libraries that contain only small protein domains (DUALSYSTEMS BIOTECH, pers. comm.).

Genome-wide Analysis of Membrane Protein Interactions in the Yeast *S. cerevisiae*

The availability of fully sequenced genomes, both of prokaryotic and eukaryotic organisms, has led to large-scale studies of gene expression (functional genomics) and more recently of the proteome. It has also been tempting to envisage large-scale studies for protein–protein interactions to complete exhaustive protein-interaction maps. Among genome-wide exploratory approaches, the two-hybrid system in yeast has outranked other techniques because it can identify pairs of proteins that physically associate with one another, and because it is simple, sensitive, and amenable to high-throughput applications. Recently, Stan Fields's group has started a genome-scale two-hybrid screen using yeast colony arrays and robotics to identify protein interactions systematically in *S. cerevisiae* (Uetz et al. 2000). Because two-hybrid interactions must occur in the nucleus to activate transcription, membrane proteins may well be misfolded and show no meaningful interactions. For that purpose, our strategy of generating ubiquitin fusions for the membrane proteins is a useful complement to the standard two-hybrid assay (J. Miller et al., unpubl.). The strategy for such a genome-wide screening using the membrane-based yeast two-hybrid system is described in Figure 5.

In theory, this will be a useful technology for comprehensive biological screenings that use the complete set of predicted open reading frames from any eukaryotic organism. Such systematic and multiple high-throughput strategies with the membrane proteins will increase our understanding of protein interactions in yeast and other eukaryotic organisms.

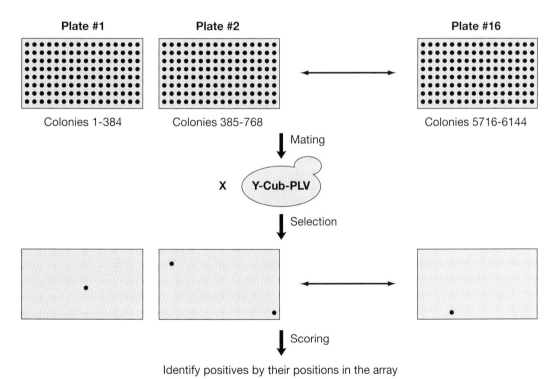

FIGURE 5. A comprehensive analysis of membrane protein interactions in yeast using the membrane-based yeast two-hybrid technology. All 6000 yeast ORFs are cloned into the NubG domain vector either as amino- or carboxy-terminal fusions and transformed into yeast to generate an array of 2 x 6000 colonies. These two different arrays (X-NubG and NubG-X) are then interrogated by mating their colonies to yeast cells of the opposite mating type expressing a membrane bait protein (Y) construct fused to Cub–PLV. Bait–prey interaction facilitates transcription of prototrophic selectable markers (*HIS3* and *URA3*) that confer growth on selective media. This process is repeated for all yeast transmembrane proteins (Y–Cub–PLV fusions) until every possible combination of bait and prey has been tested for interaction.

TABLE 2. Examples of Various Membrane Proteins Implicated in the Onset of Disease

Protein	Type	Disease	Reference
ErbB2	transmembrane protein	ovarian and breast cancer	Hynes and Stern (1994)
Androgen receptor	transmembrane protein	ovarian cancer	Ilekis et al. (1997)
Fibroblast growth factor receptor (FGFR3)	transmembrane protein	birth defect and cancer	Burke et al. (1998)
Amyloid precursor protein (APP)	transmembrane protein	Alzheimer's disease	Neve et al. (2000)
PrP^Sc, ^CtmPrp, ^NtmPrp	GPI anchor transmembrane proteins	Creutzfeldt-Jakob disease, Gerstmann-Straussler-Scheinker syndrome, scrapie and bovine spongiform encephalopathy	Jackson and Clarke (2000)
Nicotinic acetylcholine receptor (AChR)	transmembrane glycoprotein	autoimmune myasthenia gravis (MG) disease	Lindstrom et al. (1988)
GSD-1a	microsomal transmembrane protein	glycogen storage disease	Yang Chou and Mansfield (1999)
Endoglin	transmembrane protein	atherosclorose	Conley et al. (2000)
Wolframin WFS1	transmembrane protein	Wolfram syndrome	Inoue et al. (1998)
Glycine receptor α subunit	transmembrane protein	familial startle disease (STHE)	Shiang et al. (1993)

Membrane-based Yeast Two-hybrid Screening and Drug Design

Consistent with the functional importance of membrane protein–protein interactions, a number of human diseases, including a variety of different cancers and neurodegenerative diseases (Table 2), occur as a consequence of particular protein–protein association or dissociation events. For these reasons, membrane protein interactions might be considered to be important as potential drug targets. Conceptually, such protein–protein interactions can be inhibited by the use of *cis*-acting mutations in one partner or *trans*-acting molecules such as small molecules or dissociating peptides. To allow the selection for drugs and mutations that abolish the protein–protein interactions, the reverse two-hybrid system has been established (Vidal et al. 1996). Using *URA3* as a toxic reporter gene, the inhibition of the interaction between two interacting proteins leads to the loss of *URA3* expression and subsequent ability of the cells to grow on plates containing the drug 5-FOA. Our membrane-based yeast two-hybrid strategy can be combined to the reverse two-hybrid strategy to screen a library of peptides or small molecules in which preventing the interaction between two membrane protein partners provides a selective advantage for yeast (Fig. 6). In this version of the reverse membrane two-hybrid system, the interaction between two membrane proteins X and Y is toxic for the yeast cells because a toxic marker (e.g., *URA3* or *CYH2*) is used as a reporter gene (DUALSYSTEMS BIOTECH, unpubl.). Such an approach should enable identification of both interaction-defective alleles and dissociating peptides or small molecules that are able to dissociate the interaction between two membrane proteins X and Y.

ADVANTAGES AND DISADVANTAGES OF THE MEMBRANE-BASED YEAST TWO-HYBRID SYSTEM

The membrane-based yeast two-hybrid system has several advantages over the conventional yeast two-hybrid system: (1) The system can be adapted to study interactions between two membrane proteins, and/or a membrane protein and a cytosolic protein; (2) it can be applied to any transmembrane protein as a "bait" assuring that interacting modules (Cub-PLV and NubG) fused to

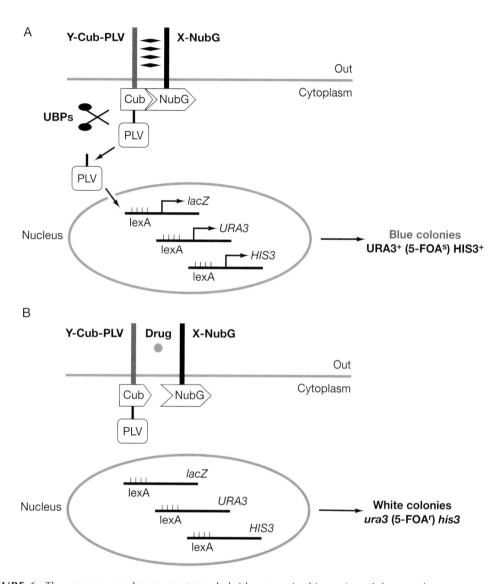

FIGURE 6. The reverse membrane yeast two-hybrid system. In this version of the membrane yeast two-hybrid system, *lacZ* and *HIS3* reporter genes are used for positive selection, whereas the *URA3* gene is used for counterselection. (*A*) Interaction between two membrane proteins Y and X results in growth on medium lacking histidine, blue coloration of the cells, but lethality on medium containing 5-fluoroorotic acid (5-FOA), a toxic metabolite of the URA3 pathway. Following the mutagenesis of Y and X proteins or screening with a complex random chemical library, the mutations or small molecules are selected that rescue the growth on 5-FOA plates.

membrane proteins to be examined are located in the cytoplasm; and (3) the system should also work in any eukaryote and is currently being adapted for mammalian hosts (DUALSYSTEMS BIOTECH, pers. comm.) (however, the yeast-based system has numerous advantages, including the ease of transformation, the convenience of retrieving plasmids, and the availability of nutritional markers for direct selection); (4) the system does not require a nuclear localization signal because it is based on diffusion of cleaved artificial transcription factor; and finally (5) the use of small ubiquitin domains is advantageous because it minimizes potential steric hindrance between the interacting partners.

The membrane-based yeast two-hybrid system cannot assay all membrane protein interactions. It cannot be applied to transmembrane proteins that multiply span the membrane and whose amino and carboxyl termini are not located in the cytoplasm. In addition, if one applies this system to study mammalian membrane protein interactions, a consideration to be taken into account is that mammalian signal sequences confirming insertion into the membrane may differ from those from yeast. In general, a number of mammalian signal sequences are not functional in yeast and should therefore be exchanged by a yeast signal peptide sequence such as those derived from yeast STE2, SUC2, or CPY proteins to target the heterologous protein of interest to the yeast membranes. However, it has been possible in some cases to express mammalian proteins in yeast with their own signal peptide (e.g., Montero-Lomeli and Okorokova Facanha 1999). Finally, the system may be difficult to apply to mammalian surface receptors (such as IGF I receptor or T-cell-antigen receptor) that rely on the cooperation of multiple subunits to facilitate optimal ligand binding and signaling function (C. Buerki and I. Stagljar, unpubl.).

In conclusion, an in vivo genetic system for monitoring the membrane protein interactions can now become a favorable technology for studying membrane protein interactions as well as a useful test for development of novel therapeutic strategies based on protein–protein interactions.

ACKNOWLEDGMENTS

We are grateful to Dr. Alcide Barberis for a critical reading of the manuscript, and to the Zuercher Krebsliga, EMDO Foundation, Walter Honegger Foundation, Bonizzi-Theler Foundation, Swiss National Foundation (Grant Nr. 31-58798.99) and Gebert Ruef Foundation for financial support of these projects. We also thank Professor Ulrich Huebscher for his support.

REFERENCES

Aronheim A., Zandi E., Hennemann H., Elledge S.J., and Karin M. 1997. Isolation of an AP-1 repressor by a novel method for detecting protein–protein interactions. *Mol. Cell Biol.* **17:** 3094–3102.

Borg J.P., Marchetto S., Le Bivic A., Ollendorff V., Jaulin-Bastard F., Saito H., Fournier E., Adelaide J., Margolis B., and Birnbaum D. 2000. ERBIN: A basolateral PDZ protein that interacts with the mammalian ERBB2/HER2 receptor. *Nat. Cell Biol.* **2:** 407–414.

Broder Y.C., Katz S., and Aronheim A. 1998. The ras recruitment system, a novel approach to the study of protein–protein interactions. *Curr. Biol.* **8:** 1121–1124.

Burke D., Wilkes D., Blundell T.L., and Malcolm S. 1998. Fibroblast growth factor receptors: Lessons from the genes. *Trends Biochem. Sci.* **23:** 59–62.

Chervitz S.A., Aravind L., Sherlock G., Ball C.A., Koonin E.V., Dwight S.S., Harris M.A., Dolinski K., Mohr S., Smith T., Weng S., Cherry J.M., and Botstein D. 1998. Comparison of the complete protein sets of worm and yeast: Orthology and divergence. *Science* **282:** 2022–2028.

Chien C.T., Bartel P.L., Sternglanz R., and Fields S. 1991. The two-hybrid system: A method to identify and clone genes for proteins that interact with a protein of interest. *Proc. Natl. Acad. Sci.* **88:** 9578–9582.

Conley B.A., Smith J.D., Guerrero-Esteo M., Bernabeu C., and Vary C.P. 2000. Endoglin, a TGF-beta receptor-associated protein, is expressed by smooth muscle cells in human atherosclerotic plaques. *Atherosclerosis* **153:** 323–335.

Ehrhard K.N., Jacoby J.J., Fu X.Y., Jahn R., and Dohlman H.G. 2000. Use of G-protein fusions to monitor integral membrane protein-protein interactions in yeast. *Nat. Biotechnol.* **18:** 1075–1079.

Fields S. and Song O. 1989. A novel genetic system to detect protein-protein interactions. *Nature* **340:** 245–246.

Goffeau A., Barrell B.G., Bussey H., Davis R.W., Dujon B., Feldmann H., Galibert F., Hoheisel J.D., Jacq C., Johnston M., Louis E.J., Mewes H.W., Murakami Y., Philippsen P., Tettelin H., and Oliver S.G. 1996. Life with 6000 genes. *Science* **546:** 563–567.

Hershko A. and Ciechanover A. 1992. The ubiquitin system for protein degradation. *Annu. Rev. Biochem.* **61:** 761–807.

Hubsman M., Yudkovsky G., and Aronheim A. 2001. A novel approach for the identification of protein-protein interaction with integral membrane proteins. *Nucleic Acids Res.* **29:** e18.

Hynes N.E. and Stern D.F. 1994. The biology of erbB2/neu/HER2 and its role in cancer. *Biochim. Biophys. Acta* **1198:** 165–184.

Ilekis J.V., Connor J.P., Prins G.S., Ferrer K., Niederberger C., and Scoccia B. 1997. Expression of epidermal growth factor and androgen receptors in ovarian cancer. *Gynecol. Oncol.* **66:** 250–254.

Inoue H., Tanizawa Y., Wasson J., Behn P., Kalidas K., Bernal-Mizrachi E., Mueckler M., Marshall H., Donis-Keller H., Crock P., Rogers D., Mikuni M., Kumashiro H., Higashi K., Sobue G., Oka Y., and Permutt M.A. 1998. A gene encoding a transmembrane protein is mutated in patients with diabetes mellitus and optic atrophy (Wolfram syndrome). *Nat. Genet.* **20:** 143–148.

Jackson G.S. and Clarke A.R. 2000. Mammalian prion proteins. *Curr. Opin. Struct. Biol.* **10:** 69–74.

Johnsson N. and Varshavsky A. 1994a. Split ubiquitin as a sensor of protein interactions in vivo. *Proc. Natl. Acad. Sci.* **91:** 10340–10344.

———. 1994b. Ubiquitin-assisted dissection of protein transport across membranes. *EMBO J.* **13:** 2686–2698.

Keegan K. and Cooper J.A. 1996. Use of the two hybrid system to detect the association of the protein-tyrosine-phosphatase, SHPTP2, with another SH2-containing protein, Grb7. *Oncogene* **12:** 1537–1544.

Laser H., Bongards C., Schüller J., Heck S., Johnsson N., and Lehming N. 2000. A new screen for protein interactions reveals that the *Saccharomyces cerevisiae* high mobility group proteins Nhp6A/B are involved in the regulation of the GAL1 promoter. *Proc. Natl. Acad. Sci.* **97:** 13732–13737.

Lindstrom J., Shelton D., and Fugii Y. 1988. Myasthenia gravis. *Adv. Immunol.* **42:** 233–284.

Montero-Lomeli M. and Okorokova Facanha A.L. 1999. Expression of a mammalian NA+/H+ antitransporter in *Saccharomyces cerevisiae. Biochem. Cell Biol.* **77:** 25–31.

Neve R.L., McPhie D.L., and Chen Y. 2000. Alzheimer's disease: A dysfunction of the amyloid precursor protein(1). *Brain Res.* **886:** 54–66.

Ozenberger B.A. and Young K.H. 1995. Functional interaction of ligands and receptors of the hematopoietic superfamily in yeast. *Mol. Endocrinol.* **9:** 1321–1329.

Shiang R., Ryan S.G., Zhu Y.Z., Hahn A.F., O'Connell P., and Wasmuth J.J. 1993. Mutations in the alpha 1 subunit of the inhibitory glycine receptor cause the dominant neurologic disorder, hyperekplexia. *Nat. Genet.* **5:** 351–358.

Stagljar I. and te Heesen S. 2000. Detecting interactions between membrane proteins in vivo using chimeras. *Methods Enzymol.* **327:** 190–198.

Stagljar I., Korostensky C., Johnsson N., and te Heesen S. 1998. A genetic system based on split-ubiquitin for the analysis of interactions between membrane proteins in vivo. *Proc. Natl. Acad. Sci.* **95:** 5187–5192.

Uetz P., Giot L., Cagney G., Mansfield T.A., Judson R.S., Knight J.R., Lockshon D., Narayan V., Srinivasan M., Pochart P., Qureshi-Emili A., Li Y., Godwin B., Conover D., Kalbfleisch T., Vijayadamodar G., Yang M., Johnston M., Fields S., and Rothberg J.M. 2000. A comprehensive analysis of protein-protein interactions in *Saccharomyces cerevisiae. Nature* **403:** 623–627.

Varshavsky A. 1996. The N-end rule: Functions, mysteries, uses. *Proc. Natl. Acad. Sci.* **93:** 12142–12149.

Vidal M., Brachmann R.K., Fattaey A., Harlow E., and Boeke J.D. 1996. Reverse two-hybrid and one-hybrid systems to detect dissociation of protein-protein and DNA-protein interactions. *Proc. Natl. Acad. Sci.* **93:** 10315–10320.

Yang Chou J. and Mansfield B.C. 1999. Molecular genetics of type 1 glycogen storage diseases. *Trends Endocrinol. Metab.* **10:** 104–113.

23 Protein Interactions in Live Cells Monitored by β-Galactosidase Complementation

Bruce T. Blakely, Fabio M.V. Rossi, Thomas S. Wehrman,
Carol A. Charlton, and Helen M. Blau

*Department of Molecular Pharmacology, Stanford University School of Medicine,
Stanford, California 94305-5175*

INTRODUCTION

The characterization of protein interactions is important to the understanding of signal transduction pathways and cellular processes. Here we describe a method utilizing β-galactosidase (β-gal) complementation that can monitor protein interactions in live mammalian cells (Rossi et al. 1997, 2000; Blakely et al. 2000). In brief, the method involves expressing chimeric proteins consisting of two potentially interacting proteins of interest fused to complementing β-gal deletion mutants. When the two proteins of interest interact, the β-gal mutants complement, reconstituting an active β-gal enzyme. Some of the advantages of β-gal complementation are: (1) It is a direct assay, that is, a signal is generated at the site of the interaction, and activation of a reporter gene is not required; (2) it is a sensitive assay, due to enzymatic amplification of the product; (3) overexpression is not required—interactions can be detected at physiological levels of expression; (4) β-gal enzyme activity can be measured using a variety of substrates that permit quantitative assays that can measure β-gal activity in single live cells, or that are amenable to high-throughput screening technologies.

A variety of biochemical methods is available for demonstrating the interaction of two proteins, the most common of which is immunoprecipitation followed by identification by immunoblotting. Such biochemical methods have proven reliable, but can be limited by the inability of some complexes to survive cellular disruption and immunoprecipitation. Complexes can be stabilized by chemical cross-linking before the cell is disrupted, but this step can introduce artifacts by cross-linking adjacent, but noninteracting, proteins. The yeast two-hybrid system has been the most powerful method for identifying novel protein interactions (Fields and Song 1989; Bai and Elledge 1996). However, proteins in this system must be capable of interacting in the yeast environment and must then translocate to the nucleus and activate transcription of a reporter gene.

An ideal system for studying protein interactions would generate a signal upon the interaction of the proteins of interest. This signal would be specific, and would be detectable in the cellular compartment in which it is generated, either directly or with minimal additional steps, such as addition of a substrate. Ideally, the signal would be detected in live cells, or the signal must be able to survive disruption of the cells. Several recent developments have come close to this goal. Fluorescence resonance energy transfer (FRET) has been used to study protein interactions via the in vitro labeling of two proteins of interest with fluorescent tags (Adams et al. 1991; Gadella and Jovin 1995; Chapter 10). However, the difficulties of introducing fluorescently labeled proteins into cells at sufficiently high concentrations to detect a signal can limit the utility of this method. An alternative FRET methodology involves expressing chimeric proteins that incorporate one partner of an interacting protein pair and a fluorescent protein, such as green fluorescent protein (GFP). By using two different GFPs, FRET analysis of the interacting chimeras is possible, but it remains to be seen whether this method will work well with a variety of interacting proteins (Miyawaki et al. 1997; Pollok and Heim 1999). Other non-FRET methods using interacting chimeric proteins depend on the complementation of two mutant protein fragments to reconstitute a functional protein. Complementation of dihydrofolate reductase (DHFR) has been used to examine protein interactions in mammalian cells, but this system is dependent on either a nonenzymatic substrate-binding assay that is quantitative only in cell lines that lack endogenous DHFR, or a nonquantitative enzymatic survival assay that only works in cells lacking endogenous DHFR (Pelletier et al. 1998; Remy et al. 1999). The complementation of bacterial β-gal, discussed in detail below, generates an enzymatically amplified signal that can be detected quantitatively by a variety of assays in many different cell types without overexpression of the proteins (Rossi et al. 1997, 2000; Blakely et al. 2000).

BACKGROUND

Intracistronic β-gal complementation is a phenomenon first observed by Jacob and Monod, in which two mutants of the bacterial enzyme β-gal that have inactivating deletions in different critical domains recreate an active enzyme by sharing the intact domains (Ullmann et al. 1965, 1967). β-Gal complementation has been used for decades as a marker for molecular cloning experiments in bacteria (blue-white colony selection). The basis of this system is the expression of a truncated inactive β-gal protein (ΔM15) by the host bacterium, which can be complemented by expression of a peptide (α peptide) from a plasmid cloning vector, but only if the α-peptide coding sequence is not disrupted by a cloned cDNA. β-gal was successfully complemented in mammalian cells using three different complementing peptides (Fig. 1): Δα, similar to the ΔM15 mutant used in bacteria, has a deletion near the amino terminus of the protein (amino acids 11–41); Δω, trun-

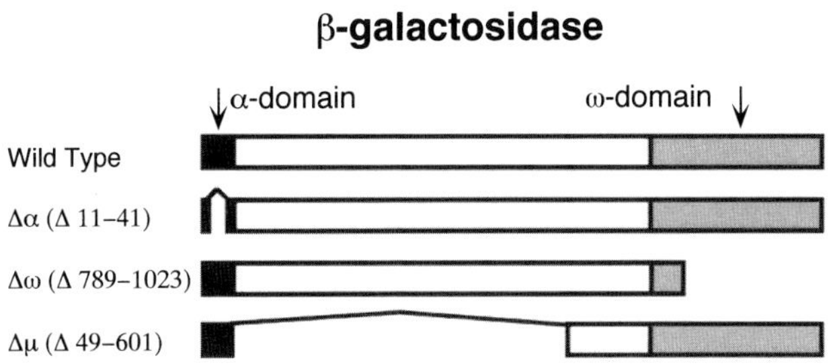

β-galactosidase

FIGURE 1. Schematic diagram of the β-gal deletion mutants. The α and ω domains, as defined by Jacobson et al. (1994), are represented by the black and gray boxes at the amino and carboxyl termini of the protein, respectively. Δα has a small deletion in the α domain, Δω has a large deletion of the carboxyl terminus, and Δμ has a large deletion in the region of the protein between the α and ω domains.

cated at amino acid 788, lacks the carboxyl terminus of the protein; and Δμ has a deletion (amino acids 49–601) in the middle of the protein (Mohler and Blau 1996). The amino-terminal (α) and carboxy-terminal (ω) regions of the protein are critical to forming a functional enzyme (Villarejo et al. 1972; Jacobson et al. 1994). Δω functions as an "α donor" (analogous to but much longer than the α peptide used in bacterial complementation), Δα functions as an "ω donor," and Δμ can provide either region but lacks the middle region that is important for the structure of the active enzyme. Any two of these mutant proteins, when expressed in the same cell, result in formation of an active β-gal enzyme (Mohler and Blau 1996). Although native β-gal forms a functional enzyme as a homotetramer, it has not been definitively established whether the active complemented enzyme complex in mammalian cells requires eight mutant proteins.

β-Gal complementation in mammalian cells was first used to study the process of cell fusion in myoblast differentiation (Mohler and Blau 1996). Each of the deletion mutants was expressed in separate populations of myoblasts. When two of the different populations were cocultured in differentiation-inducing conditions, the myoblasts fused into multinucleated syncytia, or myotubes, and complementation of β-gal was observed. Measurement of β-gal activity provided a simple quantitative assay for myoblast fusion and thus provided a rapid method to measure the effects of genetic mutations and culture conditions of myoblast differentiation (Charlton et al. 1997).

By constructing chimeric proteins incorporating one of the β-gal deletion mutants, interactions between the non-β-gal components of the chimeras can be detected (Rossi et al. 1997). Coexpression of any two of the three different deletion mutants in the same cell results in β-gal complementation, but lower activity was usually observed when Δα was paired with Δω, compared to either of these mutants with Δμ (Mohler and Blau 1996). In a complementation assay for protein interactions, the interaction should be driven by the proteins of interest, rather than the β-gal mutants; therefore, the weakest complementing pair, Δα and Δω, were used for constructing chimeras for protein interaction assays. In the initial test of this system, Δα and Δω were linked to the rapamycin-binding proteins FRAP and FKBP12 (Rossi et al. 1997). β-Gal activity was very low in the absence of rapamycin but increased significantly upon addition of rapamycin. β-Gal activity was dependent on the dose of rapamycin and continued to increase throughout the time measured (Fig. 2, left).

Membrane receptor dimerization can also be quantitated using β-gal complementation (Blakely et al. 2000). By expressing chimeric proteins that link Δα or Δω to a truncated epidermal growth factor (EGF) receptor, EGF-induced receptor dimerization resulted in increased β-gal activity. β-Gal activity initially increased more rapidly than was observed with rapamycin-binding proteins; however, after increasing about eightfold in the first hour, the activity plateaued, as would be expected with membrane receptor dimerization (Fig. 2, right). Thus, the kinetics of β-

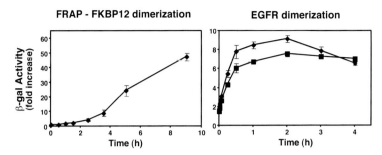

FRAP - FKBP12 dimerization

EGFR dimerization

FIGURE 2. The kinetics of chimeric β-gal protein complementation depend on the non-β-gal portion of the chimera. The rapamycin-binding proteins FRAP and FKBP12, or the extracellular and transmembrane domains of the epidermal growth factor receptor (EGFR), are fused to β-gal deletion mutants and expressed in C2C12 cells. The left panel shows that β-gal activity increases over a 9-hour period following the addition of rapamycin to a population of cells expressing the FRAP and FKBP12–β-gal chimeras. The right panel shows that β-gal activity increases rapidly over the first 1–2 hours and then plateaus after the addition of EGF to two different clones expressing the EGFR–β-gal chimeras. The β-gal activity in each experiment reflects the dimerization kinetics of the wild-type rapamycin-binding proteins and EGFR, respectively.

gal activity reflect the kinetics of the non-β-gal components of the chimeric proteins. β-Gal complementation was used to characterize fully the effects of an anti-EGF receptor antibody on receptor dimerization, which previously had not been possible using biochemical methods. Furthermore, the levels of expression of the chimeric receptors used in these studies did not exceed the levels of the endogenous receptor in normal cells.

Finally, β-gal complementation, in addition to measuring cytoplasmic protein interactions and membrane protein interactions, can also measure the interaction of a cytoplasmic protein and a membrane receptor (e.g., TGF-β receptor 1 and FKBP12; T.S. Wehrman et al., unpubl.). Thus, β-gal complementation can measure three types of interactions that are key components of most signal transduction pathways (Fig. 3).

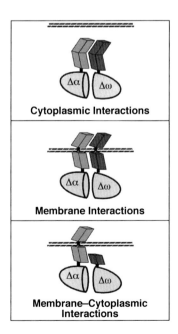

Cytoplasmic Interactions

Membrane Interactions

Membrane–Cytoplasmic Interactions

FIGURE 3. β-Gal complementation can detect protein interactions in the cytoplasm or at the membrane. Two chimeric proteins containing different β-gal deletion mutants (labeled Δα or Δω) and proteins of interest (shaded rectangular boxes in figure) are expressed in cells. If the proteins of interest interact, the β-gal mutants will complement, and β-gal enzyme activity serves as a measure of the interaction.

β-Gal complementation can be used to confirm protein interactions and to characterize known interactions. A variety of assays is available to measure β-gal activity, further increasing the utility of this method. Of particular interest are a flow cytometry assay using a fluorogenic substrate for β-gal for the measurement of β-gal activity in single live cells and a chemiluminescent assay that enables the analysis of hundreds of samples in microtiter plates using high-throughput screening technologies. The simple, rapid assays available for β-gal make it possible to test the effect of a variety of conditions (e.g., inducers, inhibitors, concentrations, antibodies, cell backgrounds). β-Gal complementation will also be useful for screening for novel inducers and inhibitors of protein interactions using high-throughput screening methods (e.g., receptor agonists and antagonists) and potentially could be used as a "mammalian two-hybrid" screen for novel interactions.

OUTLINE OF PROCEDURE

The first step is constructing two chimeras each containing one of the β-gal deletion mutants and the protein of interest. These constructs must then be expressed in cells. We use retroviral vectors to express the chimeras to rapidly obtain cells that stably express the constructs while limiting the copy number of the construct and thus the expression levels. Once cells that express both constructs are obtained, the cells can be tested for β-gal activity in the presence and absence of an inducer of the desired interaction. In some cases, cloning the cells that express the chimeras may be desirable. Several different assays for β-gal activity are presented here.

Strategy for Construction and Expression of the Chimeric Proteins

Although our initial experiments were carried out with the β-gal deletion mutant linked to the carboxyl terminus of the protein of interest, we have subsequently found that, at least in the case of cytoplasmic rapamycin-binding proteins, complementation works equally well with the β-gal mutants at either end of the chimeric protein (T.S. Wehrman et al., unpubl.). In the case of membrane proteins, the β-gal mutant has only been placed at the carboxyl terminus, so that complementation occurs in the cytoplasm. The use of PCR and standard cloning techniques is required to place the protein of interest accurately in frame with the β-gal deletion mutant.

The use of retroviral vectors simplifies the expression of the chimeric proteins. We use two retroviral vectors, which contain either Δα or Δω under the transcriptional control of the viral long terminal repeat (LTR) (Fig. 4) (Rossi et al. 1997; Blakely et al. 2000). An intracistronic ribosome entry site (IRES) and a drug-resistance marker are located downstream of the β-gal mutant. After the protein of interest is cloned in frame with the deletion mutant, the plasmid containing the retroviral vector is transiently transfected into the Phoenix-E packaging cell line using FuGENE 6 (Roche Molecular Biochemicals, Indianapolis, Indiana) according to the manufacturer's instructions. Although calcium phosphate transfection protocols can sometimes result in transfection of a higher percentage of the cells, we consistently obtain higher viral titers using FuGENE 6, possibly because of the absence of toxic side effects. Virus supernatant is used immediately to infect the target cells, or it can be frozen at –70°C with a resulting small drop in titer. Although many cell lines can be infected with nearly 100% efficiency, any uninfected cells can be eliminated through the use of the selectable marker in the vector.

Although one can infect cells with both vectors (the Δα and Δω constructs) simultaneously, we usually infect the cells sequentially. It can be useful to have cells on hand expressing only one of the vectors in case future experiments call for changing one of the interacting proteins. Furthermore, if there are problems with expression of the chimeric proteins, it may be easier to determine which of the constructs is defective if they are singly expressed, especially if the two

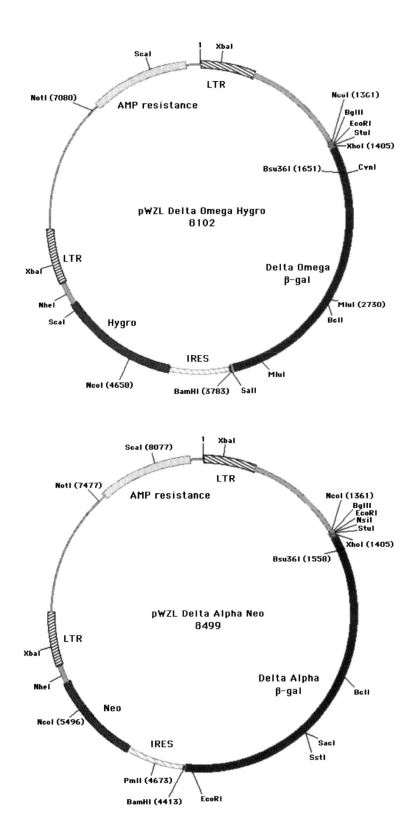

FIGURE 4. Maps of the β-gal complementation vectors. Using the plasmids shown, chimeric proteins can be engineered by using PCR and restriction digests to clone cDNAs of interest in frame with either the Δα or Δω mutants of bacterial β-gal. When transfected into packaging cell lines, retroviruses are produced (lacking the ampicillin-resistance gene and other bacterial sequences between the LTRs) that express the chimeric protein and a selectable marker (resistance to G418 or hygromycin).

chimeric proteins are similar in size. Selection with antibiotics such as G418 and hygromycin should be carried out at the lowest dose that effectively kills uninfected cells, because selection with higher doses can lead to overexpression of the constructs.

Assays for β-Gal

Chemiluminescent Assay for β-Gal

The chemiluminescent assay for β-gal is the quickest and simplest method for assaying a large number of samples. In brief, cells are plated and treated with reagents that will induce the protein interaction of interest. Gal-Screen reagent is added to the cells, resulting in the prolonged production of photons ("glow" kinetics, not "flash" kinetics) in the presence of active β-gal. The light emitted by the sample is then measured in a luminometer. Cells are plated on a 96-well plate, a format that permits the use of high-throughput screening instrumentation if desired. This format greatly simplifies the assay because cells are plated and assayed on the same dish. However, if a microplate luminometer is not available, the samples can be transferred to a tube for use in a tube luminometer.

The microtiter plate used in this assay should be of the type recommended by the manufacturer of the luminometer. For many instruments, this will be a white plate with either a white (opaque) or clear bottom. However, some instruments may specify the use of black plates. Although opaque plates are ideal, clear-bottomed plates, which permit observation under a microscope, can be used until the experimenter is comfortable that he is obtaining a subconfluent, evenly plated culture. Chemiluminescence can spill from one well to another through the clear bottom of the plate, which will be noticeable if a well producing thousands of induced light units is adjacent to a negative control or uninduced sample.

We use a Tropix TR717 microplate luminometer, although other instruments also work well, including the following instruments that we have tested: EG&G Wallac Berthold LB96V (which is nearly identical to the Tropix instrument), Turner Designs Reporter, and the EG&G Wallac MicroBeta Plus (current version is MicroBeta Trilux; MicroBeta units with the automatic injector may have reduced sensitivity). One disadvantage of the MicroBeta is that it takes 5–10 minutes to count a plate, compared to less than 2 minutes for the others. Some instruments may not have sufficient sensitivity, and this should be considered if β-gal activity is not detected.

Flow Cytometry Assay for β-Gal

By assaying the product of the fluorogenic β-gal substrate fluorescein di-β-D-galactopyranoside (FdG) using a flow cytometer, β-gal activity can be measured in single live cells. This is useful for determining changes in β-gal activity in a population of cells and thus determining whether or not cloning of the cells is necessary to achieve a uniform response. Furthermore, like the chemiluminescent assay, the FdG assay is quantitative and can be used to study the kinetics and dose response of the interaction. The fluorescence-activated cell sorter (FACS) can also be used to isolate subpopulations of cells or to clone cells that exhibit the desired response.

The protocol given here is based on that published by Nolan et al. (1988) and has been modified largely to accommodate the processing of large numbers of samples. Using this method, one person can readily assay 50–100 samples, and with two persons working together, as many as 200 samples can be assayed within a few hours. If only a few samples are to be analyzed, the assay can be completed in less than half an hour. The FdG substrate used in this assay is not cell-permeable and must be introduced into the cells by a hypotonic shock. Cleavage by active β-gal produces free fluorescein, which is also unable to cross the plasma membrane and remains inside the cells that have complemented β-gal. Following the analysis, most cells remain viable and can be returned to culture. In populations or clones that exhibit homogeneous changes in β-gal activity, the mean fluorescence of each sample is a reliable indicator of the interaction.

X-Gal and Fluor-X-Gal Assays for β-Gal

The X-gal assay is a simple chromogenic assay that requires no specialized instrumentation. In the presence of active β-gal, the substrate 5 bromo-4-chloro-3-indolyl β-D-galactopyranoside (X-gal), and a potassium ferricyanide buffer, a blue precipitate that can be observed by microscopy forms in fixed cells. This assay is not quantitative and is not particularly sensitive, but it can be used to detect β-gal complementation if the induction of complementation is sufficiently robust (greater than about fivefold) (Rossi et al. 1997).

Because the X-gal assay product quenches fluorescence, the Fluor-X-gal assay was developed so that β-gal activity could be assayed in cells that had been immunofluorescently labeled for other proteins (Mohler and Blau 1996). This assay combines an azo dye, Fast Red Violet LB, with either X-gal or 5-bromo-6-chloro-3-indolyl β-D-galactopyranoside (5-6 X-gal) to form a fluorescent precipitate in the presence of active β-gal. Fluor-X-gal is more sensitive than X-gal, but like X-gal, is not quantitative and requires fixation of the cells.

Preparation of Retrovirus and Infection of Target Cells

Retroviral vectors are used here to express the chimeras and obtain cells that express the constructs.

MATERIALS

Buffers and Reagents

Dulbecco's modified Eagle medium (DMEM), high glucose formulation (450 g/liter)
Fetal bovine serum (FBS; HyClone)
Polybrene (8 mg/ml) (1000x) in H_2O, filter-sterilized (Sigma)

Vectors

FuGENE 6 (Roche Molecular Biochemicals)
> The retroviral vector should not exceed 9 kb in length to ensure efficient packaging of the vector. This limit does not include the portions of the plasmid required for plasmid propagation which are not incorporated into the retrovirus.

Plasmids and Cells

Phoenix-E cells
> For information on obtaining these cells, see Dr. Garry Nolan's Web site at http://www.stanford.edu/group/nolan/mtas.html).
Plasmids containing β-gal deletion mutants fused in frame to the protein of interest

Special Equipment

Benchtop centrifuge with microplate carriers

METHOD

1. Transfect the plasmid containing the retroviral construct into the Phoenix-E packaging cell line using FuGENE 6. (The procedure is outlined here, but it is strongly recommended that the manufacturer's instructions be followed.)

   a. Plate 1.5×10^6 to 2×10^6 Phoenix-E cells per 60-mm dish in DMEM containing 450 g/liter glucose (high glucose formulation) and 10% FBS the day before transfection (or 3×10^6 cells on the day of transfection). Cells should be 50–80% confluent at transfection.

   b. Mix the following components in a microcentrifuge tube in the order given: serum-free DMEM, 6 μl of FuGENE 6 (vortex stock tube first, then add reagent directly to the medium, not to the side of the tube), 2 μg of plasmid DNA. The total volume should be 100 μl. Mix the components by gently flicking the tube; do not vortex the tube. Incubate at room temperature for 15–30 minutes. The volumes can be scaled for multiple plates receiving the same plasmid, or for different-sized plates.

   c. Add the 100-μl FuGENE–DNA mixture dropwise to the plate of Phoenix-E cells. It is not necessary to change the medium first or to use serum-free medium on the cells.

2. Incubate overnight (18–24 hours), and then refeed the cells with 2–4 ml of DME (high glucose) + 10% FBS.

3. At least 30 hours after transfection and at least 6 hours after refeeding, remove the medium (viral supernatant) from the dish and filter through a 0.45-μm syringe filter.

4. Add Polybrene to a final concentration of 8 μg/ml to the viral supernatant, or immediately freeze the viral supernatant at –80°C.

5. Aspirate medium from a subconfluent dish of target cells and replace with sufficient viral supernatant containing Polybrene to cover the surface of the dish. If the target cells cannot survive in the Phoenix-E media, dilute the Phoenix-E media 1:1 or more in target cell media.

6. *Optional:* To increase viral infection efficiency, centrifuge the plates (up to 100 mm in diameter) containing the target cells in a benchtop centrifuge equipped with microplate platforms at 2500 rpm (Beckman GS-6 centrifuge or equivalent) for 30 minutes.

   Place dishes carefully in the center of the microplate carrier and make sure that the rotor is balanced. Sealing the plate with Parafilm will help the dish "stick" to the center of the carrier.

7. *Optional:* Continue to harvest retrovirus from the producer cells every 6–12 hours up to 72 hours after transfection, and then discard them. Either use the harvested supernatant for additional rounds of infection or freeze immediately at –80°C.

8. Refeed the target cells 6 hours after adding retrovirus, or, to increase infection efficiency, add another aliquot of viral supernatant containing Polybrene as above. A third round of infection may be carried out after another 6 hours if desired.

   Although centrifugation and multiple rounds of infection can increase the infection frequency, it is not desirable to have multiple copies of the retrovirus in a cell, as this can lead to overexpression. To ensure single-copy transduction in most of the cells, an infection efficiency of 15% is desired. Thus, the viral supernatant may have to be diluted. Infection efficiency can be measured using any retroviral vector expressing β-gal, GFP, or any readily observable marker. However, many of the experiments described here were accomplished without precise titration of the virus.

9. Refeed the target cells with medium containing the appropriate antibiotic to select for infected cells 24 hours after the final round of infection. The time required for uninfected cells to die depends on the cell line used. Cells should continue to be cultured in the selective antibiotics throughout the experiment to prevent a loss of expression of the chimeric proteins.

Protocol 2

Chemiluminescent Assay for β-Gal

This protocol is the quickest and simplest method for assaying a large number of samples.

MATERIALS

Buffers and Reagents

        Cells from Protocol 1
        Gal-Screen reagent (Tropix)
        Phosphate-buffered saline (PBS)

Special Equipment

        Microplate luminometer
        Plates (96-well) appropriate for the microplate luminometer used

METHOD

1. Plate cells in 96-well plates the day before the assay at a density of 10,000 cells per well (number established for C2C12 or C2F3 mouse myoblasts) in a volume of 100 µl in medium appropriate for the cell line used. Cells should be subconfluent at the time of the assay.

2. Treat cells with reagents appropriate to induce or modify the interaction under study. Typically, each treatment is carried out on triplicate or quadruplicate wells.

3. At the end of the treatment period, aspirate the medium from the wells and add 200 µl of Gal-Screen reagent prepared according to the manufacturer's instructions (substrate is diluted 1:25 immediately before use with Gal-Screen buffer B equilibrated to room temperature) but additionally diluted 1:1 with 1x PBS.

   Removing the medium is not recommended by the manufacturer's instructions, but this is necessary if, because of kinetic analysis, some of the wells were refed with fresh medium at different time points. Removal of the medium also leads to a slight improvement in the signal-to-noise ratio. The Gal-Screen reagent is additionally diluted 1:1 with PBS to compensate for the removal of the medium from the plate.

4. Incubate the plate for 45 minutes to 1 hour at 26–28°C. Alternatively, incubate at room temperature for a period of time sufficient for the Gal-Screen reaction to reach a plateau.

5. Read the plate using a microplate luminometer, measuring each well for 1 second or for a time appropriate for the instrument used.

6. Analyze data using appropriate software.

Protocol 3

Flow Cytometry Assay for β-Gal

Flow cytometric assay of FdG measures β-gal activity in live cells so that a determination can be made of whether or not cloning of the cells is needed. In addition, the FdG assay is quantitative and can be used to study the kinetics and dose response of the interaction.

MATERIALS

CAUTION: See Appendix for appropriate handling of materials marked with <!>.

Buffers and Reagents

Dimethylsulfoxide (Sigma) <!>
Fluorescein di-β-D-galactopyranoside (FdG, Molecular Probes) <!>
Phosphate-buffered saline (PBS) with 5% fetal bovine serum (FBS)
Propidium iodide (10 mg/ml in H$_2$O) (Sigma) <!>

Special Equipment

Benchtop centrifuge with adapters for holding 5-ml tubes
Flow cytometer
Pipetman P-1000
Polystyrene tubes (5-ml round-bottomed; Falcon 2058)

METHOD

1. Plate cells in appropriate medium the day before the assay in 24-well plates at a density such that the culture will be subconfluent at the time of the assay (50,000 cells per well in 0.5 ml of medium for C2C12 or C2F3 mouse myoblasts).

2. Treat cells with reagents appropriate to induce or modify the interaction under study. Typically, different treatments are carried out on duplicate or triplicate samples.

3. At the end of the treatment period, aspirate the medium from the wells and rinse once with PBS.

4. Add two drops of trypsin solution and incubate (room temperature or 37°C) for 5 minutes or until a firm tap on the side of the dish dislodges most cells.

5. Add 1 ml of PBS + 5% FBS to each well. Using a Pipetman P-1000 or equivalent, triturate the cell suspension and rinse the bottom of the well to remove all cells. Transfer cell suspension to 5-ml clear polystyrene tubes (Falcon 2058) appropriate for use on a flow cytometer.

418

6. Pellet cells by spinning in a benchtop centrifuge at about 1500 rpm for 5 minutes.

7. Remove the supernatant by inverting tubes (if possible, invert entire centrifuge bucket adapter assembly with the tubes still in it). While tubes are still inverted, aspirate the remaining drop of liquid from the lip of each tube.

   > Beckman sells an insert that can be added to their adapter for 5-ml tubes that prevents the tubes from falling out when the adapter is inverted.

8. Add 100 µl of PBS + 5% FBS (room temperature) to each tube. Vortex to resuspend the cells (vortex the entire centrifuge bucket adapter assembly with tubes).

9. Prepare a 100× stock of substrate by adding 100 µl of dimethylsulfoxide to a 100-µg vial of FdG. Dilute appropriate amount of substrate in sterile deionized H_2O.

   > After preparation, the stock can be stored for several weeks at –20°C protected from light.

10. Add 100 µl of substrate to each tube of cells (hypotonic shock). Incubate for 3 minutes at room temperature.

11. Stop the uptake of FdG by adding 2.0–2.5 ml of ice-cold PBS + 5% FBS + 1 µg/ml propidium iodide to each tube. Pellet cells at about 1500 rpm for 5 minutes.

    > Propidium iodide is a fluorescent compound that accumulates in dead cells. Thus, cells with propidium iodide fluorescence can be excluded from the data analysis and sorting.

12. Remove most of the supernatant by gently inverting the tubes. If possible, invert the entire centrifuge bucket adapter assembly with tubes still in it. DO NOT shake tubes or allow liquid that clings to lip of tube to come out. Return the tubes to an upright position and vortex the entire centrifuge bucket adapter assembly with tubes. Sufficient liquid should have been retained to resuspend the cells in a volume appropriate for use on the flow cytometer (about 200 µl).

13. Place tubes on ice. Analyze the cells by flow cytometry.

X-Gal Assay for β-Gal

This assay, which is not quantitative or especially sensitive, can be used to detect β-gal complementation if the induction of complementation is sufficiently robust.

MATERIALS

CAUTION: See Appendix for appropriate handling of materials marked with <!>.

Buffers and Reagents

Dimethylformamide (Sigma)
Magnesium chloride ($MgCl_2$) <!>
Paraformaldehyde (4%) <!> in PBS
Potassium ferricyanide ($K_3Fe(CN)_6$) <!>
Potassium ferrocyanide ($K_4Fe(CN)_6$) <!>
X-Gal (Sigma)

METHOD

1. Plate the cells on tissue culture dishes or on glass coverslips at subconfluent densities.

2. Treat the cells with reagents appropriate to induce or modify the interaction under study.

3. Fix the cells for 4 minutes with cold (4°C) 4% paraformaldehyde in PBS. Rinse with PBS twice for 5 minutes.

4. Prepare the substrate by diluting a stock solution of X-gal (40 mg/ml in dimethylformamide, stored at –20°C protected from light) to a final concentration of 1 mg/ml in 5 mM $K_3Fe(CN)_6$, 5 mM $K_4Fe(CN)_6$, and 2 mM $MgCl_2$ in PBS.

5. Add the diluted X-gal to cells (sufficient volume to cover cells). Incubate overnight at 37°C (shorter times are sufficient for high levels of β-gal activity; monitor the reaction under a microscope).

6. Examine the dish or coverslip microscopically for blue cells.

Fluor-X-gal Assay for β-Gal

This assay determines β-gal activity in cells that were immunofluorescently labeled for other proteins.

MATERIALS

CAUTION: See Appendix for appropriate handling of materials marked with <!>.

Buffers and Reagents

4′,6-Diamidino-2-phenylindole dihydrochloride hydrate (DAPI; Sigma) <!>
Fast Red Violet LB (Sigma)
Paraformaldehyde (4%) <!> in PBS
5-6 X-gal (Fluka)

Special Equipment

Epifluorescence microscope

METHOD

1. Plate the cells on glass coverslips at subconfluent densities.

2. Treat the cells with reagents appropriate to induce or modify the interaction under study.

3. Fix the cells for 4 minutes with cold (4°C) 4% paraformaldehyde in PBS. Rinse with PBS twice for 5 minutes.

4. If cells are to be immunofluorescently labeled with antibody as well as Fluor X-gal, carry out all immunolabeling procedures at this point at 4°C.

5. Prepare the reagent by diluting into PBS a stock solution of Fast Red Violet LB (50 mg/ml in dimethylformamide and store at –20°C. The substrate will not completely dissolve at this concentration) to a final concentration of 100 μg/ml, and a stock solution of 5-6 X-gal (50 mg/ml in dimethylformamide; store at –20°C. The solution will change from pale blue to yellow after exposure to light, but this does not appear to affect activity) to a final concentration of up to 25 μg/ml (decrease the concentration of 5-6 X-gal if β-gal activity is strong). Filter through a 0.45-μm syringe filter to remove any precipitate.

6. Add the mixture of diluted Fast Red Violet LB and 5-6 X-gal to cells (sufficient volume to cover cells). Incubate for 60–90 minutes at 37°C.

7. Rinse in PBS for 30 minutes at room temperature.

8. *Optional:* Nuclei may be stained by diluting DAPI in PBS to a final concentration of 100 ng/ml and incubating cells for 10 minutes at room temperature, followed by two rinses in PBS for 5 minutes.

9. Mount the coverslips in PBS and seal with nail polish.

10. Detect Fluor X-gal staining with either the fluorescein (FITC) or rhodamine (TRITC) filter sets of an epifluorescence microscope. The FITC channel gives a better signal-to-background ratio for weak signals, but strong signals appear to be quenched. Therefore, Fluor X-gal stain is best viewed with TRITC filters.

In some experiments, the initial population of cells expressing the chimeric constructs will have low β-gal activity in the absence of the interaction of interest and will exhibit a severalfold increase in β-gal activity when an inducer of the interaction is present. However, in other experiments, no induction or very poor induction may be observed. Low β-gal activity accompanied by a lack of induction can be caused by a failure of one of the constructs to properly express the chimeric protein. Constitutive β-gal activity can be caused by overexpression of one or both of the constructs.

No β-gal Activity Is Observed

If no β-gal activity is observed, the expression of the chimeric constructs must be confirmed. This is best accomplished by immunoblotting cell extracts using antibodies to either β-gal or the non-β-gal component of the chimera. If an immunoreactive protein of the appropriate size is observed, the chimera is being expressed. No immunoreactive band indicates a lack of expression, and a band that is too small indicates that the protein is likely terminated at the wrong residue. Such errors can often be identified by sequencing the cDNA for the chimeric protein as well. If antibodies for immunoblotting are not available for the non-β-gal component of the chimera, commercial antibodies for β-gal are available. We have had mixed success at immunoblotting β-gal chimeras expressed in mammalian cells using anti-β-gal antibodies. The best immunoblots have been obtained using a cocktail of the following antibodies: Calbiochem (La Jolla, California) anti-β-gal monoclonal antibody (OB02-100); Sigma (St. Louis, Missouri) anti-β-gal polyclonal antisera (G4644); Sigma anti-β-gal monoclonal (G6282); each was diluted 1:1000. However, if good antibodies to the non-β-gal protein are not available, we recommend adding a peptide tag to the chimeric protein.

Constitutive β-gal Activity Is Observed

Inducible interactions will ideally have low β-gal activity in the absence of the inducer or ligand. If constitutive β-gal activity is observed, several steps can be taken to reduce the uninduced levels of β-gal activity. First, it is important to avoid overexpressing the β-gal constructs. One significant advantage of this system is that the signal is enzymatically amplified and therefore the constructs do not have to be overexpressed (Blakely et al. 2000). Overexpression can result from selection at excessively high concentrations of drug. For example, we have observed higher levels of expression in C2C12 cells selected at 1 mg/ml of G418 or hygromycin rather than at 300 μg/ml, which is the minimum amount of drug required for selection in this cell line. The use of retroviral vectors limits the number of copies of the construct to a few per cell, which can also help to avoid overexpression. To ensure that most cells receive only a single copy, it is necessary to titrate the virus so that only 15% of cells are infected, according to the Poisson distribution.

In cases where there is a high level of uninduced β-gal activity or poor induction (less than a severalfold increase) of β-gal activity, it is helpful to examine the population of cells on a flow cytometer by assaying for β-gal activity using the fluorogenic substrate FdG (Blakely et al. 2000). If uninduced β-gal activity in the population is not uniform (both high- and low-activity cells are observed), a subpopulation or clones of inducible cells can be isolated that have low background activity. However, steps must be taken to ensure that selection for low β-gal activity does not also select for cells that cannot be induced (see below). Alternatively, the constructs should be reintroduced into cells, taking steps to reduce copy number and expression as described above.

Poor Induction of β-gal Activity Is Observed

When induction fails to occur, flow cytometry is again useful to characterize the population. Low induction observed in a mass assay may be due to only a small part of the population responding. The FACS can be used to select for cells that have a high induced β-gal activity, but this criterion alone may result in cells that have constitutive β-gal activity. Ideally, cells will be sorted sequentially for a subpopulation or clones that have low β-gal activity in the absence of inducer, and higher β-gal activity in the presence of inducer (Blakely et al. 2000). Although a FACS simplifies this procedure, such selection is also possible simply by plating a large number of clones and then screening the clones for inducible β-gal activity.

Constitutive β-gal Activity Is Observed with Membrane Proteins

In our experience, constitutive β-gal activity is more likely to occur when the chimeric proteins are both membrane proteins. We suspect that this is because the membrane proteins, limited to the two-dimensional space of the membrane, have a higher effective concentration of complementation partners at a given level of expression than cytoplasmic proteins. Fortunately, membrane proteins offer an opportunity for another level of selection, because antibodies to the extracellular domain of the membrane protein of interest or to an extracellular domain tag can be used to select for chimera expression in live cells using the FACS (Blakely et al. 2000). Thus, instead of selecting only for low β-gal expression in the absence of inducer (ligand), which may also select for cells that fail to express the chimera, these cells can be simultaneously selected for low β-gal expression using fluorescein fluorescence and modest expression of the chimera using an antibody to the protein of interest and a secondary antibody that fluoresces at a different wavelength. We have also observed that high uninduced β-gal activity can be a problem when using a full-length receptor with a large cytoplasmic domain. In some cases, truncating the receptor so that the β-gal mutant is closer to the transmembrane domain increases the fold induction of β-gal activity primarily by reducing the background level of uninduced β-gal activity. In the case of some receptors, such as the EGF receptor, this has the added benefit of blocking ligand-induced internalization of the receptor (Blakely et al. 2000).

Continued Lack of β-gal Activity after Troubleshooting

In some cases, a lack of β-gal activity or constitutive β-gal activity may be difficult to correct. Some interactions will inevitably be inhibited by the attachment of the interacting proteins to the β-gal deletion mutant. In other cases, the interaction may occur, but the β-gal mutants may not be oriented in a manner that permits complementation. Increasing the distance between the interacting protein and the β-gal mutant using peptide linkers may help. For cytoplasmic proteins, moving the β-gal mutant from one end of the protein to the other may also improve complementation. Although a large majority of the interactions that we have tested work with this system, there is no definite way of predicting which ones will work well. However, given two proteins known to interact, it should be possible to generate a range of chimeric proteins in vitro and select those that display the best characteristics.

REFERENCES

Adams S.R., Harootunian A.T., Buechler Y.J., Taylor S.S., and Tsien R.Y. 1991. Fluorescence ratio imaging of cyclic AMP in single cells. *Nature* **349:** 694–697.

Bai C. and Elledge S.J. 1996. Gene identification using the yeast two-hybrid system. *Methods Enzymol.* **273:** 331–347.

Blakely B.T., Rossi F.M., Tillotson B., Palmer M., Estelles A., and Blau H.M. 2000. Epidermal growth factor receptor dimerization monitored in live cells. *Nat. Biotechnol.* **18:** 218–222.

Charlton C.A., Mohler W.A., Radice G.L., Hynes R.O., and Blau H.M. 1997. Fusion competence of myoblasts rendered genetically null for N-cadherin in culture. *J. Cell Biol.* **138:** 331–336.

Fields S. and Song O. 1989. A novel genetic system to detect protein-protein interactions. *Nature* **340:** 245–246.

Gadella T.W., Jr. and Jovin T.M. 1995. Oligomerization of epidermal growth factor receptors on A431 cells studied by time-resolved fluorescence imaging microscopy. A stereochemical model for tyrosine kinase receptor activation. *J. Cell Biol.* **129:** 1543–1558.

Jacobson R.H., Zhang X.J., DuBose R.F., and Matthews B.W. 1994. Three-dimensional structure of β-galactosidase from *E. coli*. *Nature* **369:** 761–766.

Miyawaki A., Llopis J., Heim R., McCaffery J.M., Adams J.A., Ikura M., and Tsien R.Y. 1997. Fluorescent indicators for Ca2+ based on green fluorescent proteins and calmodulin. *Nature* **388:** 882–887.

Mohler W.A. and Blau H.M. 1996. Gene expression and cell fusion analyzed by lacZ complementation in mammalian cells. *Proc. Natl. Acad. Sci.* **93:** 12423–12427.

Nolan G.P., Fiering S., Nicolas J.F., and Herzenberg L.A. 1988. Fluorescence-activated cell analysis and sorting of viable mammalian cells based on β-D-galactosidase activity after transduction of *Escherichia coli* lacZ. *Proc. Natl. Acad. Sci.* **85:** 2603–2607.

Pelletier J.N., Campbell-Valois F.X., and Michnick S.W. 1998. Oligomerization domain-directed reassembly of active dihydrofolate reductase from rationally designed fragments. *Proc. Natl. Acad. Sci.* **95:** 12141–12146.

Pollok B.A. and Heim R. 1999. Using GFP in FRET-based applications. *Trends Cell Biol.* **9:** 57–60.

Remy I., Wilson I.A., and Michnick S.W. 1999. Erythropoietin receptor activation by a ligand-induced conformation change. *Science* **283:** 990–993.

Rossi F.M., Blakely B.T., and Blau H.M. 2000. Interaction blues: protein interactions monitored in live mammalian cells by β-galactosidase complementation. *Trends Cell Biol.* **10:** 119–122.

Rossi F., Charlton C.A., and Blau H.M. 1997. Monitoring protein-protein interactions in intact eukaryotic cells by β-galactosidase complementation. *Proc. Natl. Acad. Sci.* **94:** 8405–8410.

Ullmann A., Jacob F., and Monod J. 1967. Characterization by in vitro complementation of a peptide corresponding to an operator-proximal segment of the β-galactosidase structural gene of *Escherichia coli*. *J. Mol. Biol.* **24:** 339–343.

Ullmann A., Perrin D., Jacob F., and Monod J. 1965. Identification par complémentation *in vitro* et purification d'un segment de la β-galactosidase d'*Escherichia coli*. *J. Mol. Biol.* **12:** 918–923.

Villarejo M., Zamenhof P.J., and Zabin I. 1972. β-galactosidase: *in vivo* complementation. *J. Biol. Chem.* **247:** 2212–2216.

24
Identification of Protein Single-chain Antibody Interactions In Vivo Using Two-hybrid Protocols

Antje Pörtner-Taliana, Karen F. Froning, Isolde Kusser, and Marijane Russell

Invitrogen Corporation, Carlsbad, California 92008

INTRODUCTION

As naturally occurring protein–protein interactions become increasingly well defined, it is of interest to develop a means of regulating the interactions of specific proteins in order to achieve a desirable biological or clinical effect. One means to do this is to develop a targeted agent that can bind a protein of interest, and alter its biological activity. Naturally occurring antibodies have long been known to bind to a great diversity of target proteins with high affinity. Antibody engineering efforts have built on the great wealth of structural knowledge concerning antibody–antigen recognition, to build enhanced reagents that can work not only in the extracellular milieu within which antibodies normally function, but also for specific intracellular applications. The goal of this chapter is to describe issues related to the use of modified antibodies as intracellular protein-targeted agents and to provide a protocol using a modified dual-bait two-hybrid system to generate such agents.

425

Single-chain antibodies (sFvs) are genetically engineered antibodies that consist of the variable domain of a heavy chain at the amino terminus joined to the variable domain of a light chain by a flexible peptide linker. They are generated by PCR and preserve the affinity of the parent antibody. Neutralizing sFvs against antigens of interest enable researchers to study protein function within the cell (Cattaneo and Biocca 1997; Rondon and Marasco 1997). Richardson et al. (1995) expressed a sFv against the α-subunit of the high-affinity human interleukin-2 receptor and achieved a phenotypic knockout of the receptor. Cochet et al. (1998) showed that expression of a Ras-specific sFv was able to inhibit Ras signaling pathways in *Xenopus laevis* oocytes and NIH-3T3 fibroblasts and specifically promoted apoptosis in human cells. Such neutralizing antibodies can be used to study protein function within cells and have potential in gene therapy.

sFvs are usually generated by PCR from hybridoma cell lines that express monoclonal antibodies (mAbs) with known target specificity (Nicholls et al. 1993), or they are selected by phage display from sFv libraries isolated from spleen cells or lymphocytes (Coloma et al. 1991, 1992; Hoogenboom et al. 1991; Marks et al. 1991). Neither method guarantees the isolation of sFvs that function in antigen recognition in vivo. The formation of disulfide bonds in the reducing environment of the cytoplasm and nucleus (Hwang et al. 1992) is hindered (Martineau et al. 1998), resulting in low expression levels and a limited half-life of the antibodies. Therefore, not all mAbs can be converted into sFvs and maintain their function within the cell. However, some sFvs do not require this bond for antigen recognition (Proba et al. 1998). In search of a protocol to identify intracellular binders, several groups have applied the yeast two-hybrid technology (Chapter 7) to evaluate sFv–protein interaction. This work has proven that the two-hybrid system is useful to predict whether or not a sFv will be able to recognize its target protein in vivo (Visintin et al. 1999; De Jaeger et al. 2000; Pörtner-Taliana et al. 2000). All groups cloned a known sFv or sFv libraries into the prey vector and expressed them as fusions to an activation domain (AD). Visintin et al. (1999) analyzed several sFvs with known antigen specificity. They studied a human immunodeficiency virus (HIV) integrase sFv that was derived from a mAb and exhibits neutralizing function when expressed in human cells. The sFv was able to bind to its antigen in a yeast two-hybrid system also. Next, they investigated the interaction of different sFvs isolated by phage display against the antigens Syk (a tyrosine kinase) or p21-ras using the yeast two-hybrid system. From a set of different antibodies none or only a few sFvs were able to target their antigen in a yeast two-hybrid in vivo assay. Therefore, the isolation of sFvs from phage display libraries is an inadequate criterion for their subsequent use as interacting intracellular antibodies.

The application of the yeast two-hybrid technology was also suitable for identifying intracellular binders from a set of in vitro binding sFvs. De Jaeger et al. (2000) present data of dihydroflavonol-4-reductase interacting with specific sFvs isolated by phage display. Their results correlated with previous expression analysis of the same sFvs in plant cytosol (De Jaeger et al. 1999). In a model screen, Visintin et al. (1999) isolated a sFv with known target specificity that was diluted with DNA from a library encoding nonrelevant sFv-VP16 fusion proteins. Pörtner-Taliana et al. (2000) were able to isolate transcription-factor-specific sFvs from a library using the yeast two-hybrid system and showed that the sFv isolated using a full-length bait was able to recognize this protein in mammalian cells.

In the last decade, yeast two-hybrid systems have been used to isolate interactors of many proteins of interest (see Chapter 7). However, screens with some baits result in the isolation of false positives, whereas in other cases, no interactor can be isolated. When the technology is applied to study sFv–protein interactions, the same problems may occur. Additionally, some researchers may want to screen sFv libraries with truncated bait proteins. The sFv isolated with truncated baits are sometimes not able to bind the native protein in mammalian cells (Pörtner-Taliana et al. 2000). This could be due to the inaccessibility of the fragment in the native protein. In the last years, effort has been spent on improving the yeast two-hybrid system, especially to reduce the number of false positives. The dual-bait interaction trap system (Serebriiskii et al. 1999) is a versatile expansion of the interaction trap system originally developed by Gyuris et al. (1993). A second bait plasmid was added coding for a different DNA-binding domain (DBD) that binds upstream of an additional set of unique reporter genes. The system allows testing for cross-reactivities with an unrelated or related bait protein in the same cell and for library screens with two baits at the same time. This approach has been used to test sFv–antigen interactions. Initial tests showed that the B42 AD was not strong enough to show interaction of sFvs that were isolated in a yeast two-hybrid screen that used VP16 AD. Therefore, VP16 AD was used for additional screens. This modified system was able to distinguish between two different sFv–protein interactions and can be applied for library screens (unpublished results).

Another approach to testing sFv–protein interaction in vivo is a mammalian two-hybrid system. Although mammalian systems do not offer easy library screens, this approach has the advantage that specific sFvs are tested in an environment in which they are most likely used to study protein function. A thorough titering of bait-to-prey ratio is required when characterizing interactions (Pörtner-Taliana et al. 2000). Two groups have used the mammalian two-hybrid approach and compared their results with the yeast two-hybrid approach (Visintin et al. 1999; Pörtner-Taliana et al. 2000). In contrast to the yeast two-hybrid system, Visintin et al. (1999) were only able to verify one sFv–antigen interaction with a mammalian two-hybrid system. They believe this is due to a low sensitivity of their reporter construct. Pörtner-Taliana et al. (2000) used a mammalian two-hybrid bait and prey fusion compatible with the yeast system and a sensitive luciferase reporter. The reporter plasmid contains a splice intron in the polyadenylation signal that increases reporter transcription. They showed that the sFv isolated in a yeast two-hybrid screen was also capable of recognizing the antigen in mammalian cells.

The Yeast Two-hybrid System to Test for sFv–Antigen Interactions In Vivo

Two different yeast two-hybrid systems have been used to study sFv–protein interactions. De Jaeger et al. (2000) describe studies using a Gal4-based yeast two-hybrid system using a Gal4 AD-fused sFv and a Gal4 DBD-fused bait. The laboratories of Cattaneo (Visintin et al. 1999) and Invitrogen (Pörtner-Taliana et al. 2000) used the same modified LexA-based yeast two-hybrid system originally developed by Vojtek et al. (1993), shown in Figure 1. The protein of interest is expressed as a LexA DBD fusion. The LexA DBD binds to the lexA-operator (LexA-op) upstream of the reporter. The prey proteins are sFvs fused to the VP16 AD. If a sFv binds to the bait protein, the transcription of the two reporter genes *His3* and *lacZ* is induced. The bait and prey vectors of the system are shown in Figure 2. The bait vector pBTM116 contains a *TRP1* gene for selection in yeast and the *LexA* DBD with a downstream polylinker to generate LexA-bait fusion proteins. The prey plasmid pVP16* (Pörtner-Taliana et al. 2000) is derived from the vector pVP16 (Vojtek et al. 1993). The vector pVP16 contains a *LEU2* selectable marker and was modified by inserting an ATG, followed by a nuclear localization signal and recognition sequences for *Sfi*I and *Not*I to allow cloning of sFvs (Pörtner-Taliana et al. 2000).

To study sFv–protein interactions in a two-hybrid context, the bait plasmid containing the protein of interest cloned in frame with the LexA DBD, and the prey plasmid coding for the sFv–VP16 fusion, are transformed into the yeast strain L40 (*MATa his3Δ200 trp1-901 leu2-3112 ade2 LYS2::[4lexAop-HIS3]*) (Vojtek et al. 1993). The strain has integrated *His3* and *lacZ* reporter genes. Both reporters have eight or nine *lexA* operators upstream. The transformants are plated and patched on YC-HLUW plates. Leu and Trp are selectable markers for the plasmids. An in vivo interaction is detected by growth on plates lacking histidine, and by qualitative β-galactosidase assay. The interaction is truly positive if both reporter genes are expressed and the following control transformations show no reporter activity: (1) empty bait plasmid with sFv prey plasmid—the sFv should not be able to interact with the LexA DBD; (2) bait plasmid with control protein (e.g., laminin) with sFv prey plasmid—a negative result will verify that the sFv interacts specifically with the protein of interest and not unspecifically with any protein; (3) bait plasmid with protein of interest and empty prey plasmid—the expression bait fusion and the VP16 AD should not result in the activation of the reporter genes; (4) bait plasmid with protein of interest and no prey plasmid (add Leu [L] to medium)—the result of this test will show whether or not the LexA

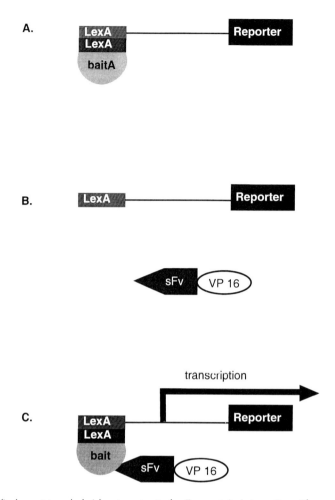

FIGURE 1. Modified yeast two-hybrid system to study sFv–protein interaction. The protein of interest (bait) is expressed as fusion to the LexA DNA-binding domain (DBD) that binds to the *lexA* op upstream of a reporter gene (*A*). The sFv is fused to the activation domain (AD) VP16 and cannot activate transcription on its own (*B*). If an sFv binds to the bait protein, the transcription factor function is reconstituted and the reporter gene is expressed (*C*).

bait fusion activates the transcription of the reporter genes alone; (5) sFv prey plasmid with no bait plasmid (add Trp [W] to medium)—a negative result will exclude the possibility that the sFv will interact unspecifically with DNA upstream of the reporter; (6) prey plasmid with nonbinding control sFv and bait plasmid containing protein of interest—the sFv should not bind to the protein of interest, and therefore, cells should not grow on medium lacking His (H) and show no β-gal activity.

If available, a positive control transformation can be performed, using a sFv/protein pair, which is known to interact in a yeast two-hybrid format. Alternatively, an expression plasmid containing a LexA–VP16 fusion could be used to obtain reporter readouts. Reporter gene expression in transformants containing a LexA–VP16 fusion is usually stronger than that obtained from experiments testing interaction of a LexA protein fusion with a sFv prey VP16 fusion.

sFv–Library Screen Applying the Yeast Two-hybrid Approach

To isolate novel sFvs that bind to a protein of interest in vivo, a bait strain is first constructed. The protein of interest is cloned in frame with the LexA DBD into the bait plasmid pBTM116. The

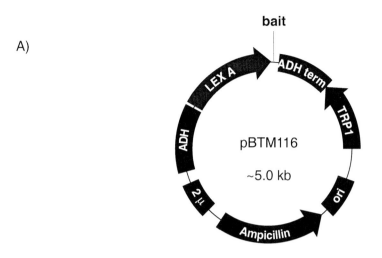

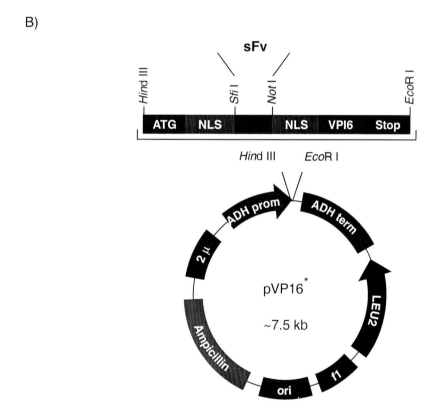

FIGURE 2. Yeast two-hybrid bait (*A*) and modified prey vector (*B*).

plasmid is transformed into the L40 yeast strain and streaked on YC-UW plates to select transformants. Expression of the bait fusion protein is verified by western blot analysis. The fusion proteins can be detected with bait-specific antibodies or with commercially available LexA antibodies. The bait fusion by itself should not be able to activate the reporter genes alone. Therefore, the bait strain is patched on a YC-HUW plate. The cells should not be able to grow on plates lacking His (H).

The prey vector pVP16* allows an easy transfer of most existing sFv libraries with *Sfi*I and *Not*I restriction enzymes. sFv libraries are derived from mRNA isolated from human and murine spleen cells or peripheral blood lymphocytes (Marks et al. 1991; Coloma et al. 1992), and the mRNA is used as a template for synthesis of a single-stranded cDNA. The variable light-chain (V_L) and heavy-chain (V_H) regions are amplified by PCR. The PCR primers are designed such that the immunoglobulin variable regions of heavy and light chain can be directly amplified without prior knowledge of their sequence (Coloma et al. 1991). The linking of V_H and V_L can occur in two ways. The 3′ primer of the V_H region and the 5′ V_L primer contain additional sequences that overlap in a recombinant PCR, and the overlapping sequence codes for the linker peptide. A second approach is to purify the V_L and V_H fragments and to reamplify them with the same set of primers containing restriction sites for cloning into an expression vector. The heavy chain is cloned 5′ of the peptide linker sequence, and the light chain 3′ of the linker sequence. The libraries used for screens with the yeast two-hybrid systems were cloned from phagemid vectors using *Sfi*I and *Not*I restriction sites.

Successful library screens with this system have been performed using the large-scale lithium acetate transformation protocol from Schiestl and Gietz (1989), Hill et al. (1991), Gietz et. (1992), and Mount et al. (1996). Experiments with other two-hybrid systems have shown that small-scale transformation is also sufficient to isolate interactors (Golemis et al. 1996). To measure primary transformation efficiency, different dilutions are prepared and plated on medium lacking Leu and Trp (YC-LUW) to select transformants that contain bait and prey plasmid. The residual transformation is plated on medium lacking additionally His (YC-HLUW). The *His* gene is under control of lexA-op and can only be transcribed when the bait protein interacts with a sFv antibody. His-positive clones generally show up after 2–3 days. The clones are patched on duplicate plates (YC-HLUW). After 1–2 days, β-gal activity is measured using a filter lift (Breeden and Nasmyth 1985) or overlay assay (Gleeson et al. 1998; Golemis and Serebriiskii 1998).

For further characterization, the sFv prey plasmid DNA has to be isolated from double-positive clones and transferred into *E. coli*. Both bait and prey plasmids contain ampicillin as a selectable marker in bacteria. Therefore, the plasmid DNA needs to be analyzed by restriction digest for the presence of sFv-pVP16* plasmid with *Sfi*I and *Not*I. The specificity of the isolated sFv is tested as described above for the interaction of characterized sFvs with their antigens (a–f). Only yeast transformations containing the bait plasmid with the protein of interest and the isolated sFv should grow on plates lacking His (H) and show β-gal activity.

A Dual-bait Yeast Two-hybrid System That Can Distinguish between Two Different sFv–Protein Interactions in a Single Step

The original yeast two-hybrid system has been modified and improved in multiple ways. The interaction trap system developed by Gyuris et al. (1993), for example, was expanded by a second bait vector (Serebriiskii et al. 1999). This vector codes for a second bait-DBD (cI) fusion binding upstream of a unique second set of reporters. The system allows testing for cross-reactivity of a prey protein with another bait in the same cell. This second bait can be any other protein of interest, a mutant, or a protein related to bait 1. To study sFv-binding sites, two different truncated mutants could serve as bait proteins. Additionally, the dual-bait system allows a library screen with two baits at the same time and therefore saves time and reagents.

A modified version of the interaction trap dual-bait system is shown in Figure 3 (unpublished results). Within a yeast cell, the two different bait proteins can be expressed: one as a LexA DBD fusion and the other as a cI DBD fusion. Each fusion has two unique reporter genes to identify interaction. The LexA DBD binds upstream of a *LEU2* and *LACZ* reporter, the cI DBD upstream of a *LYS2* and *GUS* (glucuronidase) reporter. Experiments have shown that the B42 AD is not suitable to examine sFv–protein interaction, but that an amino-terminal VP16–sFv fusion results in a stronger activation of reporter genes. Therefore, the sFvs are expressed as carboxy-terminal

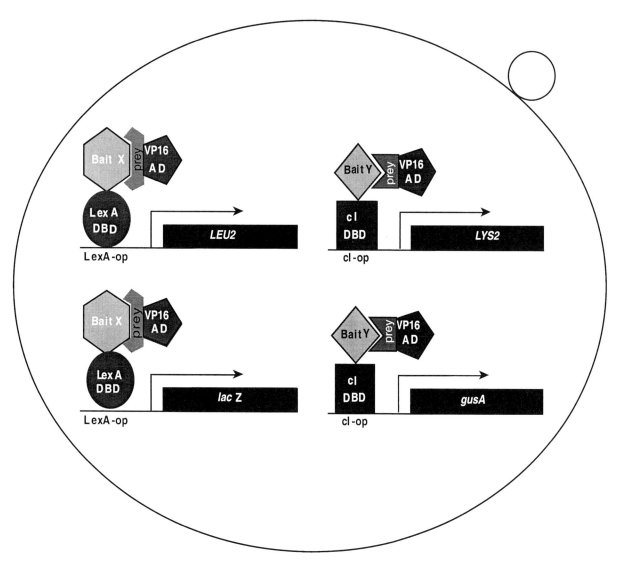

FIGURE 3. Modified dual-bait yeast two-hybrid system to study sFv–protein interaction. A VP16 activation domain-fused sFv interacts with a LexA-fused bait to drive transcription of lexA-op-responsive *LEU2* and *LacZ* reporters. Another sFv interacts with bait 2 fused to the *cl* DBD and activates transcription of cl-op-responsive *LYS2* and *GUS* reporters. This system can be exploited to screen sFv libraries with two baits simultaneously. It also allows testing for cross-reactivity of a sFv with a second bait in the same cell.

VP16 fusions. Compared to the first yeast two-hybrid system described above, the expression of the prey fusion is inducible from a *Gal1* promoter that guarantees a higher representation of the library in the yeast.

The dual-bait expression plasmids are shown in Figure 4. The bait 1 plasmid pHybLex/Zeo contains a LexA DBD upstream of a multiple cloning site (MCS) and a Zeocin marker for selection in yeast and bacteria. The bait 2 plasmid contains a *cl* DBD of the bacteriophage λ, a MCS downstream, a kanamycin gene for selection in bacteria, and a *HIS3* gene for selection in yeast. The prey plasmid pYesTrp was modified by replacing the B42 AD with VP16 and introducing an *Sfi*I and *Not*I downstream to clone sFvs easily. It contains a TRP1 marker for selection in yeast and an ampicillin marker for selection in bacteria. The dual-reporter plasmid pLacGus codes for the lexA-op-responsive *lacZ* and cl-op-responsive *GUS* reporter genes. The *Saccharomyces cerevisiae*

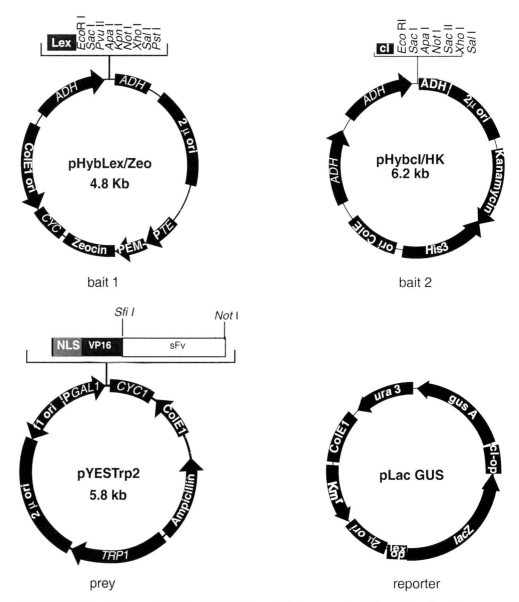

FIGURE 4. Dual-bait two-hybrid bait 1, bait 2, modified prey, and dual reporter plasmids.

strain SKY 48 (*Matα ura3 trp1 his3 6lexAop-LEU2, 3cIop-LYS2*) has integrated lexA-op-responsive *LEU2* and cI-op-responsive *LYS2* reporters (Serebriiskii et al. 1999).

The system can be applied for testing sFv–protein interaction in vivo. After transformation, the cells are first plated on glucose/Zeocin plates, which lack the amino acids His, Trp, and Ura, to select for transformants containing the plasmids bait 1, bait 2, prey, and dual reporter. The prey protein expression is controlled by a galactose-inducible promoter and induced by transferring the cells on galactose medium. The medium lacks additionally Leu (L) or Lys (K) to obtain the first reporter readout for both baits. Positive clones are patched on duplicate –L or –K plates. After a maximum of 2 days' incubation, overlay or filter lift assays are performed to detect β-gal or GUS activity, respectively. Controls are required as described for the single-bait two-hybrid system. The bait fusion proteins as well as the VP16–sFv fusion should not be able to activate reporter transcription on their own. The cross-reactivity of the sFvs with cI and LexA needs to be examined, also.

For library screens, a dual-bait strain has to be created transforming competent SKY48 cells containing the reporter plasmid pLacGUS with the two bait plasmids. Bait protein expression can

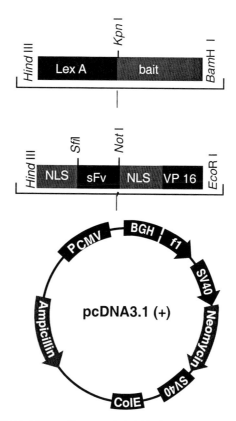

FIGURE 5. Mammalian two-hybrid bait and prey constructs.

be detected by western blot with bait-specific or LexA- and cI-specific antibodies. At the same time, it should be tested whether or not the bait fusion proteins can activate reporter transcription on their own. Either the large-scale or small-scale library transformation can then be applied. Transformation efficiency is measured by plating a sequence of 1:10 dilutions of the transformation on YC-UWH + Zeo plates. The transformants are harvested and prey protein expression is induced by incubating the cells in galactose-containing liquid medium. Different volumes are plated on galactose plates containing Zeocin, lacking UWH, and L or K. Leu[+] and Lys[+] colonies are tested for β-gal or GUS activity, respectively.

The recovery of prey plasmid DNA from double-positive clones is simplified due to the different bacterial selection markers on prey, bait, and dual reporter plasmid (cf. Golemis and Serebriiskii 1998).

A Mammalian Two-hybrid System to Evaluate sFv–Protein Interaction In Vivo

The mammalian two-hybrid system can serve as an alternative to test sFv–antigen interaction in vivo, but it cannot be applied for library screens (Visintin et al. 1999; Pörtner-Taliana et al. 2000). The system has the advantage that the sFv–protein interaction is directly assessed within a mammalian cell. Both the LexA DBD bait and the sFv–VP16 fusions are cloned into mammalian expression vector pcDNA3.1 (Fig. 5) (Pörtner-Taliana et al. 2000). Thorough titering of bait/prey ratios is required to ensure detection of interaction. A good strategy is to start with a low-bait plasmid concentration and titer the sFv–VP16 expression plasmid. The reporter readout will reach an optimum at a certain sFv–VP16 plasmid concentration and decrease with higher prey concentrations. In control experiments, it should be tested whether or not the bait fusion or the sFv–VP16 fusion is able to activate reporter transcription on its own.

In addition, the system requires a reporter with high sensitivity (Visintin et al. 1999). A luciferase reporter plasmid applied in the study of Pörtner-Taliana et al. (2000) was created by cloning 8 lex-op sequences and the minimal thymidine kinase (TK) promoter upstream of the luciferase gene in pGL-Basic (Promega). A splice intron in the polyadenylation signal increases reporter transcription. Sensitive luciferase detection assays facilitate the detection of sFv–protein interaction.

To normalize for transfection efficiency, a plasmid with a different reporter gene under control of a constitutive promoter should be cotransfected. This reporter will be a good internal control for nonspecific effects of the VP16 AD. The system described was successfully applied in CHO and COS cells with sFvs selected in yeast two-hybrid screens (Pörtner-Taliana et al. 2000). Optimal results were obtained with 25 ng of bait DNA and 250 ng of sFv–VP16 plasmid using 2 μg of reporter plasmid. Plasmid was added to a final DNA concentration of 4 μg in a 35-mm TC plate.

OUTLINE OF PROCEDURE

This chapter describes three related two-hybrid approaches to test for sFv–protein interactions in vivo. Studies with the first yeast two-hybrid system showed that the method is a valuable tool to predict whether or not a specific sFv recognizes its antigen in an in vivo environment. The second method represents an expansion of the traditional yeast two-hybrid system with a second bait and a second set of reporters. It has the advantage that cross-reactivities of sFvs with a second bait can be studied in one yeast cell, and sFv–library screens can be performed with two baits at the same time. In contrast to the first system, the prey protein expression is inducible. Therefore, a two-step selection protocol should be chosen for library screens. The first step selects for transformants containing bait, prey, and reporter plasmids. This guarantees that most of the genes in the library are represented in the population of transformants. Prey protein expression is induced by transferring the transformants to galactose-containing medium, and interaction can be assessed by examining reporter expression. The two-step selection is more time-consuming and requires more material, but it increases the probability of isolating sFvs that interact with the protein of interest. The third method describes a mammalian two-hybrid approach which has the advantage that sFv–antigen interaction can be studied in a mammalian host cell in which a neutralizing sFv is most likely used.

Yeast Two-hybrid Test for sFv–Protein Interaction

The two-hybrid approach described here can be used to predict whether or not a specific sFv recognizes its antigen in an in vivo environment. The protocol uses a second bait and a second set of reporters so that the cross-reactivation of sFv can be studied in one yeast cell; in addition, sFv–library screens can be performed with two baits at the same time. This protocol is related in concept and materials to Chapter 7 and should be cross-checked to this chapter.

MATERIALS

CAUTION: See Appendix for appropriate handling of materials marked with <!>.

Buffers and Solutions

Cracking buffer
  8 M urea
  5% SDS <!>
  40 mM Tris-HCl (pH 6.8)
  0.1 mM EDTA
  1% β-mercaptoethanol <!>
  0.4 mg/ml bromophenol blue <!>
    Store at 4°C or –20°C.
Dimethylsulfoxide (DMSO) <!>
Glycerol solution
  25 mM Tris-HCl (pH 8.0)
  0.1 M magnesium sulfate ($MgSO_4$)<!>
  65% (v/v) glycerol
    Autoclave the solution and store at room temperature.
1x LiAc
  100 mM lithium acetate (pH 7.5)
10x LiAc
  1 M lithium acetate (pH 7.5)
    Adjust pH to 7.5 using diluted glacial acetic acid <!>, filter-sterilize, and store at room temperature.
1x LiAc/1x TE
  100 mM lithium acetate (pH 7.0)
  10 mM Tris (pH 7.5)
  1 mM EDTA
    Mix together 10 ml of 10x LiAc and 10 ml of 10x TE. Add deionized H_2O to 100 ml, filter-sterilize, and store at room temperature.
1x LiAc/0.5x TE
  100 mM lithium acetate (pH 7.5)
  5 mM Tris (pH 7.5)
  0.5 mM EDTA
    Filter-sterilize and store at room temperature.

435

1x LiAc/40% PEG-3350/1x TE
   100 mM lithium acetate (pH 7.5)
   40% (w/v) PEG 3350 <!>
   10 mM Tris-HCl (pH 7.5)

   Mix together 20 ml of 10x LiAc, 20 ml of 10x TE, and 80 g of PEG 3350 <!>. Add deionized H₂O to 200 ml and dissolve the PEG. The solution may have to be heated. Autoclave (15 psi) for 20 minutes at 121°C and store at room temperature.

1x TE
   10 mM Tris (pH 7.5)
   1 mM EDTA

   Filter-sterilize.

10x TE
   100 mM Tris (pH 7.5)
   10 mM EDTA

   Filter-sterilize.

X-Gal (5-bromo-4-chloro-3-indolyl-β-D-galactose)
   25 mg or 50 mg/ml X-Gal /ml in *N,N*-dimethylformamide (DMF)<!>

X-Gluc (5-bromo-4-chloro-3-indolyl-β-D-glucuronic acid)
   25 mg of X-Gluc /ml in DMF<!>

YC medium and plates (YC is minimal defined medium for yeast)

1. Dissolve 0.12% w/v Yeast nitrogen base (without either amino acids or ammonium sulfate), 0.5% w/v ammonium sulfate, <!>, 1% w/v succinic acid, 0.6% w/v NaOH <!>, 2% w/v glucose, 0.01% w/v (adenine, arginine, cysteine, leucine [L], lysine [K], threonine, tryptophan [W], uracil [U]), and 0.005% w/v (aspartic acid, histidine [H], isoleucine, methionine, phenylalanine, proline, serine, tyrosine, valine).

   The amino acids with the one-letter code are those that need to be omitted to make selective plates, depending on the genotype of the host, plasmid markers, and reporters.

2. Add 2% w/v agar after dissolving the reagents above, if you are making plates.

3. Autoclave at 15 psi for 20 minutes at 121°C.

4. Cool to 50°C and add 10% v/v of filter-sterilized 20% w/v glucose.

   Add Zeocin (Zeo) at this point (if needed) to a final concentration of 200 µg/ml. For plates that contain galactose and raffinose, add 10% v/v 20% w/v galactose and 5% v/v 20% w/v raffinose instead of glucose.

5. Pour plates and allow to harden. Invert the plates and store at +4°C. Plates are stable for 6 months unless they contain Zeo. Plates containing Zeo are stable for about 1 month.

Yeast extract peptone dextrose medium

1. Dissolve 1% w/v yeast extract, 2% w/v peptone.

2. Add 2% w/v agar, if making plates.

3. Autoclave for 20 minutes on liquid cycle.

4. Add 2% w/v dextrose (D-glucose).

5. If desired, cool the solution to <50°C and add 200 µg/ml Zeo. Store medium at room temperature. Store medium containing Zeo at 4°C protected from exposure to light. The shelf life is ~1–2 months.

Yeast lysis buffer
   50 mM Tris-HCl (pH 8.0)
   2.5 M lithium chloride (LiCl)<!>
   4% (v/v) Triton X-100
   62.5 mM EDTA

   Adjust the pH if necessary with NaOH or HCl and bring the volume to 100 ml. Store at room temperature.

YPD ± Zeocin

Z buffer
   60 mM sodium hydrogen phosphate (Na$_2$HPO$_4$) <!>
   40 mM sodium hydrogen diphosphate (NaH$_2$PO$_4$)<!>
   10 mM potassium chloride (KCl) <!>
   1 mM magnesium sulfate (MgSO$_4$) (pH 7.0)<!>
Zeocin (Zeo) concentration and selective medium

Bacteria and Yeast

Escherichia coli: 25 µg/ml in low-salt LB medium
   Efficient selection requires that the concentration of NaCl be no more than 5 g/liter (<90 mM).
Saccharomyces cerevisiae (SKY48/pLacGUS): 200 µg/ml in YPD or other selective medium

Plasmids and Yeast Strains

Plasmids and strains for modified yeast two-hybrid system developed originally by Vojtek
   et al. (1993)
   pVP16*
   pBTM116
   L40 yeast strain (*MATα his3D200 trp1-901 leu2-3112 ade2 LYS::9lexAop-HIS3*)
Plasmids and strains for the modified dual-bait yeast two-hybrid system
   pHybLex/Zeo
   pHybcI/HK
   pYesTrp/VP16
   SKY 48 (*MATα trp1 ura3 his3 6lexop-LEU2 3cIop-LYS2*) pretransformed with pLacGus

METHOD

Small-scale Yeast Transformation (Schiestl and Gietz 1989; Hill et al. 1991; Gietz et al. 1992; Mount et al. 1996)

This protocol can be used to test whether or not sFvs isolated from libraries by phage display or mAbs are able to target their antigen in vivo. It is also applicable to create bait strains for a library screen and to control for true positive interactions after sFvs are isolated from a yeast two-hybrid library screen.

Bait strains are constructed by transforming the bait plasmid(s) containing the gene of interest in frame with the DNA-binding domain into the appropriate bait strain. These bait strains should be used to test whether or not the baits are able to activate reporter transcription on their own; the expression of the bait fusion protein should be tested by western blotting.

1. Inoculate 10 ml of YPD with a colony of SKY48/pLacGUS and shake overnight at 30°C.

2. Determine the OD$_{600}$ of your overnight culture. Dilute culture to an OD$_{600}$ of 0.4 in 50 ml of YPD and grow an additional 2–4 hours.

3. Pellet the cells at 2500 rpm and resuspend the pellet in 40 ml of 1x TE.

4. Pellet the cells at 2500 rpm and resuspend pellet in 2 ml of 1x LiAc and 0.5x TE.

5. Incubate the cells at room temperature for 10 minutes.

6. For each transformation, mix together 1 µg of plasmid DNA and 100 µg of denatured sheared salmon sperm DNA with 100 µl of the yeast suspension from step 5.

7. Add 700 μl of 1x LiAc, 40% PEG-3350, 1x TE and mix well.

8. Incubate solution for 30 minutes at 30°C.

9. Add 88 μl of DMSO, mix well, and heat shock for 7 minutes at 42°C.

10. Centrifuge in a microcentrifuge for 10 seconds and remove supernatant.

11. Resuspend the cell pellet in 1 ml of 1x TE and re-pellet.

12. Resuspend the pellet in 50–100 μl of 1x TE and plate on an appropriate selective plate.

Expression of Bait Fusion Proteins (Golemis et al. 1996; Golemis and Serebriiskii 1998)

This protocol describes how to prepare cell lysates from your L40 bait strain (Trp⁺ and Ura⁺) or His⁺, Zeo-resistant SKY48 pLacGUS bait strain for western blot analysis. Test several transformants in case of heterogeneity in LexA and cI fusion expression levels. Run lysates of untransformed L40 or SKY48/pLacGUS as negative controls.

1. Inoculate 10 ml of YC-UW (L40) or YC-UH Zeo200 (SKY48 pLacGUS) with a single colony of the bait strain. Inoculate 10 ml of YC-U with L40 or SKY48/pLacGUS as a negative control. Grow overnight, with shaking, at 30°C.

2. Streak a sample of each culture onto a fresh plate. After confirmation of bait expression, return to this plate and use it as a subsequent source of the bait strain.

3. Pellet the cells in step 1 by centrifuging at 2500 rpm for 5 minutes at room temperature. Decant the medium.

4. Transfer the cell pellets to a –80°C freezer for 10 minutes.

5. Thaw the cell pellet in 100 μl of prewarmed (60°C) cracking buffer and resuspend by pipetting the cell pellet in the buffer.

6. Transfer the cell suspension to a 1.5-ml microcentrifuge tube containing 100 μl of glass beads.

7. Incubate the solution for 10 minutes at 70°C.

8. Vortex the solution for 1 minute.

9. Centrifuge at 14,000 rpm for 5 minutes at room temperature and transfer the supernatant to a new tube.

10. Add SDS-PAGE sample buffer and boil sample for 5 minutes. Use 30–50 μl for immunoblot analysis. Detect LexA and cI fusions using antibodies to the proteins of interest or LexA and cI antibodies (available from Invitrogen).

The calculated molecular weights of the LexA and cI protein expressed from pHybLex/Zeo and pHybcI/HK, respectively, are listed below. The calculated molecular weight of each protein includes additional amino acids encoded by the multiple cloning sites. The table also lists the observed migration of each protein on an SDS polyacrylamide gel.

Protein	Calculated molecular mass	Observed molecular mass
LexA	26 kD	32 kD
cI	29 kD	36 kD

Library Transformation

SFv libraries are isolated from spleen cells or peripheral blood lymphocytes. The mRNA is isolated and the cDNA is generated. The amplifications of the V_H and V_L chains are performed with primers that do not require the knowledge of the variable domain sequences. The V_L chain is linked to the carboxyl terminus of the V_H chain by using overlapping PCR. The overlapping sequences at the V_H 3′ and V_L 5′ ends are designed to code for a $([Gly]_4Ser)_3$ linker (Marks 1991), or the purified chains are reamplified with a set of primers containing overhangs with restriction sites to clone them into an expression vector (mostly phagemids). The V_H chain is cloned amino-terminally of the linker coding sequence, and the V_L chain carboxy-terminally of that linker sequence (Finnern et al. 1997). Existing libraries can be easily cloned into the modified prey vectors of both systems using *Sfi*I and *Not*I restriction enzymes.

The large-scale protocol was applied for the single-bait system using an L40 bait strain (Schiestl and Gietz 1989; Hill et al. 1991; Gietz et al. 1992; Mount et al. 1996). The dual-bait yeast two-hybrid screen was performed using the small-scale protocol (Golemis et al. 1996; Golemis and Serebriiskii 1998). The procedures are generally interchangeable. The user should consider that the two-step selection is recommended for the dual-bait system, and the large-scale protocol should be appropriately adapted. The selection of transformants will change accordingly.

Large-scale Library Transformation for Single-bait Two-hybrid System into the L40 Bait Strain

1. Grow 5 ml of an overnight culture of the L40 bait strain in selective yeast medium lacking W and U.

2. Inoculate an aliquot of the overnight culture into 100 ml of the same medium and grow overnight at 30°C with constant shaking.

3. On the following day, dilute the culture to a final OD_{600} of 0.3 in 1 liter of YPAD (YPD with 40 µg/ml adenine). Grow for 3 hours at 30°C with constant shaking.

4. Pellet at room temperature (RT) by centrifugation at 2500 rpm for 10 minutes in a fixed-angle rotor for a medium-speed centrifuge.

5. Wash in 500 ml of 1x TE, resuspend in 20 ml of 100 mM LiAc and 0.5x TE, and transfer to a sterile 1-liter flask.

6. Add a mixture of 1 ml of 10 mg/ml denatured salmon sperm DNA, 500 µg library DNA, and 140 ml of 100 mM LiAc, 40% PEG-3350, 1x TE to the cell suspension. Incubate for 30 minutes at 30°C.

7. Swirl the mixture with 17.6 ml of DMSO, heat shock for 6 minutes at 42°C, and immediately dilute with 400 ml of YPA (YPAD without dextrose). Cool rapidly to room temperature in a water bath.

8. Pellet the cells at 2500 rpm for 10 minutes. Wash with 500 ml of YPA, resuspend in 1 liter of YPAD, and incubate for 1 hour at 30°C with constant shaking.

9. Centrifuge the suspension at 2500 rpm for 10 minutes, wash the pellet, and incubate with 1 liter of selective medium lacking W, U, and L for 16 hours with shaking.

10. Repeat the centrifugation, wash with YPA, and incubate for 1 hour with 1 liter of YPAD.

11. Plate aliquots from the 1-liter suspension on selective yeast medium lacking U, W, and L to measure primary transformation efficiency.

12. Pellet and wash the cells twice with selective medium lacking W, U, L, and H, and resuspend the final pellet in 10 ml of this selective medium. Plate aliquots of 5 µl, 10 µl, 25 µl, and 50 µl on 40 selective plates (10 plates for each aliquot) lacking W, L, U, and H. Incubate 2–3 days until colonies appear.

13. Patch plate His[+] colonies onto duplicate plates lacking W, U, L, and H, and test one plate after 2 days for β-gal activity.

Small-Scale Library Transformation for Dual-bait Screen (SKY48/pLacGUS bait strain)

In the following example, interactors are being identified in parallel for a LexA-fused and a cI-fused bait. See also Chapter 7.

1. Inoculate 20 ml of YC-UH Zeo200 with SKY48/pLacGUS containing the baits expressed in plasmids pHybLex/Zeo and pHybcI/HK. Grow overnight at 30°C.

2. The next day, dilute the culture into 300 ml of YC-UH Zeo200 to 2 × 10E6 cells/ml (OD_{600} = ~0.10). Incubate at 30°C until the culture reaches 2 × 10E7 cells/ml (OD_{600} = 1).

3. Centrifuge for 5 minutes at 1000–1500g in a low-speed centrifuge at room temperature to harvest cells. Resuspend in 30 ml of sterile H_2O and transfer to a 50-ml conical tube.

4. Centrifuge for 5 minutes at 1000–1500g. Decant the supernatant and resuspend the cells in 1.5 ml of 1× LiAc, 1× TE.

5. Add 1 μg of library DNA and 50 μg of high-quality sheared salmon sperm carrier DNA to each of 30 sterile 1.5-ml microcentrifuge tubes. Add 50 μl of the resuspended yeast solution from step 4 to each tube. The total volume of library and salmon sperm DNA added should be <20 μl and preferably <10 μl.

6. Add 300 μl of sterile 1× LiAC, 40% PEG-3350, 1× TE to each tube, and invert to mix thoroughly. Incubate for 30 minutes at 30°C.

7. Add DMSO to 10% (~40 μl per tube) and invert to mix. Heat-shock for 10 minutes in a 42°C heating block.

8. Take 28 of the 30 tubes from step 7 and plate the complete contents of one tube onto a 150-mm YC-UHW Zeo200 plate and incubate for 1–2 days at 30°C.

9. For the two remaining tubes, plate 360 μl from each tube onto separate 150-mm YC-UHW Zeo200 plates. Use the remaining 40 μl from each tube to make a series of 1:10 dilutions in sterile H_2O. Plate dilutions on 100-mm YC-UHW Zeo200 plates. Incubate all plates for 2–3 days at 30°C until colonies appear.

   The dilution series gives an idea of the transformation efficiency and allows an accurate estimate of transformants obtained.

Collect Primary Transformants

Conventional replica plating does not work well in the selection process because so many cells are transferred to new plates that very high background levels inevitably occur. Instead, the procedure described below creates a slurry in which cells derived from >10^6 primary transformants are homogeneously dispersed. A precalculated number of these cells are plated for each primary transformant.

10. Cool all of the 150-mm plates containing transformants from step 8 for several hours at 4°C to harden agar.

11. Wearing gloves and using a sterile cell scraper, gently scrape yeast cells off the plate. Be careful not to damage the agar. Pool the cells from the 30 plates into one or two sterile 50-ml conical tubes.

   This is the step where contamination is most likely to occur. Be careful, and if possible use a sterile hood.

12. Wash cells by resuspending the transferred cells into an equal volume of sterile TE buffer or H_2O. Centrifuge at 1000–1500g for ~5 minutes at room temperature, and discard the supernatant. Repeat the wash.

13. Resuspend the pellet in 1 volume of glycerol solution, mix well, and store up to 1 year in 1-ml aliquots at –80°C.

Determine Replating Efficiency

14. Thaw an aliquot of frozen transformed yeast (step 12, above) and dilute 1:10 with YC-UHW Zeo200 Gal/Raff medium. Shake for 4 hours at 30°C to induce the *GAL1* promoter to express the library.

    > Raffinose (Raff) is not required for growth, but it helps the cells to grow faster without diminishing transcription from the GAL1 promoter.

15. Make serial dilutions of the culture using the YC-UHW Zeo200 Gal/Raff medium. Plate on 150-mm YC-UHW Zeo200 Gal/Raff plates and incubate for 2–4 days at 30°C until colonies are visible.

16. Count the colonies and determine the number of colony-forming units (cfu) per aliquot of transformed yeast. If the harvest is done carefully, viability will generally be >90%.

    > Some researchers perform this step simultaneously with plating on leucine- or lysine-deficient selective medium.

Screening for Interacting Proteins

Not all transformants will contain interacting proteins. Therefore, plating should be done on leucine- or lysine-deficient medium. It is desirable that for actual selection, each primary colony obtained from the transformation be represented on the selection plate by 3–10 individual yeast cells. This will, in some cases, lead to multiple isolations of the same cDNA; however, because the slurry is not perfectly homogeneous, it will increase the likelihood that all primary transformants are represented by at least one cell on the selective plate.

It is easiest to visually scan for Leu$^+$ or Lys$^+$ colonies using cells plated at ~10^6 cfu per 150-mm plate. Plating at higher density can contribute to cross-feeding between yeast, resulting in spurious background growth. Thus, for a transformation in which 3 x 10^6 colonies are obtained, plate ~1 x 10^7 cells on a total of 10 selective plates.

17. Thaw the appropriate quantity of transformed yeast based on the plating efficiency (calculated on previous page), dilute 1:10 with YC-UHW Zeo200 Gal/Raff medium, and incubate as in step 14.

18. Centrifuge at 1000–1500g for 5 minutes at room temperature and resuspend the pellet in 1 ml of YC-UHW Zeo200 Gal/Raff medium.

19. Plate 50 μl on each of 10 YC-UHWL Zeo200 Gal/Raff plates and 10 YC-UHWK Zeo200 Gal/Raff plates. Incubate for 2–3 days at 30°C until colonies appear. Carefully pick appropriate Leu$^+$ or Lys$^+$ colonies and patch on new YC-UHWL Zeo200 Gal/Raff or YC-UHWK Zeo200 Gal/Raff master plates. Incubate for 2 days at 30°C until colonies appear, and perform LacZ and GUS assays to obtain secondary reporter readouts. Note, true interactors should have a positive phenotype only on Gal/Raff plates, not on Glu plates.

β-*Galactosidase Filter Assay*

The filter assay is applicable for a qualitative GUS assay also, but generally requires a lower substrate concentration than β-galactosidase (~20%).

1. Lay dry nitrocellulose filter onto the yeast colonies plated on selective medium.

   > Remove the filter and float it colony side up in an aluminum foil boat in a thin layer of liquid nitrogen for 30 seconds. Then immerse the filter for 5 seconds in the liquid nitrogen.

2. Remove the filter and place it at room temperature, colony side up, until thawed.

3. Prepare a petri dish for the reaction. Place 1.5 ml of Z buffer containing 15 μl of a 50 mg/ml X-gal in the lid, and lay one #1 Whatman filter circle in the Z buffer, followed by the nitro-

cellulose filter with colonies facing up.

4. Use the bottom of the dish to cover and incubate the dish at 30°C (if longer incubations are required for positive signals to be detected, place the petri dish in a humidified chamber).

> Strong interactions yield detectable color in less than 30 minutes.
>
> Alternatively, assay β-galactosidase or β-glucuronidase activity with an overlay assay.

Overlay Assay (Duttweiler 1996; Golemis and Serebriiskii 1998; Gleeson et al. 1998)

1. Add 1 g of low-melt agarose (LMA) to 100 ml of 100 mM potassium phosphate <!> buffer (pH 7.0). Dissolve the LMA by heating for 3–5 minutes in the microwave. Do not overheat the agarose because the solution will boil over.

2. Allow the agarose solution to cool to 65°C.

3. Prepare X-Gluc/DMF or X-Gal/DMF solution by adding the following amount of X-Gluc or X-Gal solution to DMF:

   For X-Gluc, add 100 µl of freshly prepared 25 mg/ml X-Gluc to 8 ml of DMF.

   For X-Gal, add 800 µl of freshly prepared 25 mg/ml X-Gal to 8 ml of DMF.

4. Mix the 8 ml of X-Gluc/DMF or X-Gal/DMF solution from step 3 with 12 ml of the dissolved LMA solution to make an X-Gluc/LMA or X-Gal/LMA solution with a final concentration of 40% DMF, 0.6% LMA. The total volume will be 20 ml.

5. Incubate the X-Gluc/LMA or X-Gal/LMA solution for 5 minutes at 65°C.

6. Carefully overlay the patched plates with the following amount of X-Gluc/LMA or X-Gal/LMA solution:

   For 100-mm plates, use 5–6 ml per plate.

   For 150-mm plates, use 12–15 ml per plate.

7. Let the plates sit at room temperature for 5–10 minutes until the X-Gluc/LMA or X-Gal/LMA solution solidifies. Do not disturb the plates during the solidification process. To prevent exposure to light, keep the plates covered with aluminum foil.

8. Incubate the plates for up to 1 hour at 30°C in the dark, but monitor the color development regularly by eye during this time.

> Colonies that are positive for β-glucuronidase or β-galactosidase activity will turn blue. Monitor the color development on the overlay plates carefully within the first hour. We suggest that you check the degree of color development every 15 minutes. This is particularly important for the β-glucuronidase assay because X-Gluc degrades easily, and long incubations can lead to high background that might be interpreted as a false-positive result.
>
> In general, the intensity and length of time that it takes for the color signal to develop should provide an indication of the strength of your positive bait–prey interaction. For a strong positive interaction between your bait and prey, you may see intense blue color develop within 15 minutes. For a weak positive interaction, the blue color may take up to an hour (GUS) or longer (β-gal) to develop.

Isolation of pVP16 sFv from His^+ β-Gal^+ clones*

The prey plasmid DNA from positive yeast clones can be recovered by two alternative methods. The first method was described by Ward (1990). The yeast cells are lysed with lysis buffer and glass beads, and the lysate is extracted with phenol/chloroform. An aliquot of the aqueous phase is transformed into *E. coli* and selected on LB plates with ampicillin. Alternatively, although more expensive to use, the yield and quality of DNA obtained using a Miniprep kit is generally higher than that isolated with the method detailed above. We have used the S.N.A.P. Invitrogen Miniprep kit; other suppliers are providing specially designed kits to isolate DNA from yeast (e.g., Qiagen).

Method developed by Ward (1990)

1. Grow 5 ml of a yeast overnight culture in selective medium lacking W, L, and H.

2. Pellet and resuspend cells in 300 µl of yeast lysis buffer.

3. Mix the suspension with 150-µl glass beads (0.45–0.50 mm) and 300 µl of phenol/chloroform. Vortex vigorously for 1 minute.

4. Spin the beads and phenol/chloroform, and transfer the aqueous phase to a new tube.

5. Precipitate plasmid DNA in the aqueous phase twice with ethanol, and resuspend in 25 µl of TE.

6. Transform *E. coli* cells with 1–2 µl of DNA and select on LB plates with ampicillin.

Isolation of Plasmid DNA from Yeast Using a MiniPrep Kit

1. Inoculate 5 ml of YC-W with a single positive colony selected as H$^+$ or K$^+$ and incubate overnight at 30°C with shaking. The culture should be in stationary phase (OD$_{600}$ = 1–2) before proceeding further.

2. Pellet the cells in a clinical centrifuge at 2500 rpm for 5 minutes.

3. Resuspend the cell pellet in 1 ml of 1× TE and repellet the cells.

4. Resuspend the cell pellet in 1 ml of 1× TE. Add 1 µl of β-mercaptoethanol and 1.5 µl of zymolyase (3 mg/ml in H$_2$O [Seikagaku]). Incubate for 1 hour at 30°C.

5. Centrifuge at 1000g for 4 minutes at room temperature to pellet the cells gently.

6. Remove the supernatant and resuspend the cell pellet in 150 µl of Resuspension Buffer containing RNase A.

7. Add 150 µl of 1% SDS. Mix gently by inversion. Incubate at 65°C for 10 minutes.

8. Place on ice for 3 minutes.

9. Add 150 µl of ice-cold precipitation salts. Mix by inverting.

10. Centrifuge at 14,000g for 10 minutes.

11. Remove supernatant to a new microcentrifuge tube and add 600 µl of binding buffer. Mix by inverting five or six times. Apply the entire solution onto the S.N.A.P. MiniPrep Column/Collection Tube.

12. Centrifuge the S.N.A.P. MiniPrep Column/Collection Tube at 1000–3000g for 30 seconds at room temperature. Discard the column flowthrough.

13. Add 900 µl of Wash Buffer. Centrifuge as in step 12.

14. Discard the column flowthrough. Centrifuge the S.N.A.P. MiniPrep Column/Collection Tube at maximum speed for 2 minutes at room temperature to dry the resin.

15. To elute the plasmid DNA, place the S.N.A.P. MiniPrep Column into a new sterile microcentrifuge tube and add 70 µl of 1× TE or sterile H$_2$O directly to the resin.

16. Incubate for 2 minutes at room temperature.

The prey plasmid DNA can be isolated from ampicillin-resistant colonies. If the first L40-based yeast two-hybrid system was used for the library screen, the plasmid DNA needs to be analyzed by restriction digest because the bait plasmid pBTM116 also contains an ampicillin-resistance gene. The prey plasmid DNA should be retransformed into L40 or SKY48 containing the dual reporter plasmid using the small-scale transformation protocol to confirm whether or not the sFv is interacting specifically with the bait protein. The control transformations should be set up as described under the section The Yeast Two-hybrid System to Test for sFv-Antigen Interactions In Vivo, p. 427.

If true positive clones are identified, the target specificity of the sFv should be investigated by western blotting. For this purpose, the sFv can be cloned into a bacterial expression vector in frame with an expression tag, e.g., myc-tag. The sFv fusion can be expressed in bacteria and the periplasmic preparation can be used to detect the antigen on western blots, detecting the epitope tag with a specific horseradish peroxidase (HRP)-conjugated antibody (Pörtner-Taliana et al. 2000). The in vivo targeting capabilities can also be verified with a mammalian two-hybrid assay (Pörtner-Taliana et al. 2000). If neutralizing sFvs are needed, the inhibiting function of the sFv should be tested with assays specific for the proteins of interest.

17. Centrifuge the S.N.A.P. MiniPrep Column/Collection Tube at maximum speed for 2 minutes at room temperature. The plasmid DNA is now eluted from the column. Remove and discard the column.

18. Transform competent *E. coli* with 10 μl of the DNA suspension and plate out the whole transformation on LB plates containing 50–100 μg/ml ampicillin.

Protocol 2

Mammalian Transient Transfection

This assay allows the sFv–antigen interaction to be studied in a mammalian host cell. This protocol was adapted from Griffiths et al. (1997) and Pörtner-Taliana et al. (2000).

MATERIALS

Buffers and Reagents

β-Galactosidase (RSV)
Growth medium
Lipofectin or Lipofectamine (LTI)
Luciferase assay (Promega)
OptiMEM
Perfect Lipid (pFx-7 for CHO cells; pFx-1 for COS cells) (Invitrogen)
Reporter lysis buffer (Promega)
Tropix β-gal assay (Bedford, Massachusetts)

Cells

CHO and COS cells

Plasmids

Lex-op TK pGL Basic luciferase plasmid
PcDNA3.1

Special Equipment

Incubator
Luminometer (Berthold Lumat LB9501)
Plates (6-well)

METHOD

1. The day prior to transfection, seed CHO and COS cells with 1×10^5 cells/well in a 6-well plate.

2. For each well, mix 75 µl of OptiMEM (LTI), 12 µl of Perfect Lipid (pFx-7 for CHO cells or pFx-1 for COS cells) with 1 µg of lex-op TK luciferase plasmid, 250 ng of RSV β-gal, and a low but constant bait plasmid concentration (e.g., 25 ng). The transfection reagents can be replaced by similar transfection reagents such as Lipofectin or Lipofectamine.

3. Add to increasing amounts of sFv-VP16 expression plasmid/well and add empty (e.g., bacterial plasmid DNA) to a final DNA concentration/well of 4 μg. Prepare duplicates or triplicates of every sample.

4. Mix well and incubate for 15 minutes at room temperature.

5. Replace the growth medium with serum-free medium or OptiMEM.

6. Add 75 μl of mixture per well.

7. Incubate for 4 hours in a 37°C incubator.

8. Replace the medium with growth medium.

9. Harvest 44 hours posttransfection by scraping cells into 150 μl of reporter lysis buffer.

10. Use 5–20 μl of sample for the luciferase assay using a kit and 10 μl of sample for the β-gal assay. Obtain readouts for both enzymes with a luminometer.

11. Normalize luciferase readouts to β-gal readouts.

> For the success of the experiment, it is essential to determine the optimal bait/prey ratio. We had good results starting with a low constant bait plasmid concentration and titering the prey plasmid concentration (compare to 2). Under these conditions, the normalized luciferase activity can increase 5–20-fold. This can vary depending on the cell line, bait prey ratio, and the strength of the particular protein–sFv interaction.

ACKNOWLEDGMENTS

This work was funded by a National Institutes of Health SBIR grant DK51418. We thank S. Hollenberg for providing the yeast prey and bait plasmids, and I. Serebriiskii and E.A. Golemis for providing the starting reagents of the dual-bait interaction trap system. The human sFv libraries were kind gifts from J. Marks and R. Finnern.

REFERENCES

Breeden L. and Nasmyth K. 1985. Regulation of the yeast HO gene. *Cold Spring Harbor Symp. Quant Biol.* **50:** 643–650.

Cattaneo A. and Biocca S. 1997. *Intracellular antibodies: Development and applications.* Springer, New York.

Cochet O., Kenigsberg M., Deluneau I., Virone-Oddos A., Multonm M.C., Fridman W.H., Schweighofer F., Teillaud J.L., and Tocque B. 1998. Intracellular expression of an antibody fragment-neutralizing p21 Ras promotes tumor regression. *Cancer Res.* **58:** 1170–1176.

Coloma M.J., Hastings A., Wims L.A., and Morrison S.L. 1992. Novel vectors for the expression of antibody molecules using variable regions generated by polymerase chain reaction. *J. Immunol. Methods* **152:** 89–104.

Coloma M.J., Larrick J.W., Ayala M., and Gavilondo-Cowley J.V. 1991. Primer design for the cloning for the immunoglobulin heavy-chain leader-variable regions from mouse hybridoma cells. *BioTechniques* **11:** 152–156.

De Jaeger G., Fiers E., Eeckhout D., and Depicker A. 2000. Analysis of the interaction between single-chain variable fragments and their antigen in a reducing intracellular environment using the two-hybrid system. *FEBS Lett.* **467:** 316–320.

De Jaeger G., Buys E., Eeckhout D., De Wilde C., Jacobs A., Kapila J., Angenon G., Van Montagu M., Gerats T., and Depicker A. 1999. High level accumulation of single-chain variable fragments in the cytosol of transgenic Petunia hybrida. *Eur. J. Biochem.* **259:** 426–434.

Duttweiler H.M. 1996. A highly sensitive and non-lethal beta-galactosidase plate assay for yeast. *Trends Genet.* **12:** 340–341.

Finnern R., Pedrollo E., Fisch I., Wieslander J., Marks J.D., Lockwood C.M., and Ouwehand W.H. 1997. Human autoimmune anti-proteinase 3 scFv from a phage display library. *Clin. Exp. Immunol.* **107:** 269–281.

Gietz D., St Jean A.S., Woods R.A., and Schiestl R.H. 1992. Improved method for high-efficiency transformation of intact yeast cells. *Nucleic Acid Res.* **20:** 1425–1431.

Gleeson M.A.G., White C.E., Meininger D.P., and Komives E.A. 1998. Generation of protease-deficient strains and their use in heterologous protein expression. *Methods Mol. Biol.* **103:** 81–94.

Golemis E.A. and Serebriiskii I. 1998. Two hybrid systems/ interaction trap. In *Cells: A laboratory manual,* Vol. 1. *Culture and biochemical analysis of cells* (ed. D.L. Spector et al.), pp. 69.1–69.40. Cold Spring Harbor Laboratory Press, Cold Spring Harbor, New York.

Golemis E.A., Gyuris J., and Brent R. 1996. Interaction Trap/two-hybrid system to identify interacting proteins. In *Current protocols in molecular biology* (ed. F.M. Ausubel et al.), pp. 20.1.1–20.1.28. Greene Publishing Associates and Wiley-Interscience, New York.

Griffiths T., Russell M., Froning K.F., Brown B.D., Scanlon S.M., Almazan M., Marcil R., and Hoeffler J.P. 1997. The PerFect Lipid optimizer kit for maximizing lipid-mediated transfection of eukaryotic cells. *BioTechniques* **22:** 982–987.

Gyuris J., Golemis E., Chertkov H., and Brent R. 1993. Cdi1, a human G1 and S phase protein phosphatase that associates with Cdk2. *Cell* **75:** 791–803.

Hill J., Donald K.A., and Griffiths D.E. 1991. DMSO-enhanced whole cell yeast transformation. *Nucleic Acid Res.* **19:** 5791–5796.

Hoogenboom H.R., Griffiths A.D., Johnson K.S., Chiswell D.J., Hudson P., Winter G. 1991. Multisubunit proteins on the surface of filamentous phage methodologies for displaying antibody (Fab) heavy and light chains. *Nucleic Acid Res.* **19:** 4133–4137.

Hwang C., Sinskey A.J., and Lodish H.F. 1992. Oxidized redox state of glutathione in the endoplasmic reticulum. *Science* **257:** 1496–1502.

Marks J.D., Hoogenboom H.R., Bonnert T.P., McCafferty J., Griffiths A.D., and Winter G. 1991. By-passing immunization. Human antibodies from V-gene libraries displayed on phage. *J. Mol. Biol.* **222:** 581–597.

Martineau P., Jones P., and Winter G. 1998. Expression of an antibody fragment at high levels in the bacterial cytoplasm. *J. Mol. Biol.* **280:** 117–127.

Mount R.C., Jordan B.E., and Hadfield C. 1996. Transformation of lithium–treated yeast cells and the selection of auxotrophic and dominant markers. *Methods Cell Mol. Biol.* **53:** 139–145.

Nicholls P.J., Johnson V.G., Blanford M.D., and Andrew S.M. 1993. An improved method for generating single-chain antibodies from hybridomas. *J. Immunol. Methods* **165:** 81–91.

Pörtner-Taliana A., Russell M., Froning K.F., Budworth P.R., Comiskey J.D., and Hoeffler J.P. 2000. In vivo selection of single-chain antibodies using a yeast two-hybrid system. *J. Immunol. Methods* **238:** 161–172.

Proba K., Worn A., Honegger A., and Plückthuhn A. 1998. Antibody scFv fragments without disulfide bonds made by molecular evolution. *J. Mol. Biol.* **275:** 245–253.

Richardson J.H., Sodroski J.G., Waldmann T.A., and Maresco W.A. 1995. Phenotypic knockout of the high-affinity human interleukin 2 receptor by intracellular single–chain antibodies against the α subunit of the receptor. *Proc. Natl. Acad. Sci.* **92:** 3137–3141.

Rondon I.J. and Marasco W.A. 1997. Intracellular antibodies (intrabodies) for gene therapy of infectious diseases. *Annu. Rev. Microbiol.* **51:** 267–283.

Schiestl R.H. and Gietz R.D. 1989. High efficiency transformation of intact yeast cells using single stranded nucleic acids as carrier. *Curr. Genet.* **16:** 339–346.

Serebriiskii I., Khazak V., and Golemis E.A. 1999. A two-hybrid dual bait system to discriminate specificity of protein interactions. *J. Biol. Chem.* **27:** 17080–17087.

Visintin M., Tse E., Axelson H., Rabbitts T.H., and Cataneo A. 1999. Selection of antibodies for intracellular function using a two-hybrid in vivo system. *Proc. Natl. Acad. Sci.* **96:** 11723–11728.

Vojtek A.B., Hollenberg S.M., and Cooper J. A. 1993. Mammalian Ras interacts directly with serine threonine kinase RAF. *Cell* **74:** 205–214.

Ward A.C. 1990. Single-step purification of shuttle vectors from yeast or high frequency back transformation into *E. coli. Nucleic Acid Res.* **18:** 5319.

25 Protein Interactions and Library Screening with Protein Fragment Complementation Strategies

Ingrid Remy,[1] Joelle N. Pelletier,[2] André Galarneau,[1] and Stephen W. Michnick[1]

[1]Département de Biochimie and [2]Département de Chimie, Université de Montréal, Montréal, Québec, H3C 3J7, Canada

INTRODUCTION

A first step in defining the function of a novel gene is to determine its interactions with other gene products in an appropriate context. Because proteins make specific interactions with other proteins as part of functional assemblies, an appropriate way to examine the function of the product of a novel gene is to determine its physical relationships with the products of other genes. This is the basis of the highly successful yeast two-hybrid system, which has been demonstrated to be effective for specific interactions or genome-wide screening for interacting proteins (Fields and Song 1989; Evangelista et al. 1996; Drees 1999; Vidal and Legrain 1999; Uetz et al. 2000; Walhout et al. 2000).

The central problem with two-hybrid screening is that detection of protein–protein interactions occurs in a fixed context, the nucleus of *Saccharomyces cerevisiae*, and the results of a screening must be validated as biologically relevant using other assays in appropriate cell, tissue, or organism models. Although this would be true for any screening strategy, it would be advantageous if library screening with tests for biological relevance could be combined into a single strategy, thus tentatively validating a detected protein as biologically relevant and eliminating false-positive interactions immediately. It was with these challenges in mind that our laboratory developed *p*rotein fragment *c*omplementation *a*ssays (PCA). In these assays, the gene for an enzyme is rationally dissected into two pieces. Fusion proteins are constructed with two proteins that are thought to bind to each other, each of which is fused to one of two probe fragments. Folding of the probe protein from its fragments is catalyzed by the binding of the test proteins to each other, and is detected as reconstitution of enzyme activity. We have demonstrated that the PCA strategy has the following capabilities: (1) It allows the detection of protein–protein interactions in vivo and in vitro in any cell type; (2) it allows the detection of protein–protein interactions in appropriate subcellular compartments or organelles; (3) it allows the detection of induced versus constitutive protein–protein interactions that occur in developmental, nutritional, environmental, or hormone-induced signals; (4) it allows the detection of the kinetic and equilibrium aspects of protein assembly in these cells; and (5) it allows the screening of protein-protein interactions in any cell type.

In addition to the specific capabilities described above, PCA has special qualities that make it appropriate for screening of molecular interactions. (1) PCAs are not a single assay but a series of assays; thus, an assay can be chosen because it works in a specific cell type appropriate for studying interactions of some class of proteins. (2) PCAs are inexpensive, requiring no specialized reagents beyond those necessary for a particular assay in the series and off-the-shelf materials and technology. (3) PCAs can be automated, allowing high-throughput screening. (4) PCAs are designed at the level of the atomic structure of the enzymes used; because of this, there is additional flexibility in designing the probe fragments to control the sensitivity and stringencies of the assays. (5) PCAs can be based on enzymes for which the detection of protein–protein interactions can be determined differently, including by dominant selection or production of a fluorescent or colored product. We have already developed six different PCA strategies based on survival-selection, colorimetric, or fluorescent outputs. Here, we discuss the most well-developed PCA, based on the enzyme murine dihydrofolate reductase (mDHFR), plus other assays based on TEM β-lactamase, green fluorescent protein (GFP), neomycin phosphotransferase, and hygromycin B phosphotransferase.

OUTLINE OF PROCEDURE

Basic Concepts

The selection of enzymes and design of PCAs have been discussed in detail previously (Michnick et al. 2000); here we review only the most basic ideas. Polypeptides have evolved to code for all of the chemical information necessary for them to fold spontaneously into a stable, unique, three-dimensional structure (Anfinsen et al. 1961; Gutte and Merrifield 1971; Anfinsen 1973). It logically follows that the folding reaction can be correctly driven by the interaction of two peptides that together contain the entire sequence in the appropriate order. This possibility was demon-

A) Interaction-directed folding from protein fragments (PCA)

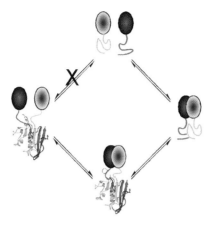

B) Weakly associating subunits

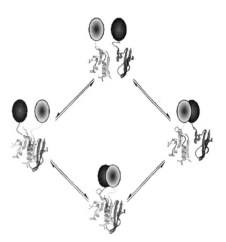

FIGURE 1. Two alternative strategies to achieve complementation. (A) The PCA strategy requires that unnatural peptide fragments be chosen that are unfolded prior to association of fused interacting proteins. This prevents spontaneous association of the fragments (pathway X) that can lead to a false signal. (B) Naturally occurring subunits that are already capable of folding can be mutated to interact with lower affinity. However, to some extent, this will always occur, requiring the selection of cells that express protein partner fusions at low enough levels that background is not detected.

strated in the classic experiments of Richards (1958) and Taniuchi and Anfinsen (1971). In practice this does not work easily; the major driving force for protein folding is the hydrophobic effect, which is also the driving force for nonspecific aggregation. However, correct folding can be favored over any other nonproductive process if soluble oligomerization domains are added to fragments that, by interacting, increase the effective concentration of the fragments (Johnsson and Varshavsky 1994; Pelletier and Michnick 1997; Pelletier et al. 1998). If the protein that folds from constitutive fragments is an enzyme whose activity can be detected in vivo, then the reconstitution of its activity can be used as a measure of interaction of the oligomerization domains (Fig. 1A). Furthermore, this binary, all-or-none folding event provides a very specific measure of protein interactions dependent not on mere proximity, but on the absolute requirement that the peptides must be organized precisely in space to allow for folding of the enzyme from the polypeptide chain. Thus, PCA probe proteins are dissected into fragments that are not capable of spontaneously folding along with their complementary fragment into a functional and complete protein in the absence of a heterologously provided dimerization function. These facts distinguish the PCA strategy from the complementation of naturally occurring and weakly associating subunits of enzymes (Rossi et al. 1997), in which some spontaneous assembly occurs, as illustrated in Figure 1B.

Applications

The PCA strategy is general in the sense that it is not represented by a single enzyme reporter, and has several different features. The question is not whether an assay can be used, but which assay or set of assays will help to address a specific question. Specific features and applications of PCAs are given in Table 1. An outline of the PCA strategy is presented in Figure 2. In some cases, a specific PCA can be used in different modes. For example, the DHFR PCA can be used as a simple survival-selection assay, or it can be used as a fluorescent assay allowing quantitative detection of

TABLE 1. Existing Protein-fragment Complementation Assays (PCA) and Their Applications

Enzyme	Molecular weight	Function	Assays	Demonstrated applications	Organism restrictions
DHFR	21 kD/ monomeric	reduces dihydrofolate to tetrahydrofolate	fluorescence survival selection	survival selection localization of protein interactions in living cells quantitation of induced associations in vivo translocation	none (universal)
β-Lactamase	27 kD/ monomeric	hydrolyzes β-lactam antibiotics (e.g., cephalosporin, ampicillin)	survival selection in bacteria fluorescence colorimetric	in vivo fluorescence (e.g., with CCf2/AM) in vitro colorimetric assay (nitrocefin)	none (universal)
GFP	28 kD/ monomeric	spontaneously fluorescent protein	intrinsic fluorescence	localization of protein interactions in living cells quantitation of induced associations in vivo translocation	none (universal)
GAR transformylase	30 kD/ monomeric	transfer of formyl group to glycineamide ribonucleotide	survival selection in bacteria	survival selection	restricted to bacterial auxo-trophs lacking GARTase
Aminoglycoside phosphotransferase	35 kD/ monomeric	phosphorylation of aminoglycosides (e.g., neomycin/G418) antibiotic	survival selection in many cell types (bacteria, yeast, mammalian, etc.)	survival selection	none (universal)
Hygromycin B phosphotransferase	35 kD/ monomeric	phosphorylation of hygromycin B	survival selection in many cell types (bacteria, yeast, mammalian, etc.)	survival selection	none (universal)

protein interactions and determination of the cellular location of protein interactions (Remy and Michnick 1999, 2001). Alternatively, the β-lactamase assay can be used as a very sensitive in vivo or in vitro quantitative detector of protein interactions as, unlike DHFR and GFP, one measures the continuous conversion of substrate to colored or fluorescent product. However, it should be noted that generation of a product by an enzyme does not always guarantee that the signal-to-background obtained is superior to that of fixed fluorophore reporters like GFP and fluorescein-conjugated methotrexate (fMTX) bound to DHFR. Observable signal-to-background depends, for example, on the quantum yield of the fluorophore, retention of fluorophore by a cell, the optical properties of the cells used, and the extent to which fluorophores are retained in individual cellular compartments. For instance, despite no enzymatic amplification, a DHFR fluorescence assay requires only between 1000 and 3000 molecules of reconstituted DHFR to distinguish a positive response clearly from background.

DHFR PCA

We have developed two types of assays based on DHFR PCA in mammalian cells: a survival-selection assay and a fluorescence assay (Remy and Michnick 1999). The principle of the survival DHFR PCA assay is that cells simultaneously expressing complementary fragments of DHFR

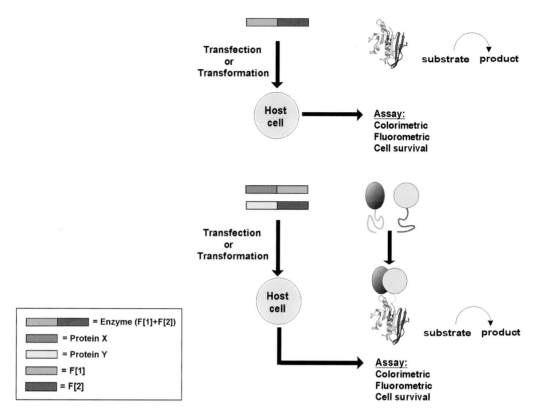

FIGURE 2. Protein fragment complementation assays (PCAs). The gene coding for an enzyme can be transformed/transfected into a host cell and its activity detected by an in vivo assay. Oligomerization domains (proteins X and Y) are fused to amino- or carboxy-terminal fragments of the enzyme. Cotransformation/transfection of oligomerization domain-fragment fusions results in reconstitution of enzyme activity by oligomerization domain-assisted reassembly of the enzyme. Reassembly of enzyme will not occur unless oligomerization domains interact.

(F[1,2] and F[3]) fused to interacting proteins or peptides will survive in media depleted of nucleotides. Reconstitution of DHFR activity can be monitored in vivo by cell survival in DHFR-negative cells (CHO-DUKX-B11, for example) grown in the absence of nucleotides. Alternatively, recessive selection can be achieved in DHFR-positive cells by using DHFR PCA fragments containing one or more of several mutations that render the refolded DHFR resistant to the antifolate drug methotrexate. The cells are then grown in the absence of nucleotides with selection for methotrexate resistance. The survival DHFR PCA is an extraordinarily sensitive assay. In mammalian cells, survival is dependent only on the number of molecules of DHFR reassembled, which we have determined to be ~25 molecules of DHFR per cell (Remy and Michnick 1999). Thus, the assay is sensitive enough to detect proteins expressed at extremely low levels. We have also demonstrated that cells containing interacting clones can be detected after being diluted in a background of 1 in 10^6 cells. Thus, this approach is particularly useful for simple screening procedures, where the goal is merely to ask the question of whether there is an interaction between proteins or not, and, in particular, for library screening, where a simple and robust assay is crucial.

The second approach, the fluorescence DHFR PCA assay, is based on the detection of fMTX binding to reconstituted DHFR. The principle of this assay is that complementary fragments of DHFR fused to interacting proteins, when expressed and reassembled in cells, will bind with high affinity ($K_d = 540$ pM) to fMTX in a 1:1 complex. fMTX is retained in cells by this complex, whereas the unbound fMTX is actively and rapidly transported out of the cells (Israel and Kaufman 1993; Kaufman et al. 1978). In addition, binding of fMTX to DHFR results in a 4.5-fold increase

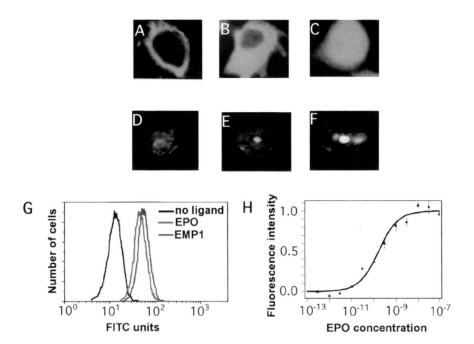

FIGURE 3. Applications of the DHFR PCA to detecting the localization of protein complexes and quantitating protein interactions. (*A–C*) Different protein pairs showing plasma membrane (*A*), cytosol (*B*), and whole-cell localization (*C*) in transiently transfected COS cells. (*D–F*) Cytosol and nuclear localization in potato protoplasts. Cytosol (*D*), nuclear localized (*E*), and DAPI (*F*) costaining of *E*. (*G*) FACS results of DHFR PCA. CHO cells expressing receptor for erythropoietin fused to complementary DHFR fragments. Receptor activation (conformation change) induced by erythropoietin (EPO) or a peptide agonist (EMP1) leads to an increase in fluorescence. (*H*) Dose–response curve for Epo-induced fluorescence as detected by FACS results in *G*.

in quantum yield. Bound fMTX and, by inference, reconstituted DHFR, can then be monitored by fluorescence microscopy, fluorescence-activated cell sorting (FACS), or spectroscopy (Remy and Michnick 1999, 2001; Remy et al. 1999). It is important to note that although fMTX binds to DHFR with high affinity, it does not induce DHFR folding from the two expressed PCA fragments. This is because the folding of DHFR from its fragments is obligatory for fMTX binding. The association of the fused oligomerization domains, by inducing the folding of the DHFR fragments, allows the creation of a binding site for fMTX. Therefore, the number of complexes observed as measured by number of fMTX molecules retained in the cell is a direct measure of the equilibrium number of protein complexes formed, independent of binding of fMTX (Remy and Michnick 1999, 2001; Remy et al. 1999). The other obvious application of the fluorescent assay is in determining the location in the cell of interactions, as illustrated in Figure 3. This is a feature shared with the GFP PCA assay. These are the only assays that allow visualization of the subcellular location of protein interactions in living cells. The protocols reported here describe the use of fMTX as the probe of DHFR reassembly. However, it should be noted that other fluorescent probe–MTX conjugates have recently become commercially available and may be simpler to use than fMTX (see step 5, Protocol 4).

Neomycin- and Hygromycin B-based PCAs

Although the DHFR survival PCA using an appropriate MTX-resistant mutant will work in virtually any cell type, one could imagine cases where perturbation of the purine synthesis pathway might be undesirable. We mention here two alternative survival PCAs that can be applied universally to any cell type. Aminoglycoside phosphotransferases (APH(3′)-IIIa) and hygromycin B phosphotransferase (APH(4)-Ia) genes, corresponding to neomycin- and hygromycin-B-resistance genes, respectively, are commonly used as dominant selectable markers in the selection of stable mammalian cell lines. Neomycin, kanamycin, and similar compounds, such as G418, inhibit protein synthesis in prokaryotes and eukaryotes. G418, for example, is an aminoglycoside antibiotic that blocks protein synthesis in eukaryotic cells by interfering with the function of 80S ribosomes (Bar-Nun et al. 1983; Eustice and Wilhelm 1984). The bacterial aminoglycoside phosphotransferases can inactivate these antibiotics by phosphorylation. Hygromycin B, another aminoglycoside antibiotic, has been reported to interfere with translocation (Cabanas et al. 1978; Gonzalez et al. 1978) and to cause mistranslation (Singh et al. 1979). The hygromycin B phosphotransferase gene also confers resistance to hygromycin B by coding for a kinase that inactivates the antibiotic through phosphorylation. We have developed two PCAs based on APH(3′)-IIIa and APH(4)-Ia that can be used as dominant selection assays in any cell types. The procedure is described below in the section Neomycin and Hygromycin B Survival Assays (Protocol 8, p. 472).

TEM β-Lactamase PCA

TEM-1 β-lactamase meets all of the essential criteria to be an excellent candidate for a PCA strategy. It is relatively small, monomeric, well-characterized structurally and functionally, can be readily expressed, and is not toxic in prokaryotes and eukaryotes. Of additional value, β-lactamase is a bacterial enzyme and has been deleted genetically from many standard *E. coli* strains. No orthologs or paralogs of β-lactamase exist in eukaryotes. Thus, a β-lactamase PCA can be used universally in eukaryotic cells and many prokaryotes without any intrinsic background. Second, assays are based on catalytic turnover of substrates with rapid accumulation of product. This enzymatic amplification allows relatively weak molecular interactions to be observed. Finally, the assay can be performed simultaneously or serially in a number of modes, including as an in vitro colorimetric assay, an in vivo fluorescence assay, or a bacterial survival assay.

Two substrates have been shown to work well with the β-lactamase-based PCA. The first one is the cephalosporin nitrocefin (O'Callaghan and Morris 1972). This substrate is used in an in vitro colorimetric assay. β-Lactamase has a k_{cat}/k_m of 1.7×10^4 mM^{-1}·s^{-1}. Substrate conversion can be easily observed by eye; the substrate is yellow in solution, whereas the product is a distinct ruby red color. The rate of hydrolysis can be monitored quantitatively with any spectrophotometer by measuring the appearance of red at 492 nm. As described below, this assay at present is performed with whole-cell lysates, as nitrocefin is not membrane-permeable. For in vivo fluorometric assay, we used the substrate CCF2/AM (Fig. 4) (Zlokarnik et al. 1998). Although it is not as good a substrate as nitrocefin (k_{cat}/k_m of 1260 mM^{-1}·s^{-1}), CCF2/AM contains butyryl, acetyl, and acetoxymethyl esters, allowing diffusion across the plasma membrane. Once internalized, cytoplasmic esterases catalyze the hydrolysis of its ester functionality, releasing the polyanionic (4 anions) β-lactamase substrate CCF2. Because of the negative charge of CCF2, the substrate becomes trapped in the cell. A fluorescence resonance energy transfer (FRET) can occur between a coumarin donor and fluorescein acceptor pair covalently linked to the cephalosporin core. The coumarin donor can be excited at 409 nm with emission at 447 nm, which is within the excitation envelope of the fluorescence acceptor (maximum around 485 nm), leading to remission of green fluorescence at 535 nm. When β-lactamase catalyzes hydrolysis of the substrate, the fluorescein moiety is eliminated as a free thiol. Excitation of the coumarin donor at 409 nm then emits blue fluorescence at 447 nm, whereas the acceptor (fluorescein) is quenched by the free thiol.

Standard Controls for a PCA Study

To assure that spontaneous fragment complementation does not occur, a set of controls can be performed with any pair of interacting proteins. These controls could include:

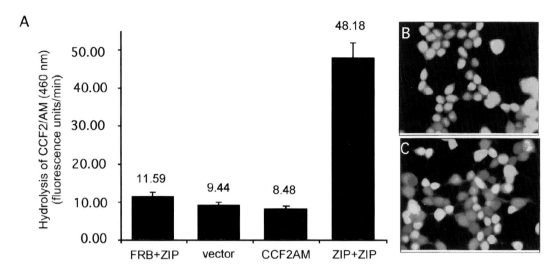

FIGURE 4. β-Lactamase PCA using the fluorescent substrate CCF2/AM. (A) The ZIP (GCN4 leucine zipper-forming sequences) are tested in HEK 293 cells as described in the text. FRB (rapamycin-FKBP binding domain of FRAP) is used as a negative control. "Vector" means the expression vector alone and ZIP + ZIP is the positive control. Data recorded in white microtiter plates on a Perkin Elmer HTS 7000 plate reader. (B,C) Fluorescent micrographs of cells expressing β-lactamase PCA showing negative response (*green* cells) (B, FRB + ZIP) or positive response (*blue* cells) (C, ZIP +ZIP).

1. *Noninteracting proteins.* A PCA response should not be observed if a protein that is not known to interact with either of two interacting proteins being tested is used as a PCA partner, nor should overexpression of this protein alone compete for the known interaction.

2. *Partner protein interface mutations.* Point or deletion mutation of a partner that is known to disrupt an interaction should also prevent a PCA response. These are the most crucial controls for demonstrating that an observed PCA response is due to specific association of two proteins.

3. *Competition.* A PCA response should be diminished by simultaneous overexpression of one of the other interacting proteins on another vector and not fused to a complementary PCA fragment.

4. *Fragment swapping.* An observed interaction between oligomerization domains should occur regardless of whether partner proteins are attached to one or the other PCA fragments.

Application of the DHFR PCA to In Vivo Library versus Library Screening for Interacting Proteins in Bacteria

The bacterial mutant DHFR PCA is, to date, the only strategy for which large-scale screening of library-versus-library has been reported (Pelletier et al. 1999; Arndt et al. 2000). Other PCAs are likely to be equally useful for robust library-versus-library screening. Currently, the main limitation is the size of the combinatorial space that can be represented, and this is limited by the efficiency of transformation. This strategy could have broad applications to protein design, to exploring the details of interactions between proteins or protein subdomains, and to a simple cDNA library screening strategy.

DHFR turnover is an absolute requirement for growth of all cells in the absence of complex nutrients, as the product of the reaction (tetrahydrofolate) is necessary for biosynthesis of (minimally) purines, thymidylate, serine, and methionine. A difference in trimethoprim sensitivity between the mammalian and the bacterial forms of DHFR provides the basis for metabolic selection, as the bacterial activity is selectively inhibited by concentrations of 1 µg/ml trimethoprim whereas the reconstituted murine enzyme remains fully active. Details of library and experimental design and statistical analysis are given in previously published work (Pelletier et al. 1999; Arndt et al. 2000).

As with any screening technique, it is necessary to introduce stringency of some sort and to select for maximal affinity and/or specificity. We have devised three selection strategies for the DHFR PCA in *E. coli*, each type having different levels of stringency. In the lowest stringency case, we screen two expressed libraries against each other beginning with a single-step selection, thereby identifying all interacting partners. In the second, we increase the stringency of the selection by using a mutant DHFR fragment (Ile114Ala) (see below), which reduces the stability of reassembled DHFR and should thus require more efficiently interacting partners to produce enough reconstituted enzyme for survival. Finally, we introduce competitive metabolic selection, where clones obtained in the second type of screening are pooled and passaged through several rounds of competition selection to enrich for optimally interacting partners.

By simultaneously screening two libraries against each other, we have demonstrated the advantages of screening a large, combinatorial sequence space in identifying stably heterodimerizing pairs. We partially sampled a sequence space of 1.72×10^{10} combinations to select novel leucine zipper pairs with characteristics consistent with stable and specific heterodimerization (Pelletier et al. 1999; Arndt et al. 2000).

Varying the Stringency of Selection by Use of the Ile114 Mutants

The interface between the two fragments of DHFR is composed in part of a hydrophobic β-sheet formed mostly by fragment-[3] but which includes a β-strand from fragment-[1,2]. We designed the mutations Ile114Val, Ile114Ala, and Ile114Gly, which sequentially reduce the volume of a side chain that is directly at the fragment interface but distant from the active site, according to the structure of the highly conserved avian enzyme (Volz et al. 1982). We hypothesized that these mutations should decrease the association of the fragments by destabilizing the hydrophobic contacts at the interface, with an increasingly severe effect (Val>Ala>Gly). Indeed, the mutations impair growth of bacteria expressing reconstituted mDHFR in the predicted fashion. In conjunction with a dimerization function provided by a homodimerizing GCN4 coiled-coil, there was no significant difference in growth rate with the valine mutant versus the Ile wild type. However, the rate was 2.5 times slower with the alanine mutant (as quantitated in liquid culture), whereas no growth was observed with the glycine mutant (Pelletier et al. 1998). Thus, the severity of the effect correlates with the difference in side-chain volume.

We hypothesized that to restore the bacterial growth rate to maximal levels when using a mutant fragment, it would be necessary to provide partner proteins that interact more stably and heterospecifically. Thus, the increased interaction between the partner-proteins should improve the reconstitution of the destabilized mDHFR. We tested this hypothesis in the selection of coiled-coil partner-proteins from semirandomized libraries. We immediately observed a 50-fold increase in selection stringency when using the Ile114Ala mutant. The features in the resulting coiled-coils were consistent with selection for more stable pairs with a higher requirement for parallel, in-register heterodimerization than in selection using the wild-type fragments (Pelletier et al. 1999; Arndt et al. 2000). Thus, the wild-type and Ile114 mutant resulted in differing selection stringencies, where destabilization of the mDHFR fragment interface was compensated by more stably and specifically interacting partners. The Ile114Val mutant could also be applied to selection strategies, and should yield a stringency intermediate between the wild type and the Ile114Ala mutant. In principle, it should also be possible to apply the Ile114Gly mutant to obtain yet higher selection stringency. However, as we have not yet observed enzymatic activity when using this mutant with a variety of partner proteins, it is possible that the interface is too perturbed in this mutant to allow fragment reconstitution.

Application of Metabolic Competition

The identification of the best-performing enzymes from a population by functional selection is a straightforward process using a similar set of reagents. We have used this approach in the selection of the best-performing reconstituted mDHFRs. As a result of the absolute requirement of cells for the product of DHFR substrate to product turnover, a limiting level of DHFR activity determines the growth rate. In a mixed clonal population where the mDHFR in each clone is reconstituted by a different interacting partner pair, the most efficiently interacting partners should allow higher DHFR turnover, leading to a higher growth rate for this clone. Therefore, in liquid culture, one can enrich for the fastest growing bacteria, which express the most efficiently interacting pairs. We have demonstrated a direct link between the rate of clonal enrichment and stability and specificity of partner association by analyzing the results of such a metabolic "competition selection" of coiled-coil partner-proteins from semirandomized libraries (Pelletier et al. 1999; Arndt et al. 2000). We observe a rapid convergence toward predominant populations of clones. Practically, we could screen 2.0×10^6 of these, and at the final passage (P12), there emerged a predominant clone representing 82% of the coiled-coil pairs sequenced. The features in the resulting coiled-coils were consistent with selection for more stable pairs than before competition, with a large increase in hetero- versus homospecificity.

Bacterial DHFR PCA Survival Assay

Before discussing library screening, one should become acquainted with the system using known, interacting proteins.

MATERIALS

CAUTION: See Appendix for appropriate handling of materials marked with <!>.

Buffers and Reagents

Ampicillin (100 µg/ml)
Isopropylthiogalactoside (1 mM) (IPTG) <!>
Kanamycin (50 µg/ml)
M9 minimal medium and rich medium (LB Broth)
Trimethoprim (1 µg/ml) (Sigma)

Vectors and Bacteria

Bacterial expression vector pQE-32 (Qiagen)
Escherichia coli strain BL21 harboring the lacIq plasmid pREP4 (Qiagen)

Special Equipment

Bacterial petri dishes

METHOD

1. Cotransform BL21 bacterial cells with vectors containing the DHFR fragment fusions X-F(1,2) and Y-F(3), where X and Y are two specific target proteins, into the pQE-32 bacterial expression vector. To obtain maximal yields in transformation, electroporate the bacteria. Then place the bacteria immediately into rich medium (LB broth) for 30–45 minutes, incubate, and wash once with M9 minimal medium to reduce carryover of complex nutrients.

2. Plate the washed cells onto selective medium consisting of M9 minimal medium with agar + 1 µg/ml trimethoprim, 1 mM IPTG, 100 µg/ml ampicillin, and 50 µg/ml kanamycin. Incubate the plates at 30°C. Colonies are observed after ~24–48 hours of incubation. The colonies can be directly picked and individually propagated for DNA sequencing.

In Vivo Library versus Library Screening for Interacting Proteins: Metabolic Competition Selection in Bacteria

It is important to determine the signal-to-background obtained, which depends on factors such as the quantum yield of the fluorophore, retention of the fluorophore by the cell, the optical properties of the cells used, and the extent to which fluorophores are retained in individual cellular compartments.

MATERIALS

CAUTION: See Appendix for appropriate handling of materials marked with <!>.

Buffers and Reagents

Ampicillin (100 µg/ml)
Isopropylthiogalactoside (1 mM) (IPTG) <!>
Kanamycin (50 µg/ml)
M9 minimal medium and rich medium (LB Broth)
Trimethoprim (1 µg/ml) (Sigma)

Vectors and Bacteria

Bacterial expression vector pQE-32 (Qiagen)
Escherichia coli strain BL21 harboring the lacIq plasmid pREP4 (Qiagen)

Special Equipment

Bacterial petri dishes

METHOD

1. Pool the colonies arising from the bacterial survival assay (given in Protocol 1; where X and Y correspond to two designed peptide libraries) by scraping the colonies into a small volume of medium. Incubate the resulting liquid with shaking (250 rpm) for 10 minutes to disrupt clumps. One portion will be used for selection; a second portion is plated for sequence analysis of individual clones arising from single-step selection (passage 0, or P0). Store the remaining volume with 15% glycerol at –80°C.

2. Calculate the volume of selective medium (described above) and pooled cells required for the selection, according to (a) the number of viable cells present and (b) the number of unique clones that should be represented in the competition.

Example: If the OD_{600} represents 4×10^8 viable cells/ml for the strain in use, and there are 4×10^5 unique clones that should be represented, then 1 ml of a pool at OD_{600} must be used to represent each clone once (on average), or 20 ml used to overrepresent each clone by a factor of 20. For the starting OD of the competition selections to be 0.001, the volume of medium used is 20 ml. Determine the OD_{600} of the pooled colonies and use an appropriate aliquot to inoculate the medium.

3. Propagate the cells until OD_{600} is between 0.1 and 1.0. Take an aliquot (passage 1, or P1) and use it to inoculate fresh medium, again taking into account the number of clones to be represented. Plate a second portion for analysis of individual clones and store the remainder at –80°C. The final OD of each passage should be no higher than 1.0, because too high a cell density results in a high enough concentration of complex metabolites in the medium for cells to propagate without the requirement for DHFR turnover. We routinely observe clonal enrichment within the first two to four passages when the final OD is approximately 0.5. The colonies can be directly picked and individually propagated for DNA sequencing.

> We find that, during the first one or two passages, the OD of inoculation should be no lower than 0.001. Otherwise, the bacteria are unable to propagate in the amount of time during which the antibiotics remain effective. However, in subsequent passages, we could decrease the OD of inoculation to 0.0005 or 0.0001, allowing a more rapid clonal enrichment. In principle, if the number of clones to be represented is small, competition selection can be performed in very small volumes. However, one may use a larger volume to allow direct sequencing or restriction analysis of the DNA from the passage. If keeping the volume low is an important consideration, a small volume of each passage can be plated and the resulting colonies analyzed.

Protocol 3

Mammalian DHFR PCA Survival Assay

This assay determines that cells simultaneously expressing complementary DHFR fused to interacting proteins or peptides will survive in media depleted of nucleotides.

MATERIALS

Buffers and Reagents

Adenosine (Sigma)
Desoxyadenosine (Sigma)
Dialyzed fetal bovine serum (FBS) (10%) (Hyclone)
Lipofectamine reagent (Life Technologies)
Minimum essential medium: α medium without ribonucleosides and deoxyribonucleosides (α-MEM) (Life Technologies)
Thymidine (Sigma)
Trypsin-EDTA (Life Technologies)

Special Equipment

Cloning cylinders (Scienceware)
Plates (6-well and 12-well tissue-culture-treated) (Corning Costar)

METHOD

1. Split CHO DUKX-B11 cells (DHFR-negative; could also be done in other DHFR-deficient cells) 24 hours before transfection at 1×10^5 cells per well in 12-well plates in α-MEM medium enriched with 10% dialyzed FBS and supplemented with 10 µg/ml each of adenosine, desoxyadenosine, and thymidine.

   Alternatively, recessive selection can be achieved in other cell lines by using DHFR fragments containing one or more of several mutations (e.g., an F31S mutation; see below) that reduce the affinity of refolded DHFR to the antifolate drug methotrexate, and growing cells in the absence of nucleotides with selection for methotrexate resistance.

2. Cotransfect cells with two mammalian expression plasmids (pCDNA3.1, Invitrogen) containing the PCA fusion partners using Lipofectamine reagent according to the manufacturer's instructions.

   The best orientations of the fusions for the DHFR PCA are: protein A-DHFR[1,2] + protein B-DHFR[3] or DHFR[1,2]-protein A + protein B-DHFR[3], where proteins A and B are the proteins to test for interaction. We typically insert a 10-amino-acid flexible polypeptide linker consisting of (Gly.Gly.Gly.Gly.Ser)$_2$ between the protein of interest and the DHFR fragment (for both fusions). DHFR[1,2] corresponds to amino acids 1–105, and DHFR[3] corresponds to amino acids 106–186 of murine DHFR. The DHFR[1,2] fragment that we use also contains a phenylalanine-to-serine mutation at position 31 (F31S), rendering the reconstituted DHFR resistant to methotrexate (MTX) treatment. We usually use pcDNA3.1 as a mammalian expression vector (Invitrogen).

3. Forty-eight hours after the beginning of the transfection, split the cells at ~5 × 10^4 cells per well in six-well plates in selective medium consisting of α-MEM enriched with dialyzed FBS but without addition of nucleotides.

   It is crucial that cell density is kept to a minimum and cells are well separated when split, to avoid cells "harvesting" nutrients from adjacent cells on dense plates. Otherwise, colonies might appear to be forming from clumps of cells that were not sufficiently separated during the splitting procedure.

   The choice of dialyzed FBS manufacturer is crucial. Cells need very little nucleotide in the medium to propagate, and this will result in false positives. Hyclone dialyzed FBS has proven a particularly reliable source.

4. Change the medium every 3 days. The appearance of distinct colonies usually occurs after 4–10 days of incubation in selective medium. Colonies are observed only for clones that simultaneously express both interacting proteins fused to one or the other complementary DHFR fragments. Only cells expressing interacting proteins will be able to achieve normal cell division and colony formation.

For further analysis of the interacting protein pair:

5. Isolate 3–5 colonies per interacting partners by trypsinization (trypsin-EDTA) using cloning cylinders and grow them separately.

6. Select the best-expressing clone by immunoblot (western blot) or using the DHFR PCA fluorescence assay (see Protocol 4). Amplify the expressed gene using methotrexate resistance, if desired, to obtain clones with increased expression (Kaufman 1990).

7. Carry out functional analysis of the clone stably expressing your interacting proteins pair fused to the complementary DHFR fragments by using the DHFR PCA fluorescence assay (Protocol 4).

Mammalian DHFR PCA Fluorescence Assays

This protocol is based on the detection of fMTX binding to reconstituted DHFR. The number of complexes observed, measured by the number of fMTX molecules retained in the cell, is a direct measure of the equilibrium number of protein complexes formed independent of binding of fMTX. This protocol can also be used to determine the location in the living cell of protein interactions.

MATERIALS

CAUTION: See Appendix for appropriate handling of materials marked with <!> .

Buffers and Reagents

Bio-Rad protein assay (Bio-Rad)
Cosmic calf serum (Hyclone)
Dialyzed fetal bovine serum (Hyclone)
Dimethyl formamide (DMF) <!>
Dulbecco's modified Eagle medium (DMEM) (Life Technologies)
Dulbecco's phosphate-buffered saline (PBS) (Life Technologies)
Fluorescein-conjugated methotrexate (fMTX) (Molecular Probes) <!>
Geltol aqueous mounting medium (Immunon)
Lipofectamine Plus reagent (Life Technologies)
Minimum essential medium: α medium without ribonucleosides and deoxyribonucleosides (α-MEM) (Life Technologies)
Trypsin-EDTA (Life Technologies)

Cells

COS
CHO DUKX-B11

Vectors

pCDNA3.1 mammalian expression plasmid (Invitrogen)

Special Equipment

Black microtiter plates (96-well) (Dynex no. 7805, VWR Scientific)
Micro cover glasses (18-mm circles, no. 2) (VWR Scientific)
Microscope slides (glass, 25 x 75 x 1.0 mm)
Microtiter plate reader (Perkin-Elmer HTS 7000 Series BioAssay Reader)
Plates (12-well tissue-culture-treated) (Corning Costar)

METHOD

Fluorescence Microscopy

For transiently transfected cells:

1. Split COS cells (this assay can be used with any other cell line) 24 hours before transfection to create a seed density of 1×10^5 cells per well on 18-mm circle glass coverslips placed in 12-well plates in DMEM medium enriched with 10% Cosmic calf serum.

   The fluorescence DHFR PCA assay is universal and in theory can be used in any cell type or organism. This assay has already been shown to work in several mammalian cell lines as well as in plant cells and insect cells.

2. Cotransfect cells transiently with mammalian expression plasmids (pCDNA3.1, Invitrogen) containing the PCA fusion partners using Lipofectamine reagent according to the manufacturer's instructions.

   The best orientations of the fusions for the DHFR PCA are: protein A-DHFR[1,2] + protein B-DHFR[3] or DHFR[1,2]-protein A + protein B-DHFR[3], where proteins A and B are the proteins to test out for interaction. We typically insert a 10-amino-acid flexible polypeptide linker consisting of (Gly.Gly.Gly.Gly.Ser)$_2$ between the protein of interest and the DHFR fragment (for both fusions). DHFR[1,2] corresponds to amino acids 1–105, and DHFR[3] corresponds to amino acids 106–186 of murine DHFR. The DHFR[1,2] fragment that we use also contains a phenylalanine-to-serine mutation at position 31 (F31S), rendering the reconstituted DHFR resistant to methotrexate (MTX) treatment. We usually use pcDNA3.1 as a mammalian expression vector (Invitrogen).

3. The next day, change medium and add fMTX to the cells to a final concentration of 10 μM.

   A stock solution of 1 mM fMTX should be prepared as follows: Dissolve 1 mg of fMTX in 1 ml of DMF. To facilitate the dissolution, incubate 15 minutes at 37°C and mix by vortexing every 5 minutes. Protect the tube from light. Keep at –20°C.

For stable cell lines:

For CHO DUKX-B11 cells (or other cell line) stably expressing PCA fusion partners, seed cells to ~2×10^5 cells per well on 18-mm glass coverslips in 12-well plates in α-MEM medium enriched with 10% dialyzed FBS. The next day, add fMTX to the cells at a final concentration of 10 μM.

4. Incubate cells with fMTX 22 hours at 37°C, remove the medium, and wash the cells with 1× PBS. Re-incubate for 15–20 minutes at 37°C in the culture medium to allow for efflux of unbound fMTX. Remove the medium and wash the cells four times with cold 1× PBS on ice and finally mount the coverslips on microscope glass slides with an aqueous mounting medium.

   Complementary fragments of DHFR fused to interacting protein partners, when expressed and reassembled in cells, will bind with high affinity (K_d = 540 pM) to fMTX in a 1:1 complex. fMTX is retained in cells by this complex, while the unbound fMTX is actively and rapidly transported out of the cells.

5. Perform fluorescence microscopy on live cells.

   All of the work reported to date has been performed in live cells (Remy and Michnick 1999, 2001; Remy et al. 1999). Although cells can be fixed, there is a significant reduction in observable fluorescence.

   Microscopic visualization must be performed within 30 minutes after the wash procedure. If the negative control (untransfected cells treated with fMTX) is too fluorescent, modify the wash procedure.

Particular attention must be given to optimizing the fMTX load and "wash" procedures. Important variables include the time of loading, temperatures at which each wash step is performed, the number and length of wash steps, and the time between washing and detection. Too little washing will mean that background cannot be distinguished from a positive result. Scrutinize the relevant parameters in the same sense as for, say, a western blot. Results may also vary with the way the cells are plated and the types of cells used. Generally, as in other fluorescent microscopy procedures, variations in the shape of cells and the localization of the fluorophore will result in better or worse results. For stable cell lines, the intensity of fluorescence will also depend on the levels of expression of the fusion proteins.

6. fMTX bleaches very rapidly (within 15 seconds); therefore, do not spend too much time scanning the cells before images are captured. Rather, quickly capture images of cells with between 0.5-second and 4-second exposures as soon as a green-fluorescing cell is identified.

Other fluorescent dye–MTX conjugates have recently become commercially available but have not been tested. Conditions for loading and washing cells must also be optimized for these dyes. Alexa Fluor-, BODIPY-, and Texas Red-MTX conjugates are sold by Molecular Probes Inc. Although they have lower quantum yields than fluorescein, these dyes fluoresce with considerably more persistence, and thus cells can be studied by eye at a more leisurely pace.

Flow Cytometry Analysis

Preparation of cells for fluorescence-activated cell sorting (FACS) analysis is the same as described for fluorescence microscopy, except that following the last 1× PBS wash at step 4 (just two times in this case), cells are gently trypsinized (trypsin-EDTA), suspended in 500 µl of cold 1× PBS, and kept on ice prior to flow cytometric analysis within 30 minutes. Data are collected on a FACS analyzer with stimulation with an argon laser tuned to 488 nm with emission recorded through a 525-nm bandwidth filter.

Fluorometric Analysis

Preparation of cells for fluorometric analysis is the same as described for fluorescence microscopy, except that following the last 1× PBS wash at step 4 (just two times in this case), cells are gently trypsinized (trypsin-EDTA). Plates are put on ice and 100 µl of cold PBS is added to the cells. The total cell suspensions are transferred to 96-well black microtiter plates and kept on ice prior to fluorometric analysis. The assay can be performed on any fluorescence microtiter plate reader; we use a Spectra MAX GEMINI XS (Molecular Devices) Reader in the fluorescence mode. The excitation and emission wavelengths for the fMTX are 497 nm and 516 nm, respectively. Afterward, the data are normalized to total protein concentration in cell lysates (Bio-Rad protein assay).

GFP PCA Fluorescence Assay

This assay allows visualization of the subcellular location of protein interactions in living cells.

MATERIALS

Buffers and Reagents

> Bio-Rad protein assay
> Cosmic calf serum (Hyclone)
> Dulbecco's modified Eagle medium (DMEM) (Life Technologies)
> Dulbecco's phosphate-buffered saline (PBS) (Life Technologies)
> Geltol aqueous mounting medium (Immunon)
> Lipofectamine reagent (Life Technologies)
> Trypsin-EDTA (Life Technologies)

Special Equipment

> Black microtiter plates (96-well) (Dynex no. 7805, VSW Scientific)
> FACS analyzers
> Fluorescence microscopy equipment
> Micro cover glasses (18-mm circles, no. 2) (VWR Scientific)
> Microscope slides (glass, 25 × 75 × 1.0 mm)
> Microtiter plate reader (Perkin-Elmer HTS 7000 Series BioAssay Reader)
> Plates (12-well, tissue-culture-treated) (Corning Costar)

METHOD

All procedures described for the DHFR PCA fluorescence assays are the same for GFP PCA, except that there is no use of fMTX and there are no washing steps. The wash procedure is obviously irrelevant in the case of the GFP PCA, where the folded/reassembled protein is a fluorophore itself.

In Vitro β-Lactamase PCA Colorimetric Assay

β-Lactamase PCA can be used universally in eukaryotic cells and many prokaryotic cells without any intrinsic background. Because these arrays are based on catalytic turnover with rapid accumulation of product, relatively weak molecular interactions can be observed.

MATERIALS

Buffers and Reagents

Cosmic calf serum (10%) (Hyclone)
Dulbecco's modified Eagle medium (DMEM) (Life Technologies)
Dulbecco's phosphate-buffered saline (100 mM) (Life Technologies)
Fugene 6 transfection reagent (Roche Diagnostics)
Nitrocefin (10 nM) (Becton Dickinson Microbiology Systems)
Trypsin-EDTA (Life Technologies)

Plasmids and Cells

COS or HEK293 cells
pCDNA3.1 mammalian expression plasmid (Invitrogen)

Special Equipment

Microtiter plate reader (Perkin-Elmer HTS 7000 Series BioAssay Reader)
Plates (12-well tissue-culture-treated) (Corning Costar)
Plates (96-well) (Corning Costar)

METHOD

1. Split COS or HEK 293 cells (this assay can be used with essentially any cell line) 24 hours before transfection at 1×10^5 cells per well in 12-well plates in DMEM medium enriched with 10% Cosmic calf serum.

2. Transiently cotransfect cells with mammalian expression plasmids (pCDNA3.1, Invitrogen) containing the PCA fusion partners to β-lactamase using Fugene 6 Transfection reagent according to the manufacturer's instructions.

    The best orientations of the fusions for the β-lactamase PCA are: protein A-BLF[1] + protein B-BLF[2] or BLF[1]-protein A + protein B-BLF[2], where proteins A and B are the proteins to test out for interaction. We typically insert a 15-amino-acid flexible polypeptide linker consisting of (Gly.Gly.Gly.Gly.Ser)$_3$ between the protein of interest and the β-lactamase fragment (for both fusions). BLF[1] corresponds to amino acids 26–196 (Ambler numbering), and BLF[2] corresponds to amino acids 198–290 of TEM-1 β-lactamase. We usually use pcDNA3 as a mammalian expression vector (Invitrogen).

3. At 48 hours after transfection, wash the cells three times with cold PBS, resuspend in 300 μl of cold PBS, and keep on ice. Then centrifuge the cells for 30 seconds at 4°C, discard the supernatant, and resuspend the cells in 100 μl of 100 mM cold PBS (pH 7.4) (β-lactamase reaction buffer).

4. Freeze cell samples in dry ice/ethanol for 10 minutes and thaw in a water bath at 37°C for 10 minutes. Repeat this freeze and thaw cycle two more times. Remove cell membrane and debris by centrifugation (10,000*g*) for 5 minutes at 4°C. Collect the supernatant whole-cell lysate and store at –20°C until assays are performed.

5. Perform assays in 96-well microtiter plates. For testing β-lactamase activity, allocate 100 μl of 100 mM PBS (pH 7.4) into each well. To this add 78 μl of H_2O and 2 μl of 10 mM Nitrocefin (final concentration of 100 μM). Finally, add 20 μl of unfrozen cell lysate (step 4) (final buffer concentration of 60 μM).

   The assay can be performed on any microtiter plate reader.

   We use a Perkin-Elmer HTS 7000 Series Bio Assay Reader in the absorption mode with a 492-nm measurement filter.

Protocol 7

In Vivo β-Lactamase PCA Enzymatic Assay

This assay is a very sensitive quantitative detector of protein interactions.

MATERIALS

CAUTION: See Appendix for appropriate handling of materials marked with <!> .

Buffers and Reagents

CCF2-AM: 1 mM stock solution in DMSO <!> (Aurora)
Cosmic calf serum (Hyclone)
Dulbecco's modified Eagle medium (DMEM) (Life Technologies)
Dulbecco's phosphate-buffered saline (PBS) (Life Technologies)
Fugene 6 transfection reagent (Roche Diagnostics)
Normal saline
   140 mM NaCl
   5 mM KCl <!>
   2 mM $CaCl_2$
   10 mM HEPES
   6 mM sucrose
   10 mM glucose (pH 7.35)
Physiological saline solution
   10 mM HEPES
   6 mM sucrose
   10 mM glucose
   140 mM NaCl
   5 mM KCl
   2 mM $MgCl_2$
   2 mM $CaCl_2$ (pH 7.35) <!>
Trypsin-EDTA (Life Technologies)

Plasmids and Cells

COS or HEK293 cells
pCDNA3.1 mammalian expression plasmid (Invitrogen)

Special Equipment

Fluorescence microscopy equipment (see note following step 7)
Glass coverslip (5-mm)
Microtiter plate reader (Perkin-Elmer HTS 7000 Series BioAssay Reader)
Plates (12-well tissue-culture-treated) (Corning Costar)
White microtiter plates (96-well) (Dynex no. 7905, VWR Scientific)

470

METHOD

1. Split COS or HEK293 cells 24 hours before transfection to 1×10^5 cells per well in 12-well plates in DMEM enriched with 10% Cosmic calf serum.

2. Cotransfect cells transiently with mammalian expression plasmids (pCDNA3.1, Invitrogen) containing the PCA fusion partners using Fugene 6 Transfection Reagent according to the manufacturer's instructions.

   The best orientations of the fusions for the β-lactamase PCA are: protein A-BLF[1] + protein B-BLF[2] or BLF[1]-protein A + protein B-BLF[2], where proteins A and B are the proteins to test out for interaction. We typically insert a 15-amino-acid flexible polypeptide linker consisting of (Gly.Gly.Gly.Gly.Ser)$_3$ between the protein of interest and the β-lactamase fragment (for both fusions). BLF[1] corresponds to amino acids 26–196 (Ambler numbering), and BLF[2] corresponds to amino acids 198–290 of TEM-1 β-lactamase. We usually use pcDNA3 as a mammalian expression vector (Invitrogen).

3. At 24 hours after transfection, split the cells again to assure 50% confluency (1.5×10^5). (The maximum loading efficiency of CCF2-AM is observed at 50% confluence.) Split the cells either onto 12-well plates for suspension enzymatic assay or onto 15-mm glass coverslips for fluorescence microscopy.

4. At 48 hours after transfection, wash the cells three times with PBS to remove all traces of serum.

   Serum may contain esterases that can destroy the substrate.

5. Load the cells with 1 μM of CCF2/AM diluted into a physiologic saline solution for 1 hour.

For in vivo enzymatic assay:

6. Wash the cells twice with the physiologic saline solution. Resuspend the cells into the same solution and aliquot 1×10^6 cells into a 96-well fluorescence white plate.

   The assay can be performed on any fluorescence microtiter plate reader; we use a Spectra MAX GEMINI XS (Molecular Devices) reader in the fluorescence mode. Blue fluorescence is detected with a 409-nm excitation filter and a 465-nm emission filter.

For fluorescence microscopy:

7. Wash the cells twice with the physiologic saline. Examine the cells directly under the microscope.

   We perform fluorescence microscopy on live HEK 293 or COS cells with an inverse Nikon Eclipse TE-200 (objective plan fluor 40× dry, numerically open at 0.75). Images were taken with a digital CCD cooled (–50°C) camera, model Orca-II (Hamamatsu Photonics; exposure for 1 second, binning of 2 × 2 and digitalization 14 bits at 1.25 MHz). Source of light is a Xenon lamp Model DG4 (Sutter Instruments). Emission filters are changed by an emission filter switcher (model Quantoscope) (Stranford Photonics). Images are visualized with ISee software (Inovision Corporation) on an O2 Silicon Graphics computer. The following selected filters are used: Filter set #31016 (Chroma Technologies); Excitation filter: 405 nm (passing band of 20 nm); Dichroic Mirror: 425 nm DCLP; Emission filter #1: 460 nm (passing band of 50 nm); Emission filter #2: 515 nm (passing band of 20 nm).

Neomycin and Hygromycin B Survival Assays

These PCAs are based on APH(3′)-IIIa and APH(4)-Ia and can be used as dominant selection assays in any cell types.

MATERIALS

Buffers and Reagents

F-12 medium (Life Technologies)
Fetal bovine serum (FBS) (Hyclone)
G418 (400 µg/ml)
Hygromycin B (250 µg/ml)
Lipofectamine reagent (Life Technologies)
Trypsin-EDTA (Life Technologies)

Plasmids and Cells

CHO cells
pCDNA3.1 mammalian expression plasmid (Invitrogen)

Special Equipment

Cloning cylinders (Scienceware)
Plates (6-well and 12-well tissue-culture-treated) (Corning Costar)

METHOD

1. Split CHO cells (or other cell lines) 24 hours before transfection at 1×10^5 cells per well in 12-well plates in F-12 medium enriched with 10% FBS.

2. Cotransfect cells with mammalian expression plasmids (pCDNA3.1, Invitrogen) containing the PCA fusion partners using Lipofectamine reagent according to the manufacturer's instructions.

   The best orientations of the fusions for the neomycin and hygromycin B PCA are: F[1]-protein A + protein B-F[2], where proteins A and B are the proteins to test out for interaction. We typically insert a 10-amino-acid flexible polypeptide linker consisting of (Gly.Gly.Gly.Gly.Ser)$_2$ between the protein of interest and the PCA fragment (for both fusions). For the design of the PCA fragments, we dissected the two enzymes between the ATP-binding and catalytic domains, corresponding to Gly-99 for the neomycin PCA and Glu-108 for the hygromycin B PCA. We usually use pcDNA3.1 as a mammalian expression vector (Invitrogen).

3. At 48 hours after the beginning of the transfection, split cells at ~5 × 10^4 cells per well in 6-well plates in selective medium consisting of F-12/FBS with addition of 400 μg/ml G418 or 250 μg/ml hygromycin B for the neomycin and hygromycin PCA, respectively.

4. Change medium every 3 days. The appearance of distinct colonies usually occurs after 10–14 days of incubation in selective medium. Colonies are observed only for clones that simultaneously express both interacting proteins fused to one or the other complementary PCA fragments.

For further analysis of the interacting protein pair:

5. Isolate three to five colonies per interacting partners by trypsinization (trypsin-EDTA) using cloning cylinders, and grow them separately.

6. Select the best-expressing clone by immunoblot (western blot). Amplify afterward, if desired, using any standard protocol, to obtain clones with increased expression.

REFERENCES

Anfinsen C.B. 1973. Principles that govern the folding of protein chains. *Science* **181:** 223–230.

Anfinsen C.B., Haber E., Sela M., and White Jr., F.H. 1961. The kinetics of formation of native ribonuclease during oxidation of the reduced polypeptide chain. *Proc. Natl. Acad. Sci.* **47:** 1309–1314.

Arndt K.M., Pelletier J.N., Muller K.M., Alber T., Michnick S.W., and Pluckthun A. 2000. A heterodimeric coiled-coil peptide pair selected in vivo from a designed library-versus-library ensemble. *J. Mol. Biol.* **295:** 627–639.

Bar-Nun S., Shneyour Y., and Beckmann J.S. 1983. G-418, an elongation inhibitor of 80 S ribosomes. *Biochim. Biophys. Acta* **741:** 123–127.

Cabanas M.J., Vazquez D., and Modolell J. 1978. Dual interference of hygromycin B with ribosomal translocation and with aminoacyl-tRNA recognition. *Eur. J. Biochem.* **87:** 21–27.

Drees B.L. 1999. Progress and variations in two-hybrid and three-hybrid technologies. *Curr. Opin. Chem. Biol.* **3:** 64–70.

Eustice D.C. and Wilhelm J.M. 1984. Mechanisms of action of aminoglycoside antibiotics in eucaryotic protein synthesis. *Antimicrob. Agents Chemother.* **26:** 53–60.

Evangelista C., Lockshon D., and Fields S. 1996. The yeast two-hybrid system: Prospects for protein linkage maps. *Trends Cell Biol.* **6:** 196–199.

Fields S. and Song O. 1989. A novel genetic system to detect protein-protein interactions. *Nature* **340:** 245–246.

Gonzalez A., Jimenez A., Vazquez D., Davies J.E., and Schindler D. 1978. Studies on the mode of action of hygromycin B, an inhibitor of translocation in eukaryotes. *Biochim. Biophys. Acta* **521:** 459–469.

Gutte B. and Merrifield R.B. 1971. The synthesis of ribonuclease A. *J. Biol. Chem.* **246:** 1922–1941.

Israel D.I. and Kaufman R.J. 1993. Dexamethasone negatively regulates the activity of a chimeric dihydrofolate reductase/glucocorticoid receptor protein. *Proc. Natl. Acad. Sci.* **90:** 4290–4294.

Johnsson N. and Varshavsky A. 1994. Split ubiquitin as a sensor of protein interactions in vivo. *Proc. Natl. Acad. Sci.* **91:** 10340–10344.

Kaufman R.J. 1990. Selection and coamplification of heterologous genes in mammalian cells. *Methods Enzymol.* **185:** 537–566.

Kaufman R.J., Bertino J.R., and Schimke R.T. 1978. Quantitation of dihydrofolate reductase in individual parental and methotrexate-resistant murine cells. Use of a fluorescence activated cell sorter. *J. Biol. Chem.* **253:** 5852–5860.

Michnick S.W., Remy I., C.-Valois F.-X., V.-Belisle A., and Pelletier J.N. 2000. Detection of protein-protein interactions by protein fragment complementation strategies. *Methods Enzymol.* **328:** 208–230.

O'Callaghan C. and Morris A. 1972. Inhibition of beta-lactamases by beta-lactam antibiotics. *Antimicrob. Agents Chemother.* **2:** 442–448.

Pelletier J.N. and Michnick S.W. 1997. A protein complementation assay for detection of protein-protein interactions *in vivo*. *Protein Eng.* **10:** 89.

Pelletier J.N., Campbell-Valois F., and Michnick S.W. 1998. Oligomerization domain-directed reassembly of active dihydrofolate reductase from rationally designed fragments. *Proc. Natl. Acad. Sci.* **95:** 12141–12146.

Pelletier J.N., Arndt K.M., Plückthun A., and Michnick S.W. 1999. An *in vivo* library-versus-library selection of optimized protein-protein interactions. *Nat. Biotechnol.* **17:** 683–690.

Remy I. and Michnick S.W. 1999. Clonal selection and in vivo quantitation of protein interactions with protein fragment complementation assays. *Proc. Natl. Acad. Sci.* **96:** 5394–5399.

———. 2001. Visualization of biochemical networks in living cells. *Proc. Natl. Acad. Sci.* **98:** 7678–7683.

Remy I., Wilson I.A., and Michnick S.W. 1999. Erythropoietin receptor activation by a ligand-induced conformation change. *Science* **283:** 990–993.

Richards F.M. 1958. On the enzymatic activity of subtilisin-modified ribonuclease. *Proc. Natl. Acad. Sci.* **44:** 162–166.

Rossi F., Charlton C.A., and Blau H.M. 1997. Monitoring protein-protein interactions in intact eukaryotic cells by β-galactosidase complementation. *Proc. Natl. Acad. Sci.* **94:** 8405–8410.

Singh A., Ursic D. and Davies J. 1979. Phenotypic suppression and misreading *Saccharomyces cerevisiae*. *Nature* **277:** 146–148.

Taniuchi H. and Anfinsen C.B. 1971. Simultaneous formation of two alternative enzymically active structures by complementation of two overlapping fragments of staphylococcal nuclease. *J. Biol. Chem.* **216:** 2291–2301.

Uetz P., Giot L., Cagney G., Mansfield T.A., Judson R.S., Knight J.R., Lockshon D., Narayan V., Srinivasan M., Pochart P., Qureshi-Emili A., Li Y., Godwin B., Conover D., Kalbfleisch T., Vijayadamodar G., Yang

M., Johnston M., Fields S., and Rothberg J.M. 2000. A comprehensive analysis of protein-protein interactions in *Saccharomyces cerevisiae*. *Nature* **403:** 623–627.

Vidal M. and Legrain P. 1999. Yeast forward and reverse 'n'-hybrid systems. *Nucleic Acids Res.* **27:** 919–929.

Volz K.W., Matthews D.A., Alden R.A., Freer S.T., Hansch C., Kaufman B.T., and Kraut J. 1982. Crystal structure of avian dihydrofolate reductase containing phenyltriazine and NADPH. *J. Biol. Chem.* **257:** 2528–2536.

Walhout A.J., Sordella R., Lu X., Hartley J.L., Temple G.F., Brasch M.A., Thierry-Mieg N., and Vidal M. 2000. Protein interaction mapping in *C. elegans* using proteins involved in vulval development. *Science* **287:** 116–122.

Zlokarnik G., Negulescu P.A., Knapp T.E., Mere L., Burres N., Feng L., Whitney M., Roemer K., and Tsien R.Y. 1998. Quantitation of transcription and clonal selection of single living cells with beta-lactamase as reporter. *Science* **279:** 84–88.

26 A Bacterial Two-hybrid System Based on a Cyclic AMP Signaling Cascade

Gouzel Karimova, Agnes Ullmann, and Daniel Ladant

Unité de Biochimie Cellulaire, CNRS URA 2185, Institut Pasteur, 75724 Paris CEDEX 15, France

INTRODUCTION

For decades, two main methods have been used to identify molecular partners of a given protein: (1) the biochemical approach, which relies on the isolation of complexes between the protein of interest and its putative partners either by coimmunoprecipitation or by using affinity chromatography; (2) the molecular genetic approach, which relies on selection of extragenic suppressor mutations of a given mutation in the gene encoding the protein of interest.

In 1989, Stanley Fields and colleagues introduced the yeast two-hybrid system, a powerful in vivo genetic technique to detect interactions between two polypeptides (Fields and Song 1989). This technique is based on the coexpression in the same cell of two hybrid proteins that, upon interaction, restore a phenotypic and/or selective trait. The proteins of interest are genetically fused to a DNA-binding domain and to a transcription activation domain from a yeast transcription factor. Association of these chimeric proteins restores a functional activator that can activate the transcription of reporter genes. Over the last decade, this technique has been improved and diversified. It has been used extensively to identify novel partners of given proteins,

477

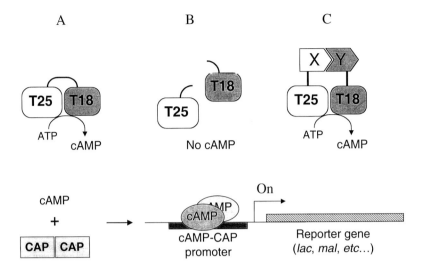

FIGURE 1. Principle of the bacterial two-hybrid system. The T25 and T18 boxes represent the two fragments of the catalytic domain of *B. pertussis* adenylate cyclase, X and Y the interacting polypeptides fused to T25 and T18, and CAP the catabolite activator protein.

to characterize the site of interaction between known proteins, and to explore the network of interactions between all putative proteins encoded by a given genome (for review, see Vidal and Legrain 1999; Legrain and Selig 2000; Uetz and Hughes 2000).

Similar two-hybrid systems using bacteria as host cells have been elaborated recently. They differ in the type of phenotypic readout that is used to detect functional interactions between the two hybrid proteins (for review, see Hu et al. 2000; Ladant and Karimova 2000; Legrain and Selig 2000). This chapter describes the bacterial two-hybrid system we have developed, which is based on the functional complementation between two fragments of an adenylate cyclase (AC) to reconstitute a cAMP signaling cascade in *Escherichia coli*.

Principle of Bacterial Adenylate Cyclase-based Two-hybrid System

The bacterial adenylate cyclase-based two-hybrid (BACTH) system that we have developed (Fig. 1) is based on the functional complementation between two complementary fragments of the catalytic domain of the AC toxin from *Bordetella pertussis* (Karimova et al. 1998), the causative agent of whooping cough (see box and Weiss and Hewlett 1986; Ladant and Ullmann 1999).

THE ADENYLATE CYCLASE FROM *BORDETELLA PERTUSSIS*

B. pertussis, the causative agent of whooping cough, secretes various toxins, among them a calmodulin-activated adenylate cyclase (CyaA) that is one of the major virulence factors of this organism. This toxin is secreted by the virulent bacteria and has the ability to enter eukaryotic cells where, upon activation by the eukaryotic protein calmodulin (CaM), it synthesizes supraphysiologic amounts of cAMP that, in turn, alter cellular physiology (for review, see Weiss and Hewlett 1986; Ladant and Ullmann 1999).

The CyaA toxin, encoded by the *cyaA* gene, is a 1706-residue-long bifunctional protein endowed with both AC and hemolytic activities (Glaser et al. 1988). The catalytic domain is located in the first 400 amino acids, and the carboxy-terminal 1306 residues are responsible for the binding of the toxin to eukaryotic cells and for its entry into these target cells. In addition, this part of the molecule has the ability to form cation-selective channels in biological membranes. Indeed, the carboxy-terminal part of CyaA exhibits several features that are characteristic of a family of bacterial cytolysins known as RTX (for *repeat in toxin*), the prototype of which is the α-hemolysin of *E. coli* (Ladant and Ullmann 1999).

We took advantage of two properties of this enzyme:

1. *AC produces a regulatory molecule, cAMP, that is a pleiotropic regulator of gene transcription in* E. coli. In *E. coli*, cAMP is a key signaling molecule (see box, below) that binds to the transcriptional activator, CAP (catabolite activator protein). The cAMP–CAP complex controls the expression of a large number of genes (Ullmann and Danchin 1983), including genes (*lac* or *mal*) involved in the catabolism of carbohydrates such as lactose or maltose (Fig. 1, lower part). *E. coli* strains deficient in their endogenous adenylate cyclase (*cya*) are unable to ferment these carbohydrates, in contrast to Cya$^+$ bacteria. Cya$^-$ and Cya$^+$ cells can be easily distinguished either on indicator media or on selective media (Ullmann and Danchin 1983; Miller 1992).

---

cAMP, A PLEIOTROPIC REGULATOR OF GENE EXPRESSION IN *E. COLI*

cAMP identified in *E. coli* cells by Makman and Sutherland in 1965 was shown by Ullmann and Monod in 1968 to overcome catabolite repression. Further studies on the mechanisms of cAMP action in *E. coli* led to the discovery of the cAMP receptor protein, also called the catabolite activator protein (CAP), encoded by the *crp* gene, and to the identification of the *cya* gene encoding adenylate cyclase, the enzyme responsible for cAMP synthesis. It was subsequently established that the cAMP–CAP complex promotes gene expression of catabolic operons by binding to sites near the transcription starts of several catabolic genes or operons and activates their expression. CAP is a dimer and has one binding site for cAMP per monomer of 209 residues. The crystal structure of the CAP dimer complexed with two molecules of cAMP was solved by Steitz and coworkers (Schultz et al. 1991). CAP has a modular structure: The amino-terminal domain binds cAMP and the carboxy-terminal domain carries a helix-turn-helix motif that binds DNA. cAMP–CAP activates transcription in *E. coli* at more than 100 promoters by binding at specific DNA sites located upstream of the RNA polymerase (RNAP) binding site. Transcription activation involves protein–protein interaction between CAP and the RNAP α subunit (for reviews, see Ullmann and Danchin 1983; Kolb et al. 1993; Busby and Ebright 1999).

---

2. *The AC catalytic domain has a modular structure made of two complementary fragments that are both required for catalytic activity.* The AC catalytic domain is encoded by the first 400 codons of the *B. pertussis cyaA* gene (Glaser et al. 1988). It exhibits a high catalytic activity (k_{cat} = 2000–5000 s^{-1}) in the presence of its activator, the eukaryotic protein calmodulin (CaM), and a residual but detectable activity (k_{cat} = 1–2 s^{-1}) in the absence of this activator (Ladant et al. 1989). The catalytic domain has a modular structure: It consists of two complementary fragments, T25 and T18, originally identified by limited proteolysis studies (Ladant 1988). The T25 and T18 fragments can associate in vitro with CaM in a fully active ternary complex. T25 (residues 1–224) carries the catalytic site, whereas T18 (residues 225–399) contains the main CaM-binding domain. Both fragments are necessary to reconstitute a fully active enzyme (Fig. 1A) (Ladant et al. 1989).

The AC catalytic domain is functional in *E. coli*. Although this bacterium, like most prokaryotes, does not produce calmodulin, the residual CaM-independent activity of AC is nevertheless sufficient to catalyze cAMP synthesis. Consequently, when expressed in an *E. coli cya* strain, AC can confer a Cya$^+$ phenotype (Fig. 1A) (Ladant et al. 1992).

When the two AC fragments, T25 and T18, are coexpressed in *E. coli cya* as separate entities, they are unable to recognize each other, and they cannot reconstitute a functional enzyme (Fig. 1B) (Karimova et al. 1998). However, when the T25 and T18 fragments are fused to peptides or proteins that can interact, heterodimerization of these chimeric polypeptides results in a functional complementation and, therefore, in cAMP synthesis (Fig. 1C) (Karimova et al. 1998). Bacteria that express chimeric proteins able to heterodimerize will exhibit a Cya$^+$ phenotype that can be easily detected.

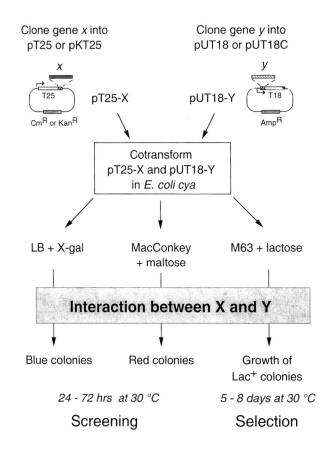

Clone gene *x* into
pT25 or pKT25

Clone gene *y* into
pUT18 or pUT18C

x

T25

pT25-X

CmR or KanR

pUT18-Y

y

T18

AmpR

Cotransform
pT25-X and pUT18-Y
in *E. coli cya*

LB + X-gal

MacConkey
+ maltose

M63 + lactose

Interaction between X and Y

Blue colonies

Red colonies

Growth of
Lac+ colonies

24 - 72 hrs at 30 °C

5 - 8 days at 30 °C

Screening

Selection

Measure cAMP and/or β-galactosidase activity in liquid culture

FIGURE 2. Analysis of protein–protein interactions using the bacterial two-hybrid system. Cloning strategy and screening/selection procedures are depicted. For detailed explanations, see text.

OUTLINE OF PROCEDURE

In the AC-based bacterial two-hybrid system, interaction between the two hybrid proteins results in a functional complementation between the two fragments T25 and T18 that leads to the synthesis of cAMP. This, in turn, activates transcription of catabolic operons. Hence the detection of protein–protein interaction with the BACTH system simply requires coexpression of the proteins of interest as fusions with the T25 and T18 fragments in an *E. coli cya* strain, and detection of the resulting Cya+ phenotype on appropriate indicator media (Karimova et al. 1998).

The basic procedure for analyzing in vivo interactions between two proteins of interest (say X and Y) is outlined in Figure 2, and is as follows:

1. The genes encoding the two proteins of interest, X and Y, are cloned into two compatible vectors, one expressing the T25 fragment, the other expressing the T18 fragment.

2. The resulting plasmids encoding the hybrid protein constructs (T25–X and Y–T18) are cotransformed in competent *E. coli cya* cells. The cotransformants are plated on indicator plates (LB/X-gal or MacConkey/maltose; see below) and incubated for 1–3 days at 30°C. A

characteristic Cya⁺ phenotype of transformants will indicate that the two hybrid proteins interact.

3. Alternatively, the cotransformants can be plated on a selective medium (minimal medium plus lactose or maltose, see below). Functional complementation between the hybrid proteins will allow the cells to grow and form colonies within 5–8 days of incubation at 30°C.

4. The efficiency of complementation can be further quantified by measuring cAMP levels and β-galactosidase activities in liquid cultures. The hybrid proteins then can be expressed in *E. coli* in order to characterize the complex by biochemical approaches.

Expression Vectors for Hybrid Proteins

Genetic screening of protein interaction requires coexpression of the two hybrid proteins. For this purpose, two compatible plasmids must be propagated within the same recipient *cya* bacteria, one expressing the T25 fusion protein and the other the T18 hybrid protein (Fig. 3) (Karimova et al. 1998, 2001).

1. Two vectors encoding the T25 fragment are available. pT25 is a derivative of the low-copy-number plasmid pACYC184, expressing a chloramphenicol-resistance selectable marker; pKT25 is a derivative of the low-copy-number plasmid pSU40, expressing a kanamycin-resistance selectable marker. Both plasmids encode the T25 fragment (encompassing the first 224 codons of *B. pertussis cyaA* gene) under the transcriptional and translational control of a *lac* promoter. A multicloning sequence is present at the 3′ end of T25 to create in-frame fusion at the carboxy-terminal end of T25 (Fig. 3).

2. Similarly, two vectors encoding the T18 fragment, pUT18 and pUT18C, are available. Both plasmids are derivatives of pUC19, expressing an ampicillin-resistance selectable marker. They encode the T18 fragment (codons 225–399 of the *cyaA* gene) under the transcriptional and translational control of a *lac* promoter, and in frame with the pUC19 multicloning sequence. In pUT18, the multicloning sequence is upstream of the T18 open reading frame, whereas in pUT18C, the multicloning sequence is downstream. The same heterologous polypeptide can therefore be fused either to the amino-terminal end or to the carboxy-terminal end of T18 (Fig. 3).

Genes encoding the proteins of interest are usually amplified by PCR using appropriate primers and subcloned into the BACTH vector by standard molecular biology techniques (Sambrook et al. 1989; Sambrook and Russell 2001, *Molecular Cloning*, Chapter 8). Vectors and recombinant plasmids are propagated at 30°C in standard *E. coli* K12 strains such as XL1-Blue, DH5a, HB101, etc. Plasmid DNA is routinely purified with the QIAprep Spin miniprep system according to the manufacturer's instructions (QIAprep Miniprep Handbook 1999).

Bacterial Strains

In vivo protein interactions with the BACTH system are assayed using two different *E. coli cya* strains, DHM1 (F⁻ *recA1, endA1, gyrA96 (Nal*ʳ*), thi1, hsdR17, spoT1, rfbD1, glnV44(AS), cya-857*) and BTH101 (F⁻, *araD139, galE15, galK16, rpsL1 (Str*ʳ*), hsdR2, mcrA1, mcrB1, cya-99*). Functional complementation between hybrid proteins is somewhat more efficient in BTH101 than in DHM1. Other known *E. coli cya* strains could, in principle, be tested as well, although we have found that the efficiency of complementation between hybrid proteins can vary greatly, depending on the background of the *cya* strain used (possibly, because of differing stability of the chimeric proteins).

pT25 / pKT25

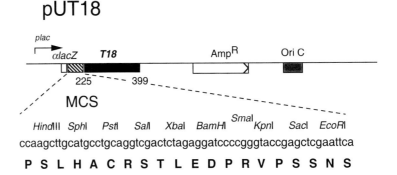

pUT18

pUT18C

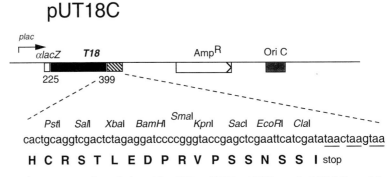

FIGURE 3. Schematic representation of plasmids pT25, pKT25, pUT18, and pUT18C used for hybrid protein expression. Black rectangles represent the open reading frames of T25 and T18 fragments. Open boxes indicate the antibiotic selectable marker, and gray boxes the plasmid origin of replication. The hatched boxes represent the multicloning sequences (MCS) that allow insertion of foreign genes. Some unique restriction sites are displayed above the nucleotide sequence, and the encoded polypeptide sequences are shown below.

Transformation in Test Bacteria

BTH101 and DHM1 bacteria can be transformed by standard CaCl$_2$ techniques (Sambrook and Russell 2001, *Molecular Cloning*, pp. 1.116–1.118) or by electroporation (see Sambrook et al. 1989; Sambrook and Russell 2001, *Molecular Cloning*, pp. 1.119–122, 16.54–16.57 for precise protocols). To prepare competent cells, first restreak the strains on MacConkey/maltose or LB-X-gal media to select a fresh Cya⁻ colony (white phenotype on these media) for seeding the liquid preculture. Competent cells can be cotransformed simultaneously with both plasmids expressing the T25 and T18 fusion proteins, especially if one wants to assay for functional interaction on indicator medium (see below). Alternatively, in the case of library screening, the *E. coli cya* cells should first be transformed with one type of plasmid (encoding the "bait" hybrid protein) and then competent cells should be prepared from these bacteria to increase the efficiency of the second transformation and, as a result, maximize the number of cotransformants.

Screening the Interaction

To test for a functional interaction between two defined hybrid proteins, the easiest and fastest approach is to plate the cotransformants on indicator media. Cya$^+$ colonies can be detected after 24–72 hours of incubation at 30°C, depending on the efficiency of functional complementation. Note that functional complementation is less efficient at 37°C than at 30°C, probably because of greater instability of the hybrid proteins at higher temperature.

Two types of indicator media can be used (Miller 1992); see the Materials section (p. 484) for precise protocols to prepare these media.

1. *LB-X-gal medium:* In *E. coli*, the expression of the *lacZ* gene encoding β-galactosidase is positively controlled by cAMP/CAP. Therefore, the bacteria that express interacting hybrid proteins will form blue colonies on rich Luria-Bertani (LB) medium in the presence of the chromogenic substrate X-gal (5-bromo-4-chloro-3-indolyl-β-D-galactopyranoside), whereas cells expressing noninteracting proteins will remain white. Isopropyl-β-D-thiogalactopyranoside (IPTG) can be included in the medium to increase β-galactosidase expression.

2. *MacConkey medium:* AC-deficient bacteria are unable to ferment lactose or maltose: They form white colonies on MacConkey indicator media containing lactose or maltose (1%), whereas Cya$^+$ bacteria form red colonies on these media (fermentation of the added sugar results in the acidification of the medium which is revealed by a color change of the dye phenol red). If hybrid proteins interact, the colonies should give a red phenotype on MacConkey/maltose, whereas all colonies should be colorless if no interaction occurs.

Selection of Cells Expressing Interacting Hybrid Proteins

To identify polypeptides that potentially interact with a given "bait" protein among a collection of noninteracting polypeptides, it is convenient to use a selection procedure. Because Cya$^+$ cells are Lac$^+$, they are able to grow on a minimal medium supplemented with lactose as a unique carbon source. Therefore, bacteria that express interacting hybrid proteins can be selected on a standard synthetic medium, M63 (plus vitamin B1), supplemented with 0.4% lactose. After transformation with the plasmids of interest, the cells should be washed at least three times with M63 medium to remove all traces of the rich medium used in the transformation procedure. Up to 10^6 cells can be plated on a single dish of M63/B1/lactose (plus appropriate antibiotics). Growth of Lac$^+$ colonies will be detected after 5–8 days of incubation at 30°C. The X-gal substrate can also be added to the medium to facilitate the early detection of Lac$^+$ colonies. The growing Lac$^+$ colonies are, in fact, of two different types: truly positive cells expressing interacting hybrid proteins and false positive cells having acquired a Lac$^+$ phenotype due to endogenous (in general promoter-up) mutations. These latter occur at a frequency of 10^{-7} to 10^{-8}.

Quantification of Functional Complementation

Quantification of functional complementation between chimeric proteins can be obtained by measuring bacterial cAMP content and β-galactosidase activities in liquid cultures, as detailed below. cAMP concentrations are measured in boiled bacterial cultures by an enzyme-linked immunosorbent assay (ELISA) (Karimova et al. 1998), and β-galactosidase measurements are performed on permeabilized cells (either exponential or overnight cultures) using o-nitrophenol-β-galactoside as a substrate (Miller 1992; Karimova et al. 1998, 2000). β-Galactosidase activity is usually expressed in units (one unit corresponds to 1 nmole of ONPG hydrolyzed per minute at 28°C) per milligram of bacterial dry weight. Under routine conditions, when no interaction occurs, the DHM1 strain expresses about 150 units of β-galactosidase/mg of bacterial dry weight (BTH101 about 80). When hybrid proteins associate, β-galactosidase activities range between 700 and 7000 units/mg, depending on the efficiency of functional complementation.

MATERIALS

CAUTION: See Appendix for appropriate handling of materials marked with <!>.

Bacterial Strains <!>

DHM1 (F⁻, *recA1, endA1, gyrA96 (Nal^r), thi1, hsdR17, spoT1, rfbD1, glnV44(AS), cya-854*)
BTH101 (F⁻, *araD139, galE15, galK16, rpsL1 (Str^r), hsdR2, mcrA1, mcrB1, cya-99*)

BACTH Vectors

Construction of these vectors has been described by Karimova et al. (1998, 2001).

Expression vectors for T18 fusions: pUT18, pUT18C
Expression vectors for T25 fusions: pT25, pKT25

Growth Media

Antibiotics
    ampicillin, final concentration of 100 µg/ml (diluted from a 100 mg/ml stock solution in sterile H₂O)
    chloramphenicol, final concentration of 30 µg/ml (diluted from a 30 mg/ml stock solution in ethanol)
    kanamycin, final concentration of 50 µg/ml (diluted from a 50 mg/ml stock solution in sterile H₂O)
Luria-Bertani (LB) broth (used for routine growth of bacteria [Miller 1992])
    10 g of NaCl
    10 g of tryptone
    10 g of yeast extract
    NaOH <!>
    H₂O (deionized)
    15 g of agar

Mix NaCl, tryptone, and yeast extract, adjust pH to 7.0 with NaOH, add deionized H_2O to a final volume of 1 liter, and autoclave. To prepare LB plates, add 15 g of agar per liter of LB broth and autoclave. Allow the medium to cool to less than 45°C; then add the antibiotics and pour the plates.

LB X-gal medium (an indicator medium used to detect functional complementation [Cya⁺/Lac⁺ bacteria form blue colonies])

    Appropriate antibiotics

    IPTG, 0.5 mM final concentration, from a 100 mM stock solution in H_2O, sterilized by filtration <!>

    X-gal, 40 µg/ml final concentration, from a 20 mg/ml stock solution in *N,N*-dimethylformamide (DMF)<!>

    Add to the melted LB agar prepared as above, just before pouring plates.

MacConkey/maltose medium (an indicator medium used to detect functional complementation [Cya⁺/Lac⁺ bacteria form red colonies as a result of sugar fermentation])

    40 g of MacConkey base (Difco)

        Pour into 1 liter of H_2O and autoclave.

    Lactose or maltose (final concentration of 1%, from a 20% stock solution in H_2O sterilized by filtration)

    Antibiotics

    IPTG (0.5 mM final concentration) <!>

    Add to the melted MacConkey medium (temperature less than 45°C) and pour 25–35 ml per dish (9-cm diameter).

M63 synthetic medium

    Standard synthetic medium M63 (Miller 1992) supplemented with 0.2–0.4% lactose is used to select for functional interaction: Only Cya⁺/Lac⁺ bacteria can grow on this medium.

    For 5x M63:

    10 g of $(NH_4)_2SO_4$ <!>
    68 g of KH_2PO_4 <!>
    2.5 mg of $FeSO_4·7H_2O$
    5 mg of vitamin B1

    Add deionized H_2O to a final volume of 1 liter, adjust pH to 7.0 with KOH <!>, and autoclave. To prepare plates, autoclave 15 g of agar in 800 ml of H_2O; add 200 ml of sterile 5x M63 medium, 0.2–0.4% lactose, and the appropriate antibiotics at half the usual concentrations (i.e., ampicillin, 50 µg/ml, kanamycin 25 µg/ml, chloramphenicol 15 µg/ml) just before pouring plates.

PM2 buffer

    70 mM $Na_2HPO_4·2H_2O$ <!>
    30 mM $NaHPO_4·H_2O$ <!>
    1 mM $MgSO_4$ <!>
    0.2 mM $MnSO_4$, pH 7.0 <!>

    Supplement with 100 mM β-mercaptoethanol <!> just before use.

Additional Reagents

    Bovine serum albumin (BSA)
    Dimethylformamide (DMF) <!>
    N-Ethoxy-carbonyl-2-ethoyl-1,2-dihydroquinoline (EEDQ; Sigma)
    HEPES-Na (pH 7.5)
    $O^{2'}$-Monosyccinyl adenosine 3′: 5′-cyclic monophosphate ($O^{2'}$-Suc-cAMP; Sigma)
    NaCl (100 mM)
    Na_2CO_3 (0.1 M, pH 9.5) <!>
    Na_2CO_3 (1 M)
    ONPG
    5′-*para*-nitrophenyl phosphate
    Radioimmunoassay or ELISA kit
    Toluene <!>

METHODS

Analytical Assays

β-Galactosidase assays are carried out on exponentially growing cells or on overnight cultures (for details, see Karimova et al. 2000).

1. To permeabilize bacteria, add 1 drop of toluene and 1 drop of a 0.1% SDS solution per 2–3 ml of cell suspension.

2. Vortex the tubes for 10 seconds and place in a shaker, lightly plugged with cotton to allow the toluene to evaporate, for 30 minutes at 37°C.

3. Add aliquots of the permeabilized cells (0.1–1 ml) to the assay medium PM2 buffer to a final volume of 2 ml.

4. Equilibrate the tubes at 28°C and start the reaction by adding 0.5 ml per tube of the substrate ONPG (4 mg/ml in PM2 buffer without β-mercaptoethanol). After sufficient yellow color has developed, stop the reaction by adding 1 ml of a 1 M Na_2CO_3 solution.

5. Record the optical density at 420 nm for each tube. The reaction is linear up to an absorbance of 1.6.

The reading at 420 nm is a combination of absorbance by the *o*-nitrophenol and light scattering of the cell debris. This latter can be neglected if small volumes of cell suspensions are used and the 420-nm reading is above 0.3. At lower absorbance, the light scattering can be corrected for by obtaining the absorbance at 600 nm from the same reaction mixture and using a correction factor $OD_{420} - 1.5 \times OD_{600}$, which then compensates for light scattering. Enzymatic activities are calculated using a molar absorption coefficient of 5 for *o*-nitrophenol at pH 11. One unit of β-galactosidase activity corresponds to 1 nmole of ONPG hydrolyzed per minute at 28°C. Results are generally given as units/mg of dry weight bacteria. One milligram of dry weight bacteria corresponds to about 2.10^9 bacteria/ml, determined from the reading at 600 nm. Other methods are described in Sambrook et al. (1989), Sambrook and Russell 2001, *Molecular Cloning*, pp.17.97–17.98, and Miller (1992).

cAMP measurements are done by an ELISA.

1. Coat a cAMP–BSA conjugate (see below) on ELISA plates and block nonspecific protein-binding sites with BSA.

2. Add boiled bacterial cultures, followed by diluted rabbit anti-cAMP antiserum in 50 mM HEPES (pH 7.5), 150 mM NaCl, and 0.1% Tween-20 (HBST buffer) containing 10 mg/ml BSA. Incubate overnight at 4°C, and then wash plates extensively with HBST.

4. Add goat anti-rabbit IgG coupled to alkaline phosphatase (AP) and incubate for 1 hour at 30°C.

5. Wash and then use 5′-*para*-nitrophenyl phosphate to reveal the AP activity.

cAMP concentrations are calculated from a standard curve established with known concentrations of cAMP diluted in LB medium.

Prepare the cAMP–BSA conjugate as follows:

1. Add 75 μl of a solution of EEDQ at 100 mg/ml in DMF to 500 μl of a solution of $O^{2'}$-Suc-cAMP in 100 mM HEPES-Na (pH 7.5).

2. Incubate for 45 minutes at room temperature, and then add 500 μl of a solution of BSA (1 mg/ml in 200 mM HEPES-Na, pH 7.5).

3. Incubate overnight at room temperature, and then dialyze the mixture against 10 mM HEPES-Na (pH 7.5), 100 mM NaCl.

4. For coating, dilute the dialyzed solution 5,000–10,000 times (optimal dilution should be determined experimentally) in coating buffer (0.1 M Na_2CO_3, pH 9.5) and add 50 µl in each well of the ELISA plate.

Commercial radioimmunoassays or ELISA kits to assay cAMP can also be purchased from various manufacturers.

CONCLUSIONS

This system has been validated by analyzing a large variety of different interactions between peptides or proteins both by us and by different independent groups (Karimova et al. 1998, 2000; Moreno et al. 2000). The system should be useful for studying structure/function relationships of proteins and for identifying potential partners of a given polypeptide. It is an easy-to-use genetic system: To test in vivo interaction between two polypeptides requires only a few genetic constructions, and results can be obtained in a few days. Once two hybrid proteins interact, it is then possible to use the system to characterize the protein–protein interface at the molecular level. We have shown that this system can be used to identify amino acid residues critical for a given interaction (Karimova et al. 2001). Identification of such mutations may shed light on the molecular basis of a particular interaction and provide experimental tools to study the physiological implications of the studied protein–protein association.

One potential drawback of this technique is that in some cases colonies will exhibit a Cya⁻ phenotype despite the fact that they express truly interacting hybrid proteins (so-called false negative). Because functional complementation depends on the spatial proximity between the AC fragments, it is expected that, in certain cases, the steric constraints imposed by the polypeptide grafted to T25 and T18 can decrease or even abolish the AC enzymatic activity of the reconstituted complex (e.g., if the fused T25 and T18 are held too far apart in space to interact properly). Indeed, we have observed large differences in complementation efficiency between hybrids that were made of the same interacting polypeptide but fused to either end of the T18 fragments (Karimova et al. 2001). Hence, the absence of a functional complementation between two hybrid proteins does not necessarily mean that the two proteins do not associate. This limitation in fact is intrinsic to all two-hybrid methodologies. Future work will give more insights into the flexibility and tolerance of this system.

We anticipate that this bacterial two-hybrid system can become a good complementary tool to the yeast two-hybrid system and will find applications in functional genomics to characterize the network of interactions among the gene products on a genome-wide scale. It should also be applicable in drug discovery and in high-throughput screening of combinatorial libraries to identify compounds that can interfere with given protein–protein interactions.

REFERENCES

Busby S. and Ebright R.H. 1999. Transcription activation by catabolite activator protein (CAP). *J. Mol. Biol.* **293:** 199–213.

Fields S. and Song O. 1989. A novel genetic system to detect protein-protein interactions. *Nature* **340:** 245–246.

Glaser P., Ladant D., Sezer O., Pichot F., Ullmann A., and Danchin A. 1988. The calmodulin-sensitive adenylate cyclase of *Bordetella pertussis:* Cloning and expression in *Escherichia coli. Mol. Microbiol.* **2:** 19–30.

Hu J. C., Kornacker M.G., and Hochschild A. 2000. *Escherichia coli* one- and two-hybrid systems for the analysis and identification of protein-protein interactions. *Methods* **20:** 80–94.

Karimova G., Ullmann A., and Ladant D. 2000. A bacterial two-hybrid system that exploits a cAMP signaling cascade in *Escherichia coli. Methods Enzymol.* **328:** 59–73.

———. 2001. Protein-protein interaction between *Bacillus stearothermophilus* tyrosyl-tRNA synthetase subdomains revealed by a bacterial two-hybrid system. *J. Mol. Microbiol. Biotechnol.* **3:** 73–82.

Karimova G., Pidoux J., Ullmann A., and Ladant D. 1998. A bacterial two-hybrid system based on a reconstituted signal transduction pathway. *Proc. Natl. Acad. Sci.* **95:** 5752–5756.

Kolb A., Busby S., Buc H., Garges S., and Adhya S. 1993. Transcriptional regulation by cAMP and its receptor protein. *Annu. Rev. Biochem.* **62:** 749–795.

Ladant D. 1988. Interaction of *Bordetella pertussis* adenylate cyclase with calmodulin. Identification of two separated calmodulin-binding domains. *J. Biol. Chem.* **263:** 2612–2618.

Ladant D. and Karimova G. 2000. Genetic systems for analyzing protein-protein interactions in bacteria. *Res. Microbiol.* **151:** 711–720.

Ladant D. and Ullmann A. 1999. *Bordetella pertussis* adenylate cyclase: A toxin with multiple talents. *Trends Microbiol.* **7:** 172–176.

Ladant D., Glaser P., and Ullmann A. 1992. Insertional mutagenesis of *Bordetella pertussis* adenylate cyclase. *J. Biol. Chem.* **267:** 2244–2250.

Ladant D., Michelson S., Sarfati R., Gilles A. M., Predeleanu R., and Barzu O. 1989. Characterization of the calmodulin-binding and of the catalytic domains of *Bordetella pertussis* adenylate cyclase. *J. Biol. Chem.* **264:** 4015–4020.

Legrain P. and Selig L. 2000. Genome-wide protein interaction maps using two-hybrid systems. *FEBS Lett.* **480:** 32–36.

Miller J.H. 1992. *A short course in bacterial genetics*, Cold Spring Harbor Laboratory Press, Cold Spring Harbor, New York.

Moreno M., Audia J.P., Bearson S.M., Webb C., and Foster J.W. 2000. Regulation of sigma S degradation in *Salmonella enterica var typhimurium:* In vivo interactions between sigma S, the response regulator MviA(RssB) and ClpX. *J. Mol. Microbiol. Biotechnol.* **2:** 245–254.

QIAprep Miniprep Handbook. 1999. QIAGEN, Valencia, California.

Sambrook J. and Russell D.W. 2001. *Molecular cloning: A laboratory manual*, 3rd edition. Cold Spring Harbor Laboratory Press, Cold Spring Harbor, New York.

Sambrook J., Fritsch E.F., and Maniatis T. 1989. *Molecular cloning: A laboratory manual*, 2nd edition. Cold Spring Harbor Laboratory Press, Cold Spring Harbor, New York.

Schultz S.C., Shields G.C., and Steitz T.A. 1991. Crystal structure of a CAP-DNA complex: The DNA is bent by 90 degrees. *Science* **253:** 1001–1007.

Uetz P. and Hughes R.E. 2000. Systematic and large-scale two-hybrid screens. *Curr. Opin. Microbiol.* **3:** 303–308.

Ullmann A. and Danchin A. 1983. Role of cyclic AMP in bacteria. *Adv. Cyclic Nucleotide Res.* **15:** 1–53.

Vidal M. and Legrain P. 1999. Yeast forward and reverse 'n'-hybrid systems. *Nucleic Acids Res.* **27:** 919–929.

Weiss A.A. and Hewlett E.L. 1986. Virulence factors of *Bordetella pertussis. Annu. Rev. Microbiol.* **40:** 661–686.

27 Peptide Aptamers to Study Proteins and Protein Networks

Russell L. Finley, Jr., Aleric H. Soans, and Clement A. Stanyon

Center for Molecular Medicine & Genetics and Department of Biochemistry and Molecular Biology, Wayne State University School of Medicine, Detroit, Michigan 48201

INTRODUCTION

The use of peptides isolated from random peptide libraries has emerged as a powerful approach to study protein function and to identify and validate new drug targets. Much of the promise of peptides lies in their potential ability to inactivate the function of specific proteins in vivo. Peptides that bind specifically to a target protein, for example, may interfere with the protein's ability to interact with other molecules. Such peptides can be expressed under controlled conditions in vivo or injected into cells to inactivate protein function dominantly. Reverse genetic approaches such as this, in which the phenotypic consequences of functionally inactivating a gene are examined, have become particularly important for the thousands of genes that have been identified but whose functions cannot be accurately predicted from their sequence. Many of the genes needing characterization will turn out to encode proteins that function in biological processes involving networks of interacting proteins. A complete understanding of the genes and the processes in which they participate will require the ability to inactivate specific members of the networks and to disrupt specific connections between members. Recent studies have shown that peptides will be useful tools for such analyses.

Two very different strategies have been used to isolate useful peptides. In one strategy, random peptide libraries are screened for peptides that promote some biological outcome or phenotype (Caponigro et al. 1998; Geyer et al. 1999; Norman et al. 1999; Blum et al. 2000). In this context, the peptides have been referred to as perturbagens because they perturb some biological process (Caponigro et al. 1998). In the second strategy, peptides are isolated on the basis of their capacity to bind directly to a specific target protein. These specific protein-binding peptides have been called peptide aptamers (Colas et al. 1996). The definitions of the terms "aptamer "and "perturbagen" are not mutually exclusive, because an aptamer can be a perturbagen and vice versa. Some peptide aptamers, for example, may be capable of blocking the function of their target proteins and thus could be used as dominant-acting perturbagens. Conversely, a peptide initially isolated as a perturbagen may be considered an aptamer once the specific protein that it binds is identified. It is often possible to identify the in vivo target of a peptide perturbagen, for example, by finding cellular binding proteins. In this chapter, we discuss strategies for isolating peptide aptamers that may be used as perturbagens, with particular emphasis on using the yeast two-hybrid system to identify, and then characterize, the aptamers.

PEPTIDE APTAMERS AS DOMINANT REAGENTS

One of the best ways to understand the function of a gene or the protein it encodes is to characterize an organism bearing loss-of-function mutations in the gene. This requires a genetic model system in which mutations in a gene can be readily obtained. For this reason, studies with loss-of-function mutants have been primarily limited to a handful of model organisms, including yeast, *Drosophila*, and *Caenorhabditis elegans*, and, with more difficulty, mouse and cultured vertebrate cells. An alternative to obtaining loss-of-function mutations is to introduce into a wild-type cell or organism an agent that can dominantly inhibit gene function. Phenotypes that result from dominant inactivation of genes often mimic loss-of-function phenotypes (Herskowitz 1987).

In recent years, several effective methods for inactivating the function of specific genes dominantly have been developed, including antisense nucleic acids, ribozymes, and RNA interference or RNAi (Izant and Weintraub 1984; Cech 1987; Zhao and Pick 1993; Fire et al. 1998). The aim of expressing or injecting one of these reagents into a living cell is to reduce or eliminate the expression of a specific protein. For example, an antisense RNA or oligonuceotide blocks expression of the corresponding gene, presumably by base-pairing specifically with an mRNA and either blocking translation or targeting the message for RNase cleavage (Izant and Weintraub 1984). RNAi, on the other hand, consists of a double-stranded RNA molecule that induces rapid and prolonged degradation of the homologous mRNA (Fire et al. 1998; Carthew 2001). RNAi inactivation can spread from one tissue to another and even from one generation to the next (Fire et al. 1998; Grishok et al. 2000). Although this might be an advantage for some applications, it may also limit the ability to control gene inactivation spatially or temporally.

The aim of using dominant inhibitors such as RNAi or antisense is to allow selective inactivation of individual members of a biological pathway. The consequence is usually a complete loss-of-function or null phenotype, which can provide an indication of the role a gene product plays in the pathway. An individual protein, however, can have multiple distinct functions. A protein that functions in a pathway or complex, for example, will contact one or several other proteins, and each of these contacts may make a unique contribution to the overall function of the protein.

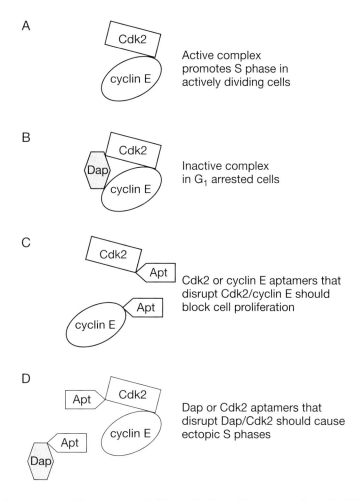

FIGURE 1. Protein interactions with aptamers. Cdk2, cyclin E, and Dap are well-studied *Drosophila* cell-cycle regulatory proteins that have homologs in many other organisms, including vertebrates (de Nooij et al. 1996; Edgar and Lehner 1996; Lane et al. 1996). Cdk2 is a kinase that must be active for cells to enter S phase. (*A*) Cdk2 must bind to the positive regulatory subunit cyclin E, for the kinase to be active. (*B*) Dap binds to both Cdk2 and cyclin E in G_1-arrested cells and inhibits Cdk2 activity. Genetic inactivation of Dap results in ectopic S phases. (*C*) Peptide aptamers that bind to Cdk2 or cyclin E and that inhibit the Cdk2–cyclin E interaction would be predicted to inhibit cell proliferation. (*D*) In contrast, peptide aptamers that bind to cyclin E or Cdk2 and block their ability to interact with Dap would be predicted to cause ectopic S phases. This example illustrates that different dominant-acting aptamers that bind to the same protein (e.g., Cdk2) could produce distinct phenotypes and reveal the functions of specific protein–protein interactions.

Ultimately, it will be useful to dissect the unique contributions of each protein–protein interaction in a pathway or complex. The functions of individual interactions could be studied with reagents that dominantly disrupt specific protein contacts, leaving other interactions intact. This might be achieved with specific binding reagents such as aptamers that bind to one surface of a protein and block intermolecular interactions involving that surface. As illustrated in Figure 1, such dominant-acting disrupters of specific protein–protein interactions could provide more detailed information about protein and pathway function than methods that ablate an entire protein.

Two types of aptamers, peptide and RNA, have been described, either of which may eventually prove useful for intracellular disruption of protein–protein interactions. An RNA aptamer is an RNA molecule selected for its ability specifically to recognize and bind to another molecule, such as a protein (Ellington and Szostak 1990). Like an antibody or a peptide aptamer, an RNA aptamer capable of targeting a protein could inhibit the protein's function. Such inhibitory RNA aptamers (iaRNA) are promising reagents for inactivating proteins in vivo. They share some of the advantages of peptide aptamers, including the fact that they can inhibit specific proteins and can be expressed intracellularly (Shi et al. 1999). One possible advantage of peptide aptamers, however, is that it may be possible to target them to a wider range of subcellular locations than an RNA molecule. Peptides could also be expressed fused to domains that give them other novel properties (Colas et al. 2000).

ISOLATING PEPTIDE APTAMERS USING THE YEAST TWO-HYBRID SYSTEM

Peptide aptamers isolated with the yeast two-hybrid system are highly specific and can be used to target specific proteins and protein interactions in vivo. The two-hybrid system was originally developed to detect protein–protein interactions and to screen cDNA libraries (Fields and Song 1989; Chien et al. 1991). In a two-hybrid screen, the target protein, or "bait," is expressed in yeast cells as a fusion to the DNA-binding domain (DBD) from a transcription factor such as yeast Gal4 or bacterial LexA. Library proteins are expressed fused to a transcription activation domain (AD). Interaction between the library protein and the bait protein results in expression of a reporter gene, which contains upstream binding sites for the DBD. The reporter gene is usually an auxotrophic marker such as yeast *LEU2, HIS3, URA3,* or *LYS2.* Yeast expressing one of these reporters can be selected by growing on media lacking the appropriate nutrient. Most two-hybrid systems also use a second reporter gene whose activity can be quantified, such as *lacZ* or green fluorescent protein (GFP). The yeast two-hybrid system has been adapted for screening libraries of random peptides fused to an AD to isolate peptides that bind to a bait protein (Yang et al. 1995; Colas et al. 1996).

Other methods for detecting protein–protein interactions may also be useful for isolating peptide aptamers (see, e.g., Smith 1985; Bunker and Kingston 1995; Roberts and Szostak 1997; Broder et al. 1998; Karimova et al. 1998; Joung et al. 2000; Zhang et al. 2000); however, the yeast two-hybrid system has several unique advantages. First, the system is well characterized and easy to work with. It has been used extensively to screen cDNA libraries, and numerous library-screening protocols are available, most of which are applicable to peptide libraries. Second, the yeast two-hybrid system is particularly well suited for detecting interactions that occur in intracellular environments. The assay takes place inside a yeast cell, which is likely to resemble more closely a variety of intracellular environments than the conditions that can be achieved in a single in vitro assay. This makes it the method of choice for identifying aptamers directed at proteins that are normally intracellular. Third, most yeast two-hybrid systems provide a means for estimating the interaction affinity of the peptide for its target; the level of reporter gene activation is usually proportional to the interaction affinity of the DBD and AD (Estojak et al. 1995). Fourth, the range of reporter sensitivities available in yeast two-hybrid systems allows detection of protein–peptide interactions over a broad range of affinities. Finally, conducting the interaction screen inside yeast cells provides a selection for peptides that are not generally toxic to eukaryotic cells.

The first report of a two-hybrid screen of a random peptide library came from Fields and coworkers (Yang et al. 1995). They constructed a library expressing random peptides of 16 amino acids fused to the AD from Gal4. To test the library, they introduced it into yeast expressing retinoblastoma protein (pRb) fused to the Gal4 DBD and screened for activation of two reporter genes, *HIS3* and *lacZ*. From ~3×10^6 library transformants, they isolated six peptides that interacted with pRb. The peptide that resulted in the highest reporter gene activity was measured for pRb-binding affinity by surface plasmon resonance using purified recombinant pRb and synthetic peptides. The measured affinity was 13–23 μM K_d (Yang et al. 1995). Remarkably, all six pRb-binding peptides contained an LXCXE, a motif found in many cellular and viral pRb-binding proteins. This suggested that natural binding partners could be identified by comparing the sequences of peptides isolated from the library with protein sequence databases.

Brent and coworkers developed a two-hybrid approach to isolate peptide aptamers displayed from a platform molecule (Colas et al. 1996). They expressed peptides as a constrained loop on the surface of *E. coli* thioredoxin, TrxA. TrxA is a small (~11.6 kD) soluble protein with a compact globular structure (Holmgren 1985). It normally participates in intracellular redox reactions through reversible oxidation of an active center dithiol, which consists of two cysteines separated by Gly-Pro. Previous work had shown that polypeptides can be inserted between the active center cysteines of TrxA without interfering with the solubility or stability of the protein (LaVallie et al. 1993). This suggested that peptides located in the TrxA active center would be conformationally constrained, because they would be anchored to the intact, structurally rigid TrxA. Further experiments showed that peptides inserted in the active center were on the surface of the molecule and capable of binding to other proteins in vitro (LaVallie et al. 1993; Lu et al. 1995). Brent and colleagues reasoned that TrxA could be used as a stable platform to display conformationally constrained peptides intracellularly, and that the two-hybrid system could be used to identify TrxA–peptide aptamers that bind to specific proteins in vivo. A TrxA–peptide library was constructed by inserting an oligonucleotide encoding a random 20 amino acids into a natural *Rsr*II site found between the codons for the two active center cysteines (Colas et al. 1996). The resulting library encoded over 10^9 different 20-mer peptides.

To test whether the TrxA–peptide library could be used in a two-hybrid screen to isolate specific high-affinity peptide aptamers, Colas et al. (1996) introduced the library into yeast expressing LexA fused to the human cyclin-dependent kinase Cdk2 and screened for activation of the two-hybrid reporters. From 6.0×10^6 library transformants, they isolated 14 TrxA–peptides that interacted with LexA–Cdk2. They demonstrated the specificity of the Cdk2 peptide aptamers by mating strains expressing them with strains expressing a number of different LexA fusions, including other highly related cyclin-dependent kinases. They also performed affinity measurements on six of the aptamers and found that they bound Cdk2 with an affinity range of K_d values from 38 to 112 nM.

Several conclusions can be drawn from these first two-hybrid experiments with TrxA–peptides. First, peptide aptamers have the capacity to be remarkably specific. Several of the Cdk2 aptamers did not interact with any of eight other cyclin-dependent kinases tested, including Cdk1, which is 66% identical to Cdk2. This level of specificity has been further demonstrated in subsequent experiments with other target proteins (Kolonin and Finley 1998, 2000; Fabbrizio et al. 1999; Butz et al. 2000), including one study showing that a TrxA–peptide aptamer could distinguish between allelic variants of Ras differing at only one amino acid (Xu et al. 1997). Second, the random peptide library contains a wide variety of recognition motifs for a given target. The 14 Cdk2 aptamers did not resemble each other, suggesting that they represented 14 different recognition motifs. This also suggests that the size of a Cdk2 recognition motif is 5 amino acids or larger. A recognition motif consisting of 4 amino acids would exist at a frequency of $\sim 10^{-5}$ in the library. Each such motif would have been isolated over 10 times from the 6.0×10^6 random peptides that were screened. Recognition motifs of 5 amino acids or larger are present at frequencies less than 10^{-6} in the library. Third, the Cdk2 aptamers had significantly higher affinities for their target than the pRb-binding peptides isolated by Fields and coworkers (Yang et al. 1995). It is likely that the conformational constraints on the TrxA-displayed peptides contribute to their high interaction affinities (see below). Fourth, the TrxA–peptide library appears to contain novel protein recognition motifs not found in nature. None of the 14 Cdk2 aptamers had significant similarity to any of the numerous natural Cdk-interacting proteins that have been identified. In other studies, only one out of six E2F-binding TrxA–aptamers resembled a known E2F-binding protein (Fabbrizio et al. 1999), and none of eight cyclin J TrxA–aptamers resembled several known cyclin J-binding proteins (Kolonin and Finley 2000).

It is interesting to note that this last conclusion is in stark contrast with the previous study in which all six of the isolated Rb-binding peptides had similarity to natural Rb-binding proteins (Yang et al. 1995). The difference between the studies with the TrxA–aptamers and the Rb-binding peptides may lie in the differences between the two-hybrid systems used in each study. However, a screen of a TrxA–peptide library adapted for the Gal4 two-hybrid system also yielded no aptamers strongly resembling natural binding proteins (Butz et al. 2000). Thus, at least in two-hybrid screens, TrxA–aptamers rarely have homology with other proteins that bind to the target protein. An explanation for this may lie in the fact that the TrxA–peptides are conformationally constrained, but it should be noted that, in other systems, TrxA-displayed peptide aptamers can resemble other binding partners. For example, TrxA–peptide libraries have been screened to identify antibody recognition epitopes, but only in an in vitro binding assay where the Trx–peptides were expressed on *E. coli* flagella (Lu et al. 1995). Nevertheless, TrxA–aptamers isolated in two-hybrid screens may not be useful for finding cellular binding partners by homology.

Peptide aptamers have the potential to inhibit specific intermolecular interactions in vivo. This application has yet to be fully explored; however, recent results with TrxA–peptide aptamers isolated in two-hybrid screens have been encouraging. Colas et al. (1996), for example, showed that a Cdk2 aptamer could competitively inhibit Cdk2 kinase activity in vitro, suggesting inhibition of Cdk2 interaction with a substrate. This same aptamer was able to inhibit cell cycle progression when expressed in Saos-2 cells, and it blocked eye development when expressed in *Drosophila*, a phenotype expected for inhibition of Cdk2 activity (Cohen et al. 1998; Kolonin and Finley 1998). In another study, a Trx–aptamer specific for the E2F transcription factor was able to block E2F binding to DNA, most likely by interfering with the ability of E2F to form heterodimers with DP1 (Fabbrizio et al. 1999). Injection or expression of the E2F aptamer in a fibroblast cell line blocked E2F-dependent transcription and cell cycle progression. Finally, a TrxA–aptamer directed at *Drosophila* cyclin J blocked interaction with Cdk2 in vitro and had inhibitory effects on the cell cycle when injected into *Drosophila* embryos (Kolonin and Finley 2000). Although each of these studies lacked definitive experiments to show that the aptamers blocked intermolecular interactions in vivo, they clearly demonstrated the potential for using TrxA–aptamers in such a manner to probe protein function. To maximize the chances of realizing this potential, screens must focus on isolating the highest-affinity aptamers. In addition, assay systems to detect peptide aptamers that can disrupt protein–protein interactions must be developed. As described below, these goals may be addressed with modified two-hybrid systems.

RANDOM PEPTIDE LIBRARIES FOR TWO-HYBRID SYSTEMS

Several random peptide libraries have been constructed for yeast two-hybrid screening (Yang et al. 1995; Colas et al. 1996; Butz et al. 2000). New libraries are relatively easy to construct, and thus might be modified, for example, to expand the number of possible peptides available for screening, to adapt a different version of the two-hybrid system for aptamer isolation, or to introduce a new platform. The methods that have been used to construct random peptide libraries for expression in yeast are all similar and take advantage of strategies used to construct random peptide libraries for phage display systems (Scott and Smith 1990; Colas et al. 1996). Commonly, oligonucleotide sequences are synthesized containing the desired number of random codons consisting of repeats of NNK, where N is A, T, C, or G, and K is either G or T. Using G or T at the third position of every codon minimizes the frequency of stop codons by permitting only amber (UAG). The random portion of the oligonucleotide is flanked on either end by a constant region of known sequence containing one or more restriction enzyme recognition sites. The single-stranded oligonucleotide is converted to double-stranded DNA by Klenow using a primer complemen-

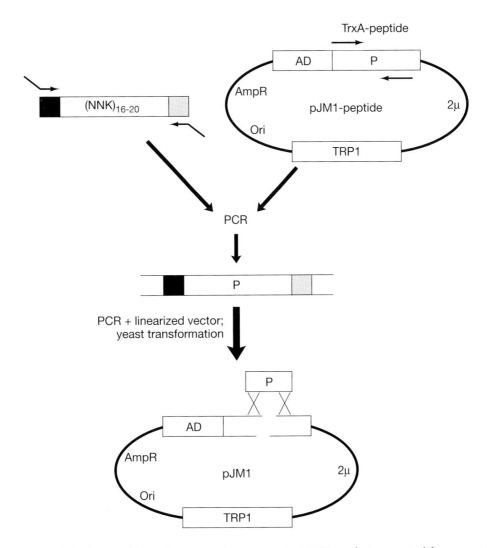

FIGURE 2. Subcloning peptide-coding regions by gap repair. A PCR product generated from a random oligonucleotide (see text) or from a TrxA–peptide vector such as pJM1 (pJM1–peptide), can be inserted into a new vector (pJM1) by gap repair in yeast. The PCR product has 20- to 40-bp ends with homology to pJM1 on either site of a restriction enzyme site. In the case of pJM1, the site is *Rsr*II, which lies between the codons for the active site cysteines in TrxA. Co-transformation of yeast with the PCR product and linearized vector results in yeast with recombinant plasmids that contain the PCR product. Yeast can be transformed using the standard LiOAc protocols along with 20–100 ng of linearized vector and ~100 ng of PCR product.

tary to the 3′-flanking sequence. Typically, the DNA fragment is then subcloned into the vector of choice by restriction enzyme digestion, ligation into the vector, and transformation of *E. coli*. The enzyme sites used in the constant regions of the oligonucleotide are chosen so that the random codons can be inserted in the NNK orientation. This can be achieved with two different restriction recognition sites on either end of the oligonucleotide, or by using a restriction enzyme that recognizes an asymmetric site, such as *Rsr*II.

An alternative approach to the cut-and-paste procedure is to use the gap repair machinery in yeast (Ma et al. 1987) to construct the peptide library. This approach is illustrated in Figure 2 for cloning into the TrxA vector, pJM1 (Colas et al. 1996). In this approach, the oligonucleotide is amplified by PCR using primers that are complementary to the constant regions of the oligonucleotide and that carry an additional 20–40 bases complementary to the vector cloning site. The

PCR product is mixed with an aliquot of the vector that has been linearized at the cloning site (e.g., at *Rsr*II in pJM1), and the mixture is used to transform yeast. Of the resulting yeast transformants, 80–90% will contain a recombinant vector that has been repaired by insertion of the PCR product. A major advantage of this approach is that it results in a library that is already in yeast and therefore ready to screen with any number of baits, as described below.

One decision to be made prior to constructing a library, or to screening an existing library, is whether or not to use a platform molecule for peptide display. Studies with TrxA and other platforms have shown that a platform can provide in vivo stability and solubility for in vitro experiments (Colas et al. 1996; Abedi et al. 1998; Norman et al. 1999). Furthermore, the conformational stability of peptides displayed from a platform may contribute to their binding energies (Ladner 1995). Presumably, a given peptide must adopt one or a very limited number of conformations to bind its target molecule. Adopting these binding conformations would impose a greater entropic cost to an unconstrained peptide than to a conformationally constrained peptide. Consistent with this, binding assays with Cdk2 and E2F1 aptamers have shown that the TrxA-displayed peptides bind their targets with 10–1000 times higher affinity than the corresponding free peptides (Fabbrizio et al. 1999; Colas et al. 2000). Interestingly, of the more than 50 reported TrxA–aptamers, none has contained a stop codon in the peptide-coding region. This is surprising, because more than half of the library members should contain a stop codon resulting in AD–TrxA–peptide fusions with truncated peptides unconstrained at the carboxy-terminal end. This suggests that constrained peptides were favored over nonconstrained peptides in the interaction screens.

Another consideration in choosing or constructing a random library is its size, both in terms of the length of the peptides and the total number of different peptides, or sequence diversity. Results with unconstrained peptides have suggested that the minimal length of an interaction motif can be relatively small (Yang et al. 1995). As noted above, results from a number of screens with TrxA-fused random peptides suggest that an interaction motif in an aptamer must be greater than four amino acids. One could argue that, if the minimum necessary size were large, aptamers would be rare in a random peptide library. The fact that aptamers have been isolated by screening 10^4–10^5 different peptides suggests that the minimum length for an interaction motif is not much greater than four amino acids. The TrxA-displayed peptide libraries that have been reported thus far have contained 16 or 20 random amino acids (Colas et al. 1996; Butz et al. 2000). Results with these libraries have shown that a peptide with 16 random amino acids is sufficient to confer specific binding.

The often-cited general rule for any library is that it should possess as many independent clones as possible. The plasmid-based random peptide libraries that have been constructed for two-hybrid screening generally have 10^7–10^9 independent members. This is far fewer than the total possible number of random peptides, which for a 16-mer or 20-mer library is 6.5×10^{20} or 1.1×10^{26}, respectively. Nevertheless, 10^7–10^9 different 16- or 20-mers appear to provide sufficient sequence diversity to contain peptide aptamers for a variety of targets. For example, from our experience and other published results, a library of 20 random amino acids displayed from TrxA will contain peptide aptamers specific for a particular target protein at a frequency of ~10^{-5}–10^{-6} (Colas et al. 1996; Fabbrizio et al. 1999; Kolonin and Finley 2000; R.L. Finley et al., unpubl.). Moreover, some of the aptamers that have been isolated have inhibitory activity toward their target protein. For example, among the 17 E6-binding aptamers isolated from ~2×10^6 TrxA–peptides, two aptamers were active inhibitors of E6 function in cells (Fabbrizio et al. 1999). In other studies, at least four of the several aptamers directed at various cell cycle proteins were able to inhibit the cell cycle in vivo (Cohen et al. 1998; Kolonin and Finley 1998, 2000). Interestingly, these frequencies are similar to those for finding perturbagen peptides in random libraries of conformationally constrained peptides (Caponigro et al. 1998; Geyer et al. 1999; Norman et al. 1999).

TWO-HYBRID STRATEGIES FOR SCREENING RANDOM PEPTIDE LIBRARIES

The TrxA–peptide library constructed by Brent and colleagues (Colas et al. 1996) is designed to be used with the LexA, or interaction trap version of the two-hybrid system (Gyuris et al. 1993). In this two-hybrid system, the DBD is LexA and the AD is an acidic polypeptide encoded by *E. coli* DNA. The AD is fused to the free amino-terminal end of TrxA, and the resulting AD–TrxA fusion is expressed from the yeast *GAL1* promoter, which is repressed in glucose and induced in galactose. The ability to control expression of the AD fusion with glucose and galactose helps identify and eliminate many of the false positives that arise during a library screen. It also allows isolation of peptides that may be somewhat toxic to yeast, which is particularly useful when screening for aptamers directed at yeast proteins. The reporters for this two-hybrid system include the yeast *LEU2* gene, required for growth in the absence of leucine, and *E. coli lacZ*. A number of different *LEU2* and *lacZ* reporters have been made that differ in their sensitivities to transcriptional activation; each has a different number of upstream LexA-binding sites or operators (Golemis et al. 1999–2000). Because the level of reporter activation correlates with the affinity of the two-hybrid interaction (Estojak et al. 1995), the range of interaction affinities obtained in a library screen can be controlled by the choice of reporter sensitivity.

Several detailed protocols are available for library screening with the LexA two-hybrid system (Finley and Brent 1995; Finley et al. 1997–2000; Golemis et al. 1999–2000; Kolonin et al. 2000). These protocols were designed for screening cDNA libraries but work just as well for screening random peptide libraries. In the protocol that we use (Finley et al. 1997–2000), the DBD and AD proteins are expressed in different haploid yeast strains, which are mated to bring together the AD and DBD. First, the AD library is introduced into a yeast strain and the transformants are frozen in aliquots. To conduct an interactor hunt, an aliquot is mated with a yeast strain expressing LexA fused to the protein of interest. For convenience, the library strain contains one reporter (*LEU2*) and the LexA strain contains the other reporter (*lacZ*). After mixing the two strains together and letting them mate by growing overnight on rich medium (e.g., YPD plates), cells from each strain fuse to form single diploid cells that contain both hybrids and both reporters. The diploids are then plated on a medium containing galactose to induce the AD–TrxA–peptide, and lacking leucine, which selects for cells expressing *LEU2*. The Leu$^+$ colonies are then assayed for *lacZ* activity by plating on media containing X-gal. Finally, the Leu$^+$ lacZ$^+$ colonies are tested to be sure that reporter gene activation depends on expression of the AD fusion by plating them onto glucose and galactose media; the Leu$^+$lacZ$^+$ phenotype should depend on galactose. The next several steps of a typical screen are designed to show that the AD plasmid encodes a fusion that interacts specifically with the LexA fusion used in the hunt. To do this, the AD plasmid is isolated from positive yeast and reintroduced into naive yeast, which is then mated with various strains expressing different LexA fusions.

The efficiency of the protocols can be optimized for conducting a peptide library screen by considering several strategic differences with cDNA library screening. First, the goal in screening for peptide aptamers is to isolate a number of interacting peptides, but not all possible interacting peptides, as is usually the goal in a cDNA library screen. With this philosophy in mind, it is possible to streamline and even automate the screening process. For example, positives that fail to carry from one step to the next—due, for example, to failed plasmid rescue or yeast transformation—can be discarded. Second, measures to reduce or eliminate redundancy, such as are employed in a cDNA library screen, are not necessary when screening a random peptide library. All clones in a typical random peptide library are unique, so little or no redundancy is expected among positive yeast. Third, for most applications, the goal of isolating peptide aptamers is to identify those with the highest affinity for their target proteins. Thus, the sensitivity of the screen need not be as high as in a cDNA library screen, where often the goal is to identify all interactors regardless of affinity. To focus a screen on obtaining high-affinity aptamers, a two-hybrid screen should employ reporters with the lowest sensitivity. In the LexA system, these include a chromosomal *LEU2* reporter with a single upstream LexA operator, as in strain EGY191, and a chromosomally integrated, single-copy *lacZ* reporter (Golemis et al. 1999–2000). The concentration of X-gal employed in media for the detection of *lacZ* activity may also be reduced at least fourfold, allowing differentiation between otherwise indistinguishably strong positives.

STREAMLINING THE PEPTIDE APTAMER HUNT

On the basis of the above considerations, the protocol for cloning and characterizing peptide aptamers can be streamlined, as depicted in Figure 3. The first part of this approach is the same as a cDNA library screen and involves mating a pretransformed library strain with a strain expressing the bait protein, DBD-X. The resulting diploids are placed on selection plates, and colonies in which the reporters are most active are tested for galactose-dependence. At this point, the protocol deviates from the standard cDNA library screen. The peptide-coding regions are amplified directly from the positive yeast by PCR using primers corresponding to the vector 20–40 bp upstream and downstream of the peptide coding region. The resulting PCR products can then be subcloned into a new AD vector by gap repair in yeast (Fig. 2). The gap repair approach eliminates the most time-consuming part of two-hybrid library screening: cloning the yeast plasmids through bacteria. The PCR product is mixed with linearized vector and used to cotransform yeast. The resulting yeast transformants are then used to confirm the two-hybrid phenotypes and to test the interaction specificity. Groups of transformants are picked and cultured in an array format, then mated on YPD plates with yeast expressing DBD hybrids (Fig. 2). Specific aptamers will interact with the original DBD but not with other DBD hybrids (see Fig. 3, Specificity Matings). The outcome of the specificity matings determines which aptamers are chosen for further analysis, such as assays for disruption of protein interactions. Three such assays are described in more detail below.

MODULATOR AND DISRUPTER ASSAYS

Three different two-hybrid-based assays for detecting the disruption of protein–protein interactions are illustrated in Figure 4. All three assays employ cotransformed strains containing fusion proteins that are known to interact (Fig. 4, X and Y). These strains are crossed with aptamer-bearing strains, and changes in reporter gene activation are measured. Disruption of the known interaction due to competition between the aptamer and one of the hybrid proteins is detected by a reduction in reporter gene activity. The first two approaches are more correctly called modulator assays, because the aptamer (or other molecule) will either enhance or inhibit the interaction between protein Y and protein X, and these effects can be detected as increases or decreases in reporter activation, respectively.

The first approach (Fig. 4A) is particularly well suited for instances where only one or a few X–Y interaction pairs are to be targeted (Kolonin et al. 2000). The example shown is for assessing aptamers that were cloned using a DBD-X strain. To assess disruption, a strain is constructed in which the original bait protein is expressed as an AD fusion (Fig. 4A, AD-X) and its interacting protein is expressed as a DBD fusion (Fig. 4A, Lex-Y). The strain is then mated with the strains expressing aptamers, and the diploids are assessed for aptamer-dependent reduction of reporter activity, indicating disruption of the X–Y interaction. Note that this approach does not require any additional cloning of the aptamer construct. The aptamer-expressing strains are the same as those created during the original screen, in the same arrayed format if desired (Fig. 2). This allows a large number of aptamers to be rapidly tested for their ability to disrupt one or a set of interactions.

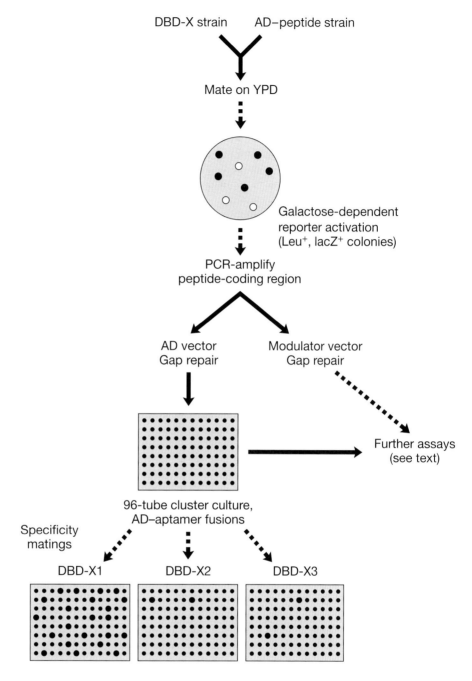

DBD-X strain   AD–peptide strain

Mate on YPD

Galactose-dependent
reporter activation
(Leu+, lacZ+ colonies)

PCR-amplify
peptide-coding region

AD vector
Gap repair   Modulator vector
Gap repair

Further assays
(see text)

96-tube cluster culture,
AD–aptamer fusions

Specificity
matings

DBD-X1   DBD-X2   DBD-X3

FIGURE 3. Aptamer hunt outline. In a typical hunt for aptamers, two pretransformed strains are mated on YPD. One strain expresses a DNA-binding domain (DBD) fused to a protein of interest (X) and the other expresses random peptides fused to an activation domain (AD–peptide). After mating, the resulting diploids are titrated and screened for *LEU2* and *lacZ* reporter activity. The Leu+lacZ+ phenotypes are tested for their dependence on galactose, to demonstrate a requirement for the AD–peptide. To test the specificity of the aptamers, the peptide-coding regions from the positive yeast are PCR-amplified and then reintroduced into the AD vector by gap repair (Fig. 2). The resulting strains are arrayed in a 96-well format and mated with various DBD strains. Highly specific aptamers will result in activation of the two-hybrid reporters when mated with the original DBD-X strain but not when mated with strains expressing other DBD fusions. The arrayed strains can also be used in a disrupter assay to detect aptamers that disrupt specific protein inter-actions with X (Fig. 4A). The PCR products can also be used for inserting the peptide-coding regions into other vectors by gap repair, such as a vector for the modulator assay shown in Fig. 4B.

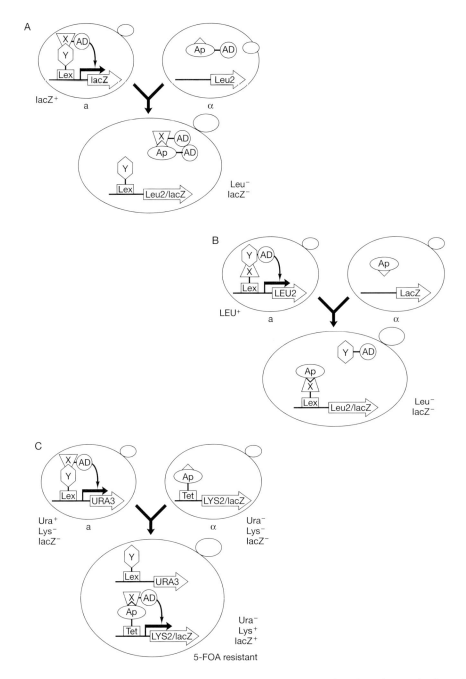

FIGURE 4. Selection strategies for disruptive aptamers. Three assays are outlined: Both *A* and *B* have been successfully implemented, whereas *C* is drawn from work with dual-hybrid systems, as described below and in the text. In the first strategy (*A*), strains expressing AD–aptamers directed at protein X are mated with strains containing an AD–X fusion and a DBD fusion to a protein that interacts with X (Lex-Y). Interaction between the AD–aptamer and AD–X may block the X–Y interaction and thus result in decreased *LEU2* and *lacZ* activity. In the second approach (*B*), the aptamers directed at protein X are expressed without an AD and tested for their ability to disrupt the interaction between DBD–X and AD–Y. Again, the disruption is detected by a decrease in *LEU2* and *lacZ* expression. A scheme is presented in *C* that would allow direct selection of aptamers that compete for an AD–X fusion in a dual-bait system. In this case, the aptamer is fused to a DBD (Tet), rather than an AD domain. Competition between the LexA–Y and Tet–aptamer for the X–AD fusion may be established by simultaneous positive selection for the X–aptamer interaction and negative selection for X–Y interactions. Aptamers that bind to X will activate LYS2 expression and allow growth in the absence of lysine (Lys+). If the same aptamer blocks X–Y interaction, it will block expression of *URA3* and allow cells to grow on 5-FOA medium.

The second assay (Fig. 4B) may be more appropriate when the protein of interest is involved in many protein interactions, all of which are to be targeted by aptamers. In this assay the cotransformed yeast strain carries the DBD-X and AD-Y fusions, and the aptamer is expressed from a vector that lacks an AD. The aptamer-coding regions can be subcloned into the non-AD vector by gap repair using the same PCR products that were generated during the library screen (Fig. 2). When the two strains are mated, competition between AD-Y and an aptamer for the DBD-X hybrid results in reduction of reporter gene activity. Although this approach requires the aptamers to be cloned into an additional vector, an advantage is that the target protein is the original DBD-X fusion used to clone the aptamer and thus is already known to be an effective DBD fusion. Moreover, the AD-Y fusions may in some cases be cDNA-encoded proteins isolated in a two-hybrid screen of a cDNA library, and thus would not require subcloning. Strains and plasmids for this modulator assay are available (Finley et al. 1997–2000).

The disrupter and modulator assays are of great utility when a limited number of DBD-X and AD-Y pairs of interactions are being studied. However, comprehensive proteome-wide protein–protein interaction maps for model genetic organisms and for humans will soon be available (Evangelista et al. 1996; Fromont-Racine et al. 2000; Ito et al. 2000; McCraith et al. 2000; Stanyon and Finley 2000; Uetz et al. 2000; Walhout et al. 2000). For even a small field of study, the number of interactions may thus number in the scores or hundreds. Such a profusion of interaction data requiring functional validation necessitates an efficient and unambiguous method for cloning aptamers that disrupt individual protein interactions. In Figure 4C we propose a combined disrupter assay and cloning approach, which is designed to allow selection of aptamers that specifically inhibit protein interactions by virtue of interacting with the AD hybrid protein in the standard two-hybrid system.

The assay depicted in Figure 4C is based on the concept of dual hybrid systems in which the two bait proteins are fused to two different DNA-binding domains (Jiang and Carlson 1996; Xu et al. 1997; Serebriiskii et al. 1999). It also takes advantage of the concept of reverse two-hybrid systems, in which it is possible to select against protein–protein interactions (Vidal et al. 1996). To detect simultaneous interaction and disruption, different promoters recognized by each DBD are fused to reporter genes that allow positive and negative selective pressure to be placed on the yeast. In the example shown, the Lex operator drives *URA3* expression, while a Tet operator is fused upstream of both *LYS2* and *lacZ*. A strain is then constructed to express an interacting pair of proteins, X and Y, as LexA-Y and AD-X; interaction of X and Y will activate the *URA3* gene. This strain is then mated with a strain expressing fusions of random peptides with the Tet DBD. In the diploid, Tet-peptides that interact with X and that block X–Y interaction will result in activation of one set of reporters (*LYS2* and *lacZ*) and decreased expression of the other reporter (*URA3*). Cells containing such peptides can be selected on media lacking lysine and containing 5-fluoro-orotic acid (5-FOA), which is toxic to cells expressing *URA3*. Dual hybrid systems such as this would also allow direct isolation of aptamers that recognize one protein but not another, perhaps even to the extent of recognizing different isoforms or alleles of proteins as demonstrated by Xu et al. (1997).

EVOLVING MORE EFFECTIVE APTAMERS

As discussed above, the frequency of aptamers that bind to a specific protein is in the order of 10^{-5} to 10^{-6}. However, whether these aptamers will be such keenly honed tools as we wish—capable of producing biological effects, for example—is another question. Clearly, among the aptamers that have been isolated there are aptamers that inhibit protein functions. It is likely, however, that they are not the most potent possible inhibitors. The fact that aptamers isolated from screens of 10^{6}–10^{7} library members do not resemble each other indicates that such screens are only capable

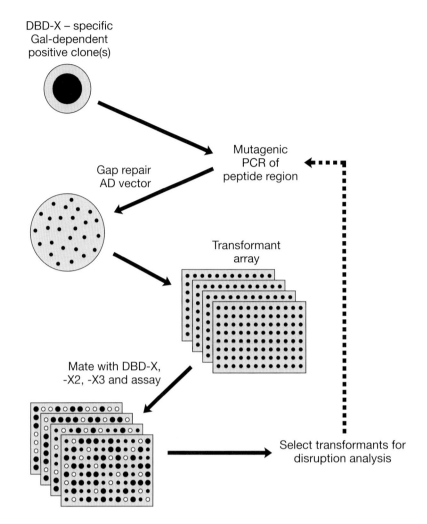

FIGURE 5. Isolating aptamer variants. In this process, aptamers with selected binding characteristics—reporter gene activation levels, modulatory patterns, etc.—are modified to allow for greater or lesser affinity for their targets, or for more specific modulatory effects. Mutagenic PCR is employed to amplify the aptamer-coding regions with low fidelity. The PCR products are then transformed into yeast with a linearized vector, with selection for transformants to create and amplify a mutagenized library (by gap repair; see Fig. 2). Crossing this library with the appropriate single or cotransformed strain and subsequent steps, as in Figs. 3 and 4, results in selection of aptamers with more desirable characteristics.

of identifying one member of any binding class. Variants from the same class may bind the target with higher affinity or greater specificity. They may also be more effective at disrupting specific protein–protein interactions. To identify such variants, a library of mutagenized aptamers could be screened for aptamers with altered binding properties (Colas et al. 2000).

Libraries containing modifications to the DNA sequence encoding just the aptamer can readily be constructed by mutagenic PCR followed by gap repair (Fig. 5). Testing all variations of a 20-mer aptamer, where each residue is varied to each of the 19 other residues, would require only 19 x 20 = 380 mutagenized sequences. This number of clones is readily obtained from a single gap-repair transformation and can be picked into arrays. Mating these arrays with various DBD hybrids, as in a standard aptamer hunt, would establish changes in affinity and specificity of the mutated aptamers. Additional rounds of selection could further enhance the effectiveness of the aptamers, and together with sequence information from many variant clones, this approach could identify the size and character of the binding motifs.

VERIFYING THE SPECIFICITY OF PEPTIDE-INDUCED PHENOTYPES

A challenge in using any dominant reagent in vivo is to prove that the observed phenotypic effects result from targeting a specific molecule. Although an aptamer can be shown to be highly specific in a yeast or in vitro assay, possible effects on other molecules cannot be easily ruled out when the peptide is expressed or injected into cells. Nevertheless, several types of experiments can be done to increase the confidence that a particular aptamer-induced phenotype is due to inactivation of the target protein. First, a biochemical interaction assay such as immunoprecipitation could be used to demonstrate that the injected or expressed aptamer interacts with the target protein in cells. Second, a peptide can be mutated (Fig. 5) to show that loss of interaction with its target protein correlates with loss of the phenotype (Blum et al. 2000). Counter-mutation of the target protein might be useful as well; mutant versions of the target protein that do not interact with the peptide may be resistant to inhibition with the peptide. Third, suppression of an aptamer-induced phenotype by overexpression or injection of the target protein would indicate that the aptamer specifically recognizes its target in vivo (Kolonin and Finley 1998; Blum et al. 2000). Such an approach requires that overexpression of the target protein does not itself cause overt phenotypes that could be confused with suppression. A caveat to this approach is that overexpression of the target protein may bind and block the aptamer's ability to inhibit some other molecule. Fourth, in cases where the aptamer binds to one protein and disrupts a particular protein–protein interaction, a second aptamer could be isolated that targets the other protein but that blocks the same interaction as the first aptamer. If the two aptamers cause similar phenotypes, it would suggest that they do so by disrupting the same specific interaction. Fifth, if an aptamer disrupts an interaction in vivo, overexpression of the non-target protein involved in the disrupted interaction would be expected to suppress the aptamer-induced phenotype. In Figure 1, for example, if a Cdk2 aptamer blocked the Cdk2-Dap interaction, the resulting phenotype (ectopic S phases) might be overcome by expression of more Dap. Because the aptamer does not bind to Dap, it is unlikely that Dap expression is competing with some unknown molecule for the aptamer.

CONCLUSIONS

Reverse genetic approaches for isolating mutants starting with cloned genes often involve significant time and cost to attempt, and do not always work for every gene. This problem is becoming more significant as more cloned genes with unknown functions become available, for example, through sequencing projects and two-hybrid interaction mapping. Additional rapid methods are needed to help determine the function of cloned genes. The use of peptide aptamers to disrupt protein interactions dominantly in vivo has the potential to help solve this problem. Inactivation of proteins with aptamers may also be able to address the functions of individual contacts made by proteins. This approach will be particularly valuable when analyzing protein interaction maps that represent many of the possible regulatory networks in a cell. The phenotypes that result from inactivation of a protein or an interaction can also be used to validate the protein or interaction as potential new drug targets. The yeast two-hybrid system has been a useful method for isolating conformationally constrained peptide aptamers that can function inside cells. Modified two-hybrid systems are making isolation and characterization of useful peptides ever more efficient and may eventually contribute to large-scale functional genomics efforts.

REFERENCES

Abedi M.R., Caponigro G., and Kamb A. 1998. Green fluorescent protein as a scaffold for intracellular presentation of peptides. *Nucleic Acids Res.* **26:** 623–630.

Blum J.H., Dove S.L., Hochschild A., and Mekalanos J.J. 2000. Isolation of peptide aptamers that inhibit intracellular processes. *Proc. Natl. Acad. Sci.* **97:** 2241–2246.

Broder Y.C., Katz S., and Aronheim A. 1998. The ras recruitment system, a novel approach to the study of protein-protein interactions. *Curr. Biol.* **8:** 1121–1124.

Bunker C.A. and Kingston R.E. 1995. Identification of a cDNA for SSRP1, an HMG-box protein, by interaction with the c-Myc oncoprotein in a novel bacterial expression screen. *Nucleic Acids Res.* **23:** 269–276.

Butz K., Denk C., Ullmann A., Scheffner M., and Hoppe-Seyler F. 2000. Induction of apoptosis in human papillomaviruspositive cancer cells by peptide aptamers targeting the viral E6 oncoprotein. *Proc. Natl. Acad. Sci.* **97:** 6693–6697.

Caponigro G., Abedi M.R., Hurlburt A.P., Maxfield A., Judd W., and Kamb A. 1998. Transdominant genetic analysis of a growth control pathway. *Proc. Natl. Acad. Sci.* **95:** 7508–7513.

Carthew R.W. 2001. Gene silencing by double-stranded RNA. *Curr. Opin. Cell. Biol.* **13:** 244–248.

Cech T.R. 1987. The chemistry of self-splicing RNA and RNA enzymes. *Science* **236:** 1532–1539.

Chien C.-T., Bartel P.L., Sternglanz R., and Fields S. 1991. The two-hybrid system: A method to identify and clone genes for proteins that interact with a protein of interest. *Proc. Natl. Acad. Sci.* **88:** 9578–9582.

Cohen B.A., Colas P., and Brent R. 1998. An artificial cell-cycle inhibitor isolated from a combinatorial library. *Proc. Natl. Acad. Sci.* **95:** 14272–14277.

Colas P., Cohen B., Ferrigno P.K., Silver P.A., and Brent R. 2000. Targeted modification and transportation of cellular proteins. *Proc. Natl. Acad. Sci.* **97:** 13720–13725.

Colas P., Cohen B., Jessen T., Grishina I., McCoy J., and Brent R. 1996. Genetic selection of peptide aptamers that recognize and inhibit cyclin-dependent kinase 2. *Nature* **380:** 548–550.

de Nooij J.C., Letendre M.A., and Hariharan I.K. 1996. A cyclin-dependent kinase inhibitor, Dacapo, is necessary for timely exit from the cell cycle during *Drosophila* embryogenesis. *Cell* **87:** 1237–1247.

Edgar B.A. and Lehner C.F. 1996. Developmental control of cell cycle regulators: A fly's perspective. *Science* **274:** 1646–1652.

Ellington A.D. and Szostak J.W. 1990. In vitro selection of RNA molecules that bind specific ligands. *Nature* **346:** 818–822.

Estojak J., Brent R., and Golemis E.A. 1995. Correlation of two-hybrid affinity data with in vitro measurements. *Mol. Cell. Biol.* **15:** 5820–5829.

Evangelista C., Lockshon D., and Fields S. 1996. The yeast two-hybrid system: Prospects for protein linkage maps. *Trends Cell Biol.* **6:** 196–199.

Fabbrizio E., Le Cam L., Polanowska J., Kaczorek M., Lamb N., Brent R., and Sardet C. 1999. Inhibition of mammalian cell proliferation by genetically selected peptide aptamers that functionally antagonize E2F activity. *Oncogene* **18:** 4357–4363.

Fields S. and Song O. 1989. A novel genetic system to detect protein-protein interactions. *Nature* **340:** 245–246.

Finley Jr., R.L. and Brent R. 1995. Interaction trap cloning with yeast. In *DNA cloning, expression systems: A practical approach* (ed. B.D. Hames and D.M. Glover), pp. 169–203. Oxford University Press, United Kingdom.

Finley Jr., R.L., Stanyon C.A., Zhong J., Kolonin M.G., Zhang H., and et al. 1997–2000. Finley Lab Home page at http://cmmg.biosci.wayne.edu/rfinley/lab.html.

Fire A., Xu S., Montgomery M.K., Kostas S.A., Driver S.E., and Mello C.C. 1998. Potent and specific genetic interference by double-stranded RNA in *Caenorhabditis elegans. Nature* **391:** 806–811.

Fromont-Racine M., Mayes A.E., Brunet-Simon A., Rain J.C., Colley A., Dix I., Decourty L., Joly N., Ricard F., Beggs J.D., and Legrain P. 2000. Genome-wide protein interaction screens reveal functional networks involving Sm-like proteins. *Yeast* **17:** 95–110.

Geyer C.R., Colman-Lerner A., and Brent R. 1999. "Mutagenesis" by peptide aptamers identifies genetic network members and pathway connections. *Proc. Natl. Acad. Sci.* **96:** 8567–8572.

Golemis E.A., Serebriiskii I., Finley Jr., R.L., Kolonin M.G., Gyuris J., and Brent R. 1999–2000. Interaction trap/two-hybrid system to identify interacting proteins. In *Current protocols in molecular biology* (ed. F.M. Ausubel et al.), pp. 13.1–13.5. John Wiley, New York.

Grishok A., Tabara H., and Mello C.C. 2000. Genetic requirements for inheritance of RNAi in *C. elegans* (see comments). *Science* **287:** 2494–2497.

Gyuris J., Golemis E., Chertkov H., and Brent R. 1993. Cdi1, a human G1 and S phase protein phosphatase that associates with Cdk2. *Cell* **75:** 791–803.

Herskowitz I. 1987. Functional inactivation of genes by dominant negative mutations. *Nature* **329:** 219–222.

Holmgren A. 1985. Thioredoxin. *Annu. Rev. Biochem.* **54:** 237–271.

Ito T., Tashiro K., Muta S., Ozawa R., Chiba T., Nishizawa M., Yamamoto K., Kuhara S., and Sakaki Y. 2000. Toward a protein-protein interaction map of the budding yeast: A comprehensive system to examine two-hybrid interactions in all possible combinations between the yeast proteins. *Proc. Natl. Acad. Sci.* **97:** 1143–1147.

Izant J.G. and Weintraub H. 1984. Inhibition of thymidine kinase gene expression by anti-sense RNA: A molecular approach to genetic analysis. *Cell* **36:** 1007–1015.

Jiang R. and Carlson M. 1996. Glucose regulates protein interactions within the yeast SNF1 protein kinase complex. *Genes Dev.* **10:** 3105–3115.

Joung J.K., Ramm E.I., and Pabo C.O. 2000. A bacterial two-hybrid selection system for studying protein-DNA and protein-protein interactions. *Proc. Natl. Acad. Sci.* **97:** 7382–7387.

Karimova G., Pidoux J., Ullmann A., and Ladant D. 1998. A bacterial two-hybrid system based on a reconstituted signal transduction pathway. *Proc. Natl. Acad. Sci.* **95:** 5752–5756.

Kolonin M.G. and Finley Jr., R.L. 1998. Targeting cyclin-dependent kinases in *Drosophila* with peptide aptamers. *Proc. Natl. Acad. Sci.* **95:** 14266–14271.

———. 2000. A role for cyclin J in the rapid nuclear division cycles of early *Drosophila* embryogenesis. *Dev. Biol.* **227:** 661–672.

Kolonin M.G., Zhong J., and Finley Jr., R.L. 2000. Interaction matting methods in two-hybrid systems. *Methods Enzymol.* **328:** 26–46.

Ladner R.C. 1995. Constrained peptides as binding entities. *Trends Biotechnol.* **13:** 426–430.

Lane M.E., Sauer K., Wallace K., Jan Y.N., Lehner C.F., and Vaessin H. 1996. Dacapo, a cyclin-dependent kinase inhibitor, stops cell proliferation during *Drosophila* development. *Cell* **87:** 1225–1235.

LaVallie E.R., DiBlasio E.A., Kovacic S., Grant K.L., Schendel P.F., and McCoy J.M. 1993. A thioredoxin gene fusion expression system that circumvents inclusion body formation in the *E. coli* cytoplasm. *Bio/Technology* **11:** 187–193.

Lu Z., Murray K.S., Van Cleave V., LaVallie E.R., Stahl M.L., and McCoy J.M. 1995. Expression of thioredoxin random peptide libraries on the *Escherichia coli* cell surface as functional fusions to flagellin: A system designed for exploring protein-protein interactions. *Bio/Technology* **13:** 366–372.

Ma H., Kunes S., Schatz P.J., and Botstein D. 1987. Plasmid construction by homologous recombination in yeast. *Gene* **58:** 201–216.

McCraith S., Holtzman T., Moss B., and Fields S. 2000. Genome-wide analysis of vaccinia virus protein-protein interactions. *Proc. Natl. Acad. Sci.* **97:** 4879–4884.

Norman T.C., Smith D.L., Sorger P.K., Drees B.L., O'Rourke S.M., Hughes T.R., Roberts C.J., Friend S.H., Fields S., and Murray A.W. 1999. Genetic selection of peptide inhibitors of biological pathways. *Science* **285:** 591–595.

Roberts R.W. and Szostak J.W. 1997. RNA-peptide fusions for the in vitro selection of peptides and proteins. *Proc. Natl. Acad. Sci.* **94:** 12297–12302.

Scott J.K. and Smith G.P. 1990. Searching for peptide ligands with an epitope library. *Science* **249:** 386–390.

Serebriiskii I., Khazak V., and Golemis E.A. 1999. A two-hybrid dual bait system to discriminate specificity of protein interactions. *J. Biol. Chem.* **274:** 17080–17087.

Shi H., Hoffman B.E., and Lis J.T. 1999. RNA aptamers as effective protein antagonists in a multicellular organism. *Proc. Natl. Acad. Sci.* **96:** 10033–10038.

Smith G.P. 1985. Filamentous fusion phage: Novel expression vectors that display cloned antigens on the virion surface. *Science* **228:** 1315–1317.

Stanyon C.A. and Finley Jr., R.L. 2000. Functional genomics: Progress and potential of *Drosophila* protein interaction maps (PIMS). *Pharmacogenomics* **1:** 417–431.

Uetz P., Giot L., Cagney G., Mansfield T.A., Judson R.S., Knight J.R., Lockshon D., Narayan V., Srinivasan M., Pochart P., et al. 2000. A comprehensive analysis of protein-protein interactions in *Saccharomyces cerevisiae*. *Nature* **403:** 623–627.

Vidal M., Brachmann R.K., Fattaey A., Harlow E., and Boeke J.D. 1996. Reverse two-hybrid and one-hybrid systems to detect dissociation of protein-protein and DNA-protein interactions. *Proc. Natl. Acad. Sci.* **93:** 10315–10320.

Walhout A.J., Boulton S.J., and Vidal M. 2000. Yeast two-hybrid systems and protein interaction mapping projects for yeast and worm. *Yeast* **17:** 88–94.

Xu C.W., Mendelsohn A.R., and Brent R. 1997. Cells that register logical relationships among proteins. *Proc. Natl. Acad. Sci.* **94:** 12473–12478.

Yang M., Wu Z., and Fields S. 1995. Protein-peptide interactions analyzed with the yeast two-hybrid system. *Nucleic Acids Res.* **23:** 1152–1156.

Zhang Z., Zhu W., and Kodadek T. 2000. Selection and application of peptide-binding peptides. *Nat. Biotechnol.* **18:** 71–74.

Zhao J.J. and Pick L. 1993. Generating loss-of-function phenotypes of the fushi tarazu gene with a targeted ribozyme in *Drosophila*. *Nature* **365:** 448–451.

28 Generation of Protein Fragment Libraries by Incremental Truncation

Marc Ostermeier,[1] Stefan Lutz,[2] and Stephen J. Benkovic[2]

[1]*Department of Chemical Engineering, Johns Hopkins University, Baltimore, Maryland 21218;*
[2]*Department of Chemistry, The Pennsylvania State University, University Park, Pennsylvania 16802*

INTRODUCTION

Protein fragmentation and domain swapping are valuable methods for the study of inter- and intra-domain and subdomain interactions in proteins. Building on classic examples of protein fragment complementation (Ullman et al. 1967; Kato and Afinsen 1969), protein fragmentation has been used in biophysical studies, particularly in the elucidation of protein folding mechanisms (Tasayco and Carey 1992; Ladurner et al. 1997). More recently, protein engineering using the recombination of protein fragments has gained popularity (for reviews, see Lutz and Benkovic 2000; Ostermeier and Benkovic 2000).

Various methodologies for protein fragmentation have been successfully implemented. On the protein level, target enzymes are usually hydrolyzed by chemical and proteolytic means.

507

Although these fragmentation methods are relatively simple, the resulting complex product mixtures require extensive purification and characterization of the various products. Although this problem is solved by limiting the product "libraries" to a dozen or fewer variants, observed biases toward more accessible regions of a target structure such as surface loops cannot easily be overcome.

Alternatively, protein fragments can be generated by genetic approaches. Using rational design, specific PCR primers can amplify predetermined gene fragments that, upon cloning and transformation into a suitable expression host, allow the isolation of a desired protein segment. Although very efficient on an individual basis, the approach rapidly becomes laborious for multiple fragments. Furthermore, it depends heavily on the availability of structural information to choose the fragments to study. In contrast, incremental truncation is a rapid method for generating a comprehensive gene library that encodes all possible fragment lengths of a target protein and does so in the absence of detailed structural information (Ostermeier et al. 1999c). One important caveat of this method is that a genetic selection system or a high-throughput screen may be needed to analyze the library if the number of possible constructs is large.

The incremental truncation methodology can be applied in a number of different ways (Ostermeier et al. 1999a). First, a library of amino-terminal or carboxy-terminal truncations of a single gene can be generated. Such protein fragments are useful in defining relationships between sequence and structure and function (Jasin et al. 1983), including the detection of regions responsible for protein–protein interactions. Furthermore, the random pairing of members of such libraries can be used to identify protein fragments that can non-covalently associate to form functional proteins (Ostermeier et al. 1999c). This method for protein fragment complementation can also be viewed as a strategy for the conversion of a monomeric protein into a heterodimer. Finally, incremental truncation can be used to create comprehensive fusion libraries between fragments of two genes (Ostermeier et al. 1999b). This method has application in protein engineering, particularly in the generation of combinatorial fusion libraries between genes with low levels of homology.

OUTLINE OF PROCEDURE

Basic Concepts

Incremental truncation is a method for creating a combinatorial library containing every one-base-pair deletion of a gene or gene fragment. As shown in Figure 1, the substrate for incremental truncation is linear double-stranded DNA. The substrate is prepared by digesting closed circular plasmid DNA with restriction enzymes such that a blunt end or 5′ overhang lies near the DNA end to be truncated, whereas a 4-base 3′ overhang on the opposite end protects the remaining vector from degradation. Exonuclease III (Exo III), a 3′-to-5′ exonuclease used to create the truncations, uses 3′ recessed ends or blunt ends as a substrate but cannot digest certain 4-base 3′ overhangs (see Table 1). Alternatively, the protected end can be prepared by filling a 5′ overhang with α-phosphothioate nucleotides (Putney et al. 1981).

The key step in the creation of an incremental truncation library is the time-dependent sampling during Exo III digestion. Over the course of the DNA digestion, small aliquots are removed frequently and the reaction is quenched in a low-pH, high-salt buffer. Temperature and buffer conditions are chosen such that the rate of hydrolysis is balanced with the rate of sampling. For example, if the rate of Exo III is controlled to 10 bases per minute and small aliquots are removed every 20 seconds, the average amount of truncation will increase by 3.3 bases for every sample. Because Exo III digests DNA at a substantially uniform and synchronous rate (Wu et al. 1976) but with a standard deviation of truncation length of 0.22 times the average number of bases truncated (Hoheisel 1993), a population of DNA is created that contains every possible integer trun-

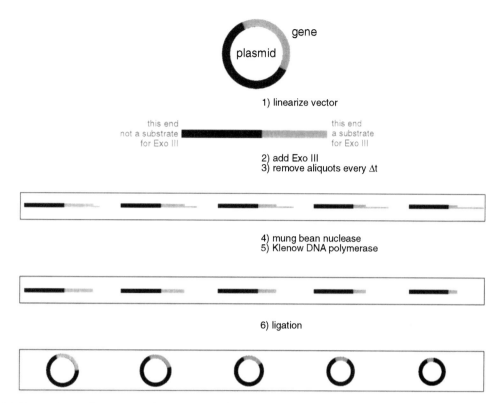

FIGURE 1. General incremental truncation schematic. Incremental truncation is performed on a linear piece of DNA (containing the segment of DNA to be truncated, shown in gray) that has one end protected from digestion and the other end susceptible to digestion. This is easily accomplished, for example, by digestion of plasmid DNA with two restriction enzymes: one that produces a 3′ overhang (which is not a substrate for Exo III) and the other which produces a 5′ overhang (which is a substrate for Exo III). Digestion with Exo III proceeds when the digestion rate is slow enough that the removal of aliquots at frequent intervals results in a DNA library with every one-base deletion. The ends of the DNA can be blunted by treatment with mung bean nuclease and Klenow DNA polymerase so that unimolecular ligation results in the desired incremental truncation library. For some applications, additional DNA manipulations are required before recircularizing the vector.

TABLE 1. Resistance to Exo III Digestion of 4-base 3′ Overhangs Created by Commercially Available Restriction Enzymes

Restriction enzyme	Is 3′ overhang resistant to Exo III?			Successfully used to protect ends for incremental truncation?
	Ref. 1[a]	Ref. 2[b]	Ref. 3[c]	
*Aat*II	yes	yes	yes	–
*Apa*I	no	no	yes	–
*Ban*II[d]	no	yes	yes	–
*Bgl*I[d]	–	yes	yes	–
*Bsp1286*I[d]	no	-	–	–
*Bst*XI[d]	–	yes	yes	–
*Hae*II[d]	no	yes	yes	–
*Kpn*I	no	–	yes	–
*Nsi*I	yes	yes	yes	yes
*Pst*I	yes	no	yes	yes
*Sac*I	–	yes	yes	yes
*Sph*I	yes	yes	yes	yes
*Sst*I	yes	–	–	–

[a]Hoheisel (1993).

[b]New England BioLabs (1998).

[a]Promega (1995).

[d]These enzymes contain ambiguous bases in their cut sites. Their resistance may depend on the exact sequence.

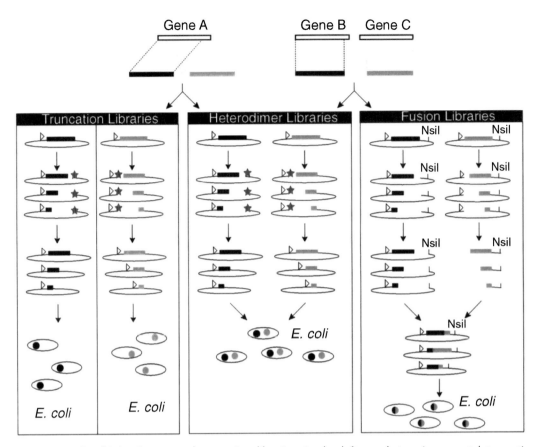

FIGURE 2. Combining incremental truncation libraries. In the left panel, two incremental truncation libraries derived from the 5´ and 3´ fragments of the same gene are prepared separately. Members of the individual library are ligated, introducing the necessary start or stop codons as indicated by the stars. In the middle panel, the two libraries, resulting from a single gene A or two different genes B and C, are randomly paired by cotransformation into *E. coli* cells. Each *E. coli* cell expresses a 5´ fragment and a 3´ fragment, allowing the possible detection fragment complementation. In the right panel, the two truncation libraries are fused on the DNA level such that each *E. coli* cell receives a fusion of a randomly paired 5´ fragment of gene B and a 3´ fragment of gene C.

cation of the starting DNA. Following the time-dependent truncation with Exo III, the single-stranded DNA tails are removed by digestion with a single-strand nuclease, preferably mung bean nuclease, and the ends are blunted using a DNA polymerase. Subsequent steps depend on the intended use of the library.

Three possible uses of libraries are presented in Figure 2. When applied to a single gene (Fig. 2, left), vectors that contain 3´-truncation libraries must carry stop codons in all three reading frames that, upon ligation, align with the library to terminate translation of the protein fragment. Vectors containing 5´-truncation libraries, on the other hand, require a start codon, fused to the 5´ end of the library upon ligation. As shown in Figure 2 (center), truncations can be performed in opposite directions on two overlapping fragments of the same gene A or for two different genes B and C. The two libraries are kept on separate plasmids and can be transformed into the same cells to randomly pair members of the two truncation libraries together. Finally, fusion libraries of the 5´ and 3´ fragments, termed ITCHY libraries, can be prepared from genes B and C as shown in the right-hand panel of Figure 2.

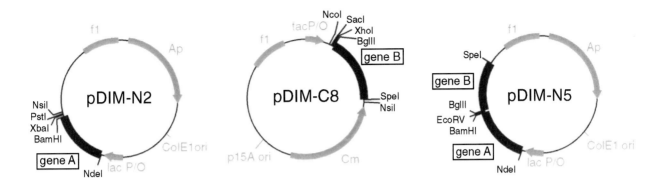

b)

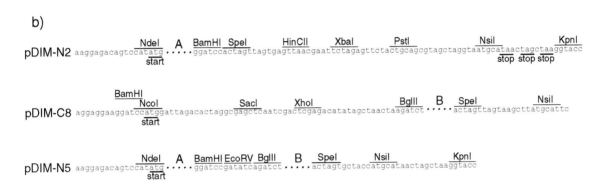

FIGURE 3. Vectors for incremental truncation. Circular maps (*a*) and sequences of cloning (*b*) and trunca-tion regions of incremental truncation vectors. All vectors have the filamentous phage origin of replication (f1) as well as plasmid origins of replication (ColE1 ori and p15A ori). In pDIM-N2 (Ostermeier et al. 1999a), the amino-terminal gene to be truncated is cloned between the *Nde*I and *Bam*HI sites downstream from an IPTG-inducible *lac* promoter (lac P/O). The vector also has an antibiotic-resistance gene (ampicillin, Ap). In pDIM-C8 (Ostermeier et al. 1999b), the carboxy-terminal gene to be truncated is cloned between the *Bgl*II and *Spe*I sites downstream from an IPTG-inducible trp/lac hybrid promoter (tac P/O). The vector also has an antibiotic-resistance gene (chloramphenicol, Cm). In pDIM-N5, both gene fragments to be truncated are cloned downstream from an IPTG-inducible *lac* promoter (lac P/O) between the indicated restriction enzyme sites (*Nde*I/*Bam*HI and *Bgl*II/*Spe*I). Between the two gene fragments is located a unique restriction site that produces blunt ends (*Eco*RV).

Description of Vectors for Incremental Truncation

Vectors designed for performing incremental truncation are shown in Figure 3. Single-gene, het-erodimer, or fusion libraries can be prepared using vectors pDIM-N2 and pDIM-C8 for the amino- and carboxy-terminal fragments, respectively. Vector pDIM-N5 is used solely for making fusion libraries. The salient features of these vectors include (1) different antibiotic-resistance genes and origins of replication in pDIM-N2 and pDIM-C8, (2) an f1 phage origin of replication, allowing the possibility of packaging into phage for efficient transfer of vector DNA into *E. coli*, (3) unique restriction enzyme sites for cloning the DNA to be truncated, (4) unique restriction enzyme sites for preparing the vectors for incremental truncation, (5) unique restriction enzyme sites for creating incremental truncation fusion libraries (ITCHY libraries), and (6) *lac*-based pro-moters to express fusion proteins.

Incremental Truncation

This protocol is divided into the following stages: Preparation of Plasmid DNA, Preparation of DNA for Truncation, and Incremental Truncation. Detailed comments are included at the beginning of each stage.

MATERIALS

CAUTION: See Appendix for appropriate handling of materials marked with <!>.

Buffers and Solutions

Ammonium acetate (7.5 M) <!>
> Store at room temperature.

100x Bovine serum albumin (10 mg/ml) (BSA)
> Store at –20°C.

Buffer EB (Qiagen, proprietary)
  10 mM Tris-HCl (pH 8.5)
> Store at room temperature.

Buffer PB
> Store at room temperature.

Buffer PE (Qiagen, proprietary)
> Store at room temperature.

dNTP mix
  0.125 mM dATP
  0.125 mM dCTP
  0.125 mM dGTP
  0.125 mM dTTP
> Store at –20°C.

Ethanol (100%) <!>
> Store at –20°C.

10x Exo III buffer
  660 mM Tris-HCl (pH 8.0)
  6.6 mM $MgCl_2$ <!>
> Store at 4°C.

7.4x Exo III stop buffer
  0.3 M potassium acetate (pH 4.6)
  2.5 M NaCl
  10 mM $ZnSO_4$
  50% glycerol
> Concentrations are approximate because this buffer can be difficult to prepare. The following protocol should be used. Mix the following together: 5.5 ml of 3 M potassium acetate, pH 4.6 (prepared by dissolving 2.94 g of potassium acetate in 2 ml of H_2O and adjusting the pH to 4.6 with 4–5 ml of glacial

acetic acid <!> followed by adjusting the volume to 10 ml), 25 ml of 5 M NaCl, 0.27 g of $ZnSO_4 \cdot 7H_2O$, and 25 ml of 100% glycerol.

   Store at $-20°C$.

1× Klenow buffer

   20 mM Tris-HCl (pH 8.0)

   100 mM $MgCl_2$ <!>

      Store at 4°C.

LB medium (per liter)

   10 g of tryptone

   5 g of yeast extract

   10 g of NaCl

10× Ligase buffer

   500 mM Tris-HCl (pH 7.5)

   100 mM $MgCl_2$ <!>

   100 mM dithiothreitol <!>

   10 mM ATP

   250 μg/ml bovine serum albumin

      Store at $-20°C$.

10× Mung bean buffer

   500 mM sodium acetate (pH 5.0)

   300 mM NaCl

   10 mM $ZnSO_4$

      Store at 4°C.

NaCl (1 M)

      Store at room temperature.

10× NEB1 buffer

   100 mM Bis Tris propane-HCl (pH 7.0)

   100 mM $MgCl_2$ <!>

   10 mM dithiothreitol <!>

      Store at 4°C.

10× *Nsi*I buffer

   100 mM Tris-HCl (pH 8.4)

   1 M NaCl

   10 mM dithiothreitol

      Store at 4°C.

PEG (50%) <!>

      Store at room temperature.

2× TY (per liter)

   16 g of tryptone

   10 g of yeast extract

   5 g of NaCl

Enzymes

Exonuclease III (~200 units/μl) (Promega)

      Store at $-20°C$.

Klenow mix

   20 μl of 1× Klenow buffer

   1 μl of Klenow (5 units) (Promega)

      Store at $-20°C$. Prepare fresh daily.

Ligase mix
   320 µl of H_2O
   40 µl of 10x ligase buffer
   40 µl of 50% PEG
   18 Weiss units of T4 DNA Ligase (Promega)
      Store at –20°C. Prepare fresh daily.
Mung bean nuclease (~10 units/µl)
   *Nco*I (~10 units/µl)
   *Nsi*I (~10 units/µl)
      Store at –20°C.

Additional Equipment and Reagents

Ampicillin
Agarose gel electrophoresis system (TAE buffer)
Chloramphenicol <!>
Heat blocks at 15°C, 22°C, 30°C, 37°C, and 72°C
Reagent kit (Qiagen Midi Prep kit)

METHOD

Stage 1: Preparation of Plasmid DNA

For incremental truncation, it is important to prepare plasmid DNA free of single-stranded nicks. Exo III can digest from single-stranded nicks in double-stranded DNA, leaving single-stranded gaps. These single-stranded gaps will be digested by mung bean nuclease and result in the undesired effect of random deletions throughout the entire plasmid.

We routinely isolate plasmid DNA of pDIM-N vectors from *E. coli* strain DH5α using commercial plasmid-prep kits. For pDIM-C vectors, we have found that the fraction of nicked molecules and total yield of plasmid DNA are highly dependent on growth conditions. However, if the following protocol for growth is followed, we routinely prepare pDIM-N and pDIM-C vectors with <10% nicked DNA. For high yield of supercoiled DNA, use the following growth conditions:

1. Grow an overnight inoculum of DH5α containing vectors in 10 ml of LB medium, 0.2% glucose, and the appropriate antibiotic (100 µg/ml ampicillin for pDIM-N2 and pDIM-N5; 50 µg/ml chloramphenicol for pDIM-C8) in a 25-ml test tube at 37°C.

2. For pDIM-N2, inoculate 100 ml of 2x TY medium, 2.0% glucose, and 100 µg/ml ampicillin with 2 ml of overnight culture and grow in 500-ml shake flask overnight at 37°C. For pDIM-C8 inoculate 500 ml of LB and 50 µg/ml chloramphenicol with 10 ml of overnight culture and grow overnight in a 2-liter baffled shake flask at 37°C.

3. Isolate plasmid DNA using a standard reagent kit (Qiagen Midi Prep kit). A typical yield of plasmid DNA is ~1 µg/ml culture for pDIM-N2 and pDIM-N5 and 0.2 µg/ml culture for pDIM-C8.

   Should production of sufficiently pure non-nicked DNA prove difficult, methods to purify non-nicked from nicked plasmid DNA include CsCl-ethidium bromide gradients (Sambrook et al. 1989), acid-phenol extraction (Zasloff et al. 1978), or removal of the nicked DNA by enzymatic digestion (Gaubatz and Flores 1990). Alternatively, treatment of nicked DNA with T4 DNA ligase should presumably repair the nicks.

Stage 2: Preparation of DNA for Truncation

Another important factor to consider in preparing non-nicked DNA for incremental truncation is the restriction enzyme digestion to linearize the DNA in preparation for truncation. Restriction enzymes from suppliers may have nuclease contamination. In addition, restriction enzymes may have single-stranded nicking activity at a high enzyme-to-DNA ratio. For this reason, we recommend digesting the DNA with the minimum amount of restriction enzyme necessary to digest the DNA fully and avoid conditions contributing to star activity (relaxed or altered specificity) in restriction enzymes (see manufacturer's product specifications).

We digest closed circular pDIM-N vectors with 1.5 units of *Xba*I and 1.5 units of *Nsi*I or *Pst*I per microgram of DNA in the manufacturer's suggested buffer at 37°C for 1.5–2 hours. pDIM-C vectors are digested with 15 units of *Sac*I and 20 units of *Xho*I per microgram of DNA in the manufacturer's suggested buffer for 1.5–2 hours at 37°C. *Sac*I and *Xho*I have difficulty digesting supercoiled DNA (New England Biolabs 2000), accounting for the higher level of enzyme activity required.

It is important to note that some 4-base 3′overhangs are not completely resistant to Exo III digestion. Table 1 summarizes commercially available enzymes that create 4-base 3′ overhangs and lists whether the 3′ overhangs are resistant to Exo III. Discrepancies between the sources as to whether a 3′ end is resistant to Exo III result from the criteria for designating an end as resistant (most 4-base 3′ overhangs are very weak substrates of Exo III) and, in the case of sites with ambiguous bases, the exact sequence of the site used for testing. An alternative way to protect DNA from digestion is to fill in a 5′ overhang with α-phosphothioate nucleotides (Putney et al. 1981).

1. Digest 10 μg of plasmid DNA with the appropriate restriction enzymes in a total volume of 100 μl at 37°C.

2. After 1.5–2 hours, heat-inactivate the enzymes at the manufacturer's recommended conditions and purify the DNA by ethanol precipitation with 50 μl of ammonium acetate and 0.3 ml of ethanol.

3. Redissolve the DNA pellet in 100 μl of buffer EB. It can be stored for several months at −20°C.

    Alternatively, quench the restriction digestion with 500 μl of buffer PB and purify the DNA by following the QIAquick protocol.

Stage 3: Incremental Truncation

Incremental truncation is performed using temperature and NaCl concentration to control the rate of Exo III digestion. We found the rate of truncation to vary with NaCl at 22°C by the following equation: rate (bp/min)= $47.9 \times 10^{(-0.00644 \times N)}$ where N = concentration of NaCl in mM (0–150 mM). Using this equation, the rate of Exo III digestion is ~10 bases/minute at 22°C in the presence of 100 mM salt. This rate expression is valid for a DNA concentration (1 μg/30 μl) and ratio of Exo III to DNA (100 units/μg DNA).

> In contrast to the truncation of pDIM-N2/ -C8 where one of the 3′ ends is protected from Exo III hydrolysis because it is a 3′-overhang (see above), the truncation of pDIM-N5 for making fusion libraries proceeds simultaneously on both ends of the linearized vector. Therefore, twice the amount of Exo III is required to keep the ratio of units of enzyme per moles of 3′ ends constant.

The dependence of Exo III digestion rate on temperature in the absence of NaCl (Henikoff 1987; Hoheisel 1993) and NaCl concentration at higher temperatures (Tomb and Barcak 1989; Hoheisel 1993) has been determined by other investigators.

Upon addition of Exo III to the reaction mixture, aliquots are taken at constant time intervals, usually every 20–60 seconds, depending on rate and overall length of truncation. After quenching the final aliquot, the single-stranded tails are removed by treatment with a single-stranded nuclease. Although we have successfully used S1 nuclease to hydrolyze the single-stranded overhang in the past, recent experiments using mung bean nuclease have produced better results, including larger, more evenly distributed libraries (Ostermeier and Benkovic 2001). Although mung bean nuclease requires a buffer change, this is conveniently accomplished using DNA purification spin columns such as Qiagen's QIAquick column. The use of spin columns greatly improves the yield and quality of truncated DNA compared to ethanol precipitation.

Subsequent processing of the DNA depends on the desired use. Accordingly, this section is divided into descriptions of the construction of heterodimer libraries and ITCHY fusion libraries: (1) single gene truncations and heterodimer libraries, (2) fusion libraries using pDIM-N2 and pDIM-C8, and (3) fusion libraries using pDIM-N5.

This general protocol is applicable for making incremental truncation libraries of a single gene, for making heterodimer libraries, or for making fusion libraries. It is written for truncating over a 300-bp range at a rate of 10 bp/minute. The rate of truncation can be changed by altering the NaCl concentration in step 2. The sample size and frequency of sampling in step 5 may be changed accordingly to truncate for different lengths of time.

1. Equilibrate 180 µl of 1x Exo III stop buffer on ice in a 1.5-ml tube (tube A).

2. To a second 0.65-ml tube (tube B) add:

DNA	2 µg
10x Exo III buffer	6 µl
1 M NaCl	6 µl

   Add H$_2$O to 60 µl.

3. Equilibrate tube B at 22°C.

4. At time = 0, add 200 units of Exo III to tube B and mix immediately.

5. Beginning at 30 seconds, remove 1-µl samples every 30 seconds and add to tube A. Mix tube A well. Note that all time points are removed to tube A, which is kept on ice.

   The rate of Exo III is very temperature dependent. It is preferable to leave tube B open during the sampling to avoid warming the tube by repeated handling.

6. After all samples are taken, add 1.2 ml of QIAquick buffer PB.

7. Follow the QIAquick protocol.

8. Elute DNA from column with 47 µl of buffer EB.

9. Add:

   5 µl of 10x mung bean nuclease buffer
   1 µl of mung bean nuclease (10 units)

10. Incubate for 30 minutes at 30°C.

11. Add 250 µl of QIAquick buffer PB.

12. Follow the QIAquick protocol to purify truncated DNA.

13. Follow Method A, B, or C (step 14 on) below depending on the type of library.

Single Gene Truncation and Heterodimer Libraries

When truncating 5′ and 3′ gene fragments individually, start and stop codons must be introduced in a separate step. For the 3′-truncated fragment, a stop codon must be fused to the library. Because the reading frame at the end of the truncated gene is unknown, a series of stop codons in all three frames is used. For two of the frames, this results in the addition of one to three carboxy-terminal residues absent in the original protein. However, for the third frame, no extra residues are added. Similarly, it is unknown what reading frame will provide the correct start codon for the 5′ truncated fragment. Thus, one-third of the library will be in-frame and therefore produce meaningful carboxy-terminal protein fragments.

As shown in Figure 4, the stop codon triplet and start codon can be exposed upon restriction enzyme digestion. Digestion of pDIM-N2 with *Nsi*I followed by removal of the 3′ single-stranded overhang exposes the sequence 5′-TAACTAGCTAA-3′ containing stop codons in all three frames for fusion to the truncated gene. In this case, the 3′ overhang is removed using the 3′ to 5′ exonuclease activity of Klenow DNA polymerase in the absence of dNTPs. Subsequently, dNTPs are added to fill in any truncations past the blunt end of the former *Nsi*I site and to blunt the end of the DNA that was truncated. Digestion of pDIM-C8 with *Nco*I followed by a fill-in reaction with Klenow DNA polymerase and dNTPs on the 5′ single-stranded overhang exposes an ATG start codon for fusion to the truncate gene. Both libraries are then circularized by ligation under dilute conditions so as to favor intramolecular ligation and prepared for transformation into *E. coli*.

Method A: Making an Incremental Truncation Library of a Single Gene or for Heterodimers

14. Elute the truncated DNA from the QIAquick column with 90 μl of buffer EB.

15. For restriction enzyme digestions: For pDIM-N2, add 10 μl of 10x *Nsi*I buffer, 1 μl of 100x BSA, and 15 units of *Nsi*I. For pDIM-C8, add 10 μl of 10x NEB1 buffer, 1 μl of 100x BSA, and 18 units of *Nco*I.

16. Incubate for 2 hours at 37°C.

17. Add 0.5 ml of buffer PB.

18. Follow the QIAquick protocol.

19. Elute DNA from the column with 82 μl of buffer EB.

20. Equilibrate the tube at 37°C.

21. Klenow treatment: For pDIM-N2, add 10 μl of Klenow buffer, incubate for 3 minutes at 37°C, add 10 μl of dNTP mix, and incubate for 5 minutes at 37°C. For pDIM-C8, add 10 μl of dNTP mix and 10 μl of Klenow buffer and incubate for 5 minutes at 37°C.

22. Inactivate the Klenow buffer by incubating for 20 minutes at 72°C.

23. Cool to room temperature and add 0.4 ml of ligase buffer.

24. Incubate for ≥12 hours at room temperature.

25. Concentrate by ethanol precipitation with 250 μl of ammonium acetate and 1.5 ml of ethanol.

26. Transform into the desired host.

a)

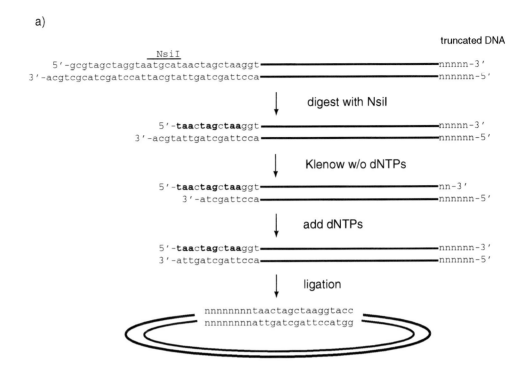

b)

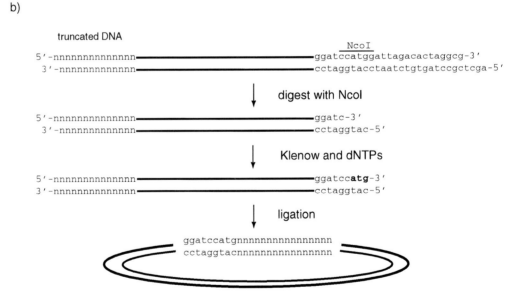

FIGURE 4. Fusion of truncation libraries to stop or start codons. (a) Truncations from the 3´ end of a gene need to be fused to a stop codon for the appropriate fragment to be expressed. Digestion with *Nsi*I exposes a series of three stop codons (shown in bold) in all three frames. The 3´-to-5´ exonuclease activity of Klenow is used to remove the 3´ overhang next to the first stop codon so that fusion to the truncated gene can be done in a seamless manner. Thus, if the gene fragment is in frame with this first stop codon, no additional carboxy-terminal residues will be added to the truncated protein. The other two frames result in one to three additional carboxy-terminal residues. (b) Truncations from the 5´ end of a gene need to be fused to a start codon in order for the appropriate fragment to be expressed. This is conveniently achieved by digestion with *Nco*I followed by a fill-in reaction with Klenow and dNTPs. The ATG start codon (shown in bold) is fused seamlessly to the truncated gene fragment, one-third of which will be in frame.

Method B: Making Fusion Libraries Using pDIM-N2 and pDIM-C8

After the mung bean nuclease treatment and purification using the QIAquick column according to the manufactuer's instructions, the DNA is blunt-ended with Klenow fragment DNA polymerase and dNTPs. Following enzyme inactivation by heat denaturation, the DNA is digested with *Nsi*I to prepare pDIM-N2 to receive the truncated library constructed in pDIM-C8. The *Nsi*I-digested DNA is subjected to agarose gel electrophoresis, and the desired range of DNA sizes for pDIM-N2 with its truncation library and the truncation library from pDIMC-8 is isolated. We obtain our largest libraries when we use electroelution to recover the DNA from the agarose, followed by ethanol precipitation and ligation in a volume of 20 μl or less.

14. Elute the truncated DNA from the QIAquick column with 72 μl of buffer EB.

15. Equilibrate the tube at 37°C.

16. Klenow treatment: add 10 μl of dNTP mix and 10 μl of Klenow buffer. Incubate for 5 minutes at 37°C.

17. Inactivate the Klenow buffer by incubating for 20 minutes at 72°C.

18. For restriction enzyme digestions: add 10 μl of 10x *Nsi*I buffer, 1 μl of 100x BSA, and 15 units of *Nsi*I.

19. Incubate for 2 hours at 37°C.

20. Incubate for 20 minutes at 72°C.

21. Isolate the large fragment from the pDIM-N2 digestion and the small fragment library from pDIM-C8 digestion by agarose gel electrophoresis.

    We have had the most success in creating large libraries by electroeluting the DNA from the gel slice using the S&S Elutrap electro-separation system (Schleicher & Schuell, Keene, New Hampshire 03431) followed by concentration by ethanol precipitation with ammonium acetate into 17 μl of H$_2$O.

22. Ligation:

    17 μl of DNA
    2 μl of ligase buffer
    6 Weiss units of T4 DNA ligase

23. Incubate for ≥12 hours at 15°C.

24. Transform into the desired host.

Method C: Making Fusion Libraries Using pDIM-N5

In this variation, after the mung bean nuclease step and purification using the QIAquick column, the DNA is treated with Klenow fragment DNA polymerase and dNTPs to fully blunt the truncated gene. After enzyme inactivation by heat denaturation, the DNA is digested with *Nsi*I to uncouple the synchronized truncation libraries. No gel purification is required, and ligation will randomly fuse members of the two libraries creating the ITCHY library.

The avoidance of the gel purification step saves time and increases the yield of DNA for subsequent ligation. However, this comes at the expense of being unable to select only those truncations that occur in the desired size range, which (1) may be important in some applications and (2) eliminates the low level of abhorrent truncation products that appear due to the small fraction of nicked DNA in the starting material and for other reasons (Henikoff 1984).

14. Elute the truncated DNA from the QIAquick column with 72 µl of buffer EB.

15. Equilibrate the tube at 37°C.

16. Klenow treatment: add 10 µl of dNTP mix and 10 µl of Klenow buffer and incubate for 5 minutes at 37°C.

17. Inactivate the Klenow buffer by incubating for 20 minutes at 72°C.

18. For restriction enzyme digestions: add 10 µl of 10x *Nsi*I buffer, 1 µl of 100x BSA, and 15 units of *Nsi*I.

19. Incubate for 2 hours at 37°C.

20. Incubate for 20 minutes at 72°C.

21. Ethanol-precipitate into 17 µl of H_2O.

22. Ligation:

    17 µl of DNA
    2 µl of ligase buffer
    6 Weiss units of T4 DNA ligase

23. Incubate for ≥12 hours at 15°C.

24. Transform into the desired host.

Controls for Incremental Truncation

Simple control experiments can provide valuable information concerning the performance of Exo III during the truncation reaction. By doubling the scale of the digestion reaction but removing only the normal sample volume over the course of the digestion, half of the volume will remain after the last time point. This second half is then quenched in a separate tube and processed as normal through the mung bean nuclease digestion and subsequent purification. The size of this DNA, which corresponds to DNA that has undergone maximum truncation, can then be analyzed by agarose gel electrophoresis to determine whether truncation has proceeded normally and over the desired length. To make size determination more accurate, or in cases where the length of truncation is small in comparison to the size of the plasmid, it may be advantageous to digest this DNA with a restriction enzyme prior to electrophoresis. It is important to remember that the truncated DNA will run as a smear on the gel because the standard deviation in truncation length varies at 0.22 times the truncation length.

Verification of Library Diversity

Following transformation into a suitable host strain, the easiest and quickest method to verify the diversity of the incremental truncation library is to perform PCR on randomly selected colonies using primers that are complementary to regions outside the desired range of truncation. The size of the amplification product is determined by agarose gel electrophoresis. Alternatively, restriction analysis can be performed on plasmid DNA isolated from randomly selected library members.

TABLE 2. Potential Applications of the Incremental Truncation Technology

Amino-terminal or carboxy-terminal truncations of a single gene
    Identify subdomains or motifs responsible for dimerization or heterologous interactions
    Define minimal functional units
    Assign subdomains or motifs responsible for various protein functions
    Create proteins with altered properties (most likely loss of function)
    Identify protein fragments suitable for crystallization experiments
Combining amino-terminal and carboxy-terminal truncations of a single gene (heterodimers)
    Identify interacting fragments of the protein
    Distinguish independent folding units within a protein
    Identify possible ancestral fusion points
Fusing amino-terminal and carboxy-terminal fusions of a single gene
    Identify subdomains or motifs responsible for dimerization or heterologous interactions
    Generate functional proteins with internal deletions or tandem duplications
    Define minimal functional units
    Assign function to subdomains or motifs within the protein
    Map out flexible segments of the protein that can accommodate changes
    Create proteins with altered properties
Fusing amino-terminal and carboxy-terminal fusions of two different genes
    Force independent proteins or domains in proximity to one another
    Create hybrid protein libraries
    Optimize linker regions between proteins to be fused

CONCLUSION

The methodology outlined above provides multiple opportunities to study protein interactions. In addition, it allows the modification and alteration of existing protein properties, as well as the generation of novel activities in the context of combinatorial libraries. A list of potential applications of the incremental truncation technology is given in Table 2.

REFERENCES

Gaubatz J.W. and Flores S.C. 1990. Purification of eukaryotic extrachromosomal circular DNAs using exonuclease III. *Anal. Biochem.* **184:** 305–310.

Henikoff S. 1984. Unidirectional digestion with exonuclease III creates targeted breakpoints for DNA sequencing. *Gene* **28:** 351–359.

———. 1987. Unidirectional digestion with exonuclease III in DNA sequence analysis. *Methods Enzymol.* **155:** 156–165.

Hoheisel J.D. 1993. On the activities of *Escherichia coli* exonuclease III. *Anal. Biochem.* **209:** 238–246.

Jasin M., Regan L., and Schimmel P. 1983. Modular arrangement of functional domains along the sequence of an aminoacyl tRNA synthetase. *Nature* **306:** 441–447.

Kato I. and Anfinsen C.B. 1969. On the stabilization of ribonuclease S-protein by ribonuclease S-peptide. *J. Biol. Chem.* **244:** 1004–1007.

Ladurner A.G., Itzhaki L.S., de Prat Gay G., and Fersht A.R. 1997. Complementation of peptide fragments of the single domain protein chymotrypsin inhibitor 2. *J. Mol. Biol.* **273:** 317–329.

Lutz S. and Benkovic S.J. 2000 Homology-independent protein engineering. *Curr. Opin. Biotechnol.* **11:** 319–324.

New England BioLabs 1998. *Exo-size deletion kit instruction manual.* New England BioLabs, Beverly, Massachusetts.

———. 2000. *New England BioLabs catalog and technical reference.* New England BioLabs, Beverly, Massachusetts.

Ostermeier M. and Benkovic S.J. 2000. Evolution of protein function by domain swapping. *Adv. Protein Chem.* **55:** 29–77.

———. 2001. Construction of hybrid gene libraries involving the circular permutation of DNA. *Biotech. Lett.* **23:** 303–310.

Ostermeier M., Nixon A.E., and Benkvoic S.J. 1999a. Incremental truncation as a strategy in the engineering of novel catalysts. *Bioorg. Med. Chem.* **7:** 2139–2144.

Ostermeier M., Shim J.H., and Benkovic S.J. 1999b. A combinatorial approach to hybrid enzymes independent of DNA homology. *Nat. Biotechnol.* **17:** 1205–1209.

Ostermeier M., Nixon A.E., Shim J.H., and Benkovic S.J. 1999c. Combinatorial protein engineering by incremental truncation. *Proc. Natl. Acad. Sci.* **96:** 3562–3567.

Promega. 1995. *Erase-a-base system technical manual,* #TM006. Promega, Madison, Wisconsin.

Putney S.D., Benkovic S.J., and Schimmel P.R. 1981. A DNA fragment with an α-phosphorothioate nucleotide at one end is asymmetrically blocked from digestion by exonuclease III and can be replicated in vivo. *Proc. Natl. Acad. Sci.* **78:** 7350–7354.

Sambrook J., Fritsch E.F., and Maniatis T. 1989. *Molecular cloning: A laboratory manual,* 2nd edition. Cold Spring Harbor Laboratory Press, Cold Spring Harbor, New York.

Tasayco M.L. and Carey J. 1992. Ordered self-assembly of polypeptide fragments to form nativelike dimeric trp repressor. *Science* **255:** 594–597.

Tomb J.-F. and Barcak G.J. 1989. Regulating the 3′-5′ activity of exonuclease III by varying the sodium chloride concentration. *BioTechniques* **7:** 932–933.

Ullmann A., Jacob F., and Monod J. 1967. Characterization by in vitro complementation of a peptide corresponding to an operator-proximal segment of the beta-galactosidase structural gene of *Escherichia coli.* *J. Mol. Biol.* **24:** 339–343.

Wu R., Rube G., Siegel B., Jay E., Spielman P., and Tu C.D. 1976. Synchronous digestion of SV40 DNA by exonuclease III. *Biochemistry* **15:** 734–740.

Zasloff M., Ginder G.D., and Felsenfeld G. 1978. A new method for the purification and identification of covalently closed circular DNA molecules. *Nucleic Acids Res.* **5:** 1139–1152.

29 Catalytic Antibodies: New Characters in the Protein Repertoire

Georgy A. Nevinsky,[1] Olga O. Favorova,[2] and Valentina N. Buneva[1]

[1]Novosibirsk Institute of Bioorganic Chemistry, Siberian Division of Russian Academy of Sciences, 8, Novosibirsk 630090, Russia; [2]Russian State Medical University, Ministry of Public Health of Russian Federation, Moscow 117437, Russia

INTRODUCTION

Classically, antibodies have been characterized as proteins produced by the immune system that have the sole function of binding other molecules, called antigens, with the goal of eliciting an immune response. Most natural antigens for antibodies are proteins, and the interactions of antigens with antibodies are very strong, with reported K_d values between 0.01 and 10 nM (Karush 1978). In this classic conception, antibody function is similar to that of enzymes in specifically binding other molecules; however, antibodies are distinct from enzymes in that they do not have the ability to catalyze chemical conversions of their bound partners. For the vast majority of antibodies, this observation is correct. However, in a 1946 consideration of the catalytic function of enzymes, Linus Pauling first hypothesized that the active center of an enzyme is closely juxtaposed to a "strained configuration" of its substrate (that is, targeted against the structure of the transition state) rather than to the native conformation of the substrate molecule (Pauling 1946). On the basis of this hypothesis, it logically followed that the binding energy between enzyme and substrate could play a significant role in lowering activation energy, thereby catalyzing a chemical

523

change. This idea led Jencks in 1969 to propose that antibodies generated in an anti-hapten immune response against chemically stable analogs of the transition state of a reaction of interest could potentially display an enzymatic activity (Jencks 1969). Indeed, this prediction was confirmed in 1986 when the first anti-hapten monoclonal catalytic antibodies were obtained (see below) and termed "abzymes" (derived from *antibody* en*zyme*). At present, artificial monoclonal abzymes catalyzing more than 100 distinct chemical reactions have been obtained. Moreover, many naturally occurring abzymes have been detected in the sera of patients with several pathologies, as well as in the milk of healthy human mothers.

From the perspective of control of protein–protein interactions, and in particular, of the study of extracellular signaling, the development of abzymes holds a unique promise: the ability selectively to modify or eliminate defined interacting proteins. For example, to date, a relatively limited set of proteases has been defined that recognize specific sequences on target proteins, and subsequently induce specific cleavages. These "restriction endoproteases" include the coagulation factor Xa (Nagai and Thogersen 1987), the caspases, and calpain (for a discussion, see Wang 2000). These proteases recognize and cleave short peptide sequences that are found on one or multiple cellular targets. The ability to create antibody-based enzymes that target different, or more selective, sequences would allow the specific cleavage and modification of essentially any surface-displayed protein to which the antibody has access. In theory, given the appropriate starting hapten, abzymes could potentially be developed that target any protein sequence of interest, allowing exact control of cell signaling processes. In practice, the first amide-catalyzing antibody was generated in 1988 and shown to enhance the rate of catalysis of its substrate 250,000 times over the uncatalyzed reaction (Janda et al. 1988). Since that time, the work of a number of groups has begun to explore the ability of abzymes as small-molecule or protein-modulatory agents. This chapter reviews some of the main themes of this work and provides a description of the important issues in identifying naturally occurring polyclonal abzymes.

DESIGN AND POTENTIALITIES OF "ARTIFICIAL" ABZYMES

In 1985, a general method for generating catalytic monoclonal antibodies (mAbs) against transition-state analogs, and a way to use those antibodies to accelerate chemical reactions, was first described (Schochetman and Massey 1985). One year later, two groups were able to produce the first mAbs with catalytic properties that were generated against hapten analogs of the transition states for *p*-nitrophenylphosphorylcholine (Pollack et al. 1986) or for monoaryl phosphonate esters (Tramontano et al. 1986a,b). Since that time, abzymes catalyzing the hydrolysis of amides and esters, as well as reactions of cyclization (see, e.g., Janda et al. 1993; Li et al. 1994, 1996; Wentworth et al. 1998), decarboxylation (see, e.g., Smiley and Benkovic 1994; Barbas et al. 1997; Hotta et al. 2000), lactonization (Napper et al. 1987), peroxidation (Ding et al. 1998), photochemical thymine dimer cleavage, bimolecular amide-bond formation, and other reactions not known to be catalyzed by known enzymes have been described. A number of papers have described abzymes that perform other specific functions directed against proteins, including formation of cyclic peptides (Smithrud et al. 2000), catalysis of peptidyl-prolyl *cis-trans* isomerization in protein folding (Ma et al. 1998), and development of a novel enzymatic activity cleaving the bacterial protein HPr (Liu et al. 1998). Some abzymes have been described that require cofactors for activity, similar to standard enzymes (Iverson and Lerner 1989). The field of artificial abzymes has been amply reviewed recently (see Lerner and Tramontano 1987; Stewart and Benkovic 1993; Suzuki 1994; Martin and Schultz 1999, and references therein, for a more detailed description of the relevant reactions).

The evolution of the technology of artificial abzymes during the last 15 years has led not only to the rapid development of direct approaches for the generation of antibodies with specified properties, but also to the creation of strategies to revise the targeting specificity of individual abzymes. Such modifications of antigen-binding specificity can be achieved genetically in vitro by application of site-directed mutagenesis, genetic selection, or screening (using approaches such as phage display detailed in Chapter 8). Alternatively, modification can be induced directly on purified antibody via selective chemical modification involving direct introduction of catalytic groups into the antibody combining site. Some studies describing these approaches include Ersoy et al. (1999), Gao and Paul (1995), Miller et al. (1997), Roberts et al. (1994), Stewart et al. (1994), and references therein. As a result of application of these approaches, the substrate specificity (and/or the specific activity) of some artificial abzymes is comparable to or even higher than that of enzymes with the same catalytic activity (Gouverneur et al. 1993; Barbas et al. 1997; Janda et al. 1997). The mechanistic basis for the activity of such abzymes is becoming well understood (see, e.g., Thayer et al. 1999 and discussion below).

NATURAL CATALYTIC ANTIBODIES

In contrast to the artificially designed transition-state-directed antibodies described in the preceding section, an alternative means of developing catalytic antibodies is through study and purification of autoantibodies produced naturally in human sera. As noted by Suzuki, naturally occurring antibodies can be quickly and easily purified at low cost, possess novel catalytic activities of interest, can be informative for development of therapeutic strategies involving antibodies administered in serum, and can be useful for design of haptens for production of artificial antibodies (Suzuki 1994). Other relevant issues in comparison of artificial versus natural, and in vitro versus in vivo, abzyme generation approaches are discussed in Fastrez (1997). The remainder of this section collates rigid criteria indicating that efficient naturally occurring catalytic antibodies do exist, and describes methods of natural catalytic antibody purification.

The idea that catalytic antibodies might be involved in natural immunity first arose after the discovery by Paul and coworkers of natural catalytic antibodies specifically hydrolyzing vasoactive intestinal peptide (VIP), a 28-amino-acid neurotransmitter, in the serum immunoglobulin G (IgG) fractions of ~50% of asthmatic patients (Paul et al. 1989). As with the overwhelming majority of antibodies, these antibodies exhibit very strong binding affinity for their cognate antigen, with an affinity for VIP (K_d ~0.76 nM). Seven antibody-sensitive peptide bonds were identified, of which six were clustered between residues 14 and 22 (Paul et al. 1991). In functional studies, separated light chains of these antibodies were found to be active in the hydrolysis of VIP (Sun et al. 1994, 1997). Compared to reversible antigen binding, it was found that antigen cleavage by catalytic antibodies was a particularly potent means to achieve the neutralization of targeted antigens (Paul 1998). Significantly, the structure and function of these antibodies were inherited via a germ-line variable light (V_L) gene, providing evidence that certain antibodies possessed an innate catalytic function of certain antibodies and arguing against the idea that abzymes required artificial haptens mimicking transition states for their generation (Gololobov et al. 1999). Subsequently, monoclonal anti-VIP abzymes were obtained and analyzed (see below), yielding further insight into their function.

Following the initial discovery of anti-VIP abzymes, a number of natural catalytic antibodies with diverse activities were detected in sera of patients with different immune pathologies. Proteolytic antibodies directed at thyroglobulin (Tg), a precursor of thyroid hormone, were found in the serum IgG of patients with Hashimoto's thyroiditis and systemic lupus erythematosus (SLE) (Li et al. 1995; Paul et al. 1997). Light chains isolated from patients with multiple myeloma possessed prothrombinase activity (Thiagarajan et al. 2000). Others have reported factor VIII-cleaving alloantibodies in the sera of patients with severe hemophilia (Lacroix-Desmazes et al. 1999). The identified proteolytic antibodies cleaved their protein substrates at a small number of fixed positions, indicating specificity of action. The K_M values of Tg- and prothrombin-directed antibodies for Tg and a peptide 268–271 of prothrombin were 39 nM and 103 μM, respectively, indicating high-affinity substrate recognition. In a study of the development of catalytic specificity by abzymes, the polyreactive (nonspecific) peptidase activity of serum IgGs from healthy individuals and patients with autoimmune disease was compared, based on the extent to which these sera cleaved a synthetic protease substrate, Pro-Phe-Arg-methylcoumarinamide (Paul et al. 1997). A transition from a polyreactive proteolytic activity to autoantigen-directed activity in autoimmune disease was suggested (Paul et al. 1997). Finally, the identification of a human immunodeficiency virus gp120-cleaving antibody light chain from multiple myeloma patients has recently demonstrated that natural catalytic immunity is not restricted to autoantigenic substrates (Paul et al. 2000).

In addition to abzymes directed against proteins, DNA- and RNA-hydrolyzing antibodies have also been found in the sera of patients with SLE and studied in detail (for examples, see Shuster et al. 1992; Buneva et al. 1994; Gololobov et al. 1995; Vlassov et al. 1998a,b; Andrievskaya et al. 2000, and references therein). Subsequently, RNA- and/or DNA-hydrolyzing IgM and/or IgG were detected in the sera of patients with several autoimmune, viral, and lymphoproliferative diseases. These sources include Hashimoto's thyroiditis, polyarthritis (Vlassov et al. 1998a,b), multiple sclerosis (MS) (Baranovskii et al. 1998), B-cell lymphomas (Kozyr et al. 1998), viral hepatitis (Baranovsky et al. 1997), and acquired immunodeficiency syndrome (AIDS) (Gabibov et al. 1994). Such abzymes were not detected in sera from normal controls or patients with other disorders, including influenza, pneumonia, tuberculosis, tonsillitis, duodenal ulcer, and some types of cancer (Baranovsky et al. 1997), suggesting specificity of generation. Last, and intriguingly, in addition to IgG nucleic acid-hydrolyzing abzymes (Kanyshkova et al. 1997; Buneva et al. 1998; Semenov et al. 1998; Nevinsky et al. 2000), amylolitic abzymes (Savel'ev et al. 1999) and protein-phosphorylating sIgA- (or IgA) abzymes (Kit et al. 1995; Nevinsky et al. 1998) were found in the milk of healthy human mothers. Of these, the "kinase" abzymes were active against a number of proteins including the major milk components α-casein and β-casein, and they represent a first example of natural abzymes catalyzing a bisubstrate synthetic reaction.

Quality Control Criteria for Intrinsic Catalytic Activities of Antibodies

Because antibodies can form complexes with other proteins, and antibody-mediated catalysis is sometimes characterized by relatively low reaction rates, it is important to prove that a catalytic activity of Ig fractions is not due to a contamination with "classic" enzymes with the same specificity. For example, if the turnover number, k_{cat}, for the antibody is 1 min^{-1}, it may be a result of 0.002% admixture of antibody with an enzyme with k_{cat} of 5×10^4 min^{-1}. Application of a set of rigid criteria worked out by Paul et al. (1989) in the first articles concerning natural abzymes (for review, see Paul 1998) allowed the authors of the initial study to conclude that the observed VIP-hydrolyzing activity is an intrinsic property of IgGs from the sera of patients with asthma. We note the consideration of these criteria is also useful in evaluating the properties of artificially generated abzymes, as described above. The most important criteria used in this process by Paul and coworkers were:

1. Electrophoretic homogeneity of the IgG (based on silver staining of an SDS-PAGE gel).

2. Retention of VIP-hydrolyzing activity by Fab fragments of the purified antibodies.

3. Complete adsorption of the catalytic activity by anti-IgG Sepharose and its elution from the adsorbent with low-pH buffer.

4. Immunoprecipitation of abzymes by anti-IgG antibodies and disappearance of catalytic activity from the solution.

5. Demonstration of the VIP-hydrolyzing activity in the antibodies of only 50% of asthma patients but in no healthy controls.

6. Demonstration that gel filtration of IgG at pH 2.7, in conditions providing dissociation of any non-covalent protein complex ("acidic shock"), did not lead to the disappearance of abzyme activity, whereas the peak of activity tracked exactly with a 150-kD IgG.

7. Characterization of a low K_m value of the abzymes for VIP (38 nM) hydrolysis, which testified to a high affinity to the substrate and was comparable with the K_d values of many antibody complexes with their antigens.

8. Determination that the substrate specificity of IgG-dependent hydrolysis of VIP was different from that of any known protease.

These rigorous checks for intrinsic catalytic activities of abzymes have been used by different groups, with some modifications, as the basis for detection of natural catalytic antibodies. As the field has developed, other controls that have been added into the characterization regimen have been developed by our groups and others. For example:

1. Preservation of the catalytic activity of antibodies is demonstrated after gel filtration not only under conditions of extremely low pH, but also in other severe conditions that effectively dissociate noncovalent complexes. Examples of such conditions include use of buffers with strongly alkaline pH (pH 10–11) or neutral buffers containing 5 M thiocyanate or 6 M guanidine chloride (Kit et al. 1995; Nevinsky et al. 1998).

2. Model experiments have been used to analyze the effectiveness of separation of mixtures containing homogeneous abzymes and human enzymes of similar specificities. After chromatography on protein A–Sepharose and gel filtration in severe conditions, an effective separation of abzymes from the added enzymes should be clearly achieved (see Andrievskaya et al. 2000 and references therein; Vlassov et al. 1998a,b).

3. Detection of a catalytic activity associated with separated L- and/or H-subunits of Ig isolated by affinity chromatography after their dissociation in mild conditions (1 mM mercaptoethanol and 4 M urea) provides good evidence of intrinsic catalytic properties of antibodies (Kanyshkova et al. 1997; Buneva et al. 1998; Nevinsky et al. 1998, 2000).

4. Direct evidence that antibodies possess enzymatic activities can also be obtained by using an affinity modification method. In this approach, covalent binding of a reactive analog of the substrate to the abzyme but not to possible contaminating proteins is demonstrated (Buneva et al. 1994; Kit et al. 1995; Kanyshkova et al. 1997; Nevinsky et al. 1998, 2000; Semenov et al. 1998).

5. A further very strong criterion of the existence of antibodies as catalysts is the in situ detection of catalytic activities of antibodies versus their Fab fragments in a gel containing the corresponding substrate (Kanyshkova et al. 1997; Baranovskii et al. 1998; Buneva et al. 1998; Semenov et al. 1998; Andrievskaya et al. 2000). For example, when IgG from the milk of healthy mothers and the Fab fragments derived from this IgG are separated by SDS-PAGE in nonreducing conditions in a DNA-containing gel, a DNA-hydrolyzing activity is revealed in the bands corresponding to both IgG and Fab fragments (Fig. 1) (Buneva et al. 1998). The same results are obtained for RNA-hydrolyzing activity (Andrievskaya et al. 2000). These criteria can in some cases be supplemented by the in situ detection of catalytic activities of separated Ig chains after SDS-PAGE using reducing conditions (in the presence of 2-mercaptoethanol). For some abzymes, a DNA-hydrolyzing activity is revealed in a DNA-containing gel for individual antibody chains (Kanyshkova et al. 1997; Baranovskii et al. 1998; Andrievskaya et al. 2000). However, this criterion is not universal: After reduction of some other Igs, the in-gel assay showed the absence of DNA-hydrolyzing activity in the separated L and H chains (Semenov et al. 1998; Nevinsky et al. 2000).

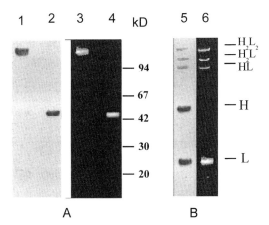

FIGURE 1. In situ gel assay of DNase activity of milk IgG (lanes *1, 3, 5, 6*) and its Fab fragments (lanes *2, 4*) by SDS-PAGE in a gel containing DNA. Before electrophoresis the samples were incubated in nonreducing (*A*) or reducing conditions (in the presence of 2-mercapthoethanol) (*B*). DNase activity was revealed as a sharp dark band on the fluorescent background after ethidium bromide staining (lanes *3, 4,* and *6*) (the negatives of the films are shown). Gels stained with Coomassie Blue show positions of IgG (lane *1*), its Fab fragments (lane *2*), and separated heavy (H) and light (L) chains (lane *5*). Arrows indicate the positions of molecular mass markers (Buneva et al. 1998).

In conjunction with Paul's original list, these tests provide strong evidence supporting the catalysis of various reactions by natural abzymes. Some of these tests, and particularly the in situ observation of catalytic activity induced by an antibody, its Fab fragment, and one of its individual separated chains, may be considered as the most rigid argument unambiguously assigning observed catalytic activity to antibodies versus contaminating proteins, while also providing the basis for subsequent structural and functional analysis of abzyme activity.

Issues in Isolation of Natural Abzymes Possessing a High Catalytic Activity

Abzymes constitute a subset of normal antibodies, and are purified and processed using similar experimental protocols (Harlow and Lane 1988). The standard scheme of antibody isolation includes as a first step a separation of a total antibody fraction consisting of many different Igs with differing affinity for various antigens, including proteins, nucleic acids, and other species, which may be tightly complexed with these antibodies. Next, monospecific preparations of antibodies against the single natural antigen are isolated. These antibodies are polyclonal in origin and may contain an extremely diverse set of Ig molecules, which include abzymes along with common antigen-binding antibodies without catalytic activity. The repertoire of antibodies differs between individual patients, and the affinity of distinct antibody fractions for an immobilized ligand may differ by several orders of magnitude (Kit et al. 1995; Nevinsky et al. 1998). To simplify problems of purification, generally only the fraction of antibodies with the highest affinity for the substrate of interest is used.

Affinity chromatography on adsorbents bearing anti-IgG, anti-IgM, and anti-IgA antibodies on protein-A Sepharose in harsh conditions is used as the first step of purification for natural abzymes. As a rule, this step yields homogeneous preparations of antibodies according to SDS-PAGE (with silver staining) and western blot with immunoprobes, and ensures the effective removal of proteins nonspecifically bound with antibodies. The next steps are chromatography on DEAE-cellulose followed by incubation of antibody preparations for 1 hour at pH 2.6 to dissociate any non-covalent complexes, followed by gel filtration on Toyopearl HW-60 or other

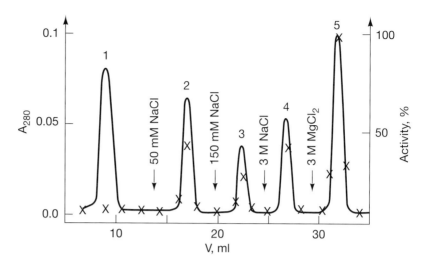

FIGURE 2. Affinity chromatography of milk sIgA with casein kinase activity on ATP-Sepharose. (–) Absorption at 280 nm; (x), relative casein kinase activity of sIgA (in % to the fraction with highest activity). Fraction 1 was eluted by 50 mM Tris-HCl (pH 6.8). (Nevinsky et al 1998).

adsorbents equilibrated with the same acidic buffer (Kit et al. 1995; Nevinsky et al. 1998 and references therein). These steps greatly reduce the likelihood that preparations of abzymes will contain any contaminating proteins: However, we note that one risk of this purification regimen is that incubation at low pH may disrupt the catalytic activity of some abzymes. The affinity chromatography of catalytic antibody preparations on an adsorbent bearing immobilized substrates (proteins, nucleic acids, nucleotides, etc.) is used as a last step and usually leads to their resolution into a set of distinct fractions demonstrating an extremely pronounced heterogeneity of monospecific catalytic antibodies, which differ from individual to individual. For example, affinity chromatography of human milk sIgA preparations on ATP-Sepharose (Fig. 2) demonstrated that these sIgA possessed a heterogeneous affinity for ATP. Usually, some fractions can only be eluted under conditions required for disruption of very stable immune complexes (3 M NaCl, 3 M MgCl$_2$) (Nevinsky et al. 1998). This step permits enrichment of antibodies with strong substrate binding.

Ideally, the final step of purification of abzymes should be their separation from noncatalytic antibodies of the same specificity. Unfortunately, at this time, no convenient method for this has been described. As a result, all described polyclonal abzyme preparations represent a mixture of catalytic and noncatalytic antibodies of similar specificity. Because of this, the relative specific activities of polyclonal enzymes may be essentially underestimated, especially as the proportion of abzymes in the high-affinity antibody fraction isolated by affinity chromatography for a given substrate may be in the order of 1–5% of the total population (G.A.Nevinsky et al., unpubl.).

MECHANISTIC STUDIES OF ABZYMES

To develop abzymes as tools for biological manipulations, it is necessary to understand the structural basis of their activity in detail. Although study of polyclonal abzymes provides important information about the potential catalytic activities of antibodies, structural studies require pure, uniform, material concentrated to high levels. Hence, these studies rely on monoclonal or recombinant sources of antibodies. A number of different catalytic antibodies have been analyzed as models for understanding the basis of catalysis. Some of this work is briefly summarized below.

The VIP-hydrolyzing abzyme, which was the first polyclonal abzyme described (Paul et al. 1989), was adapted for mechanistic analysis. The light chain of the VIP abzyme was expressed in bacteria, purified, and found to possess independent catalytic activity (Gao and Paul 1995; Tyutyulkova et al. 1996). Subsequently, single-chain Fv constructs containing the V_L domain of the anti-VIP light chain linked via a 14-residue peptide to its natural V_H domain partner were developed that possessed an increased affinity for the substrate ground state. From these and other data, a model of catalysis by the anti-VIP antibodies was proposed according to which the essential catalytic residues are located in the V_L domain and additional residues from the V_H domain are involved in high-affinity binding of the substrate (Sun et al. 1997).

Molecular modeling suggested the presence of a serine protease-like site in the light chain of VIP-hydrolyzing abzyme. This assumption was supported by inhibition of the hydrolytic activity of recombinant L chain by serine protease inhibitors, but not by inhibitors of other classes of protease. The serine protease mechanism was further supported by the observation that catalytic activity was lost following site-directed mutagenesis at a framework region residue, Asp1 (Gololobov et al. 1999), and at two complementarity determining region (CDR) residues, Ser-27a and His-93 (Gao and Paul 1995), residues forming a catalytic triad modeled to be similar to that found in serine proteases. The effect of these mutations was specific to catalysis rather than binding, and the affinity of the light chain for the substrate ground state was nearly unaffected by mutations at Ser-27a or His-93. In contrast, a Ser-26 single mutant and His-27d/Asp-28 double mutant displayed increased K_M (by about tenfold) and increased turnover (by about tenfold). Thus, two types of light-chain amino acid residues participating in catalysis were suggested: those essential for catalysis and those participating in VIP binding and indirectly limiting abzyme/substrate turnover. Of note, all three critical catalytic residues (Ser-27a, His-93, Asp-1) were present in the germ-line counterpart of the mature V_L, whereas the mature and germ-line sequences differed only by four amino acids remote from the catalytic site (Gololobov et al. 1999). Differences between the kinetic constants of the mature and germ-line light chains were marginal. These data show that catalytic activity of VIP-hydrolyzing antibodies is encoded by a germ-line VL gene, but can potentially be improved by somatic sequence diversification and pairing of the L chain with the appropriate heavy chain (for reviews, see Paul 1998; Gololobov et al. 2000).

Another intensively studied system has been catalytic antibodies with esterase-like activity. Analysis of the crystal structure of the abzyme Fab fragment has shown that the ligand *p*-nitrophenyl ester interacts with amino acid residues of both light and heavy chains of the abzyme (Golinelli-Pimpaneau et al. 1994). Two tyrosines were found to mimic the oxyanion-binding hole of serine proteases, at the catalytic core of the antibody. In a series of studies, the changes in binding of abzyme for a nitrophenyl phosphonate substrate were compared in germ-line versus affinity-matured Fab fragment (Patten et al. 1996; Wedemayer et al. 1997a,b). Intriguingly, the binding properties of the two species differed, with the germ-line abzyme showing significant changes in conformation subsequent to binding of hapten, whereas the affinity-matured Fab bound by a lock-and-key mechanism. These changes were accompanied by an increased affinity of abzyme for substrate by a factor of 10^4, reflecting a decrease in complex dissociation rate. The implication of these studies is that in vitro mutational approaches guided by detailed structural knowledge may be a useful means of generating improved novel abzyme catalytic activities targeted against the cleavage of proteins of biological interest.

SUMMARY

A growing body of data suggests that catalytic abzymes may be important mediators of immunological defense, regulation, and autoimmune dysfunction. The degree to which abzymes contribute to these biological phenomena will require continuing studies of antibody-mediated catalysis following experimental immunization and autoimmune disease, as well as mechanistic investigation of catalysis by antibodies and their subunits. From a technological point of view, the further study of both artificial and natural abzymes with the goal of understanding of structure–function relationships, to enable production of tailor-made catalysts of potential therapeutic application, is of high importance. Because the catalytic activity of certain antibodies is an innate function (Gololobov et al. 1999), catalysts with specificity for virtually any target polypeptide could potentially be developed, then improved by in vitro or in vivo affinity selection. The increasing number of available X-ray structures of catalytic antibodies shows the multiplicity of solutions to the question of how an antibody can catalyze an enzyme-like reaction. These strategies include amino acid arrangements analogous to those in enzymes (Zhou et al. 1994), but also arrangements completely different from those selected in enzymes by natural evolution (Charbonnier et al. 1995). Anti-idiotype approaches may enable the mimicry by abzymes of many useful enzymes (Kolesnikov et al. 2000), enhancing their possible functions. The phenomenon of abzyme catalysis can potentially be applied to isolate efficient catalysts suitable for passive immunotherapy of major diseases. For example, cocaine-hydrolyzing abzymes have been developed and may provide a novel approach to the problems of drug addiction (De Prada et al. 2000). Abzymes that cleave the gp120 protein of human immunodeficiency virus (HIV) may be of use in the treatment of AIDS (Paul et al. 2000). Through the rational design of abzymes with specified and novel catalytic functions allowing the selective cleavage of surface proteins, it is hoped that an unprecedented level of control over in vivo protein–protein associations may be achieved.

REFERENCES

Andrievskaya O.A., Buneva V.N., Naumov V.A., and Nevinsky G.A. 2000. Catalytic heterogenity of polyclonal RNA-hydrolyzing IgM from sera of patients with lupus erythematosus. *Med. Sci. Monit.* **6:** 460–470.

Baranovskii A.G., Kanyshkova T.G., Mogelnitskii A.S. Naumov V.A., Buneva V.N., Gusev E.I., Boiko A.N., Zargarova T.A., Favorova O.O., and Nevinsky G.A. 1998. Polyclonal antibodies from blood and cerebrospinal fluid of patients with multiple sclerosis effectively hydrolyze DNA and RNA. *Biochemistry* **63:** 1239–1248.

Baranovsky A.G., Matushin V.G., Vlassov A.V., Zabara V.G., Naumov V.A., Giege R., Buneva V.N., and Nevinsky G.A. 1997. DNA- and RNA-hydrolyzing antibodies from the blood of patients with various forms of viral hepatitis. *Biochemistry* **62:** 1358–1366.

Barbas III, C.F., Heine A., Zhong G., Hoffmann T., Gramatikova S., Bjornestedt R., List B., Anderson J., Stura E.A., Wilson I.A., and Lerner R.A. 1997. Immune versus natural selection: Antibody aldolases with enzymic rates but broader scope. *Science* **278:** 2085–2092.

Buneva V.N., Andrievskaia O.A., Romannikova I.V., Gololobov G.V., Iadav R.P., Iamkovoi V.I., and Nevinskii G.A. 1994. Interaction of catalytically active antibodies with oligoribonucleotides. *Mol. Biol.* **28:** 738–743.

Buneva V.N., Kanyshkova T.G., Vlassov A.V., Semenov D.V., Khlimankov D., Breusova L.R., and Nevinsky G.A. 1998. Catalytic DNA- and RNA-hydrolyzing antibodies from milk of healthy human mothers. *Appl. Biochem. Biotechnol.* **75:** 63–76.

Charbonnier J.B., Carpenter E., Gigant B., Golinelli-Pimpaneau B., Eshhar Z., Green B.S., and Knossow M. 1995. Crystal structure of the complex of a catalytic antibody Fab fragment with a transition state analog: Structural similarities in esterase-like catalytic antibodies. *Proc. Natl. Acad. Sci.* **92:** 11721–11725.

De Prada P., Winger G., and Landry D.W. 2000. Application of artifical enzymes to the problem of cocaine. *Ann. N.Y. Acad. Sci.* **909:** 159–169.

Ding L., Liu Z., Zhu Z., Luo G., Zhao D., and Ni J. 1998. Biochemical characterization of selenium-containing catalytic antibody as a cytosolic glutathione peroxidase mimic. *Biochem. J.* **332:** 251–255.

Ersoy O., Fleck R., Blanco M.J., and Masamune S. 1999. Design and syntheses of three haptens to generate catalytic antibodies that cleave amide bonds with nucleophilic catalysis. *Bioorg. Med. Chem.* **7:** 279–286.

Fastrez J. 1997. In vivo versus in vitro screening or selection for catalytic activity in enzymes and abzymes. *Mol. Biotechnol.* **7:** 37–55.

Gabibov A.G., Gololobov G.V., Makarevich O.I., Schourov D.V., Chernova E.A., and Yadav R.P. 1994. DNA-hydrolyzing autoantibodies. *Appl. Biochem. Biotechnol.* **47:** 293–303.

Gao Q.S. and Paul S. 1995. Site-directed mutagenesis of antibody-variable regions. *Methods Mol. Biol.* **51:** 319–327.

Golinelli-Pimpaneau B., Gigant B., Bizebard T., Navaza J., Saludjian P., Zemel R., Tawfik D.S., Eshhar Z., Green B.S., and Knossow M. 1994. Crystal structure of a catalytic antibody Fab with esterase-like activity. *Structure* **2:** 175–183.

Gololobov G., Sun M., and Paul S. 1999. Innate antibody catalysis. *Mol. Immunol.* **36:** 1215–1222.

Gololobov G., Tramontano A., and Paul S. 2000. Nucleophilic proteolytic antibodies. *Appl. Biochem. Biotechnol.* **83:** 221–232; 297–313.

Gololobov G.V., Chernova E.A., Schourov D.V., Smirnov I.V., Kudelina I.A., and Gabibov A.G. 1995. Cleavage of supercoiled plasmid DNA by autoantibody Fab fragment: Application of the flow linear dichroism technique. *Proc. Natl. Acad. Sci.* **92:** 254–257.

Gouverneur V.E., Houk K.N., de Pascual-Teresa B., Beno B., Janda K.D., and Lerner R.A. 1993. Control of the exo and endo pathways of the Diels-Alder reaction by antibody catalysis. *Science* **262:** 204–208.

Harlow E. and Lane D. 1988. *Antibodies: A laboratory manual.* Cold Spring Harbor Laboratory Press, Cold Spring Harbor, New York.

Hotta K., Lange H., Tantillo D.J., Houk K.N., Hilvert D., and Wilson I.A. 2000. Catalysis of decarboxylation by a preorganized heterogeneous microenvironment: Crystal structures of abzyme 21D8. *J. Mol. Biol.* **302:** 1213–1225.

Iverson B.L. and Lerner R.A. 1989. Sequence-specific peptide cleavage catalyzed by an antibody. *Science* **243:** 1184–1188.

Janda K.D., Shevlin C.G., and Lerner R.A. 1993. Antibody catalysis of a disfavored chemical transformation. *Science* **259:** 490–493.

Janda K.D., Schloeder D., Benkovic S.J., and Lerner R.A. 1988. Induction of an antibody that catalyzes the hydrolysis of an amide bond. *Science* **241:** 1188–1191.

Janda K.D., Lo L.C., Lo C.H., Sim M.M., Wang R., Wong C.H., and Lerner R.A. 1997. Chemical selection for catalysis in combinatorial antibody libraries. *Science* **275:** 945–948.

Jencks W. 1969. *Catalysis in chemistry and enzymology.* McGraw-Hill, New York.

Kanyshkova T.G., Semenov D.V., Vlasov A.V., Khlimankov D.I., Baranovskii A.G., Shipitsyn M.V., Iamkovoi V.I., Buneva B.N., and Nevinskii G.A. 1997. DNA- and RNA- hydrolysed antibodies from human milk and its biological role. *Mol. Biol.* **31:** 1082–1091.

Karush F. 1978. In *Immunoglobulins* (ed. G.V. Litman and R.A. Good), p. 3. Plenum Press, New York.

Kit I.I., Semenov D.V., and Nevinskii G.A. 1995. Do catalytically active antibodies exist in healthy people? (Protein kinase activity of sIgA antibodies from human milk.) *Mol. Biol.* **29:** 893–906.

Kolesnikov A.V., Kozyr A.V., Alexandrova E.S., Koralewski F., Demin A.V., Titov M.I., Avalle B., Tramontano A., Paul S., Thomas D., Gabibov A.G., and Friboulet A. 2000. Enzyme mimicry by the antiidiotypic antibody approach. *Proc. Natl. Acad. Sci.* **97:** 13526–13531.

Kozyr A.V., Kolesnikov A.V., Aleksandrova E.S., Sashchenko L.P., Gnuchev N.V., Favorov P.V., Kotelnikov M.A., Iakhnina E.I., Astsaturov I.A., Prokaeva T.B., Alekberova Z.S., Suchkov S.V., and Gabibov A.G. 1998. Novel functional activities of anti-DNA autoantibodies from sera of patients with lymphoproliferative and autoimmune diseases. *Appl. Biochem. Biotechnol.* **75:** 45–61.

Lacroix-Desmazes S., Moreau A., Sooryanarayana, Bonnemain C., Stieltjes N., Pashov A., Sultan Y., Hoebeke J., Kazatchkine M.D., and Kaveri S.V. 1999. Catalytic activity of antibodies against factor VIII in patients with hemophilia A. *Nat. Med.* **5:** 1044–1047.

Lerner R.A. and Tramontano A. 1987. Antibodies as enzymes. *Trends Biochem. Sci.* **12:** 427–438.

Li L., Paul S., Tyutyulkova S., Kazatchkine M.D., and Kaveri S. 1995. Catalytic activity of anti-thyroglobulin antibodies. *J. Immunol.* **154:** 3328–3332.

Li T., Janda K.D., and Lerner R.A. 1996. Cationic cyclopropanation by antibody catalysis. *Nature* **379:** 326–327.

Li T., Janda K.D., Ashley J.A., and Lerner R.A. 1994. Antibody catalyzed cationic cyclization. *Science* **264:** 1289–1293.

Liu E., Prasad L., Delbaere L.T., Waygood E.B., and Lee J.S. 1998. Conversion of an antibody into an enzyme which cleaves the protein HPr. *Mol. Immunol.* **35:** 1069–1077.

Ma L., Hsieh-Wilson L.C., and Schultz P.G. 1998. Antibody catalysis of peptidyl-prolyl cis-trans isomerization in the folding of RNase T1. *Proc. Natl. Acad. Sci.* **95:** 7251–7256.

Martin A.B. and Schultz P.G. 1999. Opportunities at the interface of chemistry and biology. *Trends Cell Biol.* **9:** 24–28.

Miller G.P., Posner B.A., and Benkovic S.J. 1997. Expanding the 43C9 class of catalytic antibodies using a chain- shuffling approach. *Bioorg. Med. Chem.* **5:** 581–590.

Nagai K. and Thogersen H.C. 1987. Synthesis and sequence-specific proteolysis of hybrid proteins produced in *Escherichia coli. Methods Enzymol.* **153:** 461–481.

Napper A.D., Benkovic S.J., Tramontano A., and Lerner R.A. 1987. A stereospecific cyclization catalyzed by an antibody. *Science* **237:** 1041–1043.

Nevinsky G.A., Kit Y., Semenov D.V., Khlimankov D., and Buneva V.N. 1998. Secretory immunoglobulin A from human milk catalyzes milk protein phosphorylation. *Appl. Biochem. Biotechnol.* **75:** 77–91.

Nevinsky G.A., Kanyshkova T.G., Semenov D.V., Vlassov A.V., Gal'vita A.V., and Buneva V.N. 2000. Secretory immunoglobulin A from healthy human mothers' milk catalyzes nucleic acid hydrolysis. *Appl. Biochem. Biotechnol.* **83:** 115–130; 145–153.

Patten P.A., Gray N.S., Yang P.L., Marks C.B., Wedemayer G.J., Boniface J.J., Stevens R.C., and Schultz P.G. 1996. The immunological evolution of catalysis. *Science* **271:** 1086–1091.

Paul S. 1998. Mechanism and functional role of antibody catalysis. *Appl. Biochem. Biotechnol.* **75:** 13–24.

Paul S., Johnson D.R., and Massey R. 1991. Binding and multiple hydrolytic sites in epitopes recognized by catalytic anti-peptide antibodies. *Ciba Found. Symp.* **159:** 156–167.

Paul S., Kalaga R.S., Gololobov G., and Brenneman D. 2000. Natural catalytic immunity is not restricted to autoantigenic substrates: Identification of a human immunodeficiency virus gp 120- cleaving antibody light chain. *Appl. Biochem. Biotechnol.* **83:** 71–84; 145–153.

Paul S., Volle D.J., Beach C.M., Johnson D.R., Powell M.J., and Massey R.J. 1989. Catalytic hydrolysis of vasoactive intestinal peptide by human autoantibody. *Science* **244:** 1158–1162.

Paul S., Li L., Kalaga R., O'Dell J., Dannenbring Jr., R.E., Swindells S., Hinrichs S., Caturegli P., and Rose N.R. 1997. Characterization of thyroglobulin-directed and polyreactive catalytic antibodies in autoimmune disease. *J. Immunol.* **159:** 1530–1566.

Pauling L. 1946. Molecular basis of biological specificity. *Chem. Eng. News* **24:** 1375–1377.

Pollack S.J., Jacobs J.W., and Schultz P.G. 1986. Selective chemical catalysis by an antibody. *Science* **234:** 1570–1573.

Roberts V.A., Stewart J., Benkovic S.J., and Getzoff E.D. 1994. Catalytic antibody model and mutagenesis implicate arginine in transition-state stabilization. *J. Mol. Biol.* **235:** 1098–1116.

Savel'ev A.N., Eneyskaya E.V., Shabalin K.A., Filatov M.V., and Neustroev K.N. 1999. Antibodies with amilolytic activity. *Protein Peptide Lett.* **6:** 179–184.

Schochetman G. and Massey R. 1985. International Patent no. WO85/02414.

Semenov D.V., Kanyshkova T.G., Kit Y.Y., Khlimankov D.Y., Akimzhanov A.M., Gorbunov D.A., Buneva V.N., and Nevinsky G.A. 1998. Human breast milk immunoglobulins G hydrolyze nucleotides. *Biochemistry* **63:** 935–943.

Shuster A.M., Gololobov G.V., Kvashuk O.A., Bogomolova A.E., Smirnov I.V., and Gabibov A.G. 1992. DNA hydrolyzing autoantibodies. *Science* **256:** 665–667.

Smiley J.A. and Benkovic S.J. 1994. Selection of catalytic antibodies for a biosynthetic reaction from a combinatorial cDNA library by complementation of an auxotrophic *Escherichia coli:* Antibodies for orotate decarboxylation. *Proc. Natl. Acad. Sci.* **91:** 8319–8323.

Smithrud D.B., Benkovic P.A., Benkovic S.J., Roberts V., Liu J., Neagu I., Iwama S., Phillips B.W., Smith III, A.B. and Hirschmann R. 2000. Cyclic peptide formation catalyzed by an antibody ligase. *Proc. Natl. Acad. Sci.* **97:** 1953–1958.

Stewart J.D. and Benkovic S.J. 1993. Recent developments in catalytic antibodies. *Int. Rev. Immunol.* **10:** 229–240.

Stewart J.D., Krebs J.F., Siuzdak G., Berdis A.J., Smithrud D.B., and Benkovic S.J. 1994. Dissection of an antibody-catalyzed reaction. *Proc. Natl. Acad. Sci.* **91:** 7404–7409.

Sun M., Li L., Gao Q.S., and Paul S. 1994. Antigen recognition by an antibody light chain. *J. Biol. Chem.* **269:** 734–738.

Sun M., Gao Q.S., Kirnarskiy L., Rees A., and Paul S. 1997. Cleavage specificity of a proteolytic antibody light chain and effects of the heavy chain variable domain. *J. Mol. Biol.* **271:** 374–385.

Suzuki H. 1994. Recent advances in abzyme studies. *J. Biochem.* **115:** 623–628.

Thayer M.M., Olender E.H., Arvai A.S., Koike C.K., Canestrelli I.L., Stewart J.D., Benkovic S.J., Getzoff E.D., and Roberts V.A. 1999. Structural basis for amide hydrolysis catalyzed by the 43C9 antibody. *J. Mol. Biol.* **291:** 329–345.

Thiagarajan P., Dannenbring R., Matsuura K., Tramontano A., Gololobov G., and Paul S. 2000. Monoclonal antibody light chain with prothrombinase activity. *Biochemistry* **39:** 6459–6465.

Tramontano A., Janda K.D., and Lerner R.A. 1986a. Catalytic antibodies. *Science* **234:** 1566–1570.

———. 1986b. Chemical reactivity at an antibody binding site elicited by mechanistic design of a synthetic antigen. *Proc. Natl. Acad. Sci.* **83:** 6736–6740.

Tyutyulkova S., Gao Q.S., Thompson A., Rennard S., and Paul S. 1996. Efficient vasoactive intestinal polypeptide hydrolyzing autoantibody light chains selected by phage display. *Biochim. Biophys. Acta* **1316:** 217–223.

Vlassov A., Florentz C., Helm M., Naumov V., Buneva V., Nevinsky G., and Giege R. 1998a. Characterization and selectivity of catalytic antibodies from human serum with RNase activity. *Nucleic Acids Res.* **26:** 5243–5250.

Vlassov A.V., Baranovskii A.G., Kanyshkova T.G., Prints A.V., Zabara V.G., Naumov V., Breusov A.A., Giege R., Buneva V.N., and Nevinsky G. 1998b. Substrate specificity of DNA- and RNA-hydrolyzing antibodies from blood of patients with polyarthritis and Hashimoto's thyroiditis. *Mol. Biol* **32:** 559–569.

Wang K.K. 2000. Calpain and caspase: Can you tell the difference? *Trends Neurosci.* **23:** 20–26.

Wedemayer G.J., Patten P.A., Wang L.H., Schultz P.G., and Stevens R.C. 1997a. Structural insights into the evolution of an antibody combining site. *Science* **276:** 1665–1669.

Wedemayer G.J., Wang L.H., Patten P.A., Schultz P.G., and Stevens R.C. 1997b. Crystal structures of the free and liganded form of an esterolytic catalytic antibody. *J. Mol. Biol.* **268:** 390–400.

Wentworth Jr., P., Liu Y., Wentworth A.D., Fan P., Foley M.J., and Janda K.D. 1998. A bait and switch hapten strategy generates catalytic antibodies for phosphodiester hydrolysis. *Proc. Natl. Acad. Sci.* **95:** 5971–5975.

Zhou G.W., Guo J., Huang W., Fletterick R.J., and Scanlan T.S. 1994. Crystal structure of a catalytic antibody with a serine protease active site. *Science* **265:** 1059–1064.

30 In Vitro Selection and Evolution of Protein–Ligand Interactions by Ribosome Display

Christiane Schaffitzel, Christian Zahnd, Patrick Amstutz,
Béatrice Luginbühl, and Andreas Plückthun
Biochemisches Institut, CH-8057 Zürich, Switzerland

INTRODUCTION

Ribosome display is a method for selecting and further evolving functional proteins for their properties of interaction. It is performed completely in vitro and does not involve living cells at any step (Hanes and Plückthun 1997). This method has two main advantages over most other selection methods. First, the screening of very large libraries with up to 10^{14} members is possible. In contrast, other selection methods such as phage display (Winter et al. 1994; Smith 1985), the yeast two-hybrid system (Fields and Song 1989; Chien et al. 1991), or cell-surface display methods (Georgiou et al. 1993; Boder and Wittrup 1997) all necessarily involve transformation steps in which the original library is introduced into cells. This step limits the size of the library by the transformation efficiency, which is about 10^9 to 10^{10} per microgram of DNA in *E. coli* (Dower and Cwirla 1992) and about $\sim10^7$ per microgram of DNA in yeast. Second, there is a built-in evolution during the ribosome display selection procedure. During the many PCR amplification steps that are part of the ribosome display protocol, low-fidelity DNA polymerases that introduce mutations are usually used. In most cases, the mutations are detrimental. However, in some cases, a beneficial mutation is introduced randomly that improves the stability or affinity of the displayed protein, leading to a selection advantage of this clone in the subsequent ribosome display cycle. Thus, the sequence space sampled is not limited by the initial size of the library applied to ribosome display, and diversification can be easily introduced during subsequent selection rounds if low-fidelity polymerases are used. If evolution of the selected clones is not desired, proofreading polymerases can be used, thereby virtually maintaining the original repertoire of the library (He et al. 1999). On the other hand, if in vitro evolution of a protein is desired, the diversification can be additionally enhanced by error-prone PCR (Cadwell and Joyce 1992; Zaccolo et al. 1996) and DNA shuffling (Stemmer 1994).

In comparison, all selection methods that use cells to express proteins, whether in the cytoplasm, on the cell surface, or on phages secreted from the cells, require a switch between in vitro diversification and in vivo selection in every selection round. After every in vitro diversification step, the newly generated library has to be religated and retransformed into the host organism, which is a rather time-consuming and laborious procedure. An alternative would be to use *E. coli* mutator strains (Irving et al. 1996; Low et al. 1996), but the error rate cannot be as well controlled as by in vitro methods.

The Principle of Ribosome Display

A prerequisite of protein selection is the coupling of genotype (RNA, DNA) and phenotype (protein). In ribosome display, this link is accomplished during in vitro translation by the ribosomal complexes, consisting of messenger RNA (mRNA), the ribosome, and the nascent polypeptide, which can fold correctly while still attached to the ribosome (Fig. 1) (Hanes and Plückthun 1997).

The DNA library coding for particular proteins of interest (for instance, a library of single-chain fragments of an antibody [scFv]) is transcribed in vitro. The mRNA is purified and used for in vitro translation. Because the stop codon has been removed from the protein encoding sequences in the DNA library, the ribosome stalls at the 3′ end of the mRNA during in vitro translation, giving rise to a ternary complex of mRNA, ribosome, and encoded protein (Fig. 1). The protein can fold correctly on the ribosome, because a carboxy-terminal spacer had been genetically fused to it, thus allowing the protein of interest to fold outside of the ribosomal tunnel. The ribosomal complexes are stabilized by high concentrations of magnesium ions and low temperature. After in vitro translation, the ribosomal complexes are used directly for selection either on a ligand immobilized on a surface or in solution, with the bound ribosomal complexes subsequently being captured by streptavidin-coated beads. In several washing steps, the nonbinding complexes are removed and the mRNA of the selected complexes can be eluted by dissociation of

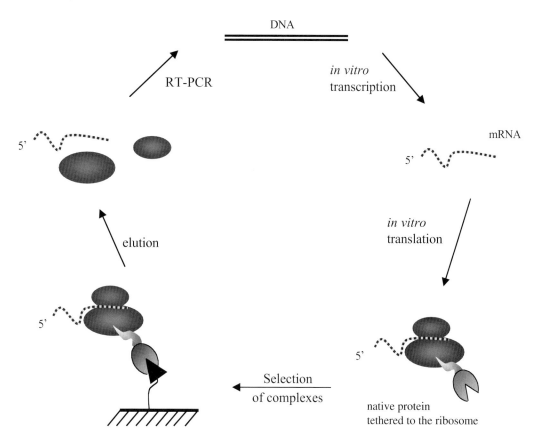

FIGURE 1. Principle of ribosome display. A DNA library encoding open reading frames lacking stop codons is transcribed in vitro. The mRNA is purified and used for in vitro translation. The ribosome stalls at the end of the mRNA. The encoded protein is not released and can fold correctly on the ribosome. The mRNA–ribosome–protein complexes are used for affinity selection on an immobilized target. After washing, the bound ribosomal complexes are dissociated. The mRNA is purified and used for reverse transcription and PCR amplification. The PCR product can be used directly for the next ribosome display selection cycle.

the ribosomal complexes with EDTA. The mRNA is purified, reverse-transcribed, and amplified by PCR. During the following PCR amplification, appropriate primers reintroduce the T7 promoter and the Shine-Dalgarno sequence. The resulting pooled DNA can either be used directly for the next selection cycle or for radioimmunoassays (RIAs), or it can be cloned in a vector for sequencing and large-scale expression of the selected binders. The enrichment of a binder after one round of ribosome display is typically 100- to 1000-fold (Hanes and Plückthun 1997).

Applications of Ribosome Display

Ribosome display was first established for the selection of peptide ligands against a protein target (Mattheakis et al. 1994; Gersuk et al. 1997). The method was further developed in our laboratory for the display of whole functional proteins that have to be correctly folded while still bound to the ribosome (Hanes and Plückthun 1997). In a model system using two distinct scFv fragments of an antibody, a 10^9-fold enrichment of a specific scFv over the nonspecific scFv was achieved by five selection cycles of ribosome display, with an average enrichment of 100 per cycle (Hanes and Plückthun 1997). Second, ribosome display was applied to the selection and simultaneous evolution of a scFv fragment binding with 40 pM affinity to a Gcn4p mutant peptide, using a library prepared from the spleen of immunized mice (Hanes et al. 1998).

Subsequently, it was demonstrated that it is possible to generate antibodies of novel specificity de novo. Starting from the human combinatorial antibody library HuCAL (Knappik et al. 2000), picomolar affinity binders to insulin were selected and evolved during ribosome display selection (Hanes et al. 2000). All antibodies selected had accumulated many mutations during the PCR amplification cycles included in the ribosome display protocol, and the affinity of the antibodies improved up to 40-fold compared to the antibodies initially present in the library. In a selection against an unusual DNA structure, namely the guanine quadruplex DNA, it was demonstrated that antibodies with high specificity could be generated by ribosome display (Schaffitzel et al. 2001). The selected anti-guanine quadruplex antibodies were applied in vivo in macronuclei of the ciliate *Stylonychia lemnae* and provided the first evidence for the existence of guanine quadruplex DNA in a physiological system (Schaffitzel et al. 2001).

Ribosome display can also be used as an in vitro evolution method starting from a single protein. In fact, by combination of ribosome display with error-prone PCR and DNA shuffling, the affinity of a scFv for its antigen was improved 30-fold, to picomolar affinity (Jermutus et al. 2001). Furthermore, ribosome display was used to improve the stability of a scFv fragment by addition of increasing amounts of dithiothreitol (DTT) during sequential rounds of in vitro translation (Jermutus et al. 2001). Using this strategy, an antibody was selected that maintained proper folding in the presence of 10 mM DTT, which is comparable to the reducing environment of the cytoplasm in which disulfide bridges usually cannot be formed. These examples suggest that ribosomal display could constitute a general and rapid method for the generation of intrabodies (antibodies that can be used in the cytoplasm). Finally, selection in the presence of a "suicide substrate" inhibitor, an active β-lactamase, was enriched over an inactive mutant with ribosome display, indicating that ribosome display can also be used for enzyme selection (P. Amstutz et al., unpubl.).

In the following protocols, we describe ribosome display selection using an *E. coli* S30 extract for in vitro translation. It should be mentioned, however, that eukaryotic in vitro translation systems such as the rabbit reticulocyte lysate and the wheat germ lysate can also be used in ribosome display (Gersuk et al. 1997; He and Taussig 1997; Hanes et al. 1999; He et al. 1999).

OUTLINE OF PROCEDURE

The Construct Used for Ribosome Display

Several features of the ribosome display construct (Fig. 2) are important for efficient ribosome display of proteins. On the DNA level, the construct needs a T7 promoter for strong in vitro transcription to generate mRNA. On the mRNA level, the construct contains the ribosome-binding site, followed by the sequence encoding the protein (library) to display. The sequence for the protein library is followed by a spacer sequence fused in-frame to the protein. The carboxy-terminal spacer tethers the nascent protein to the ribosome, and it keeps the structured part of the protein outside the ribosomal tunnel, allowing folding and interaction of the protein with ligands. One successfully applied spacer sequence is derived from gene III of filamentous phage M13, covering amino acids 130–204 (SwissProt P03622). Alternatively, spacers derived from *E. coli* genes, namely *tolA* or *tonB*, can be used as spacer. For *tolA*, the amino acids 130–204 were used (SwissProt JV0057), and for *tonB*, amino acids 62–289 (SwissProt PO2929). The open reading frame extends to the very end of the DNA used as the template for transcription, and there is no stop codon present. The presence of a stop codon would lead to the release of the protein, and thereby the ribosomal complexes would dissociate.

At both ends of the mRNA, the ribosome display construct should include stem-loops. 5′ and 3′ stem-loops are known to stabilize mRNA against RNases in vivo as well as in vitro (Hajnsdorf et al. 1996). The presence of stem-loops is important, especially in the *E. coli* ribosome display system, because at least five of 20 *E. coli* RNases have been shown to contribute to mRNA degrada-

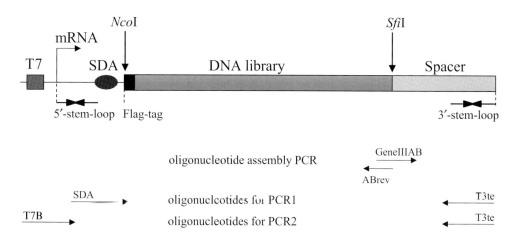

FIGURE 2. Construct used for ribosome display. A T7 promoter and a ribosome binding site (SDA) are necessary for in vitro transcription and translation. The coding sequence starts with a Flag-tag, the DNA library, and a spacer at the carboxyl terminus. The stop codon has been removed from the coding sequence. On the mRNA level, the construct is protected against ribonucleases by a 5´ and a 3´ stem-loop. Information about the oligonucleotides used in PCR is given in the text. The restriction sites used in assembly of the construct are indicated.

tion (Hajnsdorf et al. 1996), and they are probably all present in the S30 extract used for in vitro translation. The efficiency of ribosome display increased approximately 15-fold (Hanes and Plückthun 1997) when the 5´ stem-loop derived from gene 10 of phage T7 (Table 1) and the 3´ stem-loop derived from the early terminator of phage T3, slightly modified to give an open reading frame (Table 1), were introduced into the ribosome display construct (Hanes and Plückthun 1997).

TABLE 1. Oligonucleotides Used in Ribosome Display

Primer	Sequence	Remarks
T3te	5´-GGCCCACCCGTGAAGGTGAGCCTCAGTAG CGACAG-3´	introduces the 3´-stem-loop derived from the translated early terminator of phage T3; anneals to the gene III spacer
SDA	5´-AGACCACAACGGTTTCCCTCTAGAAATA ATTTTGTTTAACTTT AAG**AAGGAG**ATATAT<u>CCATGG</u>ACTACAAAGA-3´	introduces the ribosome-binding site (bold); used for the first PCR amplification step; underlined is the *Nco*I restriction site for cloning
T7B	5´-ATACGAAATT**TAATACGACTCACTATA**<u>GGG</u> AGACCACAACGG-3´	introduces the T7 promoter (bold) and the 5´-stem-loop; used for the second PCR amplification step, the transcription start is underlined.
GeneIIIAB	5´-GAGGGCGGCGGTTCTGGTT-3´	contains a 5´-overhang encoding the 3´-end of the DNA library for assembly PCR
tolAfwd	5´-TATATGGCCTCGGGGGCCGAATTCCAGA AGCAAGCTGAAG-3´	forward primer to amplify the *E. coli* gene tolA with *Eco*RI and *Sfi*I restriction sites
tolArev	5´-GGACTGAAGCTTATCACGGTTTGAAGT CCAATGGCG-3´	reverse primer to amplify tolA spacer
tonBfwd	5´-TATATGGCCTCGGGGGCCGAATTCAGCCG CCACCGGAG-3´	forward primer to amplify the *E. coli* gene tonB with *Eco*RI and *Sfi*I restriction sites
tonBtotrev	5´-CCGCACACCAGTAAGGTGTGCGGTCAGGA TATTCACCACAATCCC-3´	reverse primer to amplify tonB spacer and to introduce a 3´-stem-loop
ABrev	library dependent	reverse primer that encodes the 3´-end of the DNA library
anti-*ssrA*	5´-TTAAGCTGCTAAAGCGTAGTTTTCGTCGT TTGCGACTA-3´	inhibits the 10S-RNA peptide tagging system
anti-SDA	5´-TCTTTGTAGTCCATGGATATATCTCCTTCTT AAAGTTAAACAAAATTATTTCTAGAGGGA-3´	used for northern blotting

Generation of the Ribosome Display Construct

The ribosome display construct can be easily and rapidly prepared completely in vitro by ligation or by assembly PCR, with the advantage of not losing any diversity of the library in transformation steps. As an example, we describe the generation of a construct starting from a scFv library. As a spacer, we use the carboxy-terminal domain of the gene III protein, which can be prepared by digestion of the vector pAK200 (Krebber et al. 1997) with *Sfi*I and *Hin*dIII.

The necessary steps to generate a scFv library either from natural sources or synthetically are described elsewhere (Krebber et al. 1997; Knappik et al. 2000). The DNA library needs to encode constant regions at the amino- and carboxy-terminal ends to permit PCR amplification. In our construct, the coding sequence starts with a FLAG-tag used for detection of protein expression. The library is prepared and amplified by PCR using a primer that introduces an *Nco*I site at the amino terminus (e.g., SDA primer) and a primer that introduces an *Sfi*I restriction site at the carboxyl terminus before the stop codon of the scFv library if the library does not contain the restriction site already. After *Sfi*I digestion, the fragment encoding the library is purified by preparative gel electrophoresis. The DNA encoding the gene III spacer, which must be cut beforehand with *Sfi*I, is also purified by preparative gel electrophoresis and added in at least twofold excess to the ligation reaction with the DNA library fragment. The ligation reaction is amplified directly by PCR using the primers SDA and T3te (Table 1). The PCR product is purified by preparative gel electrophoresis or by using a PCR purification kit to remove the oligonucleotides. In a second PCR, the primers T7B and T3te (Table 1) are used.

Alternatively, assembly PCR can be used for the generation of the ribosome display construct. The advantage of assembly PCR over the ligation protocol is that handling is relatively fast and easy. Because the ligation is more efficient, however, assembly PCR is not recommended for the cloning of very complex libraries into the ribosome display format. For assembly PCR using the gene III spacer, the spacer encoding the carboxy-terminal domain of the gene III protein is PCR-amplified from the plasmid pAK200 (Krebber et al. 1997) using the primers Gene-IIIAB (containing a 5′ overhang encoding the 3′ end of the DNA library) and T3te. In parallel, the scFv library is PCR-amplified using the primers SDA and ABrev. Both PCR products are purified by gel electrophoresis. The assembly PCR is carried out by adding equimolar amounts of DNA of the scFv library and gene III spacer to the PCR without using any primers in the first few cycles, to obtain an assembled library–gene III fusion product. In the last 10–15 cycles, the primers SDA and T3te are added to the assembly PCR to amplify the desired ribosome display construct.

The *tolA* and *tonB* spacers can be directly amplified from any *E. coli* colony. For this purpose, a small amount of a colony (any *E. coli* strain; we normally use SB536; Bass et al. 1996) is picked with a toothpick and then transferred into a standard PCR mix containing primers tolAfwd/tolArev or tonBfwd/tonBtotrev, respectively. After 30 cycles of PCR, a sharp band appears that can be gel-purified. This DNA can be directly used for assembly PCR. Of course, these spacers can be amplified or digested from a plasmid containing the cloned spacers.

The quality of the PCR product is crucial for ribosome display. Thus, it should be checked on an analytical agarose gel that the PCR product contains one single strong DNA band of the expected size and no smears or by-products, which both would reduce the efficiency of ribosome display. The DNA concentration should be 20 ng/µl or more.

In Vitro Transcription

The PCR product (with the T7 promoter, the ribosome-binding site, the DNA library with the spacer fused in-frame) can be directly used for in vitro transcription using T7 RNA polymerase (Pokrovskaya and Gurevich 1994). With the given protocol, ~0.1 mg of mRNA is obtained after 2–3 hours in a 200-µl transcription reaction with 1–2 µg of PCR product as a template. The transcribed mRNA can be purified by LiCl precipitation and a subsequent ethanol precipitation.

In Vitro Translation Using *E. coli* S30 Cell Extract

For *E. coli* in vitro translation, the preparation of S30 extracts from *E. coli* MRE600 cells (Wade and Robinson 1996) is carried out following a modified protocol, based on the procedure described by Chen and Zubay (1983) and Pratt (1984). In particular, the reducing agents DTT and β-mercaptoethanol are omitted from all buffers for the display of proteins containing disulfide bridges. The *E. coli* system used for ribosome display needs to be optimized according to Pratt (1984) with respect to the concentration of Mg^{++} and K^+ ions, the amount of S30 cell extract used, and the translation time (Hanes et al. 1999). Protein synthesis follows a saturation curve reaching a plateau after ~30 minutes (Ryabova et al. 1997). At the same time, mRNA is continuously degraded. Thus, an optimal time exists, at which the concentration of intact mRNA–ribosome–protein complexes that can be used for selection is at a maximum. This optimal time for in vitro translation to be used for *E. coli* ribosome display is usually between 6 and 10 minutes after translation starts, but has to be optimized for each S30 batch. Although most proteins generally fold more efficiently at lower temperatures in vitro, we found that, at least for scFv fragments of antibodies, more ribosomal complexes containing functional protein were obtained when the reaction was carried out at 37°C, which may be attributed to the chaperone activity in the *E. coli* extract. It should be mentioned here that the synthesis of large proteins with a molecular weight >70,000 is not very efficient due to premature termination of translation (Ramachandiran et al. 2000).

An important prerequisite for efficient ribosome display is the elimination of 10Sa-RNA, a surveillance mechanism that is responsible for the release and degradation of proteins derived from mRNA lacking a stop codon (Keiler et al. 1996). To inhibit this degradation mechanism, an antisense oligonucleotide named anti-ssrA (Table 1) is added to the in vitro translation reaction, binding to the mRNA moiety of the 10Sa-RNA.

For the translation of proteins containing disulfide bridges, it is important to have an oxidizing environment. Thus, no DTT should be present in the translation reaction to allow formation of the disulfide bridges (Ryabova et al. 1997). During in vitro transcription, however, the presence of reducing agents is necessary for the stability of T7 RNA polymerase. Therefore, in vitro transcription and cell-free translation are carried out in two separate steps, requiring the isolation and purification of the mRNA. Protein disulfide isomerase (PDI), a eukaryotic chaperone that catalyzes the formation of disulfide bonds (Freedman et al. 1995), improves the efficiency of ribosome display of antibody fragments threefold (Hanes and Plückthun 1997).

Affinity Selection

The stopped diluted translation mixture can be used directly for selection experiments. The affinity selection should be performed on ice in the presence of 50 mM magnesium ions; under these conditions the ribosomal complexes are stable for at least 10 days (Jermutus et al. 2001).

If the presence of the *E. coli* proteins or other components of the translation mix is problematic in a particular experiment, ribosomal complexes can be purified by ultracentrifugation with a sucrose cushion (Mattheakis et al. 1996). When the stopped translation mix is applied to ultracentrifugation, the ribosomal complex will form a pellet and can be separated from free proteins and low-molecular-weight compounds, which stay in the supernatant. Gel filtration is a more gentle method for complex separation (P. Amstutz, unpubl.). Gel filtration columns separate ribosomal complexes efficiently from free protein and small-molecular-weight compounds. The fractionation range of the beads should be chosen from a range of 5×10^4 to 2×10^7 D (CL-4B Sepharose, Pharmacia) and the bed volume for this fractionation range should be four times the sample volume. The elution fraction containing the ribosomal complexes but no free protein can be established in a simple experiment (see below).

For affinity selection, the target protein can be immobilized on a surface. This procedure entails the risk that a protein target may be partially denatured on the surface due to hydrophobic interactions with the plastic surface, in which case the selected antibodies may not recognize the protein target in its native conformation. We found that the background (unspecific binding) is lower when selection is performed in solution. For selection in solution, the ligand has to be biotin labeled, and the ribosomal complexes binding to the ligand are captured by streptavidin-coated magnetic beads. The ligand should contain a 30 Å linker to the biotin moiety, such that the ligand is still accessible to the ribosomally displayed protein and not hidden in the streptavidin-binding pocket. To prevent the selection of streptavidin binders, alternation between streptavidin-coated magnetic beads and avidin-agarose for capture after each round of ribosome display is recommended.

Selection is performed in the presence of sterilized, debiotinylated skimmed low-fat milk (1–2%) and 0.25% (w/v) heparin to prevent nonspecific binding of ribosomal complexes to surfaces and to decrease the background signal. In addition, heparin inhibits nucleases. After several washing steps, non- or weakly bound ribosomal complexes are mostly eliminated. The mRNA of bound ribosomal complexes is recovered by dissociation of the complexes with EDTA-containing buffer, which chelates magnesium ions that are required for stabilization of the ribosomal complexes. The recovery of the mRNA by dissociation of the complexes has the advantage that the protein–ligand interaction does not have to be disrupted, and thus high-affinity binders elute as efficiently as low-affinity binders.

After dissociation of the ribosomal complexes, mRNA is isolated, reverse-transcribed using the primer T3te (Table 1), and PCR-amplified with the primers SDA (Table 1) and T3te. The product of the first PCR is purified by preparative agarose gel electrophoresis and used as the template for the second PCR amplification with the primers T7B (Table 1) and T3te. The resulting PCR product can be directly used for the next ribosome display cycle, for radioimmunoassay (RIA), or cloned in an expression vector via NcoI and SfiI.

Whether or not there is an enrichment of specific binders can be controlled by performing a selection round on a nonspecific surface, i.e., without adding antigen (background control). If the DNA pool is enriched for specific binders, the PCR signal after ribosome display selection should be higher when antigen is present during selection.

Tailoring Molecules by Adapting the Selection Pressure

Affinity is essentially determined by the off rate, as on rates fall into a relatively narrow window (Schwesinger et al. 2000, and references therein). To select for very high affinities, off-rate selection has been applied successfully (Hawkins et al. 1992; Yang et al. 1995; Boder and Wittrup 1997; Chen et al. 1999; Jermutus et al. 2001). In principle, one could also use very small amounts of antigen, but in practice, this strategy does not appear to be successful for reasons discussed elsewhere (Plückthun et al. 2000). In off-rate selection, the ribosomal complexes formed after translation are first equilibrated with biotin-labeled antigen. The concentration of the labeled antigen is chosen such that basically all complexes carrying a binding moiety are bound. After equilibration, a large excess of competitive nonlabeled antigen is added. Every complex that dissociates from its labeled antigen will be captured by the competitive antigen and therefore not be captured by the streptavidin-coated magnetic beads. There is a simple correlation between the incubation time of the complexes with competitive antigen and the mean affinity of the complexes still being bound to the labeled antigen.

This type of selection can be performed not only for affinity, but also for many other physico-chemical properties of the library. If the target protein, e.g., a scFv fragment, contains a disulfide bond needed for activity, a selection can be carried out for stability under reducing conditions (Jermutus et al. 2001). Several rounds of ribosome display with increasing concentrations of DTT present during the panning procedure lead to an increase in stability of the evolved molecules. Following this principle, selections for stability in the presence of organic solvents, detergents, or any other conditions are conceivable, provided that the ribosomal complex can be kept stable.

From Selection to Evolution

Ribosome display was first established as a very efficient method for in vitro selection (Hanes and Plückthun 1997) with enrichment factors up to 10^9 over five rounds. The complete in vitro nature of the method makes it very convenient to perform further randomization steps between different rounds of selection. The intrinsic diversification present in the original protocol, involving up to 100 cycles of PCR over all rounds, does introduce a number of mutations, especially when using nonproofreading *Taq* polymerase (Hanes et al. 1999). This effect can be enhanced by orders of magnitude by error-prone PCR and DNase I shuffling (Cadwell and Joyce 1992; Stemmer 1994; Zaccolo et al. 1996). These randomization steps are introduced after the reverse transcription (RT)-PCR.

The sequence randomization induced during error-prone PCR amplification is a well-established way to introduce mutations into a gene. Induction of errors is achieved either by addition of manganese to the PCR, which decreases the fidelity of most polymerases, or by the incorporation of dNTP analogs such as 8-oxo-guanosine or dPTP (Zaccolo et al. 1996). This leads to a mismatch base-pairing in the second-strand synthesis and, consequently, to a mutation at this position. The degree of randomization is dependent on the concentration of both manganese and dNTP-analogs and can thereby be controlled. It is worth mentioning that the number of mutations introduced is not only dependent on the polymerases used, or the concentration of additives during PCR amplification, but also on the number of PCR amplification steps.

DNase I shuffling is a highly efficient tool to introduce mutations into a gene and to recombine mutations from former selection rounds (Stemmer 1994). Because this uncouples mutations, which would otherwise be physically linked, and recombines them randomly, it is normally thought to ensure that not too many deleterious mutations are accumulated (Stemmer 1994).

After reverse transcription of the mRNA pool obtained after affinity selection, the DNA pool is PCR-amplified until a strong band appears on the analytical agarose gel. The DNA is then gel-purified and digested with DNase I until the mean fragment length is between 50 bp and 150 bp. It is worth taking a small sample after a given incubation time, shock-freezing the rest of the sample, and analyzing the degree of digestion on an agarose gel. This monitoring ensures that the fragments are not getting too small, which is undesirable and makes the assembly difficult. The digested fragments are purified over an agarose gel and isolated using a small fragment recovery kit. The fragments are then reassembled to a full-length construct using a special PCR containing Tween-20, or similar detergent, but no primers. The full-length band is isolated from an agarose gel and amplified with normal PCR to the ribosome display construct. A combination of both methods—error-prone PCR and DNase I shuffling—ensures a high degree of randomization, while decreasing the risk that the effects of beneficial mutations are hidden by deleterious ones.

RIA Analysis of the DNA Pool and of Single Clones

RIAs are performed to test for the presence of specific binders in a pool. In RIA, the in vitro translation is performed with radioactive [^{35}S]methionine, and thus the binding of radioactive protein to the immobilized ligand can be quantified. The translation time is generally about 30 minutes, which is when the protein synthesis reaches a plateau (Ryabova et al. 1997). After the translation is stopped, free ligand can be added for inhibition tests, or the translation mix is directly transferred to microtiter wells with immobilized ligand.

Several control experiments should be performed: The binding of pool-encoded proteins to a nonspecific surface (i.e., a milk-coated surface, or a neutravidin-coated surface) should be tested. Also, the RIA should be performed both in the presence and absence of free ligand as a competitor. If specific binders are present, the RIA signal should be higher on the specific surface compared to the control surface, and, particularly important, the binding should be inhibitable by the presence of free ligand that acts as a competitor.

If this is the case, the pool is clearly enriched for binders and the DNA encoding the pool is cloned in an expression vector via the *Sfi*I and *Nco*I restriction sites for identification of individual binders and for further characterization (Fig. 2; Table 1). The plasmid DNA of single clones is transcribed in vitro according to the protocol, and the mRNA of the single clones is used for RIA analysis to identify specific binders. The enriched DNA pool can also be cloned into a plasmid that adds a peptide detection tag to the protein. This allows the binders to be purified and further analyzed by enzyme-linked immunosorbent assay (ELISA), avoiding the use of radioactivity in RIA.

Optimization of Ribosome Display

Ribosome display should first be established and optimized with a defined, well-known model system. The model protein, e.g., an antibody scFv fragment, should clearly bind to its ligand in ribosome display and thus give an enrichment on a surface coated with its ligand compared to an unspecific surface. As mentioned before, the in vitro translation should be optimized for the concentration of magnesium and potassium ions, the amount of extract used, and the translation time (Pratt 1984). These parameters need to be optimized for each new preparation of *E. coli* S30 cell extract. To establish the optimal ion concentrations, it is most convenient to carry out an in vitro translation of an enzyme, such as β-lactamase or firefly luciferase, with a simple activity assay. This way, protein production under different conditions can be monitored easily. It is recommended that Mg^{++} concentration be optimized first, and then that concentration be used in all other experiments.

To determine the optimal translation time, where maximal amounts of functional ribosomal complexes are present, in vitro translation and affinity selection have to be performed using a ribosome display construct encoding a model protein, e.g., an antibody. After various incubation times at 37°C, aliquots of the in vitro translation reaction are stopped. All of the aliquots are used for affinity selection with the cognate ligand immobilized on a surface. After affinity selection, the eluted mRNA can be used either for RT-PCR or for a northern blot. Northern blotting is more sensitive in detecting small differences in the mRNA amount. In northern blot hybridization, the isolated mRNA can be monitored and quantified with a digoxigenin-labeled probe that specifically anneals to the mRNA construct (Anti-SDA, Table 1). Figure 3 shows the relationship of all of the steps in performing ribosomal display.

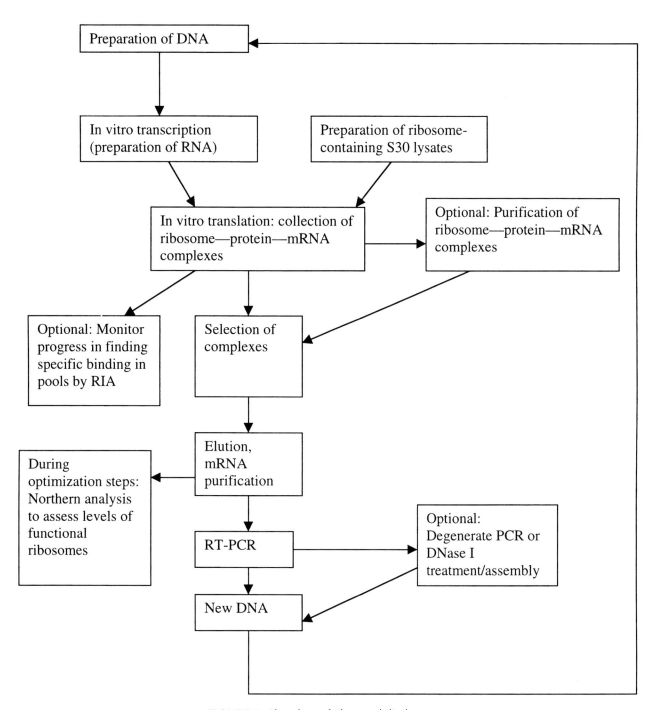

FIGURE 3. Flowchart of ribosomal display steps.

Preparation of the Ribosome Display Construct by Ligation

Efficient ribosomal display of proteins is influenced by several factors, including the presence of a T7 promoter on the DNA level and the ribosome-binding site and the sequence encoding the protein on the RNA level. Figure 2 provides a diagram of the construct elements. This protocol describes the preparation of the basic DNA template for subsequent screeening using either ligation or assembly PCR.

MATERIALS

CAUTION: See Appendix for appropriate handling of materials marked with <!>.

Buffers, Solutions, and Reagents

dNTPs, 20 mM each (Roche Diagnostics)
Dimethylsulfoxide (DMSO) (Fluka)<!>
Oligonucleotides SDA, T7B, T3te, Gene-IIIAB, ABrev (see Table 1)
QIAquick gel extraction kit (Qiagen)
Tris-HCl (10 mM, pH 8.5)

Biological molecules

*Sfi*I/*Nco*I fragment of protein encoding DNA library, e.g., scFv antibody library (150 ng,
    ~2 x 10^{11} molecules)
*Sfi*I/*Hin*dIII fragment of the gene III spacer
T4 DNA ligase (Roche Diagnostics)
Taq DNA polymerase (GIBCO BRL)

Special Equipment

Apparatus for agarose gel electrophoresis
Thermocycler

METHOD

Preparation by Ligation

1. Digest the vector pAK200 (Krebber et al. 1997) with *Hin*dIII and *Sfi*I and purify the resulting 481-bp fragment encoding the gene III spacer by agarose gel electrophoresis using the QIAquick gel extraction kit.

2. Digest the plasmid encoding the library with *Sfi*I and *Nco*I and purify it by agarose gel electrophoresis.

3. Ligate 150 ng of DNA fragment encoding the DNA library to a threefold excess of gene III spacer overnight at 16°C using 10 units of T4 DNA ligase.

4. Perform PCR amplification as described in step 5 of mRNA Purification and RT-PCR protocol (Protocol 6, p. 557) using the primers SDA and T3te. Use 5–10 μl of the ligation mix as PCR template for a 50-μl PCR. The PCR program is 4 minutes at 94°C, then 5 cycles of 30 seconds at 94°C, 30 seconds at 37°C, 2.5 minutes at 72°C, followed by 15 to 20 cycles with the same settings except with 50°C annealing temperature instead of 37°C, and finished by 10 minutes at 72°C.

5. Purify the PCR product by agarose gel extraction, and use it for the second PCR with the primers T7B and T3te (see step 7 in mRNA Purification and RT-PCR, Protocol 6, p. 557). Usually, 12–16 cycles are performed.

Preparation of the Ribosome Display Construct by Assembly PCR

1. PCR-amplify the carboxy-terminal domain of the gene III protein from the vector pAK200 with the primers Gene-IIIAB and T3te, and the DNA library with the primers SDA and ABrev.

2. Purify the PCR products by agarose gel extraction, and use equimolar amounts of each DNA for the assembly PCR. Perform the assembly PCR as follows: 4 minutes at 94°C, then 5 cycles of 30 seconds at 94°C, 30 seconds at 50°C, and 2.5 minutes at 72°C, followed by adding the primers SDA and T3te and performing an additional 12 cycles of 30 seconds at 94°C, 30 seconds at 45°C, and 2.5 minutes at 72°C.

3. Purify the PCR product of the appropriate size by agarose gel extraction, and perform the last PCR introducing the T7 promoter with the primers T7B and T3te (see step 7 in mRNA Purification and RT-PCR, Protocol 6, p. 557).

In Vitro Transcription and mRNA Purification

The DNA template from Protocol 1 is used to prepare the mRNA template by in vitro transcription. The transcribed mRNA is then purified by LiCl precipitation and ethanol precipitation.

MATERIALS

CAUTION: See Appendix for appropriate handling of materials marked with <!>.

Buffers and Solutions

Ethanol <!>
Guanidinium isothiocyanate (1 M) <!>
Lithium chloride (6 M) (LiCl) (Fluka)<!>
5x Loading buffer
   50% glycerol
   200 mM Tris-HCl (pH 8)
   100 mM acetic acid <!>
   5 mM EDTA
   bromophenol blue <!>
NTPs: ATP, CTP, GTP, TTP (50 mM each) (Sigma)
RNA denaturation buffer
   10 µl of formamide<!>
   3.5 µl of formaldehyde <!>
   2 µl of 10x MOPS buffer (0.2 M MOPS [pH 7.0], 80 mM sodium acetate, 10 mM EDTA)
Sodium acetate (3 M, pH 5.2)
5x T7 RNA polymerase buffer
   1 M HEPES-KOH <!> (pH 7.6)
   150 mm magnesium acetate
   10 mm spermidine
   0.2 mm DTT <!>
TBE buffer
   90 mM Tris-HCl
   90 mM boric acid
   10 mM EDTA

Biological Molecules

RNasin (40 units/µl, Promega, Madison, Wisconsin)
T7 RNA Polymerase (New England Biolabs, Beverly, Massachusetts)

Special Equipment

Apparatus for gel electrophoresis
Heating block
Tabletop refrigerated microcentrifuge

METHOD

1. For the in vitro transcription reaction, thaw and mix the following reagents on ice:

40 µl	5x T7 RNA polymerase buffer
28 µl	NTPs (50 mM each)
8 µl	T7 RNA polymerase (50 units/µl)
4 µl	RNasin (40 units/µl)
75 µl	RNase-free H_2O
45 µl	PCR-template (from *1*) (0.5–1 µg of DNA)
200 µl	

   Incubate for 2–3 hours at 37–38°C.

2. For mRNA purification, combine:

200 µl	in vitro transcription mixture
200 µl	RNase-free H_2O
400 µl	6 M LiCl
800 µl	

   Keep on ice for 30 minutes.

3. Centrifuge for 20–30 minutes at 14,000 rpm (4°C), discard the supernatant, and wash the pellet once with 500 µl of 70% ethanol.

4. Dry the pellet, resuspend it in 200 µl of RNase-free H_2O, and centrifuge at 14,000 rpm for 5 minutes at 4°C.

5. Mix 180 µl of the supernatant, 18 µl of 3 M sodium acetate, and 500 µl of ethanol (97%) and incubate for 30 minutes on ice.

6. Centrifuge at 14,000 rpm for 20–30 minutes at 4°C.

7. Wash the pellet with 500 µl of 70% ethanol, air-dry, and resuspend it in 40 µl of RNase-free H_2O.

8. Determine the concentration of mRNA as follows:

   a. Measure the OD at 260 nm (an OD_{260} of 1 corresponds to 40 µg/ml).

   b. Run an analytical agarose gel (1.5%): Add 10 µl of RNA denaturation buffer to 1 µg of mRNA on ice and incubate for 10 minutes at 70°C. Chill the samples on ice, mix with 2 µl of gel loading buffer and separate by 1.5% agarose gel electrophoresis in TBE buffer and in the presence of 1 M guanidinium isothiocyanate.

   > The scale of the transcription reaction can be adjusted. Care should be taken, however, that the diversity of the DNA library is conserved. Thus, the number of molecules of DNA template used should be several times higher than the diversity of the library.

Preparation of *E. coli* S30 Cell Extract ˎ

Described here is the preparation of S30 cell lysates, the source of ribosomes for subsequent in vitro translation. This protocol is based on the procedures described by Chen and Zubay (1983) and Pratt (1984).

MATERIALS

CAUTION: See Appendix for appropriate handling of materials marked with <!>.

Buffers and Solutions

Incomplete rich medium
  5.6 g of potassium dihydrogen phosphate (KH_2PO_4)<!>
  28.9 g of potassium hydrogen phosphate (K_2HPO_4)<!>
  10 g of yeast extract
  15 mg of thiamine per 1 liter of culture medium
    Autoclave the medium first and then add 25 ml of 40% (w/v) glucose, sterile-filtered.
Preincubation mix (10 ml total volume):
  3.75 ml of 2 M Tris-acetate (pH 7.5 at 4°C)
  71 µl of 3 M magnesium acetate
  75 µl of amino acid mix (10 mM of each of the 20 amino acids; Sigma)
  0.3 ml of 0.2 M ATP
  0.2 g of phosphoenolpyruvate (Sigma)
  50 units of pyruvate kinase (Sigma P 1506)
    The preincubation mix must be prepared fresh immediately before use.
S30 buffer
  10 mM Tris-acetate (pH 7.5 at 4°C)
  14 mM magnesium acetate
  60 mM potassium acetate
    Store at 4°C or chill the buffer solution before use.

Biological Molecules

Escherichia coli strain MRE600 (Wade and Robinson 1966)

Special Equipment

Baffled flask (5-liter)
Dialysis tubing with a cutoff of 6000–8000 D
French press
Refrigerated centrifuge (30,000*g*)
Shaker for bacterial culture at 25°C and 37°C

METHOD

1. Grow a 100-ml starter culture of *E. coli* MRE600 overnight at 37°C in incomplete rich medium with shaking.

2. The next day, inoculate 1 liter of incomplete rich medium in a 5-liter baffled shaker flask with 10 ml of the overnight culture and grow at 37°C.

3. Harvest the cells at OD_{550} of 1.0 (corresponding to the early exponential growth phase) by centrifugation at 3,500*g* for 15 minutes at 4°C.

4. Discard the supernatant and wash the pellet three times with 50 ml of ice-cold S30 buffer per liter of culture.

   The cell pellet can be frozen at –80°C or in liquid nitrogen and stored for up to 2 days.

5. Thaw the cell pellet on ice and wash it once again with S30 buffer.

6. Weigh the cell pellet and resuspend it in ice-cold S30 buffer at a ratio of 1.27 ml of buffer per gram of wet cells.

7. Lyse the cells by one passage through a French press using a chilled French press cell at 6000 psi.

   More than one passage of the cell suspension results in decreased translation activity of the cell extract.

8. Centrifuge the lysed cells immediately at 30,000*g* for 30 minutes at 4°C.

9. Transfer the supernatant to a clean centrifuge tube and centrifuge again at 30,000*g* at 4°C for 30 minutes.

10. Transfer the supernatant of the second centrifugation again to a clean flask and add 1 ml of preincubation mix for each 6.5 ml of S30 extract. Shake this solution slowly for 1 hour at 25°C (no foaming should occur).

    During this time, all translation of endogenous mRNA will be finished and the endogenous mRNA and DNA will be degraded by nucleases present in the cell extract.

11. Transfer the S30 cell extract to dialysis tubing and dialyze in the cold room three times against a 50-fold volume of chilled S30 buffer. Replace each dialysis solution after 1 hour.

12. Centrifuge the cell extract at 4000*g* for 10 minutes at 4°C and freeze the supernatant in aliquots of 100–500 µl in liquid nitrogen. Store it at –80°C.

    The extract can be stored for months without losing activity. It can even be frozen a second time after thawing. However, if the extract is thawed more than twice, it starts losing activity.

Protocol 4

In Vitro Translation

The actual process of in vitro translation to prepare mRNA- and protein-tethered ribosomes is described in this protocol.

MATERIALS

CAUTION: See Appendix for appropriate handling of materials marked with <!>.

Buffers and Solutions

Anti-ssrA oligonucleotide (200 μM) (see Table 1)
Methionine in H_2O (200 mM)
Magnesium acetate in H_2O (100 mM)
Potassium glutamate in H_2O (2 M)
PremixZ
    250 mM Tris-acetate (pH 7.5 at 4°C)
    1.75 mM of each amino acid except methionine
    10 mM ATP
    2.5 mM GTP
    5 mM cAMP
    150 mM acetylphosphate
    2.5 mg/ml *E. coli* tRNA
    0.1 mg/ml folinic acid
    7.5% PEG 8000 <!>
Washing buffer WBTH
    50 mM Tris-acetate (pH 7.5)
    150 mM NaCl
    50 mM magnesium acetate
    0.1% Tween-20
    2.5 mg/ml heparin (Sigma)

Biological Molecules

E. coli S30 cell extract (from Preparation of *E. coli* S30 Cell Extract protocol, p. 550)
Library mRNA (1 μg/μl) (from In Vitro Transcription and mRNA Purification protocol, p. 548)
Protein disulfide isomerase (22 μM) (bovine PDI) (Sigma) in H_2O

Special Equipment

Heating block
Tabletop refrigerated microcentrifuge

552

METHOD

1. Chill all solutions and mix the translation reaction on ice in the indicated order:

14.3	μl	RNase-free H$_2$O
11	μl	Potassium glutamate (2 M) (see note below)
7.6	μl	Magnesium acetate (0.1 M) (see note below)
1.1	μl	Methionine (200 mM)
2	μl	Anti-ssrA oligonucleotide
22	μl	PremixZ ice-cold (thaw on ice and vortex before pipetting)
40	μl	*E. coli* S30 extract
2	μl	PDI (if disulfides are present in the target protein library)
10	μl	Library mRNA (10 μg), thaw just before use and freeze the remainder immediately
110	μl	

   To optimize translation conditions, test the following conditions: for each new batch of S30, extract 11–15 mM magnesium acetate, 180–220 mM potassium glutamate, and 20–50 μl of S30 extract for a 110-μl reaction.

2. Incubate the translation reaction at 37°C for the optimized translation time (usually 6–15 minutes).

3. Stop the translation with 440 μl of ice-cold WBTH, vortex briefly, and gently place on ice.

4. Centrifuge the translation mix at 14,000*g* for 5 minutes at 4°C and transfer to a new ice-cold tube.

Affinity Selection

This procedure details the affinity selection of ribosomal complexes using either panning or a solution-based method. Variations that enable the selection of proteins with lengthened off-rate or increased stability forms of proteins are also noted.

MATERIALS

CAUTION: See Appendix for appropriate handling of materials marked with <!>.

Buffers and Solutions

Elution buffer (EB)
    50 mM Tris-acetate (pH 7.5 at 4°C)
    150 mM NaCl
    20 mM EDTA
    50 µg/ml *S. cerevisiae* RNA (Sigma)
Low-fat skim milk (sterilized) (12% in H_2O)
PBS buffer
    137 mM NaCl
    2.7 mM KCl <!>
    10 mM Na_2HPO_4 <!>
    1.8 mM KH_2PO_4 <!>
10x Washing buffer (WB)
    0.5 M Tris-acetate (pH 7.5 at 4°C)
    1.5 M NaCl
    0.5 mM magnesium acetate
1x Washing buffer (WBT)
    50 mM Tris-acetate (pH 7.5 at 4°C)
    150 mM NaCl
    50 mM magnesium acetate
    0.1% Tween-20

Biological Molecules

Avidin immobilized on agarose beads (Sigma)
Biotinylated ligand and free ligand
Streptavidin magnetic particles (Roche Diagnostics)

Special Equipment

Magnet
Microtiter plate strips or plates (Nunc)
Panning tubes (Nunc)
Rocking table or shaker

METHOD

Selection Using an Immobilized Target Protein

It is important that the buffers, microtiter plates, and pipette tips are temperature-equilibrated in the cold room before use.

1. Coat the microtiter wells overnight at 4°C with 400 ng of ligand in 100 µl of PBS.

2. Wash the coated microtiter plates or strips with PBS and block for 1 hour with 4% milk powder in PBS. As a control surface, block several noncoated microtiter wells with milk.

3. Wash the wells three times with PBS and twice with washing buffer WBT.

4. Fill the wells with 250 µl of ice-cold washing buffer WBT and put on ice.

5. For affinity selection, remove the washing buffer WBT from the ice-cold microtiter wells.

6. Supplement the in vitro translation mix (from the In Vitro Translation protocol, p. 552) with ice-cold sterilized milk in WBT to a final concentration of 2% (w/v) milk and transfer 200 µl of translation mix in each ligand-coated and milk-blocked microtiter well.

7. Gently shake the microtiter plates for 1 hour in the cold room. Pour off the translation reaction and then wash five times with WBT.

8. For elution, add 200 µl of ice-cold elution buffer EB for 5 minutes on ice and shake it gently. The eluted mRNA must be immediately purified (see step 1 in mRNA Purification and RT-PCR protocol, p. 557).

Alternative Protocol: Selection in Solution

1. Block 5-ml panning tubes with 4% (w/v) milk in PBS for 1 hour by end-over-end rotation.

2. Wash the 5-ml panning tubes three times with PBS and three times with washing buffer WBT, and then finally fill with WBT.

3. Remove biotin from the sterilized low-fat skim milk before use by end-over-end rotation of 1 ml of sterilized 12% (w/v) milk powder with 100 µl of streptavidin-coated magnetic beads for 1 hour at room temperature. Remove the streptavidin-coated beads with a magnet and discard them; transfer the milk in a new tube. Store it on ice.

4. Wash 100 µl of streptavidin-coated magnetic beads four times with ice-cold washing buffer WBT and resuspend the beads in their original volume in ice-cold WBT.

5. Empty the immunotubes and add 60 µl of sterilized, biotin-depleted low-fat skim milk to 1–2% (w/v) final concentration.

6. Add the in vitro translation mix (from mRNA Purification and RT-PCR protocol, p. 557) and 10 pmoles of biotinylated ligand to the panning tubes. Seal the immunotubes, put them into a larger tube (e.g., a 250-ml centrifuge tube that is filled with ice), and rotate end-over-end for 1 hour in the cold room.

7. For capture, add 100 µl of streptavidin-coated magnetic beads to the tubes and rotate the tube end-over-end on ice for 15 minutes in the cold room.

8. Pour off the translation reaction, then wash the magnetic beads five times with WBT, and bind them to the side of the tube with a magnet.

9. For elution, add 200 μl of ice-cold elution buffer EB for 5 minutes on ice and shake it gently. The eluted mRNA must be immediately purified (see mRNA Purification and RT-PCR, Protocol 6, p. 557).

> The capacity of the streptavidin-coated magnetic beads is dependent on the size of the biotinylated ligand. Because it is important to ensure that all biotinylated ligand can be captured by the streptavidin-coated magnetic beads, not more than 10 pmoles of biotinylated ligand should be used per 100 μl of streptavidin-coated magnetic beads, although this amount of beads can bind 100 times more free biotin. Alternating with streptavidin-coated magnetic beads, avidin-agarose should be used to avoid the selection for streptavidin-binding proteins.

Off-rate Selection

1. Perform steps 1–3 of the above protocol, Selection in Solution, to prepare the panning tubes.

2. Split the translation mix. Add to one reaction biotinylated antigen (concentration 1–100 nM; the better the binder is, the less antigen is needed).

3. Equilibrate for 2 hours to overnight.

4. Add competing antigen (1000-fold excess over the biotinylated antigen).

5. Incubate for off-rate selection time (2 hours to 15 days), depending on the expected off-rate of the best binders.

6. Wash the streptavidin-coated magnetic beads (see step 4 in the above protocol, Selection in Solution) and recover the ribosomal complexes (steps 7 and 8 in the above protocol, Selection in Solution).

Stability Selection of Disulfide Containing Proteins

1. Perform the selection as described in the Selection in Solution protocol, but during in vitro translation (step 1, In Vitro Translation protocol, p. 552) and during the affinity selection (step 6 in Selection in Solution protocol, above), add defined amounts of DTT (normally ranging from 0.5 to 10 mM).

2. Add DTT also for the analysis of the selected pools and single binders by RIA (see Radioimmunoassay, Protocol 9, p. 562).

mRNA Purification and RT-PCR

The purpose of this protocol is the purification and subsequent use for RT-PCR of mRNA from the selected ribosomal complexes. It is difficult to predict the number of PCR cycles necessary to recover the genetic information, because this depends on the amount of mRNA eluted after affinity selection. Therefore, it is useful first to perform a few PCR cycles (15–20) and then check the PCR product on an analytical agarose gel. If necessary, add more PCR cycles. Usually the fewer cycles of PCR that are needed, the more binders are present in the library. However, overamplifying the PCR product is not recommended because this results in a smeary band on an agarose gel and subsequently, after in vitro transcription, in poor-quality RNA.

MATERIALS

CAUTION: See Appendix for appropriate handling of materials marked with <!>.

Buffers, Solutions, and Reagents

dNTPs (20 mM each; Eurogentec)
Dimethylsulfoxide (DMSO) (Fluka)<!>
Dithiothreitol (DTT) (0.1 M) <!>
High Pure RNA Isolation Kit (Roche Diagnostics)
$MgCl_2$ (50 mM) <!>
Oligonucleotide primers T3te, SDA, and T7B (see Table 1)
QIAquick gel extraction kit (Qiagen)
Tris-HCl (10 mM, pH 8.5)

Biological Molecules

Superscript reverse transcriptase
5x Superscript first-strand synthesis buffer (GIBCO BRL)
RNasin (Promega)
Taq DNA polymerase, 10x PCR buffer (GIBCO BRL)

Special Equipment

Apparatus for agarose gel electrophoresis
Heating block
Thermocycler

METHOD

1. Isolate the mRNA (from Affinity Selection protocol, p. 554) using the High Pure RNA isolation kit according to the manufacturer's instructions.

2. Elute the purified RNA in 35 μl of RNase-free H_2O and immediately denature for 10 minutes at 70°C for reverse transcription. Chill the mRNA samples for 1–2 minutes on ice after denaturation.

3. For reverse transcription, prepare a premix on ice as follows:

0.25	μl	T3te primer (100 μM)
0.5	μl	dNTP (20 mM each)
0.5	μl	RNasin (40 units/μl, Promega)
0.5	μl	Superscript reverse transcriptase (GIBCO, 200 units/μl)
4	μl	5x Superscript first-strand synthesis buffer (GIBCO)
2	μl	DTT (0.1 M)
7.75	μl	

Add 12.25 μl of denatured mRNA to this premix, mix, and centrifuge briefly at 4°C.

4. Incubate for 1 hour at 50°C.

In addition to the RT-PCR sample, a negative control should be performed without template to test the buffers and primers for contamination.

5. Set up the PCR on ice:

0.125	μl	T3te primer (100 μM)
0.125	μl	SDA primer(100 μM)
0.5	μl	dNTP (20 mM each)
0.25	μl	*Taq* Polymerase (5 units/μl, GIBCO)
5	μl	10x PCR buffer (GIBCO)
2.5	μl	DMSO
1.55	μl	$MgCl_2$ (50 mM)
32.45	μl	H_2O
7.5	μl	DNA template, directly from reverse transcription
50	μl	

The PCR is carried out for 4 minutes at 94°C, followed by 20 cycles of 30 seconds at 94°C, 30 seconds at 50°C, 2.5 minutes at 72°C, and is finished by 10 minutes at 72°C.

6. Purify the PCR product by agarose gel electrophoresis.

7. Use the purified PCR product for the second PCR amplification using the same reaction setup (as above in step 5) and the primers T3te and T7B. After initial denaturation for 4 minutes at 94°C, 10–15 PCR cycles are performed with 30 seconds at 94°C, 30 seconds at 60°C, 2.5 minutes at 72°C, and the reaction is finished by 10 minutes at 72°C.

Protocol 7

Evolution: Introducing Additional Diversity

These PCR methods are used to introduce mutations and to recombine mutations from previous selection rounds. Degenerate PCR and DNase I digestion/reassembly methods can be incorporated into the selection procedure at the RT-PCR step to increase diversity for subsequent rounds of selection.

MATERIALS

CAUTION: See Appendix for appropriate handling of materials marked with <!>.

Buffers, Solutions, and Reagents

dNTP-analogs
  2 mM 6(2-deoxy-β-D-ribofuranosyl)-3,4-dihydro-8H-pyrimido(4,5-c)(1,2)-oxazin-7-one triphosphate (dPTP, Nucleix Plus, Amersham)
  2 mM 8-oxo-2′-deoxyguanosine triphosphate (8-oxo-dGTP, Nucleix Plus, Amersham)
Triton X-100 (Fluka)
Oligonucleotide primers SDA, T7B, ABrev (see Table 1)
QIAex-II gel extraction kit (Qiagen)

Biological Molecules

DNase I (Roche Diagnostics)
10x DNase I buffer
  10 mM $MgCl_2$ <!>
  10 mM $CaCl_2$ <!>
  500 mM Tris-HCl (pH 7.6)
Taq DNA polymerase, 10x PCR buffer (GIBCO BRL)

Special Equipment

Apparatus for agarose gel electrophoresis
Heating block
Thermocycler

METHOD

Error-prone PCR

Perform a standard PCR (see step 5 in the mRNA Purification and RT-PCR protocol, p. 557), but add dNTP analogs to the reaction mixture. The final mutation rate can be varied both with the number of PCR cycles and with the concentration of dNTP analogs. With up to 85 μM 8-oxo-dGTP and 85 μM dPTP, a mutation rate of 6×10^{-2} bp^{-1} can be obtained after 25 cycles of PCR.

DNase I Shuffling and Assembly PCR

1. Take about 5 μg of purified PCR product (from step 5 in the mRNA Purification and RT-PCR protocol, p. 557) and dissolve it in the following mix:

10x DNase I buffer	10 μl
PCR product in H_2O	5 μg
H_2O	up to 100 μl

2. Add 1 μl of DNase I (0.15 unit/ml) and bring the tube to room temperature.

3. Incubate for 5 minutes at room temperature.

4. Take an aliquot of 5 μl and add 2.5 μl of standard DNA loading buffer (containing EDTA). Freeze the remaining reaction immediately in liquid nitrogen.

5. Analyze the sample on a 1.5% agarose gel. The original DNA band should have been shifted to a broad band at 50–100 bp. If this is observed, purify the remaining 95 μl on a preparative agarose gel (extract from gel with Qiaex-II, Qiagen). If this is not observed, bring the reaction again to room temperature and add another 1 μl of DNase I. Repeat until the expected size range of fragment is found.

6. Mix the following PCR, containing no primers, and add 5–15 μl of the purified fragments from step 5 (try different concentrations of the template).

dNTP	0.25 μl
$MgCl_2$ (50 mM)	0.88 μl
10x PCR reaction buffer (see step 5, mRNA Purification and RT-PCR protocol, p. 557)	2 μl
Triton X-100	0.8 μl
Taq polymerase	0.5 μl

   Add H_2O to a total volume of 20 μl.

   The PCR conditions are 4 minutes at 94°C, followed by 10 cycles of 30 seconds at 94°C, 30 seconds at 40°C, 2.5 minutes at 72°C, then followed by 32 cycles of 30 seconds at 94°C, 30 seconds at 45°C, 2.5 minutes at 72°C, and the reaction is finished by 10 minutes at 72°C.

7. Extract the band of the desired size from the agarose gel. The band will be diffuse, but it will regain its sharpness after a second PCR with specific "outside" primers (e.g., SDA and ABrev).

Protocol 8

Separation of Ribosomal Complexes from Free Protein and Small-molecular-weight Compounds

This is a simple optional method for purifying ribosomal complexes prior to use in affinity selection, if the presence of *E. coli* proteins in the translation mix is problematic (see Protocol 5, p. 554).

MATERIALS

Special Equipment

CL-4B Sepharose (Pharmacia)
Gel filtration columns (1-ml) (Qiagen)

METHOD

1. Perform an in vitro translation reaction and stop it as described above (In Vitro Translation, Protocol 4, p. 552).

2. Prepare a gel filtration column (Qiagen) with 1 ml of CL-4B Sepharose (Pharmacia) and equilibrate it with ice-cold washing buffer, WB.

3. Apply 250 µl of the stopped translation mix to the column.

4. Add 200 µl of ice-cold WBT and discard the flowthrough.

5. Add another 300 µl of ice-cold WB and collect the flowthrough, containing the ribosomal complexes, which can be used for selection or other experiments.

   The fraction containing only the ribosomal complexes and no released protein can be determined in a simple experiment. Translation of an enzyme-ribosome display-construct is stopped such that stable complexes form (WB) and, in a parallel experiment, such that no complexes form (WB without Mg^{++}). Both samples are applied to gel filtration columns, and the elution profile of the enzyme is monitored by activity measurements. By comparing the two elution profiles, the fraction containing only protein in the ribosomal complex is determined.

Protocol 9

Radioimmunoassay (RIA)

This protocol determines whether a selected pool contains specific binders to a ligand. Please note that several control experiments should be performed. For a discussion of these experiments, see page 543.

MATERIALS

CAUTION: See Appendix for appropriate handling of materials marked with <!>.

Use all of the material used for in vitro translation (see In Vitro Translation, Protocol 5, p. 552) and in addition:

Buffers and Solutions

1x PBS buffer
    10 mM Na_2HPO_4 (pH 7.4) <!>
    140 mM NaCl
    15 mM KCl <!>
1x PBST buffer
    PBS with 0.5% (v/v) Tween-20
[^{35}S]Methionine (10 mCi/ml, 1175 Ci/mmole; New England Nuclear) <!>
SDS (4%) in PBS

Special Equipment

Liquid scintillation cocktail "OptiPhase2" (Wallac, Finland)
Scintillation counter

METHOD

1. Coat the microtiter plate wells overnight at 4°C either directly with the ligand (for proteins typically 100 µl of 0.2 µM) or with neutravidin (100 µl per well, 4 µg/ml in PBS).

2. Wash the plate three times with PBS. In case of neutravidin coating, add 50 pmole of biotinylated ligand in 100 µl of PBS and incubate for 30 minutes at 25°C.

3. After washing with PBST, block the microtiter plate wells with 4% skimmed low-fat milk in PBS.

4. Carry out an in vitro translation using the RNA of the pool or of single clones as a template as described in step 1 of In Vitro Translation protocol (p. 552) (10 µg of mRNA per 110-µl reaction volume) with the following modifications: Carry out the in vitro translation for 30 minutes at 37°C. Add 2 µl of [^{35}S]methionine (0.3 µM, 50 µCi/ml final), but no cold methionine.

5. Dilute the reaction mixture fourfold with PBST after translation and centrifuge the mixture at 14,000g for 5 minutes.

6. Dilute the supernatant with the same volume of 4% milk in PBST containing either no ligand or, for inhibition studies, different concentrations of free ligand, and preincubate for 1 hour at room temperature before applying to the microtiter well.

7. Add 100 μl of the radioactive reaction mix into the microtiter well and let the binding reaction with the immobilized ligand take place for 30 minutes at room temperature, with gentle shaking.

8. Wash five times with PBST and elute with 4% SDS in PBS.

9. Prepare 5 ml of scintillation fluid, add the eluted fraction, and quantify the radioactivity in a scintillation counter.

Northern Blot

Northern blotting is a sensitive method for detecting small differences in the mRNA amount. This procedure can be used during optimization of ribosome display to characterize a new preparation of S30 cell extract.

MATERIALS

CAUTION: See Appendix for appropriate handling of materials marked with <!>.

Use all of the material used for in vitro translation (In Vitro Translation, Protocol 4, p. 552) and affinity selection (Affinity Selection, Protocol 5, p. 554) and in addition:

Buffers, Solutions, and Reagents

Chemiluminescent substrate CSPD (disodium 3-[4-methoxyspiro]1,2-dioxetane-3,2′ [5′-chloro]-tricyclo-[3.3.1.1^{3,7}]decan[-4-yl]phenyl phosphate; Roche Diagnostics)
DIG Oligonucleotide Tailing Kit (Roche Diagnostics)
Elution buffer EB5
    50 mM Tris-acetate (pH 7.5)
    150 mM NaCl
    5 mM EDTA
    50 µg/ml *S. cerevisiae* RNA (Sigma)
Guanidinium isothiocyanate (20 mM) <!>
Oligonucleotide Anti-SDA (see Table 1)
TBE
    89 mM Tris
    89 mM boric acid
    2 mM EDTA in H_2O

Special Equipment

Apparatus for agarose gel electrophoresis
Turboblotter with Nytran nylon membrane (Schleicher & Schuell)
X-ray films and a developing machine

METHOD

1. To optimize translation time, carry out in vitro translations and affinity selections with a single model mRNA construct under similar conditions as described in step 1 of the In Vitro Translation protocol. Test reaction translation times from 6 to 12 minutes.

   Longer translation times may be necessary if a longer library mRNA is used for ribosome display.

2. After in vitro translation for 5, 7, 9, and 11 minutes, carry out an affinity selection as described in Affinity Selection, Protocol 5 (p. 554) with one modification: After washing, elute mRNA with 200 μl of elution buffer EB5.

3. Precipitate the RNA immediately by addition of 600 μl of ice-cold ethanol and incubate the samples for 30 minutes on ice.

4. Centrifuge at 14,000g for 30 minutes at 4°C.

5. Remove the supernatants and dry the RNA pellets for 10–15 minutes at room temperature.

6. Dissolve the RNA pellet in 10 μl of RNA denaturation buffer on ice and incubate for 10 minutes at 70°C.

   For an estimation of the amount of recovered mRNA, several other control samples containing between 0.2 ng and 10 ng of the original model mRNA can be prepared as standards.

7. Chill the samples on ice, mix with 1 μl of gel loading buffer, and separate by 1.5% agarose gel electrophoresis in the presence of TBE and 20 mM guanidinium isothiocyanate.

8. Blot the RNA samples to a Nytran nylon membrane using a Turboblotter according to the manufacturer's recommendations.

9. Carry out hybridization for at least 4 hours at 60°C as described previously (Hanes and Plückthun 1997) with the oligonucleotide Anti-SDA, labeled by 3′-tailing with digoxigenin-11-dUTP/dATP using the DIG Oligonucleotide Tailing Kit.

10. Detect the hybridized oligonucleotide probe using the DIG DNA Labeling and Detection Kit with the chemiluminescent substrate CSPD and exposure to X-ray film.

REFERENCES

Bass S., Gu Q., and Christen A. 1996. Multicopy suppressors of prc mutant *Escherichia coli* include two HtrA (DegP) protease homologs (HhoAB), DksA, and a truncated RlpA. *J. Bacteriol.* **178:** 1154–1161.

Boder E.T. and Wittrup K.D. 1997. Yeast surface display for screening combinatorial polypeptide libraries. *Nat. Biotechnol.* **15:** 553–557.

Cadwell R.C. and Joyce G.F. 1992. Randomization of genes by PCR mutagenesis. *PCR Methods Appl.* **2:** 28–33.

Chen H.Z. and Zubay G. 1983. Prokaryotic coupled transcription-translation. *Methods Enzymol.* **101:** 674–690.

Chen Y., Wiesmann C., Fuh G., Li B., Christinger H.W., McKay P., de Vos A.M., and Lowman H.B. 1999. Selection and analysis of an optimized anti-VEGF antibody: Crystal structure of an affinity-matured Fab in complex with antigen. *J. Mol. Biol.* **293:** 865–881.

Chien C.T., Bartel P.L., Sternglanz R., and Fields S. 1991. The two-hybrid system: A method to identify and clone genes for proteins that interact with a protein of interest. *Proc. Natl. Acad. Sci.* **88:** 9578–9582.

Dower W.J. and Cwirla S.E. 1992. Creating vast peptide expression libraries: Electroporation as a tool to construct plasmid libraries of greater than 10^9 recombinants. In *Guide to electroporation and electrofusion* (ed. D.C. Chang et al.), pp. 291–301. Academic Press, San Diego.

Fields S. and Song O. 1989. A novel genetic system to detect protein-protein interactions. *Nature* **340:** 245–246.

Freedman R.B., Hawkins H.C., and McLaughlin S.H. 1995. Protein disulfide-isomerase. *Methods Enzymol.* **251:** 397–406.

Georgiou G., Poetschke H.L., Stathopoulos C., and Francisco J.A. 1993. Practical applications of engineering gram-negative bacterial cell surfaces. *Trends Biotechnol.* **11:** 6–10.

Gersuk G.M., Corey M.J., Corey E., Stray J.E., Kawasaki G.H., and Vessella R.L. 1997. High-affinity peptide ligands to prostate-specific antigen identified by polysome selection. *Biochem. Biophys. Res. Commun.* **232:** 578–582.

Hajnsdorf E., Braun F., Haugel-Nielsen J., Le Derout J., and Régnier P. 1996. Multiple degradation pathways of the rpsO mRNA of *Escherichia coli*. RNase E interacts with the 5′ and 3′ extremities of the primary transcript. *Biochimie* **78:** 416–424.

Hanes J. and Plückthun A. 1997. In vitro selection and evolution of functional proteins by using ribosome display. *Proc. Natl. Acad. Sci.* **94:** 4937–4942.

Hanes J., Jermutus L., Schaffitzel C., and Plückthun A. 1999. Comparison of *Escherichia coli* and rabbit reticulocyte ribosome display systems. *FEBS Lett.* **450:** 105–110.

Hanes J., Schaffitzel C., Knappik A., and Plückthun A. 2000. Picomolar affinity antibodies from a fully synthetic naive library selected and evolved by ribosome display. *Nat. Biotechnol.* **18:** 1287–1292.

Hanes J., Jermutus L., Weber-Bornhauser S., Bosshard H.R., and Plückthun A. 1998. Ribosome display efficiently selects and evolves high-affinity antibodies in vitro from immune libraries. *Proc. Natl. Acad. Sci.* **95:** 14130–14135.

Hawkins R.E., Russell S.J., and Winter G. 1992. Selection of phage antibodies by binding affinity. Mimicking affinity maturation. *J. Mol. Biol.* **226:** 889–896.

He M. and Taussig M.J. 1997. Antibody-ribosome-mRNA (ARM) complexes as efficient selection particles for in vitro display and evolution of antibody combining sites. *Nucleic Acids Res.* **25:** 5132–5134.

He M., Menges M., Groves M.A., Corps E., Liu H., Brüggemann M., and Taussig M.J. 1999. Selection of a human anti-progesterone antibody fragment from a transgenic mouse library by ARM ribosome display. *J. Immunol. Methods* **231:** 105–117.

Irving R.A., Kortt A.A., and Hudson P.J. 1996. Affinity maturation of recombinant antibodies using *E. coli* mutator cells. *Immunotechnology* **2:** 127–143.

Jermutus L., Honnegger A., Schwesinger F., Hanes J., and Plückthun A. 2001. Tailoring in vitro evolution for protein affinity or stability. *Proc. Natl. Acad. Sci.* **98:** 75–80.

Keiler K.C., Waller P.R., and Sauer R.T. 1996. Role of a peptide tagging system in degradation of proteins synthesized from damaged messenger RNA. *Science* **271:** 990–993.

Knappik A., Ge L., Honegger A., Pack P., Fischer M., Wellnhofer G., Hoess A., Wölle J., Plückthun A., and Virnekäs B. 2000. Fully synthetic human combinatorial antibody libraries (HuCAL) based on modular consensus frameworks and CDRs randomized with trinucleotides. *J. Mol. Biol.* **296:** 57–86.

Krebber A., Bornhauser S., Burmester J., Honegger A., Willuda J., Bosshard H.R., and Plückthun A. 1997. Reliable cloning of functional antibody variable domains from hybridomas and spleen cell repertoires employing a reengineered phage display system. *J. Immunol. Methods* **201:** 35–55.

Low N.M., Holliger P.H., and Winter G. 1996. Mimicking somatic hypermutation: Affinity maturation of antibodies displayed on bacteriophage using a bacterial mutator strain. *J. Mol. Biol.* **260:** 359–368.

Mattheakis L.C., Bhatt R.R., and Dower W.J. 1994. An in vitro polysome display system for identifying ligands from very large peptide libraries. *Proc. Natl. Acad. Sci.* **91:** 9022–9026.

Mattheakis L.C., Dias J.M., and Dower W.J. 1996. Cell-free synthesis of peptide libraries displayed on polysomes. *Methods Enzymol.* **267:** 195–207.

Plückthun A., Schaffitzel C., Hanes J., and Jermutus L. 2000. In vitro selection and evolution of proteins. *Adv. Protein Chem.* **55:** 367–403.

Pokrovskaya I.D. and Gurevich V.V. 1994. In vitro transcription: Preparative RNA yields in analytical scale reactions. *Anal. Biochem.* **220:** 420–423.

Pratt J.M. 1984. Coupled transcription-translation in prokaryotic cell-free systems. In *Current protocols* (ed. B.D. Hemes and S.J. Higgins), pp. 179–209. IRL Press, Oxford.

Ramachandiran V., Kramer G., and Hardesty B. 2000. Expression of different coding sequences in cell-free bacterial and eukaryotic systems indicates translational pausing on *Escherichia coli* ribosomes. *FEBS Lett.* **482:** 185–188.

Ryabova L.A., Desplancq D., Spirin A.S., and Plückthun A. 1997. Functional antibody production using cell-free translation: Effects of protein disulfide isomerase and chaperones. *Nat. Biotechnol.* **15:** 79–84.

Schaffitzel C., Berger I., Postberg J., Hanes J., Lipps H.J., and Plückthun A. 2001. In vitro generated antibodies specific for telomeric guanine-quadruplex DNA react with *Stylonychia lemnae* macronuclei. *Proc. Natl. Acad. Sci.* **98:** 8572–8577.

Schwesinger F., Ros R., Strunz T., Anselmetti D., Güntherodt H.J., Honegger A., Jermutus L., Tiefenauer L., and Plückthun A. 2000. Unbinding forces of single antibody-antigen complexes correlate with their thermal dissociation rates. *Proc. Natl. Acad. Sci.* **97:** 9972–9977.

Smith G.P. 1985. Filamentous fusion phage: Novel expression vectors that display cloned antigens on the virion surface. *Science* **228:** 1315–1317.

Stemmer W.P. 1994. Rapid evolution of a protein in vitro by DNA shuffling. *Nature* **370:** 389–391.

Wade H.E. and Robinson H.K. 1966. Magnesium ion-independent ribonucleic acid depolymerases in bacteria. *Biochem. J.* **101:** 467–479.

Winter G., Griffiths A.D., Hawkins R.E., and Hoogenboom H.R. 1994. Making antibodies by phage display technology. *Annu. Rev. Immunol.* **12:** 433–455.

Yang W.P., Green K., Pinz-Sweeney S., Briones A.T., Burton D.R., and Barbas C.F., 3rd. 1995. CDR walking mutagenesis for the affinity maturation of a potent human anti-HIV-1 antibody into the picomolar range. *J. Mol. Biol.* **254:** 392–403.

Zaccolo M., Williams D.M., Brown D.M., and Gherardi E. 1996. An approach to random mutagenesis of DNA using mixtures of triphosphate derivatives of nucleoside analogues. *J. Mol. Biol.* **255:** 589–603.

31 Analysis of Protein Interactions with Peptides Synthesized on Membranes

Joachim Koch,[1,3] Michael Mahler,[2,3] Martin Blüthner,[2] and Stefan Dübel[2]

[1]Forschungsstelle Hantaviren der Heidelberge Akademie der Wissenschaften, 69120 Heidelberg, Germany; [2]Universität Heidelberg, Institut für Molekulare Genetik, 69120 Heidelberg, Germany; [3]Both authors contributed equally

INTRODUCTION

Large numbers of different oligopeptides assembled from up to 20 amino acids or their derivatives can be synthesized on activated cellulose membranes, in a method called SPOT synthesis. Binding studies using these membranes are well established for the analysis of protein–protein interactions (Frank 1992), with detection of protein binding to individual spots performed in the manner of an immunoblot. This method allows the mapping of antibody epitopes (Korth et al. 1997; Kramer et al. 1997; Kneissel et al. 1999; Liu et al. 1999; Blüthner et al. 2000; Mahler et al. 2000), the analysis of protein–protein interactions (McCarty et al. 1996; Schultz et al. 1998; Hilpert et al. 1999; Piossek et al. 1999), the identification of protein–DNA interaction domains (Reuter et al. 1999), and the prediction of three-dimensional structures of epitopes at the level of resolution of single side chains (Kneissel et al. 1999; Liu et al. 1999; Blüthner et al. 2000). Moreover, solid-phase peptides can be used for affinity purification of proteins under investigation (Smith and Fisher 1984). For example, epitope-specific antibodies can be purified from polyclonal sera, thus combining the monospecificity of monoclonal antibodies (mAbs) with the ease of rabbit serum preparation.

569

The SPOT method is suited for the identification of short linear interaction domains, but it does not allow the analysis of discontinuous binding sites, except in the rare cases where two short peptides combine to form the interaction domain, but both peptides separately bind the ligand with high affinity (Reineke et al. 1999). The SPOT method can be used for the analysis of interaction domains comprising up to 100 amino acid residues, when manual pipetting is used for peptide synthesis. However, the current development of robotic handling already generates arrays of several hundred different peptide spots on membranes having the size of an enzyme-linked immunosorbent assay (ELISA) plate. Commercially available pipetting robots (e.g., ASP222, Abimed, Germany) can handle larger membranes, creating arrays of several thousand spots per synthesis (Tegge and Frank 1998; Kramer et al. 1999). This allows complete analysis of large proteins, including mutational analysis (e.g., alanine-walking) on a single membrane.

Peptide Synthesis Strategy for an Interaction Study

When large proteins are analyzed, it may be helpful to narrow down the interaction site to a subdomain by using either other experimental methods or computer prediction. Analysis of the interaction of deletion mutants or the use of gene-specific phage display (Blüthner et al. 1992; Fack et al. 1997) can provide an indication of whether the interacting amino acids are clustered within a short sequence (linear binding site, typically <20 amino acids) or are distributed over a large portion of the structure (conformational binding site). Detectable interaction of the binding partner with a protein on a blot after SDS-PAGE can provide information regarding a linear interaction motif, as can computational methods. Methods based on homology with known interaction partners are more helpful than ab initio modeling.

The analysis of an interaction domain or of an entire protein using SPOT synthesis is divided into a minimum of two steps. First, 15-mer peptides, covering the entire sequence with an offset of 3 or 4 amino acids, are synthesized. To define the size of the core interaction region, reactive areas identified during this first scan are further characterized with a second set of peptides having an offset of one amino acid. For shorter proteins or interaction domains, 15-mer peptides with one amino acid offset can be used as a starting point (Fig. 1). When polyclonal sera are analyzed, which may identify overlapping epitopes bound by different antibodies, it can be helpful to shorten the length of the peptides to distinguish between the overlapping epitopes (Blüthner et al. 2000; Mahler et al. 2000).

In a second step, the core-binding motif and contribution of single amino acid residues are analyzed. The core-binding region can be verified by employing peptides successively shortened from the carboxy- and the amino-terminal ends (Blüthner et al. 2000). However, a careful interpretation of the resulting data is required. In some cases, the results from experiments with short peptides do not match the data obtained from the "one-offset" analysis. This can be explained by hypothesizing different structural properties of peptides varying in length; for example, longer peptides may form a secondary structure that hides the binding motif. The contribution of single amino acid side chains can be analyzed by using a set of peptides with point mutations; for example, alanine or glycine walks or replacements with other amino acids (Liu et al. 1999; Blüthner et al. 2000).

The contribution of side-chain modifications (e.g., phosphorylation or glycosylation) can be analyzed by synthesizing sets of peptides that have identical sequences but differ in side-chain modifications (Mukhija et al. 1998). For binding motifs located at the amino terminus of a protein, the contribution of the amino-terminal amino group to the binding can be assessed by synthesizing peptides with amino acid residues added in front of the amino-terminal amino acid (Kneissel et al. 1999). A similar approach cannot be applied for the carboxy-terminal end because the peptides are immobilized with this residue.

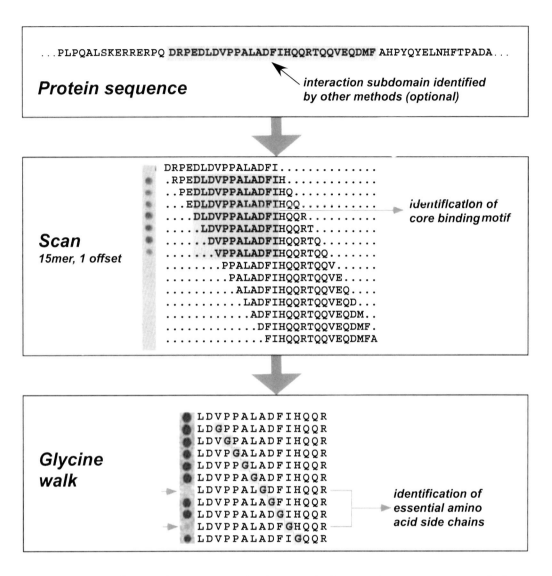

FIGURE 1. Design of a typical interaction study. The example illustrates the epitope mapping of a human serum (identification of the major epitope on the PM/Scl-100 autoantigen; Blüthner et al. 2000). The core binding area DRPED....QDMFA was identified by gene fragment phage display. The scan using 15-mer peptides with an offset of one amino acid identified the sequence VPPALADFIH as the core epitope. For the identification of the amino acid residues essential for the antibody binding (*arrows*), the epitope was subjected to a mutational analysis by successively replacing all amino acids against glycine (a glycine walk).

Detection of Bound Reaction Partners

When SPOT synthesis is used, at least one of the interaction partners must be purified and labeled for detection on a blot. We recommend biotinylation followed by detection with a streptavidin–horseradish peroxidase (HRP) complex. Alternatively, the bound protein ligand can be detected using a secondary antibody.

In epitope-mapping experiments, where an antibody is the interaction partner, secondary antibodies or labeled proteinA/G are the recommended detection systems. Controls using only the detection agents are always required, especially when enzyme-labeled animal sera are used, because unspecific binding, cross-reaction, or preimmunization may give rise to false-positive reactions (Fig. 2, Stage 2A).

In cases of weaker interactions (for instance, due to short off rates), the binding partner, after binding to a peptide, can be electrotransferred to a secondary membrane (generally nitrocellulose), where it can be immobilized, thus allowing extended detection procedures (Fig. 2, Stage 2B).

OUTLINE OF PROCEDURE

The following protocol describes the synthesis of short peptides on activated cellulose membranes coated with polyethylene glycol 500 (PEG500) spacers, and the detection of their binding partners. Peptides are synthesized using Fmoc-amino acid derivatives starting from the carboxyl terminus according to the procedure described by Frank (1992). In the first step, activated amino groups at the end of the PEG spacers are coupled to the activated carboxyl group of the Fmoc amino acid in a nucleophilic reaction, establishing the initial peptide bond. Each of the following reaction cycles is preceded by two treatments to restrict the elongation reaction to the amino terminus.

First, acetylation inactivates amino groups that did not bind to Fmoc amino acids in the previous round of synthesis. Second, the Fmoc protective side chain of the amino-terminal amino acid is chemically cleaved with piperidine to generate reactive amino groups for the coupling of the next amino acid. After completion of the synthesis, the peptides can be incubated with the potential interaction partners to identify their target sequence (see Fig. 2).

FIGURE 2. Outline of the protocols. Short peptides are synthesized on activated cellulose membranes with the auto-spot robot ASP222 (Stage I). The derived peptide spots can be probed with antibodies (Stage 2, Epitope Mapping, steps 1–3) or with protein ligands (Stage 2, Lower-affinity Protein–Protein Interaction). After detecting the bound protein, the peptide spot membranes can be regenerated to remove the compounds bound to the peptide spots and employed for another interaction study (Stage 2, Detection and Regeneration).

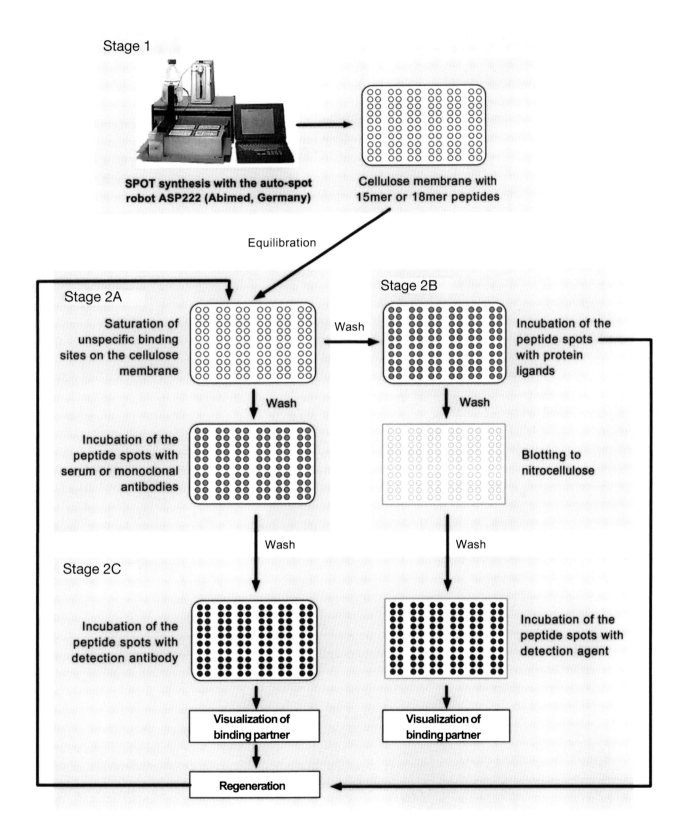

FIGURE 2. (*See facing page for legend.*)

Protocol 1

The following protocol is divided into two main stages: Synthesis of Peptides on Cellulose Membranes (Stage 1) and Immunodetection of Reactive Spots (application of the resulting membranes) (Stage 2). Stage 2 is further divided into three subprotocols: *A* describes the use of peptide spots in finding strong interaction partners with approximately nanomolar binding constants. The epitopes of antibodies are typically determined this way. *B* details the detection of weaker interactions, and *C* describes the detection of binding and membrane regeneration.

MATERIALS

CAUTION: See Appendix for appropriate handling of materials marked with <!>.

Buffers and Solutions

Acetic acid <!>
Acetic acid anhydride <!>
β-Mercaptoethanol <!>
10x Blocking buffer concentrate (SU-07-250; Genosys, Cambridge, United Kingdom)
Bromophenol blue <!>
Bromophenol blue (0.01% w/v) in *N,N*-dimethylformamide <!>
Dichloromethane <!>
N,N-Dimethylformamide (DMF) <!>
Ethanol <!>
Fmoc-L-Amino acid active esters (Genosys, Cambridge, United Kingdom, ST-20-120)
Horse serum
1-Methyl-2-pyrrolidinone (NMP) <!>
Piperidine <!>
Piperidine (20% v/v) in DMF <!>
Regeneration buffer A
   8 M urea
   1% (w/v) SDS <!>
   0.1% (v/v) β-mercaptoethanol <!>
Regeneration buffer B
   50% (v/v) ethanol <!>
   10% (v/v) acetic acid in distilled H_2O
Skim milk powder
SDS <!>
Sucrose
TBS
   10 mM Tris-Cl (pH 7.6)
   150 mM NaCl
Transfer buffer for western blotting
   25 mM Tris-Cl (pH 7.6)
   192 mM glycine
   20% methanol <!>
   0.03% SDS <!>
     When preparing from stock solutions, add methanol after H_2O has been added to avoid precipitations.

Trifluoroacetic acid (TFA) <!>
Triisobutylsilane <!>
Tween-20
Urea

Antibodies

Goat–anti-human HRP-conjugated antibody
Goat–anti-mouse HRP-conjugated antibody

Special Equipment

Amersham Lifescience Hyperprocessor (Amersham Pharmacia Biotech)
Amino-PEG membranes (Abimed, Langenfeld, Germany, APEG-UC540-10/15, 84300)
Blotting apparatus (Biometra-Fast-Blot, Göttingen, Germany, B337593)
Blotting paper (Machery-Nagel, Düren, Germany, MN 440 B)
ECL western blotting detection reagents (Amersham Pharmacia Biotech)
Hair dryer
Hypercassette (Amersham Pharmacia Biotech)
Hyperfilm ECL (Amersham Pharmacia Biotech)
Protan Nitrocellulose Transfer Membrane (Schleicher & Schuell)
Shaking platform
Spot robot (model ASP222, Abimed, Langenfeld, Germany)

All of the equipment used for synthesis and regeneration should be resistant to organic solvents. Glassware or Teflon ware should be exclusively used in all steps involving these solvents. Standard micropipetting tips (Gilson, Eppendorf) can be employed.

METHOD

Stage 1: Synthesis of Peptides on Cellulose Membranes

Preparation of the Amino Acid Derivative Solutions

With the exception of Fmoc-arginine, all amino acid derivatives can be prepared as follows:

1. Thaw the amino acid derivative powder under vacuum in a desiccator to avoid hydration of the amino acid derivatives.

    Amino acid derivative powder must be stored at –20°C.

2. Dissolve the amino acid derivatives in the appropriate volume of NMP to a final concentration of 250 mM and prepare aliquots of ~400 μl.

3. The dissolved amino acids (except arginine) can be stored for several months at –70°C. Prior to use in step 5, the amino acids should be thawed under vacuum in a desiccator.

    Fmoc-arginine is very unstable and must be prepared freshly for every cycle of the synthesis. For convenience, preweighed aliquots of arginine powder can be stored at –70°C and, directly before the cycle is started, dissolved in the appropriate amount of NMP after prewarming under vacuum as described above.

Application of Amino Acid Derivatives

On the basis of our experience with an ASP222 autospot robot, 777 peptides arrayed in 37 x 21 lanes can be conveniently synthesized on a cellulose membrane of 13 x 9 cm. All steps are carried out at room temperature, and higher temperatures should be avoided; an air-conditioned room is recommended. Steps involving the use of organic solvents should be performed under a fume hood.

4. Program the spot robot according to the manufacturer's instructions.

5. Place the tubes with the amino acid solutions and amino-PEG membranes in the appropriate positions in the robot and run the synthesis by applying 3 x 0.1 μl of amino acid dilution per spot.

   > With regard to the differences in the coupling efficiency of different amino acids (Fields and Noble 1990), we recommend three applications of the amino acids within a total incubation time of at least 2 hours.

6. Remove the membrane from the robot and proceed with the washing steps (step 7). Following the application of the last amino acid, proceed directly to step 14.

Washing of the Membranes

Following each application cycle, the membranes should be washed extensively to prepare the peptides for the next cycle of the synthesis. The protocol below is used after each cycle, except the last one. The quantities of chemicals and solutions given in this protocol are sufficient for four membranes of 13 x 9 cm, which can be processed in parallel when using the ASP222 robot.

7. Rinse the membranes in 50 ml of 4% (v/v) acetic acid anhydride in DMF for 30 seconds and again for 5 minutes in a freshly prepared solution to acetylate free amino groups of peptides that have not been elongated in the previous round of synthesis. These acetylated amino acids are no longer accessible for the binding of new Fmoc-amino acids and prevent the generation of false sequences.

8. Wash the membranes two times for 2 minutes with 50 ml of DMF to remove traces of acetic acid anhydride, which would interfere with the following steps.

9. Incubate the membranes for 5 minutes with 50 ml of 20% (v/v) piperidine in DMF to cleave the Fmoc protection groups chemically.

10. Wash the membranes ten times for 2 minutes with 50 ml of DMF to remove the piperidine quantitatively.

11. Stain the free amino groups of the peptides for 5 minutes with 0.01% (w/v) bromophenol blue in DMF to monitor the efficiency of the piperidine cleavage. Blue spots should be clearly visible, indicating accessible amino groups.

    > Due to the charge-specific staining, bromophenol blue does not bind only to amino-terminal amino groups. The side chains of other amino acids can strongly influence the staining intensity. The visible color of the peptides depends on the overall charge and thus depends on the individual amino acid sequence. Furthermore, the variable coupling efficiency of the different amino acids might influence the intensity of the staining.
    >
    > The bromophenol blue/DMF stock solution should have an intense orange color and should be discarded when the color has turned to blue-green.

12. Wash the membranes two times for 2 minutes with 50 ml of ethanol to remove unspecifically bound bromophenol blue and traces of contaminating H_2O.

13. Dry the membrane with cold air, using a regular hair dryer. The membranes are now ready for the next application cycle (return to step 5).

    Dry the membranes on the reverse side to avoid damaging the peptides. Heating of the membrane should be avoided.

    After approximately 10–12 cycles of synthesis, use a graphite pencil to mark with a grid spots that should later be incubated together. This facilitates the cutting of the membrane.

The Final Cycle

14. Remove unbound amino acids by two washes for 2 minutes with 50 ml of DMF.

15. Cleave the Fmoc protection groups by incubation with 50 ml of 20% (v/v) piperidine in DMF for 5 minutes.

16. Wash the membranes ten times for 2 minutes with 50 ml of DMF to remove the piperidine.

17. Stain the free amino groups with 50 ml of 0.01% (w/v) bromophenol blue in DMF for 5 minutes.

18. Acetylate free amino groups with two washes (30 seconds and 5 minutes) using 50 ml of 4% (v/v) acetic acid anhydride in DMF until all spots are completely destained.

    The destaining process may take longer than 5 minutes, but, according to our experience, should be completed after 10 minutes.

19. Remove traces of acetic acid anhydride by two washes for 2 minutes with 50 ml of DMF.

20. Wash the membranes two times for 2 minutes with 50 ml of ethanol to remove the DMF.

21. Dry the membrane with cold air using a regular hair dryer (as in step 13).

22. Remove all protective groups of the side chains by incubating the membranes with a freshly prepared mixture of 20 ml of dichloromethane (solvent for the chemical cleavage), 20 ml of TFA (for cleavage of the protective groups), and 1 ml of triisobutylsilane (for capture of reaction products) for 1 hour.

    Because dichloromethane is very volatile, the incubation should be carried out in a closed glass container.

> **CAUTION:** Avoid contact of the residual reaction mixture with DMF (use a separate waste container) because there is a risk of exothermal reactions (explosion!).

23. Wash the membrane four times for 2 minutes with 50 ml of dichloromethane.

24. Wash the membranes three times for 2 minutes with 50 ml of DMF.

25. Wash the membranes two times for 2 minutes with 50 ml of ethanol to remove traces of DMF.

26. Dry the membrane with a hair dryer as described in step 13.

    The membranes can either be stored at –20°C in a sealed plastic bag or directly used for equilibration with aqueous buffer, required for the interaction with the binding partner.

Stage 2: Immunodetection of Reactive Spots

A. Epitope Mapping

1. The membranes are slowly rehydrated by placing them in 50 ml of 100% ethanol. Gradually reduce the ethanol concentration by addition of 50 ml of TBS to the membranes every 5 minutes for five times.

   The white color of the spots should disappear completely as they gradually become less repellent. The membranes are now ready for immunodetection or incubation with proteins.

2. To block unspecific binding sites, incubate the membranes with blocking buffer for 2 hours at room temperature or overnight at 4°C. Depending on the antibody of interest, different blocking conditions can be compared to obtain an optimal signal-to-noise ratio. The following blocking solutions, increasing in "stringency" may be tested.

   a. 2% (w/v) skim milk powder in TBS

   b. TBS with 2% (w/v) skim milk powder, 0.2% (v/v) Tween-20

   c. TBS with 50% (v/v) horse serum, 10% (v/v) 10x blocking buffer concentrate, 0.2% (v/v) Tween-20, 150 mM sucrose.

      The blocking conditions are quite critical, and one may have to try several conditions for optimization. In our hands, blocking solution c works best for most applications; bovine fetal serum could be tried as well.

3. Wash the membranes once with TBS/0.2% (v/v) Tween-20.

4. Incubate the membranes with the antibody diluted in blocking buffer (as used in step 2). If blocking buffer (c) is used, dilute in TBS with 3% (v/v) horse serum, 10% (v/v) blocking buffer concentrate, 0.05% (v/v) Tween-20, 150 mM sucrose.

   For epitope mapping of monoclonal antibodies, use approximately 4–5 μg of purified antibody per milliliter of incubation volume. When using a polyclonal serum, we recommend a dilution of 1:100. Incubate for 90 minutes at room temperature or overnight at 4°C.

   It is not necessary to use a large volume of antibody solution for the incubation; however, make sure that the membrane is completely immersed, and prevent drying out by using a lid or wet chamber.

5. Wash the membranes three times for 5–10 minutes with TBS/0.2% (v/v) Tween-20 to remove unbound antibodies.

6. Incubate the membrane for 75 minutes with a secondary antibody diluted in the same buffer used for the first antibody. Dilutions can be used analogous to those employed in immunoblots after SDS-PAGE.

7. Wash the membranes three times for 5–10 minutes with TBS/0.2% (v/v) Tween-20, subsequently followed by washing three times for 5–10 minutes with TBS.

8. Remove excessive buffer from the membranes by placing tissue on the membrane.

   To avoid damage to the peptide coats, do not wipe or press tissue onto the membrane.

9. Detect the spots with chemiluminescence (step 1 below in Detection and Regeneration section).

B. Lower-affinity Protein–Protein Interaction

1. Equilibrate, block, and wash the peptide spot membrane as described in steps 1–4 under Epitope Mapping, above.

2. Incubate the membranes with the target protein diluted in TBS analogous to step 4, under Epitope Mapping, above. We recommend using 1–5 μg/ml.

3. Wash the membrane three times for 2 minutes with TBS.

4. Briefly equilibrate both the peptide membrane and a sheet of nitrocellulose trimmed to fit the peptide membrane in transfer buffer for western blotting.

5. Electrotransfer the proteins bound to the peptide spot membranes onto nitrocellulose for 1 hour using 0.85 mA per square centimeter. Due to denaturation by SDS, all proteins should have acquired a negative charge. Therefore, the nitrocellulose should be placed toward the positive electrode.

   Depending on the chemical properties of the protein ligands, the time required for the transfer may differ and, therefore, has to be determined empirically.

6. Block the nitrocellulose membrane with 2% skim milk in TBS for 2 hours at room temperature.

7. Continue with steps 7–9 under Epitope Mapping, above.

C. Detection and Regeneration

Detection of the reactive spots. Detection of bound protein can be achieved with a variety of conventional staining techiques. Precipitating enzyme substrates (e.g., diaminobenzidine <!> or chloronaphthol <!> for HRP-labeled detection agents) may be used. However, they interfere with the regeneration of the peptide spot membrane limiting a given membrane to only one use. A minimum of two different incubations—sample plus negative control—are recommended for each analysis; therefore, the chemiluminescent ECL system is the preferred detection agent. In case the proteins have been electrotransferred to a nitrocellulose membrane (Stage 2B), this restriction does not apply. However, given the possibility of obtaining several data sets with varying exposure time when using the ECL system, this detection method is suggested for both applications.

1. Prepare ECL detection reagent according to the manufacturer's instructions and incubate the membranes for 1 minute.

2. Place the membrane between two transparencies, apply an ECL film, and expose using a Hypercassette at room temperature.

   Start with an exposure time of 1 minute to get an impression of the signal intensity. Try other exposure times until a maximum signal-to-noise ratio is obtained.

3. Transfer the membranes to distilled H_2O and store at 4°C until they are subjected to regeneration.

   Membranes can be stored in H_2O for several days prior to regeneration.

   If no signal is seen after 30 minutes of exposure, check the detection system. If detection reagents are all right, use less stringent blocking. If no binding occurs, this may indicate a discontinuous binding site or very low affinity binding.

   If unspecific signals and a high background occur, increase the stringency of the blocking conditions and make sure that your primary binding partner and detection reagent (secondary antibody) are of high purity and are used in the highest possible dilution.

Regeneration of the membranes

4. Rinse the membranes three times for 5 minutes with distilled H_2O.

5. Wash the membranes three times for 5 minutes with DMF.

6. Rinse the membranes three times for 5 minutes with distilled H_2O.

7. Wash the membrane three times for 10 minutes with regeneration buffer A.

8. Wash the membrane three times for 10 minutes with regeneration buffer B.

9. Rinse the membranes two times for 10 minutes with ethanol.

10. Dry the membrane with a hair dryer as described in step 13 of Synthesis of Peptides on Cellulose Membranes (p. 577). The membranes can either be stored at –20°C in a sealed plastic bag or used to go on with step 1 under Epitope Mapping, p. 578.

> The membranes can usually be regenerated up to 10 times without loss of signal intensity. In very rare cases, ligands with very high affinities resist elution from the spots, and the membranes can only be used once. This can be verified by incubating a previously regenerated spot membrane directly with the detection system (steps 2 and 3, under Epitope Mapping, p. 578, then continue directly with steps 7–9 under Epitope Mapping, p. 578).

REFERENCES

Blüthner M., Bautz E.K., and Bautz F.A. 1992. Mapping of epitopes recognized by PM/Scl autoantibodies with gene-fragment phage display libraries. *J. Immunol. Methods* **198:** 187–198.

Blüthner M., Mahler M., Müller D.B., Dünzl H., and Bautz F.A. 2000. Identification of an alpha-helical epitope region on the PM/Scl-100 autoantigen with structural homology to a region on the heterochromatin p25beta autoantigen using immobilized overlapping synthetic peptides. *J. Mol. Med.* **78:** 47–54.

Fack F., Hügle-Dorr B., Song D., Queitsch I., Petersen G., and Bautz E.K. 1997. Epitope mapping by phage display: Random versus gene-fragment libraries. *J. Immunol. Methods* **206:** 43–52.

Fields G.B. and Noble R.L. 1990. Solid phase peptide synthesis utilizing 9-fluorenylmethoxycarbonyl amino acids. *Int. J. Pept. Protein Res.* **35:** 161–214.

Frank R. 1992. SPOT synthesis: An easy technique for the positionally addressable, parallel chemical synthesis on a membrane support. *Tetrahedron* **48:** 9217–9232.

Hilpert K., Behlke J., Scholz C., Misselwitz R., Schneider-Mergener J., and Höhne W. 1999. Interaction of the capsid protein p24 (HIV-1) with sequenced-derived peptides: Influence on p24 dimerization. *Virology* **254:** 6–10.

Kneissel S., Queitsch I., Petersen G., Behrsing O., Micheel B., and Dübel S. 1999. Epitope structures recognised by antibodies against the major coat protein (g8p) of filamentous bacteriophage fd (Inoviridae). *J. Mol. Biol.* **288:** 21–28.

Korth C., Stierli B., Streit P., Moser M., Schaller O., Fischer R., Schulz-Schaeffer W., Kretzschmar H., Raeber A., Braun U., Ehrensperger F., Hornemann S., Glockshuber R., Riek R., Billeter M., Wuthrich K., and Oesch B. 1997. Prion (PrPSc)-specific epitope defined by a monoclonal antibody. *Nature* **390:** 74–77.

Kramer A., Keitel T., Höhne W., and Schneider-Mergener J. 1997. Molecular basis of binding promiscuity of an anti-p24 (HIV-1) monoclonal antibody. *Cell* **91:** 799–809.

Kramer A., Reineke U., Dong L., Hoffmann B., Hoffmüller U., Winkler D., Volkmer-Engert R., and Schneider-Mergener J. 1999. Spot synthesis: Observations and optimizations. *J. Pept. Res.* **54:** 319–327.

Liu Z., Song D., Kramer A., Martin A., Dandekar T., Schneider-Mergener J., Bautz E., and Dübel S. 1999. Fine mapping of the antigen-antibody interaction of scFv217, a recombinant antibody inhibiting RNA polymerase from *Drosophila melanogaster*. *J. Mol. Recognit.* **12:** 103–111.

Mahler M., Mierau R., and Blüthner M. 2000. Fine specificity of the B-cell anti-CENP-A autoimmune response. *J. Mol. Med.* **78:** 460–467.

McCarty J., Rüdiger S., Schönfeld H.J., Schneider-Mergener J., Nakahigashi K., Yura T., and Buckau B. 1996. Regulatory region C of the *E. coli* heat shock transcription factor, sigma32, constitutes a DnaK binding site and is conserved among eubacteria. *J. Mol. Biol.* **256:** 829–837.

Mukhija S., Germeroth L., Schneider-Mergener J., and Erni B. 1998. Identification of peptides inhibiting enzyme I of the bacterial phosphotransferase system using combinatorial cellulose-bound peptide libraries. *Eur. J. Biochem.* **254:** 433–438.

Piossek C., Schneider-Mergener J., Schirner M., Vakalopolou E., Germeroth L., and Thierauch K-H. 1999. Vascular endothelial growth factor (VEGF) receptor II-derived peptides inhibit VEGF. *J. Biol. Chem.* **274:** 5612–5619.

Reineke U., Sabat R., Welfle H., Volk H.-D., and Schneider-Mergener J. 1999. A synthetic mimic of a discontinuous binding site on interleukin-10. *Nature Biotechnol.* **17:** 271–275.

Reuter M., Schneider-Mergener J., Küpper D., Meisel A., Mackeldanz P., Krüger D., and Schroeder C. 1999. Regions of endonuclease EcoRII involved in DNA target recognition identified by membrane-bound peptide repertoires. *J. Biol. Chem.* **274:** 5213–5221.

Schultz J., Hoffmüller U., Ashurst J., Krause G., Schmieder P.J., Macias M., Schneider-Mergener J., and Oschkinat H. 1998. Specific interactions between the synthrophin PDZ domain and voltage-gated sodium channels. *Nat. Struct. Biol.* **5:** 19–24.

Smith D.E. and Fisher P.A. 1984. Identification, developmental regulation, and response to heat shock of two antigenically related forms of a major nuclear envelope protein in *Drosophila* embryos: Application of an improved method for affinity purification of antibodies using polypeptides immobilized on nitrocellulose blots. *J. Cell Biol.* **99:** 20–28.

Tegge W.J. and Frank R. 1998. Analysis of protein kinase substrate specificity by the use of peptide libraries on cellulose paper (SPOT-method). *Methods Mol. Biol.* **87:** 99–106.

32 Protein Bundling to Enhance the Detection of Protein–Protein Interactions

Sridaran Natesan

Cambridge Genomics Center, Aventis Pharmaceuticals, Cambridge, Massachusetts 02139

INTRODUCTION

Many key cellular processes, including transcriptional regulation, protein degradation, and signal transduction, are the outcome of the protein–protein interactions that occur in vivo. In almost all cases, the proteins that participate in these processes exist in multiprotein complexes, the formation of which is dependent on extensive protein–protein interactions. For example, the process of transcriptional activation of eukaryotic genes is dependent on the interaction of multiple transcriptional activators with dozens of proteins that are part of several multiprotein complexes, including TFIID, RNA polymerase, and chromatin-modifying complexes (Laemmli and Tjian 1996; Kingston 1999; Lemon and Tjian 2000). A large number of protein–protein interactions occurring in many biological processes have been identified in recent years, although they represent only a small minority of the total protein–protein interactions that are thought to occur in eukaryotic cells. At present, a variety of in vitro and in vivo methods are used to study protein–protein interactions (Fields and Song 1989; Adams et al. 1991; Rossi et al. 1997; Li et al. 2001). Despite the widespread use of many of these methods, there is a growing realization that developing more sensitive and/or high-throughput methods is necessary to analyze the vast majority of protein–protein interactions that are yet to be identified. This chapter describes a method that is specifically designed to enhance the sensitivity of detection of protein–protein interactions in vivo.

Two-hybrid Interactions

The two-hybrid method is the most popular and widely used to study protein–protein interactions in vivo (Chien et al. 1991; Allen et al. 1995; Bai and Elledge 1996). The conceptual basis for the two-hybrid approach originated from the initial understanding of the mechanism of action of transcriptional activator proteins (discussed in Gill and Ptashne 1988; Mendelsohn and Brent 1994; Ptashne and Gann 1997; Keaveney and Struhl 1998). Transcriptional activators are usually composed of at least two functionally autonomous domains: a DNA-binding domain that is essential for binding to the regulatory elements in the promoter region in a sequence-specific manner and an activation domain that recruits the transcription machinery to the promoter region of a gene (Ptashne and Gann 1997). Earlier studies have shown that substituting the relatively weak activation domain of the yeast transcriptional activator protein GAL4 with the strong activation domain derived from the herpes simplex virus protein VP16 could greatly enhance its transcriptional activation potency without affecting its DNA-binding function (Sadowski et al. 1988). This simple yet elegant study showed that activation and DNA-binding domains are highly modular in nature. Importantly, studies that followed this work have shown that covalent attachment of the DNA-binding and activation domains is not essential, rather, a mere non-covalent linking of these domains is sufficient to induce the transcription of their target genes (discussed in Fields and Song 1989; Bemis et al. 1995; Estojak et al. 1995).

Separately expressed DNA-binding and activation domains can be brought together in a number of ways. As shown in a key paper by Fields and Song (1989), separately expressed activation domain and DNA-binding domain fusion proteins can be non-covalently linked by fusing them to proteins that possess an inherent affinity for each other. The presence of these fusion proteins in the same cell promotes their interaction, resulting in the reconstitution of the functional transcriptional activator and subsequent transcriptional activation of its target gene. In a different version of this method, instead of fusing the DNA-binding and activation domains to proteins that possess inherent affinity for each other, these domains were fused with proteins that are capable of interacting with a third protein or a chemical compound simultaneously (Belshaw et al. 1996; Rivera et al. 1996; Zhang and Lautar 1996; Senguptha et al. 1999). The presence of all three components within the same cell promotes the reconstruction of functional transcriptional activator molecules. These modifications of the basic two-hybrid system form the basis for the three-hybrid and dimerizer-dependent gene regulation system, respectively.

Problems in Two-hybrid Assays

Several problems associated with the expression of foreign proteins in eukaryotic cells may preclude the capture even of high-affinity interactions between proteins using the conventional two-hybrid system. It is common to find that the interaction between two-hybrid proteins cannot be detected simply because the expression of one or both hybrid proteins is lethal to the cells. Another frequently faced problem in the two-hybrid assay is that the hybrid proteins, instead of being lethal, may nevertheless be sufficiently toxic that the cells can tolerate only extremely low levels of these proteins (Tasset et al. 1990; Berger et al. 1992; Gilbert et al. 1993). In some cases, these fusion proteins could reach levels so low that the interaction between the hybrid proteins can occur only at an extremely low frequency. In this situation, even if the hybrid proteins have relatively high affinity for each other, a sufficient number of reconstituted transcriptional activators may not be formed and therefore could not be delivered to the promoter of the reporter gene to induce its transcription. Another problem that may occur in two-hybrid assays derives from the fact that many protein–protein interactions that occur in vivo are highly transient. Consequently, the complexes formed may not be stable enough to recruit the protein complexes that are required for transcriptional activation of the target gene.

Protein Bundles Enhance the Sensitivity of the Two-hybrid Assay: Focus on Mammalian Cells

Because a positive signal in the two-hybrid assay requires the assembly of a two-component transcription factor complex, both fusion proteins must be expressed at sufficient levels relative to their affinity for one another for enough of these complexes to form. In many cases, however, these fusion proteins are poorly expressed, or their affinity is below the detection threshold. These problems can be particularly acute when trying to unite a two-hybrid paradigm in mammalian cells. For example, although the interaction of the c-Src-SH3 domain and its partner c-CBL can be detected in yeast two-hybrid assays, we were unable to detect the interaction between these proteins in mammalian two-hybrid assays using conventional methods (Robertson et al. 1997; Ribon et al. 1998). Western blot assays of transfected mammalian cells suggested that the very poor expression of GAL4-cCbl (G-CBL) fusion protein caused the lack of transcriptional activation of the reporter gene. Therefore, we asked whether increasing the potency of p65 activation domain by attaching multiple copies to a c-Src-SH3 domain partner protein could overcome the negative effects of the low levels of G-CBL fusion protein on the reporter gene expression. To test this possibility, we coexpressed the G-CBL fusion protein with either SH3-S or SH3-4S (carrying a single copy or four copies of p65 activation domains, respectively) in HT1080B cells. We observed that neither combination of the hybrid proteins induced the reporter gene activity to detectable levels (Schmitz and Baeuerle 1991; Nateson et al. 1997). This finding suggested that SH3-S4 fusion proteins were either not recruited efficiently to the promoter of the target gene or, contrary to our assumption, an SH3-4S fusion protein carrying four copies of p65 activation domain is less potent than an SH3-S fusion protein that carries only a single copy of the p65 activation domain. Western blot analysis of the level of expression of hybrid proteins in the transfected cells showed that both GAL4-CBL and SH3-4S were expressed at extremely low levels and, perhaps for this reason, failed to function as potent activators of transcription in vivo (data not shown).

The inability of the hybrid proteins carrying reiterated activation domains to induce transcription robustly led us to develop an alternative method to deliver multiple activation domains to each activator-binding site in the promoter of the reporter gene. In this method, a tetramerization domain derived from the bacterial protein, lactose repressor (Friedman et al. 1995), was placed at the junction between the "target" protein and the p65 activation domain. We assumed that the presence of the tetramerization domain in the c-Src–SH3 fusion protein would allow the formation of protein "bundles" composed of four activation domains and four SH3 domains in vivo. We predicted that the bundled activation domain fusion protein, upon interacting with the G-CBL, could deliver at least a four-times-higher number of activation domains to the promoter of the responsive gene, which might result in the significant enhancement of its transcription (Fig. 1A).

To test this prediction, we coexpressed GAL4-CBL and SH3-S or its tetrameric version SH3-LS fusion protein and examined whether bundling SH3-S protein (Fig. 1B) could help to overcome the effect of very low expression of the GAL4–cCbl fusion protein. We observed that the use of bundled protein SH3-S in this assay led to an extremely strong signal from the reporter gene (Fig. 1C). This finding demonstrates that a simple modification, such as bundling the activation

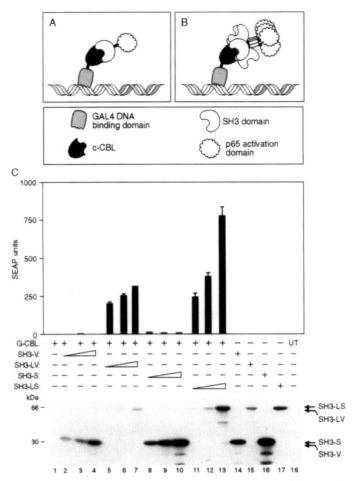

FIGURE 1. Non-covalent bundling of activation domain fusion protein enhances the detection of protein–protein interactions in mammalian cells. Diagrammatic representation of two-hybrid assays with bundled fusion protein containing the target and activation domains. (*A*) GAL4 DNA-binding domain fused to c-CBL (G-CBL in panel C) is shown interacting with its target protein SH3 fused to p65 or VP16 activation domain (SH3-S or SF3-V in panel C). (*B*) GAL4 DNA-binding domain fused to c-CBL is shown interacting with its target protein SH3 fused to lactose repressor tetramerization domain and p65 or VP16 activation domain sequences (SH3-LS or SH3-LV in panel C). (*C*) HT1080B cells carrying SEAP reporter genes placed under the control of five GAL-binding sites were transfected with 100 ng of indicated expression plasmids. The description of protein domains used in this experiment is as follows: G = Gal4 DNA-binding domain; S = p65 activation domain; V = VP16 activation domain; SH3 = SH3 domain from c-SRC protein; L = lactose repressor tetramerization domain; C = c-CBL protein. Mean values of SEAP activity secreted into the medium 24 hours after transfection are shown (± S.D.). Western blot analysis of extracts prepared from transiently transfected cells probed with anti-hemagglutinin antibody is also shown.

domain fusion protein, can improve the outcome in the two-hybrid assay very dramatically. Although it has not been tested extensively, we predict that the use of bundling strategy in general may permit the measurement of protein–protein interactions that escape detection in a conventional system.

Benefits of Bundling Hybrid Proteins in Two-hybrid Assays

The use of bundled activation domain fusion proteins in two-hybrid assays could enhance the detection of protein–protein interaction in two ways. First, unlike the conventional two-hybrid system in which an interaction between the DNA-binding domain and activation domain fusion protein could deliver only a single activation domain to the promoter, bundling allows the delivery of multiple activation domains to the promoter per interaction event. This increases the sensitivity of the assay and allows the interactions between poorly expressed proteins or proteins that have weak affinity for each other to be detected in mammalian two-hybrid assays. Second, bundled fusion proteins may create an avidity effect for the protein–protein interactions and thus may greatly enhance the interaction itself and/or increase the sensitivity of the assay.

In addition to the two-hybrid system, the bundling approach described here can also be used in other assays that are designed to detect protein–protein interactions. For example, it should be possible to employ the bundling strategy in the fluorescence resonance energy transfer (FRET) or mammalian α-complementation-based protein–protein interaction methods (Adams et al. 1991; Li et al. 2001; see also relevant chapters in this volume). In these methods, bundled proteins could enhance the interaction affinity through the avidity effect and/or significantly increase the strength of the detection signal, which, in turn, should allow the detection of interactions that may not score positively in the conventional assays.

The ability of bundled activation domain fusion proteins to induce gene expression robustly in two-hybrid assays suggested that a similar strategy could also be useful in boosting the expression of both artificially introduced and endogenous genes. To test whether bundled activation domains could induce gene expression to higher levels compared to their unbundled counterparts, we used a modified rapamycin-regulated gene expression system (see Fig. 2) (Natesan et al. 1999). The basic system is composed of a GAL4 DNA-binding domain fused to a single copy of FKBP12 and a p65 activation domain fused to the FRB domain of FKBP12-rapamycin-associated protein. An immunosuppressive drug, rapamycin, binds simultaneously to both FKBP12 and FRB; therefore, in its presence, RS (FRB domain + p65 activation domain) fusion protein can be recruited to the DNA-binding fusion protein. Under these conditions, only a single copy of the p65 activation domain can be recruited by each GAL4 monomer. To increase the number of copies of p65 activation domain recruited by each GAL4 monomer, we constructed a chimeric activator that carried a tetramerization domain between the FRB domain and a single p65 activation domain, such that each GAL4 monomer can recruit a minimum of four activation domains in the presence of rapamycin. In theory, the resulting protein should exist in cells as a tetramer, carrying four FRB domains and four p65 activation domains. In this arrangement, a single molecule of rapamycin can recruit the entire tetrameric bundle of activation domain fusion protein to each GAL4 monomer. By fusing four copies of reiterated FKBP12 moieties to the GAL4 DNA-binding domain, up to 16 copies of the p65 activation domain can be recruited to a single GAL4 monomer (Fig. 2).

To examine whether bundled activators can function as robust inducers of transcription in this system, HT1080B cells were transfected with plasmids expressing various combinations of transcription factor fusion protein and treated with 10 nM rapamycin to reconstitute functional transcriptional activators. The data from this experiment showed that delivering a single copy of the p65 activation domain to each GAL4 monomer induced the reporter gene very poorly. In contrast, delivering RLS (FRB domain + lactose tetramerization domain + p65 activation domain) fusion protein containing four copies of the p65 activation domain to each GAL4 monomer induced the reporter gene very strongly. Western blot analysis indicated that the RS and RLS fusion proteins were expressed at similar levels in the transfected cells. In this experiment, by testing various combinations of fusion proteins, it is possible systematically to vary the number of activation domains delivered to the GAL4 DNA-binding domain from 1 to 16. Under these conditions, there was an excellent correlation between the number of activation domains delivered to the promoter and induced reporter gene activity. Furthermore, these findings also suggest that increasing the number of activation domains delivered to a target promoter leads to significant increases in gene expression (Fig. 3A).

We have also found that delivering bundled activators to the promoter of the reporter gene induced its transcription to much higher levels, whereas delivering tandemly reiterated activation

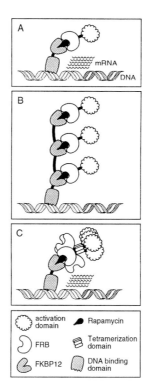

FIGURE 2. Diagram depicting the strategies used to increase the number of activation domains delivered to the promoter. (*A*) In the basic method, two fusion proteins, one containing a GAL4-DNA-binding domain fused to FKBP12 and the other composed of the p65 activation domain fused to FRB, are expressed in cells. Addition of rapamycin results in the reconstitution and subsequent recruitment of a single activation domain to each DNA-binding monomer (GF3 + RS, see panel *A*). (*B*) Fusion of multiple FKBP moieties to the DNA-binding domain allows rapamycin to recruit multiple activation domains to each DNA-binding monomer (GF3 + RS, see panel *B*). (*C*) Addition of the lactose repressor tetramerization domain to the FRB-activation domain fusion protein, producing RLS, allows rapamycin to recruit four activation domains to each FKBP fused to the DNA-binding domain. The number of activation domains recruited to the promoter can be increased by attaching more FKBP moieties to the DNA-binding domain and/or multiplying the number of binding sites for the activator. In theory, as many as 160 activation domains can be delivered to a promoter containing five GAL4-binding sites by treating the cells expressing GF4 and RLS fusion proteins with rapamycin. The actual number of activation domains delivered to the promoter of the reporter gene in vivo by using the bundling method described here has not been determined.

domains capable of delivering the same number of activation domains as the bundled activators induced the reporter gene very poorly. For example, when the DNA-binding fusion protein GF1 was expressed with either RS4 (FRB domain + four copies of p65 activation domain) or RLS and functional activators were reconstituted in vivo by adding rapamycin in the culture medium, only RLS bundles strongly induced the reporter expression (Fig. 3B). Because each delivery event is expected to bring the same number of activation domains to the promoter of the reporter gene, regardless of whether RLS or RS4 fusion protein was used in this assay, the inability of RS4 to induce gene expression is very likely due to its reduced delivery to the promoter. In support of this interpretation, western blot analysis of the extracts from transfected cells showed that RS4 fusion protein is expressed at very low levels and, perhaps for this reason, may not be delivered to the promoter efficiently. In contrast, the RLS fusion was produced at much higher levels in the transfected cells and induced the reporter gene expression very strongly. Thus, taken together, our data suggest that bundled activators are less toxic and delivered more efficiently to the promoter, leading to the strong activation of gene expression.

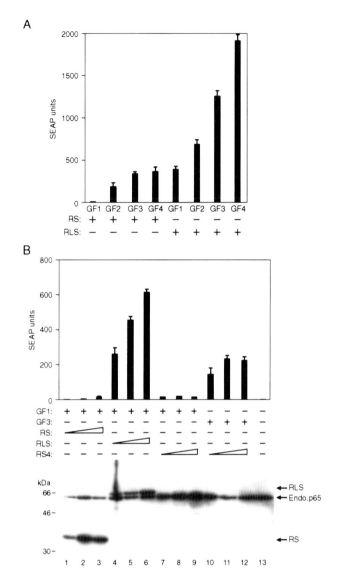

FIGURE 3. The level of expression of a stably integrated reporter gene correlates with the number and strength of the activation domains bound to its promoter. (*A*) The indicated DNA-binding domain and activation domain fusion proteins were transfected into HT1080B cells that carry a stably integrated SEAP reporter gene placed under the control of five GAL4-binding sites. In all cases, SEAP expression values are plotted for cultures receiving 100 ng of activation domain expression plasmid, which gives peak expression values in transiently transfected cells and slightly below-peak values in stably transfected cell lines. The background SEAP activity was subtracted from each value before plotting. In this experiment, expressing the GF1+RS combination of proteins produced four SEAP units above background levels in the presence of rapamycin. (*B*) The DNA-binding domain and activation domain expression plasmids were transfected into HT1080B cells. In all cases, mean values of SEAP activity secreted into the medium after the addition of 10 nM rapamycin are shown (± S.D.). Western blot analysis of the total cell lysates with anti-hemagglutinin antibody allowed an assessment of transcription factor component levels in cells.

The increased potency of bundled activator proteins has other practical applications. For example, we have found that even highly potent chimeric activator proteins such as GAL4–p65 and GAL4–VP16 fail to induce transcription of stably integrated reporter genes robustly when they are placed under the control of a single GAL4-binding site. However, stably integrated reporter genes placed under the control of a single GAL4-binding site in the promoter region can be stimulated to high levels by using bundled p65 or VP16 activation domains (S. Nateson, unpubl.). Bundled activators may also be useful in other scenarios in which the level of activator protein in the cell is too low to support target gene activation or in cell lines recalcitrant to transfection or when the messenger RNAs encoded by the target gene are highly unstable.

To test the ability of bundled activators under at least one such scenario, we generated stable reporter cell lines in which expression of the chimeric transcription factors was deliberately limited by placing the activator expression under the control a relatively weak Rous sarcoma virus promoter instead of a strong cytomegalovirus promoter (Nateson et al. 1999). One pool of stable cell lines (HT34) expressed the bundled activator, whereas the other pool (HT35) expressed the conventional activator RS. The two pools differed dramatically in their responsiveness to rapamycin; HT34 responded robustly, whereas HT35 did not respond at all. In contrast, the levels of expression of RS and RLS fusion proteins are the same in the two pools. Thus, bundled activators can robustly induce gene expression under conditions in which the conventional activators cannot (Natesan et al. 1999).

Many gene therapy applications can benefit from high-level expression of therapeutic genes. Bundled activation domain fusion proteins are well tolerated in mammalian cell types; therefore, they could induce high-level expression of therapeutic genes in many gene therapy procedures. Activation domain bundles should also allow the rapamycin-regulated system to function robustly, even when the transcriptional activators are expressed at low levels—a likely situation in many gene therapy situations. Perhaps the most important benefit of bundling in this system is that it shifts the dose response of rapamycin activation of gene expression by at least tenfold, indicating that the use of bundled activation domains could improve the practicality of regulated gene therapies by substantially reducing the level of drug required. In theory, bundled activation domains could also be used in other small-molecule-regulated gene expression systems, including the tetracycline and steroid hormone-regulated gene expression systems (Gossen and Bujard 1992; No et al. 1996; Wang et al. 1997).

The regulation of expression of an endogenous gene is dependent on the recruitment of several transcription factors to their binding sites in its promoter region. The ability to regulate the expression of endogenous genes with a single transcription factor could be highly beneficial in many gene therapy applications. At present, methods developed for this purpose use synthetic transcription factors capable of binding to specific sites in the promoter region of the gene of interest to induce or repress the expression of the endogenous gene (Beerli et al. 2000; Zhang et al. 2000). For example, it has been recently shown that it is possible to induce the expression of vascular endothelial growth factor (VEGF) in cells that normally do not produce this protein by expressing zinc-finger–VP16 activator proteins that are specifically designed to bind to the promoter of the VEGF gene (Zhang et al. 2000). It is likely that bundling these synthetic zinc-finger transcription factors could lead to a significant enhancement in the level of expression of the endogenous genes. Alternatively, the synthetic zinc fingers or the DNA-binding domains from natural transcription factors can be used in the modified rapamycin-regulated gene expression system that utilizes the bundled activation domains to regulate the expression of therapeutically relevant genes. The ability of bundled activators to induce high levels of gene expression without apparent cellular toxicity may be critical for the success of these approaches.

OUTLINE OF PROCEDURE

DNA-binding Domains and DNA-binding Domain Fusion Proteins

In theory, any sequence-specific DNA-binding domain with a modest or high affinity for its binding site can be used in two-hybrid assays. However, the vast majority of two-hybrid screens were done using DNA-binding domains derived from yeast and bacterial transcription factors, GAL4 (amino acids 1–94) and LexA, respectively. Recently, synthetic DNA-binding domains, such as ZFHD (Rivera et al. 1996), have been used in small-molecule-regulated gene expression systems; these DNA-binding domain proteins can also be used in the two-hybrid system. Plasmids carrying GAL4 or LexA coding regions are commercially available from such suppliers as Invitrogen, Stratagene, and Clontech. These sequences can easily be amplified by PCR with appropriate restriction sites, and the digested fragments can be cloned into the vector of choice. The synthetic DNA-binding domains can be obtained from either academic laboratories or biotechnology companies that make these reagents available through their Web sites (for example, Ariad Pharmaceuticals).

The hybrid protein composed of a DNA-binding domain of choice and the "bait" protein can be expressed from any episomal or viral mammalian expression vector (see Fig. 4A). It is important that the chosen expression vector contain a selectable marker such as neomycin- or zeomycin-resistance genes, so that, if necessary, stable cell lines expressing this fusion protein can be generated. It is preferable to use retroviruses to express this chimeric protein to reduce the copy number of the chimeric gene in the genome and/or toxic effects of the fusion proteins in the recipient cells. Specially designed retroviral vectors capable of simultaneously expressing multiple recombinant proteins have been published recently (Pollock et al. 2000).

Activation Domains and Activation Domain Fusion Proteins

The potency of the activation domain is a critical determinant of the outcome in the two-hybrid screens. A large number of published two-hybrid screens have used the potent activation domain derived from the herpes simplex virus protein VP16 (amino acids 410–490). We have found that the activation domain derived from the p65 subunit of NF-κB protein (amino acids 361–551) induced the transcription of the reporter gene to higher levels compared to the VP16 activation domain in a mammalian two-hybrid assay (unpublished data). Plasmids carrying the VP16 or p65 activation domain can be obtained from commercial sources such as Clontech, Invitrogen, Stratagene, and Ariad Pharmaceuticals.

Either a plasmid or viral vector can be used to express the recombinant gene encoding a fusion protein composed of the "target" protein and the activation domain (see Fig. 4B). It is not necessary that these vectors contain a selectable marker. Traditionally, plasmid-based libraries carrying cDNAs fused with the activation domain of choice have been used in the two-hybrid assays. However, a wide variety of retroviral or adenoviral vectors that can be used for this purpose are currently available from both academic and commercial institutions (Clontech and Ariad Pharmaceuticals) (Pollock et al. 2000).

Bundling Domains

In theory, any multimerization domain placed between the activation domain and the target partner protein should enhance the detection of protein–protein interactions. We have tested a limited number of multimerization domains and found that the tetramerization domain derived from the bacterial transcription factor lactose repressor (amino acids 46–360) generated the best outcome in both the two-hybrid and regulated gene expression systems. A 30-amino-acid tetramerization domain in the carboxy-terminal region of lactose repressor proteins (amino acids 330–360) appears to function as well as the entire lactose repressor protein without its DNA-binding domain (S. Natesan and E. Molinari, unpubl.). A general description of a plasmid vector carrying the coding sequences for a bundled activation domain fusion protein is shown in Figure 4C.

Reporter Genes and Reporter Plasmids

A wide variety of reporter genes can be used in two-hybrid and other types of assays for analyzing protein–protein interactions. The use of EGFP, EYFP, and CD8 reporter genes provides the

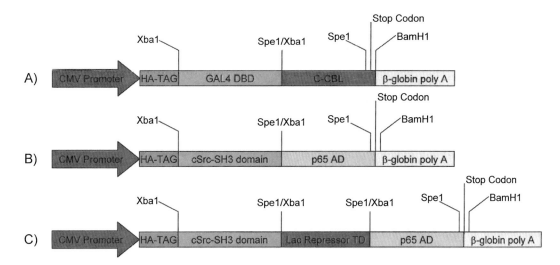

FIGURE 4. Diagram showing the components contained in the plasmids used in protein–protein interaction experiments. Plasmid pGCBL carries coding regions for the GAL4 DNA-binding domain (G) fused to c-CBL (CBL). Plasmid pSH3S carries coding sequences for the c-Src-SH3 (SH3) domain fused to p65 activation domain (S). Plasmid SH3-LS carries coding sequences for c-Src-SH3 domain (SH3) fused to the lactose repressor domain that lacks the DNA-binding domain (L) and p65 activation domain (S). The recombinant genes are under the control of CMV promoter and flanked by the hemagglutinin (HA) tag in the amino-terminal region and a poly(A) region from rabbit β-globin gene at the carboxy-terminal region.

choice of detecting and selecting cells that score positively by the fluorescence-activated cell sorting (FACS) method. If the purpose of the experiment is to map the protein–protein interaction domains, reporter genes, such as SEAP and luciferase, could be used. In general, the reporter gene is placed under the control of two or more copies of GAL4- or LexA-binding sites. For composite DNA-binding domains such as ZFHD, we have placed as many as 12 of its binding sites upstream of the reporter gene to obtain maximal transcriptional activation. However, placing multiple copies of transcription-factor-binding sites upstream of the reporter gene is not necessary when bundled p65 activator fusion proteins are utilized as they generally induce reporter genes driven by one or two copies of the transcription-factor-binding sites to high levels (unpublished data).

Like the DNA-binding and activation domain fusion proteins, the reporter gene can be delivered either in a plasmid or viral vector. The number of copies of the reporter gene integrated in the genome can be minimized or reduced to one by delivering it through retroviral vectors. Chimeric transcriptional activator proteins capable of binding to the promoter region of the reporter genes can be transiently expressed in cells to measure the degree of responsiveness of the stably integrated reporter gene.

Cell Line

The choice of cell line plays a very critical role in the outcome of the analysis of protein–protein interactions in vivo. Because many mammalian cell lines are recalcitrant to standard transient transfection protocols, the method of delivery of recombinant genes and cDNA libraries into mammalian cells needs to be optimized before undertaking the screen. If viral vectors are used for introducing recombinant genes into cells, the level of expression of the genes placed in the viral vector in the cell line of interest should be determined prior to generating custom cDNA libraries for the screen. In mammalian cells, it is not uncommon to find a steady decrease in the expression of stably integrated genes or responsiveness of the reporter genes placed under the control of reporter genes. Therefore, it is essential that the stable cell lines generated for the screen be tested periodically for their robust response. The level of reporter gene activity induced by the expression of GAL4–VP16 protein can be used to measure the responsiveness of the stably integrated reporter driven by GAL4-binding sites. The level of expression of the DNA-binding fusion protein can be measured by western blot analysis.

Controls

It is well known that the two-hybrid assays have the potential to generate a large number of false positives. Therefore, it is essential to use as many controls as possible to distinguish between the false and true positives scored in this assay. A commonly used control to eliminate the false positives is a reporter gene construct that is identical to the wild-type reporter gene, except that the binding site for the transcriptional activator protein is substituted with the mutant sites. Generally, in mammalian two-hybrid assays, we test all of the positives scored in the primary screen for their ability to induce transcription of the reporter genes driven by both the wild-type and mutant GAL4- or LexA-binding sites; only those that specifically induce the reporter gene driven by wild-type GAL4 or LexA sites are considered for further analysis. The positives isolated in this type of assay can be tested using a nonspecific bait protein attached to the DNA-binding domain used in the screen.

Methods

Analyzing the interactions between proteins and multiprotein complexes requires proficiency in several techniques, including transient transfection, stable cell line generation, reporter gene activity assays, western and immunoprecipitation assays, and general cloning procedures. See Sambrook and Russell 2001, *Molecular Cloning* for detailed protocols for all of these techniques. Also see other chapters in this volume describing two-hybrid and mammalian two-component systems for specific approaches.

ACKNOWLEDGMENT

The work described here was carried out at ARIAD Pharmaceuticals.

REFERENCES

Adams S.R., Harootunian A.T., Buechler Y.J., Taylor S.S., and Tsien R.Y. 1991. Fluorescence ratio imaging of cyclic AMP in single cells. *Nature* **349:** 694–697.

Allen J.B., Walberg M.W., Edward M.C., and Elledge S.J. 1995. Finding prospective partners in the library: The two-hybrid system and phage display find a match. *Trends Biochem Sci.* **20:** 511–516.

Bai C. and Elledge S.J. 1996. Gene identification using the yeast two-hybrid system. *Methods Enzymol.* **273:** 331–347.

Beerli R.R., Dreier B., and Barbas III, C.F. 2000. Positive and negative regulation of endogenous genes by designed transcription factors. *Proc. Natl. Acad. Sci.* **97:** 1495–1500.

Belshaw P.J., Ho S.N., Crabtree G.R., and Schreiber S.L. 1996. Controlling protein association and subcellular localization with a synthetic ligand that induces heterodimerization of proteins. *Proc. Natl. Acad. Sci.* **93:** 4604–4607.

Bemis L.T., Geske F.J., and Strange R. 1995. Use of the yeast two-hybrid system for identifying the cascade of protein interactions resulting in apoptotic cell death. *Methods Cell Biol.* **46:** 139–151

Berger S.L., Pina B., Silverman N., Marcus G.A., Agapite J., Regier J.L., Triezenberg S.J., and Guarante L. 1992. Genetic isolation of ADA2: A potential transcriptional adaptor required for function of certain acidic activation domains. *Cell* **70:** 251–265.

Chien C.T., Bartel P.L., Sternglaz R., and Fields S. 1991. The two-hybrid system: A method to identify and clone genes for proteins that interact with a protein of interest. *Proc. Natl. Acad. Sci.* **88:** 9578–9582.

Estojak J., Brent R., and Golemis E.A. 1995. Correlation of two-hybrid affinity data with in vitro measurements. *Mol. Cell. Biol.* **15:** 5820–5829.

Fields S. and Song O.K. 1989. A novel genetic system to detect protein-protein interactions. *Nature* **340:** 245–246.

Friedman A.M., Fischmann T.O., and Steitz T.A. 1995. Crystal structure of lac repressor core tetramer and its implications for DNA looping. *Science* **268:** 1721–1727.

Gilbert D.M., Heery D.M., Losson R., Chambon P., and Lemone Y. 1993. Estradiol-inducible squelching and cell growth arrest by a chimeric VP16-estrogen receptor expressed in *Saccharomyces cerevisiae:* Suppression by an allele of PDR1. *Mol. Cell. Biol.* **13:** 462–472.

Gill G. and Ptashne M. 1998. Negative effect of the transcriptional activator GAL4. *Nature* **334:** 721–724.

Gossen M. and Bujard H. 1992. Tight control of gene expression in mammalian cells by tetracycline-responsive promoters. *Proc. Natl. Acad. Sci.* **89:** 5547–5551.

Keaveney M. and Struhl K. 1998. Activator-mediated recruitment of the RNA polymerase II machinery is the predominant mechanism for transcriptional activation in yeast. *Mol. Cell* **6:** 917–924.

Kingston R.E. 1999. A shared but complex bridge. *Nature* **399:** 199–200.

Laemmli U.K and Tjian R. 1996. A nuclear traffic jam—Unraveling multicomponent machines and compartments. *Curr. Opin. Cell Biol.* **8:** 299–302.

Lemon E. and Tjian R. 2000. Orchestrated response: A symphony of transcription factors for gene control. *Genes Dev.* **14:** 2551–2569

Li H.Y., Ng E.K., Lee S.M., Kotaka M., Tsui S.K., Lee C.Y., Fung K.P., and Waye M.M. 2001. Protein-protein interaction of FHL3 with FHC2 and visualization of their interaction by green fluorescent protein (GFP) for fluorescence resonance energy transfer (FRET). *J. Cell Biochem.* **80:** 293–303.

Mendelson A.R. and Brent R. 1994. Applications of interaction traps/two-hybrid systems to biotechnology research. *Curr. Opin. Biotechnol.* **5:** 482–486.

Natesan S., Rivera V.M., Molinari E., and Gilman M.Z. 1997. Transcriptional squelching re-examined. *Nature* **390:** 349–350.

Natesan S., Molinari E., Rivera V.M., Rickles R.J., and Gilman M. 1999. A general strategy to enhance the potency of chimeric transcriptional activators. *Proc. Natl. Acad. Sci.* **96:** 13898–13903.

No D., Yao T.-P., and Evans R.M. 1996. Ecdysone-inducible gene expression in mammalian cells and transgenic mice. *Proc. Natl. Acad. Sci.* **93:** 3346–3351.

Pollock R., Isner R., Zoller K., Natesan S., Rivera V.M., and Clackson T. 2000. Delivery of a stringent dimerizer-regulated gene expression system in a single retroviral vector. *Proc. Natl. Acad. Sci.* **97:** 13221–13226.

Ptashne M. and Gann A. 1997. Transcriptional activation by recruitment. *Nature* **386:** 569–577.

Ribbon V., Printen J.A., Hoffman N.G., Kay B.K., and Saltiel A.R. 1998. Distinct classes of transcriptional activating domains function by different mechanisms. *Cell* **62:** 1177–1187.

Rivera V.M., Clackson T., Natesan S., Pollock R., Amara J.F., Keenan T., Magari S.R., Phillips T., Courage N.L., Cerasoli Jr., F., Holt D.A., and Gilman M. 1996. A humanized system for pharmacologic control of gene expression. *Nat. Med.* **2:** 1028–1032.

Robertson H., Langdon W.Y., Thien C.B., and Bowtell D.D. 1997. A c-Cbl yeast two hybrid screen reveals interactions with 14-3-3 isoforms and cytoskeletal components. *Biochem. Biophys. Res. Commun.* **240:** 46–50.

Rossi F., Charlton C.A., and Blau H. 1997. Monitoring protein-protein interactions in intact eukaryotic cells by β-galactosidase complementation. *Proc. Natl. Acad. Sci.* **94:** 8405–8410.

Sambrook J. and Russell D. 2001. *Molecular cloning: A laboratory manual,* 3rd edition. Cold Spring Harbor Laboratory Press, Cold Spring Harbor, New York.

Sadowski I., Ma J., Triezenberg S., and Ptashne M. 1998. GAL4-VP16 is an unusually potent transcriptional activator. *Nature* **335:** 563–564.

Schmitz M.L. and Baeuerle P.A. 1991. The p65 subunit is responsible for the strong transcription activating potential of NF-kappa B. *EMBO J.* **10:** 3805–3817.

Senguptha D.J., Wickens M., and Fields S. 1999. Identification of RNAs that bind to a specific protein using the yeast three-hybrid system. *RNA* **5:** 596–601.

Tasset D., Tora L., Fromental C., Scheer E., and Chambon P. 1990. Distinct classes of transcriptional activating domains function by different mechanisms. *Cell* **62:** 1177–1187.

Wang Y., DeMayo F.J., Tsai S.Y., and O'Malley B.W. 1997. Ligand-inducible and liver-specific target gene expression in transgenic mice. *Nat. Biotechnol.* **15:** 239–243.

Zhang J. and Lauter S. 1996. A yeast three-hybrid method to clone ternary protein complex components. *Anal. Biochem.* **242:** 68–72.

Zhang L., Spratt S.K., Liu Q., Johnstone B., Qi H., Raschke E.E., Jamieson A.C., Rebar E.J., Wolffe A.P., and Case C.C. 2000. Synthetic zinc finger transcription factor action at an endogenous chromosomal site. Activation of the human erythropoietin gene. *J. Biol. Chem.* **275:** 33850–33860.

33 Analysis of Genome-wide Protein Interactions Using Computational Approaches

Anton J. Enright and Christos A. Ouzounis

Computational Genomics Group, Research Programme, The European Bioinformatics Institute, EMBL Cambridge Outstation, Cambridge CB10 1SD, United Kingdom

INTRODUCTION

Recent developments in computational genomics include new methods for the functional analysis of proteins in entire genome sequences. These methods transcend the conventional approach of function prediction by sequence similarity because they take into account the genomic "context" of genes and proteins. Entire genome sequences can yield much more information than sets of proteins whose genomic context is unknown. Examples of genomic context include conservation of gene identity and position, genome-wide analysis of gene fusion, metabolic reconstruction, gene coregulation, and expression and curation of known molecular interactions for a given species. Such techniques can be used to derive information about multiple functional associations of proteins, protein complex formation, and involvement of proteins in metabolic and signaling pathways or direct physical interactions. Given the ever-expanding list of complete genome sequences, it is imperative that such methods are employed in a manner complementary to more conventional computer- and laboratory-based methods. This chapter summarizes current approaches available from computational genomics and details the application of these methods to the discovery of functional associations and physical interactions between proteins.

From Genomics to Post-genomics

Since the advent of large-scale sequencing, a number of entire genome sequences have been made available. The function of the predicted genes and proteins has been assigned on the basis of their homology with genes of known function (Andrade et al. 1999b). Yet, entire genome sequences provide a number of additional constraints that, if exploited wisely, can provide more information about the functional properties of the genes in question (Tsoka and Ouzounis 2000b). These constraints may come in the form of gene/genome structure, gene expression patterns, and involvement of gene products in biochemical networks. In other words, genomic context can be exploited to improve function prediction (Huynen et al. 2000). One of the most powerful applications of these methods is the detection of protein interactions from entire genome sequences, using a variety of approaches, as described in this chapter.

The Problem of Detecting Genome-wide Protein Interactions

Over the past few years, we have witnessed the transformation of biological research from a case-by-case approach to a systematic, large-scale, high-throughput enterprise (Nierman et al. 2000). During this time, activities such as DNA sequencing, gel electrophoresis, mass spectrometry, sequence analysis, and structure determination have all progressed to a high-throughput stage. As sequencing continued producing massive amounts of genetic information in need of interpretation, terms such as post-genomics were added to the vocabulary of the genomics revolution. The prefix in post-genomics implies analysis of the functional properties of an organism, given its entire genome sequence (Eisenberg et al. 2000).

One of the most important problems in the post-genomics era is the detection of protein interactions based on the entire genome sequence. The term protein interactions has been used in a wide variety of contexts, and it may signify more than just direct physical association of protein molecules. We have used the term "functional association" to describe the involvement of proteins in direct physical associations, multisubunit complexes, or the same metabolic/signaling pathways.

Detecting protein interactions is far more complex than detecting individual genes and their functional properties. To illustrate the difficulty, suppose that we have a genome of 70 genes and that we are interested in predicting all pair-wise interactions (Fig. 1). In addition, assume that of the 4900 possible pair-wise interactions, only 1% (49 in total) are real. The problem is much harder than detecting, for instance, genes along the chromosome, because of the large, complex space of pair-wise interactions (Fig. 1). In addition, suppose that our methods currently allow detection of protein interactions with a 50% precision rate; i.e., for every correct interaction, there is an incorrect one (Fig. 1). The problem then becomes to identify correctly the true positives from the false positives (noise), as well as to keep the number of false negatives (nondetected cases) low. It is also very hard to address the issue of true negatives (i.e., to prove that two proteins do not interact).

The challenge is to approach the problem of the accurate prediction or detection of genome-wide protein interactions using a variety of independently devised methods. These methods can be theoretical, experimental, or hybrid combinations of theory and experiment. We have divided this chapter into three main sections: (1) purely computational methods that use entire genome sequences, (2) hybrid methods where computation is essential to detect patterns in experimental information, and finally (3) database mining and curation approaches that exclusively rely on the automatic extraction or labor-intensive curation of direct experimental information (Fig. 2). The latter step is essential for the development of new prediction methods by providing a robust, well-characterized set of protein interactions. The scheme also reflects a natural sequence of events in computational biology, where predictions are first made by algorithms, some of which later are combined with experimental information, and ultimately, resources are created to store and disseminate this information to the wider community.

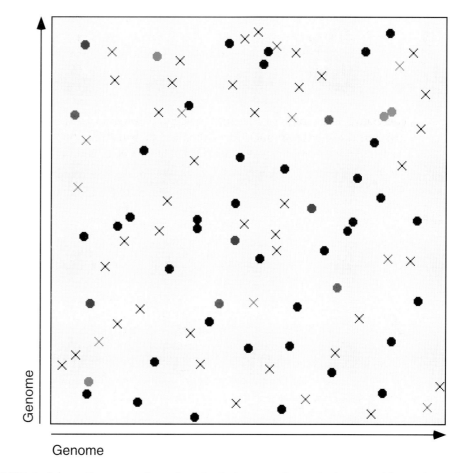

Genome

Genome

FIGURE 1. Schematic representing a hypothetical genome in quadratic space. All actual protein–protein interactions within that genome are shown (*circles*), with all cases where protein–protein interaction does not occur (*crosses*). Colors signify groupings of functionally related proteins that have been identified by a variety of methods (e.g., *red* circles and crosses signify a single set of potentially interacting proteins). For any prediction, the set of results obtained comprises true positive predictions (*blue, green,* or *purple circles*) and false-positive predictions (*blue, green,* or *purple crosses*). Depending on the accuracy and precision of any given computational approach, the number of true positives and false positives may vary considerably.

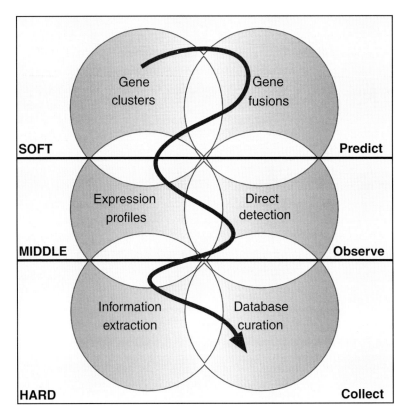

FIGURE 2. Schematic representing different approaches to computational prediction of protein interaction. Circles indicate different approaches, and they may overlap with other approaches. The curved arrow indicates a possible protocol for computational prediction of interaction.

Herein, we discuss some of the techniques used for the detection of protein interactions in entire genomes, with special emphasis on computational approaches (Fig. 2). The large-scale detection of protein interactions based on experimental approaches has been reviewed elsewhere (Chen and Han 2000; Legrain and Selig 2000).

THE SOFT WAY: PURE COMPUTATION

We call this method the soft way because it purely relies on predictions made by algorithms. Although this does not necessarily imply that the predictions are unreliable, it underlines the usual lack of experimental information to support these predictions. Most of these methods rely on the exploitation of genomic context, in the form of structural or evolutionary constraints (Marcotte 2000).

Phylogenetic Profiles

One form of genomic context is the co-occurrence of homologous genes in entire genomes. Co-occurring genes are defined as those sets of genes that are observed together in multiple species. Homologous genes may be classified as orthologs or paralogs. Orthologs are defined as genes that have arisen via speciation and have been present in the common ancestor of two species. Paralogs are defined as genes that have arisen via gene duplication after speciation. Orthologs or paralogs may (or may not) have the same function, although in most cases they share similar functional properties and usually they cannot be distinguished in the absence of speciation history

(Ouzounis 1999). Unfortunately, the term ortholog has been loosely used in genomics (Ouzounis 1999) and, to avoid further confusion, we use the following operational definition. Orthologs are defined as those genes across two species that are considered to be most similar to each other and thus are thought to perform analogous functions.

The co-occurrence of homologous genes defines a phylogenetic context; i.e., it allows the classification of genes into groups that are represented in certain species. Genes that "travel" together in evolution are assumed to be involved in similar cellular processes and thus are predicted to be functionally related. In other words, if two proteins are functionally associated, genes that encode them will tend to occur together in a given genome. The reverse of this argument may also be true; i.e., if a certain gene is absent in a given genome, it is likely that its partners will also be absent (Fig. 3).

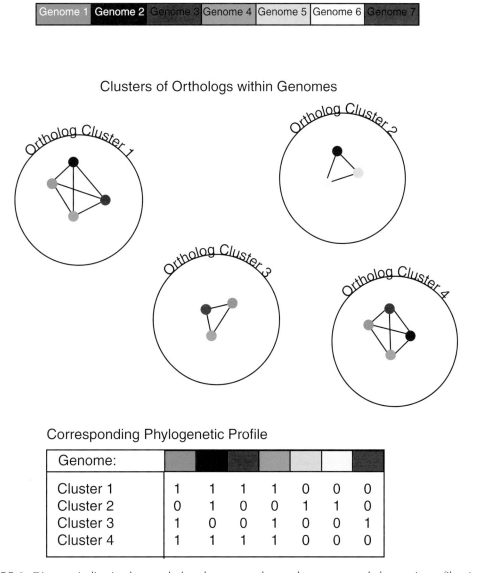

Genome Legend

| Genome 1 | Genome 2 | Genome 3 | Genome 4 | Genome 5 | Genome 6 | Genome 7 |

Clusters of Orthologs within Genomes

Ortholog Cluster 1 *Ortholog Cluster 2*

Ortholog Cluster 3 *Ortholog Cluster 4*

Corresponding Phylogenetic Profile

Genome:							
Cluster 1	1	1	1	1	0	0	0
Cluster 2	0	1	0	0	1	1	0
Cluster 3	1	0	0	1	0	0	1
Cluster 4	1	1	1	1	0	0	0

FIGURE 3. Diagram indicating how ortholog clusters may be used to generate phylogenetic profiles. Large circles indicate clusters; small, colored circles indicate a gene from a given genome within a cluster of its orthologs. The profile (*below*) indicates presence (1) or absence (0) of an ortholog from a given genome within any cluster.

To examine the evolutionary context of genomes, a phylogenetic profile is constructed. Such a profile lists classes of genes, followed by a binary representation of the presence or absence of its homologs in different genomes (Fig. 3) (Ouzounis and Kyrpides 1996; Rivera et al. 1998; Pellegrini et al. 1999). It is then possible to predict the functional association of genes that possess similar profiles (Pellegrini et al. 1999). Phylogenetic profiles have also been used to predict the cellular localization of gene products (Marcotte et al. 2000), extending the idea of interactions to a higher resolution within cell compartments.

This method becomes more powerful with an increasing number of genomes, because this allows more accurate profiles to be constructed. However, although elegant, this method suffers from a number of drawbacks. First, extensive gene duplication in a genome makes detection of orthologs very difficult, especially in the absence of speciation history (Ouzounis 1999). Second, the reliable detection of members of orthologous families is hampered by evolutionary processes such as non-orthologous gene displacement, gene loss, and horizontal gene transfer (Galperin and Koonin 2000). Third, and most important, detection of functionally associated proteins by phylogenetic profiles is a hypothesis that remains to be tested and may only be applicable for a small number of highly correlated groups of genes.

Non-orthologous gene displacement generally occurs between two genes that are unrelated, yet possess a similar or analogous function. In these cases, one of the genes displaces the role of the other protein with its own. The displaced gene may then be lost, or evolve a new, slightly different function. By definition, the newly acquired gene is not homologous to genes performing that function in other genomes where displacement has not occurred. Lineage-specific gene loss is another related problem with evolutionary context analysis. In this case, a member of a gene family in a specific lineage is lost either by deletion or by accelerated evolution to a new function. Horizontal gene transfer occurs when a gene is transferred from one lineage to another, possibly distant, lineage. These evolutionary mechanisms confound the construction of phylogenetic profiles by making the detection of a member of a gene family in a given genome very difficult. However, given the ever-increasing number of available complete genomes, the fidelity of these predictions is expected to improve over time.

Gene Clusters

Another valuable method for exploring the genomic context of genes is through the idea of colocalization, or gene neighborhood (Fig. 4A). This exploits the notion that genes which interact or are functionally associated will be kept in physical proximity to each other on the genome (Tamames et al. 1997; Dandekar et al. 1998; Overbeek et al. 1999).

The most apparent case of this phenomenon occurs with bacterial and archaeal operons, where genes that work together are generally transcribed on the same polycistronic mRNA. Although this is *generally* not the case in eukaryotic systems, it is possible to infer functional association of proteins in a genome without operons. This involves detecting homologs of these genes contained in an operon in other, noneukaryotic, genomes. This procedure has the inherent advantage that operon structures tend to vary considerably between species, providing additional cases that can suggest functional association of genes in genomes lacking operons. However, because it still remains difficult to predict operons, the amount of data available for this kind of functional inference is rather limited. The lack of operons in eukaryotic species also means that information is most valuable for bacterial genome analysis containing homologous gene families that span both prokaryotic and eukaryotic domains. This may provide little or no information about families limited to the eukaryal domain (Ouzounis and Kyrpides 1996). Yet, there have been some reports of operons (Zorio et al. 1994) and polycistronic transcription in eukaryotes (Blumenthal 1998), whose extent currently remains unknown.

This method is not limited to studies of genes that occur in operons. It is also possible to predict functional association of a pair of genes if their homologs tend to be close in physical prox-

A) Co-localization of Genes

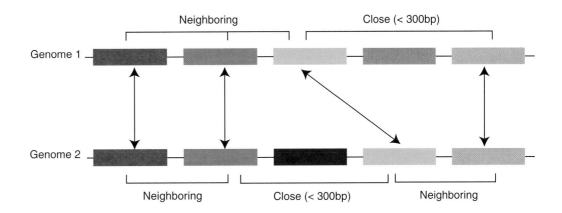

B) Gene Fusion Events

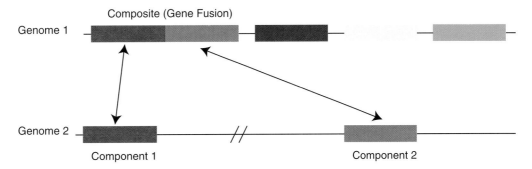

FIGURE 4. (*A*) Diagram showing a schematic representation of conservation of gene neighborhood. The two lines show separate genomes, and the colored boxes indicate genes contained within these genomes. Genes of the same color are orthologs, and arrows represent intergenomic similarity relationships. (*B*) Diagram showing a gene fusion event. The upper line indicates a genome where two genes have fused. The lower line indicates another genome where the genes are separate, and distant. Boxes of the same color indicate orthologous proteins and/or domains.

imity in many genomes (e.g., <300 bp) (Fig. 4A). This method has been used successfully to detect missing members of metabolic pathways in a number of species (Dandekar et al. 1998). The power of the method becomes more apparent as more complete genomes become available. For example, the first such report did not result in the detection of conserved gene clusters (Mushegian and Koonin 1996), a result that was later proven to be false (see above). This method is also complementary to the analysis of gene fusion, which represents the ultimate form of gene proximity, complete fusion of two genes into one single unit.

Gene Fusion

Gene fusion represents another aspect of the structure and evolution of genes that has been used to detect functionally associated or directly interacting proteins. Gene fusion involves the merging of individual gene products into multifunctional, multidomain proteins (Fig. 4B). This event is most likely to occur between two genes whose products are interacting or functionally associated, possibly due to the selective advantage of decreased regulational load (Enright et al. 1999). The detection of gene fusions in entire genomes provides an elegant way computationally to predict interactions and associations between proteins (Enright et al. 1999; Marcotte et al. 1999b). In this way, proteins that are fused into "composites" in one genome provide a template for the prediction of interaction between their nonfused homologous "component" proteins in other genomes (Fig. 4B) (Enright et al. 1999). This method is very precise and powerful: Even a single gene fusion in one species is sufficient to provide evidence for the functional association of homologous components for other species.

This approach is, however, limited in scope because it can only provide predictions for functional associations of proteins that have been fused during the course of evolution. Evidence also suggests that genes encoding metabolic enzymes are more likely to be involved in these events (Tsoka and Ouzounis 2000a). Once again, the increasing number of complete genomes available will render this method more powerful. The chance of finding a fused composite protein for a given pair of proteins of interest increases with each new genome that is made available. The method appears to work almost as well with eukaryotic species as it does with prokaryotes (Enright and Ouzounis 2001), but it requires highly accurate gene prediction, which is still an open question (Lewis et al. 2000). Care must be taken, however, to avoid the detection of so-called promiscuous domains (Marcotte et al. 1999b), such as DnaJ, cystathionine β synthase (CBS), and helix-turn-helix (HTH) domains. Such domains are particularly abundant in higher eukaryotes and need to be filtered out because they dramatically increase the false-positive rate (noise). It has been shown, however, that a low false-positive rate can be achieved by careful treatment of these cases (Enright et al. 1999). Multiple cases of components (i.e., extensive degree of paralogy) make some of these predictions less accurate because it is not possible to predict pair-wise interactions for multiple component partners.

THE MIDDLE WAY: COMPUTATION AND EXPERIMENT

We call this section the middle way because although the approaches listed here rely on robust experimental measurements, they invariably use computation to validate the relevance of experimental information. Such measurements include genome sequences, metabolic pathways, and transcription profiles. Most of these methods involve some human intervention to generate robust and meaningful predictions/observations for protein interactions.

Metabolic Reconstruction

Using annotation from entire genome sequences as a basis, many attempts have been made to reconstruct the metabolism of a given species (Karp et al. 1996; Tatusov et al. 1996; Bono et al. 1998; Selkov et al. 2000). The general principle relies on the fact that most components of central metabolism are usually conserved through evolution (for some interesting exceptions see Huynen et al. 1999). Thus, known enzymes that participate in the biosynthesis of amino acids, nucleotides, and lipids are used as the template against which the genome sequence of a species is compared (Fig. 5).

Proteins significantly similar to these known enzymes are predicted to be the corresponding enzymes in the query species. The detection is performed by sequence comparison against the reference species (e.g., *Escherichia coli*) or the full protein database from which annotations are derived. The former method may provide more reliable annotations but not necessarily sufficient coverage of the possible metabolic functions for the query species (Karp et al. 1996). The latter method provides less reliable annotations because of the increased amount of noise from sequence databases, but guarantees the maximum possible coverage for likely functions that may be absent from any given reference species.

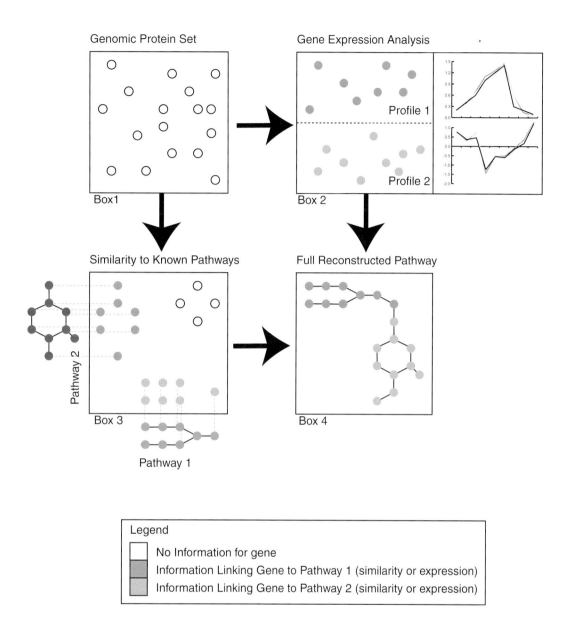

FIGURE 5. A schematic representation of how metabolic/regulatory pathway information and gene expression data may be used for reconstruction of pathways. Box 1 illustrates the initial collection of genes. Box 2 shows two gene expression profiles, and in which profile each initial gene is present (*orange* or *green circles* on the left). Box 3 shows which of the initial genes (*colored circles*) show similarity (*dashed lines*) to either of two known pathways (shown outside the box). Genes similar to the first pathway are shown in orange, and genes similar to the second pathway are shown in green. Box 4 shows the reconstruction of two pathways coming together, based on the evidence from boxes 2 and 3 (e.g., all genes coding for enzymes in these pathways may be under positive or negative control by a set of transcription factors—*orange* or *green circles*, respectively).

Metabolic pathways are then reconstructed on the basis of the presence or absence of the predicted enzymes (Bono et al. 1998). The groups of enzymes that participate in a single pathway may be considered as a functionally related group, where protein interactions may take place. Indeed, we have recently shown that the metabolic enzymes of *E. coli* appear to be involved much more frequently in gene fusion than any randomly chosen set of proteins and that most of these proteins participate in enzyme complexes (Tsoka and Ouzounis 2000a), suggesting that there may be considerable overlap between these two independent approaches. In fact, almost half of the enzyme subunits in the known metabolic complement of *E. coli* participate in protein complexes (Ouzounis and Karp 2000). Therefore, metabolic reconstruction represents another way of predicting functional associations of proteins without directly relying on homology-based prediction (Fig. 5) (Marcotte 2000).

A natural extension of metabolic reconstruction is the development of database systems that capture information on signaling pathways, for example, the Cell Signaling Networks DataBase (CSNDB), covering information for >2000 biological macromolecules (Igarashi and Kaminuma 1997). These are less well-defined than metabolic pathways and involve the presence of multiple proteins at every single step with unspecified reactions. However, signaling pathways are key components of the biochemical network that determines cell function and usually involve multiple protein interactions. At its most general incarnation, the assignment of proteins into functional classes (Tamames et al. 1996; Andrade et al. 1999a) may also be considered to provide some indication for the general involvement of proteins in some related cellular functions.

Transcription Profiles

It is now possible to generate gene expression information for an entire genome in a given state using microarray technology. A DNA chip containing all open reading frames (ORFs) from a genome such as *Saccharomyces cerevisiae* is constructed. Then mRNA from this organism is extracted from cells at a given state (e.g., diauxic shift) and used to create a fluorescently labeled cDNA population. Another population of cDNAs is constructed from mRNAs taken from a reference state. The second population is also fluorescently labeled, but with a different label. The population of labeled cDNAs should represent the amount of mRNA expression of genes in both states. These cDNAs can then be hybridized to their corresponding ORF on the DNA chip. Measurement of the fluorescence of both labels for each ORF on the chip then allows an estimate to be made for the relative expression of the genes of a cell in a particular state compared to the reference.

These expression ratio measurements represent a valuable way of detecting coexpressed genes (DeRisi et al. 1997). Genes whose patterns of expression are similar across many such experiments are possibly coregulated (Fig. 5). The corresponding gene products of such genes are then candidates for functional association and possibly interaction. Because analysis can be carried out on cells in different states (normal versus diseased) (Ross et al. 2000) and different tissues (e.g., human neurons), the amount of data that can be obtained is potentially immense.

Computational analysis of gene expression data involves the normalization of data across and within experiments, generally followed by some form of clustering operation (Eisen et al. 1998). Normalization involves modeling a statistical profile of the data, then replacing raw expression ratios with a significance score. Clustering of these normalized data points can be done in many ways. For successful clustering, a distance measure needs to be used (Eisen et al. 1998). This is a numerical value that indicates the similarity of the expression profiles of two genes. A simple Euclidean distance measure can be used, or, alternatively, a correlation coefficient (such as Pearson correlation). This allows pairs of genes to be grouped using standard clustering techniques, such as K-means. Clusters generated through these methods have been shown to group coexpressed genes remarkably well.

Gene expression data are not without flaw, however. It is possible that similar genes will exhibit cross-hybridization on the DNA chip, giving misleading results. The expression ratios generated are quite noisy, as it is generally only possible to measure strong induction or repression of a gene's expression at any given time. Subtle changes in gene expression are therefore very difficult to detect. With many experiments and many time points, noise levels can be reduced. With accurate normalization and robust clustering, it is therefore possible to generate good candidates for functional association. Transcription profiles have also been used in association with gene fusions (Marcotte et al. 1999a) in an attempt to predict protein interactions for the yeast genome.

THE HARD WAY: PURE EXPERIMENT

We call this section the hard way because the approaches in this section all rely on experimental information of protein interactions, usually verified independently and reproducibly. In other words, the compilation of information described here is textbook knowledge, for exemplary cases of protein interactions. It is also the hard way because it is the most labor-intensive of the above, involving extraction of information from the literature and curation of this information into database systems. These databases are expected to form the basis upon which novel methods will be built and provide a "gold standard" for further developments in the computational detection of protein interactions.

Curated Databases

In recent years, there have been attempts to build databases that store information about protein interactions. Previously, this type of experimentally derived information was not explicitly stored in major molecular biology databases. Two examples of protein interaction databases include the Database of Interacting Proteins (DIP) (University of California at Los Angeles) and INTERACT (University of Manchester). These databases provide mechanisms for the representation, storage, querying, and browsing of information on protein interactions. DIP also provides information about the experimental conditions and procedures that were used to determine any given interaction stored in the database. Although still at a rudimentary stage, these databases contain a few thousand well-known and documented cases of protein interactions. As the database size grows, these resources will provide the basis for further development of new algorithms and validation of existing ones (see also the section on Computational Resources, p. 606).

Text Mining

The task of curating biological databases in general and databases on protein interactions in particular is daunting. This activity involves extensive scanning of the literature to identify and extract relevant information. Thanks to the manual nature of database curation, the extracted information mostly refers to well-documented, reproducible results.

As a matter of fact, vast amounts of information pertaining to protein–protein interaction and protein function are already available in published articles. Biological abstracts are collected and maintained by the U.S. National Library of Medicine (NLM) in the MEDLINE database. Because MEDLINE only offers search and retrieval based mainly around keywords, it is very difficult to search directly for complex experimental information such as whether two proteins interact. Because of this difficulty, retrieving complex information is laborious and involves wading through countless abstracts before the desired information can be obtained. Recently, a new activity in computational genomics based on text mining has arisen around this concept (Craven and Kumlien 1999).

The principal idea is the use of natural language processing (NLP) technology to extract relevant pieces of information and include them in a database according to a predefined schema. This procedure is called information extraction and may involve sophisticated processing algorithms for the identification of complex syntactic and semantic structures in text. Attempts to detect and process textual information pertinent to protein–protein interactions have already been reported (Blaschke et al. 1999; Thomas et al. 2000). Other procedures in text mining include document clustering (Stapley and Benoit 2000; Iliopoulos et al. 2001), which generates groups of related documents. These document collections may contain information that can be very specific on a single subject, for instance, protein interactions.

Document clustering systems do not attempt to "understand" documents, but instead group similar documents based on their sharing of similar terms. From a computational perspective, document clustering is simpler than information extraction. Document clustering of biological abstracts should locate abstracts that share the same concept or topic, such as a metabolic pathway or developmental process. This can be achieved by statistical discovery of keywords in abstracts, with subsequent clustering based on the sharing of these terms. For this particular problem, protein names that frequently occur in abstracts from the same document cluster may represent molecules that are functionally related and possibly interacting. Using document clustering, it is very hard to determine the exact nature of this relationship, because one cannot query for specific information such as "Protein A interacts with Protein B."

Information extraction involves the analysis of abstracts or full-text articles using computational systems designed to "understand" abstracts. Due to the complex nature of scientific text and the inexact nature of human language, these algorithms suffer from significant noise levels. In addition, multiple synonyms for a gene or protein name and an ever-changing scientific vocabulary make this a very challenging task. This approach can potentially produce a listing of interacting proteins by attempting to understand biological abstracts and search for phrases implying interaction. A simple query would be to discover phrases such as "Protein A binds Protein B." Even with a complex understanding of grammar and synonym information, these methods have, at the moment, a high false-positive rate and poor recall.

Although it is too early to assess the utility of such systems in the construction of databases containing information on protein interactions or other kinds of functional information, it is possible that data availability will drive the development of text mining. With the recent initiatives for public repositories of biological publications, this technology will probably rapidly evolve and have a significant impact on the speed and coverage of protein interaction information from the biological literature.

It appears that computation of protein interactions using the variety of approaches described herein is not such an impossible task as it appeared a few years ago. Using algorithms that exploit certain strong constraints from gene and genome structure, combined with functional data such as metabolic pathways or transcription profiles and coupled with solid information stored in high-quality, curated databases, it is now possible to obtain indications about the possible interaction of genes or proteins. These approaches will mature in the foreseeable future and are expected to play a major role in the post-genomics revolution. We hope that functional genomics information from high-throughput biological experiments specifically designed to detect protein interactions will also be incorporated into algorithms and databases that address this issue so that computing with molecular interactions will produce significant results with positive impact on human health, agricultural production, and environmental control.

COMPUTATIONAL RESOURCES FOR THE BIOLOGIST

This section details computational resources available for the detection and analysis of protein–protein interactions (see Table 1). Many of the methods listed here are quite complex and require the use of multiple software components. For this reason, simple protocols are given for most of the computational interaction prediction techniques. More detailed complete algorithm descriptions are available from the references cited. Practically all of these computational approaches have already been performed on large numbers of complete genomes, and the results from these computations are available for analysis and download across the World Wide Web. These methods hence provide an excellent resource for biologists interested in protein interaction detection. References to well-known databases can be found in the annual database issue of *Nucleic Acids Research*.

TABLE 1. Computational Resources

	Type of Resource	URL	Reference
COGs	orthology, phylogenetic profiles	http://www.ncbi.nlm.nih.gov/COG/	Tatusov et al. (1997)
WIT	orthology, phylogenetic profiles, gene colocalization	http://wit.mcs.anl.gov/WIT2/	Overbeek et al. (2000)
STRING	gene colocalization	http://www.bork.embl-heidelberg. de/STRING/	Snel et al. (2000)
AIIFUSE	gene fusions	http://www.ebi.ac.uk/research/cgg/allfuse/	Enright and Ouzounis (2001)
KEGG	metabolic/regulatory pathway analysis and reconstruction	http://www.genome.ad.jp/keeg/	Kanehisa and Goto (2000)
ECOCYC	metabolic pathway analysis	http://ecocyc.pangeasystems.com/ecocyc/	Karp et al. (2000)
Expression Profiler	gene expression profile analysis	http://ep.ebi.ac.uk/EP/	Brazma and Vilo (2000)
Brown Lab	gene expression data	http://genome-www4.stanford.edu/MicroArray/SMD	
Church Lab	gene expression data	http://twod.med.harvard.edu/ExpressDB/	
DIP	database of protein interactions	http://dip.doe-mbi.ucla.edu/	Xenarios et al. (2000)
INTERACT	database of protein interactions	http://bioinf.man.ac.uk/interactpr.htm	Eilbeck et al. (1999)
Flybase	species-specific database for *D. melanogaster*	http://fly.ebi.ac.uk/	Gelbart et al. (1997)
YPD	species-specific database for *S. cerevisiae*	http://www.proteome.com/	Hodges et al. (1999)
WormPD	species-specific database for *C. elegans*	http://www.proteome.com/	Costanzo et al. (2000)

Pure Orthology and Phylogenetic Profiles

As described earlier, phylogenetic profiles can be a powerful tool for the prediction of functional association and possible interaction between proteins. The methods required for the generation of phylogenetic profiles are quite complex. The National Center for Biotechnology Information (NCBI) has a comprehensive catalog of orthologous proteins in complete genomes. This collection is called Clusters of Orthologous Groups, or COGs (Fig. 6). Recently, information pertaining to the phylogenetic profiles of these clusters has been added to the COG database. This allows a comprehensive analysis of phylogenetic profiles and orthology for protein sequences within complete genomes. The COG database provides an excellent framework for phylogenetic profile analysis and the computational prediction of functional association of proteins using this approach.

The COG database is assembled as follows:

1. A complete self-comparison of all complete genome protein sequences to be analyzed is performed with a sequence similarity searching algorithm such as BLAST.

2. To delineate orthologs, the Bi-directional Best Hit measure is used. This defines an ortholog as follows:

   For two proteins in two different genomes to be orthologous, they must represent the highest-scoring mutual hits when those two genomes are compared as in step 1.

   Paralogs may then be defined as two proteins within the same genome that are each other's highest-scoring similarity hit when that genome is compared against itself using BLAST.

3. The above criteria for defining these relationships are then used to cluster proteins from all genomes into families of orthologs and paralogs.

4. The initial clusters are analyzed using a multiple-alignment procedure to confirm that a conserved motif is present in each member of a given cluster. Clusters are also inspected for the presence of multidomain proteins, and if these are found, clusters are split into subgroups corresponding to each domain.

5. A phylogenetic profile for each COG may then be constructed by listing the presence or absence of a member of that COG in every complete genome.

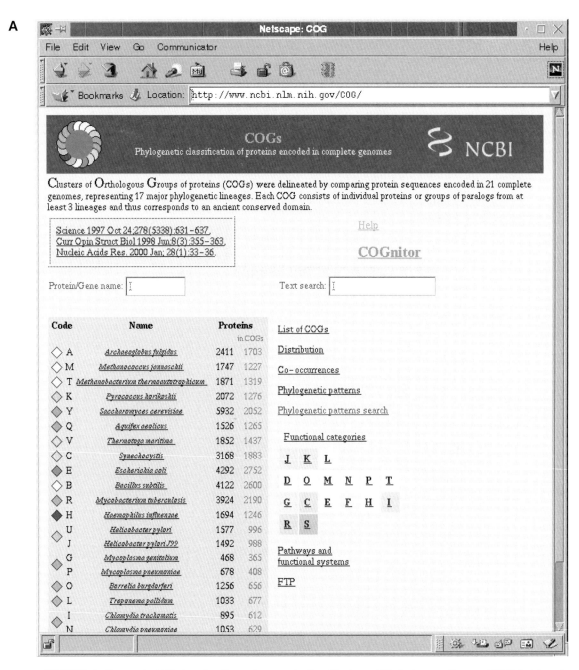

FIGURE 6. Screenshot (*A*) showing the entry page to the Clusters of Orthologous Groups (COGs) data set, and another screenshot (*B*) showing various clusters contained within the database, and their associated phylogenetic profiles.

This collection of orthologous protein families and representative phylogenetic profiles may be used for protein interaction and association analysis. For a protein of interest in cluster A, take the phylogenetic profile of this cluster and compare it to all other clusters. When comparing profiles, one can search for an exact match or look for close matches using a correlation measure (such as Pearson correlation) or a Euclidean distance measure. Proteins in clusters with very similar profiles to cluster A are hence possibly functionally associated to our protein of interest, and possibly interacting. It must be stressed once again, however, that these profiles work best with large collections of complete genomes, so a profile across 30 genomes is more informative than a profile based on fewer species.

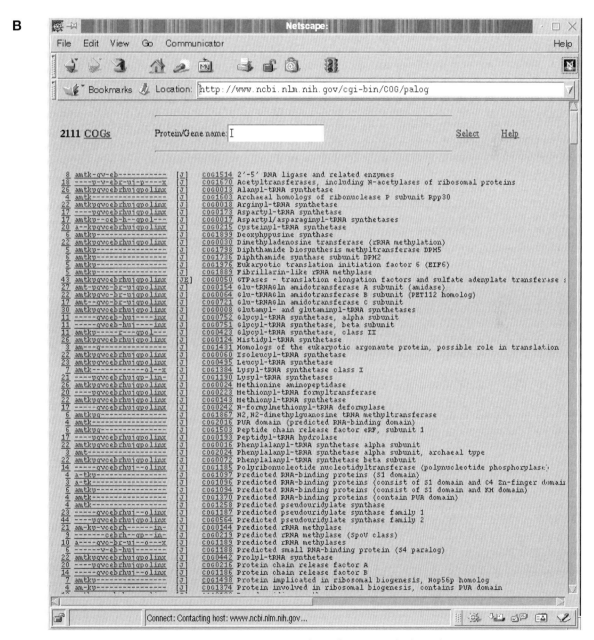

B

FIGURE 6. (*Continued, see facing page for legend.*)

Colocalization of Genes

Studies of correlated gene locations can also give insight into both functional association and possible protein interaction. Some resources exist that provide exhaustive information pertaining to gene clusters and gene neighborhood. One such resource is WIT (What Is There?) (Overbeek et al. 2000), which combines sequence clustering, metabolic pathway reconstruction, and gene colocalization information. Another such resource is STRING (Search Tool for Recurring Instances of the Neighborhood of Genes) (Snel et al. 2000), which is solely dedicated to gene colocalization (Fig. 7). Both of these resources gather information on gene colocalization in similar manners.

1. The concept of the Bi-directional Best Hit is extended to Pairs of Close Bi-directional Best Hits (PCBBHs).

   A PCBBH occurs when a pair of close genes on one genome are the bidirectional best hits of another close pair on a different genome. "Close" in this instance indicates that genes are direct neighbors or within 300 bp of each other on the DNA strand. Typically these genes are located on the same strand. The PCBBH concept allows accurate discovery of runs of genes that frequently occur in the same gene neighborhood in different genomes. The constraint of the Bidirectional Best Hit implies that these genes usually represent orthologs.

2. A less-strict constraint for locating conserved gene clusters between genomes is the concept of Pairs of Close Homologs (PCH). This allows detection of genes that frequently occur in

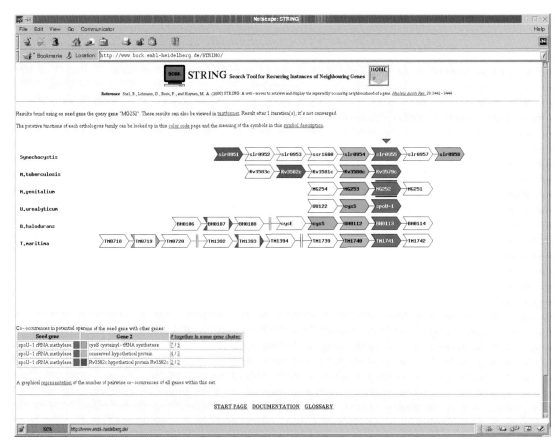

FIGURE 7. Screenshot from the STRING server at EMBL. This screenshot shows gene neighborhood information for the *Mycoplasma genitalium* gene MG252. Orthologous genes are shown in the same color in this representation.

the same neighborhood, but do not have to be each other's bidirectional best hits; the only requirement is that they are homologous above some arbitrary similarity threshold. This criterion is less strict, but allows conserved gene clusters to be located for genes that perform similar functions and are not necessarily directly orthologous.

3. Runs of genes (PCH or PCBBH) that occur in the same gene neighborhood in multiple genomes are then scored for significance. One of the most important scoring classifications depends on the phylogenetic distance between the genomes of interest. Shared gene clusters between phylogenetically similar genomes are much more likely to be detected by chance than from phylogenetically distant ones. A scoring system measures the phylogenetic distance between two genomes based on the 16S rRNA phylogenetic tree.

4. A coupling score is used to determine whether enough evidence is available to call two genes functionally linked on the basis of their common colocalization in different genomes. This generally involves the addition of all the scores from PCH and PCBBH data that link the two genes in question.

5. Finally, genes that have high coupling scores are candidates for functional association and possibly encode proteins that physically interact.

Using the STRING or WIT resource, it is relatively easy to extract from a list of genes of interest all those candidate genes that show evidence of colocalization in different genomes. This evidence may then easily be used in conjunction with phylogenetic profile and gene-fusion data to generate lists of gene product pairs that potentially interact. As with the phylogenetic profile method, this method benefits greatly from the availability of many complete genomes. As more genomes become available, the resolution of gene colocalization and phylogenetic profiles is expected to increase significantly.

Gene Fusion Analysis

Gene colocalization studies rely on the fact that evolution sometimes decides to keep physically interacting or associated genes in very close proximity on the genome. The ultimate form of this occurs in the actual fusion of these two genes into a single multifunctional protein. Many such cases of gene fusion are known to occur in nature, such as the genes involved in aromatic amino acid biosynthesis in *S. cerevisiae*. In this case, all five genes in the pathway are fused into a single unit that is capable as a single protein of performing all pathway steps. It seems quite a reasonable hypothesis to assume that when these events occur, the proteins in question are very likely to be physically interacting or functionally associated. The problem is accurate detection of these cases from complete genome databases. The AllFUSE resource is an excellent place to obtain information about genes of interest for which evidence for gene fusion exists (Fig. 8). This resource was generated from a systematic analysis within 24 complete genomes for evidence of gene fusion using a fusion detection algorithm. The process for detecting gene fusion is complex, and is performed in the following manner.

1. A query genome and a reference genome are selected. The reference genome is the genome in which fused composite proteins are sought. The query genome is where unfused proteins are detected by virtue of the fact that they form components of a fused protein in the reference genome.

2. All protein sequences from the query genome are compared against the reference genome sequences using a sequence similarity searching tool such as BLAST. The query genome is also compared against itself in a self-comparison, once again using BLAST.

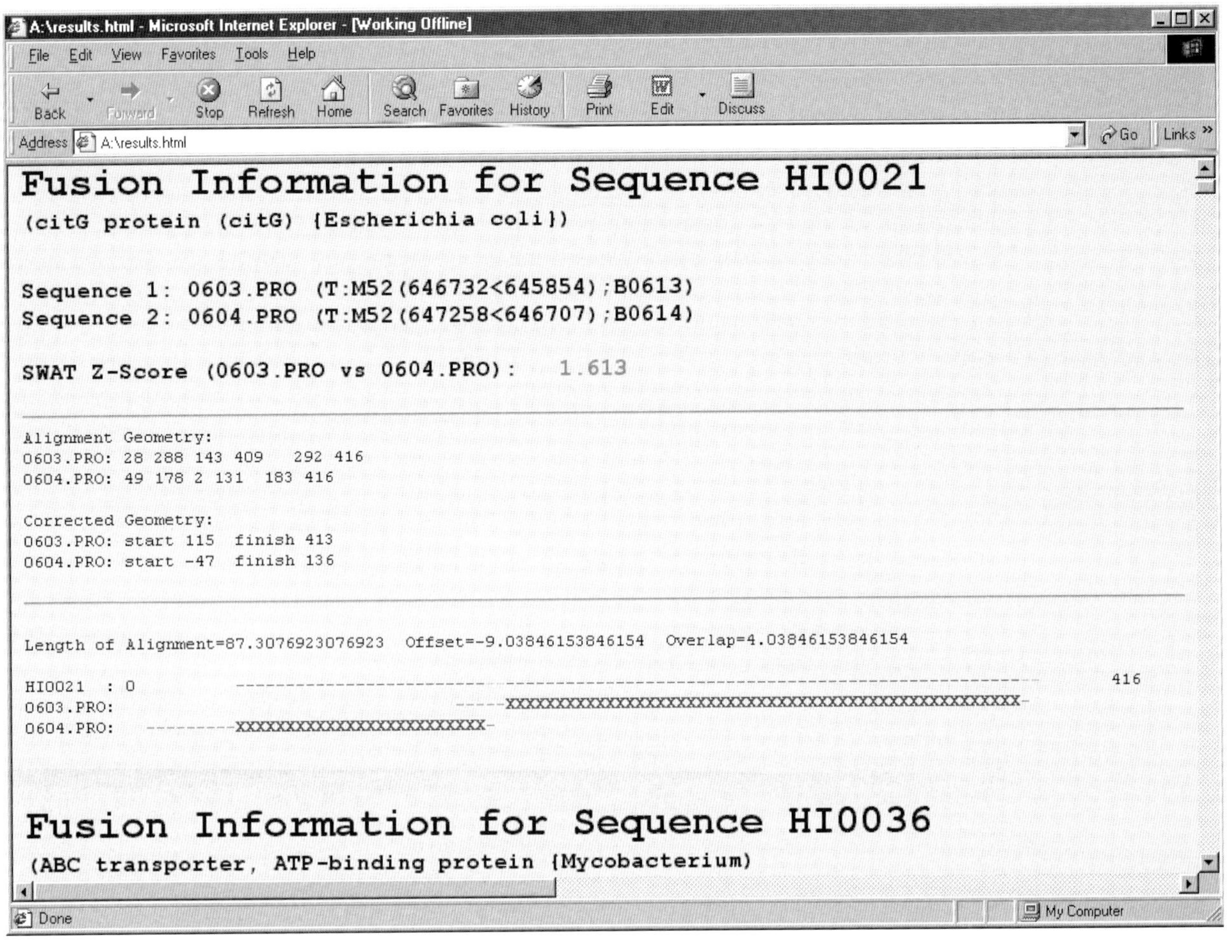

FIGURE 8. Screenshot from the AllFUSE server at EMBL-EBI. This shows two genes from the *E. coli* genome, aligned with corresponding fused protein from *H. influenzae*.

3. All query-versus-reference and query-versus-query similarities above a threshold are stored in sequence similarity matrices using a binary representation. The query-versus-query matrix is examined for symmetry (i.e., if A hits B, does B hit A ?). Nonsymmetrical hits are resolved by rechecking similarity using a very sensitive search algorithm, such as the Smith-Waterman algorithm.

4. The query-versus-reference and query-versus-query matrices are then used to detect cases where a protein A in the reference genome hits proteins B and C in the query genome. The transitivity of this relationship is then checked by looking for cases where protein A hits B and A hits C, but B and C do not hit each other. When this occurs, another round of Smith-Waterman algorithm is used to confirm that there is no significant similarity between B and C. When no similarity can be found, A is marked as a possible gene fusion and B and C as possible interactors.

5. A minimum overlap criterion is used on these candidate fusion events to filter out cases where there is more than 5% overlap of the two component proteins when they are aligned against the reference composite sequence.

6. High paralogy in the query genome may make actual assignment of possible interactors difficult, so a subsequent clustering step may then be used to classify fusion events further.

This method appears to be robust enough to detect gene fusion events with high accuracy (>90%). As such, this represents an excellent method for the detection of proteins that potentially interact. As is the case with both previously described methods, the accuracy and sensitivity improve dramatically when more complete genomes are used in this type of analysis. Two genes that appear to have fused orthologs in only one genome could be considered a sequencing error or a possible false positive, but when detected in more than one genome, the fidelity of prediction is greatly improved. This method readily detects well-known interacting proteins (e.g., tryptophan synthase α and β subunits) and many proteins known to form complexes. This leads to the conclusion that the predictions between proteins of unknown function may be very robust. It is difficult to validate the predictions from large-scale experiments such as the AllFUSE project, because large-scale experimental protein–protein interaction data are not readily available. However, comparison with gene-expression data sets from yeast indicate significant overlap (Enright and Ouzounis 2001). For the biologist looking for computational methods to determine protein interactions, AllFUSE represents a readily available resource with tens of thousands of functional associations and interactions predicted in the above manner.

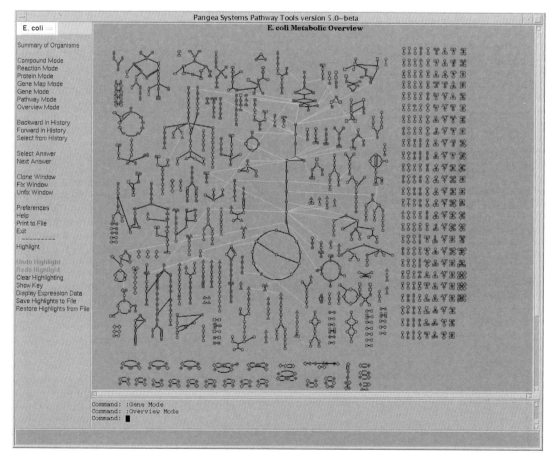

FIGURE 9. Screenshot showing the metabolic genome overview for *E. coli*, from the EcoCyc database of metabolic pathways. (Reprinted from Pangea Systems TM, Oakland, California.)

Metabolic Networks and Regulatory Pathways

Gene network analysis is a rapidly growing field in bioinformatics. Studies of how genes are interconnected in the cell in terms of genetic regulation and metabolic pathway participation can provide direct information regarding protein–protein interaction. Metabolic pathway analysis and reconstruction generally involve networks of indirect protein interaction (i.e., an enzymatic relationship). Analysis of regulatory networks of genes provides evidence for direct protein interaction, such as phosphorylation. Metabolic pathway databases have been around for some time. Examples of these databases are EcoCyc (Fig. 9) (Karp et al. 2000) and KEGG (Fig. 10) (Kanehisa and Goto 2000). KEGG (Kyoto Encyclopaedia of Genes and Genomes) has been dramatically extended to take into account not just metabolic pathway information, but also regulatory network information, orthology between genomes, and gene-expression information.

Both KEGG and EcoCyc have been developed using knowledge base technology. Although the underlying representations are quite different, one common theme is the complex and rich database schema. Knowledge bases are essentially object-oriented databases with some additional

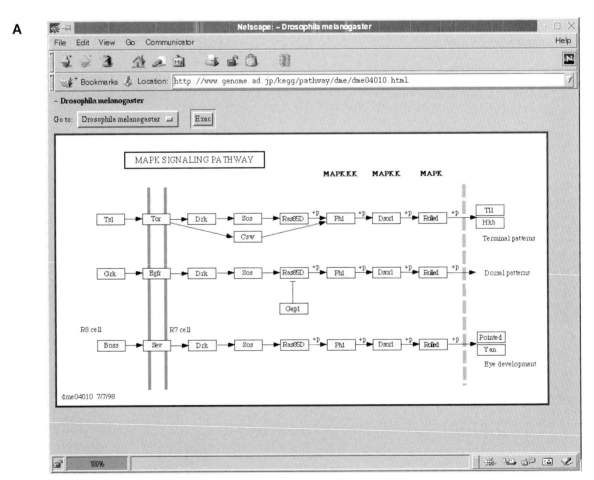

FIGURE 10. Screenshot from the Kyoto Encyclopaedia of Genes and Genomes (KEGG), showing two pathways contained within the database. (*A*) Representation of the MAP kinase regulatory signaling pathway; (*B*) the pentose phosphate metabolic cycle.

capabilities that are specific to a particular representation formalism employed. The clarity and high fidelity of representations and a powerful query capability are some advantages of knowledge bases over traditional relational databases. EcoCyc offers a variety of object types, but the most important one in the issue of protein interactions is the protein-complex type, through which homo- and hetero-polymeric complexes are documented (Karp 2000). KEGG contains a set of known metabolic and regulatory pathways (such as apoptosis and cell cycle). Both databases also link to other molecular biology databases such as SwissProt (Bairoch and Apweiler 2000). KEGG attempts to describe metabolic and regulatory pathways in a complete genome by mapping genes in that genome to known pathways and to orthologous genes in other genomes. In addition, gene expression information can be handled by both databases, and allows the systematic analysis of coregulation of genes in the context of the corresponding metabolic or regulatory pathways.

From a protein interaction point of view, it is relatively straightforward to examine genes of interest using these resources. Proteins in the same regulatory or metabolic pathway are candidates for functional association and possible interaction. The regulatory pathway prediction system is perhaps the best place to locate genes that interact physically as they are predicted to, related by mechanisms such as phosphorylation.

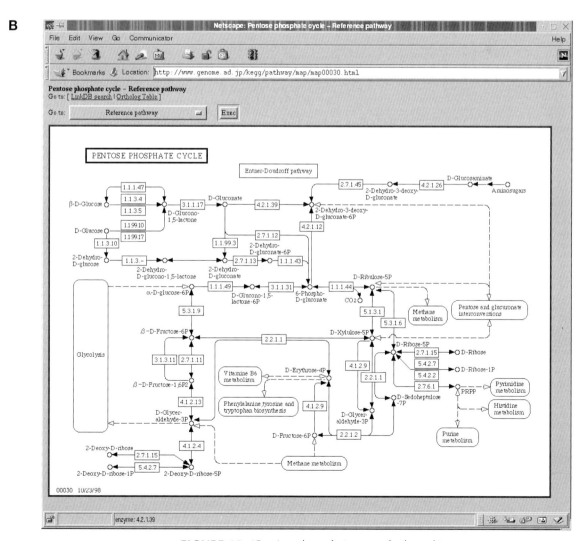

FIGURE 10. (*Continued, see facing page for legend.*)

Gene Expression Analysis Methods

Analysis of the coregulation of genes via gene expression profiling is an increasingly common way to locate functionally associated proteins. DNA chip methods now permit all of the open reading frames (ORFs) from a given genome to be placed on a single chip, allowing many gene expression experiments to be carried out very quickly. Unfortunately, at the moment, there is no standard repository for gene expression results. Some proposals for a standard system to store and use this

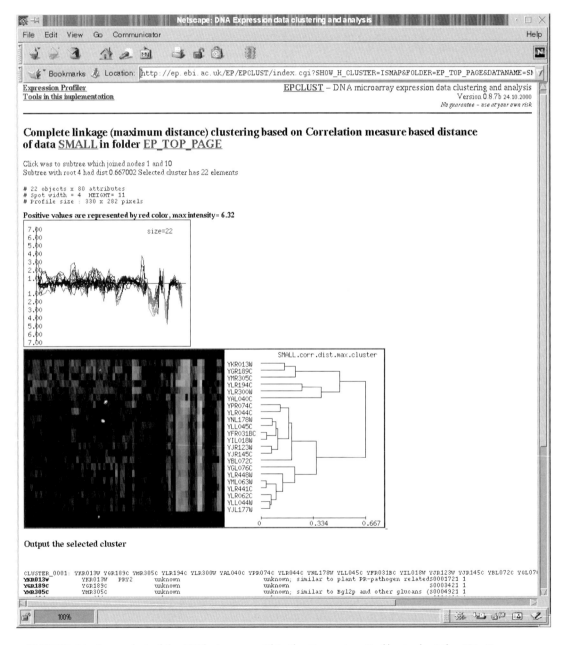

FIGURE 11. Screenshot of the EPClust server within the Expression Profiler tools at the EBI. Gene expression profiles (*top graph*) are clustered according to their similarity, and resulting clusters may then be analyzed (*bottom image and tree*).

information have been announced; for example, from the Microarray Gene Expression Database Group (http://www.mged.org). It looks hopeful that in the near future such a system will exist. It is currently possible, however, to obtain much of this information directly from the lab that produced it. Laboratories such as those at Stanford University and Harvard University are excellent sources of gene expression data for organisms such as *S. cerevisiae*.

The MicroArray group at the European Bioinformatics Institute provides some publicly available gene expression data and also the Expression Profiler (Fig. 11). This is an excellent resource for visualization, analysis, and clustering of gene expression data. It allows selection of a set of gene-expression profiles from different experiments, followed by clustering to located genes with similar expression profiles. Many different clustering methods may be used here (such as Euclidean Distance, Correlation, and K-means). Hence, it is possible to obtain both partition-based and hierarchical cluster information.

Once clustering is complete, the resulting clusters are displayed visually as normalized graph and color plots of expression ratios. This is accompanied by listings of genes that have been placed in a particular cluster on the basis of similarity between their expression profiles. Other utilities are available for the retrieval of the representative sequence of each protein and also for analysis of upstream regions of these genes at the DNA level to identify possible regulatory regions. Sophisticated pattern analysis of genes is also possible.

Using this resource, it is possible to obtain robust clusters of genes exhibiting expression profile correlation. Proteins whose genes exhibit very similar patterns of expression may then be considered candidates for functional association and possibly direct physical interaction. Gene expression analysis becomes much more reliable with more expression data. For example, genes that have high correlation across ten experiments are much more likely to be related functionally than genes correlating across two experiments. Gene expression data are still relatively susceptible to noise, and great care must be taken to minimize and filter this from the analysis. These data can be very powerful when combined with analyses involving regulatory network reconstruction, and with other methods of detection of functional association and interaction of proteins.

DATABASES CONTAINING INTERACTION INFORMATION

Database of Interacting Proteins

The Database of Interacting Proteins (DIP) is an attempt to set up a curated database of proteins that are known experimentally to physically interact (Fig. 12). Users may submit an entry to the database. The database details at present almost 3500 proteins that are known to interact from approximately 80 species. Articles and experiments that confirm these interactions are also stored in the database, making it an invaluable resource for protein interaction information. It is hoped that increasing numbers of biologists will eventually submit interactions and additional evidence to this database, making it a comprehensive guide to obtaining interaction information for any organism from a single source.

Specialist Organism Databases

Many specialist databases exist that contain information from many sources for a single organism. Many of these databases now include information regarding cellular localization, gene expression, and interaction of proteins. For *S. cerevisiae* two such databases exist: SGD (*Saccharomyces* Genome Database) (Ball et al. 2000) and YPD (Yeast Proteome Database)

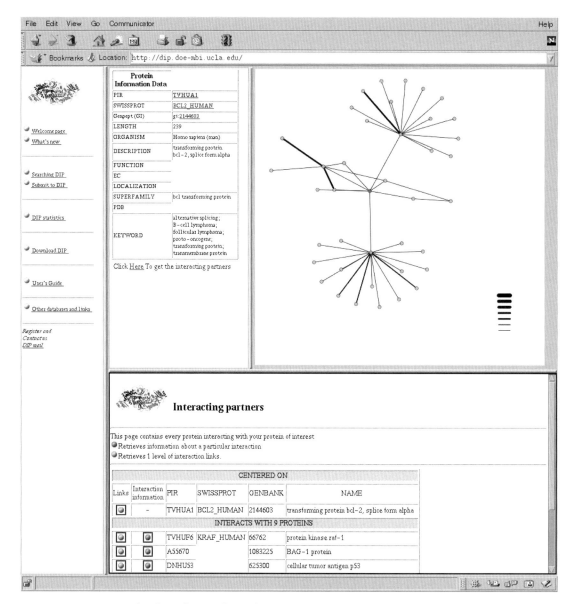

FIGURE 12. Screenshot from the Database of Interacting Proteins (DIP) server at UCLA. Information is displayed for the human Bcl-2 protein (SwissProt identifier: BCL2_HUMAN), showing a graph of interactions between this protein and other proteins. A further text index of its interacting partners is also displayed.

(Hodges et al. 1999). These databases are highly organized and provide exhaustive information regarding the genes and proteins of yeast. Information on protein interactions for yeast is available and easy to access. Another example of a specialist database is FlyBase (Gelbart et al. 1997), which is a comprehensive database for *Drosophila melanogaster*. Like YPD, FlyBase also contains a considerable amount of information regarding protein–protein interaction in *D. melanogaster*.

These rich, species-specific database resources are a valuable tool when seeking specific information regarding protein–protein interactions for particular organisms. Moreover, some of this information may be valid for other organisms that are less well documented.

REFERENCES

Andrade M.A., Ouzounis C., Sander C., Tamames J., and Valencia A. 1999a. Functional classes in the three domains of life. *J. Mol. Evol.* **49:** 551–557.

Andrade M.A., Brown N.P., Leroy C., Hoersch S., de Daruvar A., Reich C., Franchini A., Tamames J., Valencia A., Ouzounis C., and Sander C. 1999b. Automated genome sequence analysis and annotation. *Bioinformatics* **15:** 391–412.

Bairoch A. and Apweiler R. 2000. The SWISS-PROT protein sequence database and its supplement TrEMBL in 2000. *Nucleic Acids Res.* **28:** 45–48.

Ball C.A., Dolinski K., Dwight S.S., Harris M.A., Issel-Tarver L., Kasarskis A., Scafe C.R., Sherlock G., Binkley G., Jin H., Kaloper M., Orr S.D., Schroeder M., Weng S., Zhu Y., Botstein D., and Cherry J.M. 2000. Integrating functional genomic information into the *Saccharomyces* genome database. *Nucleic Acids Res.* **28:** 77–80.

Blaschke C., Andrade M.A., Ouzounis C., and Valencia A. 1999. Automatic extraction of biological information from scientific text: protein-protein interactions. *Intelligent Systems for Molecular Biology* **7:** 60–67.

Blumenthal T. 1998. Gene clusters and polycistronic transcription in eukaryotes. *Bioessays* **20:** 480–487.

Bono H., Ogata H., Goto S., and Kanehisa M. 1998. Reconstruction of amino acid biosynthesis pathways from the complete genome sequence. *Genome Res.* **8:** 203–210.

Brazma A. and Vilo J. 2000. Gene expression data analysis. *FEBS Lett.* **480:** 17–24.

Chen Z. and Han M. 2000. Building a protein interaction map: Research in the post-genome era. *Bioessays* **22:** 503–506.

Costanzo M.C., Hogan J.D., Cusick M.E., Davis B.P., Fancher A.M., Hodges P.E., Kondu P., Lengieza C., Lew-Smith J.E., Lingner C., Roberg-Perez K.J., Tillberg M., Brooks J.E., and Garrels J.I. 2000. The yeast proteome database (YPD) and *Caenorhabditis elegans* proteome database (WormPD): Comprehensive resources for the organization and comparison of model organism protein information. *Nucleic Acids Res.* **28:** 73–76.

Craven M. and Kumlien J. 1999. Constructing biological knowledge bases by extracting information from text sources. *Intelligent Systems for Molecular Biology* **7:** 77–86.

Dandekar T., Snel B., Huynen M., and Bork P. 1998. Conservation of gene order: A fingerprint of proteins that physically interact. *Trends Biochem. Sci.* **23:** 324–328.

DeRisi J.L., Iyer V.R., and Brown P.O. 1997. Exploring the metabolic and genetic control of gene expression on a genomic scale. *Science* **278:** 680–686.

Eilbeck K., Brass A., Paton N., and Hodgman C. 1999. INTERACT: An object oriented protein-protein interaction database. *Intelligent Systems for Molecular Biology* **7:** 87–94.

Eisen M.B., Spellman P.T., Brown P.O., and Botstein D. 1998. Cluster analysis and display of genome-wide expression patterns. *Proc. Natl. Acad. Sci.* **95:** 14863–14868.

Eisenberg D., Marcotte E.M., Xenarios I., and Yeates T.O. 2000. Protein function in the post-genomic era. *Nature* **405:** 823–826.

Enright A.J. and Ouzonis C.A. 2001. Functional associations of proteins in entire genomes via exhaustive detection of gene fusion. *Genome Biol.* (in press).

Enright A.J., Iliopoulos I., Kyrpides N.C., and Ouzounis C.A. 1999. Protein interaction maps for complete genomes based on gene fusion events. *Nature* **402:** 86–90.

Galperin M.Y. and Koonin E.V. 2000. Who's your neighbor? New computational approaches for functional genomics. *Nat. Biotechnol.* **18:** 609–613.

Gelbart W.M., Crosby M., Matthews B., Rindone W.P., Chillemi J., Twombly S., Russo, Emmert D., Ashburner M., Drysdale R.A., Whitfield E., Millburn G.H., de Grey A., Kaufman T., Matthews K., Gilbert D., Strelets V., and Tolstoshev C. 1997. FlyBase: A *Drosophila* database. The FlyBase consortium. *Nucleic Acids Res/* **25:** 63–66.

Hodges P.E., McKee A.H., Davis B.P., Payne W.E., and Garrels J.I. 1999. The Yeast Proteome Database (YPD): A model for the organization and presentation of genome-wide functional data. *Nucleic Acids Res.* **27:** 69–73.

Huynen M.A., Dandekar T., and Bork P. 1999. Variation and evolution of the citric-acid cycle: A genomic perspective. *Trends Microbiol.* **7:** 281–291.

Huynen M., Snel B., Lathe W., and Bork P. 2000. Exploitation of gene context. *Curr. Opin. Struct. Biol.* **10:** 366–370.

Igarashi T. and Kaminuma T. 1997. Development of a cell signaling networks database. *Pac. Symp. Biocomput.* **2:** 187–197.

Iliopoulos I., Enright A.J., and Ouzounis C.A. 2001. TextQuest: document clustering of MedLine abstracts for concept discovery in molecular biology. *Pac. Symp. Biocomput.* **8:** 6384–6395.

Kanehisa M. and Goto S. 2000. KEGG: Kyoto encyclopedia of genes and genomes. *Nucleic Acids Res.* **28:** 27–30.

Karp P.D. 2000. An ontology for biological function based on molecular interactions. *Bioinformatics* **16:** 269–285.

Karp P.D., Ouzounis C., and Paley S. 1996. HinCyc: A knowledge base of the complete genome and metabolic pathways of *H. influenzae. Intelligent Systems for Molecular Biology* **4:** 116–124.

Karp, P.D., Riley M., Saier M., Paulsen I.T., Paley S.M., and Pellegrini-Toole A.. 2000. The EcoCyc and MetaCyc databases. *Nucleic Acids Res.* **28:** 56–59.

Legrain P. and Selig L. 2000. Genome-wide protein interaction maps using two-hybrid systems. *FEBS Lett.* **480:** 32–36.

Lewis, S., Ashburner M., and Reese M.G. 2000. Annotating eukaryotic genomes. *Curr. Opin. Struct. Biol.* **10:** 349–354.

Marcotte E.M. 2000. Computational genetics: Finding protein function by nonhomology methods. *Curr. Opin. Struct. Biol.* **10:** 359–365.

Marcotte E.M., Xenarios I., van Der Bliek A.M., and Eisenberg D. 2000. Localizing proteins in the cell from their phylogenetic profiles. *Proc. Natl. Acad. Sci.* **97:** 12115–12120.

Marcotte E.M., Pellegrini M., Thompson M.J., Yeates T.O., and Eisenberg D. 1999a. A combined algorithm for genome-wide prediction of protein function. *Nature* **402:** 83–86.

Marcotte E.M., Pellegrini M., Ng H.L., Rice D.W., Yeates T.O., and Eisenberg D. 1999b. Detecting protein function and protein-protein interactions from genome sequences. *Science* **285:** 751–753.

Mushegian A.R. and Koonin E.V. 1996. Gene order is not conserved in bacterial evolution. *Trends Genet.* **12:** 289–290.

Nierman W.C., Eisen J.A., Fleischmann R.D., and Fraser C.M. 2000. Genome data: What do we learn? *Curr. Opin. Struct. Biol.* **10:** 343–348.

Ouzounis C. 1999. Orthology: Another terminology muddle. *Trends Genet.* **15:** 445.

Ouzounis C.A. and Karp P.D. 2000. Global properties of the metabolic map of *Escherichia coli. Genome Res.* **10:** 568–576.

Ouzounis C. and Kyrpides N. 1996. The emergence of major cellular processes in evolution. *FEBS Lett.* **390:** 119–123.

Overbeek R., Fonstein M., D'Souza M., Pusch G.D., and Maltsev N. 1999. The use of gene clusters to infer functional coupling. *Proc. Natl. Acad. Sci.* **96:** 2896–2901.

Overbeek R., Larsen N., Pusch G.D., D'Souza M., Selkov Jr., E., Kyrpides N., Fonstein M., Maltsev N., and Selkov E. 2000. WIT: Integrated system for high-throughput genome sequence analysis and metabolic reconstruction. *Nucleic Acids Res.* **28:** 123–125.

Pellegrini M., Marcotte E.M., Thompson M.J., Eisenberg D., and Yeates T.O. 1999. Assigning protein functions by comparative genome analysis: Protein phylogenetic profiles. *Proc. Natl. Acad. Sci.* **96:** 4285–4288.

Rivera M.C., Jain R., Mooer J.E., and Lake J.A. 1998. Genomic evidence for two functionally distinct gene classes. *Proc. Natl. Acad. Sci.* **95:** 6239–6244.

Ross D.T., Scherf U., Eisen M.B., Perou C.M., Rees C., Spellman P., Iyer V., Jeffrey S.S., Van de Rijn M., Waltham M., Pergamenschikov A., Lee J.C., Lashkari D., Shalon D., Myers T.G., Weinstein J.N., Botstein D., and Brown P.O. 2000. Systematic variation in gene expression patterns in human cancer cell lines. *Nat. Genet.* **24:** 227–235.

Selkov E., Overbeek R., Kogan Y., Chu L., Vonstein V., Holmes D., Silver S., Haselkorn R., and Fonstein M. 2000. Functional analysis of gapped microbial genomes: Amino acid metabolism of *Thiobacillus ferrooxidans. Proc. Natl. Acad. Sci.* **97:** 3509–3514.

Snel B., Lehmann G., Bork P., and Huynen M.A. 2000. STRING: A web-server to retrieve and display the repeatedly occurring neighbourhood of a gene. *Nucleic Acids Res.* **28:** 3442–3444.

Stapley B.J. and Benoit G. 2000. Biobibliometrics: Information retrieval and visualization from co-occurrences of gene names in Medline abstracts. *Pac. Symp. Biocomput.* **7:** 529–540.

Tamames, J., Casari G., Ouzounis C., and A. Valencia. 1997. Conserved clusters of functionally related genes in two bacterial genomes. *J. Mol. Evol.* **44:** 66-73.

Tamames J., Ouzounis C., Sander C., and Valencia A. 1996. Genomes with distinct function composition. *FEBS Lett.* **389:** 96–101.

Tatusov R.L., Koonin E.V., and Lipman D.J. 1997. A genomic perspective on protein families. *Science* **278:** 631–637.

Tatusov R.L., Mushegian A.R., Bork P., Brown N.P., Hayes W.S., Borodovsky M., Rudd K.E., and Koonin E.V.

1996. Metabolism and evolution of *Haemophilus influenzae* deduced from a while-genome comparison with *Escherichia coli. Curr. Biol.* **6:** 279–291.

Thomas J., Milward D., Ouzounis C., Pulman S., and Carroll M. 2000. Automatic extraction of protein interactions from scientific abstracts. *Pac. Symp. Biocomput.* **7:** 541–552.

Tsoka S. and Ouzounis C.A. 2000a. Prediction of protein interactions: Metabolic enzymes are frequently involved in gene fusion. *Nat. Genet.* **26:** 141–142.

———. 2000b. Recent developments and future directions in computational genomics. *FEBS Lett.* **480:** 42–48.

Xenarios I., Rice D.W., Salwinski L., Baron M.K., Marcotte E.M., and Eisenberg D. 2000. DIP: The database of interacting proteins. *Nucleic Acids Res* **28:** 289-291.

Zorio, D.A., Cheng N.N., Blumenthal T., and Spieth J. 1994. Operons as a common form of chromosomal organization in *C. elegans. Nature* **372:** 270–272.

34 Visualization and Integration of Protein–Protein Interactions

Peter Uetz,[1]* Trey Ideker,[2] and Benno Schwikowski[2,3]

[1]*Departments of Genetics, University of Washington, Seattle, Washington 98195;* [2]*The Institute for Systems Biology, Seattle, Washington 98105;* [3]*Department of Computer Science and Engineering, University of Washington, Seattle, Washington 98195*

*Present address: Institut für Genetik, Forschungszentrum Karlruhe, 76021 Karlsruhe, Germany.
E-mail addresses: peter.uetz@itg.fzk.de; tideker@systemsbiology.org; benno@systemsbiology.org

INTRODUCTION

The chemical composition of eukaryotic cells is remarkably well understood: We know the DNA sequences of many organisms more or less completely and can deduce many of their RNA and protein products. In the past, proteins have received the most attention from biochemists because they are the most complicated and among the most important molecules in a cell. Accordingly, proteins and their interactions have been studied in great detail, resulting in the identification of thousands of protein–protein interactions over the past 30–50 years. In addition to these classic studies, more recent large-scale proteomics projects are contributing huge amounts of systematic data relevant to the understanding of protein interactions. These large data sets also harbor information that is not immediately obvious without integrative analysis. Although integration and analysis have traditionally been carried out by humans, the sheer amount of data now calls for computer assistance.

Of course, cells consist not only of proteins, but also of a significant number of other molecules, ranging from small ions to high-molecular-weight carbohydrates and nucleic acids. Most of these nonproteinaceous compounds interact with at least one protein, because many represent the product of enzymatic reactions and, by default, associate with the enzyme that generated them (e.g., pyruvate with pyruvate kinase; see Fig. 1). Although knowledge of these interactions is essential for a complete understanding of a cell, the number of such interactions is thought to be small compared to the number of interactions among proteins or between proteins and nucleic acids. Moreover, the study of protein–DNA interactions has been a more popular starting point for interaction network assembly than the study of protein–protein interactions. DNA behaves more predictably than proteins and is subject to fewer variables, such as chemical modification, that may confer significant structural change. Furthermore, DNA-binding domains in many transcription factors are generally better characterized than protein–protein interaction domains, and extensive structural information for such modules exists. In contrast, interactions between proteins, or between proteins and RNA, are often harder to characterize due to their perceived inherent variability.

This chapter presents some computerized means to visualize and integrate protein–protein interaction data with data from DNA chip experiments and other sources, to create a starting point for future methods and discoveries.

Why Do We Need Visualization?

For most people, graphic representations of facts are much easier to understand than raw data. This is especially true for large data sets or complex situations. A long list of interacting proteins or a table of protein pairs falls short of capturing what happens in a cell, which is a dynamic process that occurs in at least four dimensions (including time). Instead, the use of graphics suits human preference for visual perception over every other sensory system. Like a road atlas, visual maps of protein interactions provide orientation for both novices and specialists. To make these maps useful for both audiences, it is desirable to generate dynamic maps that allow concealment of detail when only a rough overview is needed. Finally, protein-interaction maps stimulate the formulation of hypotheses that can be tested experimentally. For example, if a membrane protein is found to interact with a transcription factor, this result might look like a "false positive." However, such apparent incongruities have led to unexpected new insights into signal transduction, as in the cases of *notch* and *Suppressor of hairless*, Su(H) (Artavanis-Tsakonas et al. 1999) or with the SREBPs, transcription factors that are localized to the ER membrane (Edwards et al. 2000). Development of an appropriate "map" can aid in the identification of such informative anomalies.

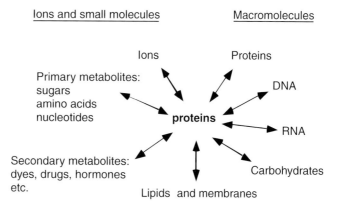

FIGURE 1. Physical interactions of proteins. Note that there are also interactions between the non-protein classes; e.g., between ions and other small molecules (such as Fe^{++} and heme). However, not many such interactions are reported. Additional interactions can be imagined with artificial molecules like drugs or synthetic ligands. In addition to physical interactions, genetic interactions (like synthetic lethality or suppression) can hint at potential physical interactions.

Protein-interaction Maps Versus Metabolic Pathways

Considerable effort has been invested in the visualization of metabolic pathways (e.g., Michal 1993, 1998), with more recent efforts using computerized systems (e.g., Küffner et al. 2000). Although the structures of metabolic pathways and protein-interaction maps are similar, there are a number of significant differences. Whereas metabolic pathways focus on the conversion of small molecules and the enzymes responsible for these conversions, protein-interaction maps (and signal transduction maps) concentrate mainly on physical contacts without obvious chemical conversions. Physical interactions are certainly of great utility when one studies single proteins or defined biological processes, but themselves do not reflect the huge amount of knowledge that has been accumulated in the biological literature. For example, Figure 5 (below) implies that the PEX proteins in yeast form a complex, but it does not provide any information about the assembly of the complex, its biological function, or its regulation. Although these shortcomings can be partly relieved by building in hyperlinks to protein databases, few such databases collate biological information in an easily accessible format.

Several groups have presented models and systems for genetic networks (Kolpakov et al. 1998; Serov et al. 1998; von Dassow et al. 2000) in addition to physical and metabolic networks. Such networks do not require physical interactions and, in fact, can suggest factors affecting control of a biological process that remain to be identified (von Dassow et al. 2000). Nevertheless, the ultimate goal of all such network models is a physical network integrated with genetic and metabolic information that is predictive of phenotype.

The biggest challenge for coherent display of physical networks remains the large amount of biological information that is available about many molecules and their interactions. Graphic networks can illustrate the complexity of biologic interactions but still fail to explain processes, mainly because important details are not visualized, such as spatial and temporal expression patterns or the conditions under which certain interactions occur.

Protein Networks, Protein Complexes, and Dynamic Protein Interactions

Currently, the most practical way to identify the components of a protein complex is mass spectrometric (MS) analysis (Yates 2000). Unfortunately, MS usually does not provide information about topology, so additional methods are required to decipher which proteins bind to which. The two-hybrid system can provide complementary information about direct interactions. However, it remains a challenge to integrate data from different experimental approaches, especially when they have been collected under different biological conditions (e.g., when cells were grown in different media). Figure 2 illustrates such nonoverlapping datasets. Protein interactions can occur in stable complexes or as transient, usually regulated, interactions. Unfortunately, most interactions are described qualitatively, and we do not know how strong the interaction really is. For example, the Database of Interacting Proteins (DIP) lists binding constants for fewer than 20 protein pairs (as of December 2000; see also Xenarios et al. 2000). Because there are hardly any quantitative data about protein interactions, protein complexes are currently difficult to analyze quantitatively based on an "informatics" approach.

Theoretically, computational analysis of protein structures should allow us to find fitting surfaces among known protein structures and thereby predict their interactions (Tsai et al. 1996; Palma et al. 2000). Of the ~6000 yeast proteins comprising the yeast proteome, about 600 have experimentally determined three-dimensional structures, whereas for up to ~2600 (~44%), the structure can be modeled on the basis of homology (http://jura.ebi.ac.uk:8765/ext-genequiz//genomes/sc/index.html; Sanchez et al. 2000; Vitkup et al. 2001; U. Pieper and M. Andrade, pers. comm.). However, limitations in experimental data and computing power still prohibit detailed and therefore successful predictions in most cases.

Protein–Protein Interactions and Associated Information

Much information is required to describe molecular interactions, especially if they are dynamic and dependent on many different parameters. For instance, once interacting proteins have been identified, one wants to know which parts or domains of the two partners interact, because this may lead immediately to further hypotheses about the specificity and nature of the interaction. The strength of interaction is also an important parameter, because it indicates whether the two proteins form a complex or interact only transiently. The strength of an interaction may also be modulated by phosphorylation or other types of modification: For example, SH2- and some WW-domain proteins only bind to phosphorylated target proteins.

Although many interactions among structural proteins are static and therefore relatively straightforward to describe, dynamic interactions are more difficult to study and visualize. For

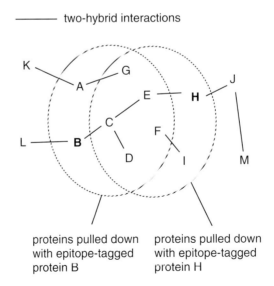

FIGURE 2. Protein complexes and networks. Mass spectrometry allows identification of proteins in a complex, but the composition may be dependent on which protein was tagged to purify the complex. Two-hybrid interactions allow reconstruction of protein networks but not "physical" complexes. Most protein-interaction diagrams ignore such contradictions.

TABLE 1. Parameters of Molecular Interactions and Complementary Sources of Biological Information

Parameter	Molecule					
	Prot.	DNA	RNA	Lipids	Carb.	Met.
Concentration	x	—	x	(x)	x	x
Localization	x	(x)	(x)	x	x	x
Covalent modifications	x	x	x	x	x	x
phosphorylation	x	—	?	x	(x)	?
acetylation	x	?	?	?	?	?
methylation	x	x	?	?	?	?
other modifications	x	?	x	x	?	?
Cleavage (degradation)	x	(x)	x	(x)	x	x
Non-covalent modification[a]	x	(x)	x	?	?	?
Logical state (ON/OFF)	x	?	?	?	?	?
Binding sites	x	x	x	(x)	(x)	(x)

For types of molecules see Fig. 1. Carb. = carbohydrates, Met. = metabolites. "x" indicates whether this paramater is relevant for the given molecule. Information about molecules and their actions can be found to various degrees in the databases listed in Table 2. Please note that many modification states or the activity of molecules is dependent on the input from other molecules, e.g., when phosphorylation activates a protein. Such actions and their conditions are usually not recorded systematically or in a standardized nomenclature, and thus are difficult to use for automated map generation. Protein-interaction maps therefore should have an option to enter some free-text annotation that can be accessed from the graphic output.

[a]Non-covalent modifications may be conformational states such as allosteric isoforms.

example, the activity of an enzyme may be regulated by several subunits whose interaction is dependent on physical or biochemical parameters such as temperature, phosphorylation, or concentration. Likewise, protein interactions may be described in terms of the actions they exert on other proteins, much like the interactions between enzymes and their substrates. Common examples are protein acetylases or proteases. Such interactions may necessarily be weak and transient to assure a high reaction rate. We have summarized parameters describing protein interactions and their conditions in Table 1. Table 2 lists databases and Web sites that have such data for large sets of proteins.

TABLE 2. Databases and Web Sites

Protein interactions

MIPS	http://www.mips.biochem.mpg.de/proj/yeast/tables/interaction/index.html, (Mewes et al. 2000)
Proteome	http://www.proteome.com/ (single interactions)
DIP	http://dip.doe-mbi.ucla.edu/
SGD (Function Junction)	http://genome-www.stanford.edu/cgi-bin/SGD/functionJunction
Myriad-Pronet	http://www.myriad-pronet.com/
Curagen	http://portal.curagen.com
BIND	http://www.binddb.org
BRITE	http://www.genome.ad.jp/brite/

Protein networks

Biocarta	http://www.Biocarta.com
Genmapp	http://gladstone-genome.ucsf.edu/introduction.asp
Kohn (1999)	http://discover.nci.nih.gov/kohnk/links.html
Signal Transduction Knowledge Environment	http://www.stke.org
Schwikowski et al.	http://depts.washington.edu/sfields/projects/YPLM/data/index.html

VISUALIZATION

Relational Visualization

Protein–protein interactions are frequently represented as a linear list of protein pairs (as in Uetz et al. 2000). In contrast, relational visualization seeks to represent entities and their relationships in a graphic form (Fig. 3). The complexity of such representations ranges from simple (Fig. 3A–C) to highly complex (Figs. 3D–H and 4). Figure 3,F and G illustrates the usefulness of graphics on a small network of protein–protein interactions in yeast.

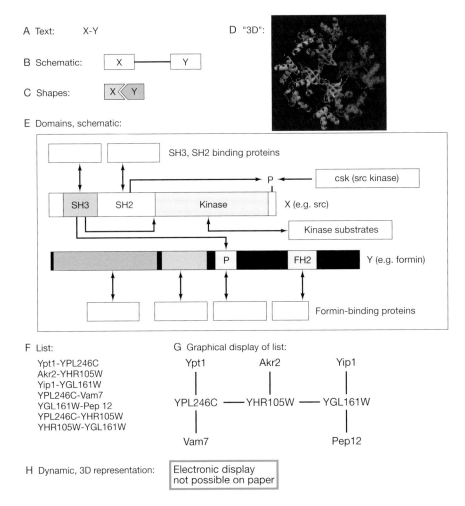

FIGURE 3. Visualization of protein interactions. For computerized display, *A* and *B* are the most common. Shapes, as in *C*, are more difficult to generate automatically by computer because topology has to be taken into account. (*D*) Computer-generated three-dimensional structure of a protein complex; such data are available only for a small set of proteins. (*E*) Scheme taking into account the domain structure of proteins and functional interactions such as phosphorylation; there are no systems available yet to generate such diagrams automatically, partly because the pertinent information is not available in databases. Boxes denote other proteins. (*F*) List of interactions as text. (*G*) Graphic representation of list in *F*. (*H*) Ultimately we want an integrated display of the aforementioned options that allows a user to look at "the big picture" of many proteins but also to zoom in to atomic detail. Such visualization tools are not available yet and will be possible only as electronic systems (i.e., not on paper). 3D structure in *D* reproduced by permission of the Protein Structure Database (http://www.rcsb.org). (Adapted, with permission, from Hargreaves et al. 1998.)

Although both representations reflect identical information, the graphic representation (frequently called layout) has fundamental advantages with respect to human perception.

1. *Localization: single versus multiple entries.* In a textual representation, the protein interactions involving a given protein X are usually spread out over different positions in the list, which requires an exhaustive search through the entire list to find all interactions involving X. In a graphic layout, X occurs exactly once.

2. *Context.* Once X has been identified in the layout, its immediate and indirect neighbors are easily identified, and their relation to X can be studied.

3. *Mental map.* A graphic representation facilitates memorization of proteins by position in a "mental map" (Eades et al. 1991). In positioning the nodes, secondary information can be employed to guide the layout; for example, proteins can be spatially grouped by localization or function. In this way, a particular arrangement of the proteins can increase the information content of the layout and facilitate its comprehension at the same time.

The magnitude of experimental data from large-scale experimental methods makes it seem impossible to visualize all protein–protein interactions in a single layout, even for relatively simple organisms such as yeast.

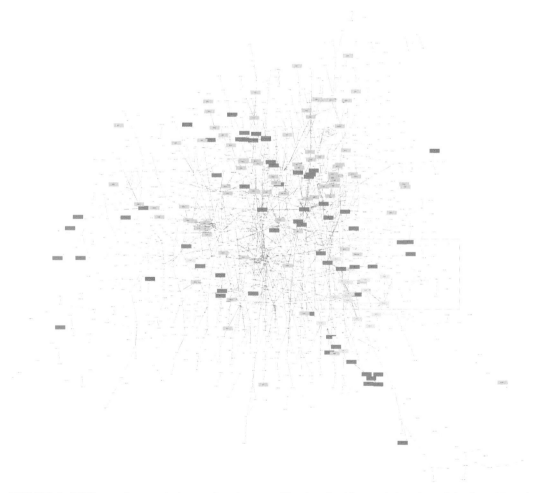

FIGURE 4. 2358 protein–protein interactions in yeast. (Reprinted, with permission, from Schwikowski et al. 2000.) Alternative wiring schemes for complex networks are shown in Kohn (1999) and Strogatz (2001).

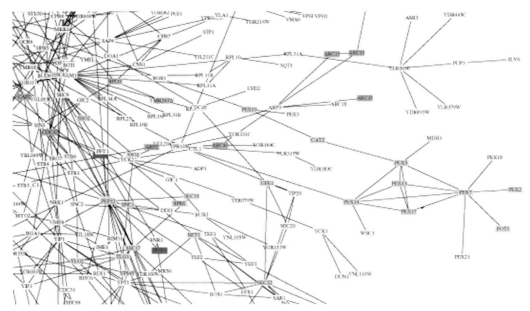

FIGURE 5. Detail of Fig. 4.

For example, Figure 4 shows a layout that contains the largest component of an experimentally determined protein–protein interaction map in yeast (as of April 2000). Specific functions (according to the YPD classification; Costanzo et al. 2000) are highlighted by color, and it becomes clear from this map that certain proteins with similar function cluster together. However, the detail in Figure 5 illustrates that, whereas proteins and their interactions appear arranged well in the peripheral regions, in central regions of the layout, the edges and name labels are drawn on top of each other, making it impossible to discern individual interactions. Finally, computer-generated interaction maps have not been designed to contain as much information as hand-drawn maps.

In summary, hand-formatted maps (such as those in Michal 1993, 1998; Kohn 1999) are usually of a higher quality, but due to the large amount of work involved to construct them, available for very limited data sets. Finally, with the greater complexity of data sets arising from more complicated genomes, even hand-formatted maps are likely to be inadequate.

Symbols and Conventions

Several authors have suggested symbols for describing protein–protein interactions. Notably, Kohn (1999) suggested some conventions for building sophisticated models of protein interactions involved in cell-cycle control and DNA repair (Figs. 6–9). Although Kohn's wiring diagrams are well worked out, they are not generated by an automated system and therefore have to be redrawn manually when larger changes need to be included. However, his symbols and conventions might also be used by a computerized system and, therefore, are reproduced here. More recently, Cook et al. (2001) suggested another system for describing complex biological systems, including protein interactions. Other projects are under way, and we refer readers to our Web site for updates (http://www.systemsbiology.org/pubs/vizprotein).

Graphs and Graph Drawing

The field of graph drawing deals with the automatic computation of maps from graphs. The abstract nature of a set of protein–protein interactions can be captured by the mathematical notion of a graph. Formally, a graph consists of nodes that represent the proteins, and edges between pairs of nodes that represent protein–protein interactions. Graphs arise in many other fields, such as sociology, project management, and software engineering, as well as in other areas of biology, such as taxonomy and biochemistry. Because of this ubiquity, there is an extensive body of graph theory that deals with mathematical properties of graphs (for an introductory text, see Bollobás 1998).

A graph is specified completely by a set of nodes and a set of node pairs as edges, but graph theory does not stipulate where its nodes and edges are to be drawn. To obtain a drawing or layout of a graph, as in Figure 3G, one needs to associate further (two- or three-dimensional) coordinates with each node and specify how the edges are drawn. Performing this task computationally is the object of graph drawing, a relatively young subfield of computer science (for an overview, see Di Battista et al. 1999). Various attempts have been made to quantify the "quality" of a two-dimensional graph layout. According to common definitions of quality, good layouts should have evenly spaced nodes: Edges should be straight lines, identical or isomorphic subgraphs should be drawn identically, and so forth. One of the most prominent criteria, the number of edge intersections in a graph drawing, has been correlated empirically with human ability to solve simple problems using that drawing (Purchase 1997).

Planar graphs are optimal in this respect; i.e., they are graphs that can, in some way, be drawn in two dimensions without any edge crossings. Planar graphs are important in applications such as the layout of electronic circuits, where different conducting paths are not allowed to cross each other. Although planar graphs usually permit many layouts without edge crossings, even efficiently testing whether a given graph is planar is not straightforward (Hopcroft and Tarjan 1974). However, most graphs, such as those that represent protein–protein interactions, are not planar.

Spring Embedder Algorithm

The most widely used algorithm for general larger two- and three-dimensional graphs is the spring embedder algorithm (Eades 1984). The layout of a graph is computed by modeling a mechanical system in which the edges of the graph correspond to springs, and nodes correspond

Symbol	Meaning
(A) ←•→ (B)	Non-covalent binding e.g. between proteins. A node represents the A-B complex itself.
(A) ←→ (B)	Asymmetric binding where protein A contributes a peptide that binds to a receptor site or pocket on protein B.
x — *z* — *y*	Z is the combination of states defined by x and y.
(A) ←•*x*→ (B) ↓ *y* (C)	Multimolecular complex: x is A-B; y is (A-B)-C.
(A) ←•→•	Formation of a homodimer. Filled circle on the right represents another copy of A. The filled circle on the binding line represents the homodimer A-A.

FIGURE 6. Kohn's symbols for describing protein–protein interactions. Please note that Kohn also suggested additional symbols for the stoichiometric conversion of A into B, degradation products, transport, etc. A complete list can be found in Kohn (1999) or on-line at http://discover.nci.nih.gov/kohnk/symbols.html. Additional symbols and conventions have been proposed by other authors such as Cook et al. (2001). (Adapted, with permission, from Kohn 1999, © by the American Society for Cell Biology.)

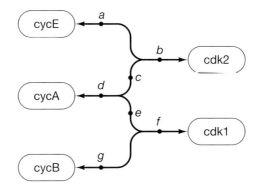

FIGURE 7. Kohn's representation of alternative binding modes. Example: heterodimers formed by Cyclins E, A, and B binding to Cdk1 or 2. Note that *a*, *c*, *e*, and *g* each lie on a unique connector line, and each represents a unique heterodimer, namely (*a*) CycE:Cdk2, (*c*) CycA:Cdk2, (*e*) CycA:Cdk1, (*g*) cycB:Cdk1. Nodes *b*, *d*, and *f*, on the other hand, represent dimer combinations, namely (*b*) Cdk2 complexed with either CycE or CycA; (*d*) CycA complexed with either Cdk2 or Cdk1; (*f*) Cdk1 complexed with either CycA or CycB. This notation simplifies the representation of multiple alternative interactions: for example, the interactions of p21, p27, or p57 with various Cyclin:Cdk dimers. A formal rule, required to avoid ambiguity, is that lines representing alternative interactions must join at an acute angle. (Reprinted, with permission, from Kohn 1999, © by the American Society for Cell Biology.)

to rings. The springs create an attracting force between the rings when they are far apart, and a repulsive force between close rings. One searches for a placement of rings that minimizes the total energy present in the system, commonly by simulating the behavior of the mechanical system over a certain period of time. Figure 4 was created using a spring embedder algorithm.

Limitations for Large Graphs

Working with layouts for very large graphs of 100 or more nodes presents certain technical limitations. First, the computer time required to execute most practical layout algorithms does not

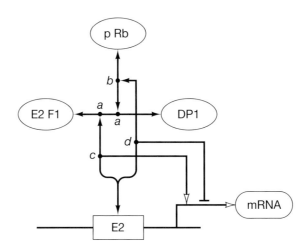

FIGURE 8. Kohn's representation of multimolecular complexes and integration of information on transcriptional regulation: stimulatory and inhibitory complexes of E2F1, DP1, and pRb. Note that the promoter element E2 can be occupied either by E2F1:DP1 or by E2F1:DP1: pRb (alternative binding represented by interaction lines joined at an acute angle). Individual complexes are (*a*) E2F1:DP1 dimer; (*b*) E2F1:DP1:pRb trimer; (*c*) E2F1:DP1 bound to promoter element E2 (transcriptional activation shown); (*d*) E2F1:DP1:pRb bound to E2 (transcriptional inhibition shown). (Reprinted, with permission, from Kohn 1999, © by the American Society for Cell Biology.)

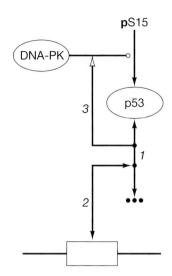

FIGURE 9. Kohn's representation of homopolymers: formation and effects of p53 homotetramer. Filled circles represent additional copies of p53. (*1*) The three additional copies of p53 monomer required to make up the tetramer are represented by three nodes placed side by side and linked to the identified p53 monomer; a node placed internally on this line represents the homotetramer itself. (*2*) p53 tetramer can bind to promoter element. (*3*) Tetramerization stimulates (or is required for) phosphorylation of p53 Ser-15 by DNA PK. (Reprinted, with permission, from Kohn 1999, © by the American Society for Cell Biology.)

scale linearly, but rather with the square of the graph size (at least). Many layout objectives even translate into NP-hard problems. For this class of problems, there probably exist no polynomial-time algorithms at all, which means that, for large problem instances, extraordinary amounts of computation are unavoidable (Garey et al. 1979, 1983) and take even more time to achieve. Second, in an interactive system with thousands of nodes or more, just drawing a dynamically changing graph, even though this is a linear operation, can take an unacceptably long time. Although faster computers may eventually mitigate the above restrictions, even the best drawings of large graphs under any of the above quality criteria may not be aesthetically pleasing or practically usable.

Extension to Three Dimensions

The increase in available computing power and the advancement of graphic displays and software standards have inspired work on three-dimensional graph layout. For graphs in three dimensions, the edge-crossing criterion is no longer helpful in selection of good drawings. Every graph can be drawn in three dimensions without edge crossings in many ways (Fary 1948). For displaying protein interactions in three dimensions, a variation of a spring embedder algorithm has been suggested (Basalaj and Eilbeck 1999).

Techniques for Visualization

There are several techniques to relieve the above-described problems with the visualization of large graphs.

Zoom and Pan

A common approach is "zoom and pan." This is the same technique that is used by Web browsers. Instead of showing a Web page from beginning to end, only part of it is shown at every given moment, and the user can continuously scroll through the Web page by means of a scroll bar.

Focus and Context Techniques

Zoom and pan has the disadvantage that zooming makes certain regions of the layout invisible: One creates a "focus," but the "context" is lost. Focus and context techniques avoid loss of context by compressing a layout toward the edges of the window, instead of hiding part of it. One example of a such a technique is the well-known fisheye effect. Note that focus and context techniques are complementary to zoom and pan—they can be used together.

Collapsing Protein Classes

A third, complementary, technique to simplify a layout collapses groups of proteins (classes) into single nodes. Figure 10 was generated from 2709 protein–protein interactions in yeast (Schwikowski et al. 2000) on the basis of the functional classification of the involved proteins according to the YPD (Costanzo et al. 2000). Each node represents a functional class. Proteins that have been assigned multiple functions thus contribute to multiple classes. The aggregated information is summarized in several ways: The number on each edge *A-B* indicates the number of protein interactions between proteins of function *A* and proteins of function *B*. This number is also reflected in the thickness of the edge. The numbers in parentheses indicate the number of intraclass interactions and the number of proteins in the class, respectively.

Available Tools for Visualization

Table 3 lists some currently available software packages that visualize arbitrary protein interactions or can be customized for that task. The databases listed in Table 4 visualize predefined, limited sets of protein interactions.

INTEGRATING PROTEIN-INTERACTION NETWORKS WITH SUPPLEMENTAL DATA

Simultaneous Display of Complementary Data Types

Proteins are not the only molecules of interest to biologists, nor do protein-interaction networks function in isolation to govern cellular processes or to influence phenotypes. On the contrary, suc-

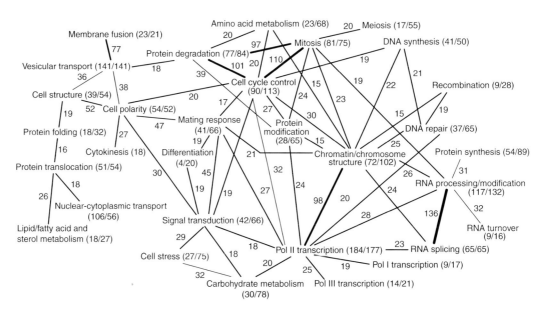

FIGURE 10. Protein classes by functional classification. For details, see above, Collapsing Protein Classes. (Modified, with permission, from Schwikowski et al. 2000.)

TABLE 3. Visualization Tools for General Networks

The LEDA library/GraphWin (http://www.mpi-sb.mpg.de/LEDA)
C++ library for efficient data structures and algorithms; contains graph drawing demo application.

Platforms: Linux-PC, Sun, Silicon Graphics, HP, Windows 95/NT (commercial)
Availability: Commercial, free license for academic users
Reference: Mehlhorn and Näher (1999)

Y-Files (http://www-pr.informatik.uni-tuebingen.de/yfiles)
Extensible, programmable graph editor, with graph algorithms. Extensions to generation of biochemical pathway diagrams are under way.

Platforms: PC, Macintosh
Availability: Free binaries/source code for academic purposes
Reference: Himsolt (1997)

Graphlet (http://www.infosun.fmi.uni-passau.de/Graphlet/)
Extensible, programmable graph editor, with graph algorithms. Extensions to generation of biochemical pathway diagrams are under way.

Platforms: Windows NT/98 or higher, Solaris, Linux
Availability: Free binaries/source code for academic purposes
Reference: Himsolt (1997)

XGvis (http://www.research.att.com/areas/stat/xgobi/index.html)
Interactive visualization system for proximity data, graphs, and networks.

Platforms: Linux, Solaris, other UNIX systems
Availability: Free, incl. source code
Reference: Buja et al. (1998)

Tom Sawyer Software (http://www.tomsawyer.com/)
Cross-platform software library and tools for drawing general graphs.

Platform: Macintosh, many UNIX versions, standard WWW browsers
Platforms: Most major operating systems, incl. Windows, Apple Macintosh
Availability: Commercial

CUtenet (http://genome6.cpmc.columbia.edu/~tkoike/cutenet/)
Interactive graphic editor for signal-transduction pathways and protein interactions.

Platform: Most major operating systems (Java application)
Availability: ?
Reference: Koike and Rzhetsky (2000)

cessful visualization of protein-interaction networks leads almost immediately to questions such as "How do these proteins interact with DNA, substrates, and other cellular components? What impact does protein interaction have on the properties and behaviors of each interacting protein?"

In fact, it is not hard to envision incorporating a number of supplemental data types into the basic network display. Sources of supplemental data generally fall into one of four categories: new types of interactions, new types of molecules, new information on existing interactions, and new information on existing molecules (see also Table 1). For instance, we may wish to visualize not only interactions between pairs of proteins, but also interactions between these proteins and ligands or other small molecules. Similarly, proteins that function as transcription factors may form complexes that bind particular DNA sequences, as well as interact with a variety of protein cofactors, and annotation or organization taking these properties into account may be useful. Moreover, it may be useful to place the protein-interaction network in the context of indirect evidence such as genetic interactions, or ultimately to annotate each protein with its instantaneous expression pattern or each interaction with its relative strength of binding as these points become known.

To date, numerous articles and textbooks have included figures displaying different types of molecules and interactions between them. However, these figures usually invoke a limited num-

TABLE 4. Visualization of Specific Data Sets

DIP (http://dip.doe-mbi.ucla.edu)
Visualization of protein–protein interactions in DIP database. Static images depict neighborhoods 2 and 3 steps away from center protein.
Platforms: Web browser
Availability: Free
Reference: Xenarios et al. (2001)

ProNet (http://pronet.doubletwist.com/)
Interactive visualization of protein–protein interactions in the ProNet database.
Platforms: Standard WWW browser
Availability: Free

GeneNet (http://wwwmgs.bionet.nsc.ru/systems/mgl/genenet/)
Interactive visualization of protein interactions in GeneNet database, based on a number of predefined diagrams. GeneNet includes entries for DNA, RNA, protein, and cellular interactions.
Platforms: Standard Web browser
Availability: Free
Reference: Kolpakov et al. (1998)

PIMRider (http://pim.hybrigenics.com)
Interactive visualization of protein–protein interactions with different viewers.
Platform: Standard Web browser
Availability: Commercial, free license for academic users
Reference: Rain et al. (2001)

BindDB (http://www.binddb.org)
Interactive visualization of protein interactions in BindDB database. BindDB contains general biomolecular interactions.
Platform: Standard Web browser
Availability: Free
Reference: Bader et al. (2001)

---

ber of components to describe an isolated biochemical process or signaling pathway, are carefully tailored to illustrate a predetermined concept, and rely heavily on accompanying textual descriptions (Pirson et al. 2000). In contrast, there is a pressing need for visual representations that can systematically present and organize the extremely large amounts of protein-interaction and expression data rapidly accumulating in the wake of two-hybrid screens, DNA microarray technology, and high-throughput proteomics. Such displays are not hand-tailored to illustrate a foregone conclusion, but should ideally stimulate the discovery of new protein functions and biological relationships. As the raw data become increasingly complex with each type of supplemental information, tools that are both visual and interactive become increasingly important for emphasizing and extracting the key features.

An Example: Integrated Networks to Study Galactose Metabolism

Here we illustrate one method to create and display an integrated interaction network systematically, as described by Ideker et al. (2001). Suppose that we are interested in viewing the molecular interactions that govern a particular cellular process—that of galactose utilization in yeast. Our specific goals might be to assess the impact of these interactions on expression of the galactose-utilization (*GAL*) genes and to understand how the process of galactose utilization interacts with other metabolic processes in yeast.

To begin, we construct a database representing all known protein–protein and protein–DNA interactions in the yeast *Saccharomyces cerevisiae*. Protein interactions may be drawn from any of the sources listed in Tables 2 and 4: In this example, we utilize the 2709 protein–protein interactions compiled by Schwikowski et al. (2000). Similarly, we obtain all of the 317 protein–DNA interactions present in either of two publicly accessible on-line databases (as of July 2000): TRANSFAC (Wingender et al. 2000) or the *Saccharomyces cerevisiae* Promoter Database (SCPD) (Zhu and Zhang 1999). These sources link known transcription factors (proteins) to the genes they regulate (DNA).

Next, we use a program based on GraphWin (Mehlhorn and Näher 1999) to display these physical interactions as a graph structure or network, as discussed above. Because several types of interactions are now involved, but all of them reflect physical binding events, we refer to this network as a physical-interaction network. As shown in Figures 11–13, each node represents a gene and is labeled with its corresponding gene name. An arrow, or directed edge, from one node to another signifies that the protein encoded by the first gene can influence the transcription of the second by DNA binding (a protein→DNA interaction), whereas a line, or undirected edge, between two nodes signifies that the proteins encoded by each gene can physically interact (a protein–protein interaction). Network layout is performed using the spring embedder algorithm included with GraphWin, so that proteins with related functions or that are involved in the same molecular pathway often end up in the same region of the display. Thus, the region shown in Figure 11 corresponds to the process of galactose utilization, whereas the regions shown in Figures 12 and 13 correspond to amino acid biosynthesis and glycogen synthesis, respectively.

According to the network, Gal4p is a transcription factor that binds to the promoters of many other GAL genes, thereby regulating their transcription through protein–DNA interactions (Fig. 11). The network also shows clearly that the activity of Gal4p may be influenced by protein–protein interactions with Gal80p and Gal11p. Note that the network specifies only that a particular protein–DNA interaction takes place: It does not dictate whether the interaction activates or represses transcription, whether the effect on transcription is rapid or gradual, or, in the case that multiple interactions affect a gene, how these interactions should be combined to produce an overall level of transcription. Because these levels of information are not encoded in the protein–DNA databases, they are also absent from the network display. Similarly, the protein–protein databases do not specify whether the Gal80p–Gal4p protein interaction, as shown in the figure, results in these proteins forming a functional complex, or whether one protein modifies another. All of this information is known outside of the databases: Classic genetic and biochemical experiments (Johnston and Carlson 1992; Lohr et al. 1995) have determined that Gal4p is a strong transcriptional activator, and that Gal80p can bind to Gal4p to repress this function.

Superimposition of mRNA- and Protein-expression Changes on the Network

To understand better how the physical-interaction network regulates genes, it can be extremely effective to augment the network with information about gene expression. As described previously (Ideker et al. 2001), we [T. Ideker] measured global changes in gene expression over 20 genetic and environmental perturbations to the GAL pathway. Wild-type (wt) and 9 genetically altered yeast strains, each with a complete deletion of a different *GAL* gene (*gal1Δ, 2Δ,3Δ, 4Δ, 5Δ, 6Δ, 7Δ, 10Δ,* or *80Δ*) were examined. These 10 strains were perturbed environmentally by growth to steady state in the presence (+gal) or absence (–gal) of 2% galactose. Because many of the deletion strains cannot grow in galactose, 2% raffinose was also provided in both media as an alternate supply of sugar. In each of these 20 perturbation conditions, we monitored changes in mRNA expression over the approximately 6200 nuclear yeast genes using a whole-yeast-genome microarray.

For any particular perturbation, we can integrate, i.e., graphically superimpose, the resulting changes in mRNA expression on the network. Although a number of visual representations are possible, an obvious choice is to use node color to represent a change in expression of the corresponding gene. For example, Figure 11 shows the expression changes resulting from the perturbation *gal4Δ*+gal, a deletion of the *GAL4* gene in the presence of galactose.

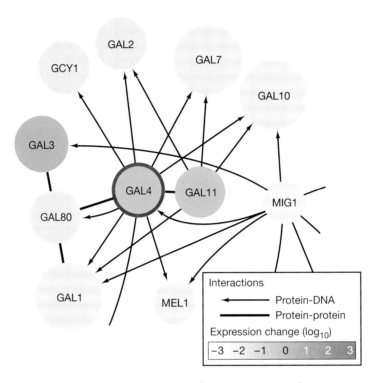

FIGURE 11. Sample region of an integrated physical-interaction network containing genes involved in galactose metabolism. Each node represents a gene, an arrow directed from one node to another represents a protein–DNA interaction, and an undirected line between nodes represents a protein–protein interaction. The gray-scale intensity of each node indicates the change in mRNA expression of its corresponding gene, with medium gray representing no change and darker or lighter spots representing an increase or decrease in expression, respectively. (To draw attention to genes with large expression changes, node diameter also scales with the magnitude of change.) To signify that the expression level of *GAL4* has been perturbed by external means, it is highlighted with a red border. According to the interactions in the network, Gal4 regulates expression of many other GAL genes through protein–DNA interactions, whereas Gal4's activity is impacted by protein–protein interactions with Gal80p and Gal11p. (See text for details.)

When protein-expression data are available, they too can be superimposed on the network display. For example, we measured changes in both mRNA and protein levels in wild-type cells grown in the presence or absence of galactose (Ideker et al. 2001). Using a procedure based on isotope-coded affinity tags (ICAT) and tandem mass spectrometry (Gygi et al. 1999), we detected a total of 289 proteins and quantified their expression-level changes between these two conditions. Figure 12 illustrates the addition of this information to the visual display, focusing on the region of the network corresponding to amino acid biosynthesis. By comparing the mRNA- and protein-expression responses displayed on each node, one can visually assess whether the mRNA and protein data are correlated and quickly spot genes for which they are remarkably discordant.

Integrated Networks Enable New Biological Insights

In the previous example, we have integrated at least four types of data into the same graphic display: protein–protein interactions, protein–DNA interactions, mRNA-expression changes, and protein-expression changes. In addition, because interconnected groups of genes tend to have related functions, the display also confers information about cellular function or process. What exactly is gained from this level of integration? Is superimposing data from multiple, complex sources on the same graphic display really worth all of the added clutter?

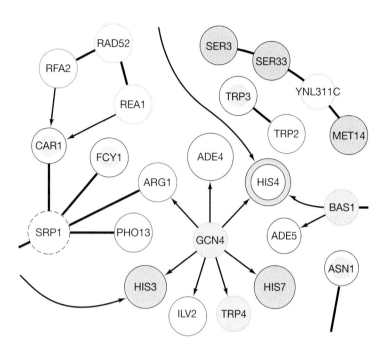

FIGURE 12. Integration of protein-expression response in the region corresponding to amino acid biosynthesis. Nodes and interactions appear as in Fig. 11, with a solid gray-scale intensity representing the change in mRNA expression. Nodes for which protein data are also available contain two distinct regions: an outer circle, or ring, representing the change in mRNA expression, and an inner circle representing the change in protein expression. Both mRNA and protein intensity scales are identical to those used in Fig. 11.

Generally, the integrated network display is useful because it provides a lucid means of summarizing existing biological knowledge about molecular behavior. Although individual researchers may amass a great deal of knowledge about the molecular interactions underlying one particular pathway, no single biologist can be familiar with the extremely complex and large number of interactions in an entire cell. A computer database, however, tracks all of these, provided the proper representation is available to allow a biologist to access, display, and interpret the information. Moreover, because changes to the database are automatically reflected in the graphic display, the integrated network is continually up to date. In short, the physical-interaction databases and the graphic display together constitute an expert system, providing knowledge about the molecular makeup of the cell that can be queried and viewed by a biologist.

In our case, study of galactose utilization, the network display engenders at least three types of biological insights. First, it provides plausible cause-and-effect explanations for numerous changes in gene expression observed in response to each of the 20 perturbations to the GAL pathway. Note that when *GAL4* is deleted in Figure 11, expression levels of *GAL1, 7, 10*, and several other genes decrease dramatically, consistent with *GAL4*'s known role as a transcriptional activator. Similarly, in Figure 12 we see that the increase in expression of *HIS3, HIS4, HIS7, ADE4, ARG1*, and *ILV2* could be controlled by the *GCN4* transcription factor. Interestingly, *GCN4* itself does not change perceptibly in mRNA expression. However, because 7 of the 8 genes it regulates *do* change, we think intuitively that *GCN4* is somehow involved. A subsequent literature search on *GCN4* reveals that this gene is in fact regulated translationally, not transcriptionally (McCarthy 1998), a detail not represented by the network display because we have not yet measured protein-expression changes for *GCN4*. Thus, through a rapid visual scan, we can determine which gene-expression changes could be caused by known protein–DNA interactions, and which changes

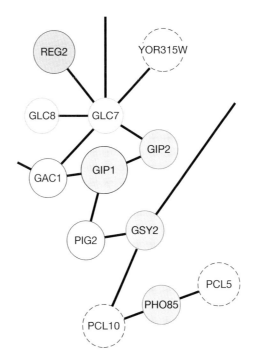

FIGURE 13. Coregulation and inverse regulation of interacting proteins. The integrated physical-interaction network is shown as in Fig. 11, for the region corresponding to glycogen synthesis. Changes in mRNA expression are due to deletion of the *GAL4* gene (a *gal4Δ* versus wild-type strain, galactose present). In this and many other perturbation conditions, the Gac1-Gip1-Pig2-Gsy2 enzyme complex increases in mRNA expression, whereas Pcl10 (which functions to inactivate Gsy2) shows a corresponding expression decrease.

require further research to identify the particular transcription factors involved. Going a step further, one can then seek explanations for how each transcription factor is itself controlled, either by protein–DNA interactions with still other transcription factors, or through protein–protein interactions with cofactors or signaling proteins.

Second, the network graph highlights groups of physically interacting proteins that display joint increases or decreases in expression level across many experimental conditions. These coordinate changes suggest that the proteins are controlled by one or more common transcription factors. For instance, the genes *GAC1*, *GIP1*, *PIG2*, and *GSY2*, shown in Figure 13, are not only involved in protein–protein interactions with each other, but also display concomitant increases in expression. Moreover, the expression levels of these genes are highly correlated over the 20 perturbation conditions. Examples of inverse regulation are also abundant among physically interacting proteins, where often one protein is known to inactivate the other. For example, we observe an increase in expression of Gsy2p, a glycogen synthase, and a corresponding decrease in expression of Pcl10p, a protein that interacts with and inactivates Gsy2p (Wilson et al. 1999). Thus, the integrated network suggests that glycogen synthesis is controlled by up-regulating an enzyme *and* down-regulating the enzyme's inhibitor.

Finally, the network graph may be used to confirm newly discovered or controversial interactions. Consistent coregulation or inverse regulation between two physically interacting proteins, over many perturbation conditions, provides strong evidence that the interaction occurs in vivo and is not an artifact of the particular assay originally used to determine the interaction. This confirmation is especially useful given that some experimental techniques for establishing a physical interaction, such as the two-hybrid screen, may return a substantial number of false-positive interactions.

Choice of Graphic Representation

Of course, it is not necessary to implement the same structural or color conventions used in the example. Nonetheless, due to the overwhelming amount of information to be loaded onto a single visual display, a clear, efficient, and consistent graphic representation remains extremely important. In the words of Edward Tufte (1983): "Graphical excellence is that which gives to the viewer the greatest number of ideas in the shortest time with the least ink in the smallest space."

In constructing our integrated interaction graphs, we have relied on a few short, common-sense guidelines:

1. High visual density is to be desired, not avoided. Often, clutter and confusion are failures of design, not complexity (Tufte 1983). Many different attributes can be varied on the same graph, each conveying a different type of information: node size and color; node border width and color; node-label font, font size, and font color; edge directionality, width, and color. We have varied only a small subset of these attributes in the visual displays of the previous example.

2. When including many different types of data in the same display, use the smallest number of colors required to represent each type. The key to managing visual complexity is to make each type of information use the fewest graphic resources possible. Because the eye is *particularly drawn to changes in color*, color should be used judiciously to emphasize only the most important features of the display. For example, although changes in gene expression are often displayed with two color scales (e.g., red for increases in expression and green for decreases in expression), these data are fundamentally one-dimensional; if necessary, they can be displayed with a single color or gray scale (e.g., Fig. 11), freeing up a larger range of colors to encode other types of data. Moreover, contrasts of red and green are particularly ill suited to convey information to the ~8% of men who are fully or partially red–green color-blind (Passarge 1995). In short, careful use of color is the key to representing highly dimensional data sets.

3. Display new types of data only if there is a clear biological goal that relies on these data. The choice of which information to display, out of all information accessible from the databases, must always be driven by biological inquiry. For instance, if the goal is to understand which protein–DNA interactions can cause a particular gene-expression pattern, information such as the amino acid sequences of each protein or the genomic location of the corresponding gene would be of low interest. Alternatively, each of these data types may be assigned to a distinct visual layer, which can be temporarily hidden when the data in the layer are not directly relevant.

From Visual Representation to a Predictive Model: Early Steps toward Gene-expression Modeling Using Physical-interaction Networks

Beyond their use as graphic displays, physical-interaction networks can function as predictive models of the cell. For instance, with a few added formalisms that we discuss below, the physical-interaction network developed in our example can predict all of the changes in gene expression that could be caused by a particular perturbation. Such predictions are highly informative when compared to their true, observed values measured in laboratory experiments: For any gene-expression level that changes in experiment but not in simulation, we may conclude that one or several physical interactions are unknown and/or absent from the network model.

Predictions are obtained by running simulations. To see how, recall that the network represents two types of interactions, protein–DNA and protein–protein. These interactions can produce very different effects: A protein–DNA interaction can affect the expression level of a gene, whereas a protein–protein interaction can cause a protein to become active or inactive with regard to its biological function. However, a protein–protein interaction by itself cannot elicit a change in gene expression (without the aid of an associated protein–DNA interaction), just as a pro-

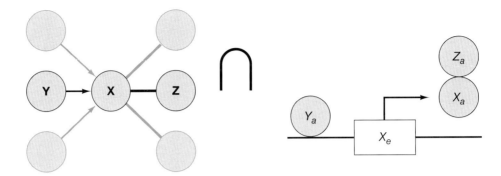

X_e *may depend on* Y_a *for all protein -DNA interactions* Y ♦ X

X_a *may depend on* X_e *and* Z_a *for all protein -protein interactions* Z → X

FIGURE 14. Rules governing the effects of physical interactions on gene expression and protein activity. Each node X in the physical-interaction network (*left*) has an associated expression level X_e and activity level X_a (*right*). These levels may change in response to (i.e., may depend on) changes at adjacent nodes Y or Z according to rules highlighted in the box. Rules are qualified with the word "may" because the network does not specify in what conditions the interactions occur or how multiple interactions should be combined to produce an overall expression or activity level.

tein–DNA interaction usually affects protein activity only indirectly, by influencing whether the protein is expressed.

These interaction types imply at least two types of information associated with each node X: a gene expression level X_e and a protein activity X_a. A perturbation to the network may elicit a change in X_e through an incoming protein–DNA interaction from a node Y (Y→X), if Y also undergoes a change in activity Y_a. In contrast, X_a may change either if the perturbation causes a change in X_e, or if X is involved in a protein–protein interaction with a node Z, which undergoes a corresponding change in activity Z_a. Figure 14 summarizes these rules.

Along with the network model, these rules are sufficient to predict possible changes in gene expression resulting from perturbation of any particular node in the network. It then becomes straightforward to perform these simulations automatically, by implementing these rules directly in software. Because the network model does not specify precisely how a node combines the relevant input interactions to determine its expression level or protein activity, it is not possible to state definitively whether a change actually occurs: We only know if a change may occur. However, it is possible to predict which nodes are not, under any circumstances, affected by a particular perturbation to the network.

Prediction of Expression Changes Resulting from Particular Perturbations to the Galactose-utilization Network

As an example, consider Figure 15A, which once again displays the region of the physical-interaction network corresponding to galactose use. In this case, the network has been perturbed by deletion of the *GAL3* gene in the presence of galactose. The resulting changes in gene expression, as observed by microarray experiment, are superimposed on this graph. To perform the corresponding simulation, we reason that deletion of *GAL3* is likely to affect $GAL3_e$ and $GAL3_a$. In turn, a change in $GAL3_a$ may affect $GAL80_a$, which then may affect $GAL1_a$ or $GAL4_a$, as mediated through protein–protein interactions. Although a change in $GAL1_a$ has no further impacts on the network, a change in $GAL4_a$ can affect $GAL7_e$, $GAL10_e$, and many other expression levels through

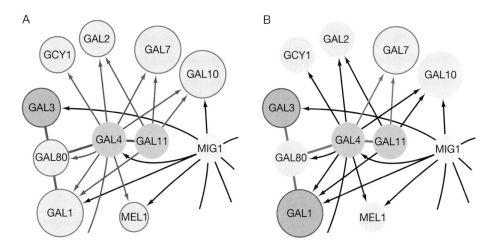

FIGURE 15. Predicting changes in gene expression in the region of the integrated network corresponding to galactose metabolism. The gray-scale intensity of each node reflects the experimentally observed change in mRNA expression for a *gal3Δ* versus wild-type strain in galactose. (*a*) Forward simulation, starting from the perturbed gene *GAL3* (highlighted in *red*). Red edges denote interactions that may transmit a change, either in expression or activity, from one node to another (according to the rules described in Fig. 14). Nodes highlighted in blue denote genes whose expression levels may change as a result. Experimentally observed expression changes in these blue genes are consistent with the simulation. (*b*) Reverse simulation, tracing backward from *GAL7* (highlighted in *blue*), whose expression state has changed in response to perturbation of the network. Here, blue edges denote interactions that may transmit a change, leading to nodes that are highlighted in red if their corresponding genes were observed to change significantly in expression. These red nodes are possible "causes" of the change observed at *GAL7*.

protein–DNA interactions. Thus, experimentally observed expression changes in *GAL7* and *GAL10* are consistent with those predicted by the network.

For observed changes in gene expression that are not predicted by the network model, a legal path between the perturbed and affected gene does not exist. However, it is possible that a segment of this path is present in the network, offering at least a partial explanation for the observed change. One approach to finding this partial path is to start at an affected gene and work backward toward the perturbed gene. Figure 15B gives an example of this type of simulation. Here, we start at *GAL7*, which exhibits a clear decrease in gene expression under this perturbation condition. Working backward, we see that a change in $GAL7_e$ could be explained by incoming protein–DNA interactions implicating $GAL4_a$ or $GAL11_a$. $GAL4_a$ could likewise be affected by $GAL80_a$, and $GAL80_a$ affected by $GAL1_a$ or $GAL3_a$. Since dramatic changes in gene expression were observed for *GAL1* and *GAL3*, they become possible "causes" of the change observed at *GAL7*.

Although our simulation implicates *GAL3* as a possible cause of the expression change at *GAL7*, the perturbed node may not always be reachable through a backward path. However, in performing the simulation, we hope to identify upstream nodes that are one or several steps closer to it.

FUTURE DIRECTIONS

In the future, the available methods for data integration and network visualization should be extended in a number of important directions. First, there is a need for more complex integration schemes than have been presented here. For instance, although small molecules such as metabolites, drugs, or hormones are known to directly influence the expression of many genes and proteins, they do not appear in the network graph. One could represent these compounds as nodes in the graph and define a new type of physical interaction to represent the enzymatic transformation of one metabolite to another.

Second, we also need better algorithms for automated layout. Although the spring embedder and similar algorithms draw the graph so that strongly connected subsets of nodes are grouped in two dimensions, an improved layout algorithm would not only group interacting proteins, but also could attempt to explicitly group nodes that are of similar biological function or subcellular localization, or that have similar gene-expression responses to perturbation.

Finally, it would also be extremely useful to display a much wider range of information about existing nodes and interactions. For example, the improved display might supply information about the structural or functional implications of a protein–protein interaction. Are the interacting proteins subunits of a larger complex, or does the interaction instead result in covalent modification of one of the proteins? Alternatively, when using the network graph as a guide to explain experimental observations, one might like to know how much confidence to place in each interaction. For example, was the interaction predicted computationally or determined experimentally, and is it supported by corroborating evidence? We anticipate that these types of data will become increasingly available as annotation of the public databases becomes more systematic and complete. The challenge will then be to integrate the new data in such a way as to increase our understanding of the underlying biological processes, not obscure them in convoluted figures or excessive detail. Ultimately, these added layers of information will make the network even more powerful as a model on which simulations may be performed to predict experimental outcomes.

REFERENCES

Artavanis-Tsakonas S., Rand M.D., and Lake R.J. 1999. Notch signaling: Cell fate control and signal integration in development. *Science* **284:** 770–776.

Bader G.D., Donaldson I., Woting,C., Ouellette B.F.F., Pawson T., and Hogue C.W.V. 2001. BIND—The biomolecular interaction network database. *Nucleic Acids Res.* **29:** 242–245.

Basalaj W. and Eilbeck K.. 1999. Straight-line drawings of protein interactions. In *Graph Drawing 1999* (ed. J. Kratochvil), pp. 259–266. Springer-Verlag, Heidelberg.

Bollobás B. 1998. *Modern graph theory.* Springer, New York.

Buja A., Swayne D.F., Littman M., and Dean N. 1998. XGvis: Interactive data visualization with multidimensional scaling. Available at http://www.research.att.com/ ~andreas/xgobi/

Cook D.L., Farley J.F., and Tapscott S.J. 2001. A basis for a visual language for describing, archiving and analyzing functional models of complex biological systems. *Genome Biol.* **2:** RESEARCH0012. (http://genomebiology.com/2001/2/4/research/0012)

Costanzo M.C., Hogan J.D., Cusick M.E., Davis B.P., Fancher A.M., Hodges P.E., Kondu P., Lengieza C., Lew-Smith J.E., Lingner C., Roberg-Perez K.J., Tillberg M., Brooks J.E., and Garrels J.I. 2000. The yeast proteome database (YPD) and *Caenorhabditis elegans* proteome database (WormPD): Comprehensive resources for the organization and comparison of model organism protein information. *Nucleic Acids Res.* **28:** 73–76.

Di Battista G., Eades P., Tamassia R., and Tollis I.G. 1999. *Graph drawing—Algorithms for the visualization of graphs.* Prentice Hall, Upper Saddle River, New Jersey.

Eades P. 1984. A heuristic for graph drawing. *Congressus Numerantium* **42:** 149–160.

Eades P., Lai W., Misue K., and Sugiyama K. 1991. Preserving the mental map of a diagram. In *First International Conference on Computational Graphics and Visualization Techniques*, pp. 34–43. Elsevier Science Publishers, Amsterdam.

Edwards P.A., Tabor D., Kast H.R., and Venkateswaran, A. 2000. Regulation of gene expression by SREBP and SCAP. *Biochim Biophys Acta.* **15:** 103–113.

Eilbeck K., Brass A., Paton N., and Hodgman, C. 1999. INTERACT: An object oriented protein-protein interaction database. In *The Proceedings of Seventh International Conference on Intelligent Systems for*

Molecular Biology, pp. 87–94. Heidelberg, Germany.

Fary I. 1948. On straight lines representation of planar graphs. *Acta Sci. Math. Szeged* **11:** 229–233.

Garey M.R. and Johnson D.S. 1979. *Computers and intractability: A guide to the theory of NP-completeness*. W.H. Freeman, New York.

———. 1983. Crossing number is NP-complete. SIAM *J. Algebraic Discrete Methods* **4:** 312–316.

Gygi S.P., Rist B., Gerber S.A., Turecek F., Gelb M.H., and Aebersold R. 1999. Quantitative analysis of complex protein mixtures using isotope-coded affinity tags. *Nat. Biotechnol.* **17:** 994–999.

Hargreaves D., Rice D.W., Sedelnikova S.E., Artymiuk P.J., Lloyd R.G., and Rafferty J.B. 1998. Crystal structure of *E. coli* RuvA with bound DNA Holliday junction at 6 Å resolution. *Nat. Struct. Biol.* **5:** 441–446.

Himsolt M. 1997. The Graphlet System (system demonstration). In *Symposium on Graph Drawing GD '96 Proceedings* (ed. S. North), pp. 233–240. Springer, Berkeley.

Hopcroft J. and Tarjan R.E. 1974. Efficient planarity testing. *J. ACM* **21:** 549–568.

Ideker T., Thorsson V., Ranish J.A., Christmas R., Buhler J., Eng J.K., Bumgarner R., Goodlett D.R., Aebersold R., and Hood, L. 2001. Integrated genomic and proteomic analysis of a systematically-perturbed metabolic network. *Science* **292:** 929–934.

Johnston M. and Carlson M. 1992. Regulation of carbon and phosphate utilization. In *The molecular and cullular biology the yeast* Saccharomyces (ed. E. Jones et al.), pp. 193–281. Cold Spring Harbor Laboratory Press, Cold Spring Harbor, New York.

Kolpakov F.A., Ananko E.A., Kolesov G.B., and Kolchanov N.A. 1998. GeneNet: A gene network database and its automated visualization. *Bioinformatics* **14:** 529–537.

Kohn K.W. 1999. Molecular interaction map of the mammalian cell cycle control and DNA repair systems. *Mol. Biol. Cell* **10:** 2703–2734.

Koike T. and Rzhetsky A. 2000. A graphic editor for analyzing signal-transduction pathways. *Gene* **259:** 235–244.

Küffner R., Zimmer R., and Lengauer T. 2000. Pathway analysis in metabolic databases via differential metabolic display (DMD). *Bioinformatics* **16:** 825–836.

Lohr D., Venkov P., and Zlatanova J. 1995. Transcriptional regulation in the yeast GAL gene family: A complex genetic network. *FASEB J.* **9:** 777–787.

McCarthy J.E.G. 1998. Posttranscriptional control of gene expression in yeast. *Microbiol. Mol. Biol. Rev.* **62:** 1492–1553.

Mehlhorn K. and Näher S.. 1999. *LEDA. A platform for combinatorial and geometric computing*. Cambridge University Press, United Kingdom.

Mewes H.W., Frishman D., Gruber C., Geier B., Haase D., Kaps A., Lemcke K., Mannhaupt G., Pfeiffer F., Schuller C., Stocker S., and Weil B. 2000. MIPS: A database for genomes and protein sequences. *Nucleic Acids Res.* **28:** 37–40.

Michal G. 1993. *Biochemical pathways* (Poster). Boehringer Mannheim GmbH.

———. 1998. On representation of metabolic pathways. *Biosystems* **47:** 1–7.

Palma P.N., Krippahl L., Wampler J.E., and Moura J.J. 2000. BiGGER: A new (soft) docking algorithm for predicting protein interactions. *Proteins* **39:** 372–384.

Passarge E. 1995. *Color atlas of genetics*. Thieme, Stuttgart.

Pirson I., Fortemaison N., Jacobs C., Dremier S., Dumont J.E., and Maenhaut C. 2000. The visual display of regulatory information and networks. *Trends Cell Biol.* **10:** 404–408.

Purchase H.C. 1997. Which aesthetic has the greatest effect on human understanding? In *Proceedings of Symposium on Graph Drawing GD '97*, pp. 248–261. Berlin.

Rain J.-C., Selig L., De Reuse H., Battaglia V., Reverdy C., Simon S., Lenzen G., Petel F., Wojcik J., Schächter V., Chemama Y., Labigne A., and Legrain P. 2001. The protein–protein interaction map of *Helicobacter pylori*. *Nature* **409:** 211–215.

Sanchez R., Pieper U., Mirkovic N., de Bakker P.I., Wittenstein E., and Sali A. 2000. MODBASE, a database of annotated comparative protein structure models. *Nucleic Acids Res.* **28:** 250–253.

Schwikowski B., Uetz P., and Fields S. 2000. A network of protein-protein interactions in yeast. *Nat. Biotechnol.* **18:** 1257–1261.

Serov V.N., Spirov A.V., and Samsonova M.G. 1998. Graphical interaface to the genetic network database GeNet. *Bioinformatics* **14:** 546–547.

Strogatz S.H. 2001. Exploring complex networks. *Nature* **410:** 268–276.

Tsai C.J., Lin S.L., Wolfson H.J., and Nussinov R. 1996. Protein-protein interfaces: Architectures and interactions in protein-protein interfaces and in protein cores. Their similarities and differences. *Crit. Rev. Biochem. Mol. Biol.* **31:** 127–152.

Tufte E.R. 1983. *The visual display of quantitative information*. Graphics Press, Cheshire, Connecticut.

Uetz P., Giot L., Cagney G., Mansfield T.A., Judson R.S., Knight J.R., Lockshon D., Narayan V., Srinivasan M., Pochart P., Qureshi-Emili A., Li Y., Godwin B., Conover D., Kalbfleisch T., Vijayadamodar G., Yang

M., Johnston M., Fields S., and Rothberg J.M.. 2000. A comprehensive analysis of protein-protein inter-actions in *Saccharomyces cerevisiae*. *Nature* **403:** 623–627.

Vitkup D., Melamud E., Moult J., and Sander C. 2001. Completeness in structural genomics. *Nat. Struct. Biol.* **8:** 559–566.

von Dassow G., Meir E., Munro E.M., and Odell G.M. 2000. The segment polarity network is a robust devel-opmental module. *Nature* **406:** 188–192.

Wilson W. A., Mahrenholz A. M., and Roach P. J. 1999. Substrate targeting of the yeast cyclin-dependent kinase Pho85p by the cyclin Pcl10p. *Mol. Cell. Biol.* **19:** 7020–7030.

Wingender E., Chen X., Hehl R., Karas H., Liebich I., Matys V., Meinhardt T., Pruss M., Reuter I., and Schacherer F. 2000. TRANSFAC: An integrated system for gene expression regulation. *Nucleic Acids Res.* **28:** 316–319.

Xenarios I., Fernandez E., Salwinski L., Duan X.J., Thompson M.J., Marcotte E.M., and Eisenberg D. 2001. DIP: The database of interacting proteins: 2001 update. *Nucleic Acids Res.* **29:** 239–241.

Yates J.R. 3rd. 2000. Mass spectrometry. From genomics to proteomics. *Trends Genet.* **16:** 5–8.

Zhu J. and Zhang M.Q. 1999. SCPD: A promoter database of the yeast *Saccharomyces cerevisiae*. *Bioinformatics* **15:** 607–611.

35 Modulating Protein–Protein Interactions to Develop New Therapeutic Approaches

Valérie Jérôme and Rolf Müller

Institute of Molecular Biology and Tumor Research, Philipps University, 35033 Marburg, Germany

INTRODUCTION

The manipulation of protein–protein interactions, which are involved in nearly all biological functions, is a potentially powerful tool for developing new therapeutic strategies. Several approaches have already been applied to develop improved promoters for gene therapy or to alter the tropism of viral vectors. This chapter describes some representative examples in this area of research.

TRANSCRIPTIONAL TARGETING

In the context of cancer gene therapy, transcriptional targeting allows the expression of a transgene in a controlled manner (Nettelbeck et al. 2000). In an ideal system, basal expression of the gene to be regulated should be low or undetectable but induced to high absolute levels by specific transcription factors affecting only targeted cells. For example, tumor cells, but frequently not the cells in the tissue of tumor origin, are proliferating. On the basis of this observation, we developed a strategy combining cell-cycle-regulated (Müller 1995) and cell-type-specific gene expression (Sikora 1993) in the same promoter. The basic idea of the strategy established in our laboratory is to drive transcription of a transgene by an artificial heterodimeric transcription factor, whose DNA-binding subunit is expressed from a tissue-specific promoter, whereas the *trans*-activating subunit is transcribed from a cell-cycle-regulated promoter (Fig. 1) (Müller et al. 1996). The construction of this heterodimeric transcription factor is based on the modular structure of the transcription factor, which allows the combination of DNA-binding and *trans*-activation domains derived from different proteins. The critical features of such a heterodimeric transcription factor are (1) a DNA-binding domain that binds with high affinity to a sequence that is not present in the mammalian genome and is not recognized by endogenous transcription factors; (2) an activation domain that, when recruited to a promoter, induces high levels of transcription of the target gene; and (3) a target gene promoter that has little or no basal activity but promotes transcription to high levels on binding of the heterodimeric transcription factor. We have called this approach a *dual specificity chimeric transcription factor* (DCTF) system. Apart from the chosen promoters, the DCTF system is critically dependent on specific protein dimerization interfaces that cannot bind or be bound by endogenous factors.

To avoid any interference of the DCTF system with endogenous proteins, we first sought to identify a setting that would allow the assembly of an efficient *trans*-activating complex using nonmammalian proteins. We first tested the bacterial LexA protein as a transcriptional DNA-binding domain (LexA-DBD amino acids 1–202; Brent and Ptashne 1984; Lech et al. 1988) and the activation domain of the herpesvirus VP16 protein (amino acids 411–455; Triezenberg et al. 1988) in combination with dimerization interface of the yeast transcription factors Gal4p (amino acids 820–900) and Gal80p (amino acids 1–435) (Chasman and Kornberg 1990); of the heterodimerizing yeast proteins Ino2p (amino acids 213–305) and Ino4p (amino acids 1–151) (Nikoloff et al. 1992), or of yAP-1 (amino acids 89–136) (Moye-Rowley et al. 1989), which can form homodimers. As a reporter gene, we used the firefly luciferase gene. However, none of these combinations gave rise to detectable expression of the transgene, presumably due to insufficient interactions of the yeast transcription factors in the context of mammalian cells.

In a separate approach, we made use of the dimerization interface of human lymphoid proteins CD4 and p56^{Lck}, which have been shown to associate also in nonlymphoid cells (Simpson et al. 1989). The CD4/LCK dimerization system has an advantage over other mammalian proteins in that the former molecules are normally associated with the plasma membrane and expressed only in a very restricted number of cell types (Shaw et al. 1989; Turner et al. 1990), making interference (i.e., competition) with the heterodimer formation in the nucleus unlikely. CD4 is a 55-kD T-lymphocyte membrane glycoprotein mediating the interaction of T cells with antigen-presenting cells (Biddison et al. 1982; Swain et al. 1983). It is composed of an amino-terminal extracellular domain of 372 amino acids, a transmembrane domain of 23 amino acids, and

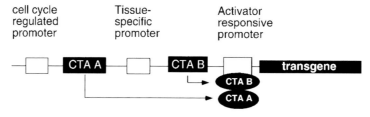

FIGURE 1. The dual-specificity chimeric transcription factor (DCTF) system. A DNA-binding subunit expressed from a tissue-specific promoter and a *trans*-activating subunit expressed from a cell-cycle-regulated promoter interact to form a heterodimeric chimeric transcriptional activator (CTA). The expression of a functional (i.e., heterodimeric) CTA will, therefore, be restricted to proliferating cells of a certain tissue type. The CTA binds and activates an activator-responsive promoter, thus leading to the expression of the transgene.

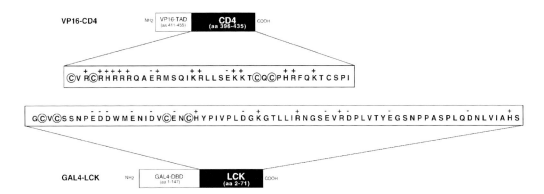

FIGURE 2. Structure of the CD4 and LCK chimeric transcription factors and a schematic representation of the GAL4–LCK and VP16–CD4 fusion proteins. The dimerization domains have been enlarged to show the sequence of the region. The acidic and basic residues that potentially make electrostatic interactions between subunits are indicated by a (–) or (+). The cysteine residues are circled. (+) Basic; (–) acidic.

a highly conserved cytoplasmic tail of 38 amino acids (Maddon et al. 1985; Littman et al. 1988). p56^{Lck}, a member of the src family of protein tyrosine kinases, is found predominantly in T lymphocytes (Marth et al. 1985; Voronova and Sefton 1986; Veillette et al. 1987). Structure–function analysis of mutant and chimeric molecules containing fragments of CD4 and p56^{Lck} showed that the cytoplasmic tail of CD4 is necessary and sufficient for interaction with 32 amino-terminal residues of p56^{Lck} (Shaw et al. 1989). Complex formation is critically dependent on the presence of free cysteines in the cytoplasmic tails of CD4 and in the amino-terminal domain of p56^{Lck} but does not imply disulfide bonds (Rudd et al. 1988; Shaw et al. 1990). In addition, the charged amino acids surrounding the cysteine motif are predominantly basic in CD4 (+-+-X-Cys-X-Cys-Pro), where X is a nonconserved residue and + is a basic residue), and acidic in p56^{Lck} (Cys-X-X-Cys), suggesting that salt bridges may stabilize an interaction centered on the cysteine residues (see Fig. 2).

Using CD4/LCK as the interaction interface, we demonstrated that the GAL4 DNA-binding domain (DBD; amino acids 1–147) (Webster et al. 1988; Chasman and Kornberg 1990) and VP16 *trans*-activation domain (amino acids 411–455; Lech et al. 1988) in combination with the interaction domains of CD4 (amino acids 396–435) and p56^{Lck} (LCK; amino acids 2–71) (Shaw et al. 1989; Turner et al. 1990) were far superior to the yeast interaction domains mentioned above. The basic system was further optimized by the amino-terminal linkage of a nuclear localization signal (NLS) from SV40 T antigen (Kalderon et al. 1984) to both transcription factors. The addition of the NLS sequence to both transcription factors increased the overall activity of the system fivefold. Moreover, the insertion of ten instead of five GAL4-binding sites (Webster et al. 1988) into the luciferase vector also increased promoter activity fivefold.

To establish a regulatory mechanism that selectively operates in proliferating melanoma cells, the expression of the chimeric transcription factors was driven by the *cyclin A* promoter for S/G$_2$-phase-specific transcription (Zwicker et al. 1995) and a tyrosinase promoter/enhancer construct for melanocyte-specific expression (Shibata et al. 1992). The constructs used for the subsequent studies are shown in Figure 3. The tissue-specific transcription of the DCTF system was demonstrated by cotransfection experiments in proliferating tumor cell lines originating from three different carcinoma types. The tyrosinase promoter driving LCK gives rise to 10- to 20-fold higher levels of luciferase activity in melanoma cell lines compared to prostate and hepatoma cell lines. In addition, deletion of the CD4 interaction domain results in levels of expression representing the background activity of the system in all cases. Furthermore, we tested the DCTF system for simultaneous tissue-specific and cell-cycle-regulated expression. Tumor cells were synchronized in G$_1$ by methionine deprivation and cotransfected with the constructs described above.

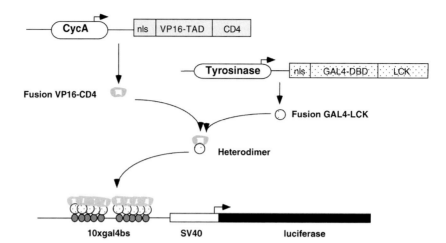

FIGURE 3. Outline of the experimental strategy. One subunit of an artificial transcription factor (CD4), consisting of the transcriptional activation domain of the herpes simplex virus protein VP16 (amino acids 411–455) and the cytoplasmic domain of the human CD4 (amino acids 396–435), is expressed from the cell-cycle-regulated cyclin A promoter. The second subunit (LCK), consisting of the GAL4 DNA-binding domain (amino acids 1–147) and the amino-terminal domain of the human p56^{Lck} (LCK; amino acids 2–71), which interacts with the CD4 cytoplasmic domain, is expressed from the tissue-specific tyrosinase promoter. Only in proliferating cells originating from this tissue will both subunits be expressed and thus be able to form a complex through the CD4–LCK interaction. Binding of this heterodimeric transcription factor to an effector/reporter gene construct containing GAL4-binding sites (gal4bs) and the minimal SV40 promoter (nucleotides –56 to +40) will then lead to transcriptional activation through the strong VP16 activation domain. (CycA) Cyclin A; (nls) nuclear localization signal; (TAD) *trans*-activation domain.

Determination of luciferase activities in growing and G$_1$-arrested cells clearly demonstrated a preferential expression in the proliferating melanoma cells. Cell-type specificity was in the range of 10-fold and cell cycle regulation approximately 5-fold (Jérôme and Müller 1998). The extent of regulation was further increased (1.5-fold) by cloning the two expressing cassettes into the same plasmid, a prerequisite for future testing of the DCTF system in animals. Although these results demonstrated the suitability of the DCTF system for directing gene expression to proliferating tumor cells, the extent of the regulation had to be improved further for a potential application in cancer gene therapy. Because the CD4–LCK interaction is weak, and potential interactions with endogenous molecules might result in the sequestration of the chimeric transcription factor used in the DCTF system, we reasoned that the replacement of the heterodimerization interface domains might enhance the overall performance of the system.

Leucine zippers (LZiPs) represent strong interaction domains commonly found in transcription factors (Busch and Sassone-Corsi 1990). They have been shown to be functional in a heterologous context and form particularly strong interactions (Kouzarides et al. 1989; Neuberg et al. 1989; Schmidt-Dörr et al. 1991). Dimerization through LZiPs is mediated by regularly spaced leucines (heptad repeats) in parallel α-helices through hydrophobic interactions, whereas the choice of the dimerization partner is determined by other amino acids, mainly charged residues forming salt bridges (Kouzarides et al. 1988; O'Shea et al. 1989, 1992; Schuermann et al. 1989, 1991; Cohen and Parry 1990; Oas et al. 1990). For the Fos and Jun transcription factors, the preferential formation of the heterodimer over either of the homodimers is brought about by the positively charged amino acids in the Jun LZiP and the negatively charged residues in the Fos counterpart, leading to strong self-repulsion of both proteins (O'Shea et al. 1992; Glover and Harrison 1995). The strength and specificity of the interaction between Fos and Jun, as well as their preferential heterodimerization, suggested that they could be useful for the improvement of the DCTF system. Both Jun and Fos family members are expressed to variable extents in all cells. To make

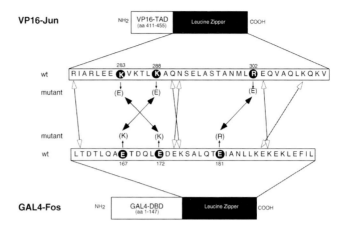

FIGURE 4. Structure of the chimeric transcription factors. Schematic representation of the GAL4–fos and VP16–jun fusion proteins. The leucine zipper structures have been enlarged to show the sequence of the region. Arrows connect the residues making electrostatic interactions between subunits. Black arrows link the amino acids switched from one zipper to the other, and the mutants generated are indicated in brackets. (Reprinted, with permission, from Jérôme and Müller 2001.)

the Jun/Fos LZiPs suitable for the DCTF system, it was therefore necessary to prevent interactions between the chimeric transcription factors and the endogenous Jun and Fos proteins. Using the resolved crystal structure (Glover and Harrison 1995), we introduced three acidic amino acids from c-Fos into the c-Jun LZiP (mJun) and three corresponding basic amino acids from c-Jun into the c-Fos LZiP (mFos), as depicted in Figure 4, thereby creating mixed-charged structures. Analysis of the oligomerization properties in an in vitro translation/association assay showed that these mutated zippers interact specifically with each other, but not with the wild-type LZiPs. After insertion into the transcription factors used in the DCTF system in place of the CD4 and p56^{Lck} sequences, these new heterodimerization domains enhanced promoter activity >100-fold to a level that makes this system suitable for applications in gene therapy. Moreover, the analysis of the fos/jun DCTF system in an in vivo setting of experimental melanoma showed a specific expression that was almost 4-fold higher than the expression of a SV40 construct after intratumoral injection of naked DNA (Jérôme and Müller 2001).

The DCTF system opens up the possibility of designing a particularly efficient system for *gene-directed enzyme prodrug therapy* (GDEPT). A promoter could be restricted to the proliferating cells of a defined tissue type (i.e., tumor cells) to direct the expression of an enzyme that cleaves a prodrug to a toxic drug that is not cell-cycle-dependent and gives rise to a strong bystander effect. This should lead to the destruction, not only of the expressing tumor cell, but also of all adjacent cells irrespective of their proliferative activity. In contrast, the commonly used thymidine kinase/ganciclovir system leads only to a selective destruction of the proliferating cells in the tumor (Conners 1995; Rigg and Sikora 1997), which is a major drawback in view of the fact that a substantial fraction of tumor cells is in a resting state or a prolonged G$_1$ phase at any given time. Moreover, instead of a prodrug-activating system it might also be possible to use transgenes encoding secreted cytotoxic proteins or cytokines stimulating a tumor-directed immune response. In addition, the mJun/mFos LZiP domains should also be useful for addressing other problems, as, for example, directing partner proteins to a specific subcellular location (Dang et al. 1991), to increase the activity of multienzyme systems by virtue of a physical link between the enzymes (Davidson et al. 1993; Gontero et al. 1993) or to improve the function of heterologous proteins in a gene therapeutic setting by enabling oligomerization (Walczak et al. 1999).

a) In the absence of rapamycin

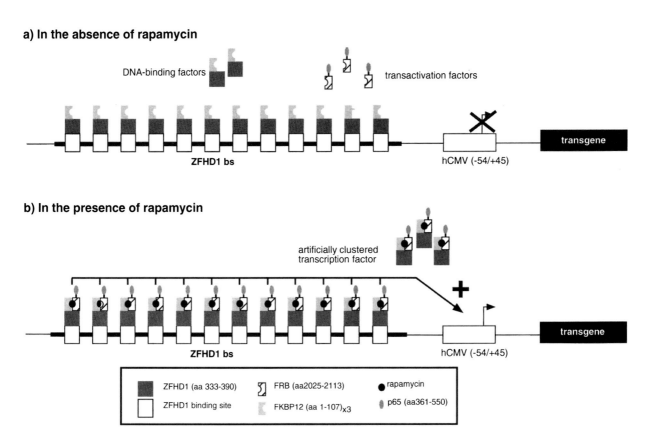

b) In the presence of rapamycin

FIGURE 5. The *r*apamycin-*i*nducible *h*umanized *s*ystem (RIHS). (a) In the absence of rapamycin, the two components of the *trans*-activator do not interact with each other and transcription does not proceed. (b) In the presence of rapamycin, the FKBP12-rapamycin binding (FRB) domain interacts with FKBP12. Rapamycin-mediated association of the domains results in a fully functional transcription factor that binds to and activates expression of a target gene containing binding sites for ZFHD1.

INDUCIBLE TRANSCRIPTION

In a clinical setting, a system that can be switched on and off at will to allow the timing of delivery and the optimization of gene product dosage would be highly desirable. This is particularly important for studies involving genes encoding toxic or otherwise highly bioactive proteins. Several systems regulatable by small molecules have been described, including promoter systems that are inducible by tetracycline (Baron and Bujard 2000), ecdysone (No et al. 1996), RU486/ mifepristone (Wang et al. 1997), or FK506/rapamycin (Rivera et al. 1996). Here, we focus on the latter, because this is the most advanced system with respect to gene therapy.

The goal of a "gene switch" is to place the expression of a gene of interest under control of a transcription factor, the activity of which can be regulated with a small-molecule drug (Rivera et al. 1996). The cyclophilin and immunophilin (FKBP12-rapamycin-associated protein; FRAP) protein families and the FK506-binding proteins (FKBPs) contain chemical-binding domains that can be fused to DNA-binding and *trans*-activation domains. The dimerization of these chimeric transcription factors can be achieved via chemical inducers of dimerization (CIDs), which are the immuno-suppressors FK506, rapamycin, and cyclosporin A (CsA) (Fig. 5) (Spencer et al. 1993; Klemm et al. 1998). These cell-permeable molecules display pharmacological properties that favor their use in vivo. They are used to control the interaction of a transcription activation domain and a DNA-binding domain and to allow the level of expression of a target gene to be regulated in a dose-responsive manner by the concentration of dimerizer independently of normal cellular physiology.

One of the most promising systems is a humanized version (Rivera et al. 1996) of the previously published system (Belshaw et al. 1996; No et al. 1996). A DNA-binding function is provided by a chimeric DNA-binding domain (ZFHD1; amino acids 333–390 of Zif268 fused to amino acids 378–439 of Oct-12 and spaced by 2 Gly residues) that is not recognized by endogenous transcription factors in the experimental context (Pomerantz et al. 1995). Three tandemly repeated copies of human FKBP12 (amino acids 1–107; Bierer et al. 1990) were fused to the ZFHD1 DNA-binding domain. A subdomain of p65 NF-κB (amino acids 361–550; Schmitz and Baeuerle 1991) is used as a transcriptional activator moiety and the small molecule binding function by a domain derived from the human FRAP (amino acids 2025–2113) (Brown et al. 1994). These two chimeric transcription factors have no inherent affinity to one another in the absence of rapamycin.

Using this system, a precise control of circulating protein levels in vivo from implanted, stably transfected cells secreting human growth hormone (hGH) in response to stimulation by rapamycin could be achieved. The combination of genetically engineered cells and the oral administration of a therapeutic compound to which the cells are responsive should permit the safe and prolonged delivery of therapeutic proteins (Rivera et al. 1996). Recently, the exploitation of the *rapamycin-inducible humanized system* (RIHS) to express the murine erythropoietin in mice and nonhuman primates after adeno-associated virus (AAV) transfer demonstrated the ability to titrate in vivo expression of the transgene with the dose of rapamycin (Ye et al. 1999).

To summarize, the RIHS displays the following advantages: (1) It is, in principle, applicable to human gene therapy because it is composed completely of human proteins; (2) there is very low background activity in vitro and in vivo, and a high induction ratio independent of host physiology and cell-type-specific factors; (3) it has a small DNA-coding capacity required for the regulatory proteins (<2.5 kb); and (4) the system's modularity allows each component to be optimized and engineered independently. Furthermore, in contrast to bacterial repressors (Gossen et al. 1992), which rely on allosteric intramolecular interactions to control DNA-binding activity, the RIHS can be applied to virtually any DNA-binding and activation domain. Nonetheless, some drawbacks might be the slow kinetics of de-induction and the growth-inhibitory and immunosuppressive effects of rapamycin. A potential solution to the latter problem may be the use of new nonimmunosuppressive analogs of rapamycin (Liberles et al. 1997; Harvey and Caskey 1998). Thus, the RIHS may become a method of choice for applications in human subjects.

RETARGETING OF VIRAL VECTORS

Viral vectors derived from retroviruses, adenoviruses, adeno-associated viruses, and herpesvirus are frequently used as gene delivery vehicles because of their capacity to carry foreign genes and their ability to deliver and express these genes efficiently (for review, see Walther and Stein 2000). Recently, efforts have been made to improve viral targeting based on control of protein–protein interactions, which are likely to play an essential role in promoting efficient and tolerable gene therapy protocols. The following section gives a short overview of the progress made in recent years within this field.

Retroviruses often suffer from a low efficiency of infection (Wang et al. 1991). One way to remedy this problem could be to alter the cell tropism of these viruses by incorporating receptor ligands or single-chain antibodies into the envelope protein (Young et al. 1990; Russell et al. 1993; Kasahara et al. 1994; Valsesia-Wittman et al. 1994; Cosset et al. 1995; Han et al. 1995; Somia et al. 1995; Marin et al. 1996; Schnierle et al. 1996; Ager et al. 1996; Chu et al. 1997). Although retargeting could be demonstrated, viral titers were generally very low, which represents a serious obstacle with respect to their applicability. One explanation might be that only specific cell-surface molecules on target cells are able to mediate efficient infection. Another approach consists of the introduction of protease target sequences or spacer peptides between the Env protein and the ligand introduced for retargeting (Valsesia-Wittman et al. 1997; Peng et al. 1997, 1998, 1999). Although this retargeting approach was more successful, numerous technical problems still remain to be solved before retargeted retroviral vectors can be used in a clinical setting of gene therapy.

Replication-deficient adenoviruses have several key attributes that make them potentially useful for clinical gene therapy. These include their large packaging capacity, the relative ease of virus production at high titers, and their ability to infect both dividing and nondividing cells. There are, however, also important limitations, including its widespread tropism, short-term transgene expression, and stimulation of inflammatory and immune responses.

Adenovirus infection is initiated by high-affinity binding of the fiber protein "knob" domain to the coxsackie-adenovirus receptor (CAR) (Bergelson et al. 1997; Tomko et al. 1997). The virion subsequently binds via its penton capsid protein to a cellular integrin receptor (via a RGD motif) and internalizes via the receptor-mediated endocytosis pathway (Bilbao et al. 1998). A low level of CAR expression on target cells has emerged as a key limiting factor in adenovirus gene delivery to certain neoplastic (Dmitriev et al. 1998; Miller et al. 1998; Blackwell et al. 1999; Kasono et al. 1999) and nonneoplastic (Zabner et al. 1997; Kaner et al. 1999; Walters et al. 1999) cells. Additionally, expression of CAR in non-target tissues may lead to undesired transgene expression with risk of toxicity (Yee et al. 1996). Therefore, it is important to target the delivery and/or expression of adenovirus-encoded transgenes to the appropriate set of cells. Moreover, sequestration of adenovirus vectors in the liver is a key limiting factor to the use of these agents via the systemic route (Van der Eb et al. 1998).

On the basis of the well-understood mechanism of adenoviral cell entry, several groups are working on approaches toward an "immunologic" or a "genetic" retargeting of adenoviral vectors (Douglas and Curiel 1995; Curiel 1999). Immunologic retargeting strategies are based on the use of bispecific conjugates (antibody directed against a component of the virus and a targeting antibody or ligand) and require both the abolition of native targeting and the introduction of a new tropism (Fig. 6). For example, a folate-conjugated neutralizing antibody against the knob region of adenovirus enabled the targeting of several tumor types (Weitman et al. 1992). Nettelbeck et al. (2001) developed a bispecific single-chain diabody directed against endoglin and the adenovirus knob domain, which restricts and enhances the adenoviral transduction to endothelial cells. Another approach made use of a fusion protein between a neutralizing anti-knob scFv and the epidermal growth factor (EGF). This "adenobody" was successfully used to retarget an adenoviral vector to EGF receptor-positive tumor cells (Watkins et al. 1997). Interestingly, this pathway of infection appears to bypass the need for penton base/integrin interaction for internalization of the virus. More recently, Reynolds and coworkers published the first in vivo data demonstrating that an immunological retargeting approach can modify the tropism of a systemically injected adenovirus (Reynolds et al. 2000).

Genetic retargeting involves the modification of the capsid proteins at the level of the viral genome (for review, see Bilbao 1998; Curiel 1999). Genetic modifications can be directed to the

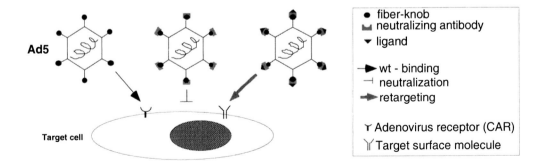

FIGURE 6. Immunologic retargeting strategy. An antibody directed against the knob domain can neutralize the adenoviral binding of the fiber-knob to the CAR receptor. The fusion of a targeting protein or ligand to this anti-knob neutralizing antibody allows a selective binding of an adenovirus to the target tissue.

fiber, to the penton base, or to capsid proteins (Wickham et al. 1995). Wickham et al. (1996) showed that incorporating a high-affinity peptide ligand for α_v integrin into the adenoviral fiber protein can enhance adenoviral transfection efficiency in endothelial cells in vitro and in vivo (Wickham et al. 1997a,b; McDonald et al. 1999). Retargeting can also improve the transducibility of certain tissues, as shown for Kaposi's sarcoma (Goldman et al. 1997). Importantly, retargeting should allow a substantial reduction of the viral dose required to achieve a therapeutic effect, thus potentially reducing toxic side effects, which are a major concern with respect to adenoviral vectors (Schulick et al. 1995a,b).

SUMMARY

Manipulations of protein–protein interactions to develop new therapeutic strategies have been extensively developed within the past few years. The work of several groups provides new tools, based on transcriptional targeting, that allow restricted expression of a transgene to a defined tissue type (DCTF system; Jérôme and Müller 1998, 2001), or a controlled expression of a transgene via cell-permeable molecules (RIHS; Rivera et al. 1996). Other groups successfully modified the adenoviral tropism via immunologic or genetic retargeting (for review, see Curiel 2000). In the near future, one can predict that researchers will focus their work on improving these existing tools further, and combining transcriptional targeting with adenoviral retargeting; thus achieving a tightly controlled and localized transgene expression in vivo, making these systems suitable for human gene therapy. Above all, the main goal of cancer gene therapy is to provide tools to cure cancer avoiding the side effects of the currently used protocols.

REFERENCES

Ager S., Nilson B.H., Morling F.J., Peng K.W., Cosset F.L., and Russell S.J. 1996. Retroviral display of antibody fragments; interdomain spacing strongly influences vector infectivity. *Hum. Gene Ther.* **7:** 2157–2164.

Baron U. and Bujard H. 2000. Tet repressor-based system for regulated gene expression in eukaryotic cells: Principles and advances. *Methods Enzymol.* **327:** 401–421.

Belshaw P.J., Ho S.N., Crabtree G.R., and Schreiber S.L. 1996. Controlling protein association and subcellular localization with a synthetic ligand that induces heterodimerization of proteins. *Proc. Natl. Acad. Sci.* **93:** 4604–4607.

Bergelson J.M., Cunningham J.A., Droguett G., Kurt-Jones E.A., Krithivas A., Hong J.S., Horwitz M.S., Crowell R.L., and Finberg R.W. 1997. Isolation of a common receptor for Coxsackie B viruses and adenoviruses 2 and 5. *Science* **275:** 1320–1323.

Biddison W.E., Rao P.E., Talle M.A., Goldstein G., and Shaw S. 1982. Possible involvement of the OKT4 molecule in T cell recognition of class II HLA antigens. Evidence from studies of cytotoxic T lymphocytes specific for SB antigens. *J. Exp. Med.* **156:** 1065–1076.

Bierer B.E., Mattila P.S., Standaert R.F., Herzenberg L.A., Burakoff S.J., Crabtree G., and Schreiber S.L. 1990. Two distinct signal transmission pathways in T lymphocytes are inhibited by complexes formed between an immunophilin and either FK506 or rapamycin. *Proc. Natl. Acad. Sci.* **87:** 9231–9235.

Bilbao G., Contreras J.L., Gomez-Navarro J., and Curiel D.T. 1998. Improving adenoviral vectors for cancer gene therapy. *Tumor Targ.* **3:** 59–79.

Blackwell J.L., Miller C.R., Douglas J.T., Li H., Peters G.E., Carroll W.R., Strong T.V., and Curiel D.T. 1999. Retargeting to EGFR enhances adenovirus infection efficiency of squamous cell carcinoma. *Arch. Otolaryngol. Head Neck Surg.* **125:** 856–863.

Brent R. and Ptashne M. 1984. A bacterial repressor protein or a yeast transcriptional terminator can block upstream activation of a yeast gene. *Nature* **312:** 612–615.

Brown E.J., Albers M.W., Shin T.B., Ichikawa K., Keith C.T., Lane W.S., and Schreiber S.L. 1994. A mammalian protein targeted by G1-arresting rapamycin-receptor complex. *Nature* **369:** 756–758.

Busch S.J. and Sassone-Corsi P. 1990. Dimers, leucine zippers and DNA-binding domains. *Trends Genet.* **6:** 36–40.

Chasman D.I. and Kornberg R.D. 1990. GAL4 protein: Purification, asssociation with GAL80 protein, and conserved domain structure. *Mol. Cell. Biol.* **10:** 2916–2923.

Chu T.H. and Dornburg R. 1997. Toward highly efficient cell-type-specific gene transfer with retroviral vectors displaying single-chain antibodies. *J. Virol.* **71:** 720–725.

Cohen C. and Parry D.A. 1990. Alpha-helical coiled coils and bundles: How to design an alpha-helical protein. *Proteins* **7:** 1–15.

Conners T.A. 1995. The choice of prodrugs for gene directed enzyme prodrug therapy. *Gene Ther.* **2:** 702–709.

Cosset F.L., Morling F.J., Takeuchi Y., Weiss R.A., Collins M.K., and Russell, S.J. 1995. Retroviral retargeting by envelopes expressing an N-terminal binding domain. *J Virol.* **69:** 6314–6322.

Curiel D.T. 1999. Strategies to adapt adenoviral vectors for targeted delivery. *Ann. N.Y. Acad. Sci.* **886:** 158–171.

———. 2000. Rational design of viral vectors based on rigorous analysis of capsid structure. *Mol. Ther.* **1:** 3–4.

Dang C.V., Barrett J., Villa-Garcia M., Resar L.M., Kato G.J., and Fearon E.R. 1991. Intracellular leucine zipper interactions suggest c-Myc hetero-oligomerization. *Mol. Cell. Biol.* **11:** 954–962.

Davidson J.N., Chen K.C., Jamison R.S., Musmanno L.A., and Kern C.B. 1993. The evolutionary history of the first three enzymes in pyrimidine biosynthesis. *BioEssays* **15:** 157–164.

Dmitriev I., Krasnykh V., Miller C.R., Wang M., Kashentseva E., Mikheeva G., Belousova N., and Curiel D.T. 1998. An adenovirus vector with genetically modified fibers demonstrates expanded tropism via utilization of a coxsackievirus and adenovirus receptor-independent cell entry mechanism. *J. Virol.* **72:** 9706–9713.

Douglas J.T. and Curiel D.T. 1995. Targeted gene therapy. *Tumor Targ.* **1:** 67–84.

Glover J.N. and Harrison S.C. 1995. Crystal structure of the heterodimeric bZIP transcription factor c-Fos-c-Jun bound to DNA. *Nature* **373:** 257–261.

Goldman C.K., Rogers B.E., Douglas J.T., Sosnowski B.A., Ying W., Siegal G.P., Baird A., Campain J.A., and Curiel D.T. 1997. Targeted gene delivery to Kaposi's sarcoma cells via the fibroblast growth factor receptor. *Cancer Res.* **57:** 1447–1451.

Gontero B., Mulliert G., Rault M., Giudici-Orticoni M.T., and Ricard J. 1993. Structural and functional properties of a multi-enzyme complex from spinach chloroplasts. 2. Modulation of the kinetic properties of enzymes in the aggregated state. *Eur. J. Biochem.* **217:** 1075–1082.

Gossen M. and Bujard H. 1992. Tight control of gene expression in mammalian cells by tetracycline-responsive promoters. *Proc. Natl. Acad. Sci.* **89:** 5547–5551.

Han X., Kasahara N., and Kan Y.W. 1995. Ligand-directed retroviral targeting of human breast cancer cells. *Proc. Natl. Acad. Sci.* **92:** 9747–9751.

Harvey D.M. and Caskey C.T. 1998. Inducible control of gene expression: prospects for gene therapy. *Curr. Opin. Chem. Biol.* **2:** 512–518.

Jérôme V. and Müller R. 1998. Tissue-specific, cell cycle-regulated chimeric transcription factors for the targeting of gene expression to tumor cells. *Hum. Gene Ther..* **9:** 2653–2659.

———. 2001. A synthetic leucine zipper-based dimerization system for combining multiple promoter specificities. *Gene Ther..* **8:** 725–729.

Kalderon D., Roberts B.L., Richardson W.D., and Smith A.E. 1984. A short amino acid sequence able to specify nuclear location. *Cell* **39:** 499–509.

Kaner R.J., Worgall S., Leopold P.L., Stolze E., Milano E., Hidaka C., Ramalingam R., Hackett N.R., Singh R., Bergelson J., Finberg R., Falck-Pedersen E., and Crystal R.G. 1999. Modification of the genetic program of human alveolar macrophages by adenovirus vectors in vitro is feasible but inefficient, limited in part by the low level of expression of the coxsackie/adenovirus receptor. *Am. J. Respir. Cell. Mol. Biol..* **20:** 361–370.

Kasahara N., Dozy A.M.. and Kan Y.W. 1994. Tissue-specific targeting of retroviral vectors through ligand-receptor interactions. *Science* **266:** 1373–1376.

Kasono K., Blackwell J.L., Douglas J.T., Dmitriev I., Strong T.V., Reynolds P., Kropf D.A., Carroll W.R., Peters G.E., Bucy R.P., Curiel D.T., and Krasnykh V. 1999. Selective gene delivery to head and neck cancer cells via an integrin targeted adenoviral vector. *Clin. Cancer Res.* **5:** 2569–2571.

Klemm J.D., Schreiber S.L., and Crabtree G.R. 1998. Dimerization as a regulatory mechanism in signal transduction. *Annu. Rev. Immunol.* **16:** 569–592.

Kouzarides T. and Ziff E. 1988. The role of the leucine zipper in the fos-jun interaction. *Nature* **336:** 646–651.

———. 1989. Leucine zippers of fos, jun and GCN4 dictate dimerization specificity and thereby control DNA binding. *Nature* **340:** 568–571.

Lech K., Anderson K., and Brent R. 1988. DNA-bound Fos proteins activate transcription in yeast. *Cell* **52:** 179–184.

Liberles S.D., Diver S.T., Austin D.J., and Schreiber S.L. 1997. Inducible gene expression and protein translocation using nontoxic ligands identified by a mammalian three-hybrid screen. *Proc. Natl. Acad. Sci.* **94:** 7825–7830.

Littman D.R., Maddon P.J., and Axel R. 1988. Corrected CD4 sequence. *Cell* **55:** 541.

Maddon P.J., Littman D.R., Godfrey M., Maddon D.E., Chess L., and Axel R. 1985. The isolation and nucleotide sequence of a cDNA encoding the T cell surface protein T4: A new member of the immunoglobulin gene family. *Cell* **42:** 93–104.

Marin M., Noel D., Valsesia-Wittman S., Brockly F., Etienne-Julan M., Russell S., Cosset F.L., and Piechaczyk M. 1996. Targeted infection of human cells via major histocompatibility complex class I molecules by Moloney murine leukemia virus-derived viruses displaying single-chain antibody fragment-envelope fusion proteins. *J. Virol.* **70:** 2957–2962.

Marth J.D., Peet R., Krebs E.G., and Perlmutter R.M. 1985. A lymphocyte-specific protein-tyrosine kinase gene is rearranged and overexpressed in the murine T cell lymphoma LSTRA. *Cell* **43:** 393–404.

McDonald G.A., Zhu G., Li Y., Kovesdi I., Wickham T.J., and Sukhatme V.P. 1999. Efficient adenoviral gene transfer to kidney cortical vasculature utilizing a fiber modified vector. *J. Gene Med.* **1:** 103–110.

Miller C.R., Buchsbaum D.J., Reynolds P.N., Douglas J.T., Gillespie G.Y., Mayo M.S., Raben D., and Curiel D.T. 1998. Differential susceptibility of primary and established human glioma cells to adenovirus infection: Targeting via the epidermal growth factor receptor achieves fiber receptor-independent gene transfer. *Cancer Res.* **58:** 5738–5748.

Moye-Rowley W.S., Harshman K.D., and Parker C.S. 1989. Yeast YAP1 encodes a novel form of the jun family of transcriptional activator proteins. *Genes Dev.* **3:** 283–292.

Müller R. 1995. Transcriptional regulation during the mammalian cell cycle. *Trends Genet.* **11:** 173–178.

Müller R., Lührmann R., and Sedlacek H.H. 1996. Nukleinsaürekonstrukte mit Genen kodierend für transportsignale. *Patent Appl.* DE 19617851.7.

Nettelbeck D.M., Jérôme V., and Müller R. 2000. Gene therapy: designer promoters for tumour targeting. *Trends Genet.* **16:** 174–181.

Nettelbeck D.M., Miller D.W., Jérôme V., Zuzarte M., Watkins S.J., Hawkins R.E., Muller R., and Kontermann R.E. 2001. Targeting of adenovirus to endothelial cells by a bispecific single-chain diabody directed against the adenovirus fiber knob domain and human endoglin (CD105). *Mol. Ther.* **3:** 882–891.

Neuberg M., Adamkiewicz J., Hunter J.B., and Müller R. 1989. A Fos protein containing the Jun leucine zipper forms a homodimer which binds to the AP1 binding site. *Nature* **341:** 243–245.

Nikoloff D.M., McGraw P., and Henry S.A. 1992. The INO2 gene of *Saccharomyces cerevisiae* encodes a helix-loop-helix protein that is required for activation of phospholipid synthesis. *Nucleic Acids Res.* **20:** 3253.

No D., Yao T.P., and Evans R.M. 1996. Ecdysone-inducible gene expression in mammalian cells and transgenic mice. *Proc. Natl. Acad. Sci.* **93:** 3346–3351.

O'Shea E.K., Rutkowski R., and Kim P.S. 1992. Mechanism of specificity in the Fos-Jun oncoprotein heterodimer. *Cell* **68:** 699–708.

O'Shea E.K., Rutkowski R., Stafford W.F.D., and Kim P.S. 1989. Preferential heterodimer formation by isolated leucine zippers from fos and jun. *Science* **245:** 646–648.

Oas T.G., McIntosh L.P., O'Shea E.K., Dahlquist F.W., and Kim P.S. 1990. Secondary structure of a leucine zipper determined by nuclear magnetic resonance spectroscopy. *Biochemistry* **29:** 2891–2894.

Peng K.W., Morling F.J., Cosset F.L., and Russell S.J. 1998. Retroviral gene delivery system activable by plasmin. *Tumor Targ.* **3:** 112–120.

Peng K.W., Vile R., Cosset F.L., and Russell S. 1999. Selective transduction of protease-rich tumors by matrix-metalloproteinase-targeted retroviral vectors. *Gene Ther.* **6:** 1552–1557.

Peng K.W., Morling F.J., Cosset F.L., Murphy G., and Russell S.J. 1997. A gene delivery system activatable by disease-associated matrix metalloproteinases. *Hum. Gene Ther.* **8:** 729–738.

Pomerantz J.L., Sharp P.A., and Pabo C.O. 1995. Structure-based design of transcription factors. *Science.* **267:** 93–96.

Reynolds P.N., Zinn K.R., Gavrilyuk V.D., Balyasnikova I.V., Rogers B.E., Buchsbaum D.J., Wang M.H., Miletich D.J., Grizzle W.E., Douglas J.T., Danilov S.M., and Curiel D.T. 2000. A targetable, injectable adenoviral vector for selective gene delivery to pulmonary endothelium in vivo. *Mol. Ther.* **2:** 562–578.

Rigg A. and Sikora K. 1997. Genetic prodrug activation therapy. *Mol. Med. Today* **3:** 359–366.

Rivera V.M., Clackson T., Natesan S., Pollock R., Amara J.F., Keenan T., Magari S.R., Phillips T., Courage N.L., Cerasoli Jr., F., Holt D.A., and Gilman M. 1996. A humanized system for pharmacologic control of gene expression [see comments]. *Nat. Med.* **2:** 1028–1032.

Rudd C.E., Trevillyan J.M., Dasgupta J.D., Wong L.L., and Schlossman S.F. 1988. The CD4 receptor is complexed in detergent lysates to a protein-tyrosine kinase (pp58) from human T lymphocytes. *Proc. Natl. Acad. Sci.* **85:** 5190–5194.

Russell S.J., Hawkins R.E., and Winter G. 1993. Retroviral vectors displaying functional antibody fragments.

Nucleic Acids Res. **21:** 1081–1085.

Schmidt-Dörr T., Oertel-Buchheit P., Pernelle C., Bracco L., Schnarr M., and Granger-Schnarr M. 1991. Construction, purification and characterization of a hybrid protein comprising the DNA binding domain of the LexA repressor and the Jun leucine zipper: A circular dichroism and mutagenesis study. *Biochemistry* **30:** 9657–9664.

Schmitz M.L. and Baeuerle P.A. 1991. The p65 subunit is responsible for the strong transcription activating potential of NF-kappa B. *EMBO J.* **10:** 3805-3817.

Schnierle B.S. and Groner B. 1996. Retroviral targeted delivery. *Gene Ther.* **3:** 1069–1073.

Schuermann M., Hunter J.B., Hennig G., and Müller R. 1991. Non-leucine residues in the leucine repeats of Fos and Jun contribute to the stability and determine the specifity of dimerization. *Nucleic Acids Res.* **19:** 739–746.

Schuermann M., Neuberg M., Hunter J.B., Jenuwein T., Ryseck R.P., Bravo R., and Müller R. 1989. The leucine repeat motif in Fos protein mediates complex formation with Jun/AP-1 and is required for transformation. *Cell* **56:** 507–516.

Schulick A.H., Newman K.D., Virmani R., and Dichek D.A. 1995a. In vivo gene transfer into injured carotid arteries. Optimization and evaluation of acute toxicity. *Circulation* **91:** 2407–2014.

Schulick A.H., Dong G., Newman K.D., Virmani R., and Dichek D.A. 1995b. Endothelium-specific in vivo gene transfer. *Circ. Res.* **77:** 475–485.

Shaw A., Amrein K.E., Hammond C., Stern D.F., Sefton B.M., and Rose J.K. 1989. The lck tyrosine protein kinase interacts with the cytoplasmic tail of the CD4 glycoprotein through its unique amino-terminal domain. *Cell* **59:** 627–633.

Shaw A.S., Chalupny J., Whitney J.A., Hammond C., Amrein K.E., Kavathas P., Sefton B.M., and Rose J.K. 1990. Short related sequences in the cytoplasmic domains of CD4 and CD8 mediate binding to the amino-terminal domain of the p56lck tyrosine protein kinase. *Mol. Cell. Biol.* **10:** 1853–1862.

Shibata K., Muraosa Y., Tomita Y., Tagami H., and Shibahara S. 1992. Identification of a cis-acting element that enhances the pigment cell-specific expression of the human tyrosinase gene. *J. Biol. Chem.* **267:** 20584–20588.

Sikora K. 1993. Gene therapy for cancer. *Trends Biotechnol.* **11:** 197–201.

Simpson S.C., Bolen J.B., and Veillette A. 1989. CD4 and p56lck can stably associate when co-expressed in NIH3T3 cells. *Oncogene.* **4:** 1141–1143.

Somia N.V., Zoppe M., and Verma I.M. 1995. Generation of targeted retroviral vectors by using single-chain variable fragment: An approach to in vivo gene delivery. *Proc. Natl. Acad. Sci.* **92:** 7570–7574.

Spencer D.M., Wandless T.J., Schreiber S.L., and Crabtree G.R. 1993. Controlling signal transduction with synthetic ligands. *Science* **262:** 1019–1024.

Swain S.L. 1983. T cell subsets and the recognition of MHC class. *Immunol. Rev.* **74:** 129–142.

Tomko R.P., Xu R., and Philipson L. 1997. HCAR and MCAR: The human and mouse cellular receptors for subgroup C adenoviruses and group B coxsackieviruses. *Proc. Natl. Acad. Sci.* **94:** 3352–3356.

Triezenberg S.J., Kingsbury R.C., and McKnight S.L. 1988. Functional dissection of VP16, the trans-activator of herpes simplex virus immediate early gene expression. *Genes Dev.* **2:** 718–729.

Turner J.M., Brodsky M.H., Irving B.A., Levin S.D., Perlmutter R.M., and Littman D.R. 1990. Interaction of the unique N-terminal region of tyrosinase p56lck with cytoplasmic domains of CD4 and CD8 is mediated by cysteine motifs. *Cell* **60:** 755–765.

Valsesia-Wittmann S., Morling F.J., Hatziioannou T., Russell S.J., and Cosset F.L. 1997. Receptor co-operation in retrovirus entry: Recruitment of an auxiliary entry mechanism after retargeted binding [published erratum appears in *EMBO J.* 1997 Jul 1;16(13):4153]. *EMBO J.* **16:** 1214–1223.

Valsesia-Wittmann S., Drynda A., Deleage G., Aumailley M., Heard J.M., Danos O., Verdier G., and Cosset F.L. 1994. Modifications in the binding domain of avian retrovirus envelope protein to redirect the host range of retroviral vectors. *J. Virol.* **68:** 4609–4019.

Van der Eb M.M., Cramer S.J., Vergouwe Y., Schagen F.H., van Krieken J.H., van der Eb A.J., Rinkes I.H., van de Velde C.J., and Hoeben R.C. 1998. Severe hepatic dysfunction after adenovirus-mediated transfer of the herpes simplex virus thymidine kinase gene and ganciclovir administration. *Gene Ther.* **5:** 451–458.

Veillette A., Foss F.M., Sausville E.A., Bolen J.B., and Rosen N. 1987. Expression of the lck tyrosine kinase gene in human colon carcinoma and other non-lymphoid human tumor cell lines. *Oncogene Res.* **1:** 357–374.

Voronova A.F. and Sefton B.M. 1986. Expression of a new tyrosine protein kinase is stimulated by retrovirus promoter insertion. *Nature* **319:** 682–685.

Walczak H., Miller R.E., Ariail K., Gliniak B., Griffith T.S., Kubin M., Chin W., Jones J., Woodward A., Le T., Smith C., Smolak P., Goodwin R.G., Rauch C.T., Schuh J.C., and Lynch D.H. 1999. Tumoricidal activity of tumor necrosis factor-related apoptosis-inducing ligand in vivo. *Nat. Med.* **5:** 157–163.

Walters R.W., Grunst T., Bergelson J.M., Finberg R.W., Welsh M.J., and Zabner J. 1999. Basolateral localization of fiber receptors limits adenovirus infection from the apical surface of airway epithelia. *J. Biol.*

Chem. **274:** 10219–10226.

Walther W. and Stein U. 2000. Viral vectors for gene transfer: A review of their use in the treatment of human diseases. *Drugs* **60:** 249–271.

Wang H., Kavanaugh M.P., North R.A., and Kabat D. 1991. Cell-surface receptor for ecotropic murine retroviruses is a basic amino-acid transporter. *Nature* **352:** 729–731.

Wang Y., DeMayo F.J., Tsai S.Y., and O'Malley B.W. 1997. Ligand-inducible and liver-specific target gene expression in transgenic mice. *Nat. Biotechnol.* **15:** 239–243.

Watkins S.J., Mesyanzhinov V.V., Kurochkina L.P., and Hawkins R.E. 1997. The 'adenobody' approach to viral targeting: Specific and enhanced adenoviral gene delivery. *Gene Ther.* **4:** 1004–10012.

Webster N., Jin J.R., Green S., Hollis M., and Chambon P. 1988. The yeast UASG is a transcriptional enhancer in HeLa cells in the presence of the GAL4 *trans*-activator. *Cell* **52:** 169–178.

Weitman S.D., Lark R.H., Coney L.R., Fort D.W., Frasca V., Zurawski V.R., and Kamen B.A. 1992. Distribution of the folate receptor GP38 in normal and malignant cell lines and tissues. *Cancer Res.* **52:** 3396–3401.

Wickham T.J., Carrion M.E., and Kovesdi I. 1995. Targeting of adenovirus penton base to new receptors through replacement of its RGD motif with other receptor-specific peptide motifs. *Gene Ther.* **2:** 750–756.

Wickham T.J., Lee G.M., Titus J.A., Sconocchia G., Bakacs T., Kovesdi I., and Segal D.M. 1997a. Targeted adenovirus-mediated gene delivery to T cells via CD3. *J. Virol.* **71:** 7663–7669.

Wickham T.J., Segal D.M., Roelvink P.W., Carrion M.E., Lizonova A., Lee G.M., and Kovesdi I. 1996. Targeted adenovirus gene transfer to endothelial and smooth muscle cells by using bispecific antibodies. *J. Virol.* **70:** 6831–6838.

Wickham T.J., Tzeng E., Shears 2nd, L.L., Roelvink P.W., Li Y., Lee G.M., Brough D.E., Lizonova A., and Kovesdi I. 1997b. Increased in vitro and in vivo gene transfer by adenovirus vectors containing chimeric fiber proteins. *J. Virol.* **71:** 8221–8229.

Ye X., Rivera V.M., Zoltick P., Cerasoli Jr., F., Schnell M.A., Gao G., Hughes J.V., Gilman M., and Wilson J.M. 1999. Regulated delivery of therapeutic proteins after in vivo somatic cell gene transfer. *Science* **283:** 88–91.

Yee D., McGuire S.E., Brunner N., Kozelsky T.W., Allred D.C., Chen S.H., and Woo S.L. 1996. Adenovirus-mediated gene transfer of herpes simplex virus thymidine kinase in an ascites model of human breast cancer. *Hum. Gene Ther.* **7:** 1251–1257.

Young J.A., Bates P., Willert K., and Varmus H.E. 1990. Efficient incorporation of human CD4 protein into avian leukosis virus particles. *Science* **250:** 1421–1423.

Zabner J., Freimuth P., Puga A., Fabrega A., and Welsh M.J. 1997. Lack of high affinity fiber receptor activity explains the resistance of ciliated airway epithelia to adenovirus infection. *J. Clin. Invest.* **100:** 1144–1149.

Zwicker J. and Müller R. 1995. Cell cycle-regulated transcription in mammalian cells. *Prog. Cell Cycle Res.* **1:** 91–99.

Appendix: Cautions

GENERAL CAUTIONS

The following general cautions should always be observed.

- **Become completely familiar with the properties of substances used** before beginning the procedure.

- **The absence of a warning** does not necessarily mean that the material is safe, since information may not always be complete or available.

- **If exposed to toxic substances,** contact your local safety office immediately for instructions.

- **Use proper disposal procedures** for all chemical, biological, and radioactive waste.

- **For specific guidelines on appropriate gloves,** consult your local safety office.

- **Handle concentrated acids and bases** with great care. Wear goggles and appropriate gloves. A face shield should be worn when handling large quantities. Do not mix strong acids with organic solvents, as they may react. Sulfuric acid and nitric acid especially may react highly exothermically and cause fires and explosions. Do not mix strong bases with halogenated solvent, as they may form reactive carbenes that can lead to explosions.

- **Never pipette solutions using mouth suction.** This method is not sterile and can be dangerous. Always use a pipette aid or bulb.

- **Keep halogenated and nonhalogenated solvents separately** (e.g., mixing chloroform and acetone can cause unexpected reactions in the presence of bases). Halogenated solvents are organic solvents such as chloroform, dichloromethane, trichlorotrifluoroethane, and dichloroethane. Some nonhalogenated solvents are pentane, heptane, ethanol, methanol, benzene, toluene, N,N-dimethylformamide (DMF), dimethyl sulfoxide (DMSO), and acetonitrile.

- **Laser radiation,** visible or invisible, can cause severe damage to the eyes and skin. Take proper precautions to prevent exposure to direct and reflected beams. Always follow manufacturers' safety guidelines and consult your local safety office. See caution below for more detailed information.

- **Flash lamps,** due to their light intensity, can be harmful to the eyes. They also may explode on occasion. Wear appropriate eye protection and follow the manufacturers' guidelines.

- **Photographic fixatives and developers** also contain chemicals that can be harmful. Handle them with care and follow manufacturers' directions.

- **Power supplies and electrophoresis equipment** pose serious fire hazard and electrical shock hazards if not used properly.

- **Microwave ovens and autoclaves** in the lab require certain precautions. Accidents have occurred involving their use (e.g., to melt agar or Bacto-agar stored in bottles or to sterilize). If the screw top is not completely removed and there is not enough space for the steam to vent, the bottles can explode and cause severe injury when the containers are removed from the microwave or autoclave. Always completely remove bottle caps before microwaving or autoclaving. An alternative method for routine agarose gels that do not require sterile agar is to weigh out the agar and place the solution in a flask.

- **Use extreme caution when handling cutting devices** such as microtome blades, scalpels, razor blades, or needles. Microtome blades are extremely sharp! Use care when sectioning. If you are unfamiliar with their use, have someone demonstrate proper procedures. For proper disposal, use the "sharps" disposal container in your lab. Discard used needles *unshielded*, with the syringe still attached. This prevents injuries (and possible infections; see Biological Safety) while manipulating used needles, since many accidents occur while trying to replace the needle shield. Injuries may also be caused by broken Pasteur pipettes, coverslips, or slides.

GENERAL PROPERTIES OF COMMON CHEMICALS

The hazardous materials list can be summarized in the following categories:

- Inorganic acids, such as hydrochloric, sulfuric, nitric, or phosphoric, are colorless liquids with stinging vapors. Avoid spills on skin or clothing. Spills should be diluted with large amounts of water. The concentrated forms of these acids can destroy paper, textiles, and skin, as well as cause serious injury to the eyes.

- Inorganic bases such as sodium hydroxide are white solids that dissolve in water and under heat development. Concentrated solutions will slowly dissolve skin and even fingernails.

- Salts of heavy metals are usually colored powdered solids that dissolve in water. Many are potent enzyme inhibitors and therefore toxic to humans and to the environment (e.g., fish and algae).

- Most organic solvents are flammable volatile liquids. Avoid breathing the vapors, which can cause nausea or dizziness. Avoid skin contact.

- Other organic compounds, including organosulfur compounds such as mercaptoethanol and organic amines, can have very unpleasant odors. Others are highly reactive and should be handled with appropriate care.

- If improperly handled, dyes and their solutions can stain not only your sample, but also your skin and clothing. Some of them are also mutagenic (e.g., ethidium bromide), carcinogenic, and toxic.

- All names ending with "ase" (e.g., catalase, β-glucuronidase, zymolase) refer to enzymes. There are also other enzymes with nonsystematic names like pepsin. Many of them are provided by manufacturers in preparations containing buffering substances, etc. Be aware of the individual properties of materials contained in these substances.

- Toxic compounds are often used to manipulate cells. They can be dangerous and should be handled appropriately.

- Be aware that several of the compounds listed have not been thoroughly studied with respect to their toxicity. Handle each chemical with the appropriate respect. Although the toxic effects of a compound can be quantified (e.g., LD_{50} values), this is not possible for carcinogens or mutagens where one single exposure can have an effect. Realize that dangers related to a given compound may also depend on its physical state (fine powder vs. large crystals/diethylether vs. glycerol/dry ice vs. carbon dioxide under pressure in a gas bomb). Anticipate under which circumstances during an experiment exposure is most likely to occur and how best to protect yourself and your environment.

HAZARDOUS MATERIALS

Acetic acid (concentrated) must be handled with great care. It may be harmful by inhalation, ingestion, or skin absorption. Wear appropriate gloves and goggles and use in a chemical fume hood.

Acetic anhydride is extremely destructive to the skin, eyes, mucous membranes, and upper respiratory tract. It may be harmful by inhalation, ingestion, or skin absorption. Wear appropriate gloves and safety glasses, and use in a chemical fume hood.

Acetone causes eye and skin irritation and is irritating to mucous membranes and upper respiratory tract. Do not breathe the vapors. It is also extremely flammable. Wear appropriate gloves and safety glasses.

Acetonitrile is very volatile and extremely flammable. It is an irritant and a chemical asphyxiant that can exert its effects by inhalation, ingestion, or skin absorption. Treat cases of severe exposure as cyanide poisoning. Wear appropriate gloves and safety glasses and use only in a chemical fume hood. Keep away from heat, sparks, and open flame.

Acrylamide (unpolymerized) is a potent neurotoxin and is absorbed through the skin (the effects are cumulative). Avoid breathing the dust. Wear appropriate gloves and a face mask when weighing powdered acrylamide and methylene-bisacrylamide. Use in a chemical fume hood. Polyacrylamide is considered to be nontoxic, but it should be handled with care because it might contain small quantities of unpolymerized acrylamide.

3-Amino-1,2,4-triazole (ATA) is a carcinogen. It may be harmful by inhalation, ingestion, or skin absorption. Wear appropriate gloves, safety glasses, and other protective clothing. Avoid breathing vapors. Use only in a chemical fume hood.

Ammonium bicarbonate, NH_4HCO_3, may be harmful by inhalation, ingestion, or skin absorption. Wear appropriate gloves and safety glasses and use in a chemical fume hood.

Ammonium persulfate, $(NH_4)_2S_2O_8$, is extremely destructive to tissue of the mucous membranes and upper respiratory tract, eyes, and skin. Inhalation may be fatal. Wear appropriate gloves, safety glasses, and protective clothing and use only in a chemical fume hood. Wash your hands thoroughly after handling.

Ammonium sulfate, $(NH_4)_2SO_4$, may be harmful by inhalation, ingestion, or skin absorption. Wear appropriate gloves and safety glasses.

Aprotinin may be harmful by inhalation, ingestion, or skin absorption. It may also cause allergic reactions. Exposure may cause gastrointestinal effects, muscle pain, blood pressure changes, or bronchospasm. Wear appropriate gloves and safety glasses and use only in a chemical fume hood. Do not breathe the dust.

Bacterial strains (shipping of): The Department of Health, Education, and Welfare (HEW) has classified various bacteria into different categories with regard to shipping requirements (see Sanderson and Zeigler, *Methods Enzymol. 204:* 248–264 [1991]). Nonpathogenic strains of *E. coli* (such as K12) and *B. subtilis* are in Class 1 and are considered to present no or minimal hazard under normal shipping conditions. However, *Salmonella, Haemophilus,* and certain strains of *Streptomyces* and *Pseudomonas* are in Class 2. Class 2 bacteria are "Agents of ordinary potential hazard: agents which produce disease of varying degrees of severity...but which are contained by ordinary laboratory techniques." For detailed regulations regarding the packaging and shipping of Class 2 strains, see Sanderson and Ziegler (*Methods Enzymol. 204*:248–264 [1991]) or the instruction brochure by Alexander and Brandon (*Packaging and Shipping of Biological Materials at ATCC* [1986]) available from the American Type Culture Collection (ATCC), Rockville, Maryland.

Biotin may be harmful by inhalation, ingestion, or skin absorption. Wear appropriate gloves and safety glasses and use in a chemical fume hood.

Bisacrylamide is a potent neurotoxin and is absorbed through the skin (the effects are cumulative). Avoid breathing the dust. Wear appropriate gloves and a face mask when weighing powdered acrylamide and methylene-bisacrylamide.

Bromophenol blue may be harmful by inhalation, ingestion, or skin absorption. Wear appropriate gloves and safety glasses and use in a chemical fume hood.

CaCl$_2$, *see* **Calcium chloride**

Calcium chloride, CaCl$_2$, may be harmful by inhalation, ingestion, or skin absorption. Wear appropriate gloves and safety glasses and use in a chemical fume hood.

C$_6$H$_5$CH$_3$, *see* **Toluene**

C$_2$H$_4$INO, *see* **Iodoacetamide**

CHCl$_3$, *see* **Chloroform**

CH$_3$CH$_2$OH, *see* **Ethanol**

C$_6$H$_5$CH$_2$SO$_2$F, *see* **Phenylmethylsulfonyl fluoride**

C$_7$H$_7$FO$_2$S, *see* **Phenylmethylsulfonyl fluoride**

Chloramphenicol may be harmful by inhalation, ingestion, or skin absorption and is a carcinogen. Wear appropriate gloves and safety glasses and use in a chemical fume hood.

Chloroform, CHCl$_3$, is irritating to the skin, eyes, mucous membranes, and respiratory tract. It is a carcinogen and may damage the liver and kidneys. It is also volatile. Avoid breathing the vapors. Wear appropriate gloves and safety glasses and always use in a chemical fume hood.

4-Chloro-1-naphthol is irritating to the eyes, skin, mucous membranes, and respiratory tract. Handle with care. Wear appropriate gloves and safety glasses.

Cyanogen bromide is extremely toxic and is volatile. It may be fatal by inhalation, ingestion, or skin absorption. Do not breathe the vapors. Wear appropriate gloves and always use in a chemical fume hood. Keep away from acids.

Cycloheximide may be fatal if inhaled, ingested, or absorbed through the skin. Wear appropriate gloves and safety glasses and use in a chemical fume hood.

DAB, *see* **3,3´-Diaminobenzidine tetrahydrochloride**

DAPI, *see* **4´,6-Diamidine-2´phenylindole dihydrochloride**

DCM, *see* **Dichloromethane**

4´,6-Diamidine-2´phenylindole dihydrochloride (DAPI) is a possible carcinogen. It may be harmful by inhalation, ingestion, or skin absorption. It may also cause irritation. Avoid breathing the dust and vapors. Wear appropriate gloves and safety glasses and use in a chemical fume hood.

3,3′-Diaminobenzidine tetrahydrochloride (DAB) is a carcinogen. Handle with extreme care. Avoid breathing vapors. Wear appropriate gloves and safety glasses and use in a chemical fume hood.

Dichloromethane (DCM), CH$_2$Cl$_2$, (also known as **Methylene chloride**) is toxic if inhaled, ingested, or absorbed through the skin. It is also an irritant and is suspected to be a carcinogen. Wear appropriate gloves and safety glasses and use in a chemical fume hood. Do not breathe the vapors.

***N,N*-Dimethylformamide (DMF), HCON(CH$_3$)$_2$,** is irritating to the eyes, skin, and mucous membranes. It can exert its toxic effects through inhalation, ingestion, or skin absorption. Chronic inhalation can cause liver and kidney damage. Wear appropriate gloves and safety glasses and use in a chemical fume hood.

Dimethyl sulfoxide (DMSO) may be harmful by inhalation or skin absorption. Wear appropriate gloves and safety glasses and use in a chemical fume hood. DMSO is also combustible. Store in a tightly closed container. Keep away from heat, sparks, and open flame.

Dithiothreitol (DTT) is a strong reducing agent that emits a foul odor. It may be harmful by inhalation, ingestion, or skin absorption. When working with the solid form or highly concentrated stocks, wear appropriate gloves and safety glasses and use in a chemical fume hood.

DMF, *see* **N,N-Dimethylformamide**

DMSO, *see* **Dimethyl sulfoxide**

DTT, *see* **Dithiothreitol**

EDC, *see* **N-ethyl-N′-(dimethylaminopropyl)-carbodiimide**

Ethanol, CH$_3$CH$_2$OH, may be harmful by inhalation, ingestion, or skin absorption. Wear appropriate gloves and safety glasses.

Ethanolamine, HOCH$_2$CH$_2$NH$_2$, is toxic and harmful by inhalation, ingestion, or skin absorption. Handle with care and avoid any contact with the skin. Wear appropriate gloves and goggles and use in a chemical fume hood. Ethanolamine is highly corrosive and reacts violently with acids.

1-Ethyl-3-[3-dimethylaminopropyl] carbodiimide (EDC), *see* **N-Ethyl-N′-(dimethylaminopropyl)-carbodiimide**

***N*-Ethyl-*N*′-(dimethylaminopropyl)-carbodiimide (EDC)** is irritating to the mucous membranes and upper respiratory tract. It may be harmful by inhalation, ingestion, or skin absorption. Wear appropriate gloves and safety glasses. Handle with care.

FITC, *see* **Fluorescein isothiocyanate**

Fluoroscein may be harmful by inhalation, ingestion, or skin absorption. Wear appropriate gloves and safety glasses and use in a chemical fume hood.

Fluorescein isothiocyanate (FITC), may be harmful by inhalation, ingestion, or skin absorption. Wear appropriate gloves and safety glasses.

Formamide is teratogenic. The vapor is irritating to the eyes, skin, mucous membranes, and upper respiratory tract. It may be harmful by inhalation, ingestion, or skin absorption. Wear appropriate gloves and safety glasses and always use a chemical fume hood when working with concentrated solutions of formamide. Keep working solutions covered as much as possible.

Formic acid, HCOOH, is highly toxic and extremely destructive to tissue of the mucous membranes, upper respiratory tract, eyes, and skin. It may be harmful by inhalation, ingestion, or skin absorption. Wear appropriate gloves and safety glasses (or face shield) and use in a chemical fume hood.

β-**Galactosidase** is an irritant and may cause allergic reactions. It may be harmful by inhalation, ingestion, or skin absorption. Wear appropriate gloves and safety glasses.

β-**Glucuronidase (GUS)** may be harmful by inhalation, ingestion, or skin absorption. Wear respirator, appropriate gloves, and safety glasses.

Gonadotropin is a possible teratogen and poses a risk of irreversible effects. it may be harmful by inhalation, ingestion, or skin absorption. Wear appropriate gloves and safety goggles and use in a chemical fume hood. Do not breathe in dust.

Guanidine hydrochloride is irritating to the mucous membranes, upper respiratory tract, skin, and eyes. It may be harmful by inhalation, ingestion, or skin absorption. Wear appropriate gloves and safety glasses. Avoid breathing the dust.

Guanidine thiocyanate may be harmful by inhalation, ingestion, or skin absorption. Wear appropriate gloves and safety glasses.

Guanidinium isothiocyanate, *see* **Guanidine thiocyanate**

Guanidinium thiocyanate, *see* **Guanidine thiocyanate**

GUS, *see* β-**Glucuronidase**

HCl, *see* **Hydrochloric acid**

H$_3$COH, *see* **Methanol**

HCOOH, *see* **Formic acid**

H$_2$O$_2$, *see* **Hydrogen peroxide**

HOCH$_2$CH$_2$NH$_2$, *see* **Ethanolamine**

HOCH$_2$CH$_2$SH, *see* β-**Mercaptoethanol**

H$_3$PO$_4$, *see* **Phosphoric acid**

H$_2$SO$_4$, *see* **Sulfuric acid**

Hydrochloric acid, HCl, is volatile and may be fatal if inhaled, ingested, or absorbed through the skin. It is extremely destructive to mucous membranes, upper respiratory tract, eyes, and skin. Wear appropriate gloves and safety glasses and use with great care in a chemical fume hood. Wear goggles when handling large quantities.

Hydrogen peroxide, H$_2$O$_2$, is corrosive, toxic, and extremely damaging to the skin. It may be harmful by inhalation, ingestion, and skin absorption. Wear appropriate gloves and safety glasses and use only in a chemical fume hood.

N-**Hydroxysuccinimide** is an irritant and may be harmful by inhalation, ingestion, or skin absorption. Wear appropriate gloves and safety glasses.

Imidazole is corrosive and may be harmful by inhalation, ingestion, or skin absorption. Wear appropriate gloves and safety glasses and use in a chemical fume hood.

Iodoacetamide, C$_2$H$_4$INO, can alkylate amino groups in proteins and can therefore cause problems if the antigen is being purified for amino acid sequencing. It is toxic and harmful by inhalation, ingestion, or skin absorption. Wear appropriate gloves and safety glasses and use only in a chemical fume hood. Do not breathe the dust.

IPTG, *see* **Isopropyl-β-D-thiogalactopyranoside**

Isoamyl alcohol (IAA) may be harmful by inhalation, ingestion, or skin absorption and presents a risk of serious damage to the eyes. Wear appropriate gloves and safety goggles. Keep away from heat, sparks, and open flame.

Isopropyl-β-D-thiogalactopyranoside (IPTG) may be harmful by inhalation, ingestion, or skin absorption. Wear appropriate gloves and safety glasses.

KCl, *see* **Potassium chloride**

K$_3$Fe(CN)$_6$, *see* **Potassium ferricyanide**

K$_4$Fe(CN)$_6$·3H$_2$O, *see* **Potassium ferrocyanide**

KH$_2$PO$_4$/K$_2$HPO$_4$/K$_3$PO$_4$, *see* **Potassium phosphate**

KOH, *see* **Potassium hydroxide**

Laser radiation, both visible and invisible, can be seriously harmful to the eyes and skin and may generate airborne contaminants, depending on the class of laser used. High-power lasers cause permanent eye damage and can burn exposed skin, ignite flammable materials, and activate toxic chemicals that release hazardous by-products. Avoid eye or skin exposure to direct or scattered radiation. Do not stare at the laser and do not point the laser at anyone. Wear appropriate eye protection and use suitable shields that are designed to offer protection for the specific type of wavelength, mode of operation (continuous wave or pulsed), and power output (watts) of the laser being used. Avoid wearing jewelry or other objects that may reflect or scatter the beam. Some nonbeam hazards include electrocution, fire, and asphyxiation. Entry to the area in which the laser is being used must be controlled and posted with warning signs that indicate when the laser is in use. Always follow suggested safety guidelines that accompany the equipment, and contact your local safety office for further information.

> **Ion lasers** present a hazard due to high-voltage high-current power supplies. Always follow manufacturers' suggested safety guidelines.

> **Ultraviolet lasers** present a hazard due to invisible beam, high-energy radiation. Always use beam traps, scattered light shields, and fluorescent beamfinder cards.

> **Blue-green lasers** present a hazard due to photothermal coagulation. Blue and green wavelengths are readily absorbed by blood hemoglobin.

Leupeptin (or its **hemisulfate**) may be harmful by inhalation, ingestion, or skin absorption. Wear appropriate gloves and safety glasses and use in a chemical fume hood.

LiCl, *see* **Lithium chloride**

Lithium acetate may be harmful by inhalation, ingestion, or skin absorption. Wear appropriate gloves and safety glasses and use in a chemical fume hood. Do not breathe the dust.

Lithium chloride, LiCl, is an irritant to the eyes, skin, mucous membranes, and upper respiratory tract. It may be harmful by inhalation, ingestion, or skin absorption. Wear appropriate gloves, safety goggles, and use in a chemical fume hood. Do not breathe the dust.

Magnesium chloride, MgCl$_2$, may be harmful by inhalation, ingestion, or skin absorption. Wear appropriate gloves and safety glasses and use in a chemical fume hood.

Magnesium sulfate, MgSO$_4$, may be harmful by inhalation, ingestion, or skin absorption. Wear appropriate gloves and safety glasses and use in a chemical fume hood.

β-Mercaptoethanol (2-Mercaptoethanol), HOCH$_2$CH$_2$SH, may be fatal if inhaled or absorbed through the skin and is harmful if ingested. High concentrations are extremely destructive to the mucous membranes, upper respiratory tract, skin, and eyes. β-Mercaptoethanol has a very foul odor. Wear appropriate gloves and safety glasses and always use in a chemical fume hood.

Methanol, H$_3$COH, is poisonous and can cause blindness. It may be harmful by inhalation, ingestion, or skin absorption. Adequate ventilation is necessary to limit exposure to vapors. Avoid inhaling these vapors. Wear appropriate gloves and goggles and use only in a chemical fume hood.

Methotrexate (MTX) is a carcinogen and a teratogen. It may be harmful by inhalation, ingestion, or skin absorption. Exposure may cause gastrointestinal effects, bone marrow suppression, and liver or kidney damage. It may also cause irritation. Avoid breathing the vapors. Wear appropriate gloves and safety glasses and always use in a chemical fume hood.

N-**Methyl-2-pyrrolidinone (2-methyl-2-pyrrolidinone)** may be harmful by inhalation, ingestion, or skin absorption. Wear appropriate gloves and safety glasses and use in a chemical fume hood. Do not breathe the vapors. Keep away from heat, sparks, and open flame.

MgCl$_2$, *see* **Magnesium chloride**

MgSO$_4$, *see* **Magnesium sulfate**

3-(*N*-Morpholino)-propanesulfonic acid (MOPS) may be harmful by inhalation, ingestion, or skin absorption. It is irritating to mucous membranes and upper respiratory tract. Wear appropriate gloves and safety glasses and use in a chemical fume hood.

MOPS, *see* **3-(*N*-Morpholino)-propanesulfonic acid**

MTX, *see* **Methotrexate**

NaF, *see* **Sodium fluoride**

Na$_2$HPO$_4$, *see* **Sodium hydrogen phosphate**

NaH$_2$PO$_4$/Na$_2$HPO$_4$/Na$_3$PO$_4$, *see* **Sodium phosphate**

NaN$_3$, *see* **Sodium azide**

NaOH, *see* **Sodium hydroxide**

Na$_3$VO$_4$, *see* **Sodium orthovanadate**

(NH$_4$)$_2$SO$_4$, *see* **Ammonium sulfate**

Paraformaldehyde is highly toxic. It is readily absorbed through the skin and is extremely destructive to the skin, eyes, mucous membranes, and upper respiratory tract. Avoid breathing the dust. Wear appropriate gloves and safety glasses and use in a chemical fume hood. Paraformaldehyde is the undissolved form of formaldehyde.

PEG, see **Polyethyleneglycol**

Pepstatin A may be harmful by inhalation, ingestion, or skin absorption. Wear appropriate gloves and safety glasses and use in a chemical fume hood.

Phenol is extremely toxic, highly corrosive, and can cause severe burns. It may be harmful by inhalation, ingestion, or skin absorption. Wear appropriate gloves, goggles, protective clothing, and always use in a chemical fume hood. Rinse any areas of skin that come in contact with phenol with a large volume of water and wash with soap and water; do not use ethanol!

Phenylmethylsulfonyl fluoride (PMSF), C$_7$H$_7$FO$_2$S or **C$_6$H$_5$CH$_2$SO$_2$F,** is a highly toxic cholinesterase inhibitor. It is extremely destructive to the mucous membranes of the respiratory tract, eyes, and skin. It may be fatal by inhalation, ingestion, or skin absorption. Wear appropriate gloves and safety glasses and always use in a chemical fume hood. In case of contact, immediately flush eyes or skin with copious amounts of water and discard contaminated clothing.

Phosphoric acid, H$_3$PO$_4$, is highly corrosive and may be harmful by inhalation, ingestion, or skin absorption. Wear appropriate gloves and safety glasses.

Piperidine is highly toxic and is corrosive to the eyes, skin, respiratory tract, and gastrointestinal tract. It reacts violently with acids and oxidizing agents and may be harmful by inhalation, ingestion, or skin absorption. Do not breathe the vapors. Keep away from heat, sparks, and open flame. Wear appropriate gloves and safety glasses and use in a chemical fume hood.

PMSF, *see* **Phenylmethylsulfonyl fluoride**

Polyethyleneglycol (PEG) may be harmful by inhalation, ingestion, or skin absorption. Avoid inhalation of powder. Wear appropriate gloves and safety glasses.

Potassium chloride, KCl, may be harmful by inhalation, ingestion, or skin absorption. Wear appropriate gloves and safety glasses.

Potassium ferricyanide, $K_3Fe(CN)_6$, may be fatal by inhalation, ingestion, or skin absorption. Wear appropriate gloves and safety glasses and always use with extreme care in a chemical fume hood. Keep away from strong acids.

Potassium ferrocyanide, $K_4Fe(CN)_6 \cdot 3H_2O$, may be fatal by inhalation, ingestion, or skin absorption. Wear appropriate gloves and safety glasses and always use with extreme care in a chemical fume hood. Keep away from strong acids.

Potassium hydroxide, KOH and KOH/methanol, can be highly toxic. It may be harmful by inhalation, ingestion, or skin absorption. Solutions are caustic and should be handled with great care. Wear appropriate gloves.

Potassium phosphate, $KH_2PO_4/K_2HPO_4/K_3PO_4$, may be harmful by inhalation, ingestion, or skin absorption. Wear appropriate gloves and safety glasses. Do not breathe the dust. *$K_2HPO_4 \cdot 3H_2O$ is dibasic and KH_2PO_4 is monobasic.*

Propidium iodide may be harmful by inhalation, ingestion, or skin absorption. It is irritating to the eyes, skin, mucous membranes, and upper respiratory tract. It is mutagenic and possibly carcinogenic. Wear appropriate gloves, safety glasses, protective clothing, and always use with extreme care in a chemical fume hood.

Radioactive substances: When planning an experiment that involves the use of radioactivity, include the physicochemical properties of the isotope (half-life, emission type and energy), the chemical form of the radioactivity, its radioactive concentration (specific activity), total amount, and its chemical concentration. Order and use only as much as really needed. Always wear appropriate gloves, lab coat, and safety goggles when handling radioactive material. **X-rays** and **gamma rays** are electromagnetic waves of very short wavelengths either generated by technical devices or emitted by radioactive materials. They may be emitted isotopically from the source or may be focused into a beam. Their potential dangers depend on the time period of exposure, the intensity experienced, and the wavelengths used. Be aware that appropriate shielding is usually of lead or other similar material. The thickness of the shielding is determined by the energy(s) of the X-rays or gamma rays. Consult your local safety office for further guidance in the appropriate use and disposal of radioactive materials. Always monitor thoroughly after using radioisotopes. A convenient calculator to perform routine radioactivity calculations can be found at:

http://www.graphpad.com/calculators/radcalc.cfm

SDS, *see* **Sodium dodecyl sulfate**

Sodium acetate, see **Acetic acid**

Sodium azide, NaN_3, is highly poisonous. It blocks the cytochrome electron transport system. Solutions containing sodium azide should be clearly marked. It may be harmful by inhalation, ingestion, or skin absorption. Wear appropriate gloves and safety goggles and handle with great care. Sodium azide is an oxidizing agent and should not be stored near flammable chemicals.

Sodium dodecyl sulfate (SDS) is toxic, an irritant, and poses a risk of severe damage to the eyes. It may be harmful by inhalation, ingestion, or skin absorption. Wear appropriate gloves and safety goggles. Do not breathe the dust.

Sodium fluoride, NaF, is highly toxic and causes severe irritation. It may be fatal by inhalation, ingestion, or skin absorption. Wear appropriate gloves and safety glasses and use only in a chemical fume hood.

Sodium hydrogen phosphate, Na$_2$HPO$_4$, (sodium phosphate, dibasic) may be harmful by inhalation, ingestion, or skin absorption. Wear appropriate gloves and safety glasses and use in a chemical fume hood.

Sodium hydroxide, NaOH, and solutions containing NaOH are highly toxic and caustic and should be handled with great care. Wear appropriate gloves and a face mask. All other concentrated bases should be handled in a similar manner.

Sodium orthovanadate, Na$_3$VO$_4$, may be harmful by inhalation, ingestion, or skin absorption. Wear appropriate gloves and safety glasses and use in a chemical fume hood.

Sodium phosphate, NaH$_2$PO$_4$/Na$_2$HPO$_4$/Na$_3$PO$_4$, is an irritant to the eyes and skin. It may be harmful by inhalation, ingestion, or skin absorption. Wear appropriate gloves and safety goggles. Do not breathe the dust.

Sulfuric acid, H$_2$SO$_4$, is highly toxic and extremely destructive to tissue of the mucous membranes and upper respiratory tract, eyes, and skin. It causes burns, and contact with other materials (e.g., paper) may cause fire. Wear appropriate gloves, safety glasses, and lab coat, and use in a chemical fume hood.

TCA, *see* **Trichloroacetic acid**

TEMED, *see N,N,N´,N´-*Tetramethylethylenediamine

*N,N,N´,N´-*Tetramethylethylenediamine (TEMED) is highly caustic to the eyes and mucous membranes and may be harmful by inhalation, ingestion, or skin absorption. Wear appropriate gloves and tightly sealed safety goggles.

Toluene, C$_6$H$_5$CH$_3$, vapors are irritating to the eyes, skin, mucous membranes, and upper respiratory tract. Toluene can exert harmful effects by inhalation, ingestion, or skin absorption. Do not inhale the vapors. Wear appropriate gloves and safety glasses and use in a chemical fume hood. Toluene is extremely flammable. Keep away from heat, sparks, and open flame.

Trichloroacetic acid (TCA) is highly caustic. Wear appropriate gloves and safety goggles.

Urea may be harmful by inhalation, ingestion, or skin absorption. Wear appropriate gloves and safety glasses.

UV light and/or **UV radiation** is dangerous and can damage the retina. Never look at an unshielded UV light source with naked eyes. Examples of UV light sources that are common in the laboratory include hand-held lamps and transilluminators. View only through a filter or safety glasses that absorb harmful wavelengths. UV radiation is also mutagenic and carcinogenic. To minimize exposure, make sure that the UV light source is adequately shielded. Wear protective appropriate gloves when holding materials under the UV light source.

Xylene is flammable and may be narcotic at high concentrations. It may be harmful by inhalation, ingestion, or skin absorption. Wear appropriate gloves and safety glasses and use only in a chemical fume hood. Keep away from heat, sparks, and open flame.

Xylene cyanol, *see* **Xylene**

Zymolase may be harmful by inhalation, ingestion, or skin absorption. Wear appropriate gloves and safety glasses.

Appendix: Suppliers

With the exception of those suppliers listed in the text with their addresses, all suppliers mentioned in this manual can be found in the BioSupplyNet Source Book and on the Web site at:

http://www.biosupplynet.com

If a copy of the BioSupplyNet Source Book was not included with this manual, a free copy can be ordered by any of the following methods:

- Complete the Free Source Book Request Form found at the Web site at: http://biosupplynet.com
- E-mail a request to info@biosupplynet.com
- Fax a request to 1-919-659-2199

Index